VIRTUAL MODELING AND RAPID MANUFACTURING

ADVANCED RESEARCH IN VIRTUAL AND RAPID PROTOTYPING

BALKEMA – Proceedings and Monographs
in Engineering, Water and Earth Sciences

PROCEEDINGS OF THE 2ND INTERNATIONAL CONFERENCE ON ADVANCED RESEARCH AND RAPID PROTOTYPING, LEIRIA, PORTUGAL, 28 SEPTEMBER–1 OCTOBER, 2005

Virtual modeling and rapid manufacturing

Advanced Research in Virtual and Rapid Prototyping

Paulo Jorge Bártolo
Artur Jorge Mateus
Fernando da Conceição Batista
Henrique Amorim Almeida
Joel Correia Vasco
Mário António Correia
Nuno Fernandes Alves

Nuno Carpinteiro André
Paulo Parente Novo
Paulo Lima
Pedro Conceição Custódio
Pedro Gonçalves Martinho
Rui Adriano Carvolho

Polytechnic Institute of Leiria, Portugal

Taylor & Francis
Taylor & Francis Group

LONDON/LEIDEN/NEW YORK/PHILADELPHIA/SINGAPORE

Published by: Taylor & Francis/Balkema
P.O. Box 447, 2300 AK Leiden, The Netherlands
e-mail: Pub.NL@tandf.co.uk
www.balkema.nl, www.tandf.co.uk, www.crcpress.com

ISBN 0 415 39062 1

Printed in Great Britain

Virtual modeling and rapid manufacturing – Bártolo (eds)
© 2005 Taylor & Francis Group, London, ISBN 0 415 39062 1

Table of Contents

CAD and 3D data acquisition technologies

Materials

Rapid tooling and manufacturing

Advanced rapid prototyping technologies and nanofabrication

Virtual environments and concurrent engineering

Applications

Virtual modeling and rapid manufacturing – Bártolo (eds)
© 2005 Taylor & Francis Group, London, ISBN 0 415 39062 1

Preface

Virtual Modeling and Rapid Manufacturing contains papers presented at the 2nd International Conference on Advanced Research in Virtual and Physical Prototyping (VR@P2005), held by the School of Technology and Management of the Polytechnic Institute of Leiria, Portugal. This event was designed to be a major forum for the scientific exchange of multi-disciplinary and inter-organisational aspects of virtual and rapid prototyping and related areas, making a significant contribution for further development of these fields. It joined participants from more than 20 countries. Such diversity was parallel to the various multi-disciplinary contributions to the conference, whose subjects enclosed a wide range of topics like making a significant contribution for further development of these fields like biomanufacturing, materials for rapid prototyping, advanced rapid prototyping technologies, reverse engineering, etc. This research community is strongly engaged in the development of innovative solutions to solve industry's problems, this way contributing to a more pleasant and healthy life for everyone.

I am deeply grateful to authors, participants, reviewers, the international scientific committee, session chairs, student helpers and administrative assistants, for contributing to the success of this conference. This conference was endorsed by:

- The Polytechnic Institute of Leiria (IPL)
- The School of Technology and Management (ESTG)
- The Global Alliance of Rapid Prototyping Associations (GARPA)
- The Portuguese Rapid Prototyping Association (ANPR)

Finally, I would like to express my gratitude to ROLAND who sponsored this book.

Paulo Jorge Bártolo
Leiria, September 2005

Virtual modeling and rapid manufacturing – Bártolo (eds)
© 2005 Taylor & Francis Group, London, ISBN 0 415 39062 1

International Scientific Committee

Wei Sun, Drexel University, USA
Rui Vilar, IST, Portugal
Tamás Várady, Hungarian Academy of Sciences, Hungary
Terry Wohlers, Wohlers Associates, USA
Yongnian Yan, Tsinghua University, China

Keynotes

Virtual modeling and rapid manufacturing – Bártolo (eds)
© 2005 Taylor & Francis Group, London, ISBN 0 415 39062 1

New trends and developments in additive fabrication

Terry Wohlers

Wohlers Associates, Inc., Fort Collins, Colorado, USA

ABSTRACT: The use of 3D printers for concept models, and even refined prototypes, continues to gain appeal and momentum. In 2004, unit sales growth of this class of additive fabrication grew an astounding 90.9%, according to research conducted by Wohlers Associates. Even the smallest organizations are now considering the purchase of a machine. Meanwhile, companies are discovering ways to apply additive processes to the manufacture of finished production parts in quantities of one to several thousand. This approach to manufacturing is allowing companies to introduce new products that before were not feasible due to tooling costs, long lead times, and risk.

1 INTRODUCTION

The two hottest growth areas of additive fabrication are 1) 3D printing for concept modeling and design validation and 2) the use of the technology for the manufacture of series production parts. With the machine prices less than $25,000, even the smallest organizations are purchasing equipment. This is dramatic change from years ago when only large organizations and service providers acquired machines. Meanwhile, high-end equipment is getting faster and building better quality parts, making it feasible to use it for finished production parts. A growing number of companies are now using technology, such as laser sintering, to manufacture parts for a wide range of industries such as aerospace, motor sports, military, dental, and medicine.

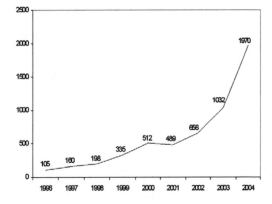

Figure 1. Source: Wohlers Report 2005.

2 3D PRINTERS

Wohlers Associates estimates that Stratasys, Z Corp., 3D Systems, Objet Geometries, Envisiontec, and Solidimension sold $74.6 million worth of 3D printers in 2004. In unit sales, the combined total was an estimated 1,970 machines. The estimate represents an unprecedented increase over the year before.

The graph in Figure 1 shows the growth of 3D printer sales from 1996 through 2004. 3D printers now represent 37.9% of all additive fabrication systems installed worldwide, up from 30.7% in 2003 and 25.8% in 2002.

In new product development, communication is critically important to success. Engineering drawings provide necessary information such as dimensions, section views, and details. Shaded renderings offer a view of parts and assemblies that most people appreciate, especially when the design is complex. However, when the design is described through prints and renderings, people can misinterpret the design and often do not fully understand how the product will look, feel, and function. While developers of CAD systems have created impressive design aids, these tools are no substitute for the tactile and visual feedback provided by touching and studying a physical model of a design.

Constar, a plastic packaging design and manufacturing company, uses its Z Corp. machine to clearly depict the finite element analysis (FEA) results to its customers. Before using it, Constar attempted to describe the FEA results through words and 2D color prints. This method often left Constar's clients confused and disappointed since they could not fully

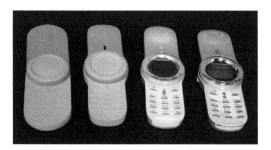

Figure 3. Unfinished (left) and finished (right) ski boot buckle parts, courtesy of Techologia & Design.

Figure 2. 3D printed concept models (left), courtesy of Motorola.

understand the need for design changes. With color models from its Z Corp. system, Constar has an effective method of communicating both the problem and the solution. The 3D representation of the FEA results allows the client to fully understand the need for design modifications, and in the end, results in a stronger customer relationship and a better product. Go to http://wohlersassociates.com/constar.html to see color examples.

During the design phase of its V70 cell phone, Motorola used 3D printer technology to cut development time and cost. Early in the design phase, when various concepts were under consideration, the company produced concept models for review by the design, engineering, and marketing departments. Motorola found that the physical models allowed everyone involved in the review to better understand the designs. In turn, the company got better feedback that resulted in a better product. When compared to previous projects that did not use additive fabrication, Motorola found that the development time was reduced by 50%. Working with team members around the globe, 3D printing allowed the company to achieve real-time, global collaboration around the clock.

3 RAPID MANUFACTURING

Rapid manufacturing (RM) has a promising future, with a compelling list of potential benefits. With RM, tooling is eliminated, thus reducing substantial time and cost. However, there are other powerful advantages that result from the absence of tooling, including increased design freedom, heterogeneous materials, custom products, just-in-time production, and decentralization of production.

The principal advantage of additive fabrication is the ability to construct models and prototypes of virtually any complexity without the need for tooling or machining. When this principal is applied to manufacturing processes, the opportunities for product design and manufacturing are immense.

Rapid manufacturing can be applied to more than just end-use parts. In fact, stereolithography (SL) and fused deposition modeling (FDM) have both been applied to the creation of parts for the manufacturing process. Northhrop Grumman has made more than 700 tools (hand tools, not molds or dies) to support the manufacturing, repair, and servicing of aircraft. In one case, the company produced an aircraft repair kit that included metal shaping tools, guides, and gauges. Built entirely from SL, this field kit allowed the reshaping of an inaccessible component without tearing down a section of the aircraft.

At an FDM user's manufacturing plant, a pulley on a production line belt sander cracked, taking the belt sander out of commission. The cracked pulley was made of aluminum, but to avoid any more downtime, the fabrication manager decided to try a polycarbonate part from the FDM Titan. After a month of operation, the pulley was still going, and the fabrication manager was in no hurry to replace it with its aluminum counterpart.

Techologia & Design of Italy used laser sintering to manufacture small production runs of a ski boot buckle. The design required repeated loading and impact while maintaining high aesthetic standards. The organization produced the parts in PA 2200 (Nylon 12) on an EOSINT P 380 machine from EOS.

The plastic buckles passed impact testing, as well as fatigue testing of 2,000 cycles at a load of 70 kg. Polishing was automated using abrasive tumbling. The finished parts are indistinguishable from parts manufactured using the conventional process.

The cost to produce one set of buckles was €80, which sounds like a lot, until you compare it to the cost of tooling and molding. Tooling cost would have been €48,000 and molding cost would have been €2 per set. The breakeven point was 600 sets. In other words, it was less expensive to use laser sintering to produce fewer than 600 sets of buckles. If the objective was to produce more than 600 sets, tooling and molding would have been less expensive.

The industry is currently in a transitional phase. Additive fabrication, in spite of its limitations, is being used for low-volume production and custom parts. For RM to succeed and flourish across a much wider spectrum of applications, the limitations of additive fabrication must be addressed—limitations such as speed, surface finish, repeatability, material properties, and material cost.

4   THE FUTURE

The industry is on the brink of monumental change. No longer just a technical curiosity, additive processes have become ingrained in product development around the world. It is a rare company that has not employed the technology and enjoyed its benefits. Rather than reaching a plateau of maturity, the industry seems ready to surge forward with new processes, new materials, and new systems that will lead to new customers and new applications.

The technology has developed into three basic categories: 3D printing for concept modeling; mainstream rapid prototyping for fit and function applications, as well as master patterns; and the rapid manufacture of finished series production parts. As these categories and "subindustries" develop, the machines and their manufacturers will become much more specialized and sophisticated.

In the years to come, 3D printing will capture a significant portion of its potential user base and will eventually experience the slowed growth that comes with maturity. Meanwhile, RM will experience double-digit growth. Rapid prototyping will be caught in the middle, as 3D printing and RM systems on both sides perform the prototyping function. As developments in medicine, microsystems, art, and science grow, new classes of additive processes will emerge.

REFERENCE

Selected sections of this paper were taken from *Wohlers Report 2005*, a 250-page worldwide progress report on the rapid prototyping, tooling, and manufacturing state of the industry. Visit http://wohlersassociates.com to learn more.

Virtual modeling and rapid manufacturing – Bártolo (eds)
© *2005 Taylor & Francis Group, London, ISBN 0 415 39062 1*

Rapid Prototyping: A review

I. Gibson

The University of Hong Kong, Hong Kong

ABSTRACT: This paper discusses the current status of layer-based manufacturing Rapid Prototyping (RP) technology and how it is currently being implemented as a tool for product development (PD). A discussion on RP for PD is given, focusing on the limitations of existing technology. Future trends for RP development are then discussed with further consideration for software issues in future applications and how the technology is being accepted worldwide.

1 INTRODUCTION

Rapid Prototyping (RP) represents a range of technologies that permit the automated fabrication of physical objects directly from virtual 3D CAD data without significant process planning related to part features and geometry. RP machines work by simplifying complex 3D problems into a series of simpler 2D problems. Thin, 2D 'layers' are combined or 'added' together to form complete 3D objects. The fact that this is an automated process removes considerable amounts of time and manual effort, whilst making it possible to deal with the demand for increasingly complex geometries and shorter product development times. This capability has fuelled considerable investment in order to improve RP technology and increase the range of applications.

This paper aims at analyzing RP technology to determine its current status in order to understand how it may develop in the future. Firstly, it is important to discuss why RP is an important product development tool in order to appreciate why it attracts such interest from manufacturers. Secondly, anyone who uses RP technology must realize that there are limitations to how it works and so these limitations are discussed in detail. By addressing these issues, it is inevitable that the range of applications for RP will expand and this paper goes on to discuss the key focus areas for RP technology development in the future. The steady expansion of this technology sector can be seen in the numerous RP associations that have sprung up around the world and the worldwide institutionalization of RP is also discussed here.

RP is a tool that has been found to be vital in speeding up the process of product development (PD). Early RP systems were developed as a result of 3D CAD being able to provide output representations of complete product designs using surfaces and solids. Whilst early 3D CAD systems could provide such output, they were cumbersome to use and the model data was still difficult to properly visualize. RP technology at the time was relatively slow and inaccurate compared with today, but it still provided an effective mechanism for models to be evaluated in the early stages of design. Now, CAD systems are increasingly easy to use and the rendering quality can be very high, obviating the need for RP for many such visualization tasks. However, whilst RP is still used for concept modeling, improvements in the technology (speed, accuracy, cost, etc.) have opened up other applications further downstream in the PD process.

The advantages in using RP as part of the PD process can best be explained in figure 1, which has been adapted from D.T. Pham's definition of Time Compression Engineering (Pham & Dimov, 2003). Obviously this is a somewhat stylized representation, focusing on the point that technological solutions can now provide you with an opportunity to perform tasks concurrently, thus saving time in the process. It is clear from this diagram that there is a need for significant levels of data and platform sharing to provide a mechanism for iteration and feedback. RP fits into this framework very well in numerous places. Firstly, RP can assist in consolidating the conceptual design stage as shown in point ① of figure 1. RP can also be used to assist in the analysis stage by providing test models ②; the results of tests using these models can be fed back into the detailed design process. Models can also assist the tool designers and fabricators by providing reference to the final parts. They can also assist in the tool-making process by providing patterns for soft tooling ③ and, in some cases, hard tooling that can be used to advance the product release date in the form of a small batch manufacturing bridge

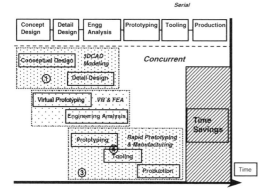

Figure 1. Time Compression Engineering, showing the contribution of RP to this process (after Pham & Dimov, 2003).

prior to full-scale manufacture ④. These additional uses for RP have resulted from improvements in the technology, moving further downstream in the product development process to reduce the time to market.

In his annual state of the industry reports, Terry Wohlers has catalogued recent development of RP technology (Wohlers, 2004). These reports have shown numerous examples of how RP has been implemented to enable the concurrent processes illustrated in figure 1. However, even with multiple uses for RP technology, it is interesting to note that the number of machines is still only around 15,000 units worldwide. Compared with the hundreds of thousands of companies developing new products around the world, this can only represent the tip of a very large iceberg in terms of the overall market potential. This is notwithstanding the possibility of completely revolutionizing the way manufacturing is carried out in what is becoming termed as Rapid Manufacture (RM). That subject will be discussed later on in this paper.

2 LIMITATIONS OF CURRENT COMMERCIAL RP TECHNOLOGY

As already mentioned, the basic approach used by all commercial RP systems is layer-based fabrication. There are many different ways in which the part layers can be made and consequently there are a number of different RP machines and manufacturers. All of these machines have limitations according to the following categories:

Layer thickness: The thinner the layers, the more accurate the part will be. However, the typical value for many machines is 0.1 mm layer thickness, which is still too thick for many applications. The best layer thickness commonly available is around 0.02 mm, but

users must be made aware that reduction in layer thickness means more expensive machines and a slower build process. All RP parts exhibit a characteristic 'stair step' texture, which is most evident on sloping surfaces and therefore manual surface finishing is required for most RP models that are made.

Part accuracy: In addition to layer thickness, there are a number of other accuracy issues that affect the building of parts. In particular one can expect an RP machine to have a minimum wall thickness that is particularly relevant for shell-type parts. Normally this will be a few millimeters. Repeatability of RP processes is generally good (in the order of a few microns), but part shrinkage due to material and process constraints can lead to tolerances in the few tenths of a millimeter range. In addition, many RP processes require overhanging and isolated regions to be supported. These supports must be removed at a later period. Normally, where the supports were in contact with the fabricated model, there is a surface roughness that again affects the overall part accuracy.

Part size: RP is often described as a geometrically independent process. Geometrical independence is only true up to a certain extent. As well as the need for supports, as previously mentioned, the working envelope of the machine restricts the maximum dimensions of a single model. Many RP machines are in the order of 300 mm cubed and the largest are around 500 mm cubed. Some applications may therefore require construction of parts in sections or for them to be made to scale in order that they may fit inside the machine.

Materials: Much of the part fabrication process in RP is dependent on the ability to bond layers together. This forces severe constraints on the materials suitable for a particular process. The majority of RP processes build models from polymeric materials since these provide satisfactory material properties for bonding and maintaining good part accuracy without the need to resort to very high temperatures and/or forces.

Part strength: Since parts are built in layers, which are then bonded together in some way, it is likely that these bonded interface regions represent weaknesses in the overall structure. Even within the material range of a particular process, it is commonly found that the mechanical strengths of parts made are slightly inferior to parts made with the same material using other manufacturing processes (e.g. injection moulding).

Speed: RP is not as 'rapid' as many people think and would like it to be. Parts generally take a matter of hours, to perhaps a couple of days to fabricate, depending on the chosen process and size of part. Whilst this is a significant improvement on conventional model making approaches (with the addition of improved accuracy, material properties, etc.), there is always a demand for further increase in speed. Many potential users may not be prepared to wait for models to be made in this time frame.

Cost: Of course, the capacity to create models quickly, accurately and reliably using RP technology must come at a price. RP technology is still something of a novelty and machines are generally constructed in small volume production. Many of the higher end machines are over US$200,000. Having said that, the prices of all machines are steadily coming down and smaller machines that are focused more at the concept modeling sector, generally with limited part properties, are approaching a cost similar to many high-end computer products (i.e. as low as US$50,000). Many concept-based applications can be addressed using these lower-cost machines.

System integration: Currently, RP machines are generally expected to operate as stand-alone systems. Whilst most machines are networked to assist file transfer, machine setup and part removal requires a significant amount of manual labour and they are not considered to be fully automated manufacturing processes.

Work is being carried out by system manufacturers and researchers to address all of the above points so that better parts can be made to suit a wider range of applications. For example, systems are being developed that can make parts with micron-scale layer thicknesses. Other machines are focusing on producing parts with a wider variety of materials, namely ceramics, bio-materials and metals. Ink-jet printing technology has been widely used in more recent RP machines, with the major benefits of controllable precision and increased speed of build. Many of these technologies are aimed at the ultimate goal of RM.

3 FUTURE TRENDS

From figure 1, it is quite easy to see why RP technology that has aroused the interests of many manufacturers and researchers. Whilst it is a technology in common use in industry, it is still relatively new and requires a significant change in philosophy on how it is to be implemented as mainstream manufacturing technology. RP is a direct link to CAD without the need to focus specifically on the fabrication process, thus providing a mechanism to produce parts automatically. This section discusses the various ways in which RP is developing in order to overcome one or more of the limitations mentioned above.

3.1 Development of existing technology

Obviously, RP vendor companies wish to reduce the cost of their machines in order to make them more competitive and widespread. Incremental improvements of the RP machines include reduction in layer thicknesses, faster scanning speeds, and larger build volumes. Improvements in materials include the ability to withstand higher temperatures, reduce shrinkage and produce higher resolution parts. Cost reduction is primarily achieved by streamlining the production process and selling more machines. Whilst vendors have taken advantage of improvements in associated technologies (e.g. computer systems with greater data processing capabilities), the machines bought today employ the same techniques developed when RP started to emerge in the late 1990's. What might be termed 2nd generation RP technology is now starting to encroach on the commercial market. These are the results of RP research being carried out with RP technology in the last 10–15 years and are described briefly in the following sections.

3.2 Rapid Tooling (RT)

RT is an objective that RP system manufacturers have been trying to improve upon for many years. The basic rationale used by most RT approaches employs RP parts as patterns for casting. In 'soft-tooling' the pattern is used in conjunction with rubber moulding technology to provide short production runs. This has been successful commercially for a number of years and companies routinely use this approach to test new products prior to decisions on making significant financial (and time) commitment to large-scale production.

'Hard-tooling' based on RP, on the other hand, is still a relatively unfulfilled objective. A number of companies have introduced methods for creating steel (or other metal) parts, based on RP technology. An example process is Laserform (3D Systems, 2005), owned by the most well-known RP vendor; 3D Systems. In this case a metal/polymer composite is created in the RP machine and further furnace processing results in a fully dense metal part. This, and similar processes, can result in good quality parts. However, the process is considerably more involved than just using a single RP machine and requires additional capital investment.

An alternative approach appropriate to all RP technologies is to create an investment-casting pattern using RP. This approach requires the manufacturer to have access to foundry facilities with an understanding that the RP parts being used may not behave in exactly the same way as conventional wax or foam parts. Most tooling applications would like parts to be very accurate, with tight tolerances and good surface finish so that the resultant moulded parts are of high quality. Most RP parts can therefore only be near-net shape, thus requiring conventional machine-tool processing in most cases to achieve acceptable accuracy and surface finishes.

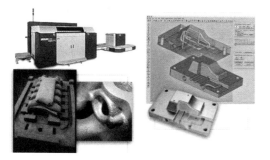

Figure 2.   The ProMetal system that can be used to directly fabricate metal parts and mould inserts (ProMetal, 2005).

More recently there have been a number of direct metal RP systems (like the ProMetal system shown in figure 2 (ProMetal, 2005)). Nearly all direct metal RP parts are made using a powder feed approach, using laser or similar beam technologies to locally melt the powder. Use of powders means there is generally no problem with supports, which would be very difficult to remove from a metal part. However, there will always be a powdery surface to the part, which will require additional machining for tight-tolerance and surface-critical features. Upward-facing surfaces are often considerably better defined than downward-facing ones. However, this is not usually a problem when using direct metal RP for making mould inserts. Direct manufacture of moulds using RP can result in what might be called 'optimized' or 'high performance' tooling. The layer-based approach makes it possible to include internal features into the mould. Most specifically, conformal cooling channels that follow the mould surface can be included. Such channels can result in more evenly cooled moulded parts, which will improve part tolerance. In addition, control over material feed during the melting process makes it possible to vary the material properties. This can lead to more durable moulds or better thermal conductivity properties. These multiple material capabilities for RP will be discussed in a later section.

RT is therefore an achievable goal for RP but requires process integration in order to make it work effectively. Soft tooling exploits the benefits of RP very well. With advancements in high-speed machining, the benefits of hard tooling may be difficult to justify in terms of cost. For RP to be of benefit in this area, the technology must provide something 'extra' in terms of overall performance.

3.3   Rapid Manufacture (RM)

Much of the future of RP lies in the ability to directly manufacture parts. In RM, the justifications resulting from changes in product development are stacked in favour of RP. Product life cycles are becoming shorter, designs are becoming more complex, and markets are demanding greater variety and customized features in many products. Polymers are the material of choice for most products, which is also the most common RP material. Decisions for RM are therefore made in terms of cost, time and volume. If a small number of complex geometry parts is wanted in a short period of time, then RP can become RM. Numerous case studies can now be found to demonstrate this fact.

A good example of RM comes from Phonak, who produce hearing aids (Phonak, 2005). The NemoTech process uses a combination of reverse engineering of ear impressions, specialized software, and the Selective Laser Sintering (SLS) RP technology. This makes it possible to produce custom hearing aids that fit inside the outer-ear canal. The fit is very comfortable for users, who claim improved, high quality performance. To make individual moulds for each hearing aid would be time consuming and cost-prohibitive, so RP technology comes out as the only viable solution for this stage in the manufacturing process. Note that this is a form of semi-customisation since most of the components (in particular the electronics) are the same. RP therefore gives this particular product an edge that allows a generic system to be suited to individual customer's needs.

Wohlers cites a number of similar cases in his report (Wohlers, 2004), including studies carried out at the University of Loughborough in the UK, which analysed the cost implications of using RP as a manufacturing method in place of conventional injection moulding. There are a numerous caveats concerning adopting this approach, but on cost alone RP can become RM for any technologically suitable production run up to a few thousand parts.

3.4   Multiple material systems

Returning to the discussion on RT, an interesting characteristic of some RP systems is the capability to produce parts with multiple material properties. With RP being a layer-based technology, it has long been the concern of system manufacturers to reduce the layered effect. Although most evident in the form of a characteristic 'stair-stepping', many RP parts also exhibit mechanical, surface and other volumetric defects that will vary according to the position, orientation and other ways in which the part was built within the machine. Obviously this is a problem, but if the behaviour can be predicted then this heterogeneity also represents a potential advantage over other manufacturing processes.

A number of RP processes, like the LENS system from Optomec (Optomec, 2005) and the SDM process developed at CMU (Liu, 2004), have the capability to vary the feed of different materials into the build region. Since parts are made by 'selectively' building up material, there is the opportunity to control how that material is added in order to produce the following effects:

Composite materials: By mixing filler materials together but without alloying them, it is possible to make composite material structures taking advantage of the constituent material properties. This can lead to stiffer, harder, or more heat resistant parts but the planar nature of the layers still represents a major source of mechanical weakness. If the layers do not represent the directionality of, for example the stiffness required using carbon fibre-reinforced components, then the performance of the part will be compromised. The development of the curved-LOM process illustrates how this can be overcome to a certain extent by curving the layers rather than using planar additive fabrication (Klosterman et al., 1999).

Functionally Gradient Materials: If a beam is built with steel on one side and copper on the other, the interface between the two materials represents a significant problem. Physical bonding of the materials plus thermal and chemical effects will lead to stress, bending, cracking and corrosion. Instead of having a clear junction between the two materials, grading them by varying the ratio across the boundary can alleviate some of these problems. Injection moulds for example can have the durability of steel at the mould surface, whilst benefiting from the heat transfer properties of the copper to assist in the cooling process. Gradient blending can reduce the possibility of delamination. Of course there are still material affinities to deal with relating to temperature, chemistry, etc. that would have to be solved and therefore not all materials can be mixed in this way.

Multiphase perfect materials: Taking the previous examples to an extreme, RP represents a technology that can produce multiphase perfect materials (Chen et al., 2004). This is a process of choosing material compositions and geometries to meet the requirements of a particular performance-related task. For example, animal bone has a combination of materials combined in specific ways to solve the task for supporting the body. There are dense fibrous cortical bone regions to provide support and stiffness in particular load directions. There are less dense cancellous bone regions that assist the mechanical behaviour of the bone with other features that provide nutrients to the system. Furthermore there are layers of cartilage at the joint interfaces that provide smooth and wear-resistant motion. RP systems do not have such capabilities as yet, but the capability to specify the material properties in different regions can be exploited to this effect.

Development of multiple material capabilities represents a significant area of current RP research. As RP systems develop in this way, there could be many repercussions; in particular in terms of how parts are designed. Incorporating heterogeneity into a component can affect numerous attributes like mass, strength, stiffness, etc. Parts may therefore have to be carefully optimized to suit a specified application. Some of the consequences of this will be discussed in later sections on software and system integration.

3.5 *Micron-scale RP*

Another area of great excitement is that of micro- and nano-scale fabrication. There are many researchers who are developing technologies focused on creating objects on a nano-scale. However, we cannot really say that these are RP technologies since they require significant amounts of design and planning and technical constraints. All this goes somewhat against the RP principle of 'plug and play' where the designer does not have to worry too much about the fabrication process when creating the design. Micro-scale RP, on the other hand, is very viable.

The only RP technology that has been successfully scaled down to a micron level is stereolithography (SLA). The advantage of this process is that the liquid resin photoreactive polymer can be presented at a very low viscosity prior to polymerization. This makes it possible to generate very thin layers (around 10 microns). In-plane resolution is achieved by the use of very finely focused optic systems, which can even be in the form of masking technology (Tan & Gibson, 2004). However, it is not merely a case of reducing the size of all the machine components. For example, very small parts will have very fragile features. Mechanical spreading of the resin to ensure the layer is flat cannot be easily done for fear of damaging the part. Similarly, since it uses liquid resins, SLA is a technology that requires support structures. Removal of parts from the build platform as well as removal of the supports themselves represents a significant problem when the parts may only be 1 or 2 mm in size. The support geometry is therefore significantly different from conventional SLA.

However, there is increasing interest in this technology to assist the electronics (for connectors, waveguides, etc.), jewelry and optics industries, etc. The demand is unlikely to be large compared with the rest of the RP industry, but one can envisage a lucrative service sector; fabricating prototype micro-parts and tooling, like the ones produced by Bertsch et al. with their system developed at EPFL in Lausanne (Figure 3, taken from Bertsch, et. al., 2000).

3.6 *Non-manufacturing applications*

RP was a technology that was initially developed to solve the problems of manufacturing industry. The types of materials, machine dimensions and processing speeds are generally compatible with product

11

Figure 3. Microstereolithography parts made at EPFL (Bertsc et al, 2000).

development as it is carried out by manufacturers of typical products like consumer goods, consumer electronics, automotive, etc. However, there are a number of potential applications outside of this industry:

Medicine: Already a number of discussions in this paper have alluded to applications in medicine. Applications in medicine are not merely confined to medically-related manufactured products like the Phonak hearing aids. RP models have also proven to be extremely successful aids to diagnosis and planning for complex surgical cases. For example models are routinely used in planning the separation of conjoined twins (Christensen et al., 2004). Many applications are therefore created from individual patient data. However, it is interesting to note that whilst linking RP to medical imaging data (CT, MRI, etc.) seems a logical and natural development of this technology, there are numerous cost, speed and materials (specifically biocompatibility) issues that must be resolved before it is widely used in this application sector (Gibson et al, 2004).

Architecture: The main issue surrounding the use of RP for architecture is effective communication. Models are routinely used at various stages of the architectural design process (Gibson & Kvan, 2002). However models, by definition, can never become the final product. Buildings are too large and too complex to be made using RP (although a system is currently under development that uses RP techniques for construction of full-scale buildings (Khoshnevis, 2005)). Architects therefore make decisions on what details they wish to focus on by using models to make certain statements and assess certain aspects of the

Figure 4. Architectural models, showing different scaling, which therefore have different feature definitions.

design (for example in the scale models of figure 4). As a result, they would like an RP system that has the aesthetic capability of representing multiple material (although not necessarily mechanically). Currently there is one colour RP system commercially available from the ZCorp Company (ZCorp, 2005). Whilst this technology is improving in its ability to represent colours, multiple material representation is not just about colour, but also texture, translucency, etc.

Art: Using RP for artistic modeling (sometimes referred to as digital sculpturing (Digital Sculpturing, 2005)) is not the same as in the previous example of architecture. Here, artists are attracted to RP because it provides a different artistic media compared to others (like clay, marble, etc.). As well as the aesthetic impact of RP materials, it has been recognized for many years that the technology has the capability of creating parts that would be difficult or even impossible to fabricate any other way. Objects trapped inside other objects, fine detail and precisely defined surfaces are all relatively easy to create using RP. RP has therefore become a particularly popular tool for artists who base there designs on mathematical equations (Dickson, 2005).

Certainly there will be new applications that can benefit from RP technology in the future (for example providing spare parts for space exploration vehicles whilst on a journey to Mars). One possible conclusion of this is to see the obviation of manufacturing processes as we know them. Fabrication of new products may eventually end up being performed using RP machines in the home, where application is limited only by the imagination of the user.

4 INTEGRATION OF RP WITH OTHER TECHNOLOGIES

Obviously CAD is the entry point for any product development system involving RP. RP is driven directly from the CAD file of the product generated. This CAD representation must consist of fully-enclosed surface model data or else the resulting RP part may be compromised. In the past this caused problems for early surface modeling systems and produced best results from solid modeling packages. Nowadays, RP is an accepted output route for most CAD systems and part quality is rarely affected by the package used. With more widespread use of RP in manufacturing, attention has been directed towards effective integration with other technologies. There are numerous software solutions to assist in running RP systems as noted in (Gibson, 2002). Two of the key areas where there is considerable activity are discussed in the following sections.

4.1 *Representations*

The *de facto* standard for RP model representation is the STereoLithography (STL) file. This is a simple facetted representation that uses triangles to approximate the model surfaces. The size of triangles used can be varied according to the maximum allowable deviation from the actual surface. If this deviation is close to the RP machine accuracy limits then faceting as a result of the STL file would not normally be noticed on the resulting part. The STL file represents something of a lowest common denominator that makes it relatively easy to make good quality output regardless of the RP machine used. Slicing is easy to perform by just calculating the intersection of the appropriate layer plane with the model triangles. It is possible to use the additional information of triangle facet direction to determine the internal and external surfaces, but this is more commonly calculated by looking at the nested curves generated in 2D. RP machines have their own operating systems with slicing algorithms that are process dependent. Some RP technologies require calculation of supports as well as vectors that drive the scanning mechanisms. Whilst it may be more efficient to slice the model directly from the original CAD file, this would cause problems with model file transportability, thus supporting the widespread use of STL as an intermediate file format. However, the STL file poses some constraints that may ultimately limit the development of RP technology, for example:

Feature based systems: There are systems that can adapt some of their parameters according to the model being built. For example, some systems can vary the layer thickness so that vertical surfaces can be manufactured using thick layers whilst more complex curved and sloping surfaces can be fabricated using thinner layers. The objective here is to optimize build speed without compromising build accuracy; thick layers being quicker whilst thin layers more accurate. Other systems have been developed that

will not necessarily build the components in a purely homogeneous, layer-wise fashion (e.g. Shape Deposition Modeling (Liu et al., 2004) and Contour Crafting (Khoshnevis, 2005)). In these cases the STL file may not be the best way to determine the build process, requiring greater knowledge of the design than a simple geometric representation. Being able to break the model up into sections, using a feature-based approach may therefore benefit development of this kind of RP technology.

Heterogeneous systems: STL is a surface representation. The systems under development that have heterogeneous properties cannot use existing STL files since they are defined in terms of the solid structure. Even the colour system from ZCorp cannot properly create a coloured model using STL files alone. A modified version of STL can apply a colour tag to each triangle, but there is no real reason why triangles should be just one colour. VRML has been used and can be directly applied to the model, but so far only limited implementations exist as discussed in (Gibson & Ling, 2001), which illustrates the difficulty in this area. Most probably there is a need for a new standard to deal with heterogeneous solid modeling, some initial work on which is described in (Gibson, 2002). This new standard should have the capability to properly model the RP process within CAD. This can take advantage of the functional benefits of a heterogeneous structure and incorporate it into the design process; probably with finite element analysis of the design in this loop as well to permit feedback and optimization.

4.2 *Technology management*

RP technology is designed to operate for long periods of time unattended whilst the build process is being carried out. Totally ignoring the system can however result in dire consequences as a build can go wrong and many hours of operation can be wasted. Good control software ought to have a top level system that monitors the quality of the output regularly so that the machines are kept in good condition and parts are produced reliably. This may be expensive to achieve automatically so some operators use remote camera monitoring systems (Axis, 2005). It is also interesting to note that a fast, reliable build time estimator is still not commonly available for most machines.

Whilst RP is viewed as a form of automation, there are in fact quite a number of manual tasks that must be carried out. Machines need to be set up. Parts must be unloaded from the machines. Post-processing of parts can require a lot of manual labour if they are for presentation purposes. Similarly parts used in additional process chains generally require manual treatment. In addition, material management is often an important issue, requiring the operators to keep an eye on the part quality as a measure of how good the raw material is. Unacceptable part quality can mean a need to replace the raw material. This means that operating RP technology is a mixture of good machine skills coupled with good manual skills. Such rare individuals who possess both qualities must not only be well rewarded but kept productive. Since RP is supposed to be a versatile technology, the process planning software ought to reflect this versatility. This is issue is described in detail in (Gibson, 2002).

5 RP IN A GLOBAL PERSPECTIVE

We live in what is increasingly being referred to as the Global Marketplace. Mobile communications, cable and satellite systems, and high speed Internet means that offices need not be in the same building and people can work on the move and no longer be restricted to using the same channels of communication for every task. This has coupled with a tendency to use outsourcing, downsizing companies, customer networking systems, supply chain management, etc., to produce highly flexible ways of doing business. Suppliers need not be close by, or customers, making it possible to seek out the best quality and lowest prices for components and the best markets for products.

There is often discussion about how different parts of the world are becoming similar to each other. In fact global market development *requires* different parts of the world to be different. As well as geographical differences (climate, size, raw materials, etc.) there are political and cultural differences that affect the choice of a particular part of the world for a particular product or way of doing business. Obviously labour costs are a critical issue. Combined with the comparatively high level of training and education, the low-cost labour forces in China, for example, present an extremely lucrative region for manufacturers to exploit.

Cultural, social, and language differences represent huge problems in communication of ideas. With a new product concept, the designer may have a very clear view. However, during the detailing, testing, tool design, manufacture, etc., this view may become confused or even lost. It may be easy to rectify this if all the product development stages were under one roof, with all personnel speaking the same language and with the same cultural background. With global enterprises this will not be the case. The concept of 24-hour product development is now quite common (Hirschberg, 2005). Many companies have design offices in different parts of the world where a product may be worked on 24 hours per day. To have design in Europe, detailing in the US, and manufacture in Asia is nothing unusual. English may be the most common language to use but it will not be the first language of

many involved in the project, which in turn can lead to misinterpretation and misunderstanding.

Designers must therefore consider designing for different markets. For effective product development, they must have an understanding of local cultural differences and requirements. Communicating these concepts back to the designer is not easy. Anything that provides an effective means of communicating ideas across boundaries of distance, language and thinking will avoid mistakes that require time, effort and cost to settle. RP can assist greatly to achieve this.

5.1 *RP development worldwide*

It may be interesting to note that different RP systems are popular in different parts of the world. To understand this, it is necessary to discuss how RP technology developed worldwide. Cultural, political and economic factors have all had parts to play in this development. We can perhaps consider how RP has grown up in different key regions:

USA: Whilst there is some debate as to who invented the first RP machine, there is no doubt that the first commercial success was stereolithography developed by 3D Systems in the US. These machines gained significant market share in the early years of RP. The first users in automotive and aerospace were able to justify the capital costs in a very short time. This provided the impetus for market growth in the US for local manufacturers 3D Systems, DTM and Stratasys. This network of machines made it more difficult for the newer 3D printers to gain market share. These newer machines therefore had to be much lower in price and focus on the niche concept modeling application arena in order to gain a market foothold. Aggressive pursuance of patent rights meant the local market was dominated by US-made machines. However, this still represented a considerable diversity of technologies, the US being the major source of innovation in RP. It therefore remains the primary market, giving US manufacturers a significant time advantage in implementing new RP technology.

Europe: There were also a number of different RP technologies developed in Europe. In essence, Europe represents the greatest diversity of RP technology because there were fewer patent concerns allowing imports of machines from all parts of the world. However, the impression was that Europe was always struggling to catch up with implementation. This may be due to the relatively small scale of the manufacturers when compared with the US; many European companies preferring to develop distributor networks through SMEs rather than the huge conglomerated companies of the US. This makes implementation of new technology and capital outlay a slower process. Europe also seemed to be characterized in terms of how RP was implemented, with a specific focus on rapid tooling and manufacturing. This was generally promoted through the European R&D funding networks. Concept modelling companies of course did make use of RP, but again because the companies were generally smaller the market did not take off until the lower cost machines became more readily available.

Asia: Asia is now the second biggest market for RP technology after the US. However, the potential for market growth is huge, because the majority of the current market is in Japan. Japan has a predominance of SLA-type (photocurable resin) technologies because a number of companies developed similar machines at the same time. Japanese companies normally concentrate on the home market first and then overseas. By the time companies were ready to exploit the overseas market, 3D Systems had already covered most of the world with patent protection of SLA. China already represents a huge market with forecasted high growth in manufacturing maintaining for many years to come. To have only 6% of the total number of RP machines in the world is testament to how fast this market has grown and is of great concern to the RP manufacturing industry. Chinese manufacturers build products for a huge number of foreign designer companies. Most of these companies use digital design methods and really need to ensure their designs are being dealt with correctly. There has been considerable resistance, mainly due to the comparatively low cost of highly skilled manual labour being able to manufacture prototypes using traditional methods. However, increasing costs of living add to the increasing levels of complexity of each new product to make it increasingly important for manufacturers to implement this technology. There are a number of Chinese-made RP machines. However, the machine quality and level of support of foreign-made machines currently is far superior. Other parts of Asia are also developing quickly.

Others: RP is being used in many parts of the world and a number of other significant points are worth mentioning here:

Starting with Cubital SGC, there have been a number of systems developed in Israel. The most successful of these is currently the Objet Company. Israel appears to have a good record for funding technological innovation.
South Africa has recently developed a keen interest in applying RP. Some novel areas include architecture and reproduction of tribal artifacts in addition to the more conventional uses.
There is a very active artist community that uses RP as an artistic medium. Sculptors regularly come together from all over the world to exhibit their art at the Intersculpt exhibition (Digital Sculpturing, 2005).

Australia, New Zealand, Brazil, India, Finland, along with other countries that have a small and geographically dispersed population and industry, are all developing their own local RP communities. This serves to illustrate that RP can assist in bringing industries together to make them more effective on a global scale.

5.2 GARPA

GARPA (GARPA, 2005) was set up as an informal way for the original 12 national RP organizations in the world at the time to link together. This number has now risen to 19, with at least 2 other countries currently developing their own associations and waiting to become full GARPA members. GARPA remains informal in structure; member associations are not required to pay a fee and the only real requirement is for members to demonstrate that they are the main organization that represents RP in the country and do not actively exclude individual RP users. To explain the diversity of GARPA, it is worth noting:
US representation is through the Society of Manufacturing Engineers. This is a huge organization that involves hundreds, if not thousands of members. Similarly, in the UK the RP association is part of the IMechE. Hong Kong on the other hand has a small number of representatives from industry, distributors, vendors, and academics as a steering committee to promote RP locally and has no formalized general membership structure. This is also similar to the Chinese Laser Rapid Forming Committee.

In Germany, RP is considered as part of the NC machinery association, whilst in other countries special associations were created.
In Australia, the RP community was formed out of a consortium that involved the majority of users in the country at the time.

There are many ways to set up an organization involving RP within a country based on the number of users, the types of users, the requirements of an organization and the political and social system in place. The aim is to promote RP and GARPA assists in the following ways:

By endorsing meetings and conferences: Often such meetings have a GARPA session that shows how RP is implemented around the world.
GARPA has a web site at www.garpa.org. This is a hub for GARPA members to present activities that may be of interest to users in other parts of the world.
GARPA collects RP application case studies from member representatives to distribute in the public domain.

There is a GARPA summit meeting every year, where chairmanship of GARPA is passed from one country to another.
GARPA has a permanent secretariat that can be accessed through the web site, which keeps members informed of activities occurring throughout the world.

As the RP community grows, GARPA will grow with it. Informal methods of communication will give way to a more formalized structure as the benefits of alliance increase. GARPA already is a valuable resource to many member organizations. It is a way to lobby for local support and gain recognition at the formative stages. Small communities can have a say on the global development of RP and members can access the wealth of experience that is available around the world. Larger organizations can also benefit in terms of access to new markets and establishing new contacts.

6 CONCLUSIONS

RP is a technology that can be used for many different applications, both manufacturing and non-manufacturing based. It can enhance and optimize the product development process. Whilst there are still many outstanding technological issues surrounding development and application of RP technology, it has already proved that it can be a valuable addition to the range of automated systems available to manufacturers.

This paper has presented an overview of RP technology, followed by discussion on the contribution it can make to product development. It has then gone on to discuss various limitations of the technology and how researchers are attempting to address these. Finally, there has been a brief discussion on how software needs to be developed in order to further enhance the technology.

It can be concluded that RP has made a good introduction and has a bright future in making PD more effective and efficient. Rapid Tooling assists this in current applications, but one should maintain a watching brief on the development of the exciting area of Rapid Manufacturing, which is ultimately set to revolutionize the way we manufacture products to meet the demands of modern consumers.

It can also be concluded that RP is an accepted technology for widely distributed companies and/or companies involved in producing for a global market. By facilitating the transfer of design intent through model data it is much easier to communicate without making costly errors. RP has grown up in different ways around the world and development is now being increasingly supported by local associations of users and suppliers. Assistance in establishing and supporting these associations can be achieved through GARPA, which has already linked many countries together to help provide institutional support for this technology.

REFERENCES

3D Systems, (2005), LaserForm, http://www.3dsystems.com/products/sls/laserformoven/datasheets.asp,.

Axis, (2005), Axis 2100 network camera, http://www.axis.com/products/cam_2100/, Axis Communications.

Bertsch A., et al., (2000), Rapid prototyping of small size objects. Rapid Prototyping Journal 6 (4), pp. 259–266.

Chen K.Z., Zhu F. and Feng X.A. (2004), Generating layered manufacturing information for components made of a multiphase perfect material from their CAD models, *Proceedings of CAD04*, Pattaya Beach, Thailand, May 24–28, 2004.

Christensen A., Humphries S., Goh K. and Swift D. (2004), Advanced 'tactile' medical imaging for separation surgeries of conjoined twins, *Childs nervous system, monograph on conjoined twins*, Springer-Verlag, Heidelberg, Sept. 2004.

Dickson S. (2005), Mathematics-based sculptures, http://emsh.calarts.edu/~mathart/portfolio/SPD_portfolio_summary.html.

Digital sculpturing, (2005), http://intersculpt.org/, Intersculpt and Ars Mathmatica.

GARPA, (2005) – http://www.garpa.org/

Gibson I. (2002), *Software solutions for rapid prototyping*, Prof. Engg. Pub., ISBN 1 86058 360 1, 2002.

Gibson I. and Kvan T. (2002), The Use of Rapid Prototyping for Architectural Concept Modelling, *SME Technical paper no.PE02-222*, Society of Manufacturing Engineers, 2002.

Gibson I. and Ling W.M. (2001), Colour RP, *Rapid Prototyping Journal*, vol. 5(4), 2001, pp. 212–216.

Gibson I. et al., (2004), 'The use of Rapid Prototyping to assist medical applications', Proc. 10th Assises Europeennes du Prototypage Rapide, Paris, France, 14&15 Sept., 2004, file s5_1_aepr.pdf

Hirschberg U. et al., (2005), 24-hour design – http://faculty.washington.edu/brj/Publications/caadria99.PDF

Khoshnevis B. (2005), Countour crafting http://www-rcf.usc.edu/~khoshnev/RP/CC/Contour%20Crafting.htm,

Klosterman D., et al. (1999), Direct Fabrication of Polymer Composite Structures with Curved LOM, *Proc. Solid Freeform Fabrication Symposium Proceedings*, University of Texas at Austin, Austin, TX, August, 1999, pp. 401–409.

Liu Hao-Chih, et al. (2004), RP of Si_3N_4 burner arrays via assembly mould SDM, *Rapid Prototyping Journal*, Vol. 10, No. 2, 2004, pp. 239–346.

Optomec Co. (2005), LENS, http://www.optomec.com/,

Pham D.T. and Dimov S.S. (2003), Rapid manufacturing – technologies and applications, *Proc. Int. Conf. on Manufacturing Automation*, Prof. Engg. Pub., ISBN 9 86058 376 8, 2003, pp. 3–22.

Phonak Co. (2005), NemoTech – Digital Mechanical Processing, http://www.phonak.com/professional/productsp/nemotecheshell.htm.

Prometal Co. (2005), Prometal direct metal RP system, http://www.prometal-rt.com/

Tan W., Gibson I. (2004), 'Microstereolithography and application of numerical simulation on its layer preparation', Proc. Int. Conf. on Manufacturing Automation (ICMA), Wuhan, PRC, Oct. 26–29, 2004, pp. 675–68.

Wohlers, T.T. (2004), *Wohlers Report 2004*, Wohlers Associates, 2004, http://www.wohlersassociates.com/

ZCorp (2005), Colour 3D Printing, http://zcorp.com/

Virtual modeling and rapid manufacturing – Bártolo (eds)
© 2005 Taylor & Francis Group, London, ISBN 0 415 39062 1

Rapid prototyping in tissue engineering: A state-of-the-art report

Chee-Kai Chua, Wai-Yee Yeong & Kah-Fai Leong
School of Mechanical & Aerospace Engineering, Nanyang Technological University, Singapore

ABSTRACT: This article provides a comprehensive and state-of-the-art review of the application of rapid prototyping in the field of tissue engineering. These advanced techniques serve as a promising candidate for the production of tissue scaffold with much potential yet to be explored. Challenges of scaffold-based technology are discussed from the perspective of rapid prototyping technology.

1 INTRODUCTION

1.1 *Background*

Tissue engineering aims to produce patient-specific biological substitutes, which may circumvent the limitations of existing clinical treatments for damaged tissue or organs. These limitations include shortage of donor organs, issue of chronic rejection and cell morbidity. The main regenerative tissue engineering approaches include injection of cells alone, development of encapsulated systems and transplantation of cells in scaffold (Sonal *et al* 2001). The last approach appears to be the dominant method in the research of tissue engineering for the flexibility that permits user intervention at three levels to achieve optimal construct, namely the cells, the polymer scaffolds and the construction method (Jennifer *et al* 1998).

1.2 *Scaffold-based tissue engineering*

The three primary roles of scaffold are: (1) to serve as an substrate for cell adhesion and facilitate the delivery of cells when it is implanted, (2) to provide temporary mechanical support to the newly grown tissue and (3) to guide the development of new tissues with the appropriate function (Kim *et al* 2001).

Successful scaffolds should have the following characteristics (Leong *et al* 2003): (1) A suitable macrostructure to promote cell proliferation and matrix production. (2) An open-pore geometry with a highly porous microstructure. (3) Optimal pore size for tissue regeneration with minimum pore occlusion. (4) Suitable surface morphology and chemistry to encourage intracellular signaling and attachment of cell. (5) A predictable material degradation profile with non-toxic degradation product.

Scaffolds can be produced in a variety of ways using conventional techniques or advanced rapid prototyping (RP) methods, which is the focus of this article. The RP techniques can be classified as (1) melt-dissolution deposition techniques and (2) Particle bonding techniques.

2 MELT-DISSOLUTION DEPOSITION TECHNIQUE

2.1 *Working principles*

In this technique, an injector extrudes a filament of material while it moves across the build platform, forming a layer of the cross section of the part. Successive layer formation, one atop another, forms a complex 3D solid object (Chua *et al* 2003).

A porous structure is achieved by drawing the filament at a specific pattern. XYZ gap is produced by depositing the filament at an interval space and at an angle with respect to the previous layer. A demonstrative system using melt deposition is the Fused Deposition Modeling, FDM. Several new systems that operate under similar principle are presented in this section too.

2.2 *FDM process*

In FDM, roll of filament is fed and melted inside a heated liquefier before being extruded through a nozzle. The extruded filaments cool and solidify to the previous layer, forming a 3D structure.

Researchers have successfully fabricated functional scaffold directly using FDM. The materials adopted were polycaprolactone (PCL) (Zein *et al* 2002) and polypropylene-tricalcium phosphate (PP-TCP) (Samar *et al* 2003). In a study, human mesenchymal progenitor cells (hMPCs) were seeded on PCL and PCL-HA scaffolds fabricated by FDM (Endres *et al* 2003). Proliferation of cell toward and onto the scaffolds surfaces was detected.

The choice of material in FDM is very limited since the input material must be made into filament and melted into semi-liquid phase before extrusion. Temperature-sensitive natural biomaterial and bio-molecules is incompatible for this melting process thus limiting the biomimetic aspects of the scaffold produced. Furthermore, the solidified filaments form a dense surface with no microporosity throughout the structure; this will further reduce the effectiveness of the scaffold.

2.3 *Variants of FDM process*

In view of the above mentioned limitations, new techniques were developed to eliminate the requirement of precursor filament. Some variants of the FDM process include 3D fiber-deposition technique (Woodfield *et al* 2003), precision extruding deposition (Wang *et al* 2004) and precise extrusion manufacturing (Zhuo *et al* 2001).

2.3.1 *3D Fiber-deposition technique*
The material is in pellet or granule form that can be input into the heating liquefier directly. The feasibility of the machine was demonstrated by the fabrication of Poly(ehyelene glycol)-terephthalate-poly (butylenes terephthalate) (PEGT/PBT) block co-polymer scaffolds for articulate tissue engineering applications. The scaffolds produced contain smooth fiber approximately 250 μm in diameter and a fiber spacing ranging between 0.5 and 2.0 mm. The mechanical properties reported are similar to native articular cartilage explants. In vitro cell seeding studies showed homogenous cell distribution after 5 days of dynamic seeding (Woodfield *et al.* 2003).

2.3.2 *Precision Extruding Deposition (PED)*
Extruder of this system is equipped with a built-in heating unit to melt the feedstock material. PCL scaffold with pore size of 250 μm was fabricated. Preliminary result of biological experiment showed that the pore size of 250 μm is conducive for fibroblast cell growth without the need for a filler material (Wang *et al* 2004).

2.3.3 *Precise Extrusion Manufacturing (PEM)*
Similar to the previous described system, the sprayer of this system is equipped with a built-in heating unit to melt the feedstock material, which can be in pellet form. PLLA scaffold with controllable porous architectures from 200 to 500 μm in size was produced (Zhuo *et al* 2001).

To eliminate the elevated operating temperature in the flow of processing, researchers replaced the melting process with that of dissolution of material. Systems developed include low-temperature deposition manufacturing (Zhuo *et al* 2002), multi-nozzle deposition manufacturing (Zhuo *et al* 2003); pressure assisted microsyringe (Vozzi *et al* 2002) and robocasting (Cesarano & Calvert 2000).

2.3.4 *Low-temperature Deposition Manufacturing (LDM)*
As suggested by the name of the process, no heating process is adopted during the liquefying process of materials. The material supply is liquid form. The scaffold building cycle is performed in a low temperature environment under 0°C. PLLA/TCP pipe scaffold is produced. Macropore dimension of the scaffold produced is about 400 μm in size. Microporosity at the scale of 5 μm is created from the sublimation of solvent during freeze-drying (Zhuo *et al* 2002).

2.3.5 *Multi-nozzle Deposition Manufacturing (MDM)*
This is an improvement of LDM with enhanced material opportunity. The enhancement is achieved by incorporating more than one jetting nozzle into the system. Different liquid materials are fed into individual extrusion/ jetting nozzles, which are specially designed to compensate on the different material properties. Support structure can be built using water. This technique also allows the incorporation of biomolecules into the scaffold during building cycle. In the work by Zhuo *et al* biomolecules bone morphogenic protein (BMP) was suspended in water and being sprayed into the scaffold during its forming process. The biomolecules were released slowly from the bulk material as the scaffold degraded.

2.3.6 *Pressure Assisted Microsyringe (PAM)*
A microsyringe is used to expel the dissolved polymer under low and constant pressure across the substrate surface to form desired pattern. The resolution of this method is remarkably high at cellular scale. Vozzi developed PCL and PLLA scaffolds with line width of 20 μm. Hexagons morphology with side of 250 μm was also reported. The performance of this method is comparable to soft lithography. However, capillary with very small diameter requires careful handling to avoid tip breakage. Higher pressure is needed to expel the material out of the small orifice.

2.3.7 *Robocasting*
This patented system is able to lay down highly concentrated colloidal suspension that is pseudoplastic-like (Cesarano & Calvert 2000). Therriault *et al* fabricated 3D microvascular network by robocasting fugitive organic ink, followed by scaffold infiltration with

epoxy resin and further post processing (Therriault et al 2003).

In general, the techniques described above are meant to fabricate hard tissue scaffold. Landers and Mülhaupt has developed an aqueous system, 3D bioplotter (Landers & Mülhaupt 2000), producing hydrogel scaffold to meet the demand for soft tissue engineering. Hydrogels is a cross-linked polymer matrix that exhibits the ability to swell in water without dissolving and to retain a large amount of water within its three-dimensional structure, simulating the mechanical properties to those of many soft tissues in the human body. Ang et al developed a robotic dispenser, RPBOD, for the fabrication of chitosan scaffold (Ang et al 2002). A multi-nozzle deposition system for the construction of 3D biopolymer scaffolds was reported recently (Khalil et al 2005).

2.3.8 3D Bioplotter

The principle of this method is dispensing material into a liquid medium with matched density. The material then solidifies in the plotting medium after bonding to the previous layer. Hydraulic force in the liquid medium provides support to any overhang feature printed and hence no support structure is needed. It is possible to incorporate bioactive components such as cells, growth factors and drugs into the material during building cycle.

Landers et al fabricated hydrogel scaffolds that had interconnected pore at the scale of 200–400 μm (Landers et al 2002). However, the hydrogel presented a smooth surface, which require further surface coating to render it favorable for cell-adhesion. Besides, the scaffold had limited resolution and mechanical strength.

2.3.9 Rapid Prototyping Robotic Dispensing System (RPBOD)

This system with a one-component pneumatic dispenser was designed by Ang et al. Liquid material was dispensed into a dispensing medium through a small nozzle. Chitosan scaffolds with pore size of 400–1000 μm were produced in the preliminary study.

2.3.10 Multi-nozzle deposition system

This system is capable of depositing material containing live cells. The examples shown are based on alginate solutions and PCL. Further characterization of the construct was not reported (Khalil et al 2005).

3 PARTICLE BONDING TECHNIQUES

3.1 Working principles

In this technique, input powder particles are selectively bonded to form a 2D layer. The layers are then bonded one upon another to form a complex 3D solid object. During fabrication, the object is supported by and embedded in unprocessed powder. Therefore this technique permits the fabrication of through channels and overhanging features.

These techniques are capable of producing a porous structure with both macroporosity and microporosity. The microporosity arises from the space between the individual granules of powder. Macroporosity is achieved by manipulating the region of bonding or by mixing of porogen into the powder bed before the bonding process.

The fabricated scaffold yields a characteristic rough surface, which could be advantageous for cell growth since studies have shown that topographical cues might have significant effect upon cellular behavior (Flemming et al 1999). From the attached rough surface, receptors on the cell surface might be subjected to varying degrees of deformation that led to activation of cell signal transduction pathway.

3.2 Example of systems

Typical systems in this category include 3-Dimensional Printing™ (Lam et al 2002), TheriForm™ (Griffith & Naughton 2002) and Selective Laser Sintering (Calvert & Weiss 2001).

3.2.1 3-Dimensional Printing™ (3DP™)

In this process, a stream of adhesive droplets is expelled through an inkjet printhead, selectively bonding a thin layer of powder particles to form a solid shape (Chua et al 2003). The resolution achieved is about 300 μm. Kim et al employed 3DP™ with a particulate leaching technique to create porous scaffolds using PLGA mix with salt particles and a suitable organic solvent. Pores formed were at the scale of 45–150 μm with 60% porosity. In vitro cell culture with hepatocytes (HCs) showed in-growth of HCs into the pore space (Kim et al. 1998).

To improve the biocompatibility of the process, Lam et al formulated a blend of starch-based polymer powders that can be bonded together using distilled water (Lam et al 2002). However, post-processing step like infiltrating the structure with polymer solution was needed to enhance the mechanical property of the scaffold.

3.2.2 TheriForm™

This system is similar to 3DP™ where a printhead assembly deposits binder droplets onto selected regions of the powder, dissolving and binding the polymer powder in the printed regions.

Zeltinger et al performed a study on a TheriForm™ built PLLA scaffold with different pore size using three different cell types namely fibroblasts, vascular smooth muscle cells and microvascular epithelial cells.

Figure 1. Sintered PEEK/HA powder.

Their result suggested the existence of an optimal pore size for different cells type.

3.2.3 *Selective Laser Sintering (SLS)*

The machine uses a deflected CO_2 laser beam to selectively scan over the powder surface following the cross-sectional profiles carried by the slice data. The laser power melt increase the local temperature to glass-transition temperature, causing surfaces in contact to deform and fuse together (Chua *et al* 2003). It is the preferred forming process to produce complex porous ceramic matrices for bone implant. (Vail *et al* 1999).

Recent work done illustrates the ability to design and fabricate PCL scaffolds using SLS. The scaffolds have sufficient mechanical properties for bone tissue engineering applications (Williams *et al* 2005).

The authors' group has successfully sintered polyetheretherketone-hydroxyapatite (PEEK/HA) powder blends on a commercial SLS machine (Tan *et al* 2003). Different weight percentage compositions of physically mixed PEEK/HA powder blends were sintered by varying the machine settings (Figure 1).

4 INDIRECT SCAFFOLD-FABRICATION METHODS BY RP

RP systems can also be utilized indirectly to fabricate tissue-engineering scaffolds. These multi-step methods usually involve casting of material in a mold and then removing the mold to obtain the final scaffold. Such techniques allow the user to control both the external and also the internal morphology of the final construct. In addition, indirect methods also require less raw scaffold material while extending the choice of material. Further advantage include well-conservation of original material properties as no heating process is imposed on the scaffold material.

4.1 *Examples*

Some examples of indirect RP techniques employed include melt deposition (Bose *et al* 2003), droplet deposition (Sachlos & Czernuska 2003) photopolymerization (Chu *et al* 2001) and 3D Printing (Lee *et al* 2005).

4.1.1 *FDM*

Bose *et al* produced alumina and β-tricalcium-phosphate (β-TCP) ceramic scaffolds with pore sizes in the range of 300–500 μm and porosity of 25–45% (Bose *et al* 2003). The molds were made using standard thermoplastic polymer and removed by binder burnout. The experiment aimed to understand the influence of porosity parameters such as pore size and pore volume on mechanical and biological properties.

4.1.2 *ModelMakerII™(MMII™)*

This machine uses a single jet each for a plastic build material and a wax-like support material. The printerhead ejects droplets of the materials as they are moved in X-Y fashion. After an entire layer of the object is hardened, a milling head is passed over the layer to achieve the specified uniform thickness.

Taboas *et al* produced PLLA scaffolds with micro and macro porosity for trabecular bone tissue engineering (Taboas *et al* 2003). Channels of 500 μm wide were computationally designed while the 50–100 μm local pores were formed by porogen. PLA/PGA discrete composites were made using melt processing.

Limpanuphap and Derby fabricated TCP scaffolds with controlled internal porosity using a suspension of TCP in acrylate binder (Limpanuphap & Derby 2002). A similar route was used to produce polymer/TCP scaffold, which is believed to show more potential for cell adhesion. Wilson *et al* fabricated HA scaffolds with a defined macro-architecture (Wilson *et al* 2004). Channels achieved are approximately 350–400 μm wide. They suggested that the inherent surface texture obtained from a MMII built mold increased the surface area of the scaffold and led to higher degree of calcium and phosphate release, which is advantageous to bone formation. This technique is selected by researchers to fabricate the calcium phosphate scaffold because the building material has a very low coefficient of thermal expansion. There will be minimal risk of fracture due to thermal mismatch with the ceramic during pyrolysis.

Sachlos *et al* have successfully produced collagen scaffolds with predefined internal channels (Sachlos *et al* 2003). The smallest channel width achieved was as low as 135 μm. In a variation to Sachlos's work, the author produced chitosan-collagen scaffolds. MMII™ molds with intricate channels were built to contain chitosan-collagen solution (Yeong *et al* 2004). The resultant gel-like scaffold was capable of attaining the designed morphology (Figure 2).

Figure 2. Replication of mold morphology with flow channels (left) onto scaffold (right).

These channels can be tailored to serve as flow channels for efficient perfusion of culture medium when coupled with a customized bioreactor.

4.1.3 Stereolithography (SLA)

A UV laser traces out the model's cross-section areas of photo-curable resin, solidifying only the part area. The elevator then drops enough to introduce another layer of liquid resin. The laser traces the second layer atop the first to form a 3D part.

Chu et al have produced HA based porous implants using SLA built epoxy molds. Thermal curable HA/acrylate suspension was cast into the mold to obtain a scaffold with interconnected channels. The resolution of channel width achieved was as low as 366 μm diameter (Chu et al 2001).

In another study, investigators showed that it is possible to control the overall geometry of the regenerated bone tissue through specific architectural design of the scaffolds. An in vivo study was performed using two different architecture designs, orthogonal and radial channels (Chu et al 2002). In the orthogonal design, regenerated bone conformed to the shape of the channels while in radial design; bone formation was centered at the periphery of the central column.

4.1.4 3DP™

A multi-step process involving a printed mold and sucrose porogen was developed to produce PLGA scaffold with villi features (Lee et al 2005). The villi features produced are at the scale of 500 microm diameter and 1 mm height. Anatomically shaped zygoma scaffolds with 300–500 microm interconnected pores were produced and characterized. This method provides an alternative method to complement other direct RP fabrication methods.

5 CHALLENGES OF RP IN TISSUE ENGINEERING

5.1 Material issue

5.1.1 Material processability

Each RP technique requires a specific form of input material such as filament, powder, solid pellet or solution. Therefore, the material selected for the scaffold must be compatible with the chosen process.

5.1.2 Degradation rate

The scaffold should be remodeled by growing cells and gradually replaced by the newly formed extracellular matrix and differentiated cells. In ideal case the polymer degradation rate is synchronized with the rate of tissue ingrowths. Therefore, the degradation properties of a scaffold are of crucial importance for the success of scaffold-based tissue engineering.

The degradation-absorption mechanism is the compound result of many factors that include hydrophilicity of the polymer backbone, degree of crystallinity, presence of catalysts, volume of porosity and the amount of surface area. Balancing each of these factors will allow an implant to degrade slowly while transferring stress at an appropriate rate to surrounding tissues as they heal. The degradation-absorption mechanism of a 3D scaffold is one of the major challenges facing TE research today.

5.1.3 Degradation product

Even though degradation products of biodegradable polymers are known to be largely non-cytotoxic, little information is available regarding the rate-dependent degradation effect of the scaffold. Sunga et al found that fast degradation of the polymer affects negatively the viability and migration of cell in vitro and in vivo (Sung et al 2004). The negative impact can be explained by the rapid local acidacation due to the polymer degradation. A more systematic investigation approach is needed to classify the material degradation profile.

5.1.4 Material rigidity of scaffold

Material rigidity was shown to influence cell spreading and migration speed as demonstrated by Wong et al (Wong et al. 2003). Cells displayed preference on stiffer region and tended to migrate faster on these regions.

The rigidity of the scaffold also influences the anchorage of cells. Cells are very sensitive to the mechanical properties of the adhesion substrate. The integrin binding, assembly of focal adhesion plaques and cytoskeleton are regulated accordingly to the rigidity of the substrate (Wang et al 2001). The material should maintain an appropriate degree of rigidity to resist the tractional forces generated by the assembling cytoskeleton; as well as sufficient compliancy to allow the attached cells to reorganize and recruit the receptors into focal adhesion plaques.

5.2 Structural factor

5.2.1 Pore size

Despite numerous proof-of-concept studies exhibiting the existence of optimal range of pore size for different cell type (Ranucci et al 2000; Dalton et al 1999;

Figure 3.   Discontinuity of structure morphology.

Bignon *et al* 2003) little is known about the exact value of the optimal pore size for a specific type of cell. The current selection of pore size is governed by general empirical guidelines.

5.2.2   *Scaffold morphology*
RP fabricated scaffold generally presents many edges and groves (Figure 3). The effect of these discontinuities in topography might affect the adhesion and migration of cell as shown by Yin's work. Yin *et al* grew cardiac cell on microgrooved elastic scaffolds to investigate the topography-driven changes in cardiac electromechanics (Yin *et al* 2004). They demonstrated direct influence of the microstructure on cardiac function and susceptibility to arrhythmias via calcium-dependent mechanisms.

5.2.3   *Surface roughness and morphology*
The surface roughness of the scaffold is important in cell-matrix interaction. The rough powder surface produced from powder based RP technique may enhance cell adhesion. However, if the surface is too rough, the cells adhering to these surfaces may not be able to develop distinct focal adhesion plaques or bridge over the irregularities. Besides, the protruded surface might damage the cell mechanically. In certain RP systems such as FDM and bioplotter, further surface modification or coating is required to ensure firm cell adhesion.

5.3   *Bioactivity of RP fabricated scaffold*
The interaction of cell on the scaffold is governed by both structural and chemical signaling molecules. These molecules play a decisive role for cell adhesion and further behavior of cells after initial contact (Bacakova *et al* 2004).

The extent of initial cell adhesion decides the profile of focal adhesion plaques formed on cell membrane, which subsequently describe the size, and shape of cell spreading area. The extent of spreading is crucial for further cell dynamics such as migration, proliferation and differentiation behavior of anchorage dependent cells.

Current strategies to control the proliferation and other behavior of cells is by patterning the material surfaces with adhesive molecules or by direct-loading of biomolecules such as natural growth factors on the scaffold (Rai *et al* 2005). Some RP systems that operate at low temperature such as MDM, and bioplotter allow the incorporation of biomolecule during part building cycle.

However, further information such as the type of biomolecule, the optimal concentration and spatial control of these biomolecules is needed to produce a most favorable scaffold. Furthermore, the degradation profile of the scaffold would affect the release kinetics of the loaded molecules.

5.4   *Cell seeding and vascularization of scaffold*

Bioreactors technologies serve as a very important enabling technology in scaffold-based tissue engineering to guarantee the localization of uniformly distributed cells throughout the scaffold at an efficient density. Bioreactors are generally defined as devices in which biochemical processes develop under tightly controlled operating conditions (Martin *et al* 2004).

Current FDA approved cell seeding process, using petri-dish, is incompetent in delivering cell deep inside the scaffold with uniform distribution (Li *et al* 2001; Xiao *et al* 1999; Kim *et al* 1998). The successful cellularization of a 3D scaffold is therefore closely associated to the development of bioreactor technologies. Examples of bioreactor technologies are spinner flask, perfusion cartridge and rotary cell culture system (Goldstein *et al* 2001). Each of the system utilizes different physics principles and hence requires specific design consideration on the scaffold shape and strength.

RP systems provide a high degree of freedom in scaffold design and development. Interconnected flow channels can be designed to fit into the operation of the bioreactor as displayed by the work of Sakai *et al*.

Rapid and sufficient vascularization of the scaffold is essential to maintain adequate perfusion in engineering large tissue constructs. Current paradigm to achieve vascularization is by incorporating growth factor into the scaffold or by transplanting endothelial cells on scaffold. Experimental studies have confirmed that the vascularization of matrices is accelerated with endothelial cell transplantation (Holder *et al* 1997). The bioactivity of the scaffold plays a crucial role in this approach.

RP fabrication method offers the flexibility and capability to couple the design and development of a bioactive scaffold with the advances of cell-seeding technologies in order to ensure the success of scaffold-based tissue engineering.

6 NEW DEVELOPMENT AND POTENTIAL

6.1. *Automated design, development and characterization*

RP as a group of systematic techniques holds the potential of automating the design and fabrication of patient-specific scaffolds. In the work of Cheah *et al* automated scaffold assembly algorithm was developed to design scaffold internal architectures from a selection of open-celled polyhedral shapes (Cheah *et al* 2004). The program can be interfaced with various RP technologies to achieve automated production of scaffolds.

Since the RP process offers complete user-control on the structural features of the scaffold, it is hence possible to characterize the scaffold using computer-aided characterization. The approach can be used to estimate the effective mechanical properties of scaffolds and also to investigate the effect of design and process parameters on the structural properties of the scaffolds. Fang *et al* characterized the effective mechanical properties of porous PCL scaffolds manufactured by PED using a computational algorithm for finite element implementation and numerical solution of asymptotic homogenization theory (Fang *et al* 2005).

6.2 *Scaffold as a learning tool*

Scaffold fabricated using RP techniques can serve as a learning tool to study the cell-matrix interaction since the parts obtained are of high dimensional accuracy and high reproducibility. Effects of material rigidity, surface roughness and morphology, pore size and structure design can be investigated independently to understand cell physiological behavior.

The concept of layered manufacturing techniques also extends into the research of direct organ production. These new technologies are organ printing (Mironov *et al* 2003; Wilson & Boland 2003; Boland *et al* 2003); laser printing of cells (Ringeisen *et al* 2004) photopatterning of hydrogel (Valerie & Sangeeta 2002) and microfluidics technology (Tan & Desai 2004).

6.3 *Organ printing*

Boland *et al* developed a cell printer that is capable of printing single cells, cell aggregates and the supportive thermoreversible gel. A tubular collagen gel with bovine aortal endothelial cells was printed.

6.4 *Laser printing of cells*

A device termed matrix-assisted pulsed laser evaporation direct write (MAPLE DW), was used to deposit micron-scale patterns of pluripotent embryonal carcinoma cells onto thin layers of hydrogel. Cell viability of 95% was reported (Ringeisen *et al* 2004).

6.5 *Photopatterning of hydrogels*

Photolithographic techniques was borrowed from the silicon chip industry. In this process, a UV light is shone through a patterned template atop a thin film of polymer-cell solution, curing the exposed polymer that sets with cells inside. Different template can be used to achieve a complex 3D structure with region of different cells.

6.6 *Microfluidics technology*

A layer-by-layer microfluidic method was reported to build a 3D tissue-like structure inside microchannels (Tan & Desai 2004). This approach is an extension of the 2D cell patterning technique into the vertical axis. The technology encompasses immobilization of a cell–matrix assembly, cell–matrix contraction, and pressure-driven microfluidic delivery processes.

7 CONCLUSION

Rapid prototyping technologies hold great potential in tissue engineering especially in the context of scaffold fabrication. The technologies offer high degree of freedom in the design, fabrication and modeling of the scaffold being constructed. RP technologies are dynamic ones; new techniques and new apparatus are being developed globally in the effort to bring the technologies closer to the goal of engineering tissue.

REFERENCES

Ang, T.H. *et al* (2002) Fabrication of 3D chitosan-hydroxy-apatite scaffolds using a robotic dispensing system. *Materials Science and Engineering* C 20, 35–42

Bacakova, L. *et al* (2004) Cell Adhesion on Artificial Materials for Tissue Engineering. *Physiol. Res.* 53, S35–S45

Bignon, A. *et al* (2003)Effect of micro- and macroporosity of bone substitutes on their mechanical properties and cellular response. *J. Mater. Sci. Mater. Med.* 14, 1089–1097

Boland, T. *et al* (2003) Cell and organ printing 2: Fusion of cells aggregates in three-dimensional gels. *Anat. Rec. Part A* 272A, 497–502

Bose, S. *et al* (2003) Pore size and pore volume effects on alumina and TCP ceramic scaffolds. *Mater. Sci. and Eng.* C 23, 479–486

Calvert, J.W. and Weiss, L. (2001) Assembled scaffolds for three-dimensional cell culturing and tissue regeneration. US Patent 6,143,293

Cesarano, J. and Calvert, P. (2000) Freeforming objects with low-binder slurry, US Patent 6,027,326

Cheah, C.M. et al (2004) Automatic Algorithm for Generating Complex Polyhedral Scaffold Structures for Tissue Engineering. *Tissue Eng.* 10, 595–610

Chu, T.M.G. et al (2001) Hydroxyapatite implants with designed internal architecture. *J. Mater. Sci. Mater. Med.* 12, 471–478

Chu, T.M.G. et al (2002) Mechanical and in vivo performance of hydroxyapatite implants with controlled architectures. *Biomaterials* 23, 1283–1293

Chua, C.K. et al (2003) *Rapid prototyping principles and applications.* World Scientific

Dalton, B.A. et al (1999) Modulation of corneal epithelial stratification by polymer surface topography. *J. Biomed Mater Res.* 45, 384–394

Endres, M. et al (2003) Osteogenic Induction of Human Bone Marrow-Derived Mesenchymal Progenitor Cells in Novel Synthetic Polymer-Hydrogel Matrices *Tissue Eng.* 9, 689–702

Fang, Z. et al (2005) Computer-aided characterization for effective mechanical properties of porous tissue scaffolds. *Computer-Aided Design* 37, 65–72

Flemming, R.G. et al (1999) Effects of synthetic micro- and nano-structured surfaces on cell behavior. *Biomaterials* 20, 573–588

Goldstein, A.S. et al (2001) Effect of convection on osteoblastic cell growth and function in biodegradable polymer foam scaffolds. *Biomaterials* 22, 1279–1288

Griffith, L.G. and Naughton, G. (2002) Tissue engineering-current challenges and expanding opportunities. *Science* 298, 1009–1014

Holder, W.D. et al (1997) Increased vascularization and heterogeneity of vascular structures occurring in polyglycolide matrices containing aortic endothelial cells implanted in the rat. *Tissue Eng.* 3, 149–160

Jennifer, J.M. et al (1998) Transplantation of cells in matrices for tissue regeneration *Adv. Drug Deliv. Rev.* 33, 165–182

Khalil, S. et al 2005 Multi-nozzle Deposition for Construction of 3D Biopolymer Tissue Scaffolds, Rapid Prototyping Journal 11, 9–17

Kim, B.S. et al (1998) Optimizing seeding and culture methods to engineer smooth muscle tissue on biodegradable polymer matrices. *Biotechnol. Bioeng.* 57, 46–54

Kim, B.S. et al (2001) Development of biocompatible synthetic extracellular matrices for tissue engineering. *Trends Biotechnol.* 16, 224–229

Kim, S.S. et al. (1998) Survival and function of hepatocytes on a novel three-dimensional synthetic biodegradable polymeric scaffold with an intrinsic network if channels. *Ann. Surg.* 228, 8–13

Lam, C.X.F. et al (2002) Scaffold development using 3D printing with a starch-based polymer. *Mater. Sci. and Eng.* C 20, 49–56

Landers, R. and Mülhaupt, R. (2000) Desktop manufacturing of complex objects, prototypes and biomedical scaffolds by means of computer-assisted design combined with computer-guided 3D plotting of polymers and reactive oligomers. *Macromol Mater. Eng.* 282, 17–21

Landers, R. et al (2002) Fabrication of soft tissue engineering scaffolds by means of rapid prototyping techniques. *J. Mater. Sci.* 37, 3107–3116

Landers, R. et al (2002) Rapid prototyping of scaffolds derived from thermoreversible hydrogels and tailored for applications in tissue engineering. *Biomaterials* 23, 4437–4447

Langer, R. and Vacanti, J. (1993) Tissue engineering *Science* 260, 920–926

Lee, M. et al 2005 Scaffold fabrication by indirect three-dimensional printing. Biomaterials 26, 4281–4289

Leong, K.F. et al (2003) Solid freeform fabrication of three-dimensional scaffolds for engineering replacement tissues and organs. *Biomaterials* 24, 2363–2378

Li, Y. et al (2001) Effects of filtration seeding on cell density, spatial distribution, and proliferation in nonwoven fibrous matrices. *Biotechnol. Prog.* 17, 935–944

Limpanuphap, S. and Derby, B. (2002) Manufacture of biomaterials by a novel printing process. *J. Mater. Sci. Mater. Med.* 13, 1163–1166

Martin, I. et al (2004) The role of bioreactors in tissue Engineering. *Trends in Biotechnology* 22, 80–86

Mironov, V. et al (2003) Organ Printing: computer-aided jet-based 3D tissue engineering. *Trends Biotechnol.* 21, 157–161

Rai, B. et al (2005) Novel PCL_based honeycomb scaffolds as drug delivery systems for rhBMP-2. *Biomaterials* 26, 3739–3748

Ramakrishna, S. et al (2001) Biomedical applications of polymer-composite materials: a review. *Composite Science and technology* 61, 1189–1224

Ranucci, C.S. et al (2000) Control of Hepatocyte Function on Collagen Foams: Sizing Matrix Pores for Selective Induction of 2-D and 3-D Morphogenesis. *Biomaterials* 21, 783–793

Ringeisen, R.B. et al (2004) Laser printing of pluripotent embryonal carcinoma cells. *Tissue Eng.* 10, 483–491

Sachlos, E. and Czernuska, J.T. (2003) Making tissue engineering scaffolds work. Review on the application of solid freeform fabrication technology to the production of tissue engineering scaffolds. *European Cells and Materials* 5, 29–40

Sachlos, E. et al (2003) Novel collagen scaffolds with predefined internal morphology made by solid freeform fabrication. *Biomaterials* 24, 1487–1497

Sakai, Y. et al (2004) A novel poly-L-lactic acid scaffold that possesses a macroporous structure and a branching/joining three-dimensional flow channel network: its fabrication and application to perfusion culture of human hepatoma Hep G2 cells. *Mater. Sci. and Eng.* C 24, 379–386

Samar, J.K. et al (2003) Development of controlled porosity polymer-ceramic composite scaffolds via fused deposition modeling *Mater. Sci. and Eng.* C 23, 611–620

Sonal, L. et al (2001) Tissue engineering and its potential impact on surgery. *World Journal of Surgery* 25, 1458–1466

Sung, H. et al (2004) The effect of scaffold degradation rate on three-dimensional cell growth and angiogenesis. *Biomaterials* 25, 5735–5742

Taboas, J.M. et al (2003) Indirect SFF fabrication of local and global porous, biomimetic and composite 3D polymer-ceramic scaffolds. *Biomaterials* 24, 181–194

Tan, K.H. et al (2003) Scaffold development using selective laser sintering of polyetheretherketone-hydroxyapatite biocomposite blends. *Biomaterials* 24, 3115–3123

Tan, W. and Desai, T.A. (2004) Layer-by-layer microfluidics for biomemitic three-dimensional structures. *Biomaterials* 25, 1355–1364

Therriault, D. *et al* (2003) Chaotic mixing in three-dimensional microvascular networks fabricated by direct-write assembly *Nature Materials* 2, 265–271

Tienen, T.G.V. *et al* (2002) Tissue ingrowth and degradation of two biodegradable porous polymers with different porosities and pore size. *Biomaterials* 23, 1731–1738

Vail, N.K. *et al* (1999) Materials for biomedical applications. *Materials and Design* 20, 123–132

Valerie, A.L. and Sangeeta, N.B. (2002) Three-dimensional photopatterning of hydrogels containing living cells. *Biomedical Microdevices* 4, 257–266

Vozzi, G. *et al* (2002) Microsyringe-based deposition of two-dimensional and three-dimensional polymer scaffolds with a well-defined geometry for application to tissue engineering. *Tissue Eng.* 8, 1089–1098

Vozzi, G. *et al* (2003) Fabrication of PLGA scaffolds using soft lithography and microsyringe deposition. *Biomaterials* 24, 2533–2540

Wang, F. *et al* (2004) Precision extruding deposition and characterization of cellular poly-ε-Caprolactone tissue scaffolds. *Rapid Prototyping Journal* 10, 42–49

Wang, N. *et al* (2001) Mechanical behavior in living cells consistent with the tensegrity model. *Proc Natl Acad Sci. USA* 98, 7765–7770

Williams, J.M. *et al* 2005 Bone tissue engineering using polycaprolactone scaffolds fabricated via selective laser sintering. Biomaterials, 26, 4817–4827

Wilson, C.E. *et al* (2004) Design and fabrication of standardized hydroxyapatite scaffolds with a defined macro-architecture by rapid prototyping for bone-tissue-engineering research. *J. Biomed. Mater. Res.* 68A, 123–132

Wilson, W.C. and Boland, T. (2003) Cell and organ printing 1: Protein and cell printers. Anat. Rec. Part A 272A, 491–496

Wong, J.Y. *et al* (2003) Directed movement of vascular smooth muscle cells on gradient-compliant hydrogels. Langmuir 19, 1908–1913

Woodfield, T.B.G. *et al* (2003) Design of porous scaffolds for cartilage tissue engineering using a three-dimensional fiber-deposition technique. *Biomaterials* 25, 4149–4161

Xiao, Y.L. *et al* (1999) Static and dynamic fibroblast seeding and cultivation in porous PEO/PBT scaffolds. *J. Mater. Sci. Mater. Med.* 10, 773–777

Yeong, W.Y., Chua, C.K. Leong, K.F. & Chandrasekaran M. 2004. Rapid prototyping in tissue engineering: challenges and potential *Trends in Biotechnology.* 22, 643–652

Yin, L. *et al* (2004) Scaffold Topography Alters Intracellular Calcium Dynamics in Cultured Cardiomyocyte Networks. *Am J Physiol Heart Circ Physiol.* 287, H1276–85

Zein, I. *et al* (2002) Fused deposition modeling of novel scaffold architectures for tissue engineering applications. *Biomaterials* 23, 1169–1185

Zeltinger, J. *et al* (2001) Effect of pore size and void fraction on cellular adhesion, Proliferation and Matrix Deposition. *Tissue Eng.* 7, 557–572

Zhuo, X. *et al* (2001) Fabrication of porous poly(L-lactic acid) scaffolds for born tissue engineering via precise extrusion. *Scripta Materialia.* 45, 773–779

Zhuo, X. *et al* (2002) Fabrication of porous scaffolds for bone tissue engineering via low-temperature deposition. *Scripta Materialia.* 46, 771–776

Zhuo, X. *et al* (2003) Layered manufacturing of tissue engineering scaffolds via multi-nozzle deposition. *Mater. Lett.* 57, 2623–2628

Virtual modeling and rapid manufacturing – Bártolo (eds)
© 2005 Taylor & Francis Group, London, ISBN 0 415 39062 1

Real virtuality: Beyond the image

P.S. D'Urso
Alfred Hospital Department of Neurosurgery, Melbourne, Australia

ABSTRACT: Biomodelling allows 3D medical imaging data to be translated into an accurate solid facsimile of patient anatomy. 'Real virtuality' is the term coined to describe this visualisation medium. A study of 1000 cases in cranio-maxillofacial surgery, neurosurgery, orthopaedics, vascular surgery and prosthetics was performed. 3D CT, MR, angiography and US data were transferred via DICOM network into a workstation running ANATOMICS BIOBUILD software. The data were edited and converted into build code that directed rapid prototyping systems to manufacture biomodels. Stereolithography, selective laser sintering, thermojet printing, and polyjet printing technologies were investigated. Bureau and hospital based systems were developed and tested. Surgeons have reported that biomodelling is superior to standard imaging in complex cases for diagnosis, surgeon confidence, surgical planning, informed patient consent and surgical outcomes. Surgeons have indicated that biomodelling reduces operation time. A hospital based system has been found to significantly reduce biomodel turn-around time and expand applications into time critical conditions such as trauma. Biomodels have been found to have 5 fundamental applications in medicine. Firstly, as a diagnostic aid, secondly, as a communication tool, thirdly, by allowing surgical simulation, fourthly, by allowing manufacture of customised prostheses, and finally, by enabling stereotaxy. Real virtuality has unique advantages over images which include tactile feedback, user friendliness, implant customisation, realistic surgical simulation, versatility and enhanced patient communication. The main limitations of the technology are reimbursement and the reluctance of RP vendors to meet the needs of the medical market.

1 INTRODUCTION

Surgery is a practical art! As every patient is unique there is a need for the surgeon to attain a specific understanding of the individual's anatomy preoperatively. In modern surgical practice such detailed information is captured by medical imaging. Advances in medical imaging have created ever increasing volumes of complex data. The interpretation of such information has become a speciality in itself and the surgeon at times may be left bewildered as to how to best apply the available information to the practicalities of physical intervention. Three dimensional (3D) imaging has been developed to narrow the communication gap between radiologist and surgeon. By using 3D imaging a vast number of complex slice images can be combined into a single 3D image which can be quickly appreciated. The term 'three dimensional' however is not a truly accurate description of these images as they are still usually displayed on a radiological film or flat screen in only two dimensions. It was the yearning of the surgeon for the most realistic portrayal of data that initiated the evolution of 3D imaging and has now fuelled the development of solid biomodelling.

'Biomodelling' is the generic term that describes the ability to replicate the morphology of a biological structure in a solid substance. Specifically, biomodelling has been defined by the author as 'the process of using radiant energy to capture morphological data of a biological structure and the processing of such data by a computer to generate the code required to manufacture the structure by rapid prototyping apparatus.'[1] A 'biomodel' is the product of this process. 'Real virtuality' is the term coined to describe the creation of solid reality from the virtual imagery.[1] Virtual reality in contrast creates a computer synthesised experience for the observer without a real basis. In medicine, biomodelling has been used to create anatomical real virtuality. Biomodels are a truly remarkable and exciting tool in the practice of surgery and the applications of such a generic technology will be discussed in this chapter.

2 THE BIOBUILD PARADIGM

Anatomics BioBuild (www.anatomics.com) software was specifically developed to integrate all the functionality required to import, visualise, edit, process and export 3D medical image volumes for the production of physical biomodels by way of rapid prototyping (RP). An overriding design goal has always been compatibility with as many different medical scanners, and as many different image formats as possible. Consequently, BioBuild is able to bridge the gap between patient scans and physical biomodels in many varying environments. This has seen BioBuild used in conjunction with CT, MR, 3D ultrasound and 3D angiography data. 3D data sets can be imported into BioBuild in one of several ways. Data can be loaded directly from a series of 'digital image and communications in medicine' (DICOM) files or generic 'raw' image files on the local computer, on a network drive, or a remote DICOM server. Loading data from many different and varied data sources is one of BioBuild's strengths.

Many hospitals now use the DICOM standard for storage and transmission of patient data from multiple imaging modalities. BioBuild is able to access this patient data seamlessly across the hospital network. Patient searches are made even easier by masks, such as the first letter of the patient's name, the patient's hospital ID, or even the patient's doctor.

The general process for producing a physical biomodel from a patient scan is straight-forward. First, the patient data set is imported. The data is then converted into a volume for inspection and processing. After completing the necessary processing, a 3D surface is extracted that represents the physical biomodel. To ensure that all regions of interest have been correctly modelled it is important to first visualise and inspect the surface. Finally, the software model must be exported to a format suitable for physical biomodel production. Anatomics BioBuild software was designed to integrate each of these steps into a single, user-friendly system. Further, these steps form the basis of the BioBuild paradigm for biomodel production.

The process is complicated by the many varying data formats and vendor specific peculiarities that arise when loading data from varying scanners. There are also potentially many different biomodelling applications. Because of this, BioBuild provides the user with a vast array of options for modifying and transforming a volume. Generally however, only a few of those operations are ever likely to be performed on any single volume. The number of possible options can initially intimidate novice users: as there is often more than one way to perform any given task. However, once novice users become accustomed to the BioBuild paradigm, they quickly become comfortable with the interface.

The usual procedural steps required for processing image volumes using BioBuild can be summarised as follows.

- Import and reduce volume, and confirm orientation
- Inspect anatomy and find intensity threshold
- Edit and optimise volume
- 3D visualisation
- Patented RP build optimisation
- RP build file generation

Because BioBuild was designed with ease-of-use in mind, virtually all volume processing can be accomplished via simple point-and-click operations. Although providing advanced editing features, the majority of BioBuild's functionality is accessed through moveable toolbars, composed of simple and intuitive icons. Comprehensive help and tutorials are also available within the software. Step by step instructions and demonstrations assist in training and familiarity with the software. These features significantly reduce the learning curve for novice users, and make common editing operations completely automatic for experienced users.

We have investigated the use of a variety of RP technologies. Biomodels need to be biocompatible, safely sterilised, translucent, multi-coloured and produced quickly in an office friendly environment. Stereolithography (SL) remains the gold standard at present as it does meet most of these requirements except office friendliness. Selective laser sintering (SLS) has significant limitations which although the direct fabrication of biocompatible implants may be a niche application. Thermojet wax biomodels also have the most significant limitations not least of which is fragility. Fused deposition modeling (FDM) is limited by slow build speeds and opacity. 3D printing with extruded binder and powder has advantages of speed and low cost but is limited by opacity, difficulties in sterilization and the messy substrate. Polyjet acrylic 3D printing is a new technology that meets most of the requirements for biomodelling but can only print one colour. We have implemented a hospital based system with a polyjet printer. We believe that this system will become a platform for in house biomodelling in the future.

Our research has allowed the identification of factors that are presently limiting the development of biomodelling technology. The main factor has been the failure of RP vendors to specifically address the medical requirements of medical RP and allow integration of the technology into the hospital environment. If the issues highlighted were addressed by RP vendors we believe that the uptake of the technology would increase significantly.

Anatomics has been addressing some of these limitations by interfacing RP vendor software into the medical imaging environment with BioBuild software. It is anticipated that in future BioBuild will act as a seemless user friendly interface between DICOM and a number of suitable RP systems.

3 SURGICAL APPLICATIONS OF REAL VIRTUALITY

The author initiated research into biomodelling technology in 1991. Anatomics Pty Ltd (www.anatomics.net)

Table 1.

Speciality	Cases
Cranio-maxillofacial	
Craniofacial	16
Maxillofacial	259
Cranioplastic	285
Orthopaedic	
Spinal	105
Pelvic	18
Upper Extremity	4
Lower Extremity	6
Neurosurgical	
Skull based tumour	11
Stereotactic	9
Vascular	41
Basal Ganglia	2
Aortic	5
Ears	14
Bile ducts	2
Chest	3
Fetus	3
Natural Sciences	117
Total	**1000**

was subsequently founded as a research and development company in 1996. This paper will reflect our experience with 1000 patients in a broad range of applications. Such applications have been developed in cranio-maxillofacial surgery, neurosurgery, customised prosthetics, orthopaedics and other miscellaneous specialties. The breakdown of applications is listed in Table 1.

3.1 *Generic applications of biomodelling*

Biomodels were found to have five general applications.[2]

1. As a communication aid for professionals and patients,
2. To assist surgeons with diagnosis and surgical planning,
3. For the rehearsal and simulation of surgery,
4. For the creation of customised prosthetics,
5. As an aid for stereotactic surgery.

3.2 *Cranio-maxillofacial surgery*

The complexity of cranio-maxillofacial anatomy combined with the morphological variation encountered by the reconstructive surgeon makes cranio-maxillofacial surgery a conceptually difficult task in explanation, planning and execution. Cranio-maxillofacial surgeons have shown a high level of affinity for biomodelling, perhaps because they routinely use dental casting and the articulated models produced from such castings. The use of biomodels of the mandible and maxilla is

a natural extension to the use of these dental castings. The biomodels have been commonly used to improve the diagnostic relevance of the data and for surgical simulation.[2–4]

In craniofacial surgery biomodels have been traditionally used by surgeons to gain insight into unusual or particularly complex cases. Often in craniofacial surgery multidisciplinary a team is involved. The biomodels are often used to assist communication amongst team members, to achieve informed consent from patients and relatives and between surgeons intra-operatively. Biomodels have also been used to simulate craniofacial reconstruction. Standard surgical drills and saws can be used to fashion osteotomies. The bone fragments can be reconstructed using wire, plates and screws or glue. A technique that has been developed to assist with the reconstruction involves the manufacture of two biomodels.[2] The first biomodel is reconstructed by the surgeon and acts as an 'end-point' biomodel, illustrating the desired pre-planned reconstruction. The surgeon then marks on the second biomodel the osteotomy lines that he or she intends to use. This biomodel becomes the 'start-point' biomodel. Intra-operatively the two biomodels are used in different ways. The start-point biomodel is used to navigate the anatomy and to accurately transfer the osteotomy lines onto the exposed patient's skull. The osteotomies are then made and the pieces of bone reconstructed according to the end-point biomodel. Such reconstruction, as well as the shaping of bone grafts, can be performed by a second surgeon working at a side table whilst the other surgeon continues to operate on the patient. This technique has been reported to improve the results of surgery as well as shorten the operating time. The end and start point biomodels have also proven very useful for informed consent and team communication.

Maxillofacial surgeons have also developed some interesting specific applications of biomodelling as follows:

1. Integration of biomodels with dental castings to form articulated jaws.
2. The use of biomodels to shape prosthetic and autograph implants.
3. The use of biomodels to prefabricate templates and splints.
4. The use of biomodels in restorative prosthetics.
5. The use of biomodels to plan distraction osteogenesis.

The combination of dental casting (replicating the teeth) with biomodels (replicating the jaws) is advantageous because it removes the problem of artefact from metallic fillings and it allows the more accurate casting to be used to simulate the occlusion.[3] Such hybrid biomodels are particularly helpful in orthognathic surgery where the effects of osteotomies on the dental occlusion may be preoperatively examined.

Biomodels may be used in several ways to shape maxillofacial prosthetic and autographic implants. A simple way to do this is to use the biomodel as a template on which a graft may be directly shaped intraoperatively.[2–4] The bone graft may be harvested from the iliac crest and shaped directly on the sterilised biomodel. Once the contouring is satisfactory the surgeon places the graft in situ and fixes it. This approach can dramatically reduce operating time whilst improving the end result. This can be achieved because the surgeon can shape the graft on the biomodel while the assistant simultaneously prepares the exposure of the donor site. This technique also avoids the need for repeated 'fitting and chipping' of the graft when the patient is directly used as the template, since direct shaping is restricted by soft tissue cover and limited surgical access.

Another approach is to use acrylic, or a similar material, to pre-operatively create a master implant to serve as a guide for the shaping of the bone graft intraoperatively. This is particularly appropriate when the graft requires a somewhat complex shape. The surgeon can minimise operating time by pre-operatively moulding the acrylic to the exact shape required, using the biomodel as the template.

More recently synthetic bioabsorbable materials, such as polygalactic acid, have been introduced into cranio-maxillofacial surgery. Such materials in the form of sheets or plates can be intra-operatively shaped to fit specific anatomy. As with autograph implants such bioabsorbable implants can be shaped to a biomodel to save time and avoid difficulties with limited anatomical exposure.

Biomodels may be used to plan endosseous surgery and to create customised drill guide templates.[3] Edentulous patients may have teeth restored by mounting them on titanium pins which are implanted into the jaw. The implantation of the titanium pins however can be difficult and can be complicated by damage to the underlying dental nerve. Transparent mandibular biomodels accurately replicate the neurovascular canal through which the mandibular nerve travels. The course of this nerve may easily be displayed by passing a malleable coloured wire along the neurovascular canal or replicating it in a second colour. The biomodel can then be used to determine and rehearse the positioning and depth of the holes required to receive the titanium mounting pins using a standard dental drill. The pins can then be inserted into position and 'cold cure' acrylic moulded around them and the mandibular contour to form a relocatable drill guide. The depth of each hole can also be determined relative to the drill guide and recorded. Whilst firmly held in place the guide can be used to drill the holes with the correct positioning and depth as pre planned in the biomodel without risk of injury to the underlying mandibular nerve.

Distraction osteogenesis is a technique use to promote the remodelling and lengthening of bones. The technique uses an implantable device to slowly distract an osteotomised bone by about 1 mm per day. Biomodels have been used to plan the positioning of the distraction device as well as to determine the extent of distraction to achieve the desired result. The use of biomodels has proven extremely valuable in this regard.[5,6]

Yet another use of biomodels in maxillofacial surgery is for the prefabrication of restorative prosthetics. Surgery usually requires titanium fixative implants to be inserted on which a prosthesis is mounted. A biomodel can be used to determine the best location for these implants and to construct the overlying prosthesis. This may be performed by inserting the mounts into the biomodel, constructing the overlying wire scaffolding and then adding the plastic nasal and dental prostheses. A biomodel is therefore used not only to plan and rehearse implantation but also for the construction of the prosthesis. This improves the ability to form an accurate prostheses, which enhances the cosmetic result and shortens the operative time. Another advantage is that less time is required to shape and trial fit the prosthesis, as much of this can be performed on the biomodel.

3.3   Customised cranio-maxillofacial prosthetics

The method traditionally used by neurosurgeons, before the advent of biomodelling, to repair craniotomy defects (for which no autologous bone is available) was to shape cold cure acrylic to fit the defect. This acrylic is moulded and polymerised in situ to form the implant. The technique is limited by the artistic skill of the surgeon to achieve the desired contour of the implant within the short time before polymerisation. This moulding process can be time consuming, especially if the surgeon takes several attempts to achieve the desired contour. Longer surgical time increases infection risk. Infection is a major complication in any implant surgery as its presence will usually necessitate removal of the implant. Another disadvantage of cold cure acrylic monomer is its toxicity. If polymerisation is incomplete monomer can leach into the patients tissues with detrimental effects.

The application of real virtuality to assist in the manufacture of customised prostheses was realised at an early stage.[7] Early techniques were based on the creation of CNC milled models that could be made into a mould from which the implant could be cast.[8,9] A model could also act as a template. Wax could be moulded to create a master implant that could then be used to create a mould from which the implant could be cast.[10] Alternatively the model could act as a template over which titanium[11–13] or hydroxylapatite[14] could be moulded to generate an implant. Such techniques have also been applied to construct prostheses in maxillofacial surgery and orthopaedic surgery.[15]

Methyl methacrylate (acrylic) has been long accepted for use in cranioplasties.[16,17] Other materials that allow the ingrowth of bone, such as hydroxylap-atite,[14] ceramic,[18] and ionomeric bone cement.[19] These materials are highly expensive, somewhat difficult to mould or shape, and if infected can be difficult to remove in toto. Titanium is also used commonly[12–14,23] and has the advantage of being biologically inert, though this is offset by cost, difficulties in moulding, casting or milling, and the artefact generated in CT and MR scanning after implantation. We have used acrylic and mesh titanium for almost 300 implants now without any material related problems.

Once generated the implant is sterilised by gas or autoclaving. The implant is used by the surgeon in conjunction with the biomodel of the region to determine the exact attachment sites and means. Attachment is usually achieved with titanium miniplates and screws. Titanium miniplates and screws can also be preoperatively attatched to the implant to minimise operating time.

Some of the acrylic implants occasionally required minor trimming to achieve good contact between the bevelled edge of the implant and the bone defect. The fit after such minor trimming was consistently within 1.5 mm. The use of prefabricated implants was reported to reduce operating time compared to the traditional cold cure method. Surgeons have reported that the ability to study the implant and the biomodel preoperatively allows fixation to be exactly planned and for possible problems to be identified, e.g. the site of cranial sinuses to be determined.[10]

Some inaccuracy has been occasionally noted. The most likely cause was that the biomodel warped after being removed from the SL machine or that the surgeon failed to remove all of the soft tissues from around the craniotomy defect. Warpage has been reported as a source of error by other investigators.

The advantages of the custom biomodel implants compared to the traditional cranioplasty techniques (where autologous bone is not available) can be summarised as follows:[10]

1. Improved cosmesis and fit,
2. Non-toxic thermally polymerised acrylic or titanium mesh is used,
3. Reduced operating time and risk of infection,
4. Opportunity for surgeons to assess the defect and implant preoperatively, and improve fixation,
5. Improved patient informed consent,
6. Allows one stage resection and implantation procedures to be performed.

3.4 Vascular biomodelling

The feasibility of vascular biomodelling was first assessed by experimental work with the CT angiogram (CTA) of a patient's Circle of Willis which contained an anterior communicating artery aneurysm.[21] Various techniques have been utilised within the Anatomics BioBuild software to accurately define both the bone and the vasculature. A connectivity algorithm was used on only the larger vessels in the Circle of Willis so that unconnected smaller branches could be deleted. Smaller branches tend to become disconnected as the density of contrast within falls below the original threshold and as their diameter becomes such that partial volume effects create gaps in the data. In CTA the subtle difference in the density of the bone and vessel can make thresholding difficult. CTA does allow for both the skull and the vascualture to be generated. This is helpful for neurosurgeons to plan the craniotomy and exposure but does limit the definition of the vessel as it passes through bony structures. SL with two colour StereoCol resin is best used to highlight the internal course of a vessel through the skull base. A limitation of CTA data is that both arterial and venous vessels can be included which at first appearance may be confusing and interfere with the depiction of arterial structures.

MR angiography (MRA) data from a 1.5 Tesla General Electric scanner has been used to biomodel cerebral angiograms containing aneuysms.[21] The main difference between MRA data and that from CTA is that the MRA arterial structures are of much higher pixel intensity in the image and are thus easily segmented. The biomodels replicated fine vessels with a diameter of .8 mm. The resolution of MRA was noted to be less than that of CTA. MRA has the advantages of displaying only arterial structures, free of bony artefacts, without the need for contrast. Disadvantages include not having bony supports or bony landmarks in the biomodel. This can make the MRA biomodels delicate and difficult to orient. MRA data also has lower resolution and any biomodel generated is likely to reflect this.

More recently data from 3D spin angiography has been used to generate biomodels.[22] This imaging technique uses multiple 2D projections of subtraction angiography to reconstruct a high resolution 3D vascular model. The 3D spin angiography data has higher resolution than CT or MR and has been found to produce high quality biomodels of even the finest vessels. 3D spin angiography data is similar to MR data as the arterial structures can be easily defined without bone or venous structures. We anticipate that 3D spin angiography data will become the gold standard for vascular biomodelling in the future.

The most difficult part of biomodelling fine blood vessels was removing the supports. Great care must be taken when removing the supports to avoid damage to finer vessels.

Cerebrovascular biomodels have been found to be very useful to the surgeons for pre-operative evaluation and intra-operative reference.[21,22] Neurosurgical applications that have been identified include surgery for intracranial aneurysms and arterio-venous malformations. For these conditions the biomodels were found to accurately represent the vasculature at operation. Neurosurgeons reported that the biomodels gave a better overview of the anatomy compared to standard 2 and 3 dimensional images as the anatomy was often complex and difficult to interpret from multiple two dimensional projections which are often selected by the radiographer.[21] The biomodels allowed the surgeons to physically manipulate the data into any perspective.[21,22] Biomodels were sterilised and used intra-operatively for micro-surgical navigation. The biomodels allow the surgeon to anticipate problems and assisted in the selection of appropriate aneurysm clips. The biomodels were also very valuable for gaining consent from patients and relatives. As cerebrovascular surgery carries a considerable risk of morbidity and mortality the biomodels were seen to be particularly cost effective. It was noted that the small resin volume in most cerebrovascular biomodels meant that they could be produced at low cost.

Several biomodels of abdominal aortic aneurysms were also manufactured. These biomodels were produced to ascertain feasibility rather than for specific surgical use. Other authors have also reported the manufacture of aortic biomodels.[23] The introduction of multislice helical CT will allow longer vessels to be manufactured and it is likely that such biomodels will be useful for planning the endovascular repair of diseased aortic, iliac and renal arteries.[24]

It is anticipated that vascular biomodelling will allow the manufacture of flow phantoms that will enable scientific research into the haemodynamics of blood flow. Such research is likely to advance the simulation software for endovascular interventions.

3.5 Skull base tumour surgery

Tumours occurring in the region of the cranial floor occupy the most anatomically complex and inaccessible part of the human body. The anatomy of the cranial floor leaves little room for error and allows only limited surgical access. Resection or debulking of these tumours is immensely challenging for the surgeon and of high risk to the patient. To minimise operative morbidity and mortality, and to maximise therapeutic success, accurate surgical strategies must be tailored to each patient and must be carefully planned using the best possible anatomical information.

The biomodelling of skull base tumours is challenging because two or three (bone, tumour and blood vessels) differing tissue types need to be incorporated into the biomodel. The differing densities of the soft tissue composing the tumour and the adjacent bone must be edited from all other structures. When only biomodelling the bone and vasculature this is achieved by the relatively simple process of segmentation. The voxel threshold can be set so that only bone and vessels

remain (high threshold) removing all of the soft tissues (lower thresholds). The problem encountered when biomodelling a soft tissue mass is how to separate it accurately from surrounding structures of similar density. Fortunately tumours tend to be more vascular than the surrounding normal tissues, and contrast agents will highlight them in the image data. This enhancement in the threshold values of tumour voxels can be used to edit the tumour by segmentation, but not to the extent that it can be relied on completely to separate the abnormal from normal. In order to accurately include the tumour and bone together it is often necessary to edit the data on a slice by slice basis.[25]

The close association of the tumours to the skull base has both positive and negative effects on the biomodelling process.[25] A positive effect is that the tumour is supported by the bone during and after manufacture so that additional support structures were unnecessary. If a tumour occurred completely within surrounding tissue that was edited away it would fall to the bottom of the biomodel or build platform, unless the automatically generated support structures held it in position. It is important that these support structures are attached to other anatomy and not just the build platform, because as the biomodel is removed from the platform the relationship between structures may be destroyed. The anatomy can be rotated so that supports are generated attached to other surfaces, or supports may be manually drawn in using the pixel editor to interconnect the structures, thus maintaining the spatial relationships.

The negative effect is a result of the binary nature of the process at present. When a tumour comes into contact directly with bone it is often difficult to see the margin of tumour to bone transition. The same is, of course, true whenever two differing tissue types are modelled together in contact. This problem will be solved when multi-coloured resins are made available. One resin can be transparent, allowing the second to be clearly seen through it.

3.6 Spinal surgery

The combination of CT/MR scans with plain X-rays when planning spinal surgery yields a complex combination of data. Even with all of the standard data a true appreciation of the anatomy can remain elusive. This is particularly the case with the spine as it is constituted of 23 individual bones, each with complex anatomy and articulation. The overall 3D alignment of the bones over long segments can be particularly difficult to assimilate from standard imaging.

The author and several colleagues have investigated the utility of biomodelling in the spine.[26] Studies have been performed on children and adults with congenital, traumatic and degenerative anomalies. It was immediately obvious that the spinal biomodel gave an immediate unique and tactile overview of the anatomy. In all cases a better appreciation of the bony anatomy was demonstrated with the biomodel. In some cases the feasibility of surgery was determined on the basis of the biomodel. Patients also universally stated that the biomodel provided facilitated a better understanding of the anatomy and surgical plan.

Commonly in more complex spinal deformities the placement of pedicle screws and instrumentation is necessary to improve deformity and ensure bony fusion. The placement of such screws can be very difficult in the pathological spine due to anatomical distortion. In these patients the spinal biomodelling is particularly helpful. The author has developed a technique to assist with the placement of spinal instrumentation using the biomodel as a stereotactic aid. Pre-operatively a standard electric hand drill is use to drill screw trajectories into the biomodel at the desired locations. As the biomodel is transparent and replicates the thickness of cortical bone accurately optimal trajectories can be easily and quickly drilled into the biomodel. Metal trajectory pins can then be placed into the drill holes to visualise the trajectory as well as to directly measure the length of the screw required. Once the trajectory pins are in situ cold cure acrylic cement can be molded around the pin and the immediate conture of the biomodel to fashion a contour matching stereotactic drill guide. The template and the biomodel are sterilised and used intra-operatively to navigate and confirm anatomical relationships. The template may then be contour matched to the spine and used as a drill guide to replicate the pre-operatively planned trajectories in the patients spine. A similar technique has previously been described for cranial applications.[26]

With experience it has become evident that by having a sterile biomodel with the trajectory pins in the operative field screw trajectories could be transferred accurately by visual cue alone. The entry point of the trajectory can be readily determined by direct comparison of the biomodel anatomy (incorporating the pre-operatively determined trajectory) to the intra-operative patient anatomy. Once the entry point is established the trajectory vector can be easily determined by holding the biomodel with the trajectory pin in the same orientation as the patient and within the direct line of sight of the surgeon. The surgeon can align the pedicle finder to the same trajectory as the pin in the biomodel by direct visual comparison. The length of the screw can easily be determined by direct measure from the biomodel. In the experience of the author the direct visual alignment method is adequate for almost all cases, and is easy in preparation and execution. Not only does this technique allow for the rapid accurate placement of screws but it also reduces the need for intra-operative x-ray to assist placement. A reduction of x-ray exposure to both surgeon and patient is clearly advantageous for both health and safety and cost reasons.

Biomodels have been found to be particularly helpful for minimal invasive spinal procedures. In such cases surgery is performed via a small incision with the aid of a 'tube like' retractor. This technique limits the visible anatomy and can make orientation and navigation difficult. The surgeon compensates for this by using intra-operative x-ray. The use of biomodels in minimal invasive procedures has been found to increase confidence and accuracy of screw placement whilst reducing the use of x-ray.

More recently multislice CT has been introduced. This does significantly reduce radiation exposure whilst increasing the length of the Z axis which can be scanned. Presently it is recommended that CT scanning be performed on a multislice machine in helical mode.

As the spine is tall and narrow to build a number of vertebrae in the axial orientation many layers will be required. In RP the manufacturing time is directly related to the number of layers required as each recoating cycle takes a set amount of time between layers. In such situations the spine will take less time to manufacture when built rotated through 90 degrees to minimise the build height. The number of supports should also be taken into consideration, and optimising the build orientation can reduce build time by ensuring the supports are minimised. These considerations are important when biomodelling any structure with axial length greater than its maximum width, e.g. long bones. We recommend the use of the patented build optimisation algorthim within Anatomics BioBuild software to automatically determine the optimum build orientation.[27]

3.7 Orthopaedic biomodelling

Orthopaedic surgeons have been slow to use biomodels. This has been surprising as the skeletal anatomy commonly treated by orthopaedic surgeons is ideal for biomodelling. Applications that have been studied include, hip anomalies, particularly congenital dysplasia, acetabular trauma and revision arthroplasty procedures. Biomodels have also been used in complex revision surgery for the knee and elbow. As most orthopaedic procedures are planned on the basis of plain x-ray alone there are likely to be barriers against the introduction of CT scanning purely so that a biomodel can be manufactured. It is likely though that a steady increase in the use of multislice CT will lead to an increase use of biomodelling technology.

Another factor that limits the use of biomodelling in orthopaedics is the size of the structures imaged. A pelvis for example may easily exceed the size of the build platform of a SLA 250, and must be manufactured in parts that have to be glued together. Such a biomodel is very expensive to make and cumbersome to transport and use. The cost of such large biomodels is a deterrent to their use. A way of reducing this problem is to scale the biomodel down to reduce size. The advantage of reducing the image pixel size (x, y dimensions) by half is that it reduces the biomodel volume by a factor of eight and thus the cost of manufacture by a similar proportion. Another advantage is that for large structures, such as the pelvis, a smaller biomodel is easier to transport, handle and store.

4   CONCLUSION

Anatomical biomodelling is a technology that uses patient's medical scan data to manufacture an exact plastic replica (biomodel) of a selected anatomical structure using rapid prototyping. Biomodels were found to have five general applications:[2] to enhance communication between professionals and patients, to educate patients and improve informed consent, to assist surgeons with diagnosis and surgical planning, for the rehearsal and simulation of surgery, for the creation of customised prosthetics, and as an aid for stereotactic surgery.

The medical specialty applications of biomodelling have been growing steadily since the introduction of the technology over 10 years ago. We have identified applications in cranio-maxillofacial surgery, neurosurgery, spinal surgery, vascular surgery and orthopaedic surgery. Biomodelling appears to be a cost effective intuitive visualisation medium that improves surgeon confidence, facilitates communication, shortens procedure times and improves outcomes.

One of the main limitations of the technology is the lack of investment and insight by RP vendors to address the specific requirements of medical prototyping. We hope that improvements in RP technology will facilitate integrated, cheaper and faster hospital based biomodelling systems in the future.

Another limitation of the technology is related to difficulties reimbursing the cost of the technology. As biomodelling doesn't readily fit into existing rebate schedules clinical uptake has been limited severely by cost. It is hoped that data verifying the utility of the technology will support the inclusion of biomodelling in rebate schedules in the future.

REFERENCES

1. D'Urso PS, Thompson RG. Fetal biomodelling. Australian & New Zealand Journal of Obstetrics & Gynaecology. (1998) 38:205–7.
2. D'Urso PS, Atkinson RL, Lanigan MW, Earwaker WJ, Bruce IJ, Holmes A, Barker TM, Effeney DJ, Thompson RG. Stereolithographic (SL) biomodelling in craniofacial surgery. British Journal of Plastic Surgery. (1998) 51:7 522–530.
3. Arvier JF, Barker TM, Yau YY, D'Urso PS, Atkinson RL, Mcdermant GR. Maxillofacial Biomodelling. British Journal of Oral and Maxillofacial Surgery. (1994) 32, 276–83.
4. D'Urso PS, Barker TM, Arvier JF, Earwaker WJ, Bruce IJ, Atkinson RL, Lanigan MW, Effeney DJ. Stereolithographic (SL) biomodelling in cranio-maxillofacial surgery: a prospective trial. The Journal of Craniomaxillofacial Surgery. (1999) Feb;27 (1):30–7.
5. Yamaji KE, Gateno J, Xia JJ, Teichgraeber JF. New internal Le Fort I distractor for the treatment of midface hypoplasia. J Craniofac Surg. 2004 Jan;15(1):124–7.
6. Whitman DH, Connaughton B. Model surgery prediction for mandibular midline distraction osteogenesis. Int J Oral Maxillofac Surg. 1999 Dec;28(6):421–3.
7. D'Urso PS, Tomlinson FH, Earwaker WJ, Barker TM, Atkinson RL, Weidmann MJ, Redmond M, Hall B, M'Kirdy B, Loose S, Wakely G, Reik AT, Effeney DJ. Stereolithographic (SLA) Biomodelling in Cranioplastic Implant Surgery. Proceedings of the International Conference on Recent Advances in Neurotraumatology. Gold Coast, Australia. September 25–28 (1994) 153–6.
8. White DN. Method of forming implantable prostheses for reconstructive surgery. U.S. Patent Office, Patent 4436683, 3 June 1982.
9. Toth BA, Ellis DS, Stewart WB. Computer-designed prostheses for orbitocranial reconstruction. Plas Recon Surg 1988;81:315–324.
10. D'Urso PS, Earwaker WJ, Barker TM, Redmond MJ, Thompson RG, Effeney DJ, Tomlinson FH. Custom cranioplasty using stereolithography and acrylic. British Journal of Plastic Surgery. (2000) 53(3): 200–04.
11. Blake GB, MacFarlane MR, Hinton JW. Titanium in reconstructive surgery of the skull and face. Br J Plast Surg 1990;43:528–535.
12. Joffe JM, McDermott PJC, Linney AD, Mosse CA, Harris M. Computer-generated titanium cranioplasty: report of a new technique for repairing skull defects. Br J Neurosurg 1992;6:343–350.
13. Abbott J, Netherway D, Wingate P, Abbott A, David D. Craniofacial imaging, models and prostheses. Aust J Otolaryng 1994;1:581–587.
14. Waite PD, Morawetz RB, Zeiger HE, Pincock JL. Reconstruction of cranial defects with porous hydroxylapatite blocks. Neurosurgery 1989;25:214–217.
15. Rhodes ML, Kuo Y, Rothman SLG, Woznick C. An application of computer graphics and networks to anatomic model and prosthesis manufacturing. 1987 IEEE CG&A;2:12–25.
16. Manson PN, Crawley WA, Hoopes JE. Frontal cranioplasty: Risk factors and choice of cranial vault reconstructive material. Plastic and Reconstr Surg 1986;77:888–900.
17. Remsen K, Lawson W, Biller HF. Acrylic frontal cranioplasty. Head and Neck Surgery. 1986;Sept/Oct:32–41.
18. Kobayashi S, Hara H, Okudera H, Takemae T, Sugita K. Usefulness of ceramic implants in neurosurgery. Neurosurgery. 1987;21:751–755.
19. Ramsden RT, Herdman RCD, Lye RH. Ionomeric bone cement in neuro-otological surgery. J Laryngol Otol. 1992;106:949–953.

20. Eufinger H, Wehmoller M, Harders A, Heuser L. Prefabricated prostheses for the reconstruction of skull defects. Int J Oral Maxillofac Surg (1995) 24:104–110.

21. D'Urso PS, Thompson RG, Atkinson RL, Weidmann MJ, Redmond MJ, Hall BI, Jeavons SJ, Benson MD, Earwaker WJS. Cerebrovascular biomodelling. Surgical Neurology. (1999) 52(5):490–500.

22. Wurm G, Tomancok B, Nussbaumer K, Adelwohrer C, Holl K. Cerebrovascular stereolithographic biomodeling for aneurysm surgery. Technical note. J Neurosurg. 2004 Jan;100(1):139–45.

23. Lermusiaux P, Leroux C, Tasse JC, Castellani L, Martinez R. Aortic aneurysm: construction of a life-size model by rapid prototyping. Ann Vasc Surg. 2001 Mar;15(2):131–5.

24. Kato K, Ishiguchi T, Maruyama K, Naganawa S, Ishigaki T. Accuracy of plastic replica of aortic aneurysm using 3D-CT data for transluminal stent-grafting: experimental and clinical evaluation. J Comput Assist Tomogr. 2001 Mar–Apr;25(2):300–4.

25. D'Urso PS, Askin G, Earwaker WJS, Merry G, Thompson RG, Barker TM, Effeney DJ. Spinal Biomodelling. Spine. (1999) 24(12):1247–1251

26. D'Urso PS, Hall BI, Atkinson RL, Weidmann MJ, Redmond MJ. Biomodel guided stereotaxy. Neurosurgery. (1999) 44(5):1084–1093.

27. D'Urso P.S. Stereolithographic modelling process. Australian Patent 684 546; US Patent 5 741 215. September 1993.

Virtual modeling and rapid manufacturing – Bártolo (eds)
© 2005 Taylor & Francis Group, London, ISBN 0 415 39062 1

Positive direct-mask stereolithography with multiple-layer exposure: Layered fabrication with stair step reduction

T. Murakami

The University of Tokyo, Tokyo, Japan

ABSTRACT: In this paper, we explain a new stereolithography involving the separate use of s liquid photoinitiator and a base resin (photopolymer resin without photoinitiator). First, the base resin is supplied as a layer, and then a mask pattern is drawn onto the surface with photoinitiator by inkjet printing. When the surface is exposed to a UV lamp, only the pattern drawn with the photoinitiator is cured. In this process, the photoinitiator acts as a positive mask. Also, the idea of multiple-layer exposure, which may enable stair step reduction and fabrication time reduction, is introduced. The proposed idea and method are implemented in an experimental stereolithography system and some experiments are conducted using the system. As a result, the effectiveness and feasibility of the idea and method are confirmed.

1 INTRODUCTION

Presently, various methods of rapid prototyping or solid free-form fabrication, each of which has its own merits and characteristics, have been studied and developed. Among those methods, stereolithography is a method whose characteristics are fabrication accuracy, high resolution and material transparency.

In conventional stereolithography, objects are fabricated in liquid photopolymer resin. In contrast, the author proposed and studied a new stereolithography method of fabricating solid objects in solid resin using a sol-gel transformable photopolymer resin. This study result was commercialized as a stereolithography system by Denken Co., Ltd., Japan in June 2001.

In this paper, the author briefly explains the basic idea and process of the stereolithography using solgel transformable photopolymer resin, and then describes the results of our continuing studies as follows:

– positive direct mask as a new exposure method,
– multiple layer exposure for stair step reduction.

2 STEREOLITHOGRAPHY USING SOL-GEL TRANSFORMABLE PHOTOPOLYMER RESIN

2.1 Basic idea and fabrication process

Figure 1 shows our basic idea and the process of stereolithography for fabricating solid (photopolymerized) objects in solid (gel) photopolymer resin (Murakami

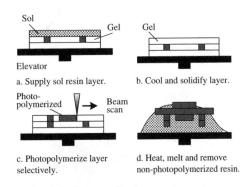

a. Supply sol resin layer.   b. Cool and solidify layer.

c. Photopolymerize layer selectively.   d. Heat, melt and remove non-photopolymerized resin.

Figure 1. Process of stereolithography using sol-gel transformable photopolymer resin.

et al. 2000a, b, c). First, a liquid (sol) photopolymer resin is supplied to form a new layer, cooled to the gel state, and then selectively photo-polymerized. After repeating this process for all layers, a photopolymerized object is fabricated in a gel resin block. The object can be obtained by heating the block, and then melting and removing only the non-photopolymerized resin.

The use of a gel layer, instead of a liquid layer as in conventional stereolithography, eliminates the need for the "support" structures of isolated or overhanging shapes.

2.2 Sol-gel transformable photopolymer resin

To achieve the process shown in Figure 1, a special sol-gel transformable photopolymer resin (UV1214-15) composed of urethane-acrylate photopolymer resin and sol-gel transformable resin was prepared by a Japanese resin company.

Figure 2 shows the relationships between temperature and viscosity of photopolymer resins: UV1214-15, commercially available resin for the conventional stereolithography process (SCR-500, JSR: Japan Synthetic Rubber Co., Ltd.), and specially prepared urethane-acrylate resin (KC1026-8, JSR) used in our previous study. In our previous experiments, we needed to cool

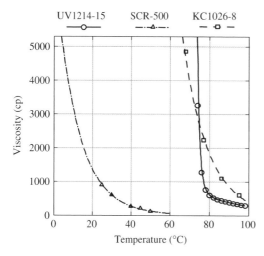

UV1214-15   SCR-500   KC1026-8

Figure 2. Relationship between temperature and viscosity of photopolymer resins.

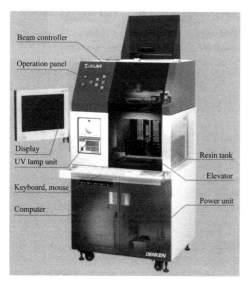

Figure 3. SolidJet SJ-200P from Denken Co., Ltd., Japan (http://www.denken-eng.co.jp/).

SCR-500 and KC1026-8 to −50 and −10°C, respectively, to make them sufficiently solid. Problems arising at such low temperatures are inactive photopolymerization reaction, frost on the resin surface, and the need for a large and complex cooling mechanism (e.g., using liquid nitrogen).

UV1214-15, however, changes sharply between sol and gel states at about 80°C, and its gel strength is 600 kPa, which has been experimentally proven to be sufficiently solid to support the fabricated objects during our stereolithography process.

After further improvement of the sol-gel transformable photopolymer resin, our research result was commercialized as a stereolithography system (Figure 3) by Denken Co., Ltd., Japan in June 2001.

3 POSITIVE DIRECT MASK: A NEW EXPOSURE METHOD

3.1 *Direct-mask exposure*

Possible additional merits of using a gel resin layer instead of liquid is that we can apply some treatments to the layers, such as drawing some mask patterns directly on the gel resin surface. If we draw mask patterns with a material that blocks UV light directly on the gel layer surface, we may be able to introduce selective-area exposure with a UV lamp into our stereolithography process, as in Figure 4a. We named this method as "direct-mask exposure".

This direct-mask exposure has the following merits.

– Direct-mask exposure can be implemented using inkjet printing of the mask material which is less expensive than using imaging devices such as liquid crystal (Hayashi 2000) and a digital micromirror.
– If there is some distance between the mask and the fabrication surface as in the solid ground curing method (Jacobs 1992), divergence of the light definitely reduces fabrication accuracy. On the other hand, the mask is drawn directly on the fabrication surface (distance is 0) in our method, and divergence of the light becomes less critical. Also, conventionally, only a specific wavelength of the light source can be used if we need parallel light because the design of the optical system depends on the wavelength of the light. On the contrary, in direct-mask exposure, all wavelengths of the light source can be used because it does not need highly parallel light, which may lead to efficient use of light power.

The direct-mask exposure, however, also has the following demerits.

1. Since the mask is drawn on the nonphotopolymerized portion, the uncured photopolymer resin is mixed with the mask material, making it difficult to

reuse. This may lead to problems of material cost and chemical material waste.

2. It is difficult to realize a mask material to satisfy all conditions of good UV blockage, highly inkjet printable, and well matched to the resin surface.

3.2 Positive direct-mask exposure

The reason for demerit 1 of direct-mask exposure is that the mask, as shown in Figure 4a, is a "negative" mask drawn on the portion not to be photopolymerized. Therefore, one possible solution of this problem is to use a "positive" mask drawn on the portion to be photopolymerized.

The photopolymer resin the author has been using is a mixture of the base resin to be photopolymerized and a photoinitiator which absorbs light and triggers photopolymerization. Here the author considers the use of these two components separately before mixture: base resin without photoinitiator and liquid photoinitiator (Murakami et al. 2004a). A sol-gel transformable resin component is mixed with the base resin. By using the photoinitiator as a mask material, positive mask exposure should be accomplished, as shown in Figure 4b.

Positive direct-mask exposure should solve the demerits and add the merits of negative direct-mask exposure as follows.

- Mask material is mixed in the cured portion and the uncured resin can be reused (solution to demerit 1).
- A liquid photoinitiator mask matches the base resin surface well, since they are usually used mixed (solution to demerit 2).
- Keeping the base resin and photoinitiator separate reduces their reactivity and leads to longer pot life of materials.

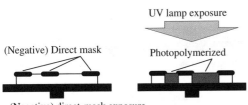

a. (Negative) direct-mask exposure.

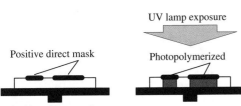

b. Positive direct-mask exposure.

Figure 4. Two types of direct-mask area exposure.

- Low viscosity and the non curability of the liquid photoinitiator itself enable greater stability and liability of the inkjet process. Also, the base resin can be of high viscosity and thus the restriction on applicable materials should be milder than in the method of inkjet-printing the photopolymer resin itself (Eden system (Objet Geometries Ltd. (http://www.objet.co.il/)); InVision system (3D Systems, Inc. (http://www.3dsystems.com/))).
- Since only the base resin without photoinitiator is heated in our sol-gel process, possible heat polymerization can be avoided.

4 IMPLEMENTATION OF POSITIVE DIRECT-MASK EXPOSURE STEREOLITHOGRAPHY SYSTEM

Figure 5 illustrates the process of stereolithography using sol-gel transformable photopolymer resin and positive direct-mask exposure. The heated sol base resin supplied as a new layer is cooled and becomes a gel layer. The liquid photoinitiator is drawn onto the gel surface as a positive direct mask by inkjet printing. When the photoinitiator permeates into the base resin,

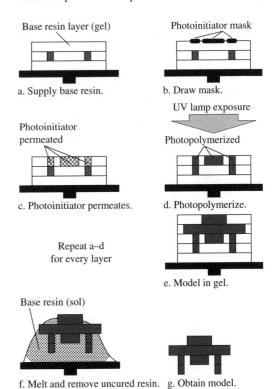

Figure 5. Stereolithography process using positive direct-mask exposure.

41

the UV lamp exposes the gel surface, and only the portion where the positive direct mask is drawn is photopolymerized.

To examine the feasibility of this process, we implemented a semi-automatic (partly manual) experimental fabrication system. Figure 6a, b and Table 1 show the appearance, configuration and specifications, respectively, of the system. We implemented C++ software to process mask data in BMP image files. Base resin is heated in the resin tank and driven by the pump through the slit next to the modeling table in the sol state. The base resin is supplied onto the modeling table by the recoater blade. At the same time, an inkjet head attached behind the recoater blade prints photoinitiator onto the resin surface as positive mask. After the resin is cooled to room temperature and becomes a gel layer, it is exposed to the UV lamp from above.

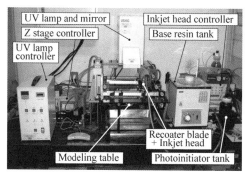

(a) Appearance

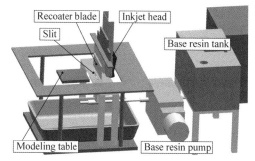

(b) Configuration

Figure 6. Stereolithography system using sol-gel transformable photopolymer resin and positive direct-mask exposure.

5 EXPERIMENTS ON POSITIVE DIRECT-MASK EXPOSURE

5.1 Layer thickness

The layer thickness in positive direct-mask stereolithography is determined by the vertical permeation of the liquid photoinitiator mask into the base resin.

To obtain the basic data, we drew a 50×50 pixel square mask (about 7×7 mm in 180 dpi) and exposed it using a UV lamp for 2 s with $130 \, \text{mW/cm}^2$ after several wait times for photoinitiator permeation. We washed the fabricated object with toluene to remove uncured resin and measured its thickness with a micrometer calliper. In order to check the influence of the photoinitiator mask on the upper resin layer, we measured the thickness of only the layer below and the layers below and above laminated (Figure 7). The results

Table 1. System specifications.

Item	Specification	
UV lamp	Ushio UV lamp HB25103BY, Exposure area diameter 60 mm, Band-pass filter 365 nm.	
Inkjet head	Xaar XJ500/180 (Resolution180 dpi, 80 pl/drop)	
Base resin	Urethane base resin LW2155 SCBS20	These are the components of sol-gel transformable photopolymer resin SJR101 used for SolidJet SJ-200P (Figure 3).
Positive mask	Solution of radical-polymerization photoinitiator I1173	

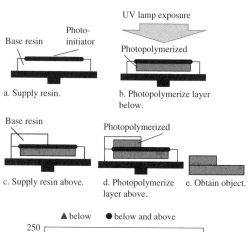

a. Supply resin.

b. Photopolymerize layer below.

c. Supply resin above.

d. Photopolymerize layer above.

e. Obtain object.

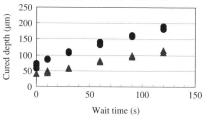

f. Relationship between wait time and cured depth.

Figure 7. Layer thickness in positive direct-mask exposure.

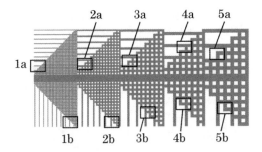

Figure 8. Lattice pattern of positive direct mask (gray portion was drawn with photoinitiator).

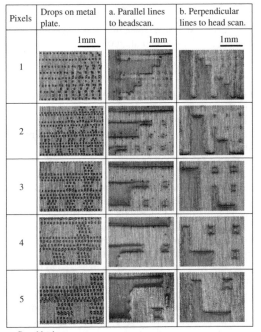

Pixels	Drops on metal plate.	a. Parallel lines to headscan.	b. Perpendicular lines to head scan.
	1mm	1mm	1mm
1			
2			
3			
4			
5			

a. Cured lattice pattern.

▲ Solid vertical width ■ Solid horizontal width
△ Hollow vertical width □ Hollow horizontal width
○ Nominal width

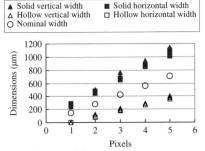

b. Cured lattice resolution.

Figure 9. Horizontal resolution of positive direct-mask exposure.

show that layer thickness can be about 50 to 200 μm depending on wait time.

5.2 Horizontal resolution

The horizontal resolution of positive direct-mask stereolithography is determined by the horizontal permeation of the liquid photoinitiator mask into the base resin.

To obtain the basic data, we conducted an experiment to photopolymerize the gel resin layer with the mask pattern shown in Figure 8. The line and interval widths of a lattice vary as 1, 2, 3, 4 and 5 pixels sequentially from left to right. We drew this 100 μm by inkjet printing (180 dpi), waited 30 s for permeation and exposed a UV lamp. Figure 9a shows inkjet drops on a metal plate without permeation and cured patterns (lighter portions) for each of 1 to 5 pixels. Figure 9b shows nominal width (calculated as 180 dpi) and measured widths of cured lines and intervals (between lines) for each pixel. This result shows that the resolutions are about 250 μm for the solid portion (minimum cured width) and about 100 μm for the hollow portion (minimum uncured width) with the inkjet head (180 dpi, 80 pl/drop) used in these experiments. Since an inkjet head with higher resolution (smaller drop) is easily available now, higher resolution should be achievable by positive direct-mask exposure.

6 LAYERED FABRICATION BY POSITIVE DIRECT-MASK STEREOLITHOGRAPHY

Based on the basic data obtained above, we conducted experiments on solid-shape fabrication by positive direct-mask stereolithography. Inkjet resolution was 180 dpi and layer thickness was 200 μm for all the following fabrication examples.

Figure 10 shows layered fabrication examples. When we first conceived the idea of positive direct-mask exposure, we thought of the possibility of photoinitiator permeating into the base resin insufficiently to form a sufficient layer thickness or in too much excess to form a sufficiently sharp horizontal shape. Although the current fabrication examples do not have sufficient accuracy or resolution for practical solid free-form fabrication, they at least show the possibility of positive direct-mask stereolithography.

Since inkjet printing produces a positive direct mask on the portion to be photopolymerized, color fabrication can easily be introduced by using a colored photoinitiator solution. Figure 11 shows the result of a fan-shaped object fabricated using transparent photoinitiator and red photoinitiator (mixed with dye). The solid fan shape was drawn with a transparent positive mask by inkjet printing for the first five and last five layers. For the intermediate 20 layers, the outer

43

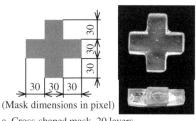

(Mask dimensions in pixel)

a. Cross-shaped mask, 20 layers.

b. Circular mask (diameter 49 pixels), 30 layers.

c. Square mask (100 × 100 pixels), 10 layers + 10 layers (position shifted by 50 pixels).

Figure 10. Examples of layered fabrication by positive direct-mask stereolithography.

(Mask dimensions in pixels)

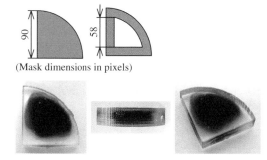

Figure 11. Examples of color fabrication by positive direct-mask stereolithography.

hollow fan shape was drawn with a transparent mask and then an inner small solid fan shape was manually drawn using a metal stamp.

7 MULTIPLE-LAYER EXPOSURE OF POSITIVE DIRECT MASK

Another possible advantage of the positive direct mask is the exposure of multiple layers simultaneously, which

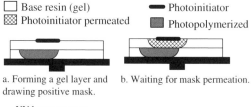

a. Forming a gel layer and drawing positive mask.

b. Waiting for mask permeation.

UV lamp exposure

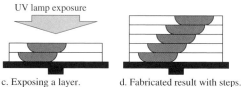

c. Exposing a layer.

d. Fabricated result with steps.

Figure 12. Fabrication process by layer-by-layer exposure.

may lead to stair step reduction and fabrication time reduction (Murakami et al. 2004b).

7.1 Stair step reduction

Figure 12 shows a fabrication process of conventional layer-by-layer exposure. In this process, the sequence of forming a new gel layer, drawing a positive direct mask, waiting for mask permeation and exposing the layer is repeated for every layer.

Here we modify the process as shown in Figure 13. In this modified process, a sequence of only forming a gel layer and drawing the positive direct mask is repeated for multiple layers, and then the multiple gel layers are exposed simultaneously. This is possible because the base resin is transparent. In this process, the liquid photoinitiator drawn between layers may permeate in every direction in multiple layers, as shown in Figure 13b. This phenomenon may lead to the stair step reduction effect in layered fabrication, as shown in Figure 13c, compared with that in Figure 12d.

To confirm this effect, an experiment on layered fabrication with this multiple-layer exposure was conducted. As for the sample shape, we prepare a slope model with 5 layers, as shown in Figure 14a. First, we fabricated the slope by conventional layer-by-layer exposure. We waited 200 s after the positive mask was drawn for every layer and exposed the layer for 2 s at 130 mW/cm^2. Figure 14b shows the result. Next, we fabricated the slope model by multiple-layer exposure. Since our experimental fabrication system takes about 5 s for forming the gel layer and drawing the positive direct mask and 9 s for lowering the table and returning the coater to the origin, the time when each positive mask was drawn is as shown in Figure 14a. After five layers were prepared, we waited 120 s, 240 s and 360 s for photoinitiator permeation and exposed the 5 layers for 2 s at 130 mW/cm^2. Figure 14c, d, e show the result. We can see the stair step reduction effect.

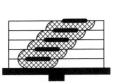

a. Forming gel layer and drawing positive mask (repeated for multiple layers).

UV lamp exposure

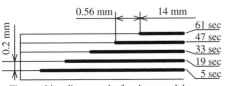

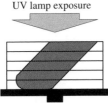

b. Photoinitiator permeation in every direction in multiple layers.

c. Fabricated result with reduced steps.

Figure 13.   Fabrication process by multiple-layer exposure.

a. Five positive direct masks for slope model.

1mm

b. Layer-by-layer exposure.

c. Multiple-layer exposure (wait time 120 s)

d. Multiple-layer exposure (wait time 240 s)

e. Multiple-layer exposure (wait time 360 s)

Figure 14.   Examples of slope fabrication by multiple-layer exposure.

7.2   Fabrication time reduction

Here, we assume the fabrication time parameters given in Table 2 and compare the total time taken to fabricate 10 layers by layer-by-layer exposure and by multiple-layer exposure (e.g., 5 layers, twice). A simple calculation yields the following result.

Table 2.   Fabrication time parameters.

Lowering table and returning coater	T_t	9 s	
Forming gel layer and drawing positive mask	T_l	5 s	
Photoinitiator permeation	T_p	30 s	(Layer thickness 100 μm)
Exposing one layer	T_o	2 s	
Exposing multiple layers	T_m	10 s	(Assumed rather long)

$$T_{layer-by-layer} = (T_t + T_l + T_p + T_o) \times 10 = 460$$

$$T_{multiple-layer} = \{(T_t + T_l) \times 5 + T_p + T_m\} \times 2 = 220$$

Although this is a very rough calculation, the possibility of fabrication time reduction by using multiple-layer exposure is suggested.

8   CONCLUSIONS

In this paper, a new stereolithography method using sol-gel transformable photopolymer resin and positive direct-mask area exposure was presented. The basic experiments showed that positive direct-mask area exposure can achieve layer thicknesses of 50–200 μm and horizontal resolution of 100–250 μm with an inkjet head resolution of 180 dpi and 80 pl/drop. Layered fabrication experiments showed that fabrication of simple shapes and multicolor fabrication are possible. Also, the idea of multiple-layer exposure was presented, and the results of experiments indicated stair step reduction and fabrication time reduction.

Since an inkjet head with higher resolution and smaller drop size is easily available, the values presented above should be able to be improved so that the described method should become applicable as a practical fabrication system.

REFERENCES

Hayashi, T. 2000. Direct 3D Forming Using TFT LCD Mask, *Proceedings of the 8th International Conference on Rapid Prototyping, Tokyo, Japan*: 172–177.
Jacobs, P.F. 1992. *Rapid Prototyping & Manufacturing: Fundamentals of Stereolithography*. Society of Manufacturing Engineers: 416–419.
Murakami, T. et al. 2000a. Refrigerative Stereolithography Using Sol-Gel Transformable Photopolymer Resin and Direct Masking. In S.C. Danforth et al. (eds), *Solid Freeform and Additive Fabrication – 2000 (Materials Research Society Symposium Proceedings Volume 625), San Francisco, California, USA*: 91–96.
Murakami, T. et al. 2000b. Refrigerative Stereolithography Using Direct Masking. *Proceedings of the*

8th International Conference on Rapid Prototyping, Tokyo, Japan: 184–189.

Murakami, T. et al. 2000c. Refrigerative Stereolithography Using Sol-Gel Transformable Photopolymer Resin and Direct Masking, *Proceedings of the ASME Manufacturing Engineering Division 2000, MED-Vol.11, Orlando, Florida, USA*: 289–296.

Murakami, T. et al. 2004a. Stereolithography Using Positive Direct-Mask Exposure, *The 26th Rapid Prototyping Symposium, Omiya, Japan*: 101–106 (in Japanese).

Murakami, T. et al. 2004b. Stereolithography Using Positive Direct-Mask Exposure, *JSME 14th Design and Systems Conference, Fukuoka, Japan*: 47–50 (in Japanese).

Virtual modeling and rapid manufacturing – Bártolo (eds)
© 2005 Taylor & Francis Group, London, ISBN 0 415 39062 1

Microfabrication of dental root implants with a porous surface layer by microstereolithography

T. Laoui, S.K. Shaik & R.F. Hall
University of Wolverhampton, RIATec, School of Engineering and Built Environment, Wolverhampton, UK

A. Schneider
Rutherford Appleton Laboratory, Central Microstructure Facility, Didcot, UK

ABSTRACT: Microstereolithography (μSL) has been utilized for the production of 3D polymeric parts with miniature features and/or micro-sized surface structures. This investigation presents the preliminary successful results obtained from a feasibility study concerning the manufacturing of dental root implant prototypes composed of a dense core and a porous surface layer using μSL process and acrylic resin and their corresponding characterization. To minimize distortions in part geometry and pore dimension, improvements were introduced into the CAD models of the implant as well as to the μSL process parameters.

1 INTRODUCTION

Micro-rapid prototyping and fabrication (μRPF) techniques, based on methods, which build up material layer-by-layer, are emerging rapidly as alternative techniques to the conventional micro-fabrication methods, which are based on material subtraction, for the production of complex 3D parts with miniature features and/or micro-sized surface structures. Among these μRPF techniques is microstereolithography (μSL) used for the production of 3D polymeric and complex microstructures with a resolution in micrometer range. μSL machines based on layer-by-layer manufacture fall within two main categories: the first is a vector-by-vector process in which a scanning laser beam is focused onto the surface of a photopolymer liquid resin; and, the second is a process whereby the entire layer is subjected to irradiation and is cured in one single step (Nakamoto et al. 1996, Maruo et al. 1998, Bertsch et al. 1999, Beluze et al. 1999). The latter process offers a clear advantage as it is a fast process and thus leads to a shorter manufacturing time. Recently an advanced μSL system called Super-IH (Integrated Hardness) process has been developed based on pinpoint solidification, avoiding the layer-by-layer technique, which is capable of achieving a resolution as high as 0.5 μm (Maruo et al. 2002). This process allows free forming of 3D microstructures by scanning a focused laser beam inside a liquid photopolymer without relying on support structures, which

are in general necessary for fabrication of SL-objects. Even a subdiffraction-limit spatial resolution of 120 nm can be achieved by using special μSL resins, which enables and facilitates two-photon absorption (Kawata et al. 2001). The disadvantage of this process is that, because of the reduced layer thickness, fabrication of objects with an overall size of several millimeters would take an order of magnitude longer in processing compared with a one-photon-absorption μSL process. Therefore, it is recommended that prototypes of dental implants be produced by a layer-by-layer projection μSL system using conventional one-photon absorption.

The objective of the present study concerns the development of an idealized structure for a dental root implant possessing a porous surface and a dense core using μSL technique. The mechanical stability of medical implants can be greatly enhanced by providing a porous surface layer as an integral part of the implant as compared to conventional techniques of roughening the surface by plasma spraying of a separate biocompatible coating material. It has been reported that bone in-growth is greater in the 3D interconnected structure of a porous surface than in the irregular geometry of the plasma-sprayed coating; as the initial matrix mineralization leading to osseointegration occurs more rapidly with the porous-surfaced implants which promotes greater attachment strength and interfacial stiffness (Simmons et al. 1999). Other coating techniques include pulsed laser deposition, which has more chemical stability than the plasma sputtering technique (Lo et al. 2000). The non-coating techniques employed for roughening the surface of medical implants are powder blasting and acid etching leading to an increase in the interfacial shear strength (Li et al. 2001).

2 EXPERIMENTAL

The schematic description of the microstereolithography (μSL) machine installed in the Central Manufacturing Facility (CMF) at the Rutherford Appleton Laboratory, Didcot (UK), is shown in Figure 1. This is a system for building 3D shapes from an AutoCAD design via projection of sliced images of these shapes layer-by-layer into a photoreactor with liquid photosensitive resin.

The system works by rapid prototyping of microstructure in photocurable acrylic resin. This layer-by-layer fabrication of microstructures gives process times of several hours per object depending on its dimensions, using a maskless process (integral process).

The μSL machine consists of a digital micro-mirror device (DMD) called the pattern generator, which is illuminated by an UV light source. The computer controlled DMD holds the image of the sliced layer, which is projected onto the surface of the resin and solidifies the resin in the illuminated area. The light beam, controlled by a shutter, illuminates the surface of the resin for a short time, seven seconds per layer. Then the platform lowers to a depth of 10 μm into the liquid resin with the Z-stage (controlled by a motor) so that the already solidified part of the object is re-coated with fresh unexposed resin for the next layer. The process height of the whole part could be few micrometers to several millimeters.

3 RESULTS AND DISCUSSION

Research and experimental work related to the development of medical implants using μRPF techniques are still in their infancy. This investigation presents the initial results obtained from a feasibility study on manufacturing dental root implants composed of a dense core and a porous surface layer using μSL integral process and acrylic resin (to be utilized as prototypes at this stage) and their corresponding preliminary characterization.

3.1 Dental root implant CAD models

An idealized geometry and CAD model of an implant was prepared using ProEngineer software. The first model of a dental root implant, with 5 mm diameter and 6 mm length, was prepared with a dense core and a porous surface consisting of channels of 50 μm diameter and 1 mm depth, as shown in Figure 2. However, difficulties were encountered in fabricating such a model due to its high aspect ratio (channel length/channel diameter: 1 mm/0.05 mm = 20) as the non-crosslinked resin could not easily flow out of the channels after removing the part from the liquid resin in the photoreactor. Thus, all further models were prepared with a lower aspect ratio (below or equal to 10). To approximate the channels to pores, the second model of the dental root implant, of 5 mm diameter and 6 mm length, was prepared with channels having 50 μm pore diameter and 25 μm depth.

3.2 Processing of dental root implants

The CAD/ProEngineer models of the dental root implant were converted into STL files and transferred

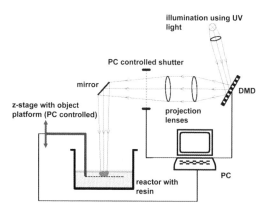

Figure 1. Schematic description of the μSL system at Rutherford Appleton Laboratory.

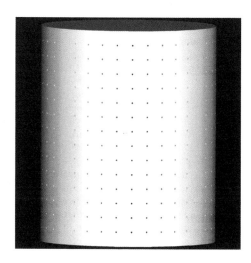

Figure 2. Model of a dental root implant with a high channel length to diameter aspect ratio.

to the μSL machine, which used dedicated software to slice the STL files into Bitmap images so that the part could be built up layer-by-layer. The pattern generator shaped the image, which was projected onto the surface of liquid resin by the DMD and the liquid solidified at the illuminated surface. The model of the solid implant, mentioned above, was built with a support, a hollow cylindrical structure with internal and external diameter of 4.6 mm and 5.0 mm respectively and a height of 0.5 mm, for easy removal of the μSL object. The building time of this model took approximately three hours. The μSL prototype had an approximate diameter of 4.82 mm instead of 5 mm, as shown in Figures 3a and 3b, and the shape of the pores was more rectangular measuring approximately 120 μm by 40 μm, instead of a spherical shape (50 μm diameter) originally designed in the CAD model.

The discrepancies in the implant shape (particularly prototype top surface) and pore (referring to channel diameter) dimensions between the implant model and the built sample (μSL prototype) were thought to be caused by various factors. Geometry/aspect-ratio of the

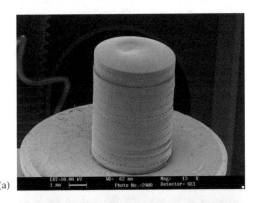

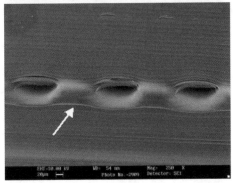

(a)

(b)

Figure 3. (a) SEM image of a μSL dental implant prototype – solid cylinder with micropores, fabricated without any specific re-coating relaxation; (b) enlarged picture of the prototype showing deformed pores (~120 μm × ~40 μm), arrow indicates warped layers.

channels, structural stability of the cross-linked resin layer in the vicinity of the pore (see waviness of layers – indicated by an arrow in Fig. 3b), μSL process parameters, mechanical stress in-built in each layer and surface tension of the liquid resin during re-coating of each layer may be only a few of these factors. To overcome the pore deformation, the distance between the layers containing the pores was reduced from 0.5 mm to 0.3 mm. Furthermore, to reduce the surface tension obtained in the relatively large lateral surface area causing its deformation during resin curing (see Figure 3a), a hollow cylindrical implant model was prepared with a cylindrical support structure.

It was also noticed that the horizontal channels in the previous model prototypes hindered the flow of resin through them and as a result the channels were tilted downwards at an angle of 30° to the horizontal plane to provide a slope for the resin to flow easily out of the channels. Optimizing this angle would provide a prototype, which has suitable pores/channels for operativeness of a dental implant.

Considering the above aspects, a modified 6 mm long hollow dental implant model was made, with external and internal diameters of 4 mm and 3 mm respectively, and channels representing the surface layer with dimensions of 100 μm diameter and 0.5 mm depth titled at 30°. This model was built on a cylindrical support structure. A μSL prototype of such model is displayed in Figures 4a and 4b. The processing of this prototype took approximately 7 hours because of a prolongation of the pre-exposure relaxation time for each re-coated layer. The dip distance for re-coating each layer was also increased in comparison with the process used for the previous prototype (Fig. 3). In Figure 4b, the improvement of the micropores is apparent. Still the opening (approximately a diameter of 89 μm) is less than expected but the shape of a pore is more regular and the waviness of the layers is reduced (most layers are parallel to each other, except in closest vicinity of the pore where a slight distortion of the layers is detected).

3.3 Some properties of the μSL implants

As discussed in the previous chapter, the geometry of the AutoCAD designed prototype has already considerable influence on spatial tolerances of the final μSL object. This is also reflected in the surface roughness of differently fabricated μSL objects. The roughness of one μSL cylindrical dental implant prototype was scanned between the pores in axial cylinder direction, perpendicular to the layers (Fig. 5). For comparison, the roughness of a similar μSL object used as a reference was also quantified with a stylus profilometer (surface A in Fig. 5). The reference object has a flat side wall. Clearly, the surface roughness of the dental implant prototype is significantly larger than that of the reference object. The surface of the dental implant has an average deviation from centerline mean roughness $R_a \sim 5.55\,\mu m$ and a maximum peak-to-valley profile height (R_{max}) of $25.09\,\mu m$ in the scanned area, whereas the surface of the reference object has an

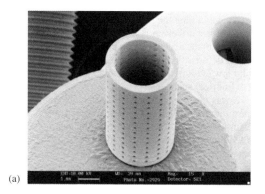

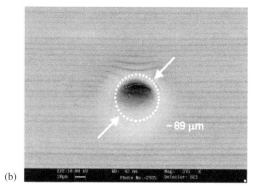

(a)

(b)

Figure 4. SEM image of a μSL dental implant – hollow cylinder with 30° tilted pores made using improved building parameters (e.g. re-coating relaxation for several seconds); (a) the whole sample, (b) enlargement of one micropore.

$R_a = 0.88\,\mu m$ and $R_{max} = 5.68\,\mu m$. These latter values indicate that the roughness of flat surfaces of μSL objects made with the system at CMF is much less than the resolution of the system (z-resolution: $10\,\mu m$ layer thickness; x/y-resolution: $10\,\mu m$ defined by pixel size of DMD). It might be possible to improve further this roughness by using an improved vibration insulation of the system. On the other hand, the dental implant shows typical steps in the surface profile, which have a step size approximately the size of the pixels projected into the resin. Presumably larger roughness in this case results from the 'slicing'-algorithm used for the STL files to produce 2D Bitmap images. Because the STL standard uses triangular faces to represent the 3D shape of a μSL object, the formation of curved surfaces lacks exact reproducibility of Euclidian geometry arced planes.

Even if the properties of the liquid resin (viscosity, surface tension, wetting contact angle, etc.) are optimized for the process; and the 'pixel errors' of arced surfaces are smoothed, it is still necessary to improve the conversion algorithm and the resolution of the projection system. Nevertheless, the surface roughness measured for the μSL prototype is adequate for medical applications such as dental implants.

The other design modification made (changing from a solid cylinder to a hollow one) is necessary because the solid μSL dental implant prototype reveals a substantial undesired curvature of the top surface of the cylinder (Fig. 3a). The reason for that is that with increasing height of a μSL object – dental implant prototypes with a height between 6 mm and 11 mm – the internal stress due to solidification in a solid structure causes the top surface to be deformed. Figure 6 shows a profile of this top surface. The curvature was simulated by a simple model and the radius (r) of the surface is approximately 2.9 mm in accordance with the simulated curve. An ideal flat surface would have a curvature radius $r = \infty$. A hollow cylinder consists of

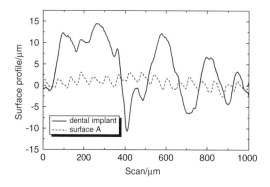

Figure 5. Surface roughness of sidewall of μSL objects: cylindrical dental implant and reference sample with a flat surface (A).

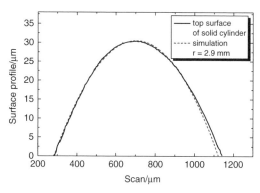

Figure 6. Top surface profile of solid μSL cylinder (including simulated profile).

less material and, as a result, the internal stress is reduced. The top surface curvature of side walls in these hollow cylinders can be neglected for this application (dental root implant). The hollow prototypes therefore produce improved results.

3.4 Future development work

Further experimental work leading to improvement of the μSL prototypes is being carried by optimizing the implant model geometry and the process parameters to obtain medical implants with desired properties and it is expected that additional results will be reported in the near future.

For instance, the next cylindrical dental implant model will be prepared with a bottom lid containing channels/pores of $100\,\mu m$ and 1 mm height. The purpose of incorporating such a lid underneath the μSL prototype is to control the outflow of non-cross linked resin in the hollow cylinder. Although the unexposed non-cross linked resin, located in the central region of the μSL cylinder, flows out slowly through the pores of the lid, a considerable amount of resin will be still present in the large cavity. For future work, it is intended to perform a flood exposure after μSL process so that the liquid resin in the hollow cylinder will rapidly solidify and be part of the final dental implant prototype.

4 CONCLUSION

Microstereolithography has been shown to be capable of manufacturing a medical implant geometry. Modifications to the dental root implant CAD model have had to be made to ensure the internal stresses during production do not distort the part geometry and pore shape (channels). Further experiments are being carried out to optimise the implant CAD models as well as the μSL process parameters to obtain medical implants with desired properties and it is expected that additional results will be reported in the future.

Clearly, for real dental implants different materials will be necessary and other geometries will be investigated which forms the next stage of the research to be performed.

ACKNOWLEDGMENTS

The authors would like to thank EPSRC for funding and CCLRC for the opportunity to carry out the Pump Priming project PCMF 042.

REFERENCES

Beluze L., Bertsch A. & Renaud P. 1999. Micro-stereolithography: a new process to build complex 3D objects, *Symposium on Design, Test and microfabrication of MEMs/MOEMs, Proceedings of SPIE* 3680 (2): 808–17.

Bertsch A., Lorenz H. & Renaud P. 1999. 3D microfabrication by combining microstereolithography and thick resist UV lithography, *Sensors and Actuators* 73: 14–23.

Kawata S., Sun H.-B., Tanaka T. & Takada K. 2001. Finer features for functional microdevices – Micromachines can be created with higher resolution using two-photon absorption, *Nature* 412: 697–698.

Li D., Ferguson S.J., Beutler T., Cochran D.L., Sittig C., Hirt H.P. & Buser D. 2001. Biomechanical comparison of the sandblasted and acid etched and the machined and acid-etched titanium surface for dental implants, *J Biomed Mater Res* 60: 325–332.

Lo W.J., Grant D.M., Ball M.D., Welsh B.S., Howdle S.M., Antonov E.N. & Bagratashvlili V.N. 2000. Physical, chemical and biological characterization of pulsed laser deposited and plasma sputtered hydroxyapatite thin films on titanium alloy, *J Biomed Mater Res* 50: 536–545.

Maruo S. & Kawata S. 1998. Two-photon-absorbed near-infrared photopolymerization for three-dimensional microfabrication, *Journal of Microelectromechanical Systems* 7 (4): 411–415.

Maruo S. & Ikuta K. 2002. Submicron stereolithograhy for the production of freely movable mechanisms by using single-photon polymerization, *Sensors and Actuators A* 100: 70–76.

Nakamoto T., Yamaguci K., Abraha P.A. & Mishima K. 1996. Manufacturing of three-dimensional micro-parts by UV laser induced polymerisation for three-dimensional microfabrication, *J. Micromech. Microeng.* 1(2): 240–253.

Simmons C.A., Valiquette N. & Pilliar R.M. 1999. Osseointegration of sintered porous-surfaced and plasma spray-coated implants: An animal model study of early post implantation healing response and mechanical stability, *J Biomed Mater Res* 47: 127–138.

Virtual modeling and rapid manufacturing – Bártolo (eds)
© 2005 Taylor & Francis Group, London, ISBN 0 415 39062 1

Virtual Engineering of complex systems: Towards robust solutions based on solid modeling

N.S. Sapidis

Department of Product and Systems Design Engineering, University of the Aegean,
Ermoupolis (Syros), Greece

ABSTRACT: Virtual Engineering systems aim at robust "Design-Interrogation Methods" which require availability of informationally-complete models for all parts of a design-project, including spatial constraints. This is the subject of the present investigation, leading to a new model for spatial constraints, the "virtual solid", which has been analyzed in a series of papers. This paper focuses on the solid-modeling aspects of the virtual-solid methodology, and studies solid-modeling problems (related to object definition and to object processing) that appear in layout-constraint analysis. It is established that "3D-constraint modeling" requires more sophisticated SM technologies compared to a CAD system for physical artifacts and assemblies.

1 INTRODUCTION

Design/analysis of complex electromechanical systems involves a large number of components each associated to various *design constraints*. Thus, the whole design-problem involves numerous constraints that must be analyzed/solved, so that the final synthesis does not violate them. Current "constraint solvers" ignore the 3D-geometrical nature of many of these constraints, and thus they are unable to treat complex design problems.

The present research builds on the papers Sapidis & Theodosiou (2001) and Theodosiou & Sapidis (2004), which propose use of Solid Modeling (**SM**) to accurately describe complex design-constraints. This paper analyzes one particular kind of constraints: Required Free Spaces (**RFSs**). These are essential, e.g., in a complex plant where many electromechanical systems require free spaces for maintenance, inspection and parts replacement. Also, some of these systems produce heat and thus RFSs must exist so that other pieces of machinery are not overheated. In other cases, RFSs define free floor-area for temporary storage of parts and for the plant personnel to efficiently perform inspection and other tasks.

This work focuses on geometric modeling of RFSs and identifies related geometric models and operations that the employed SM foundation must support. The surprising conclusions are:

- A CAD system for Required Free Spaces (in short, "*Free-Space CAD*" or **FSCAD**) involves more sophisticated SM models, methods and tools than a CAD system for physical artifacts and assemblies!

- Geometric modeling of 3D constraints involves some very interesting, *new* solid-modeling problems not addressed by the state-of-the-art in that field.

2 CAD MODELING OF SPATIAL CONSTRAINTS USING "VIRTUAL SOLIDS"

Informationally-complete representation of spatial constraints, based on "virtual solids" and SM, was introduced in Sapidis & Theodosiou (2001) and subsequently investigated in Theodosiou (2003) and Theodosiou & Sapidis (2004). Prior to this work, similar ideas, in a preliminary form, appeared in few publications: Chang & Li (1995) studied "assembly maintainability" employing user-defined "constraint volumes", modeled as spheres blended with each other by frustums of cones. Kim & Gossard (1991) solved a "packaging task" using standard mathematical optimization, yet the introductory part of that paper included an interesting comment on the "accessibility problem"; there, one reads about a proposal to define "a fictitious solid object with the shape of the desired access space". However, this idea was not explored any further by these authors.

2.1 Solid modeling for spatial constraints

A mechanical assembly includes a number of components whose position must satisfy various design constraints. Suzuki et al. (1997) considers the related 2D design problem, and classifies these constraints into three groups: *dimensional, regional and interference constraints.*

The first group includes distance and angular dimensions constraining the size, geometric form, location and orientation of components. *Regional constraints* restrict the region (area) where a component may lie, and finally, *interference constraints* specify components and regions that must not overlap. Current methods to solving these constraints adopt the following strategy (Sapidis & Theodosiou 2001):

a. All constraints are translated (exactly or approximately) into dimensional constraints.
b. Dimensional constraints are viewed as either a system of algebraic (in)equalities or as a graph, and are solved using a numerical or graph-theoretic method.

This methodology is rightly characterized as "design with dimensional constraints" and is one of the early approaches investigated by CAD pioneers. It is based on the fundamental assumption, implied by standard engineering drawings, that

*A design is fully described by a prescribed number of parameters.

This assumption, in general, is invalid, thus, it is not a surprise to anyone that, despite research efforts of twenty years (which considerably advanced the available tools for treating constraints), the scope of "design with dimensional constraints" is still limited to 2D- and trivial 3D-problems. More specifically, even for 2D problems this method has many disadvantages:

1. The assumption* is, in general, erroneous and should be applied only to simple 2D layouts.
2. This method often leads to "black box" software-tools, disallowing user involvement.

3. This method is unable to take advantage of evolving Solid-Modeling and CAD technologies.

The proposal of Sapidis & Theodosiou (Sapidis & Theodosiou 2001, Theodosiou & Sapidis 2004) aims at producing informationally complete descriptions of design constraints, and implies exactly the inverse approach:

(i) All design constraints are translated into properties (or constraints/problems) of real and/or virtual solids and relationships between these solids.
(ii) The resulting "virtual-solid modeling problems" are treated using procedures operating on these solids.

Obviously, regional and interference constraints may be directly described as "solidified constraints" according to **(i)**, thus we focus on dimensional constraints. Those describing features of a solid may be directly incorporates into the parametric description of the solid. In the case a parametric solid modeler is not available, then one faces the problem of incorporating inequality constraints like "the height H of the object X must be less than h". In this case, one must use virtual solids to translate the dimensional constraint into a regional one. For the above example, the geometric description of X and the value h uniquely define a virtual solid Y allowing statement of the above constraint as "X must be a subset of Y". Finally, dimensional constraints describing relations between components are incorporated into the SM system using either virtual solids or appropriate local-coordinate systems.

Analyzing industrial examples (Theodosiou 2003 and Theodosiou & Sapidis 2004) leads to the conclusion that the task of classifying/modeling a design constraint is not a trivial procedure, and it is doubtful that this can be fully automated. This is demonstrated by the example of Figure 1: This constraint is "in principle"

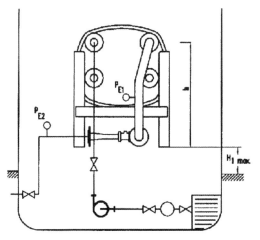 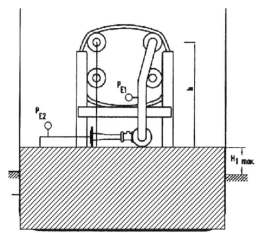

Figure 1. [Left] Manufacturer's specification regarding highest position Hlmax for the subsystem S. [Right] Equivalent solid-modeling based design-constraint employing the virtual solid V: $S \cap v \times \varnothing$.

a dimensional constraint as it refers to the position of the depicted subsystem. Thus, the obvious decision is to include it in the layout model "as is". However, it is straightforward, for a human user, to see that this is essentially a regional constraint and it can be modeled using a virtual solid and the intersection set-operation, as explained in Figure 1. Clearly, this choice is much "designer friendlier" than the first one.

3 SOLID-MODELING FOUNDATION FOR DEFINING FREE SPACES

Early research on "geometric modeling of design constraints" was relying on primitive-solids (box, cylinder, etc.) and elementary SM techniques like extrusions and set operations; see Section 2 and Chang & Li (1995), Sapidis & Theodosiou (1999). Recent work (see Sapidis & Theodosiou 2001, Theodosiou & Sapidis 2004) showed that industrial applications often involve complicated spatial-constraints whose geometric description is possible only in a state-of-the-art SM system. The analysis presented below establishes that RFSs impose severe requirements on the employed SM kernel. Indeed, modeling such constraints often employs complex SM operations and sometimes even poses new SM-problems; see details below.

3.1 Solid-model simplification for FSCAD

A vital component of FSCAD is the "2D/3D object simplification procedure", as FSCAD employs plant components and structural elements (e.g., see Figs 2–3) which are defined by final, and thus fully-detailed, CAD-models. Using these as "input information" often leads to over detailed SM descriptions of spatial constraints. Indeed, in Figure 2, this engine's "Lubricating Oil Cooler" (the component at the low-left

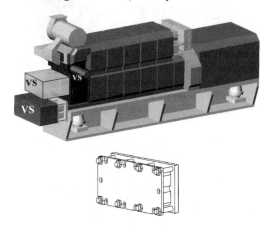

Figure 2. [Top] An engine and its RFSs. [Bottom] A component of this engine.

end of Figure 2 [Top] shown also in Figure 2 [Bottom]) corresponds to the RFS VS_C. Definition of VS_C involves faces of the "Cooler", which should be simplified (prior to construction of VS_C) so that VS_C is free of unnecessary details. In the example of Figure 3, the "exact RFS" (Fig. 3[Left]) is an extruded solid with many faces corresponding to a SAT file of 67 KB. However, the given layout-problem is affected negligibly if one replaces it with the "simplified FS" of Figure 3[Right]. This includes only five faces and is described by a SAT file of 5 KB, reducing the amount of required computer-memory by 93%!

The simplified profile in Figure 3[Right] has been produced by a "2D simplification" method which is an extension of that in (Sapidis & Theodosiou 1999). Specifically, Sapidis & Theodosiou (1999) forces the simplified profile to be inside the convex-hull of the original profile, while the present technique is free of this restriction and it extrapolates profile-edges when this extrapolation (a) does not violate the validity of the profile and (b) results in an error within the specified tolerance.

3.2 Modeling "illegal solids" in FSCAD

Often a manufacturer describes RFSs for humans and/or equipment by specifying required "floor area" or "traffic lanes" or by incompletely defining volumes using few surfaces. Here, one must deal with "volumes" that violate several of the fundamental laws of SM. For instance, in Figure 4: crew- and passenger-areas are incompletely defined using ship surfaces, 2D profiles, and planes defined by the designer.

Also, it is quite common that the designer wants some of these "volumes" to be *unbounded* in certain directions (e.g., see the two "vertical volumes" at the center of Figure 4). Finally, it is not uncommon for these models to include 2D or 3D geometric-elements (e.g., lines) "symbolically" describing accessories of RFSs.

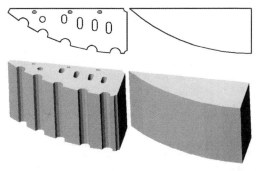

Figure 3. [Left] A volume defined by extruding a given profile. [Right] The "simplified volume" defined by extruding a "simplified profile".

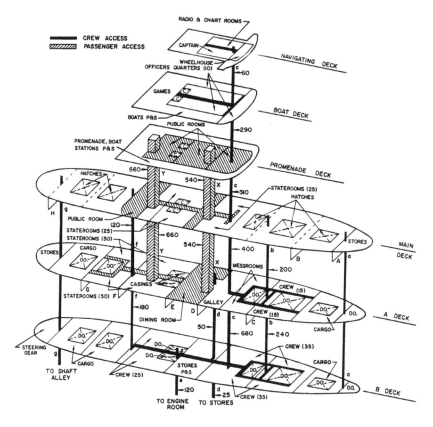

Figure 4. Access diagram for crew and passengers in a ship design.

Conclusion: a FSCAD system must be based on a robust SM kernel tolerating "illegal solids" and "non-standard modeling-operations" like those described above.

3.3 *Simultaneous handling of fully- and partially-evaluated free spaces in FSCAD*

When a designer deals with the geometric definition of a product/assembly, he/she always *fully evaluates* the specified geometric entities so that he/she can visually inspect the design, apply interrogation procedures, communicate his/her proposals to others, etc. This is not, in general, the preferred practice when one designs RFSs: here, one needs only to make sure that (a) the whole design (= "product models" + "design constraints") is valid, and (b) the RFSs are fully defined, in an appropriate form, in the design data-base. Thus, RFSs may be *unevaluated solid models*, a practice that leads to an economical description of these entities. For example a typical RFS is an extruded object defined by a planar polyline and a "height" (see Fig. 5(a)). Clearly, storing only the polyline and the "height" is

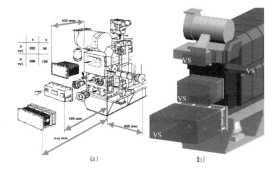

Figure 5. An engine and some of its RFSs.

much more economical than storing the resulting extruded solid (Fig. 5(b)).

Using exclusively "unevaluated Solid Models" to define RFSs is not a realistic approach, as some of the RFSs, e.g., the most complex ones, must be "detailed" as explicit 3D objects by the designer. This directly leads to the conclusion that the SM foundation of FSCAD must be able to handle 2D/3D solids with a varying degree of "explicitness" in their definition.

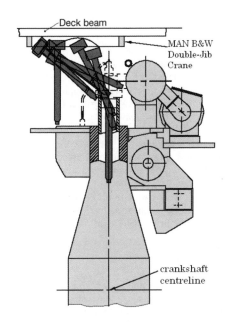

Figure 6. Tilted lift of an engine's piston using a double-jib crane: the RFS is a solid sweep.

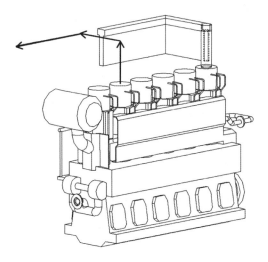

Figure 7. A "generalized assembly" including a standard assembly (ship engine) and a virtual solid (RFS) defined as a solid sweep.

3.4 2D & 3D solid-sweep in FSCAD

Many cases of RFSs, and also other spatial constraints, involve 3D Solid-Sweep in their definition; see, e.g., Figures 5–7. Although the "general solid-sweep", with an arbitrary 3D path, is still an unsolved problem, various special cases are solvable (Theodosiou 2003). For

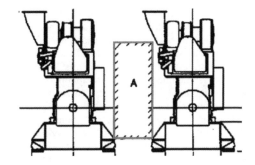

Figure 8. A "generalized assembly" including the virtual solid "A" (RFS for maintenance personnel) and two standard assemblies (ship engines).

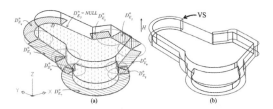

Figure 9. (a) A "generalized assembly" including a standard solid and several 2D virtual solids representing "required floor areas". (b) The RFS implied by these "required floor areas".

example, in Figure 6, the related solid-sweep problem may be solved by applying an "extrude" operation on the 2D domain defined as a "2D sweep" according to the movements of the piston-profile shown in this figure; indeed, although the 2D-sweep in Figure 6 is complex, an exact description of the outcome domain is possible, since the piston-profile undergoes only planar translations and rotations. On the other hand, in Figure 7, we have exactly the "opposite case" where the sweep problem is genuinely 3D yet still solvable as the union of a small number of instances of the moving solid (piston) and of three standard "extrusions". In conclusion, RFS modeling involves many special cases of 3D Solid-Sweep which are solvable; the corresponding tools must be included in the SM kernel supporting free-space design/analysis.

3.5 Assembly-modeling methods for artifacts and free spaces

The obvious way to combine RFS information with the corresponding assembly CAD-model is to define a "generalized assembly" including both physical objects and RFSs.

Existing assembly-modeling methods employ "mating conditions" ("against", "fit", "contact", etc) to define "relationships" between components. Often, the mating condition between an RFS and a mechanical part is indeed a standard one; see examples in Figures 5–7, where a face of the depicted RFSs and a face of the engine are related with an "against" condition. However, sometimes no standard mating-condition can describe the relationship between an RFS and a solid; see example in Figure 8.

Extending the list of mating conditions is not the only required modification of the "assembly-modeling subsystem" for it to support FSCAD. As explained in Section 3.2, often an RFS is only partially defined by "surfaces", which also must be included in the "generalized assembly"; see example in Figure 9.

4 CONCLUSION

The publications Sapidis & Theodosiou (2001), Theodosiou (2003) and Theodosiou & Sapidis (2004) have introduced a new methodology for modeling 3D design-constraints, which uses Solid Modeling to produce informationally-complete descriptions of these constraints. This approach gives rise to interesting solid-modeling problems, which have been the subject of this paper. The present study leads to the conclusion that: a *CAD system for 3D design-constraints involves more sophisticated solid- and assembly-modeling problems than a CAD system for physical artifacts and assemblies.* More specifically, Section 3 of the present paper establishes that design-constraint modeling requires:

1. A robust method to simplify solids and 2D domains
2. Solid-modeling methods for "illegal solids" and for "incompletely-described volumes"
3. Simultaneous handling of fully-evaluated and unevaluated solids and assemblies
4. Solid sweeping (at least for those many cases that a solution is known)
5. Assembly modeling with "generalized" mating-conditions

ACKNOWLEDGEMENTS

The author thanks G. Theodosiou for his contributions and an anonymous referee for a thorough examination of this paper. This work was supported by the Ministry of Development of Greece through the Research Projects "E-MERIT: An Integrated E-Collaborative Environment for Product & Process Modeling Using 3D Models and Avatars" and "Information Management and Exchange in Network-Centric Product Design: Modeling Component Knowledge in Design Repositories". This research was also supported by the Ministry of Education of Greece through the "*Heraclitus Research Grant:* Preliminary Industrial Design: Geometric/Information Models and Methods for Interactive Product-Design".

REFERENCES

Chang H. & Li T.Y. 1995, Assembly Maintainability Study with Motion Planning. Proceedings of the IEEE International Conference on Robotics & Automation, 1012–1019.
Kim J. & Gossard D. 1991, Reasoning on the Location of Components for Assembly Packaging. ASME Journal of Mechanical Design, 113 (4): 402–407.
Sapidis N. & Theodosiou G. 1999, Planar Domain Simplification for Modeling Virtual-Solids in Plant and Machinery Layout. CAD, 31 (10): 597–610.
Sapidis N. & Theodosiou G. 2001, Informationally Complete Product Models of Complex Arrangements for Simulation-Based Engineering, Modeling Design Constraints using Virtual Solids. Engineering with Computers 2000, 16 (3&4): 147–161.
Suzuki H., Ito T., Ando H., Kikkawa K. & Kimura F. 1997, Solving Regional Constraints in Components Layout Design based on Geometric Gadgets, AIEDAM, 11 (4): 343–353.
Theodosiou G. 2003, Solid Modeling of Free Spaces in Plant Layout: Application in Design of Ship's Engine Room. PhD Thesis, National Technical University of Athens.
Theodosiou G. & Sapidis N. 2004, Information Models of Layout Constraints for Product Life-Cycle Management: A Solid-Modeling Approach, CAD, 36 (6): 549–564.

Virtual modeling and rapid manufacturing – Bártolo (eds)
© 2005 Taylor & Francis Group, London, ISBN 0 415 39062 1

Femtosecond laser rapid prototyping and its applications to micro-nanodevices

Hong-Bo Sun

State Key Laboratory on Integrated Optoelectronics, College of Electronic Science and Engineering, Jilin University, China;
Department of Applied Physics, Osaka University, Suita, Osaka, Japan

Satoshi Kawata

Department of Applied Physics, Osaka University, Suita, Osaka, Japan;
JST-CREST; RIKEN, Hirosawa, Wako, Saitama, Japan

ABSTRACT: Use of femtosecond laser for rapid prototyping (RP) dramatically improves the technology, making RP a high-precision three-dimensional processing tool for fabrication of micro, nano-structures and devices. Depending on materials used, typical fabrication mechanisms include two-photon photopolymerization of resins and laser micro-explosion of transparent solids; considering the strategies of optical exposure, we have direct laser writing and multi-beam interference patterning methods. Here we present our fundamental study on direct laser micro-nanowriting based on two-photon polymerization and its application on photonic and mechanical micro-nanodevices.

1 WHY FEMTOSECOND LASERS

1.1 Laser-matter interactions

Use of femtosecond lasers for rapid prototyping (RP) leads to significant improvement of the technology from at least two aspects (Maruo et al, 1997; Sun et al, 2004): the *real three-dimensional (3D) resolution* and *high accuracy*. In order to achieve 3D resolution, we use laser wavelengths at near-infrared region, a transparent window of many macromolecules and inorganic materials. The laser thus penetrates into the bulk of a material without power dissipation, most importantly, without excitation pollution to the material along the light path so that fabrication is confined only at the laser focal spot. For attaining high spatial resolution, a femtosecond laser is indispensable. First, material excitation for fabrication is accomplished via nonlinear optical process. Since early 1990s, it was recognized that light-matter interactions for femtosecond pulses were fundamentally different from those resulting from longer pulses or CW lasers (Kieffer et al, 1989; Diels et al, 1996) because femtosecond pulses carry much larger peak power. With conventional light source the strength of light field is in the range of 1 V/cm and the resulting elongation of dipole is smaller than 10^{-16} m, much smaller than atomic or molecular diameters ($10^{-10} \sim 10^{-7}$ m). With femtosecond laser pulse irradiation, the field strength could be as intense as 10^{8} V/cm,

sufficient to induce direct bond breaking. Various nonlinear effects (Shen, 1984; Boyd, 1992) are easily launched, among which the most important for laser fabrication is multiphoton absorption (Goeppert-Mayer, 1931; Kaiser et al, 1961). Multiphoton absorption has an extremely small cross-section; it was confined to occur only in a small 3D volume around the close vicinity of the laser focus, less than the cubic wavelength (λ^{3}). Hence, a quite high 3D spatial resolution is achieved for pinpoint exposure. Secondly, when materials are irradiated with a femtosecond laser pulse, the photon energy is deposited much faster than electrons could transfer it to the lattice or molecule/atom oscillations through phonon emission, meaning that the excitation is a heat insulation process (Saeta et al, 1991; Glezer et al, 1996, 1997). This provides an ideal optical excitation means for many photochemical or photophysical reactions where thermal effect, a process difficult to localize, are not desired. Because of merits of 3D processing and high precision, femtosecond laser has now been utilized as a nanofabrication tool, playing a more and more important role in basic nano science and industry.

1.2 Various techniques of femtosecond laser-based micro-nanofabrication

Photons from a femtosecond laser could be tailored in time, spatial, and phase domains (Menzel, 2001), and usable photochemical and photo-physical processes in materials are almost infinite, meaning that the laser micro-nanofabrication is a broad field to explore. As the exposure strategy, direct focal spot scanning or called direct laser writing (DLW) and multi-beam interference patterning, and as material excitation mechanisms, two-photon induced photopolymerization of resins and laser-induced micro-explosion of transparent solids have been widely utilized. Before entering the major topic, rapid prototyping by femtosecond lasers, generally called two-photon photopolymerization, we first briefly introduce several other closely related researches to get a general idea on the field of laser micro-nanofabrication.

1.2.1 *Direct laser writing in transparent solids*

Glass is an important type of optical materials. We focused 800-nm-wavelength and 150-fs-width laser pulses into glass, it was found that if the deposited energy and the focusing condition were properly chosen, for example, use of high numerical aperture (NA = 1.4) objective lens, well-defined near-spherical spots with significant refractive index change ($\Delta n/n > 10^{-2}$) could be induced at the focal point. A broad fluorescence band associated with the irradiation-induced defects in several hundreds nanometer diameter spots were found (Watanabe et al, 1999a, 2000; Sun et al, 2000a) and was utilized for readout of 3D optical memory, by which a storage density of 10^{12} bits/cm^3 has been achieved (Watanabe et al, 1999b). Furthermore, we treated the spots as photonic atoms and arranged them to array the same way as atoms in real crystals. 3D photonic crystals (PhCs) (Yablonovitch, 1987; John, 1987) with varied Bravais lattices were therefore obtainable. The concept of PhC will be further detailed in Section 3. For example, a 3D face centered cubic (FCC) PhC with lattice constant around 1.0 μm shows a pronounced transmission dip, the fingerprint of bandgap effect, at 3490 cm^{-1} (Sun et al, 2001b). By continuous scanning the laser focus, lines were produced, which were piled into 2D triangular lattices (Sun et al, 1999a).

1.2.2 *Multi-beam interference patterning*

The idea of the technology is recording multi-beam interference created light patterns in various materials. It is nowadays mainly used for PhC fabrication. Both PhCs and holograms have periodically 3D dielectric functions and perform complete reconstruction of the electromagnetic fields (Berger et al, 1997a, 1997b). As a consequence of the periodicity of PhCs, the Fourier transform of the refractive index distribution function was well approximated by a small number of dirac functions, implying that PhCs or holograms can be recorded by a small set of plane waves. The Fourier transformation manifests the relation of PhC lattices and the laser wavevectors, i.e., the wavevectors are the

vectors of the reciprocal lattices. Therefore, PhC patterns can be designed by choosing a suitable beam number, their geometrical arrangement and polarizations (Campbell et al, 2000; Kondo et al, 2001). With the high peak power of femtosecond lasers, the interference patterns are recordable in many transparent solids or liquid media. Here we use photopolymerizable resins. If the light intensities of each beam are properly chosen, it is possible that the resin at the light intensity maxima sites is solidified, while the unpolymerized or less polymerized material at light minima volume was removed in the post exposure developing. The structure consisting of polymer skeleton and air voids is thus produced and functions as 3D PhCs.

The idea of constructing 3D PhCs by using multibeam interference was proposed in 1995 (Mei et al, 1995). Berger et al utilized this technology for 2D photoresist patterning (Berger et al, 1997b). 3D fabrication in photopolymers was first reported independently by some of the current authors (Shoji et al, 2000) and by Campbell et al (Campbell et al, 2000), respectively. The feature of our work (Shoji et al, 2003; Sun et al, 2003) is multi-step exposure, by which various lattice types and lattice constants could be arbitrarily chosen.

2 FEMTOSECOND LASER RAPID PROTOTYPING VIA TWO-PHOTON POLYMERIZATION

2.1 *The principle and setup*

Two-photon photopolymerization is one of the multiphoton absorption-induced photochemical reactions, which was experimentally observed for the first time in 1965 (Pao et al, 1965), and was utilized for micronanofabrication from 1997 (Maruo et al, 1997). It is simply realized by scanning tightly focused red or NIR laser beam according to preprogrammed patterns. Since the chemical reaction of polymerization was confined to occur solely at the close vicinity of the focal spot, pinpoint material solidification and pattern drawing can be achieved. After scanning, the sample was washed by a developer in order that unsolidified precursory liquid polymer was removed and a solid skeleton remains. Due to the requirement of high accuracy, a microscopic system (Parthenopoulos et al, 1989; Denk et al, 1990) is needed. A homemade system is schematically shown in Fig. 1.

The resin usable for the fabrication is a UV optical adhesive, which consists primarily of monomer, oligomer, initiator and photosensitizer (Odian, 1991). Monomer and oligomer, i.e., prepolymer that comprises of several monomer units, are the major components of a resin. The quantum yield of general monomers and oligomers is too low to polymerize with a reasonable rate. In order to increase the photopolymerizing

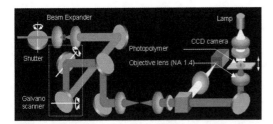

Figure 1.   Schematic two-photon photopolymerization system. We use a laser of 150-fs pulsewidth, 780-nm wavelength, and of operation under 82 MHz mode locking as the irradiation source. The laser beam is focused by a high-NA iol-immersion objective lens into the photopolymerizable resin. The 3D writing was accomplished either by mirror scanning the laser beam in the two horizontal dimensions plus vertical movement of the sample stage, or by stage translation in all 3D dimensions. The laser power was adjusted by a neutral density (ND) filter, and the optical adjustment and the entire fabrication is in-situ monitored with a coupling charged device (CCD) camera.

efficiency, one or several low-weight molecules that are more sensitive to light irradiation, called photoinitiator, are added. They form initiating species of radicals or cations by absorbing photons. Take the radical case for example, the initiation step:

$$I \xrightarrow{2h\nu} I^* \to R \cdot \tag{1}$$

where symbols denote photoinitiator (I), radical (R·) and I^*, an intermediate status of the photoinitiator after absorbing photon energy. The polymerization process is then described by the following equation:

$$R \cdot + M \to RM \cdot \xrightarrow{M} RMM \cdot \cdots \to RM_n \cdot \tag{2}$$

where M is monomer or oligomer molecules. The photo produced radicals react with monomers (or oligomers), producing monomer radicals, which are liable to combine with new monomers. Therefore the monomer radicals expand in form of chain reaction, until two radicals meet with each other. The chain propagation stops in either of the following channels:

$$RM_n \cdot + RM_m \cdot \to RM_{m+n}R \tag{3}$$

$$RM_n \cdot + RM_m \cdot \to RM_n + RM_m \tag{4}$$

In many cases, the energy collection (Eq. 1) and triggering of chain polymerization (Eq. 2) were cooperatively accomplished by more than one type of molecules. A photosensitizer is the molecule that absorbs light and then transfers the energy to a photoinitiator.

With such a scheme, the photoinitiation process is further expressed as:

$$S \xrightarrow{h\nu} S^* \cdots \xrightarrow{I} I^* \to R \cdot \tag{5}$$

where S is photosensitizer.

Compared with the general UV RP, here the process features that the initiation is fulfilled by simultaneously absorbing, by an initiator molecule, two red photons, instead of one UV photon. The above descriptions of polymerization were based on radical initiation. Another common mechanism, ionic polymerization isn't discussed because its less use in our research.

2.2   Spatial resolution

The fabrication accuracy or spatial resolution is an important issue to address. The lateral size of volume element (voxel) of photopolymerization was considered to be limited by Rayleigh's criterion arising from the optical diffraction limit (Born et al, 1999), to $1.22\lambda/NA$. For an ideal Gaussian beam of 800-nm wavelength under 1.4 NA focusing and the refractive index of the material of 1.5, the limit is approximately 460 nm. The spatial resolution could be improved by utilizing either shorter wavelength or larger NA focusing, but the diffraction limit cannot be circumvented in optical imaging due to interactions of light signals from the neighboring features in the focal plane. In laser fabrication, we concern only the focal spot. Thus only if a threshold exists in material response to light excitation, the reaction volume could be reduced smaller than that defined by the diffraction limit. The *threshold* is a level of light intensity, above which the photochemical reactions become irreversible. On this case, the diffraction limit becomes just a measure of the focal spot size; it does not put any actual restraint to voxel sizes (Tanaka et al, 2002).

For the particular radical initiated photopolymerization, the oxygen molecules dissolved in the resin savage radicals (Flory, 1952), acting as a thresholding mechanism. By tailoring the light intensity, it is possible to reach such status that light excited radicals survive and initiate polymerization only at the central portion of the focal spot where exposure energy is larger than the threshold. The intensity of the surrounding volume was low, and therefore easily clapped under the TPA threshold. Figure 2 shows the SEM image of voxels formed under different exposure durations and laser pulse energies [Fig. 2 (a)]. A lateral spatial resolution down to 120 nm, far below the diffraction limit, has been achieved. Fig. 2 (c) and (d) is a micro-bull structure (Kawata et al, 2001), which consists of smooth and tough surfaces, curvatures and sharp horns. The 10-μm long and 7-μm high bull is the smallest animals ever made artificially, and is as

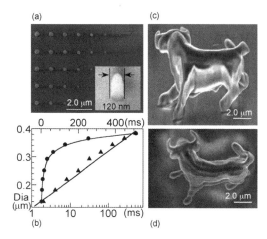

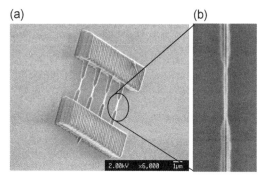

Figure 4. A two-photon photopolymerized fiber suspended between two anchors. (a) The SEM image, and (b) the magnified view of the fiber center portion, indicating a line width of ~65 nm.

Figure 2. Sub-diffraction-limited spatial resolution. (a) Voxels formed at varied exposure time and laser pulse energy, and (b) a typical dependence of voxel lateral size on the exposure time. (c) and (d) are different views of a micro-bull sculpture.

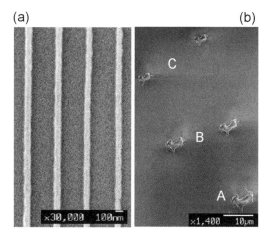

Figure 3. The spatial resolution improvement by intentionally inducing a radical quencher into the resin. (a) SEM image of 100-nm-width lines, and (b) the microbull structures of lengths: A–10 μm, B–7 μm and C–4 μm.

resin, SCR 500 except for an additional quencher. The nearly 20% improvement of the fabrication accuracy would reduce the dimensions of polymerized objects in principle by the same percentage. The SEM images of micro-bulls fabricated with the improved precision are shown in Fig. 3(b). Their size reduces from the original 10-μm length [A], to 7 μm [B] and 4 μm [C].

In order to explore how small feature may be depicted by the two-photon photopolymerization scheme, we scanned a fiber line suspended between two polymerized anchors by minimizing the laser pulse energy to a level, at which the fiber is sufficiently narrow but still robust enough to resist washing by the solvent. Figure 4 shows the SEM image of the fiber, of which the central part is 65 nm in width and the two ends are exposed heavier.

We haven't discussed the longitudinal spatial resolution yet, which is however, isn't independent from the lateral resolution. They are correlated with each other through the point spread function (Born et al, 1999). In order to characterize the 3D spatial resolution, an important problem is how to obtain isolated, complete voxels since voxel truncation happens when the laser is focused too near the substrate surface [Fig. 5 (a)], while on the other hand, floating voxels would be formed and they were flushed away during developing if the laser was focused far above the substrate.

To solve this problem, we proposed an ascending scan method (Sun et al, 2002). The laser focus was scanned along a slant line, [e.g., the arrowed straight line in Fig. 5 (b)] from below to above substrate surface, residing at series of positions. At each position, the shutter for laser was switched ON and kept for a short term, the exposure time, to get voxels at an identical exposure condition.

Figure 5 (c) shows SEM images of thus produced voxels. The left voxels (A, B, C) were truncated. They

small as the red blood cell. It is a good proof of the feasibility of depicting sub-diffraction-limit features by two-photon photopolymerization.

Any chemical species that tends to prohibit photopolymerization reaction even with a small concentration functions as a quencher. Oxygen is just one of many choices. By attentively adding a quencher solution into the resin, it is much easier to control the polymerization than the case of dissolved oxygen. Figure 3(a) shows line structures of 100-nm width that was two-photon photopolymerized with the same

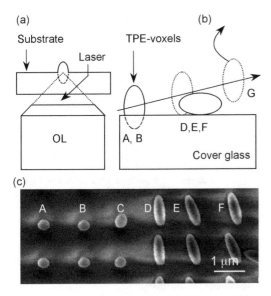

Figure 5. Schematic ascending scan method for achieving isolated and complete 3D voxels. (a) Laser beam focusing illustration, showing the substrate truncation effect, (b) voxels formed at different focusing levels, and (c) SEM image of voxels produced by scanning the laser focal spot from inside to above the substrate.

stuck to and erected on the substrate, revealing only their lateral size information. The rightmost voxels were floated away (not shown). A transition state always exists between these two regions: the edge of voxels was bordered at and weakly adhered to the substrate surface, and they were overturned during developing. That is the case of D, E, F, from which both lateral, and most importantly, the longitudinal information could be attained. The longitudinal resolution is typically 2–3 times the lateral one for high NA, for example, NA = 1.4 focusing when polymerization occurs at near-threshold level.

2.3 Surface roughness

Surface roughness (Takada et al, 2005) is another important issue to address since it significantly affects the surface forces that dominate in nanomechanical devices and micromachines (Sun et al, 2000b; Galajda et al, 2001), and as for photonic elements (Miura et al, 1997; Cumpston et al, 1999) like waveguides, mirrors, and switches, coarseness-induced scattering is the principal source of power dissipation. These devices, however complicated their designs are, are realized through point-by-point scanning. Hence, the vestige of voxels is reasonably considered as the prime factor to cause the surface roughness.

As a test model, a series of cube structures of $12\,\mu m \times 12\,\mu m$ surface area, a size range of interest

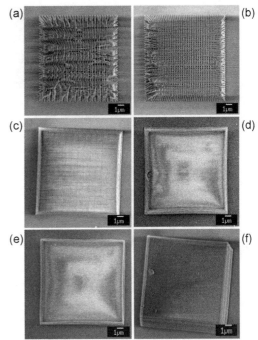

Figure 6. Cube structures for testing the surface roughness under different voxel distances or depicting pitches. (a) $400\,nm \times 400\,nm$, (b) $300\,nm \times 300\,nm$, (c) $200\,nm \times 200\,nm$, (d) $100\,nm \times 100\,nm$, (e) $50\,nm \times 50\,nm$, (f) The tilted view of the 50-nm cube in (e).

for micro-nanosystems, were designed and polymerized with different planar voxel distance (D), or pitch [Fig. 6]. It is obvious that when D is larger than a certain value, e.g., D = 100 nm for the current condition: NA = 1.4 focusing, photon flux density $p = 2.0\,MW/cm^2$, and exposure duration, $\tau = 1\,ms/voxel$, the scanning trace governs the surface quality. That is the case of Figs 6 (a–c).

In order to make quantitative characterization, we use atomic force microscope (AFM) to analyze the sample surface, for which the sampling was conducted at 256×256 points on $10\,\mu m \times 10\,\mu m$ area. Moreover, we used the definition of the surface roughness, R, as follows

$$R = \frac{1}{S_0} \iint |f(x,y) - z_0| dx dy \qquad (6)$$

where S_0 is the scanned area, $f(x,y)$ is the height at each point, and z_0 the average height. From Fig. 7 (a), we confirmed the pitch-dependent roughness, for example, R falls rapidly from 76 nm to 9 nm as the pitch decreases from 400 nm to 100 nm. It is also interesting to notice that the roughness tends to saturate, to a value

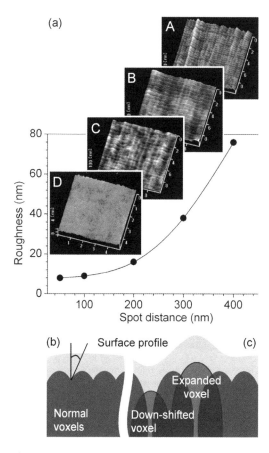

(a)

(b) Surface profile (c)

Normal voxels · Down-shifted voxel · Expanded voxel

Figure 7. A possible mechanism of surface roughness formation. (a) R-D plot. The inset are AFM images of cube surfaces with (A) D = 300 nm, (B) 100 nm, (C) 50 nm, and (D) a spincoated polymer surface. (b) An illustrative model to show the voxel-induced roughness can't propagate to the sample surface when $D < D_s$. (c) Therefore ambient and material factors should be responsible for the surface unflatness.

we may define as R_s, when the pitch is smaller than the critical value D_c. Both SEM and AFM measurement demonstrate the surfaces scanned with 100-nm [Fig. 6(d) and the inset Fig. 7 (a)-B] and 50-nm [Fig. 6(e) and Fig. 7 (a)-C] pitches are almost identical. Judging from Fig. 7 (a), $R_s \sim 8$ nm, the value of D ~ 50 nm. The existence of the critical pitch, D_c, and saturated roughness, R_s could be interpreted by examining how a surface is piled out of voxels in two dimensions. Voxels take ellipsoid shape that is associated with the NA of the focusing lens, and their size is delicately determined by the laser pulse energy and the exposure time. Therefore, the roughness dependence on the pitch could be attributed as different degrees of voxels overlapping. The gradually improved surface quality in Fig. 6 corresponds to the overlapping degrees, defined as the ratio

of voxel size to the pitch, from 61%, 81%, 122%, 243%, and 486% for (a)–(e), respectively. On the other hand, liquid resins in the concaves boarded by voxel ends [Fig. 7(b)] is difficult to remove due to the surface Gibbs free energy that is minimized by the negative curvature. As a result, the roughness propagated to the polymerized surface becomes negligible compared with those caused by other factors when the overlapping degree is high enough, or equivalently the pitch is sufficiently small. This is the origin of the critical scanning pitch, which is obviously a function of NA, laser pulse energy and exposure duration.

The roughness of a surface spin-coated with the same resin is approximately 1 nm [Fig.7 (a)-D], while surfaces scanned at varied laser wavelengths, focusing and exposure conditions were at the range of 4~11 nm. The R_s was found insensitive to the variation of the above parameters but slightly affected by the laser power, e.g., R_s = 6.7 nm, 7.5 nm, and 8.3 nm for the photon flux density of 1.7, 6.2, and 8.9 MW/cm^2, respectively.

The larger roughness may arise from (i) nonuniform surface relief occurring during the material solidification, to which the surface protrusions in Fig. 7 (a)-C, features much larger than voxel size, could be attributed; (ii) ambient factors, e.g., vibration may cause voxel displacement [Fig.7 (c)]; and (iii) jittering of the laser output induces the voxel size fluctuation [Fig.7(c)], which should be the origin of the weak power dependence.

The photopolymerized surface roughness tends to be minimized at several nanometers, or 1/50~1/100 visible or near-infrared wavelengths, which could satisfy the requirements of various photonic and optoelectronic devices to their surface and interfaces.

3 APPLICATIONS TO 3D PHOTONIC CRYSTALS

3.1 *The challenge of 3D photonic crystals*

Photonic crystals are microstructures with periodical distribution of refractive indexes. The formation of photonic bandgap (Joannopoulos et al, 1995; Noda 2000, 2001) leads to the inhibition of spontaneous emission and light propagation along a sharp turn, making PhC an ideal candidate for all-optical integration circuits (Bitwas et al, 1995). In order to fully use these unique properties of PhCs for exploring photon-matter interactions in arbitrarily tailored electromagnetic circumstance and pushing the artificial composite materials for diversified optoelectronic and photonic applications, 3D structures are ultimately needed. However, their fabrication is technically challenging. For example, theoretical work has predicted that complex lattice structures may possess stronger photonic bandgap effect, like distorted diamond (Soukoulis et al, 1995), face-centered cubic (FCC) with non-spherical atoms (Li, et al, 1998), and spiral structures (Toader et al, 2001). These geometries are not possibly or very difficult to produce by conventional technologies such as self-organization of colloidal particles, layer-by-layer packing of two-dimensional semiconductor meshes, electrochemical etching and vacuum deposition on structured substrates. To a large degree due to the technical reason, the current PhC study is mainly focused on two-dimensional structures. Furthermore, inclusion of materials of versatile characteristics to PhC structures is essential for realizing device functions like high-efficiency light emission, large photonic bandgap tunability, enhancement of nonlinear phenomena, and so forth. This is sometimes a harsh requirement to the currently existing methods since they are generally applicable to some a specific material system. This situation hinders the deep understanding of PhC physics and versatile industrial applications of the delicate novel optical material. The problems are now possibly solved by the femtosecond laser prototyping technology.

3.2 Arbitrary 3D lattices

There are at least two merits in fabricating PhCs using two-photon photopolymerization technology. First of all is the potential to produce PhCs of arbitrarily designed lattices. PhCs of varied lattice types, lattice constants, and filling factors were realizable just by scanning different CAD patterns. Secondly, the diversification of usable materials and functions. The progress of molecular material engineering has made it possible to synthesize polymers with performances similar to or better than their inorganic counterparts. By introducing functional groups to unsaturated monomer or oligomer units in molecular structure or just by doping the functional polymers into known photopolymerizable materials, optical, electronic, magnetic, and mechanical functions can be imparted to devices.

Some of the current authors proposed for the first time using two-photon photopolymerization technology for fabrication of 3D PhCs (Sun et al, 1999b, 2001a) and observed pronounced PBG effect. A commercially available resin Nopcocure 800 consisting of radical photoinitiator and acrylic acid ester was utilized. The linear absorption of the resin extends from UV to around 370 nm. Due to a low TPA cross-section, 10^{-56} cm^{4}s photon^{-1}, the fabrication was conducted using regeneratively amplified laser pulses.

Shown in Fig. 8 (a) is the schematic illustration of the log-pile structure and the fabricated PhC in Fig. 8 (b). Fourier transform infrared spectroscopic measurement shows the 20-layer logpile lattices with different in-plane rod spacing: 1.2 μm, 1.3 μm, and 1.4 μm give rise to transmittance dips under normal incidence

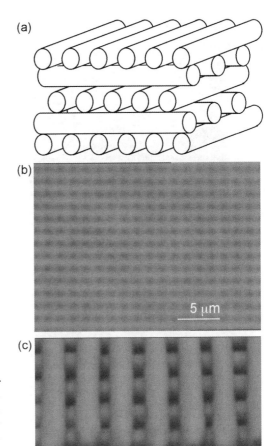

Figure 8. A logpile PhC structure. (a) an illustration and (b) the actually fabricated lattices.

at wave numbers of 2550, 2510, and 2450 cm^{-1}, respectively. The increase of the wavelength of transmission minima versus lattice constant is an expected photonic bandgap feature.

Diamond structure is one of ideal lattices to give strongest PBG effect. It is composed of two interpenetrating FCC lattices, one displaced 1/4 of a lattice constant, in each direction from the other. When the two sub-lattices are of different atoms, then the diamond lattice becomes the zincblende. Examples of electronic crystals with the diamond crystal structure are diamond, silicon and germanium. In PhCs, the crystal atoms are represented as photonic atoms. Different from the case in solid matrix (Sun et al, 2001b), inside which atoms are inlaid or, in-situ fixed where they are created, photonic atoms polymerized from liquids need

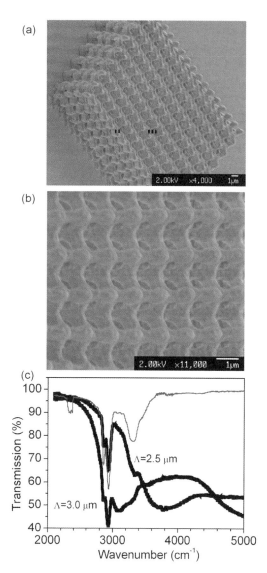

(a)

(b)

(c)

Figure 9. Diamond-lattice photonic crystal structure. (a) SEM images of the fabricated structure, and (b) the transmission spectra.

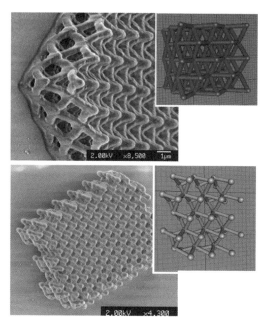

Figure 10. Atom-photonic lattices: Mg-lattice (top) and Cu-lattice (bottom). The right insets are stick-an-ball molecular models.

connection. The linking bridges, or covalent bonds in a stick-and-ball molecular model, are actually the major component in attained crystal configuration. Figure 9(a) shows a <100>-orientated diamond lattice, where the period is $\Lambda = 2.5\,\mu m$ (Kaneko et al, 2003b). The distance between the rod-ended balls, or intuitively called photonic atoms, is 1.1 μm. The rods and the photonic atoms have diameters of 500 nm and 580 nm, respectively. From the figure, it is clearly seen that a quite smooth rod surface have been created. FTIR measurement [Fig. 9(b)] shows that the transmission spectra of PhC structures of 2.5-μm and 3.0-μm lattices give rise to transmission minima at $3790\,cm^{-1}$ and $3100\,cm^{-1}$, respectively. They occur at the identical normalized frequency of 1.05, as defined by λ/Λ, showing the linear scaling performance of the diamond lattices. The single-layer attenuation is as large as 35%.

By mimicking the arrangement of atoms in other real-world crystals, more PhCs can be conveniently designed and fabricated, as an example, the Cu-lattice PhC in the top of Fig. 10 and Mg-lattice PhC in the bottom. We called this type of PhC as atom-photonic lattices.

3.3 *Materials functionalization*

As another commonly used type of femtosecond laser micro-nanofabrication material, vitreous silica glasses have very low optical nonlinearity and therefore doping is essential to realize most of desired functions. By rare-earth metallic ions doping, long-lasting phosphorescence, rewritable memory by valence states control, and ultrafast switching have been realized (Miura et al, 1997; Hirao et al, 2001). Compared with glass, polymers, in addition to their possible stronger optical nonlinear performances, exhibit excellent characteristics in homogeneity, transparency, ease of fabrication and compatibility to versatile solvents, which are comparable or better than their glass counterparts.

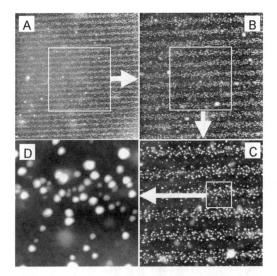

Figure 11. AFM images of the two-photon photoreduced nanoparticles array of areas $20 \times 20\,\mu m^2$ (up-left), $10 \times 10\,\mu m^2$ (up-right), $5 \times 5\,\mu m^2$ (low-right), and $1 \times 1\,\mu m^2$ (low-left).

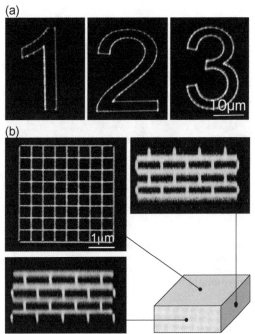

Figure 12. Laser micro-nanowriting by photoreduction of metal ions in a polymer matrix. (a) Several digits at different depths, and (b) a 3D PhC defined by nanoparticle lines.

In order to achieve nano level uniformity of particles dispersion in the matrix polymer, we used metal ions instead of nanoparticles as the precursory material (Kaneko et al, 2003a), in details, a gold ion-doped polyvinyl alcohol (PVA) film was exploited. In order to test the photoreduction of the metal ions, we split an 800-nm and 150-fs laser beam into two beams and irradiated the sample with the interference pattern. Regularly arrayed nanoparticle grating lines are created as indicated by the AFM images [Fig. 11].

From the sequentially reduced scanning areas, starting at $20\,\mu m \times 20\,\mu m$, to $10\,\mu m \times 10\,\mu m$, $5\,\mu m \times 5\,\mu m$ and $1\,\mu m \times 1\,\mu m$, it is resolved that the smallest discernible beads have diameters around 15 nm. The particles were assigned as gold nanoparticles mainly because of the emergence of the 585-nm peak in the absorption spectrum, a feature of plasmon absorption of gold nanoparticles. The distribution of particle size basically follows the light intensity distribution: larger particles appear at the bright interference fringes while at the dark region, both particle size and number density are small, agreeing with the prediction of absorption spectrum.

When a femtosecond laser beam is focused into the metal ion-polymer hybrid material, complicated gold nanoparticle patterns were created following the 3D scanning trace. Figure 12 (a) shows digits written in different depths of the film with $10\,\mu m$ layer interval. Figure 12 (b) is a 3D logpile structure depicted the same way. This work is important for two-photon fabrication of nanodevices using metal ion-doped polymers for inducing electric conductivity and enhanced mechanical strength, as well as metal-particle size dependent peculiar optical and electronic functions.

4 NANOMECHANICAL DEVICES AND NANOMECHANICS

4.1 Micro-nanomechanical devices and their 3D diagnosis

With the sub-diffraction-limited spatial resolution, a number of micro components have been fabricated, as those shown in Fig. 13.

3D micromechanical systems and micromachines need proper positioning, shaping and jointing. A pre-operation evaluation is critical for judging and optimizing designs and fabrications. The tube shown in Fig. 13 (d) should be hollow according to design. However, it is not an easy task to confirm the internal status even if it is a quite simple device. Common optical microscope doesn't have sufficient resolving power to distinguish details in three dimensions, particularly in the longitudinal directions. Electronic microscopes have high spatial resolution and high imaging quality, however, they are useful only for observing the appearance of objects. Two-photon confocal

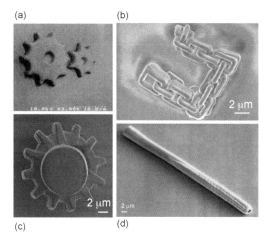

(a)   (b)

2 μm

2 μm   2 μm

(c)   (d)

Figure 13. Micro-nanomechanical components and devices fabricated by two-photon photopolymerization. (a) A micro-gearwheel pair, (b) a microchain, (c) a micro-gearwheel affixed to a shaft, and (d) a micro tube.

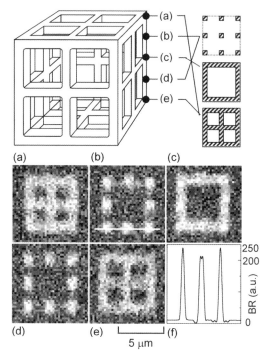

Figure 14. Fluorescence images of a dye-doping photopolymerized cubic cage (a–e), of which a schematic is illustrated by the top line drawing. The heights are (a)-0.0, (b)–1.35, (c)-2.7, (d)–4.05 and (e)–5.4 μm, respectively. A fluorescence intensity distribution is extracted from (b) and is given by (f).

microscope possesses 3D imaging feasibility with resolution better than conventional transmission or reflection optical microscope [Wilson, 1990; Gu, 1999]. To attain a high signal-to-noise ratio, fluorescence read-out is preferable. The structure is therefore required to emit fluorescence, which is not sufficiently strong in general resins. To reach this end, we proposed a fluorescence dye-labeling method (Sun et al, 2001c). Experimentally the laser dye of LD490 was induced by dissolving SCR500 resin to ethanol solution of LD490 till saturation. The mixed material was photopolymerized following the normal procedure. For two-photon confocal readout, the system as used for fabrication was configured at the reading part by further equipping with a confocal pinhole and a CCD detector.

Figure 14 shows the two-photon confocal fluorescence images of a 5.4 μm × 5.4 μm × 5.4 μm cubic cage. From the design, the top illustration, it is seen that different heights of the structure correspond to varied cross-sections. This was clearly reflected by the sliced fluorescent images [Fig. 14 (a–e)]. The fluorescence signal is more intense than that needed for reading. A signal/noise ratio more than 20 [Fig. 14(f), extracted from Fig. 14(b)] was obtained. In contrast, features of an identical cage solidified from unadulterated SCR resin can't be resolved due to a low signal-to-noise ratio.

The above result implies the feasibility of dye doping for characterization of 3D micro-nanodevices. Figure 15 shows confocal fluorescence images of axial and radial cross-sections of a tube similar to that shown in Fig. 13 (d). The absence of internal fluorescence confirms its vacancy. The 1.6 μm internal diameter agrees well with designed value. These results show the effectiveness of the 3D micro-diagnosis technology.

4.2 Device optical actuating

The fabricated micro-nanodevices should work for a certain function. The essential issue is actuating mechanism. Appropriate electric, optical, thermal, magnetic, and chemical effects should be found to reach this end. Electrically control of micro systems, the requirement of MEMS, is most desirable, however, there is a long way to go before involving conductive polymer into structures and integrating polymer devices on an IC-contained semiconductor chip or developing polymer ICs. Optical force provides a simple solution, which is the currently most practical mechanism for actuating micro-nanodevices. Here we introduce several possible mechanisms (Sun et al, 2004).

Windmills rotate when facing wind blowing, which is known since ancient time. It is not surprising that similar rotation phenomena have been frequently observed in laser-trapped particles (Ashkin, 1970; Ashkin et al, 1990). The rotation torque arises from the axial irradiation force and from the asymmetrical or rotation-symmetrical shape of the particles. In micro-machine, it is important to design a device structure

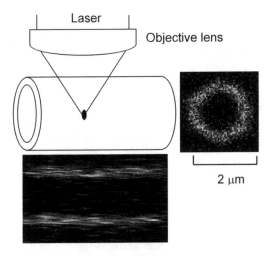

Figure 15. Internal micro-diagnosis of a 3D micro-tube. Schematic scheme (top), and axial (bottom) and radial (right) cross-sectional fluorescence images.

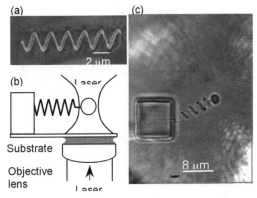

Figure 16. A two-photon polymerized micro oscillator. (a) a spring, (b) the driving scheme, and (c) a fabricated spring-contained micro-oscillator under laser trapping.

of helical shape of proper rotation symmetry so that the structure could be fixed (trapped) at a suitable position and with desired orientation for high stability and for minimizing the friction between the rotating parts and its axel if there is, whereby the translation momentum from the "photon wind" is efficiently converted to the spinning momentum of the object. The gearwheel shown in Fig. 13 (c) was rotated by this strategy.

Light itself can carry angular momentum. Another machine rotating mechanism is coupling the photon angular momentum to the object to be rotated. Microparticles rotating experiments have been carried out using (i) elliptically polarized light, (ii) laser beam with a helical phase structure interacting with absorptive particles. Changing the polarization of the light from plane to circular caused the rotation frequency to increase or decrease, depending on the sense of the polarization with respect to the helicity of the beam, (iii) by rotating the asymmetric laser beam. These mechanisms need experimentally application to micronanomachines.

A microparticle can be three-dimensionally trapped at the laser focus (Ashkin et al, 1992). Then it moves following the movement of the focal spot. The phenomenon isn't specialized to isolated particles but also applicable to a portion of an object. This implies that the entire object may be pushed or pulled in random directions only if a part of a structure is captured by a laser focus.

Comparing with the last two technologies, the push-pull method (i) doesn't need designing machine to special shapes, as needed by the windmill driving mechanism, and (ii) the actuation isn't limited to rotating, but random 3D movement. As an example, a micro-oscillator system was driven by this principle, as to be introduced in the next section.

4.3 Mechanics of nanodevice operation

Figure 16 (a) shows a micro-spring, of which the spiral radius is 150 nm, the coil radius is 1.0 μm, and the pitch 2 μm (Sun et al, 2001d). Calculation according to the mechanical formula with the above parameters gives rise to a spring constant of $k = 4.8 \times 10^{-3}$ N/m.

In order to optically operate the spring, we fixed the spring to an anchor and designed a sphere at its free end [Fig. 16(b)], expecting that sphere movement by laser trapping triggers the oscillation of the spring. The actually fabricated structure is shown in Fig. 16 (c). When the laser focus was carefully adjusted, the bead was found three-dimensionally trapped and freely manipulated. The spring was pulled by moving the trapped bead. Releasing the bead by blocking the laser initiated the oscillation. The spring was observed to be prolonged [Fig. 17 (b)] from its original length [Fig. 17 (a)], and restoring [Fig. 17 (c)] to its original state after the laser was turned off [Fig. 17 (d)]. An elongation up to 7 μm in multi-operation didn't cause any elasticity failure, as evidenced by the fact that the spring always restored to its original length.

Calculated from the damping oscillation curve, the spring constant is 1×10^{-8} N/m. A more reliable characterization using dragging experiments (Asikin et al, 1992) shows the value is 4.8×10^{-6} N/m. Preliminary analysis shows that the 3-order positive error ($10^{-6} \rightarrow 10^{-3}$ N/m) are due to the significant change of material property at sub-micron to nanoscale, for example, the effective removal of inter-network small molecules reduces the steric resistance, and therefore softens the

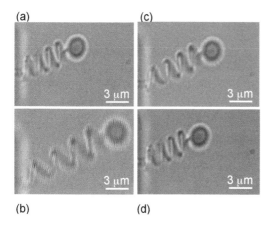

(a)   (c)

(b)   (d)

Figure 17. A working micro-oscillator system. The micro-spring is (a) in its natural state, (b) pulled by a length, (c) released, and (d) recovered to its original states 20 s after the releasing.

spring. The 2-order negative error ($10^{-6} \rightarrow 10^{-8}$ N/m) is mainly a result of neglect of the viscosity or resistance exerting to the spring surface.

Above results indicates that the behavior of micro-nanodevices is much different from that of macro scale devices, abiding by laws possibly unknown to us, although we have the knowledge that at small size, mass and gravity are less important but various surface forces dominate. Understanding of the micro-nano material characteristics, exploring the laws of nanodevices operation and investigating their origin at molecular level are important task for the future research in the filed. That is also the reason why we propose the research of nano mechanics, which is now enabled by the femtosecond laser rapid prototyping.

5 CONCLUSION

Laser micro-nanofabrication concerns production of structures of sub-micro-to-nano feature size through photon-matter interactions in light electromagnetic fields or photon force exerting to a material. It is an important content of nanophotonics research. Two-photon induced photopolymerization as introduced here is one of the currently most widely used laser nanofabrication approaches, by which novel optoelectronic devices on the basis of photonic crystals and functional micro-nanomachines are expected. It is reasonably expected that the femtosecond laser micro-nanoprototyping technology would play a role for polymer and transparent solid media in their nanodevicing as the role the planar lithography has played for semiconductors.

REFERENCES

Ashkin, A. 1970. Acceleration and trapping of particles by radiation pressure. *Physical Review Letters* 24 (4): 156–159.

Ashkin, A., Schutze, K., Dziedzic, J. M., Eutenuer, U. & Schliwa, M. 1990. Force generation of organelle transport measured invivo by an infrared laser trap. *Nature* 348 (6299): 346–348.

Ashkin, A. 1992. Forces of a single-beam gradient laser trap on a dielectric sphere in the ray optics regime. *Biophysical Journal* 61 (2): 569–582.

Berger, V., GauthierLafaye, O. & Costard, E. 1997a. Photonic band gaps and holography. *Journal of Applied Physics* 82 (1): 60–64.

Berger, V., GauthierLafaye, O. & Costard, E. 1997b. Fabrication of a 2D photonic bandgap by a holographic method. *Electronics Letters* 33 (5): 425–426.

Biswas, R., Chan, C. T., Sigalas, M., Soukoulis, C. M. & Ho, K. M. 1995. In: Soukoulis, C. M. (ed) *Photonic Band Gap Materials*. London: Kluwer Academic Press.

Born, M. & Wolf, E. 1999. *Principles of Optics*. 7th edn. Cambridge: Cambridge Univ Press.

Boyd, R. W. 1992. *Nonlinear Optics*. San Diego: Academic.

Campbell, M., Sharp, D. N., Harrison, M. T., Denning, R. G. & Turberfield, A. J. 2000. Fabrication of photonic crystals for the visible spectrum by holographic lithography. *Nature* 404 (6773): 53–56.

Cumpston, B. H., Ananthavel, S. P., Barlow, S., Dyer, D. L., Ehrlich, J. E., Erskine, L. L., Heikal, A. A., Kuebler, S. M., Lee, I. Y. S., McCord-Maughon, D., Qin, J. Q., Rockel, H., Rumi, M., Wu, X. L., Marder, S. R. & Perry, J. W. 1999. Two-photon polymerization initiators for three-dimensional optical data storage and microfabrication. *Nature* 398 (6722): 51–54.

Denk, W., Strickler, J. H. & Webb, W. W. 1990. 2-photon laser scanning fluorescence microscopy. *Science* 248 (4951): 73–76.

Diels, J. C. & Rudolph, W. 1996. *Ultrashort Laser Pulse Phenomena: Fundamentals, Techniques, and Applications on a Femtosecond Time Scale (Optics and Photonics)*. New York: Academic.

Flory, P. J. 1952. *Principles of Polymer Chemistry*. New York: Cornell University Press.

Galajda, P. & Ormos, P. 2001. Complex micromachines produced and driven by light. *Applied Physics Letters* 78 (2): 249–251.

Glezer, E. N., Milosavljevic, M., Huang, L., Finlay, R. J., Her, T. H., Callan, J.P. & Mazur, E. 1996. Three-dimensional optical storage inside transparent materials. *Optics Letters* 21 (24): 2023–2025.

Glezer, E. N. & Mazur, E. 1997. Ultrafast-laser driven microexplosions in transparent materials. *Applied Physics Letters* 71 (7): 882–884.

Goeppert-Mayer, M. 1931. Ann Phys 9: 273

Gu, M. 1999. *Advanced Optical Imaging Theory*, Heidelberg: Springer.

Hirao, K., Mitsuyu, T., Si, J. & Qiu, J. (eds.) 2001. *Active Glass for Photonic Devices: Photoinduced Structures and Their Applications*. Berlin: Springer.

Joannopoulos, J. D., Meade, R. D. & Winn, J. N. 1995. *Photonic Crystals: Modeling the Flow of Light*. Singapore: Princeton Univ Press.

John, S. 1987. Strong localization of photons in certain disordered dielectric superlattices. *Physical Review Letters* 58 (23): 2486–2489.

Kaiser, W. & Garrett, C. G. B. 1961. *Physical Review Letters* 7: 229

Kaneko, K., Sun, H. B., Duan, X. M. & Kawata, S. 2003a. Two-photon photoreduction of metallic nanoparticle gratings in a polymer matrix, *Applied Physics Letters* 83 (7): 1426–1428.

Kaneko, K., Sun, H. B., Duan, X. M. & Kawata, S. 2003b. Submicron diamond-lattice photonic crystals produced by two-photon laser nanofabrication, *Applied Physics Letters* 83 (11): 2091–2093.

Kawata, S., Sun, H. B., Tanaka, T. & Takada, K. 2001. Finer features for functional microdevices, *Nature* 412 (6848): 697–698.

Kieffer, J. C., Matte, J. P., Belair, S., Chaker, M., Audebert, P., Pepin, H., Maine, P., Strickland, D., Bado, P. & Mourou, G. 1989. Absorption of an ultrashort laser-pulse in very deep plasma-density gradients. *IEEE Journal of Quantum Electronics* 25 (12): 2640–2647.

Kondo, T., Matsuo, S., Juodkazis, S. & Misawa, H. 2001. Femtosecond laser interference technique with diffractive beam splitter for fabrication of three-dimensional photonic crystals. *Applied Physics Letters* 79 (6): 725–727.

Li, Z. Y., Wang, J. & Gu, B. Y. Full band gap in fcc and bcc photonic band gaps structure: Non-spherical atom. *Journal of Physical Society of Japan* 67 (9): 3288–3291.

Maruo, S., Nakamura, O. & Kawata, S. 1997. Three-dimensional microfabrication with two-photon-absorbed photopolymerization. *Optics Letters* 22 (2): 132–134.

Mei, D. B., Cheng, B. Y., Hu,W., Li, Z. L. & Zhan, D. H. 1995. *Optics Letters* 20 (5): 429–431.

Menzel, R. 2001. *Photonics: Linear and Nonlinear Interactions of Laser Light and Matter*. Berlin: Spriger.

Miura, K., Qiu, J. R., Inouye, H., Mitsuyu, T., Hirao, K. 1997. Photowritten optical waveguides in various glasses with ultrashort pulse laser. *Applied Physics letters*, 71 (23): 3329–3331.

Noda, S., Tomoda, K., Yamamoto, N. & Chutinan, A. 2000a. Full three-dimensional photonic bandgap crystals at near-infrared wavelengths *Science* 289 (5479): 604–606.

Noda, S., Chutinan, A. & Imada, M. 2000b. Trapping and emission of photons by a single defect in a photonic bandgap structure. *Nature* 407 (6804): 608–610.

Odian, G. 1991. *Principles of Polymerization*, 3rd edn. New York: Wiley.

Pao, Y. H. & Rentzepis, P M. 1965. Laser induced production of free radicals in organic compounds. *Applied Physics Letter* 6(5): 93.

Parthenopoulos, D. A. & Rentzepis, P. M. 1989, 3-Dimensional optical storage memory. *Science* 245 (4920): 843–845.

Saeta, P., Wang, J. K., Siegal, Y., Bloembergen, N. & Mazur E. 1991. Ultrafast electronic disordering during femtosecond laser melting of GaAs, *Physical Review Letters* 67 (8): 1023–1026.

Shen, Y. R. 1984. *The Principles of Nonlinear Optics*. New York: Wiley.

Shoji, S. & Kawata, S. 2000. Photofabrication of three-dimensional photonic crystals by multibeam laser interference into a photopolymerizable resin. *Applied Physics Letters* 76 (19): 2668–2670.

Shoji, S., Sun, H. B. & Kawata, S. 2003. Photofabrication of wood-pile three-dimensional photonic crystals using four-beam laser interference. *Applied Physics letters* 83 (4): 608–610.

Soukoulis, C. M. (ed) 1995. *Photonic Band Gap Materials, NATO Asi Series. Series E, No 315*. Dordrecht: Kluwer Academic.

Sun, H. B., Xu, Y., Matsuo, S. & Misawa, H. 1999a. Microfabrication and characteristics of two-dimensional photonic crystal structures in vitreous silica. *Optical Review* 6 (5): 396–398.

Sun, H. B., Matsuo, S. & Misawa, H. 1999b. Three-dimensional photonic crystal structures achieved with two-photon-absorption photopolymerization of resin. *Applied Physics Letters* 74 (6): 786–788.

Sun, H. B., Juodkazis, S., Watanabe, M., Matsuo, S., Misawa, H. & Nishii, J. 2000a. Generation and recombination of defects in vitreous silica induced by irradiation with a near-infrared femtosecond laser. *Journal of Physical Chemistry B* 104 (15): 3450–3455.

Sun, H. B., Kawakami, T., Xu, Y., Ye, J. Y., Matsuo, S., Misawa, H., Miwa, M. & Kaneko, R. 2000b. Real three-dimensional microstructures fabricated by photopolymerization of resins through two-photon absorption. *Optics Letters* 25 (15): 1110–1112.

Sun, H. B., Mizeikis, V., Xu, Y., Juodkazis, S., Ye, J. Y., Matsuo, S. & Misawa, H. 2001a. Microcavities in polymeric photonic crystals. *Applied Physics Letters* 79 (1): 1–3.

Sun, H. B., Xu, Y., Juodkazis, S., Sun, K., Watanabe, M., Matsuo, S., Misawa, H. & Nishii, J. 2001b. Arbitrary-lattice photonic crystals created by multiphoton microfabrication. *Optics Letters* 26 (6): 325–327.

Sun, H. B., Tanaka, T., Takada, K. & Kawata, S. 2001c. Two-photon photopolymerization and diagnosis of three-dimensional microstructures containing fluorescent dyes. *Applied Physics Letters* 79 (10): 1411–1413.

Sun, H. B., Takada, K. & Kawata, S. 2001d. Elastic force analysis of functional polymer submicron oscillators. *Applied Physics Letters* 79 (19): 3173–3175.

Sun, H. B., Tanaka, T. & Kawata, S. 2002. Three-dimensional focal spots related to two-photon excitation. *Applied Physics Letter* 80 (20): 3673–3675.

Sun, H. B., Nakamura, A., Shoji, S., Duan, X. M. & Kawata S. 2003. Three-dimensional nanonetwork assembled in a photopolymerized rod array. *Advanced Materials* 15 (23): 2011–2014.

Sun, H. B. & Kawata, S. 2004. Two-photon photopolymerization and 3D lithographic microfabrication. *Advances in Polymer Science* 170: 169–273.

Tanaka, T., Sun, H. B. & Kawata, S. 2002. Rapid subdiffraction-limit laser micro/nanoprocessing in a threshold material system. *Applied Physics Letter* 80 (2): 312–314.

Takada, K., Sun, H. B. & Kawata, S. Improved spatial resolution and surface roughness in photopolymerization-based laser nanowriting. *Applied Physics Letter* 86 (7): Art. No. 071122.

Toader, O. & John S. 2001. Proposed square spiral microfabrication architecture for large three-dimensional photonic band gap crystals. *Science* 292 (5519): 1133–1135.

Wilson, T. (ed) 1990. *Confocal Microscopy*. London: Academic.

71

Watanabe, M., Juodkazis, S., Sun, H. B., Matsuo, S. & Misawa, H. 1999a. Luminescence and defect formation by visible and near-infrared irradiation of vitreous silica. *Physical Review B* 60 (14): 9959–9964.

Watanabe, M., Juodkazis, S., Sun, H. B., Matsuo, S., Misawa, H., Miwa, M. & Kaneko, R. 1999b. Transmission and photoluminescence images of three-dimensional memory in vitreous silica. *Applied Physics Letters* 74 (26): 3957–3959.

Watanabe, M., Juodkazis, S., Sun, H. B., Matsuo, S. & Misawa, H. 2000. Two-photon readout of three-dimensional memory in silica. *Applied Physics Letters* 77 (1): 13–15.

Yablonovitch, E. 1987. Inhibited spontaneous emission in solid-state physics and electronics. *Physical Review Letters* 58 (20): 2059–2062.

Virtual modeling and rapid manufacturing – Bártolo (eds)
© *2005 Taylor & Francis Group, London, ISBN 0 415 39062 1*

Virtual reality: Engineering applications, promises and challenges

M.C. Leu
Intelligent Systems Center and Department of Mechanical and Aerospace Engineering, University of Missouri-Rolla, Rolla, Missouri, USA

X. Peng
Department of Mechanical Engineering, Prairie View A&M University, Prairie View, Texas, USA

ABSTRACT: Virtual reality (VR) techniques provide natural interactions between the virtual environment and the user. They enable the user to create, modify and manipulate 3D models intuitively, and at the same time allow the user to visualize and analyze data in an immersive virtual environment. By immersing the user in the virtual environment, virtual reality reveals spatially complex structures in a way that makes them easy to understand and study. VR has been applied to many aspects of the modern society and is becoming an increasingly important technology in a number of fields including design, manufacturing, medicine, entertainment, training, etc. This paper provides a review of virtual reality technology and its engineering applications, and also discusses the future promises and challenges of this technology.

1 INTRODUCTION

Since the first virtual reality simulator "Sensorama Simulator" was invented in 1956, the concept of virtual reality (VR) has continued to fascinate engineers, scientists, and the general public. The term virtual reality has been used by many researchers and described in various ways. According to the book "The Silicon Mirage" (Aukstakalnis & Blatner 1992), it can be broadly defined as: "Virtual Reality is a way for humans to visualize, manipulate and interact with computers and extremely complex data." Generally speaking, the features of virtual reality are: 1) it is a medium of communication, 2) it requires physical immersion, 3) it provides synthetic sensory stimulation, 4) it can mentally immerse the user, 5) it is interactive (Sherman & Craig 2003). Virtual reality is closely associated with an environment commonly known as Virtual Environment (VE). Virtual environment is an interactive graphic display enhanced by special processing and by non-visual display modalities, such as auditory and haptic feedback, to convince the users that they are immersed in a real physical space (Banerjee & Zetu 2001).

Because of the availability of the more advanced computer and other hardware and the increasing efforts of researchers, virtual reality technology and application has made considerable progress in the past decade. Growing interest in virtual reality has led to numerous applications of this technology in automobile, aircraft, entertainment, medicine, sports and other industries as well as in education and training (Leu et al. 2001, 2002, 2003, Lu et al. 1999, Maiteh et al. 2000). Based on factors including display hardware, graphics rendering algorithms, level of user involvement, and level of integration with the physical world, VR systems can be classified into different categories as follows:

Immersive VR: In an immersive virtual environment, typically a user wears a head-mounted display (HMD) or shutter glasses while viewing a computer monitor or standing inside a spatially immersive display. The concept of immersion is that the virtual environment surrounds the user, wholly or partially. For example, a CAVE™ has screens on several (commonly four) sides of a cube to provide the sense of immersion as shown in Figure 1.

Desktop VR: Non-immersive VR systems normally run on standard desktop computers, hence the term "desktop VR." Desktop VR systems use the same 3D computer graphics as immersive VR systems, but there are two key differences. First, the virtual environment does not surround the user – it is seen only on a single screen in front of the user. Second, the user typically navigates through and interacts with the environment using traditional desktop input devices such as a mouse and a keyboard (although specialized 3D input devices may also be used).

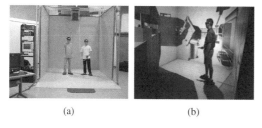

(a)   (b)

Figure 1. The CAVE™ system at the University of Missouri-Rolla and an application example.

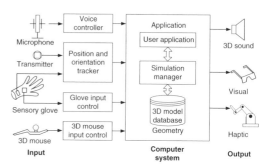

Figure 2. Virtual reality configuration.

Image-Based VR: This displays a realistic virtual world using images (instead of computer generated models). Its basic approach is to manipulate the pixels in images to produce the illusion of a 3D scene, rather than to build the 3D scene on a computer. The simplest type of image-based VR is a panorama – a series of images taken with a camera at a single position pointed in multiple directions.

Telepresence is a computer-generated virtual environment consisting of interactive simulations and computer graphics in which a human experiences presence in a remote location. An example is a pilot in a sophisticated simulator, which is used to control a real airplane 500 miles away and provides to the pilot visual and other sensory feedback as if the pilot were actually in the cockpit looking out through the windscreen and feeling the turbulence. Other applications of telepresence involve the use of remotely operated vehicles (e.g. robots) to handle dangerous conditions (e.g. nuclear accident sites) or for deep sea and space exploration.

Augmented reality (AR) is the use of a transparent HMD to overlay computer generated images onto the physical environment. Rapid head tracking is required to sustain the illusion. In the most popular AR systems, the user views the local real-world environment, but the system augments that view with virtual objects. For example, the Touring Machine system (Güven & Feiner 2003) acts as a campus information system, assisting a user in finding places and allowing him/her to pose queries about items of interest such as buildings and statues.

The aim of this paper is to give an overview of the virtual reality technology, including the hardware and software components and engineering applications. The future promises and research challenges of this technology are also discussed.

2   VIRTUAL REALITY SYSTEM COMPONENTS

Virtual reality enables the user's immersion into a sensory-rich, interactive experience. A typical virtual reality configuration is shown in Figure 2. Using input and output sensors, a virtual world is generated and manipulated under the control of both the reality engine and the participant. Software integrates various hardware elements into a coherent system that enables the user to interact with the virtual environment. Below we will briefly review the hardware and software components commonly used for the creation of virtual environments.

2.1   VR hardware

VR hardware plays an extremely important role in a virtual reality system. It provides physical devices that comprise an immersive environment such as CAVE™, HMD, etc. Input and output devices transmit the information between the participant and the virtual environment. They provide interactive ways for the information communication. This section will describe the hardware used in various VR systems including tracking devices, input devices, visual stereo displays, haptic devices and auditory displays.

2.1.1   Input devices

An input device is the bridge to connect the real world and the virtual world. It gathers the input information from the user and transmits it to the virtual world. Using different input devices provides an intuitive and interactive environment for the user to immerse in the virtual world.

A key input device is the tracking device. There are two types of tracking devices commonly used in VR systems: position tracker and body tracker. A position tracker is a sensor that reports an object's position (and possibly orientation) and maps it to the object's relative position in the virtual environment. The current position tracking devices used in VR include electromagnetic, mechanical, optical, video-metric, ultrasonic, and neural position-sensing devices (Sherman & Craig 2003). A body tracking device monitors the position and action of the participant. Commonly used body tracking techniques include tracking the head, the hand and fingers, the eyes, the torso, the feet, etc.

Other than the tracking device, 3D mice, 6-DOF mice, joy sticks, sensory gloves, voice synthesizers, force balls, etc. can be used as input devices for interaction with the user in the virtual environment. In addition to functioning as a normal mouse, the 6-DOF mouse can report height information as it is lifted in three dimensions. The force ball senses the forces and torques applied by the user in three directions. The CyberGlove® is a fully instrumented glove that uses resistive bend-sensing elements to accurately transform hand and finger motions into real-time digital joint-angle data. Voice input devices and biocontrollers are also used to provide a convenient way for the user to interact with the virtual environment.

2.1.2 Output devices

How the user perceives the virtual reality experience is based on what feedback the virtual environment can provide. Output devices are used in the virtual environment to present the user with feedback about his or her actions. Three of the human perceptual senses, i.e. visual, aural, and haptic, are commonly presented to the VR user with synthetic stimuli through output devices.

There are three general categories of visual displays. The least expensive method is using stereo graphics on a desktop monitor and a pair of LCD shutter glasses. The head-mounted display (HMD) can provide a fairly immersive experience for the user. It is a helmet worn by the user with one small monitor in front of each eye to display images as a 3D view of the virtual world. Often the HMD is combined with a head tracker so as to allow the participant a full 360-degree view. HMDs are less costly and more portable than projection systems. However, they have a lower resolution and can be cumbersome. Projection systems can be made up of a single large screen or several screens. By providing stereoscopic images and with the application of stereo glasses, the projection system gives the most realistic representation possible of graphical information. The most commonly used multi-screen projection system is CAVE™, which uses four to six large screens in the shape of a cube. It allows multiple viewers to physically immerse in the virtual environment.

Haptics broadly refers to touch sensations that occur for the purpose of perception or physical contacts with objects in the virtual world. Haptic interface devices such as the CyberTouch™ read finger contact information and output resistive forces to individual fingers. PHANToM™ is a ground-based device that provides the user a feel of force when he/she manipulates a virtual object with a physical stylus that can be positioned and oriented in six degrees of freedom.

Besides visual and haptic feedback, sound can be incorporated in the virtual environment to enhance the participant's sense of presence. Standard speakers or headphones are generally used as the auditory display hardware in the virtual environment. An application example is to generate sound to confirm the selection when an object is selected. Another example is to provide different pitches of sound to simulate the machining with different properties of material.

2.2 VR software

Software is required to integrate various hardware devices into a coherent system that enables the user to interact with the virtual environment. When deciding which VR software is best suited to a particular application, various features of the VR software need to be considered. They include features to support cross-platform, VR hardware, and importing 3D models from other systems, 3D libraries, optimization of level of detail, interactivity of virtual world, and multi-user networking.

VR implementation software can be classified into two major categories: software development kits (SDKs) and authoring tools (Isdale 1998). SDKs are programming libraries (generally in C or C++) that have a set of functions for the developer to create VR applications. Authoring tools provide the graphical user interfaces (GUIs) for the user to develop the virtual world without requiring tedious programming.

3 ENGINEERING APPLICATIONS

3.1 Concept design

In a concept design, the exact dimensions of the design part are not determined initially, and the designer is more interested in creating part shapes and features. Commercial CAD systems such as Unigraphics, Ideas, Catia, PRO/E, etc. are powerful geometric modeling tools, but they require precise data for designing objects and thus do not allow the users to implement their ideas on shape and feature design in an intuitive manner. Their user interface generally consists of windows, menus, icons, etc., which tend to prevent the users from focusing on the design intent. Another limitation of the conventional CAD system is in the use of input devices. The designers use a two-dimensional input device, usually the mouse, for the construction of three-dimensional objects. Concept design using virtual reality techniques enables the user to create, modify and manipulate solid CAD models intuitively, and at the same time allows the user to visualize CAD models in a virtual environment immersively. Some systems that provide such capabilities are: 1) 3-Draw, a system for interactive 3D shape design introduced at MIT (Sachs et al. 1991); 2) 3DM, built by researchers at the University of North Carolina (Butterworth et al. 1992); 3) JDCAD, developed by researchers at the University of Alberta, Canada (Liang & Green 1994); 4) Holosketch, created by Deering (1996) at Sun Microsystems; and 5) COVIRDS, developed by Dani & Gadh (1997) for creating shape designs in a virtual reality environment called COVIRDS (Conceptual VIRtual Design System).

Peng & Leu (2004a, b) have developed a Virtual Sculpting system for freeform design as shown in Figure 3. The virtual sculpting method is based on the metaphor of carving a solid block into a 3D freeform object workpiece like a real sculptor would do with a piece of clay, wax or wood. The VR interface includes stereo viewing and force feedback. The geometric modeling is based on the Sweep Differential Equation method (Blackmore & Leu 1992) to compute the boundary of the tool swept volume, and based on the ray-casting method to perform Boolean operations between the tool swept volume and the virtual stock in dexel data to simulate the sculpting process (Peng & Leu 2004b). Incorporating a haptic interface into the virtual sculpting system provides the user with a more realistic experience. Force feedback enables the user to feel the model creation process like actual sculpting with physical materials. The PHANToM™ manipulator is used as a device to provide the position and orientation data of the sculpting tool and at the same time to provide haptic sensation to the user's hand during the virtual sculpting process.

3.2 Manufacturing

Application of virtual reality technologies to modeling and simulation of manufacturing facilities and operations is gradually emerging. Generally speaking, virtual reality in manufacturing applications can be categorized into the following areas: process simulation, factory layout design, assembly prototyping, and part flow simulation.

Figure 3. A chair generated using the virtual sculpting system developed at the University of Missouri-Rolla (Peng & Leu 2004a).

The biggest advantage of using virtual reality for factory layout design is that it supports the user in planning space or logistical issues by allowing interactively moving and relocating the machines after the simulation has been carried out. VR-Fact! developed at University of Buffalo can be used to create digital mock-up of a real factory shopfloor for a given product mix and set of machines. An example is shown in Figure 4. By intuitively dragging and placing modular machines in the factory, designers can study issues such as plant layout, cluster formation and part flow analysis. It uses mathematical algorithms to generate independent manufacturing cells. The VR walk through environment of this software provides a unique tool for studying physical aspects of machine placements.

Chawla & Banerjee (2001, 2002) at the Texas A&M University presented a 3D virtual environment to simulate the basic manufacturing operations (unload, load, process, move, and store). It enables a 3D facility to be reconstructed using spacefilling curves from a 2D layout. The facility layout is represented in 3D as a scenegraph structure. The scenegraph structure automatically encapsulates the static and the dynamic behavior of the manufacturing system.

Kibira & McLean (2002) at the National Institute of Standards and Technology presented the design of a production line for a mechanically assembled product in a virtual reality environment. Their research mainly focused on the partitioning and analysis of the assembly operation of the prototype product into different tasks and allocation of these tasks to different assembly workstations. The manufacturing process design is constructed using three commercial

Figure 4. VR-Fact! Simulation example (Virtual Reality Laboratory, University of Buffalo).

software applications: The geometry of components is created by using AutoCAD™; IGRIP™ is used to provide the graphical ergonomic modeling of workstation operations; and QUEST™ provides the discrete event modeling of the overall production line.

Kelsick et al. (2003) presented an immersive virtual factory environment called VRFactory, which was developed with an interface to the commercial discrete event simulation program SLAM II. The virtual environment is built in a CAVE™-like projection facility. This factory allows the user to explore the effect of various product mixes, inspection schedules, and worker experience on productivity while immersing in a visual, three-dimensional space. The geometric models of the machines are created using the Pro/Engineer CAD software and the WorldUP™ modeling software and then loaded in a virtual world. Figure 5 illustrates an application example of the VRFactory.

Zachmann & Rettig (2001) at University of Bonn, Germany investigated interaction techniques to perform assembly tasks in the virtual environment. They identified that interaction metaphors for virtual assembly must be balanced between naturalness, robustness, precision and efficiency. Multimodal input techniques were utilized to achieve robust and efficient interaction with the system including speech input, gesture recognition, tracking, and menus. Precise positioning of parts was addressed by constraining interactive object motions and abstract positioning via command interfaces. A natural grasping algorithm was presented to provide intuitive interaction.

The simulation of a robotic work cell in a factory was created by the Virtual Reality and Prototyping Laboratory at University of Missouri-Rolla as shown in Figure 6. The goal was to provide a viability check for a proposed robotic work-cell at a nearby industrial manufacturing plant. The company had done 2D analysis to ensure that the process was feasible and that no problems such as interference or crashing existed. However, the only way to know what would happen when the designs were extended into the third dimension was to build the working environment in some fashion. A physical mock-up would be complex and very expensive. The virtual work cell environment was designed for realistic stereoscopic visualization and to provide the capabilities of interference detection, constraints checking, and cycle-time analysis for support of actual work cell development.

Figure 5.   VRFactory application example (Kelsick et al. 2003).

Figure 6.   Robotic work cell simulation developed at University of Missouri-Rolla.

3.3   Driving simulation

A driving simulator is a virtual reality tool that gives the user an impression that he/she drives an actual vehicle. This is achieved by taking the driver inputs from the steering wheel and pedals and feeding back the corresponding visual, motion and audio cues to the driver (Fang et al. 2001). The advantages of a driving simulator vs. driving with a real vehicle may be a wide range of possible configurations, repeatable conditions, easy to change tasks and parameters and good experimental efficiency. A driving simulator can be used to study design and evaluation of vehicles, highway systems, in-vehicle information and warning systems, traffic management systems and many more.

The National Advanced Driving Simulator (NADS) developed by National Highway Traffic Safety Administration in the United States is the most advanced driving simulator existing in the world today (Chen et al. 2001). The applications of this driving simulator have included driver crash avoidance study, evaluation of advanced in-vehicle systems and control technologies, and highway design and engineering research related to traffic safety. NADS uses a dome 24 feet in diameter with an interchangeable car cab sitting inside of the dome. Different cabs that can be used currently include Ford Taurus, Chevy Malibu, Jeep Cherokee and Freightliner. Within the dome are the projectors that provide 190° front and 65° rear field of view, as shown in Figure 7. At the same time, the

Figure 7. National Advanced Driving Simulator (NADS).

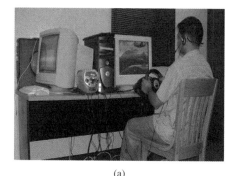

(a)

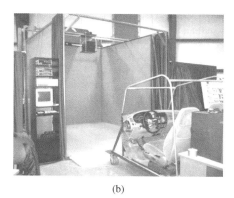

(b)

Figure 8. (a) Drunk Driving Simulator on a PC at University of Missouri-Rolla. (b) Car cab integrated with CAVE™ system used for driving simulation at University of Missouri-Rolla.

motion subsystem, on which the dome is mounted, provides 64 feet of horizontal and longitudinal travel and 330 degrees of rotation providing a total of 9 degrees of freedom. The VR effect enables the driver to feel acceleration, braking and steering cues as if he/she were driving a real car, SUV, truck, or bus. The system supports generation and control of traffic within the virtual environment.

Ford Motor Company developed a motion-based driving simulator VIRTTEX (VIRtual Test Track EXperience) to test the reactions of sleepy drivers (Grant et al. 2001). This simulator can generate forces that would be experienced by a person while driving a car. The simulator dome houses five projectors, three for the forward view and two for the rear view, that rotate with the dome and provide a 300° computer-generated view of the road. Different car cabs in the simulator are attached to a hydraulic motion platform, called the hexapod as Stewart platform, that can simulate the motion associated with more than 90 percent of the typical miles in the U.S. including spinouts. A continuation of many years of driver drowsiness research is conducted by Volvo using VIRTTEX. The new safety technology developed through this research will be integrated into the new car design.

The Drunk Driving Simulator developed at the University of Missouri-Rolla (Sirdeshmukh 2004) is the initial development of a virtual reality based driving simulator that is intended for use as a tool to educate people, especially college and high school students, about the consequences of drunk driving. The virtual environment is initially built on a PC with a desktop monitor as shown in Figure 8a. The developed system consists of hardware interface, force feedback, virtual driving environments and evaluation software. The drunk effects are simulated by implementing the control time lag and visual effect. The simulator has

been extended to the CAVE™ system to provide sufficient immersion and improved realism (Nandanoor 2004). The design and implementation of such a simulator was quite different from the single-screen driving simulator. It involved multiple projections, where the display of each projection was controlled by a specific PC and all the PCs involved with the projection have to synchronize with each other. This driving simulator uses a car cab as shown in Figure 8b.

3.4 Construction

Recently virtual reality techniques have been successfully explored in the area of civil engineering and construction. Virtual reality offers considerable benefits for many stages of the construction process planning in the following aspects: (1) Enabling a designer to evaluate the design and make modifications by immersing himself/herself in a building, (2) Allowing virtual disassembly and reassembly of the components to design the construction process, and (3) Offering a "walk through" view of the facility and experience the near-reality sense of the construction.

VIRCON (The VIrtual CONstruction Site) is a collaborative project between the University College London, Teeside University, the University of Wolverhampton and eleven construction companies in the United Kingdom (Dawood et al. 2003). The aim of this project was to provide a decision support system which allows the planners to trade off the temporal sequencing of tasks with their spatial distribution using VR technologies. The research involved both the development of a space scheduling tool called "critical space analysis", its combination with critical path analysis in a space-time broker, and the development of advanced visualisation tools for both of these analyses. This system applied visual 4D planning techniques that combine solid CAD models with the construction schedule (time) to allow not only visualizing the construction products but also visualizing the movement of objects. This provides better evaluation and communication of activity dependency in spatial and temporal aspects.

DIVERSITY (Distributed Virtual Workspace for Enhancing Communication within the Construction Industry) is a project funded by the European Union Information Society Technologies (Sarshar & Christiansson 2004). Ten universities and research institutes are participating in this project to investigate the use of VR in construction. DIVERCITY aims to develop a "shared virtual construction workspace" that allows construction companies to conduct client briefing, design reviews, simulate what-if scenarios, test constructability of buildings, communicate and coordinate design activities between teams. It allows the user to produce designs and simulate them in a virtual environment. It also allows the project teams based in different geographic locations to collaboratively design, test and validate shared virtual projects. Figure 9 illustrates a virtual environment for construction engineering generated by DIVERCITY.

3.5 Natural resources exploration

The Center for Visualization at University of Colorado has worked on exploring virtual reality applications in the oil and gas industry. Dorn et al. (2001) developed an Immersive Drilling Planner for platform and well planning in an immersive virtual environment as shown in Figure 10. The planning of a well requires a wide variety of data, geometric accuracy of the data display, interaction with the data, and collaborative efforts of an interdisciplinary team. The system can import and visualize critical surface and sub-surface data in the CAVE™ system. The data include seismic data, horizons, faults, rock properties, existing well paths and surveys, log data, bathymetry, pipeline maps, drilling hazards, etc.

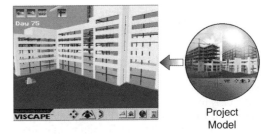

Project Model

Figure 9. DIVERCITY construction planning (Christiansson et al. 2002).

Figure 10. Immersive well path editing (BP center for visualization).

Using the data the user can design the platform and other properties of the well interactively. The Immersive Drilling Planner has been applied to several projects and has demonstrated the development planning cycle times reduced from several months to one week or less.

The Virtual Reality Laboratory at Laurentian University in Canada (Kaiser et al. 2002) offers a collaborative immersive virtual reality environment for mine planning and design. The virtual environment is built on a large spherical stereoscopic projection system with advanced earth modeling software. The system has a 9 by 22 foot curved screen and can contain up to 20 persons in the theater. Because of the constantly changing work environment and enormous overflow of data, mine design and planning is a difficult, time consuming and expensive exercise. Large volumes of data of mine geometry, geology, geomechanics and mining data can be interpreted and evaluated efficiently by utilizing a large-scale immersive visualization environment. The collaborative environment also makes the planning process quicker. The key benefits of using Virtual Reality in mine design are data fusion, knowledge transfer, technical conflict resolution, and collaboration.

3.6 *Homeland security*

Since the devastating terrorist attack at the United States on September 11, 2001, Homeland Security (HLS) has become an issue of utmost importance in the world. There is great need for developing measures for counter-terrorism and for training of military and civilian first responders engaged in HLS. The application of virtual reality is a powerful approach to planning strategies and training personnel for HLS operations. Building physical facilities would be time-consuming and costly, and the result would be of only limited utility in planning/training activities. Also, using realistic physical facilities to develop decision-making and operation skills of personnel in counter-terrorism and other HLS missions might be dangerous. The same skills can be gained through training of the personnel with synthetic reality environments that represent highly realistic simulations of physical environments, which provide great advantages in increased safety, reduced cost and time, and unlimited scenarios that can be generated with computers.

The First Responder Simulation and Training Environment (FiRSTE) project is conducted by the University of Missouri-Rolla in collaboration with Battelle Memorial Institute and U.S. Army. Its purpose is the application of virtual reality simulations for training civilian first responders to deal with Weapons of Mass Destruction (WMD) and Hazardous Material events in a zero-risk environment while applying proper procedures, techniques, and protocols (Leu et al. 2003). The software development of this project uses the Rational Unified Process. The hardware development includes a Virtual Environment Navigation Pad, an instrumented treadmill, and a mock-up Photo Ionization Detector, which are integrated with head mounted displays and computer hardware and software to form the FiRSTE system, as shown in Figure 11a. Training vignettes mirror WMD incidents involving hazardous materials inside a building as shown in Figure 11b. Analyses of the data and information collected using quantitative and qualitative measures during the training exercises have been carried out with the goal of evaluating the effectiveness of such a training system.

Other user navigation hardware interfaces for this type of training with a virtual environment have been developed. For example, the Omni-Directional Treadmill (U.S. Patent No. 6,743,154) as shown in Figure 12 is one of the locomotion devices affording a user to explore the virtual environment while doing physical walking, running, or turning. The Omni Directional Treadmill is an ideal device for navigation

(a)

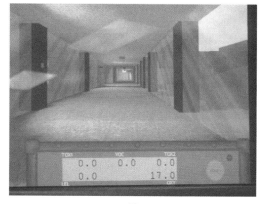

(b)

Figure 11. (a) Virtual Environment Navigation Pad (VENP) in the FiRSTE system. (b) Training vignettes in the FiRSTE system.

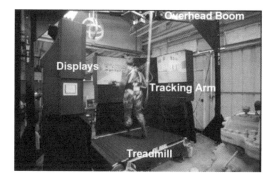

Figure 12. The Omni-Directional treadmill.

in virtual worlds as it allows natural walking movement of the user in the forward, backward and sideward direction. However, the apparatus is very expensive and is difficult to maintain and transport.

4 FUTURE PROMISES AND CHALLENGES

We have discussed the components of a virtual reality system and promising engineering applications of this emerging technology. In this section we will discuss great promises of virtual reality in the future and the research and technical challenges that need to be addressed in order to realize these promises and further advance VR technology.

4.1 Promising future applications and research needs

The applications of virtual reality are expected to continue to grow. Some of the promising future applications of this technology and the research need to realize these advanced applications are discussed below.

4.1.1 Multimodal virtual training

Using virtual reality for training provides the potential of a realistic, safe, and controlled environment for the trainees to practice real operations in a virtual environment, allowing them to make mistakes without serious consequences. VR training also offers the possibility of providing a standardized performance evaluation for the trainee. Such a system will be useful for training of special personnel like novice orthopaedic surgeons, medical first responders, etc. The key to an effective VR training system is to have the capability of providing high-fidelity dynamic graphic displays with realistic force and sound feedback during the VR training process.

Developing such a VR training system is a major undertaking and needs to develop advanced geometric modeling and physical modeling methods and multimodal rendering techniques to simultaneously achieve high-fidelity dynamic graphic displays with realistic force and sound feedback in real time during the VR training process. Because of the drastically different update rates required for simultaneous graphic, haptic and sound displays, multithreading needs to be employed to meet the different computational demands for multimodal rendering. How to represent, organize and manipulate the huge set of geometric and physical data in order to drastically reduce computations is a major research issue.

4.1.2 Networked collaborative engineering

Due to the increasing complexity of engineering artifacts and the continuous trend of globalization, a geographically distributed collaborative environment for design and manufacturing is becoming ubiquitous. Collaborative product development and realization has recently become a new paradigm for today's engineering organizations, and products have increasingly been developed through intensive collaborations that are distributed across people, organizations, and space. The future collaborative design and manufacturing environment will consist of a network of engineering applications, in which state-of-the-art multimedia tools will enhance closer collaboration among geographically distributed applications or design agents, virtual reality systems will allow visualization and simulation in a synthetic environment, and information exchange standards will facilitate seamless interoperation of heterogeneous applications.

A networked virtual environment offers the possibility of major breakthroughs in collaborative engineering, with team participants located anywhere in different geographical locations to share a virtual world. A change to the virtual environment at one site is transmitted and displayed at other sites simultaneously. The virtual artifacts can be manipulated to query "what if" scenarios from all the team members in different locations and visualized simultaneously in their respective virtual environments. This kind of collaborative environment potentially offers significant advantages over the current web-based collaborative technologies, via significantly more realistic, adaptable, and flexible simulation and communication. The development of such a highly advanced collaborative engineering system poses many technical challenges. Fundamental research is needed to enable effective conflict detection and resolution among the collaborating participants in different geographical locations and from multiple disciplines, to display a complex virtual environment with a large amount of data transmitted over the internet without significant time delay, and to devise a communication architecture for concurrency control.

4.1.3 Human workload management

By conducting experiments in a variety of VR environments, it is possible to better understand the perceptual capabilities of the human processor and to increase the effectiveness of human interaction with the actual work environment. This understanding will allow us to maximize the system's performance and to provide a set of guiding principles that enable intuitive and efficient interactions which will lead to maximizing human workload management in manufacturing and other affectively intense operations.

Research with well designed, human-centered VR environments will allow us to understand the relationship between mental workload (task demand versus available mental resources) and human performance. The research could lead to mental workload quantification through psychophysiological indices, and to creation of algorithms which merge human physiological activity with real-time immersive virtual environments and allow the VR system to increase or decrease workload demands on the performer. The objective is to regulate the workload in real time in order to increase the system's performance while maintaining a stable level of engagement.

4.2 *VR technology challenges*

Most of the tracking devices connect the user with the control boxes through cables. The cables are cumbersome and have limited the movement of the user in the virtual environment. Wireless systems have become available, reducing encumbrances on the participant. However, these wireless systems tend to employ more receivers than typically used in a VR system, making them bulkier and more costly. This has prevented the wireless tracking device from widespread use. Reduced infrastructure, improved robustness, and reduced latency are the main challenges in offering high fidelity tracking performance in a virtual environment. In the future, the wireless tracking device needs to be improved in the following categories: size reduction, higher accuracy, less latency, immune to occlusions, self-contained, robustness and lower price. Hybrid tracking combining different types (optical, ultrasonic, electromagnetic, etc.) of tracking devices has been demonstrated with advantages over a single type tracking device. Significant research efforts in data fusion are required to develop an advanced hybrid tracking system.

Two main types of visual display systems used in the virtual environment have been introduced, including head mounted displays and projection-based displays. Both of these display devices have their suitable applications. The big challenges for head mounted displays are increasing the resolution and field of view and reducing the weight of the helmet. For the projection systems, reducing the cost of the system is a major challenge. In the future, a flat panel display will likely replace a set of rear-projection screen and projector. Flat panel displays have higher resolution, require less maintenance, and take up less room than typical rear-projection systems. Another likely future trend is autostereoscopic displays, where the display provides different images when viewed from different angles. They eliminate the need for wearing special headgears.

Haptic displays in use today like the PHANToM™ require the device to be attached to an anchor or ground to which the reaction force can be applied. This limits the user's movement in the virtual environment. Self-grounded systems can be worn by the user and are thus more mobile. However, these devices tend to be cumbersome to wear. Haptic devices are costly because they are fairly complex systems involving sensors, actuators and sophisticated control software. The challenge of haptic devices is making them light and easy to wear, yet still can generate significant force feedback. Haptic devices providing both tactile and force stimuli will offer the user richer touch feedback allowing him/her to feel the surface texture, shape and softness/hardness of objects.

5 CONCLUSION

This paper first gives a brief overview of VR software and hardware. It then reviews various virtual reality applications in engineering including concept design, manufacturing, driving simulation, construction, natural resources exploration, and homeland security. The paper then discusses three promising VR applications: multimodal virtual training, networked collaborative engineering, and human workload management. The discussion includes the research and technical challenges that need to be addressed in order to realize these promises and further advance VR hardware, i.e. tracking devices, visual displays, and haptic displays.

REFERENCES

Aukstakalnis, S. & Blatner, D. 1992. *Silicon Mirage: the Art and Science of Virtual Reality.* Berkeley, CA: Peachpit Press, Inc.

Banerjee, P. & Zetu, D. 2001. *Virtual Manufacturing.* New York: John Wiley & Sons, Inc.

Blackmore, D. & Leu, M.C. 1992. Analysis of Swept Volume via Lie Groups and Differential Equations. *International Journal of Robotics Research* 11(8): 516–537.

Butterworth, J., Davidson, A., Hench, S. & Olano, M. 1992. 3DM: A Three Dimensional Modeler Using a Head-Mounted Display. *Proceedings 1992 Symposium on Interactive 3D Graphics*: 135–138.

Chawla, R. & Banerjee, A. 2001. A virtual environment for simulating manufacturing operations in 3D. *Proceedings of the 2001 Winter Simulation Conference* 2: 991–997.

Chawla, R. & Banerjee, A. 2002. An Automated 3D Facilities Planning and Operations Model Generator for Synthesizing Generic Manufacturing Operations in Virtual Reality. *Journal of Advanced Manufacturing Systems* 1(1): 5–17.

Chen, L., Papelis, Y., Watson, G. & Solis, D. 2001. NADS at the University of Iowa: A Tool for Driving Safety Research. *Proc. of the 1st Human-Centered Transportation Simulation Conference*, Iowa City, Iowa, 4–7 November.

Christiansson, P., Dalto, L.D., Skjaerbaek, J.O., Soubra, S. & Marache, M. 2002. Virtual Environments for the AEC sector – The Divercity experience. *Presentation at European Conference of Product and Process Modelling*, Portoroz, Slovenia, 9–11 September.

Dani, T.H. & Gadh, R. 1997. Creation of Concept Shape Designs via a Virtual Reality Interface. *Computer-Aided Design* 29(8): 555–563.

Dawood, N., Sriprasert, E., Mallasi, Z. & Hobb, B. 2003. Implementation of Space Planning and Visualisation in a Real-life Construction Case Study: VIRCON Approach. *Conference on Construction Applications of Virtual Reality*, Virginia Tech, Blacksburg, VA, 24–26 September.

Deering, M.F. 1996. The Holosketch VR Sketching System. *Communications of the ACM* 39(5): 54–61.

Dorn, G.A., Touysinhthiphonexay, K., Bradley, J. & Jamieson, A. 2001. Immersive 3-D Visualization Applied to Drilling Planning. *The Leading Edge* 20(12): 1389–1392.

Fang, X., Hung, A.P. & Tan, S. 2001. Driving in the Virtual World. *Proceedings of the 1st Human-centered Transportation Simulation Conference*, Iowa City, Iowa, 4–7 November.

Grant, P., Artz, B., Greenberg, J. & Cathey, L. 2001. Motion Characteristics of the VIRTTEX Motion System. *Proceedings of the 1st Human-Centered Transportation Simulation Conference*, Iowa City, Iowa, 4–7 November.

Güven, S. & Feiner, S. 2003. Authoring 3D Hypermedia for Wearable Augmented and Virtual Reality. *Proceedings of the ISWC '03 (Seventh International Symposium on Wearable Computers)*, White Plains, NY, 21–23 October: 118–226.

Isdale, J. 1998. What is Virtual Reality? *Website: http://vr.isdale.com /WhatIsVR.html.*

Kaiser, P.K., Henning, J., Cotesta, L. & Dasys, A. 2002. Innovations in Mine Planning and Design Utilizing Collaborative Immersive Virtual Reality (CIVR). *Proceedings of the 104th CIM Annual General Meeting*, Vancouver, Canada, 28 April–1 May.

Kelsick, J., Vance, J.M., Buhr, L. & Moller C. 2003. Discrete Event Simulation Implemented in a Virtual Environment. *Journal of Mechanical Design* 125(3): 428–433.

Kibira, D. & McLean, C. 2002. Virtual Reality Simulation of a Mechanical Assembly Production Line. *Proceedings of the Winter Simulation Conference*, San Diego, California: 1130–1137.

Leu, M.C., Maiteh, B.Y., Blackmore, D. & Fu, L. 2001. Creation of Freeform Solid Models in Virtual Reality. *Annals of CIRP* 50(1): 73–76.

Leu, M.C., Peng, X. & Velivelli, A. 2002. Creating Freeform Model by Carving Virtual Workpiece with Haptic Interface. *Proceedings of ASME International Mechanical Engineering Conference*, New Orleans, LA, on CD-ROM.

Leu, M.C., Hilgers M.G., Agarwal S., Hall R.H., Lambert T., Albright, R. & Nebel, K. 2003. Training in Virtual Environments for First Responders. *Proceedings of the ASEE Midwest Section Meeting*, Rolla, Missouri.

Liang, J. & Green, M. 1994. JDCAD: A Highly Interactive 3D Modeling System. *Computers and Graphics* 18(4): 499–506.

Lu, S.C-Y., Shpitanlni, M. & Gadh, R. 1999. Virtual and Augmented Reality Technologies for Product Realization. *Annals of CIRP* 48(2).

Maiteh, B.Y., Blackmore, D., Abdel-Malek, L. & Leu, M.C. 2000. Swept-Volume Computation for Machining Simulation and Virtual Reality Application. *Journal of Materials Processing and Manufacturing Science* 7: 380–390.

Nandanoor, K.R. 2004. Development and Testing of Driving Simulator with a Multi-wall Virtual Environment. *M.S. Thesis.* University of Missouri-Rolla, Rolla, MO.

Peng, X. & Leu, M.C. 2004a. Interactive Virtual Sculpting with Force Feedback. *Proceedings of 2004 Japan-USA Symposium on Flexible Automation*, Denver, Colorado, 19–21 July.

Peng, X. & Leu, M.C. 2004b. Interactive Solid Modeling in a Virtual Environment with Haptic Interface. In Ong, S.K. & Nee, A.Y.C. (eds), *Virtual and Augmented Reality Applications in Manufacturing.* London: Springer-Verlag.

Sachs, E., Roberts, A. & Stoops, D. 1991. 3-Draw: A Tool for Designing 3D Shapes. *IEEE Computer Graphics and Applications* 11(6): 18–24.

Sarshar, M. & Christiansson, P. 2004. Towards Virtual Prototyping in the Construction Industry: The Case Study of the DIVERCITY project. *Proceedings of Incite2004 – International Conference on Construction Information Technology*, Langkawi, Malaysia, 18–21 February.

Sherman, W.R. & Craig, A.B. 2003. *Understanding Virtual Reality: Interface, Application, and Design.* San Francisco: Morgan Kaufmann Publishers.

Sirdeshmukh, M. 2004. Development of Personal Computer Based Driving Simulator for Education to Prevent Drunk Driving. *M.S. Thesis.* University of Missouri-Rolla, Rolla, MO.

Zachmann, G. & Rettig, A. 2001. Natural and Robust Interaction in Virtual Assembly Simulation. *Proceedings of the 8th ISPE International Conference on Concurrent Engineering: Research and Applications*, Anaheim, California, 28 July–1 August.

Virtual modeling and rapid manufacturing – Bártolo (eds)
© 2005 Taylor & Francis Group, London, ISBN 0 415 39062 1

The effects of rapid manufacturing on virtual and physical prototyping

Phill Dickens
Loughborough University, England

ABSTRACT: Rapid Manufacturing is starting to have an impact on the way designs are formulated and then on how parts are produced. This paper discusses the wide variety of impacts on the Virtual and Physical Prototyping that is being undertaken and will be in the future. Many changes will need to occur in both of these areas with Physical Prototyping becoming less important and Virtual Prototyping become paramount.

1 INTRODUCTION

1.1 Definition of rapid manufacturing

For this paper Rapid Manufacturing (RM) is defined as the direct production of finished goods from an additive process. The technique uses additive processes to deliver finished parts directly from digital data and this then eliminates the need for tooling.

In the future RM technology may develop and the layer-based approach may be combined with subtractive (machining) operations or replaced by additive processes that use a multi-axis approach instead of a layered process. However, the definition of RM above will continue.

1.2 Benefits of rapid manufacturing

Rapid manufacturing offers many benefits in a number of areas such as in reduction of tooling costs, reduced lead times and product cost, design freedom, heterogeneous materials, custom products, just-in-time production, and decentralization of production (Wohlers, 2004). However, this paper will concentrate on the two areas of Virtual and Physical Prototyping.

2 EFFECTS OF RAPID MANUFACTURING ON PHYSICAL PROTOTYPING

The effects of Rapid Manufacturing on Physical Prototyping will be both positive and negative.

The positive effect will be a greater acceptance of the prototype as being representative of the manufactured part because they will both be made on the same machine. Conventional manufacturing processes often involve expensive tooling and so there have arisen a number of processes developed as an intermediate stage. This can be easily seen in the area of plastic parts where prototypes may be made by one of the Rapid Prototyping processes but the end-user part is to be injection moulded in a completely different material. In this case both the material and the manufacturing process are different. This will therefore lead to a reduced need for many secondary prototyping processes such as vacuum casting of urethanes and so a whole section of industry could suffer as a result of this. The best option for companies in this area is to invest over the next 10 years and migrate to small lot manufacturing with Rapid Prototyping machines.

Another positive benefit will be the greater complexity of parts which can be prototyped (Hague et al., 2003a). This ability to produce both prototypes and production parts by an additive process will mean that parts could be much more complex than before. This could be greater complexity in geometry and/or greater complexity in material composition. However the material complexity may be a problem as discussed later. At the moment, we have the ability to produce very complex parts by Rapid Prototyping but the intermediate processes like vacuum casting or the final processes such as injection moulding are very limited. The use of Rapid Manufacturing will ensure consistency from generation of concept to delivery of product.

A negative effect of Rapid Manufacturing will be the reduced need for tooling and machining and so some of the more traditional companies will suffer due to this.

However, having discussed the advantages and disadvantages above, the question should be asked whether prototyping will be really needed? This really depends on the quantities that will be manufactured by additive processes in the future. Existing work has already shown that the current Rapid Manufacturing processes are economic for tens of thousands of parts (Hopkinson & Dickens, 2003) and clearly in this case it makes sense to manufacture a physical prototype and test it. However, where parts will be made on an individual basis it will not make sense to produce a prototype and then physically test it to be followed by a single part which is sold. In this situation Virtual Prototyping will become much more important and so the majority of this paper will concentrate on this subject.

3 EFFECTS OF RAPID MANUFACTURING ON VIRTUAL PROTOTYPING

With the reduced investment costs involved in prototype tooling and production tooling then it is likely that there will be more entrepreneurial designs. Many good ideas are currently not pursued because of these investment costs so there is likely to be a growth in product variety and small companies established to exploit an inventor's idea.

However, there will be a need for more imagination from current designers to exploit the possibilities of more complex geometry. In some ways the current designers have been trained into producing simple designs which are not optimal. This is simply because manufacturing engineers have pushed the concept of having designs that are simple and easy to produce. There is also the likelihood that products will be much more integrated in terms of functionality and aesthetics due to part reduction and greater complexity. Ultimately, a hybrid designer may emerge—one that is skilled in aesthetics and mechanical design as well as Rapid Manufacturing techniques (Hague et al., 2003b).

One area that will require less Virtual Prototyping will be concerned with interfaces as parts will be more complex and there will be fewer parts to model. It is often the interfaces between parts where the greatest difficulties arise in modeling because the surface properties may be different from the bulk properties. With more complex parts there will be fewer interfaces requiring fasteners, seals, adhesives etc.

The current CAD systems are very good for modeling solid and surface geometry but they will not be appropriate for more complex parts. For example, some CAD systems cannot tolerate topologically incorrect models such as a Kline Bottle. They are also unable to model textures, porosity and material composition, all of which have the possibility of control using additive processes. In some situations we will need to wrap textures over CAD models and this texture could be variable (Sachs et al., 1994). The same variability could apply to porosity and material composition. Some of these areas are being addressed by TNO (Knoppers et al., 2004)) for graded materials and by Hague (2004).

A serious deficiency at the moment is concerned with the area of Reverse Engineering for custom fitting products. For example, if a grip is to be designed and manufactured for a customer then it is possible to capture the geometry of the hand. However, what is required is the geometry for the hand to fit into that gives the most comfortable and effective grip. As the customer grips an object then the soft tissue deforms so there has to be some intermediate geometry between the geometry of the soft tissue and that of the bones. A Virtual Prototyping system is required that will take both sets of geometry and produce the grip geometry that gives the best fit.

As the number of Rapid Manufactured parts and products will be much lower than those from conventional processes and in some cases this will be a single unit the need for Virtual Prototyping will increase and this will have an impact on the type approval process to obtain CE marking (European Commission, 2000). As many of the initial applications appear to be in safety equipment then there will need to be some work to understand the implications in this area as products may be covered by the Personal Protective Equipment Directive (European Commission, 1989). If Virtual Prototyping is to replace physical prototyping then there would need to be great reliability in the results. In this area then Virtual test standards become very important to replace or augment the existing physical test standards.

In general there will be a much greater need for Virtual Prototyping and there will be much more of a need for design optimization of structures, airflow etc. This will also lead to simultaneous optimization (Sobieszczanski-Sobieski, 1996) where for example the strength of a Formula 1 fairing is being optimized at the same time as the air flow in CFD.

There will be less need for some aspects of virtual prototyping such as assembly and clash detection due to the reduced number of parts and therefore the smaller chances of interference.

Much of the existing Virtual Prototyping is associated with consumer products, automotive, aerospace etc. However, there will also be a need for a wider range of modeling capability in other areas and an example of this could be in the modeling of bone in-growth into implants.

Of course all of this modeling is impossible if the properties of the materials are unknown and so it is important to obtain accurate material property data from the RM processes. To date there has been limited work in this area except for recent work to investigate the long term properties of Stereolithography resins and Laser Sintered nylon (Hague et al 2004). One of the main problems here is that the materials are still developing and being replaced at a rapid rate. By the time the data is obtained for a material it is often replaced by a new version!

There is also the issue of how the customer will interact with the design process in the future. With mass produced products the customer is either canvassed before for opinions or takes part in product assessment clinics. Therefore, the customer input is either opinions based on general likes and dislikes, or is opinions based on seeing the product. To have customer input for customized products then it is likely that they will have to be involved more in the actual design stages. If this involves CAD then we will have a major problem. CAD systems are not designed for this type of activity and require an extensive skill level to be operated efficiently. Even if we had a 'customer-friendly' CAD system the next issue would be how the interaction would take place? There are a number of possibilities:

- Customer uses the CAD system to design a product
- Customer uses a design system on the web
- Customer goes to local 'Customisation Centre'
- Customer goes to the company

Each of these possibilities will have a major effect on the way companies do business.

Another effect of Rapid Manufacturing will be that more resources will switch from prototyping to design and manufacturing. At the moment the early design stages account for a small proportion of the total cost. With Rapid Manufacturing and Customization there will be a very great increase in resources to the design process and possibly a small increase to manufacturing.

One of the potential problems of Rapid Manufacturing will be the use of graded structures/mixed materials. We are currently developing a number of processes that can combine materials into complex structures. However, taking them apart for recycling will be a major difficulty. This will become even more of a problem with the end of life directives for cars and electrical products.

REFERENCES

European Commission, 1989, Directive on Personal Protective Equipment (PPE) 89/686/EEC, *Official Journal L 399, 30/12/1989 P. 0018–0038,* [http://europa.eu.int/eur-lex/lex/LexUriServ/LexUriServ.do?uri = CELEX:31989L0686:EN:HTML].

European Commission, 2000, Guide to the implementation of directives based on the New Approach and the Global Approach, Luxembourg: Office for Official Publications of the European Communities, 2000, ISBN 92-828-7500-8, [http://europa.eu.int/comm/enterprise/newapproach/newapproach.htm].

Hague, R.J.M., Mansour, S. and Saleh, N., 2003a, Design opportunities with Rapid Manufacturing, *International Journal of Assembly Technology & Management*, 23(4), 346–356, ISSN 0144-5154.

Hague, R.J.M., Campbell, I., Dickens, P.M., 2003b, Implications on Design for Rapid Manufacture, *IMechE Part C, Journal of Mechanical Engineering Science*, Vol. 217, 25–30.

Hague, R.J.M., 2004, Materials Analysis & design optimisation for Rapid Manufacturing and Rapid Manufacturing of Textiles and Textures, *Proceedings of Joint NASUG/SLSUG Conference*, Apr 25–29, Anaheim, California, USA.

Hague, R.J.M., Mansour, S. and Saleh, N., 2004, Material and design considerations for Rapid Manufacturing, *International Journal of Production Research*, 42(22), 4691–4708, ISSN 0020-7543.

Hopkinson, N. and Dickens, P.M., 2003, Analysis of Rapid Manufacturing – Using Layer Manufacturing Processes for Production, *IMechE Part C, Journal of Mechanical Engineering Science*, Vol. 217, 31–39.

Knoppers, G.E., Gunnink, J.W., van de Hout, J. and van Vliet, W.P., 2004, The Reality of Functionally Graded Material Products, *Proc. of Solid Freeform Fabrication Symposium*, University of Texas in Austin, 2–4 August, 38–43, ISSN 1053-2153.

Sachs, E., Curodeau, A., Gossard, D. Ad Jee, H., 1994, Surface texture by 3D printing, *Proc. of Solid Freeform Fabrication Symposium*, University of Texas in Austin, 8–10 August, 56–64, ISSN 1053-2153.

Sobieszczanski-Sobieski, J. and Haftka, R.T., 1996, Multidisciplinary Aerospace Design Optimization: Survey of Recent Developments, **34th AIAA Aerospace Sciences Meeting and Exhibit, Reno, Nevada, AIAA Paper 96-0711.**

Wohlers, T., 2004, *Wohlers Report 2004: Rapid Prototyping, Tooling & Manufacturing State of the Industry*, Wohlers Associates, Fort Collins, USA, ISBN 0-9754429-0-2.

Biomanufacturing

Virtual modeling and rapid manufacturing – Bártolo (eds)
© *2005 Taylor & Francis Group, London, ISBN 0 415 39062 1*

The direct cell controlled assembly technology on the base of RP

H.X. Liu[1,2], Y.N. Yan[1,2], X.H. Wang[1,2], J. Cheng[1,2], Y.Q. Pan[1,2], F. Lin[1,2] & Z. Xiong[1,2]

[1]*Center of Organism Manufacturing, Department of Mechanical Engineering, Tsinghua University, Beijing, P.R. China*
[2]*Institute of Life Science & Medicine, Tsinghua University, Beijing, P.R. China*

ABSTRACT: This paper reports a technology based on rapid prototyping (RP) to assembly the cells and hydrogels mixture. A cell controlled assembler is developed, which is capable of extruding cells and hydrogels simultaneously into pre-defined three dimensional structures, which is a kind of viable geometry with high-density live cells. The assembling process is biocompatible and occurs at a digital controlled temperature. Three dimensional scaffolds with alginate-gelatin mixture or scaffolds with alginate-gelatin-cells mixture can be formed freely. The cells in the freeform structures indicated a normal morphology and the analyses of the cell survivability showed that more than 90% of the cells are viable during assembling. Furthermore, culturing *in vitro*, the cells in the structure still retain their ability to function, and the proliferation and division are normal. The RP-based cell controlled assembly technology holds a potential for creating living tissue analogs, has a promising application in the reparation of tissues, especially complex internal organs. Also, there may be a broad application in pharmacology and toxicology study.

1 INTRODUCTION

Tissue engineering cells-seeding on scaffolds to growth the new tissues for the replacement of nonfunctional tissues are relatively well known (Langer R. & Vacanti J. 1993). Due to fabricate the 3D scaffolds before guide cells to form functional tissue, maybe these tissue engineering techniques are called cell indirect controlled assembly. Nearly full-scale simple tissue rehabilitation through cell indirect controlled assembly techniques have been operated worldwide effectively (Maemura T. et al, 2003), but it is still in a dilemma as to repair the large and complex tissue, especially the multi-cell tissue (Vacanti J.P. 2003), because the above techniques may not provide enough cells adhesive mass and different cells' spatially distribution (Vacanti J.P. et al, 2001). Recently, research on forming the viable structures containing the cells commences (Xu T. et al, 2005), namely cell directly controlled assembly (Mironov V. et al, 2003), (Sun W. et al, 2003). Here, we present a novel attempt and rationale to fabricate three dimensional viable structures through assembly cell directly and provide some preliminary results. Furthermore, we review the application, advancement and future directions of cell direct controlled assembly in the design and creation of viable three dimensional structures for use in rehabilitation, biopharmaceutical manufacturing and drug test.

2 ASSEMBLER SYSTEM STRUCTURE

RP are computerized fabrication techniques that can rapidly produce highly complex three dimensional objects drove by the CAD system directly, the prototyping materials are deposited to build the final structure in a layer-by-layer process (Sachs E. et al, 1992). This makes the RP techniques very attractive for direct assembling the cells to a complex structure (Xiong Z. et al, 2002). But among the expanding number of commercial RP, very few are directly suitable for this rather special application. Stringent criteria must be met in bio-functional rapid prototyping. First, bio-functional rapid prototyping must process nontoxic, biocompatible, and biodegradable materials without exposing them to elevated temperatures, which are known to cause the denaturation of highly sensitive biopolymers such as proteins (Landers R. et al, 2002). Here, we combine the hydrogel solidification mechanics with RP technique and develop a specified machine (Fig. 1) and stepwise forming process for cell controlled assembly.

2.1 *Material extrusion unit*

The four-axial motion NC card (AT6400) is selected to drive X-Y directional platform and Z-directional elevator. The fourth axis is assigned to drive the materials extrusion unit. All of them are driven by two linear-stepping motor and controlled by leading screw.

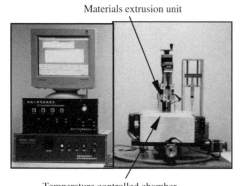

Materials extrusion unit

Temperature controlled chamber

Figure 1. The direct cell assembler I.

Droplet mode

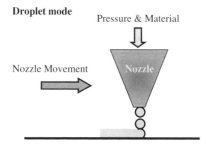

Figure 2. Droplet mode of deposition.

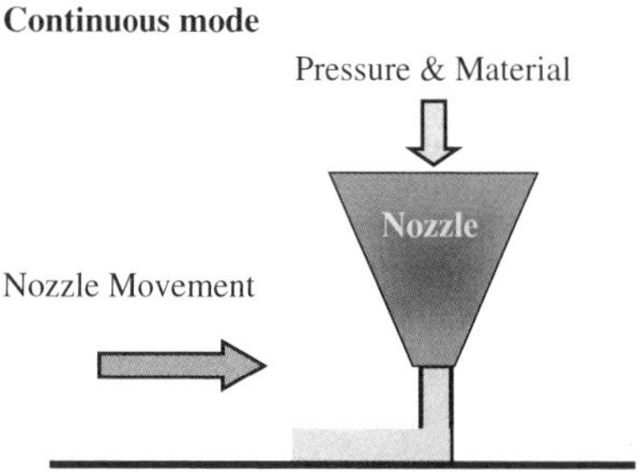

Continuous mode

Figure 3. Continuous mode of deposition.

In general, there are two modes of deposition polymer in RP process; the droplet mode (Fig. 2) and the continuous mode (Fig. 3). In the droplet mode, the materials droplet is controlled by frequency signal, the droplet mode can form a structure by depositing multiple droplets at desired locations on a substrate. This process can be repeated layer by layer to develop a freeform fabricated part. In the continuous mode, the material is extruded out of the nozzle under an applied pressure. This mode can basically lay down the material in the form of line structures to create the desired model by moving the nozzle over a substrate

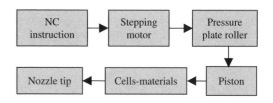

Figure 4. The material extrusion system workflow chart.

in the designed path. In the same way, this process can be repeated to form a 3D structure. Each material delivery mode has its advantages and limitations, here we choose the later for delivering the mixture because it is more efficient.

Material extrusion unit includes stepping motor, pressure plate roller, precise bar, storing barrel and piston. All input signals are controlled through computer programming. The extrusion system workflow is shown in Figure 4. Because adopting the stepping motor to drive directly, the controllability of micro extrusion flow and minimum drip can be accurately controlled, the large-scale extrusion pressure can be got through adjusting the velocity of the stepping motor. Therefore, a lot of materials can be selected as the forming primary stuff, from high viscous gel to low viscous weak solution, such as gelatin, collagen, fibrin, chitosan, PLGA, protein and growth factors etc. Here, we choose the mixture of gelatin and sodium alginate as the primary experiment materials. A stepwise forming process is achieved by controlled sol-gel phase transition between 0°C and 37°C.

2.2 Temperature control forming chamber unit

The main part of the temperature control forming chamber unit is an adiabatic envelope box fixed on the X Y platform (outline dimension 300 × 250 × 200 mm, wall thickness 30 mm), the semiconductor refrigerator pieces (DC12V) are fitted on the right and left side of box, three grade temperature (room temperature, around 15°C, around 6°C) can be get. An organic glass slip cover moves with the forming chamber, is equivalent to a barrier layer, can keep the sterilized laminar flow. The forming basement (70 × 60 mm) will be coated the biomaterials when used, often choose the same materials as extrusion materials.

2.3 Nozzle unit

We choose the disposable syringe nozzles, so can get a series changeable diameter nozzle tips, tip diameter is in the range of 0.06 mm–1.2 mm. It is easier to cater for injecting the different diameter cells and different viscosity materials. Furthermore, it is more safe and convenient for cell assembly to use the sterilized syringe nozzles.

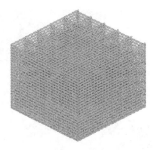

Figure 5.   CAD model.

Figure 6.   12 mm × 12 mm × 3 mm structure with SAG.

3   FORMING PROCESS AND METHOD

3.1   *Gel and cell preparation*

2 g SAG (the mixtures of gelatin (Tianjin green-island Company, China) and sodium alginate (Sigma-Aldrich, USA) at the rate of 1:1) was sterilized for 30 min in an autoclave and then dissolved in 20 ml of hot physiology saline (70°C) at sterilized environment.

Hepatocytes were isolated from a 8-week-old male rat, the obtained hepatocytes (Seglen P.O. 1976) were suspended in the SAG mixture solution at the concentration of 2×10^8 cells/mL.

3.2   *CAD model and process workflow*

A grid framework CAD model is shown in Figure 5, an object is build up by extruding layers of parallel strands in succession. The space interval designed between strands is a kind of channel to supply the nutrients and metabolism for cells, for different assembled cell, the size maybe is different. Here is 500 um to satisfy the hepatocytes mentioned above.

After one layer SAG with cells was extruded, an offset intersection was extruded for the second layer in the designed CAD path. A structure (10 mm × 10 mm × 5 mm) with live hepatic cells was formed (Fig. 7).

Figure 7.   10 mm × 10 mm × 5 mm structure with SAG and hepatic cells.

3.3   *Post-treatment*

The finished 3D structure was put into a Petri dish containing 6% $CaCl_2$ crosslinking 5 min for more firming, then moved into a well plate containing Dulbecco's Modified Eagle Medium (DMEM) and Hepato ZYME-SFM (Gibco BRL) (the volume ratio is 1:1) and cultured in a incubator, which was maintained at 37°C, 5% CO_2 in air.

4   RESULTS AND DISCUSSION

4.1   *Morphologic observation*

As shown in Figure 7, some of the macro pores showed almost round shapes or even irregular shapes, mostly due to the hydrogel diffusion before cross-linking.

Cell viability was estimated by the trypan blue staining immediately after the 3D structure was formed, More than 90% of the cells were alive. In other words, almost no cells were destroyed during the whole process including extrusion, low temperature and $CaCl_2$ cross-linking.

6 days after culture, hepatocytes survived and multiplied everywhere in the 3D structure (Fig. 9). The shape kept well and the cytoplasm membrane can be distinctly observed. Some of the hepatocytes of the inner structure were aggregated to form multicellular spheroids.

4.2   *Immunochemistry test*

It is known that albumin is a specific and the most abundant protein only synthesized by functional adult hepatocytes. To determine whether the hepatocytes in the cultured structure were functional, we used immunohistochemical staining techniques to evaluate the albumin expression. LSCM pictures after one month culture, shown as Figure10, the states of cells were much well, cell moiety phase and secreta (green color) could be clearly observed. These results

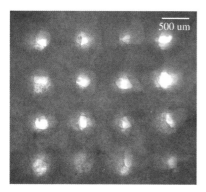

Figure 8. Light microscopy image of hepatocytes in the structure post-processing.

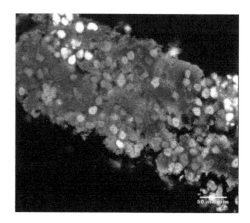

Figure 10. 9d LSCM of hepatocytes in the structure.

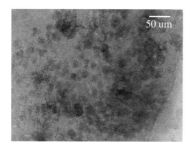

Figure 9. 6d light microscopy image of hepatocytes in the structure.

confirm that the viability and function of hepatocytes can be maintained in the cultured structure.

This research indicates that cell controlled assembly apparatus based on RPT can form the mixture of cells and hydrogel into 3D structure directly and keep the activity of cells. The advantages of this machine could be summarized as follows:

(1) The forming process is non-invasive, which keeps the material/cell active.
(2) The pore size and connectivity of formed structure could be designed and fabricated flexibility according to the different cell.
(3) The distribution and position of different kinds of cells, extracellular matrix (ECMs) and bioactive factors could be fabricated exactly.
(4) The disposable nozzle is easy to maintain and cheap, at the same time, needn't sterilize.
(5) High-throughput and high-density cells can be assembly in a short time.

Because the current works are on improving the cell viability after extrusion, there are still some limitation existed in the process and the machine, such as follow.

(1) The control system and operating interfaces are poor, while future works will be emphasized on developing them.
(2) Optimize the forming parameters and forming conditions.
(3) Research and develop the multiple nozzles system.

5 CONCLUSIONS

This is the first time that cells and ECMs are assembly into 3D analogy human environment structure using RP technology. The appearance of the cells in the structure indicated a healthy morphology. Although it is still in the early developing stage of manufacture human organs, it is the most promising way to create 3D living organ precursors with complex structure and functionality. The advances in direct cell 3D assembling technology and its integration with life science, biology and medicine will make it possible for organ regeneration manufacture, especially for the complex internal organ rehabilitation.

REFERENCES

Landers R., Hübner U., Schmelzeisen R. & Mülhaupt R. 2002. Rapid prototyping of scaffolds derived from thermoreversible hydrogels and tailored for application in tissue engineering. *Biomaterials*. 23: 4437–4447.
Langer R. & Vacanti J. 1993. Tissue engineering. *Science*. 260: 920–925.
Maemura T., Shin M., Sato M., Mochizuki H. & Vacanti J.P. 2003. A tissue-engineered stomach as a replacement of the native stomach. *Transplantation*. 76(1): 61–65.

Mironov V., Boland T., Trusk T., Forgacs G. & Markwald R.R. 2003. Organ printing: computer-aided jet-based 3D tissue engineering. *TRENDS in Biotechnology*. 21(4): 157–161.

Sachs E., Cima M., Williams P,. Brancazio D. & Cornie J. 1992. Three dimensional printing: rapid tooling and prototypes directly form a CAD model. *J Eng Ind*. 144: 481–488.

Seglen P.O. 1976. Preparation of isolated liver cells. *Method Cell Biol*. 13: 29–83.

Sun W., Darling A., Starly B. & Nam J. 2003. Computer aided tissue engineering Part I: overview, scope and challenges. *Biotechnol Appl Biochem*. 39(1): 29–47.

Xu T., Jin J., Gregory C., James J., Hickman & Boland T. 2005. Inkjet printing of viable mammalian cells. *Biomaterials*. 26: 93–99.

Vacanti J.P., Leonard J.L., Dore B., Bonassar L.J., Cao Y. & Stachelek S.J. 2001. Tissue-engineered spinal cord. *Transplant P*. 33(1): 592–598.

Vacanti J.P. 2003. Tissue and organ engineering: can we build intestine and vital organs? *J astrointest Surg*. 7(7): 831–836.

Xiong Z., Yan Y., Wang S., Zhang R. & Zhang C. 2002. Fabrication of porous scaffolds for bone tissue engineering via low-temperature deposition. *Scripta Materialia*. 46(11): 771–776.

Virtual modeling and rapid manufacturing – Bártolo (eds)
© *2005 Taylor & Francis Group, London, ISBN 0 415 39062 1*

Fabrication of soft and hard biocompatible scaffolds using 3D-Bioplotting™

C. Carvalho, R. Landers & R. Mülhaupt
Freiburg Materials Research Center, Freiburg, Germany

U. Hübner & R. Schmelzeisen
Klinik für Mund-, Kiefer- und Gesichtschirurgie, Freiburg, Germany

ABSTRACT: In Tissue Engineering and bone reconstruction, alongside the choice of materials, the scaffold design is of great importance. Three dimensional structures not only permit the tuning of chemical and mechanical properties, but they can also copy the outer form of the required bone or cartilaginous structures. While new processes that create such 3D scaffolds by means of Rapid Prototyping have been developed, they are still restricted to a limited type of materials. At the Freiburger Materialforschungszentrum, our group has developed a new process called 3D Bioplotting™. Most kinds of polymers and biopolymers can be used for the fabrication of 3D scaffolds with 3D Bioplotting™, including hydrogels (e.g. collagen, agar), polymer melts (e.g. PLLA, PGA) and two-component systems (e.g. chitosan, fibrin). Tailor made biodegradable scaffolds can be fabricated in a short time using individual computer-tomography data from the patient. In-vitro tests showed promising results and in-vivo experiments are now under observation.

1 INTRODUCTION

The creation of prototypes of a product is a necessary step in almost any industry. In the past, prototypes were painstakingly created by hand, which not only took a considerable amount of time, but also produced prototypes which were hard to reproduce. Following an increase in competition in the industry, the demand for faster methods to fabricate prototypes with higher rates of reproducibility increased by itself considerably. By using Computer Aided Design (CAD) to develop prototypes virtually on a computer, machines could be automated to fabricate any needed number of perfectly equal prototypes in a far lower amount of time. Thus Rapid Prototyping was developed, at first for the creation of prototypes which served for the visualization of ideas and complex 3D structures: Concept Modeling (Mankovich et al. 1994, Chua et al. 1998). This technology was further developed as an aid to the fabrication of machine parts, for example in the creation of molds: Rapid Tooling (Winder et al. 1999, Aung et al. 1998). With the increase in research to improve mechanical properties and decrease production time of rapid prototyped 3D objects, "ready for use" parts can be fabricated using Rapid Prototyping to create fully working models: Functional Modeling (Yang et al.

2002). A new branch of Functional Modeling is gaining in importance nowadays, specially in Europe. Cooperating with Tissue Engineering, highly customized medical implants can be fabricated. Research on the construction of titanium plates (Fisher et al. 2003), drug releasing systems (Katstra et al. 2000) and support structures for bone and nerve defects (Pfister et al. 2004) is currently being followed by several workgroups.

A great number of different Rapid Prototyping techniques have been developed in the last years (Wohlers 2000, Gebhardt 2003, Pham & Dimov 2001), which can be distinguished by the materials used to create the layers building the 3D object, as well as in the method of hardening said layers. The hardening processes are temperature changes, solvent evaporation, photopolymerisation or chemical reactions. Yet each Rapid Prototyping technique is restricted to one or two types of materials, as well as method of hardening used. Most techniques do not permit the handling of a wide range of bioactive components, which are temperature sensitive and require exclusively non-toxic chemicals. At the Freiburg Materials Research Center we were able to develop 3D-Bioplotting™, a technique that can use a large amount of different materials, including a number of biopolymers and bioactive components, as well as a wide range of

solidification methods for Rapid Prototyping purposes (Landers 2004).

2 3D-BIOPLOTTING™

3D-Bioplotting™ is a 3D dispensing technique that
uses a pressure-controlled dispenser movable in each
direction. 3D objects can thus be fabricated layer by
layer, as typical of Rapid Prototyping techniques.
Each layer is composed of single parallel strands, for
each layer the direction of these strands can be varied.
By setting a certain distance between the strands,
porous structures can be built.

The surface and surroundings onto which the
object is built is called plot medium. This defines the
solidification method used. While high-temperature
polymer melts can be plotted on a glass plate at room
temperature on air or an inert gas, many other materials will require substances that speed up the solidification process. By using liquid plot mediums which
have the same density as the plot material, buoyancy
can be provided to the plotted layers, making strands
of soft materials remain in the correct position instead
of sacking due to the gravitational pull. This makes
this the only technique that can fabricate 3D hydrogel
scaffolds in a direct plotting process, without the use
of moulds or solid support materials. Using all hardening methods available we are able to fabricate scaffolds out of polymer melts, ceramics, monomers and
even hydrogels (Landers 2002).

Our main interest lies in the field of Tissue Engineering. By using computer tomography data from each
individual patient, support structures for bone defects
can be planed using CAD software. Tailor-made 3D
implants can this way be fabricated using biocompatible or biodegradable materials (Webb 2000, Sun &
Lal 2002, Mahoney 1995), which can decrease the
duration of surgical operations and comply to the aesthetic needs of the patient. By cultivating cells from
each patient onto the scaffolds previous to implant a
decrease in the duration of the self-regeneration process
should be possible (Kuschnierz et al. in prep.). This
will also reduce the risk of rejection.

By adding drugs and growth-factors to the plotting material previous to the plotting process, the
self-regeneration process could be promoted further
and the risk of infection reduced (Yamada 1997).

3 SCAFFOLD PREPARATION

For this series of tests we decided to use all three types
of materials available to the bioplotting technique:
ceramics, polymer melts and hydrogels. For the ceramic
scaffold we used hydroxyapatite, a biomaterial found in
bones. As for the polymer melt, PLGA was chosen, one

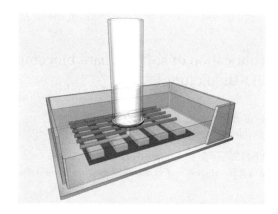

Figure 1.   Schematics of the 3D-Bioplotting™ process, building of the first layer.

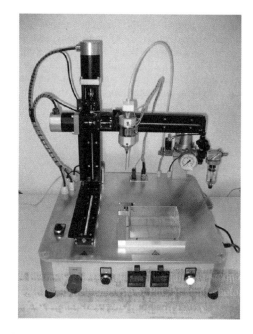

Figure 2.   The 3D-Bioplotter at the Freiburg Materials
Research Center, build by Envisiontec GmbH.

of the most often used materials in biodegradable
implants. The hydrogel was a mixture of collagen and
chitosan, two well known biomaterials with added
hydroxyapatite powder to improve mechanical stability.

3.1 *3D-Plotting of hydroxyapatite with polyvinyl alcohol*

A very fine hydroxyapatite with particle sizes ranging
between 1 and 10 μm was sieved through a 50 μm sieve

Figure 3. An hydroxyapatite scaffold.

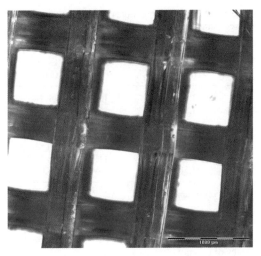

Figure 4. A PLGA scaffold.

on a sieving machine (Retsch, Haan, Germany – Model AS 200 control) producing a powder without larger aggregates. A 10% solution of polyvinyl alcohol was made by dissolving the polymer in boiling water. 37,5 g of hydroxyapatite powder were mixed in 62,5 g of the polyvinyl alcohol solution and stirred firmly for 20 minutes. A white paste with large aggregates is created, which can be pressed through a 100 μm sieve. The sieved paste was filled in a PE syringe for the plotting process.

After plotting, the scaffolds were kept on the plotting surface in a freezer at −15°C for 2 hours. The frozen scaffolds were freeze-dried (B. Braun Biotech, Melsung, Germany – Christ, Model Alpha 2) for 2 days using 0.05 mbar and −20°C. Afterwards, the dried scaffolds were inserted in an oven (Nabertherm, Lilienthal, Germany – Model HT04/17). The oven was slowly heated to 350°C using a heating rate of 25 K/h. After 5 hours at 350°C, the temperature was raised to 1150°C using a heating rate of 100 K/h, at which the scaffolds were sintered for 12 hours. The oven was cooled slowly at a rate of 50 K/h to room temperature, to avoid cracking of the ceramic scaffolds.

3.2 3D-Plotting of PLGA

The Resomer LG 824 was purchased as a white powder from Boehringer Ingelheim. This powder was filled into a glass syringe and heated to melting point. A slow degradation of the polymer during the plotting process could be seen in the gradual change of the color of the polymer from white to yellow.

During the plotting process, a reduction of the viscosity also takes place due to degradation, for this reason only small amounts of the polymer were filled into the syringe each time. Used syringes were cleaned with dichloromethane before being refilled, to remove all rests of degraded PLGA, or new syringes were used.

3.3 3D-Plotting of collagen-chitosan-hydroxyapatite

Collagen was extracted from bovine tendons. The collagen was dissolved in 0.5 M acetic acid, the inhomogeneous solution was grinded. The hydrogel was precipitated in a 5% NaCl solution, filtered and dissolved once more. The solution was filtered through a 1 mm sieve to remove insoluble collagen and then filled into dialysis tubes, which were kept in distilled water. The water was exchanged every 6 hours for two days. After dialysis the collagen/water mixture was frozen in a −70°C freezer for 3 hours, after which they were freeze-dried for 1 day at 0.001 mbar.

0,2 g of collagen were solved in 20 ml 0.5 M acetic acid, while 2,4 g chitosan were solved in the same amount acetic acid. Both gels were mixed together and grinded with a mortar into an homogeneous paste. 5 g hydroxyapatite were suspended in 20 ml 0.5 M acetic acid, to which the collagen-chitosan paste was added in small portions and grinded into an homogeneous paste. The hydrogel mixture was pressed through a 100 μm sieve and the sieved paste was filled in a PE syringe for the plotting process.

The scaffolds were plotted onto a glass plate which laid on a liquid hydrogen cooled metal plate. The viscous gel could this way be frozen directly during plotting, maintaining a very good mechanical stability. After plotting, −20°C cold ethanol was added to the scaffolds, which was changed three times during a period of 24 hours to remove all traces of acetic acid.

Table 1. E-Module for scaffolds with increasing distance between strands.

Material	Distance between strands (mm)	EMod (MPa)	E-F max (%)
HA*	1,05	125,11	11,74
HA	1,50	77,49	12,14
HA	1,75	57,63	17,25
PLGA	0,45	111,62	44,68
PLGA	0,55	132,82	50,83
PLGA	0,70	119,72	68,17
PLGA	1,50	89,35	78,07

* HA: Hydroxyapatite.

After this time the scaffolds were washed once with distilled water and twice with PBS, to remove all traces of ethanol.

3.4 Seeding of the scaffolds

The plotted scaffolds were sterilized using plasma sterilization (STERRAD 100S), after which they were washed five times using PBS (once per hour) and twice times with nutrient media (once every 24 hours). $1*10^6$ cells were seeded on top of the scaffolds and nutrient media was added, after which the scaffolds were cultivated for 7 days at 37°C under a 5% CO_2 atmosphere. The nutrient media was exchanged every two days. Before measurement with the EZ4U-test, the scaffolds were moved to another well plate, so cells on the plate were not measured.

3.5 EZ4U-test

The substrate is first dissolved in 2,5 ml activator solution at 37°C, colouring the solution yellow. Per well, 25 µl of the substrate is added to 200 µl cell culture, after which these solutions are incubated for 5 hours at 37°C. After incubation, the plate is shaken on the reader (Anthos Reader 2001) and the absorption at 450 nm (492 nm) against 620 nm is measured. The absorption of a blank value is deducted from the measured values.

4 SCAFFOLD DESIGN

Several factors have a major impact in cell proliferation on an scaffold. The used material needs to be hydrophilic, to allow adhesion of the cells onto the surface, for this reason we for example coat hydroxyapatite scaffolds with a thin collagen film.

Another major impact on cell proliferation is the interior design of the scaffold itself. The design of the scaffold is chosen at first accordingly to the needs of

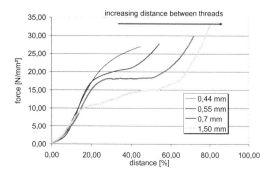

Figure 5. Compressive strength measurement on cylindrical PLGA scaffolds (d = 14 mm, h = 5 mm).

Table 2. Cell activity on PLGA scaffolds after two weeks with different distances between strands.

Gap between strands (mm)	Optical density
0,85	1,483
1,00	1,315
1,25	1,661
1,60	1,579
2,00	1,856

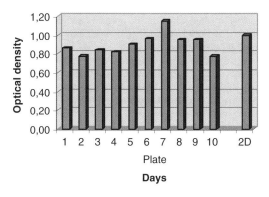

Figure 6. Cell proliferation on a PLGA scaffold.

the particular medical application regarding mechanical properties. By maintaining the same strand thickness but increasing the gap between strands the firmness of the scaffold can be reduced. An optimal support scaffold for bone regeneration would therefore have around the same firmness as the surrounding bone (Table 1).

An interesting effect was noted on PLGA scaffolds, these could be compressed nearly indefinitely, the layers would collapse on one another without breaking the scaffold itself (Figure 5). Hydroxyapatite scaffolds behaved as expected, breaking up below 10 N/mm².

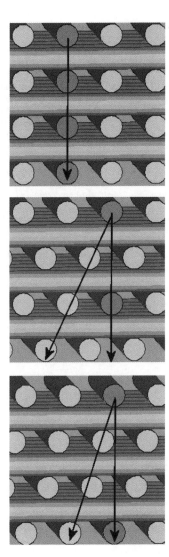

Figure 7. Scaffold design using complex offsets between layers. (From top to bottom) Offset 0%, offset 50% and off-set 25%.

Table 3. Cell proliferation on PLGA scaffolds using complex offsets between layers.

Offset (%)	Optical density
0	2,087
0.25	1,868
0.50	1,779

increasing the distance between the strands (Table 2). One week is not enough for this test, as cells tend to proliferate only on the top layers of the scaffold until completely filling these layers. Only after a certain amount of time do cells start migrating to lower layers, where they have abundant space to reproduce. After two weeks the cells have achieved an equilibrium between living space and provided nutrient media (Figure 6).

Attempts to create scaffolds with more complex interior designs, like a "staircase" building method, proved to give only poor results regarding cell proliferation. The interconnecting pores became too narrow to allow the needed amount of nutrient media through the scaffold, only on the top layers could living cells be found (Figure 7 & Table 3).

5   CONCLUSION

At the Freiburg Materials Research Center we developed a new Rapid Prototyping technique which allows for the use of a large number of materials as well as any method of hardening. Scaffolds made of ceramics, polymer melts and hydrogels have been successfully fabricated using this method. Using CT data from patients, tailor-made implants can be created, which have a porous design. The scaffolds have interconnecting pores, which allow for the flow of nutrient media, promoting not only the growth of cells on the scaffold surface, but as well inside of the scaffold. The design of the scaffolds is being optimized to increase cell proliferation mechanical stability.

By working under a cell friendly environment, this method can be used to fabricate hydrogels scaffolds which contain cells, also known as cell plotting. By adding further dispensing units to the Bioplotter, it will be possible to create scaffolds containing different cell types on different layers. The fabrication of such constructs using no solid support materials, but only a nutrient media plot medium, will be an important step towards Organ Printing.

This porous design of the scaffolds was chosen to provide enough space for cell growth while maintaining a good mechanical stability. The interconnecting pores allow the flow of nutrient media to every part of the scaffold and the swift removal of any degradation products from the proximity of the cells. The gap between strands cannot be too narrow though, or else not enough nutrient media will be able to reach all cells in the interior of the scaffold. The interconnecting pores must be large enough to allow the flow of nutrient media throughout the whole scaffold. Testing showed an increase in cell activity after two weeks by

REFERENCES

Aung, S.C., et al., 1999. Selective laser sintering: application of a rapid prototyping method in craniomaxillofacial reconstructive surgery. *Annals of the Academy of Medicine, Singapore*, 28(5): p. 739–43.

Chua, C.K., et al., 1998. Biomedical Applications of Rapid Prototyping Systems. *Automedica*, 17: p. 29–40.

Fischer, P., et al., 2003. Sintering of commercially pure titanium powder with a Nd:YAG laser source. *Acta Materialia*, 51(6): p. 1651–1662.

Gebhardt, A., 2003. *Rapid Prototyping*, 1st edition. Munich: Hanser Verlag.

Katstra, W.E., et al., 2000. Oral dosage forms fabricated by Three Dimensional Printing (TM). *Journal of Controlled Release*, 66(1): p. 1–9.

Landers, R., et al., 2002. Fabrication of soft tissue engineering scaffolds by means of rapid prototyping techniques. *Journal of Materials Science*, 37(15): p. 3107–3116.

Landers, R., 2004. *Materialdesign und bioaktive Formgebung für Anwendungen in der regenerativen Medizin*, Albert-Ludwigs-Universität Freiburg.

Mahoney, D.P., 1995. Rapid Prototyping in Medicine, in *Computer Graphics World*. p. 42–48.

Mankovich, N.J., et al., 1994. Surgical planning using three-dimensional imaging and computer modeling. *Otolaryngologic Clinics of North America*, 27(5): p. 875–89.

Pfister, A., et al., 2004. Biofunctional rapid prototyping for tissue-engineering applications: 3D bioplotting versus 3D printing. *Journal of Polymer Science, Part A: Polymer Chemistry*, 42(3): p. 624–638.

Pham, D.T. and Dimov, S.S., 2001. An overview of rapid prototyping and rapid tooling. *Gepgyartas*, 41(9–10): p. 58–68.

Sun, W. and Lal, P., 2002. Recent development on computer aided tissue engineering-a review. *Computer Methods and Programs in Biomedicine*, 67(2): p. 85–103.

Webb, P.A., 2000. A review of rapid prototyping (RP) techniques in the medical and biomedical sector. *Journal of Medical Engineering and Technology*, 24(4): p. 149–53.

Winder, J., et al., 1999. Medical rapid prototyping and 3D CT in the manufacture of custom made cranial titanium plates. *Journal of Medical Engineering and Technology*, 23(1): p. 26–8.

Wohlers, T., 2000. *Rapid prototyping & tooling state of the industry annual worldwide progress report*. Inc. Fort Collins: Wohlers Associates.

Yamada, K., et al., 1997. Potential efficacy of basic fibroblast growth factor incorporated in biodegradable hydrogels for skull bone regeneration. *Journal of Neurosurgery*, 86(5): p. 871–875.

Yang, S., et al., 2002. The design of scaffolds for use in tissue engineering. Part II. Rapid prototyping techniques. *Tissue Engineering*, 8(1): p. 1–11.

Virtual modeling and rapid manufacturing – Bártolo (eds)
© 2005 Taylor & Francis Group, London, ISBN 0 415 39062 1

Virtual and medical prototype assisted craniofacial reconstructive surgery

B. Starly[1], J.H. Piatt[2], W. Sun[1] & E. Faerber[3]
[1]Department of Mechanical Engineering, Drexel University, Philadelphia, PA, USA
[2]Section of Neurosurgery, St Christophers Children Hospital, Philadelphia, PA, USA
[3]Department of Radiology, College of Medicine, Drexel University, Philadelphia, PA, USA

ABSTRACT: Computer aided design models in the form of 3-dimensional visual computer models or in the form of actual physical prototypes has aided surgeons in the reconstructive surgical procedure although in only *qualitative* terms. The goal of this research is to harness the full power of modern computer imaging technologies for *quantitative* control of craniofacial reconstructive procedures. This research has addressed: 1) a science and technology based approach to the construction of BioCAD models from patient specific anatomical images; 2) the application of such models to craniofacial surgical planning using computer aided design technology, and 3) the marriage of BioCAD models and/or prototypes with commercially available surgical image guidance systems in the operating room. This scheme presented here can be incorporated within any hospital environment.

1 INTRODUCTION

Despite scientific advances in our understanding of the biology of the craniofacial skeleton and its associated tissues and despite many technological advances in imaging and instrumentation, the surgical correction of the deformities associated with craniosynostosis remains a craft. With each particular case, the surgeon based on his aesthetic sensibility and technical experience must make qualitative and quantitative judgments regarding which features of the deformity require correction and how much correction to attempt. Modern digital technologies have the potential, however, to put preoperative planning and surgical execution on a quantitative footing through the marriage of extraoperative simulation and intraoperative image-guidance. We report a single case that illustrates a particularly simple example of such a marriage employing a method not previously described for transfer of the simulated reconstruction into the clinical environment.

As shown graphically in Figure 1, the patient's data is obtained using any of the standard image acquisition systems such as CT/MRI. These images are then imported into a medical image processing software capable of reconstructing the patient's anatomy. This model is then used as the basis for virtual reconstruction using any of the available STL manipulation software. At this stage, both the surgeon and CAD operator would work in unison to virtually reconstruct the skull. After this has been accomplished, the STL model is then prototyped using any of the available rapid prototyping systems. After the model has been fabricated, the reconstructed prototype model is then rescanned using a CT machine with same parameters as the original CT scan. The new image dataset is then superimposed on the original data set so that differences between the two datasets can be measured and identified. The new image data set obtained from the CT machine is also imported into an image guided surgical system that would allow a real time evaluation of the procedure during surgery. The following section describes the process in detail.

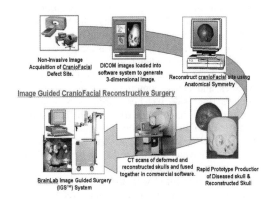

Figure 1. Image guided reconstructive surgery.

2 VIRTUAL RECONSTRUCTION

2.1 *Patient diagnosis*

A 7-year-old male child was brought to attention by his mother because of a facial deformity. There was flattening and recession of the left inferior frontal region and superior orbital margin. The orbital margin was elevated and a prominence in the left temporal region. The nose was rotated clockwise in the plane of the face as viewed by the examiner. There was no significant deformity of the right frontal or orbital regions. The clinical diagnosis was anterior plagiocephaly due to left coronal synostosis. The degree of the deformity was relatively mild, and in view of the patient's age, correction of the inferior frontal flattening with an on-lay bone graft and hydroxyapatite paste was planned.

2.2 *Deformed virtual model*

Computed tomographic (CT) scanning of the head was performed with 1.5 mm cuts and 3-dimensional reconstruction thereafter. The imaging dataset was transferred to a commercial neurosurgical image-guidance workstation (VectorVision®; BrainLAB AG, Heimstetten, Germany) over the hospital's local area network (LAN), where it was saved as a DICOM file on a Zip disc without any patient identifiers. On a workstation outside the operating room, the DICOM file was imported using medical image processing software MIMICS® (Materialise Inc) where in all images were imported, stacked upon each other and registered for its orientations. The region of interest was identified by appropriate color masks and grown into a 3D voxel representation of the deformed patient's skull. Figure 2 gives two different views of the 3D rendered voxel model.

2.3 *Virtual reconstruction of skull*

The voxelized model was converted to an STL based format and imported in the STL manipulation software MAGICS RP® (Materialise Inc) where the model was used as a starting point for the virtual reconstruction. The STL model was modified in terms of cutting out unwanted regions of the skull and smoothing out the surfaces to reduce the number of triangles and hence computer memory usage. The patient's right side has been used as template for reconstructing the left side of the skull. The surgeon identified a midline plane along the nose-bridge. The deformed left side of the re-presentation was deleted and the relatively normal right side was reflected across the midline plane to create a mirror representation, which was fused with the original right hemi-image to create a symmetrical, digitally reconstructed skull. The process is shown in Figure 3.

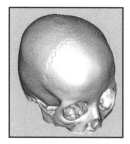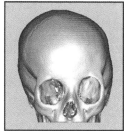

Figure 2. Different views of the voxel model.

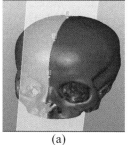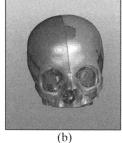

(a)   (b)

Figure 3. (a). Relatively good half is mirrored across a defined mid-plane. (b). Reconstructed model with better symmetry on the patient's forehead region.

3 QUANTITATIVE ASSESSMENT

3.1 *Prototype fabrication*

The goal of the prototype creation of the digital reconstruction was to serve as an intraoperative template for application by means of the image-guidance system to be used during surgery. However, the CAD software package did not allow for the reconstructed dataset (skull model) to be saved as a DICOM file readable by the image-guidance workstation. A physical transfer of the reconstruction was therefore undertaken. The reconstructed dataset was transferred to a three dimensional printer (ZCorp 420 3D Printer; Z Corporation, Burlington, MD), and a plaster model of the virtual reconstruction was created (Figure 4 and Figure 5). The model was fabricated in a laminar fashion by the printing of 0.076 mm axial cross-sectional layers over 16 hours. Plaster was chosen as our material in order to render the reconstructed skull prototype radio opaque. The plaster model was then subjected to CT scanning using the same protocol as the original clinical scan of the patient. The CT dataset describing the digital reconstruction was transferred to the image-guidance workstation through the hospital LAN, as had been the original clinical dataset.

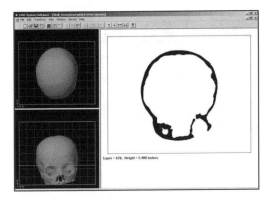

Figure 4. Fabrication layered manufacturing view. (slice parameters: 0.003 in, plaster material, build time: 16 hrs)

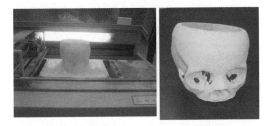

Figure 5. Prototype fabricated using Z402 system.

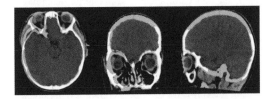

Figure 6. Fusion of original and reconstructed dataset to illustrate the difference between the 2 sets in a quantitative manner. (brighter regions indicate the data set from the reconstructed plaster model.)

3.2 Fusion of data sets

On the image-guidance workstation, the surgeon fused the data set of the clinical image with that of the dataset of the digital reconstruction. The right orbits, which were identical in both images, were superimposed (Figure 6) to create the surgical plan. In the operating room the patient's head was immobilized in a 3-point pin head-holder and registered with the image-guidance system using the soft tissue, 3-dimensional reconstruction of the clinical CT image and a laser-scanning device (Z-Touch®; BrainLAB AG, Heimstetten, Germany). When the frontal bones, the left superior and lateral orbital margins, and the

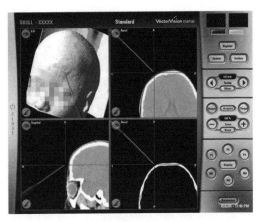

Figure 7. Real time image guidance using bayoneted probe to find out the accuracy of the reconstruction. The image on the bottom right half of the screen shows the reconstructed plaster skull data set.

left temporal bone had been exposed through a bicoronal incision, the prominent squamous temporal bone was harvested for an on-lay graft for the inferior frontal region. The graft was secured in place by means of through-and-through absorbable screws, and it was supplemented with hydroxyapatite paste. As the augmentation of the inferior frontal region and the orbital margin proceeded, it was checked against the digital reconstruction using the bayoneted image-guidance probe (Figure 7). At the conclusion of the procedure, the frontal regions and the superior orbital margins were symmetrical as judged both by the eyes of the surgeon and by the extraoperative digital surgical plan. Therefore the virtual/prototype model have been used not merely as a qualitative assessment alone but more importantly a quantitative measurement on the accuracy of the surgery performed. This therefore can be used as a real time quality control assessment for reconstructive surgeries.

4 DISCUSSION

Treatment of this older child's relatively mild facial deformity leant itself to extraoperative simulation and intraoperative control using a popular computer-assisted design package and a commercially available image-guidance system. Software incompatibility created a vexing problem with data transfer that was solved by fabrication of a plaster model of the simulated reconstruction has been described.

Routine application of this paradigm to the more severe deformities typically encountered in infancy requires a number of refinements. A safe and dependable method for immobilization of the infant skull is necessary for registration with the image-guidance system. More powerful CAD software is necessary to support complex manipulations of the craniofacial skeleton compatible, ideally, with the physical properties of bone in infancy. And this CAD software must save its work product in either a digital form or, as in the current case, a physical form that can be accessed by the neurosurgical image-guidance system. The technical challenges presented by the application of CAD methods and image-guidance in the correction of craniofacial deformities have been reviewed by Hassfeld (2001) and Vannier (1996). The individual components of this paradigm have been developed to varying degrees of sophistication by groups working predominantly in oral and maxillofacial surgery and reconstructive plastic surgery. Initial efforts focused on fabrication of stereolithographic biophysical models from clinical CT datasets for a physical qualitative assessment of the patient defect [Lambrecht (1990), D'Urso (1998), Levi (2002), Muller (2003)]. Many surgeons have found such models to be useful qualitative aids in the study of deformities and the planning of reconstructions. The models themselves can be measured, marked, cut, and reassembled to create an analogical surgical simulation. Increasingly elegant software systems have subsequently been described for quantitative digital simulation of craniofacial reconstructive surgery [Fujioka (1988), Burgielski (2002), Jans (1999), Girod (2001), Troulis (2002)]. The most sophisticated of these systems offer a set of basic surgical maneuvers, such as ablation of bone, osteotomy, translation and rotation of osseous fragments, and even bending of osseous surfaces [Mommaerts (2001), Meehan (2003)]. Haptic feedback can be incorporated as illustrated by Munchenberg (2000). Soft tissue contours can be laid onto the reconstructed craniofacial skeleton to create life-like previews of cosmetic results [Girod 1995)]. Commercially available software systems for manipulation of biomedical images have been reviewed by Vannier (1996). The problem of transfer of a digitally simulated reconstruction to the clinical setting has been solved in various ways as well. The use of photographs and the creation of sterilizable metal templates have been reported by Altobelli (1993). Unlike the MAGICS® system employed in the current case, some of the medical CAD software systems allow work products to be saved as DICOM files that, in theory, can be opened by many commercial image-guidance systems. Alternately, unlike the VectorVision® system employed in the current case, some commercial image-guidance systems apparently incorporate rudimentary CAD functionality and thus negate the transfer problem entirely for relatively simple reconstructions [Vougioukos (2004)]. So far as the authors are aware, the use of radio-opaque rapid prototyped models to transfer a digitally simulated reconstruction to a surgical image-guidance system has not been described before. The precision of the

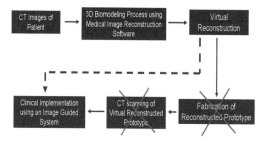

Figure 8.   STL model conversion to DICOM format presents the ideal approach to remove the need for any fabrication step. This should be a necessary feature in commonly available medical image processing software for easier clinical implementation.

rapid prototyping fabrication process has not been analyzed quantitatively, but the ease with which the shared sections of the clinical and the reconstructed images could be superimposed and "fused" on the image-guidance workstation suggests that it is sufficient for the correction of calvarial deformities. Although its dependence on access to hardware and software systems for rapid prototyping limits its general clinical applicability, the authors believe that enhancements to the commonly available commercial medical image processing software such as MIMICS can help avoid the need for any rapid prototyping system. This therefore can be used by hospitals and can be adapted to any commercially available surgical image-guidance system. Figure 8 illustrates the process that would be ideal in such a scenario. This also presents a cost effective approach to reconstructive surgeries conducted in third world countries.

The ability for surgeons to have a real time quantitative assessment will certainly aid in his/her judgment during the surgical procedure without relying on past experience. Hence such processes involve a proficiency in both technical as well as artistic creativity skills. The image guided reconstructive surgery aims at reducing the artistic component in such surgical procedures allowing them to correct deviations from the reconstructed template on a real time basis. Discrepancies between the planning performed at the CAD workstation and the actual surgical execution can be detected and corrected on the spot. The image guided reconstructive surgical procedure provides the surgeon with a tool aimed in reducing surgical errors and thereby increase the rate of successful reconstructive procedures with minimal post operative treatments.

5   DISCLOSURES

The authors have no financial tie or association with any of the commercial products used in this project or with their manufacturers.

REFERENCES

1. Altobelli DE, Kikinis R, Mulliken JB, Cline H, Lorensen W, Jolesz F: Computer-assisted three-dimensional planning in craniofacial surgery. Plast Reconstr Surg 92:576–585; discussion 586–577, 1993
2. Burgielski Z, Jansen T, von Rymon-Lipinski B, anssen N, Keeve E: Julius–a software framework for computer-aided-surgery. Biomed Tech (Berl) 47:101–103, 2002
3. D'Urso PS, Atkinson RL, Lanigan MW, Earwaker WJ, Bruce IJ, Holmes A, et al.: Stereolithographic (SL) biomodelling in craniofacial surgery. Br J Plast Surg 1:522–530, 1998
4. Fujioka M, Yokoi S, Yasuda T, Hashimoto Y, Toriwaki J, Nakajima H: Computer-aided interactive surgical simulation for craniofacial anomalies based on 3-D surface reconstruction CT images. Radiat Med 6:204–212, 1988
5. Girod S, Keeve E, Girod B: Advances in interactive craniofacial surgery planning by 3D simulation and visualization. Int J Oral Maxillofac Surg 24:120–125, 1995
6. Girod S, Teschner M, Schrell U, Kevekordes B, Girod B: Computer-aided 3-D simulation and prediction of craniofacial surgery: a new approach. J Craniomaxillofac Surg 29:156–158, 2001
7. Hassfeld S, Muhling J: Computer assisted oral and maxillofacial surgery – a review and an assessment of technology. Int J Oral Maxillofac Surg 30:2–13, 2001
8. Jans G, Vander Sloten J, Gobin R, Van der Perre G, Van Audekercke R, Mommaerts M: Computer-aided craniofacial surgical planning implemented in CAD software. Comput Aided Surg 4:117–128, 1999
9. Lambrecht JT, Brix F: Individual skull model fabrication for craniofacial surgery. Cleft Palate J 27:382–385, 1990
10. Levi D, Rampa F, Barbieri C, Pricca P, Franzini A, Pezzotta S: True 3D reconstruction for planning of surgery on malformed skulls. Childs Nerv Syst 18:705–706, 2002
11. Meehan M, Teschner M, Girod S: Three-dimensional simulation and prediction of craniofacial surgery. Orthod Craniofac Res 6:102–107, 2003
12. Mommaerts MY, Jans G, Vander Sloten J, Staels PF, Van der Perre G, Gobin R: On the assets of CAD planning for craniosynostosis surgery. J Craniofac Surg 12:547–554, 2001
13. Muller A, Krishnan KG, Uhl E, Mast G: The application of rapid prototyping techniques in cranial reconstruction and preoperative planning in neurosurgery. J Craniofac Surg 14:899–914, 2003
14. Munchenberg J, Worn H, Brief J, Kubler C, Hassfeld S, Raczkowsky J: Intuitive operation planning based on force feedback. Stud Health Technol Inform 70:220–226, 2000
15. Troulis MJ, Everett P, Seldin EB, Kikinis R, Kaban LB: Development of a three-dimensional treatment planning system based on computed tomographic data. Int J Oral Maxillofac Surg 31:349–357, 2002
16. Vannier MW, Marsh JL: Three-dimensional imaging, surgical planning, and image-guided therapy. Radiol Clin North Am 34:545–563, 1996
17. Vougioukas VI, Hubbe U, van Velthoven V, Freiman TM, Schramm A, Spetzger U: Neuronavigation-assisted cranial reconstruction. Neurosurgery 55:162–167, 2004

Virtual modeling and rapid manufacturing – Bártolo (eds)
© *2005 Taylor & Francis Group, London, ISBN 0 415 39062 1*

Mechanical properties of porous ceramic scaffolds made by 3D printing

H. Seitz, S.H. Irsen, B. Leukers, W. Rieder & C. Tille
Research center caesar, Bonn, Germany

ABSTRACT: Rapid Prototyping and especially 3D printing is well suited to generate complex porous ceramic scaffolds for bone tissue engineering. Anatomical information obtained from patients can be used to design and optimize the implant for a bone defect. 3D printing allows manufacturing of porous matrices with complex shape as well as designed internal channel network to mimic bone structures. The scaffolds are manufactured straight from 3D data using hydroxyapatite granulates. A 3D printing based fabrication process for porous ceramic scaffolds has been developed. Prototypes of scaffolds were successfully manufactured and characterized. The aim of the current study was to analyze the mechanical characteristics of ceramic scaffolds manufactured from different hydroxyapatite granulates. The compression strength of the test structures ranges between that of human spongiosa and cortical bone. It was observed that materials featuring good processability result in parts with low compression strength and vice versa.

1   INTRODUCTION

Critical-size bone defects often result from trauma or tumors and cannot be repaired by the repaired by the body's healing mechanisms. The application of synthetic bone replacement has a high clinical potential for filling such defects because bone grafts do not have to be harvested from donors or other places in the body. As a consequence, the risk of tissue rejection and disease transfer from donor tissue can be eliminated. Today different synthetic bone replacement materials are available. Normally, these materials are produced in simple geometries like blocks, pins or splines.

Rapid Prototyping and especially 3D printing is well suited to generate complex porous ceramic scaffolds for bone tissue engineering (Yang et al. 2002). Anatomical information obtained from patients can be used to design and optimize the implant for a bone defect. 3D printing allows manufacturing matrices with complex shapes as well as designed internal channel network to mimic bone structures. The scaffolds are manufactured straight from 3D data using a ceramic granulate.

The matrices generated by 3D printing can be used for bone tissue engineering using patient's cells seeded onto the scaffolds. The scaffolds serve as three-dimensional templates for initial cell attachment and subsequent tissue formation. The aim is to produce biocompatible 3D scaffolds for bone replacement with fully interconnected channel network.

Several studies have investigated the application of 3D printing technologies for direct and indirect manufacturing of scaffolds for tissue engineering (Dutta et al. 2003a), (Dutta et al. 2003b), (Lam et al. 2002), (Lee et al. 2005), (Park et al. 1998), (Wu et al. 1996). The materials used in these studies are polymers, ceramics and composites of polymers and ceramics.

The aim of the current study was to investigate the capabilities of a ceramic-based process chain for the fabrication of porous scaffolds for bone tissue engineering. The basic technology is the 3D printing technique, which is used to manufacture ceramic green parts that are sintered afterwards (Seitz et al., in press). Different ceramic materials have been tested and evaluated classifying the different powders by means of 3D printing quality as well as materials aspects (Irsen et al., submitted). Furthermore, we performed biocompatibility tests with 3D printed test structures. Good cell viability as well as good proliferation behavior was proved (Leukers et al., in press).

The aim of the current study was to analyze the mechanical properties of the ceramic scaffolds manufactured from different hydroxyapatite granulates. The compression strength of the material is an important parameter of scaffolds for bone replacement. Therefore, a test structure was designed for mechanical tests. Various specimens of the test structure were fabricated by 3D printing, sintered and finally analyzed. Results from the mechanical tests were compared to results of a previous study focusing on the processing behavior of the same powder materials in a 3D printer.

2 3D PRINTING PROCESS CHAIN

2.1 Data acquisition and preparation

Rapid Prototyping requires 3D datasets for manufacturing physical models. These data can come from various sources. In order to design an implant for a target defect, anatomical information obtained from patients acquired by computer tomography (CT) or magnetic resonance imaging (MRI) can be used.

The software tool MIMICS (Materialise, Belgium) is used to transform the 2D medical datasets derived from CT or MRI scans into 3D surface models (STL file format). Specific object regions can be extracted selecting the corresponding threshold value (Houndsfield units) for segmentation.

The software tool MAGICS (Materialise, Belgium) allows to edit the resulting STL-datasets. The damaged site and the mirrored area of the corresponding undamaged bone on the opposite side of the body can be used to calculate the shape of an implant for a bone defect.

In case of process investigations test bodies with simple geometries like blocks, cylinders etc. with or without inner channels have to be designed. The software tool MAGICS is an appropriate tool to generate such test geometries and datasets.

Furthermore, MAGICS is also used to slice the 3D datasets in a stack of two-dimensional layers and to convert these layers into bitmap-files. These files can be processed by the print control software of the 3D printer.

2.2 3D printing process

The 3D printing process is a powder-based process that builds physical 3D models directly from computer data (Sachs et al. 1992). The process starts with a three-dimensional dataset that is sliced by a computer in order to generate the print matrix of each layer. The recoating mechanism carries an amount of ceramic powder from the powder reservoir to the building box, thereby creating a thin layer of powder onto the top of the building box. Liquid binder is printed on the layer of powder according to the current cross-section of the part using a micro-dispensing valve. The ceramic powder is bonded in these selected regions. When the layer is completed, the building box piston moves down by the thickness of a layer and a new layer of powder is deposited on the first one. These process steps are repeated until the whole part is formed within the powder bed (see figure 1, step 1 to 4). The surrounding powder material supports the part during the building process. Thus,

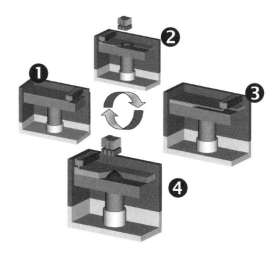

Figure 1. 3D printing process scheme; step 1 to 4 are repeated until the part is completed.

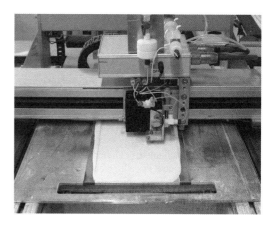

Figure 2. 3D printing test setup.

there is no need for support structures for features such as large overhangs. After completion, the part is removed from the building box and is cleaned using an air blower.

A 3D printing test setup was used for process investigations (see figure 2). The flexibility of the 3D printing setup makes it possible to investigate both new process techniques and new material combinations. The test setup is run by a machine software that imports stacks of bitmap-files and controls the complete printing process.

2.3 Materials

In this study, three different spray-dried hydroxyapatite granulates were chosen for 3D printing tests. These ceramic materials have already been tested and evaluated classifying the different powders by means of 3D printing quality as well as materials aspects (Irsen et al., submitted). It was found that HA 2 (also referred to as V5.2), HA 7 (also referred to as V10) and HA 9 (also referred to as V12) are suited for 3D printing. Consequently, these materials have been chosen to perform mechanical tests.

The main component of all tested granulates is hydroxyapatite (HA). Besides HA, the granulates are composed of polymeric additives to improve bonding and flowability. Hydroxyapatite has been chosen because the chemical composition of HA is comparable to the inorganic part of native bone. Therefore, it is a promising ceramic material to manufacture scaffolds for bone tissue engineering (Hench & Wilson 1993). The polymeric binder Schelofix was dissolved in water and used as binder fluid to be jetted by the micro-dispensing valve. All materials were obtained from the Friedrich-Baur-Research Institute for Biomaterials (Bayreuth, Germany).

2.4 Sintering

After 3D printing the specimens were sintered for 2 h at a temperature of 1250°C in an electrically heated chamber furnace (Carbolite, Germany) in air. The organic binder is removed during sintering by pyrolysis and the green body obtains its desired mechanical properties. The sintering also typically causes shrinkage of ceramic green body.

3 EVALUATION

3.1 Test body

In order to investigate the compression strength of the different materials a test structure was designed. It is a simple cylinder with a diameter of 15 mm and a height of 30 mm. A second variant of the cylindrical part with the same dimensions but with round inner channels with a diameter of 2 mm was also defined. The regular distance between the channels is 4 mm. Four channels were defined form the top to the bottom of the cylinder; two parallel rows of channels pierce the structure laterally. Both structures are shown in figure 3.

The CAD models were sliced in a regular distance of 300 μm into a stack of horizontal layers using the software MAGICS. Then the data were converted into bitmap-files. The two-dimensional bitmap files-were imported into the print control software.

3.2 Mechanical and geometrical evaluation

Physical properties of the printed bodies like compression strength and shrinkage were determined. The

Figure 3.   Three-dimensional rendering of the CAD files of the test structures for compression tests without and with channels.

compression strength of the test parts was measured using a uniaxial testing system (Zwick, Germany). The strain rate was chosen at 1 mm/min. The shrinkage was determined by measuring the diameter and length of unsintered and sintered test parts using a sliding caliper.

4 RESULTS AND DISCUSSION

A total of 24 specimens of the test parts were manufactured and analyzed. Figure 4 shows examples of a green part and a sintered specimen of the test part.

Table 1 resumes the results of the compression test. The table is supplemented by results from a previous study (Irsen et al., submitted) focusing on the processing behavior of the same powder materials in a 3D printer.

Before performing the compression tests the convex bottom surface of the test parts were cut by a saw to get coplanar planes that were needed for uniaxial compression test.

Parts made from the material HA 2 feature the highest compression strength but HA 2 causes high distortion of the printed layers during the 3D printing process. The effect of distortion is depicted in figure 5. Distortion can lead to geometrical inaccuracies and in the worst case to a process abort. The effects of distortion strongly depend on the geometry of the parts to be printed. On the other hand, the material HA 7 involves no distortion of layers during printing but the resulting parts have only low compression strength. The material HA 9 is a good compromise between processability of the material and the compression strength of the resulting parts.

In general, the compression strength of the test parts with channels is lower than the compression strength of parts without channels. When cutting the convex bottom surface of the test parts with channels made from HA 7 and HA 9 the test parts tended to break.

All tested materials can be used to fabricate scaffolds for bone tissue engineering. Since every material has its advantages and disadvantages, the appropriate material has to be chosen dependent on the desired geometrical accuracy and mechanical characteristics of the scaffold to be built.

The compression strength of the test parts lies in the range of human spongiosa and cortical bone (An 2000). On the other hand, scaffolds for bone tissue engineering with a complex channel network made from the tested materials can be less stable than the tested specimens and are not able to carry higher forces in load bearing regions in the human skeleton. In highly loaded regions, scaffolds have to be fixed by metal plates.

The shrinkage of the parts caused by sintering is almost isotropic and ranges between 18 and 22%. Since shrinkage is reproducible for each material, it can be compensated by rescaling the CAD-files of the parts to be printed by the appropriate factor.

The SEM scan in figure 6 shows the cross section of a sintered test part with two crossing channels.

The shape of the powder granules and the geometry of the channels are preserved. The particles have a good connection trough sintering. The usage of granulates leads to parts with high porosity.

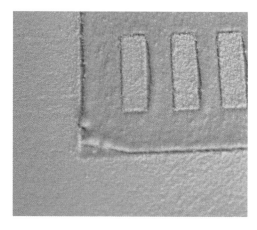

Figure 5. The distortion in the corner of a printed layer, material: HA 2.

Figure 4. Sintered specimen of the test part (left) and ceramic green part before sintering (right); material: HA 2.

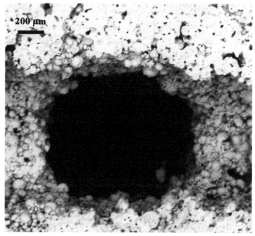

Figure 6. Cross section of a test part with two crossing channels; material: HA 2.

Table 1. Result of the compression tests. The distortion results are taken from a previous study (Irsen et al., submitted).

Material	Compression strength (test part without channels) [MPa]	Compression strength (test part with channels) [MPa]	Shrinkage after sintering (diameter/ length) [%]	Distortion
HA 2	21.2 ± 2.2	13.4 ± 1.6	18/19	High
HA 7	2.0 ± 0.4	1.5 (n = 2)	21/22	Very low
HA 9	6.6 ± 1.9	4.3 (n = 2)	20/20	Low

Generally, channel structures and high porosity of ceramic scaffolds are supposed to improve the vascularisation, the biodegradability and the cell proliferation of cells seeded onto the surface of the fabricated scaffolds (Dutta et al. 2003), (Gauthier et al. 1995).

5 CONCLUSIONS

This paper reports the mechanical characteristics of ceramic scaffolds made from different hydroxyapatite granulates. The structures were produced by a 3D printing process chain developed for manufacturing custom-made three-dimensional porous ceramic scaffolds for bone replacement The complete fabrication process, beginning from data preparation to 3D printing tests and finally sintering of the scaffolds has been presented. Test structures for mechanical characterization have been designed, manufactured using different granulates and finally characterized.

The evaluation of the results shows that all tested materials can be used to manufacture stable scaffolds for bone tissue engineering. It has been found that the powder material leading to highly mechanically stable test structures shows insufficient processing behavior like distortion of printed layers. Dependent on the application of the scaffolds and the required physical characteristics, the appropriate material has to be chosen for manufacturing.

Future work will focus on the osteogenic differentiation capability of human mesenchymal stem cells grown on 3D printed scaffolds and the resorbability of the scaffolds.

ACKNOWLEDGEMENT

The authors gratefully acknowledge the Friedrich-Baur-Research Institute for Biomaterials (Bayreuth, Germany) for a generous donation of the ceramic powder and Schelofix binder.

REFERENCES

An, Y.H. 2000. Mechanical Properties of Bone. In An, Y.H. & Draughn R.A. (ed.), *Mechanical testing of Bone and the Bone-Implant Interface:* 41–63. Boca Raton:CRC Press.

Dutta Roy, T.; Simon, J.L.; Ricci J.L.; Rekow, E.D.; Thompson, V.P. & Parsons, J.R. 2003a. Performance of degradable composite bone repair products made by three-dimensional fabrication techniques. *J Biomed Mater Res* 66A:283–291.

Dutta Roy, T.; Simon, J.L.; Ricci, J.L.; Rekow E.D.; Thompson, V.P. & Parsons, J.R. 2003b. Performance of hydroxyapatite bone repair scaffolds created via three-dimensional fabrication techniques. *J Biomed Mater Res* 67A:1228–1237.

Gauthier, O.; Bouler, J.-M.; Aguado, E.; Pilet, P. & Daculsi, G. 1995. Macroporous biphasic calcium phosphate ceramics: influence of macropore diameter and macroporosity percentage on bone ingrowth. *Biomaterials* 19:133–139.

Hench, L.L. & Wilson, J. (ed.). 1993. *An Introduction To Bioceramics.* Singapore:World Scientific.

Irsen, S.H.; Deisinger, U.; Leukers, B.; Rieder, W.; Scheler, S.; Stenzel, F.; Ziegler, G.; Tille, C. & Seitz, H. Evaluation and optimization of hydroxyapatite granules for 3D printing bone grafts. *Biomaterials* (submitted).

Lam, C.X.F.; Mo, X.M.; Teoh, S.H. & Hutmacher, D.W. 2002. Scaffold development using 3D printing with a starch-based polymer. *Mater Sci Eng C* 20:49–56.

Lee, M.; Dunn, J.C.Y. & Wu, B.M. 2005. Scaffold fabrication by indirect three-dimensional printing. *Biomaterials* 26: 4281–4289.

Leukers, B.; Gülkan, H.; Irsen, S.; Milz, S.; Tille, C.; Seitz, H. & Schieker, M. Biocompatibility of ceramic scaffolds for bone replacement made by 3D printing. *Materialwissenschaft und Werkstofftechnik* (in print).

Park, A.; Wu, B.M. & Griffith, L.G. 1998. Integration of surface modification and 3D fabrication techniques to prepare patterned poly(L-lactide) substrates allowing regionally selective cell adhesion. *J Biomater Sci Polym Edn* 9:89–110.

Sachs, E.; Cima, M.; Williams, P.; Brancazio D. & Cornie, J. 1992. Three dimensional printing: rapid tooling and prototypes directly from a CAD model. *J Eng Ind* 114:481–488.

Seitz, H.; Rieder, W.; Irsen, S.; Leukers, B. & Tille, C.: Three-dimensional printing of porous ceramic scaffolds for bone tissue engineering. *J Biomed Mater Res Part B: Appl Biomater* (in print).

Wu, B.M.; Borland, S.W.; Giordano, R.A.; Cima, L.G.; Sachs, E.M. & Cima, M.J. 1996. Solid free form fabrication of drugdelivery devices. *J Controlled Release* 40:77–87.

Yang, A.; Leong, K.F.; Du, Z. & Chua, C.K. 2002. The Design of Scaffolds for Use in Tissue Engineering. Part II. Rapid Prototyping Techniques. *Tissue Engineering* 8:1–11.

Virtual modeling and rapid manufacturing – Bártolo (eds)
© 2005 Taylor & Francis Group, London, ISBN 0 415 39062 1

Development of scaffolds for tissue engineering using a 3D inkjet model maker

Wai-Yee Yeong, Chee-Kai Chua & Kah-Fai Leong
School of Mechanical and Aerospace Engineering, Nanyang Technological University, Singapore

Margam Chandrasekaran & Mun-Wai Lee
Forming Technology Group, Singapore Institute of Manufacturing Technology, Singapore

ABSTRACT: One of the major approaches in tissue engineering is the utilization of scaffold as a temporary template to support tissue ingrowth and to guide the organization of cells. As a result, scaffold technology is critical to ensure the success of this approach. Scaffold with intricate internal morphology can be achieved by the application of Rapid Prototyping (RP). In this project, a commercial RP system was employed to produce tissue-engineering scaffolds via an indirect method. This method offers great degree of design freedom compared to conventional as well as direct fabrication method. Material opportunity is significantly enhanced with this sacrificial mold method while enabling the utilization of natural biomaterial. Flow channels can be incorporated into the design to increase the nutrient diffusion rate through the scaffold.

1 INTRODUCTION

1.1 Tissue engineering

Tissue engineering is gaining more recognition in the past decades owing to the apparent gains offered by the industry. The main regenerative tissue engineering approaches include injection of cells alone, development of encapsulated systems and transplantation of cells in scaffold. The last approach appears to be the dominant method in the research of tissue engineering (Lalan S. et al 2001). Scaffolds used in regeneration of tissues or organs serve the purpose of providing cells with a suitable environment for cells seeding and proliferation.

The solid scaffold approach in tissue engineering is often faced with the issue of optimal vascularization, which often impedes the successful cellurization of the scaffold. When cells are first transplanted on a 3-D porous scaffold, the area within the scaffold is a vascular and the transplanted cells are dependent on the diffusion of nutrients and waste for survival (Peters et al. 2002). This diffusion is adequate only if the tissue is small (<1 mm thick) or the cells' metabolic needs are low. But in the case of engineering an artificial organ, such as the liver, most cells will die soon after transplantation because of mass transport limitations (Peters et al. 2002).

1.2 Rapid prototyping in scaffolding technology

One proposed solution to overcome this limitation is to stimulate the rapid development of a vascular network within three-dimensional tissue engineering matrices, incorporating flow channels into the matrix as part of the architecture. Such matrix with intricate internal morphology can be achieved by the application of rapid prototyping (RP) (Yang et al 2002).

RP is a common name for a group of techniques that can generate physical model directly from computer-aided design data. It is an additive process in which a part is constructed in a layer-by-layer manner (Chua et al 2003). Originally created for the convenience of product designer, RP has shown to be a potential tool in tissue engineering (Yeong et al 2004).

Currently available RP technologies operate with a limited number of materials such as synthetic polymers and certain ceramics, as the standard working material (Yang et al 2002; Yeong et al 2004). If non-standard materials such as thermally sensitive natural biomaterials are to be utilized, indirect RP method is required to transfer the morphology of the fabricated part into the desired target material. This is because most RP systems use elevated temperature or chemicals to fuse the material with the purpose of producing a continuous 3-D part.

The advantages of natural biomaterials include, their closeness to mimic the natural extracellular matrix, they are non-toxic, generally have a more native surface relative to the synthetic polymers that may positively support cell adhesion and function (Liu & Ma 2004).

Based on the criteria of utilizing natural biomaterial and achievable resolution, a 3-D inkjet printer (MMII™) is selected as the vehicle in this project. This RP process provides the most significant advantage that the fabricated part, which serves as a sacrificial mold, can be removed by dissolution. The original properties of the biomaterial are well conserved as no heating process is imposed on the scaffold material.

2 DESIGN OF SCAFFOLD

2.1 Design process

As in any design problem, the development of scaffold using indirect RP is an iterative process. Three main issues in the application of indirect RP technique to fabricate tissue-engineering scaffold are: thorough understanding of the chosen RP technique, the design of the scaffold and the application of the system to fabricate scaffold (Figure 1).

Each segment requires the knowledge input of the related segments in order to achieve fruitful results. Hence, the primary motivation for this work is to understand and model these relationships so as to produce patient-specific scaffolds as well as overcoming the challenges of optimal vascularization.

2.2 Design of flow channel

In this design, interconnected channels were incorporated to act as flow channels that would increase the mass diffusion efficiency through the scaffold (Figure 2).

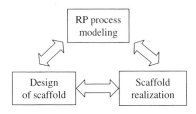

Figure 1. Interrelationships of RP and scaffold.

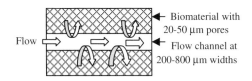

Figure 2. Flow channel.

The calculations of the channels lengths can be performed based on the estimated flow shear in the channels and the oxygen mass balance required in the scaffold. Channel geometries can then be modified according to the constraints and technical limits.

3 THEORETICAL BACKGROUND IN INKJET PRINTING

3.1 3D inkjet printing

MMII™ is a droplet-based process in which the smallest building block of a part printed is the droplet. The droplet generated also defines the minimum resolution of this process. These droplets need to be of consistent volume and eject speed as these two parameters determine momentum of the droplet. Consequently, the momentum of the droplet would determine the extent of spreading upon impact.

Important parameters that determine the momentum of the droplets are the input signal (pulse width, voltage amplitude) and the piezoelectric response of the printhead. It is also crucial to understand the phenomenon of liquid droplet formation, ejection, and impact on the substrate. With the understanding of these phenomena, it is then possible to predict and control the profile of a deposited droplet.

3.2 Modeling the height of a line

The layer thickness is an important performance index for rapid prototyping since it is a layer-by-layer manufacturing process. A model is presented in Figure 3 to relate the process parameters (printhead frequency and velocity) to the layer thickness. Surface roughness of the part produced is found to be dependent on the layer thickness as well.

It is noted that the volume of the line printed on the substrate is equivalent to the volume of droplets generated from the printhead (Gao & Sonin 1992). Assuming $A \approx w \times h$ hence,

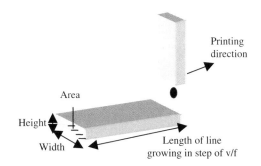

Figure 3. Modeling the height of a line.

$$h = \frac{Vol_{drop}}{d_o} \cdot \frac{f}{U} \left[\frac{We + 12}{3(1 - \cos\theta) + \frac{4We}{\sqrt{Re}}} \right]^{-\frac{1}{2}} \quad (1)$$

where A = cross sectional area; w = width; h = height; Vol_{drop} = volume of a droplet; d_o = initial diameter of droplet; f = printing frequency; U = printhead scanning speed; We = Weber number; Re = Reynold number; and θ = contact angle.

The thickness increases as printing frequency increases, or when the printhead moves at a slower speed. Higher frequency translates into higher mass transfer rate while a slower speed allows more material to be deposited at the particular spot.

The ability to predict the achieved layer thickness is important in order to guarantee the resolution of the channels and hence the scaffold design.

4 FABRICATION OF SCAFFOLD

4.1 Material and equipment

The mold was fabricated using 3D phase change inkjet printer ModelMakerII™ from Sanders Prototype. The material used in the inkjet printing process is the proprietary material of MMII™, namely ProtoBuild and ProtoSupport.

Collagen scaffold material was prepared by dispersion. The formula of dispersion is adjusted to pH 3.2, which is below 4.2 that can cause the collagen to swell and helps in its homogenization. The chosen dispersion medium is acetic acid (CH_3COOH).

4.2 Fabrication process flow

The design and development of mold design was carried out using CAD software. Data handling is carried out using ModelWorks™ application software. The CAD file was opened in this application to orientate the printing axis and to specify the layer thickness. The software also generated the model support structure automatically.

The mold was fabricated and the material in the form of dispersion was cast in. The mold-material construct was then subject to freezing at $-20°C$. Freezing the dissolution results in the production of ice crystals that grow, aggregating the collagen fiber into the space between. The kinetics of the freezing mechanism is governed by the heat transfer in the dispersion.

The construct was put into a freeze-dryer. Freeze-drying is the process of removing water from a product by sublimation and desorption. The ice crystals created inside are removed by inducing the sublimation of the ice. The mold was removed by immersing the construct into bath of ethanol. The scaffold was dried using freeze dryer to bring the scaffold to solid state at ambient pressure.

The process flow is summarized in Figure 4.

4.3 Fabricated mold

Qualitative examinations show that at larger width, the printed lines have parallel sides and the corners of the edges are sharp (Figure 5). This shows that the printheads are aligned correctly and the volume of the droplets deposited is consistent.

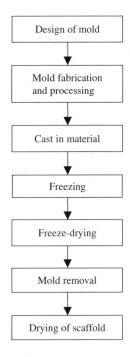

Figure 4. Process flow.

Figure 5. Fabricated mold.

Figure 6. Surface finish of a mold.

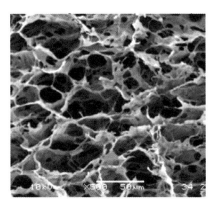

Figure 7. Freeze-dried collagen foam at 500x magnification.

The mold shown in Figure 6 exhibits characteristic burr formations on the surface as a result of the droplet-based manufacturing process. The droplets spread and flattened into a half spherical cap upon impact to the substrate. Hence, the edge defined by the material have a convex and rounded profile.

4.4 Fabricated scaffold

SEM observation reveals that highly porous networks are formed in the conventionally fabricated collagen scaffold after freeze-drying (Figure 7). The bar shown is 50 μm in length. Qualitative examination indicates that the micropore sizes are not uniform and range from 20 μm to 50 μm.

The morphology of the mold could be transferred to the collagen scaffold successfully. Figure 8 compares the mold to the morphology on the scaffold

Figure 8. Morphology of the scaffold.

(Figure 8). The outline of the morphology is shown with dotted line. The channels would ensure the interconnectivity of the macroporosity.

5 CONCLUSION

This indirect method provides the designer the opportunity to manipulate the characteristic of the scaffold at three different scales namely the macroscopic scale, intermediate scale and the cellular scale. Initial results showed that the method is able to transfer the RP part architecture to shape the internal morphology of collagen.

REFERENCES

Chua, C.K., Leong, K.F. & Lim, C.S. 2003. Rapid prototyping principles and applications. World Scientific
Gao, F. & Sonin, A.A. 1992. Precise deposition of molten microdrop: The physics of digital microfabrication. Proc. R. Soc. Lond. A 444: 533–554
Lalan, S., Pomerantseva, I. & Vacanti, J.P. 2001. Tissue engineering and Its potential impact on surgery. World J Surg. 25:1458–1466
Liu, X. & Ma, P.X. 2004. Polymeric scaffolds for bone tissue engineering. Ann Biomed Eng. 32: 477–486
Peters, M.C., Polverini, P.J. & Mooney, D.J. 2002. Engineering vascular networks in porous polymer matrices. J Biomed Mater Res. 60: 668–678
Yang, S., Leong, K.F., Du, Z. & Chua, C.K. 2002. The design of scaffolds for use in tissue engineering. Part II. Rapid prototyping techniques. Tissue Eng. 8: 1–11
Yeong, W.Y., Chua, C.K., Leong, K.F. & Chandrasekaran, M. 2004. Rapid prototyping in tissue engineering: challenges and potential Trends in Biotechnology, 22: 643–652

Virtual modeling and rapid manufacturing – Bártolo (eds)
© 2005 Taylor & Francis Group, London, ISBN 0 415 39062 1

Direct manufacture of hydroxyapatite based bone implants by Selective Laser Sintering

F. Cruz & J. Simões
Higher School of Technology of Setubal, Polytechnic Institute of Setubal, Portugal

T. Coole
Faculty of Technology, Buckinghamshire Chilterns University College, UK

C. Bocking
CRDM, Centre for Rapid Design and Manufacture, UK

ABSTRACT: This paper presents an overview of the research work carried out for the rapid manufacture of hydroxyapatite (HA) based components by means of the SLS (Selective Laser Sintering) technology, to be used for bone tissue implantation in order to replace and/or repair bone defects due to traumatized, damaged or lost bone. SLS is a Rapid Prototyping Technology (RPT) that selectively sinters powders of engineering materials by means of a CO_2 laser. HA is a bioceramic used since several years for medical applications although being mainly processed by conventional methods (cast, machined, or manually produced). For that purpose CAD/CAM, RPT and medical imaging techniques were integrated, and this enabled the customised production of anatomical models. This approach provides a direct method of producing HA based prostheses.

1 INTRODUCTION

This paper presents the results of a study, carried out from 2001 to 2004, aiming to investigate the feasibility to directly manufacture bone implants (i.e., without post-operations) from a bioceramic (HA) / biopolymer (PLLA) composite, using a Rapid Prototyping Technology (Selective Laser Sintering – SLS), rather than a conventional technology as investment casting, slip cast, pressing followed by thermal sintering, or machining.

One of the goals of the research work was to show the capability of RP to produce medical models, and thus enlarging the field of application of this emergent technology, taking advantage of its main characteristics: to manufacture parts directly from a virtual model, no need of any kind of tools, virtually independent of the geometric complexity of the parts being built, very quick processing, adequate geometric and dimensional accuracy, and a large range of materials used.

The final aim of the study was to manufacture customised bone implants, from medical scanning data captured from patients, having adequate physical and biological properties to allow their direct application in the human body, for bone grafting and bone replacement.

For this purpose the following steps had to be made:

– To convert the medical data into a 3D-CAD system (virtual model) and to generate the required files to be sent to a RP machine;
– To use powder bio-materials (bioceramic / biopolymer composite) for the manufacture of physical bone implants (physical model), by means of the SLS technology;
– To evaluate the physical and biological properties of the implant manufactured;
– To determine the field of possible applications of those implants, as a function of the results obtained.

A key requirement in many biomedical applications, including artificial bones, is a property called connected porosity. Green parts produced by the chosen SLS technology inherently have this property. While HA has been used in bone replacement and grafting for more than twenty years, this material was a logical choice for this work.

PLLA was chosen as binder because it offers a higher degradation time and a low melting temperature when compared with other acid lactic biodegradable derivates. This biopolymer acts as binder during the sintering operation, and afterwards improves the mechanical properties of the implants.

The study has take in consideration prior work made at University of Texas, USA, concerning the Selective Laser Sintering of calcium phosphate powders [Lee & Barlow, 1994; Barlow et al., 2000], although that work involved an indirect approach, unlike the direct approach presented in this paper.

2 BIOMATERIALS

Calcium phosphate based bioceramics have been in use in medicine and dentistry in the last thirty years, in several applications: dental implants, periodontal treatment, alveolar ridge augmentation, orthopaedics, maxillofacial surgery, and otolaryngology. Different phases of calcium phosphate ceramics are used depending upon whether a resorbable or bioactive material is desired.

Apatite is a natural calcium phosphate containing a little fluorine (Ca_4 (Ca F) (PO_4)$_3$) or chlorine (Ca_4 (Ca Cl) (PO_2)$_3$). It is found in granular limestone, igneous rocks and metalliferous ores.

Hydroxyapatite (HA or HAp) is a complex phosphate of calcium (Ca (PO_4)$_3$OH) that occurs as a mineral and is the chief structural element of vertebrate bone. It is a very important material for bioceramics.

One of the important research issues concerning HA is its mechanical behaviour limitation. In fact, brittleness, poor fatigue resistance, low tensile strength, and low fracture toughness value precludes HA from use in load bearing situations.

Dense, porous, or particulate forms have been prepared. However, porous HA as an implant material, is preferred. The pores (100–300 μm) allow bone to grow into implant, promoting mechanical fixation with the natural bone. Nevertheless, porosity and pore size can reduce the mechanical properties of HA ceramic. The minimum pore size of approximately 100–150 μm has been established as necessary for the continued health of bony ingrowths.

Porous hydroxyapatite has been studied for use in repairing large defects in bone. Implanted HA is slowly reabsorbed by the body over several years and replaced by bone. HA has the capability to form a direct chemical bond with hard tissues. On implantation of HA particles or porous blocks in bone, new lamellar cancellous (soft) bone forms within 4 to 8 weeks.

Porous materials are used on bone compatible implants to encourage bony ingrowths. Pore size can be of considerable biological importance. Studies have shown [Hench, 1991; Chelule, 2002] that with pore sizes ≥150 μm, in orthopaedic implants, bony ingrowths into the pores occur, and this is useful to anchor the implant. It was found experimentally that pores smaller than 75 μm did not permit the ingrowths of bone tissue. Moreover, it was difficult to maintain fully viable osteons within pores in the 75–150 μm size range. This large pore size is needed so that

capillaries can provide a blood supply to the in-growth of connective tissues. Vascular tissues do not appear if pores <150 μm. If the micro-movements occur at the interface of a porous implant, the capillaries can be cut off, leading to tissue death, inflammation and destruction of interfacial stability [Hench, 1991].

When a porous material is implanted in bone, the pores become filled first with blood, which clots, then with osteoprogenitor mesenchymal cells, then after about 4 weeks, bony trabeculae. The ingrowths bones then become remodelled in response to mechanical stress. The bony ingrowths process depends on a degree of mechanical stability in the early stages. If too much motion occurs, the ingrowths tissue will be collagenous scar tissue, not bone.

The porous and dense HA materials show excellent biocompatibility after implantation. The porous material can only de used to replace those places in the skeleton which are not loaded or are loaded mainly in compression. The dense sintered material does not convert to natural bone after implantation [Chelule, 2002].

Synthetic polymeric materials have been widely used in medical disposable supplies, prosthetic materials, dental materials, implants, dressings, extracorporeal devices, encapsulates, polymeric drug delivery system, and orthopaedic devices such as metal and ceramics substitutes.

The interest in biodegradable polymeric biomaterials for biomedical engineering has increased dramatically during the past decade, because this class of materials has two major advantages that non-biodegradable biomaterials do not have. First, these biomaterials don't elicit permanent chronic foreign-body reaction due to the fact that the human body would gradually absorb them, and they do not permanently retain trace of residuals in the implantation sites. Second, some of them have recently been found to be able to regenerate tissues (tissue engineering), through the interaction of their biodegradation with immunologic cells. Hence, surgical implants made from biodegradable biomaterials could be used as temporary scaffold for tissue regeneration.

This approach toward the reconstruction of injured, diseased, or aged tissues is considered one of the most promising fields of this century [Stevens, 2002].

The most commercially significant biodegradable polymeric biomaterials are originated from:

– Glycolic acid (e.g. polyglycolide (PG), polyglycolic acid (PGA));
– Lactic acid (e.g. polylactide (PLA), poly-L-lactide (PLLA)).

Their biomedical applications have been limited to mainly orthopaedic surgery, drug control/release devices, coating materials for suture, vascular grafts, and surgical meshes to facilitate wound healing after dental extraction.

PLLA is a biodegradable semi-crystalline polymer derived from lactic acid ($C_3H_6O_3$).

Lactic acid is a natural organic acid. Long before it became commercially available lactic acid was formed by natural fermentation in products such as cheese, yoghurt, soy sauce, meat products, pickled vegetables, beer and wine. Animal and human bodies also produce significant amounts of L (+)-lactic acid during daily activities such as walking and running.

Because metals are too stiff to prevent stress protection, polymers tend to be too flexible and to weak to meet the mechanical demands for an internal fixation device, and bioceramics (e.g. HA) are to brittle and have unfavourable mechanical properties where weight bearing is concerned, composite materials should be considered for this purpose. Recently attention has been paid to the application of HA in combination with a polymeric substance. The use is still confined to the field of the filling of bony defects and as drug carrier. Polyethylene, polybyturate and PLLA are the most frequently used polymers in such composite materials.

Composites of HA/PLLA combines bioresorption (PLLA) with bone bonding potentials (HA), resulting in a potentially bioactive and bioresorbable composite with higher strengths and stiffness than the unfilled polymer [Kesenci et al., 2000].

3 SELECTIVE LASER SINTERING

The Selective Laser Sintering (SLS) Process belongs to the solid-based Rapid Prototyping Systems, which primarily use powder as the basic medium for prototyping. The objects are built layer by layer by sintering/interfusing thin layers of powder. Each layer fuses to the previous one creating a physical object.

This technology was developed and patented at the University of Texas, Austin, USA, and commercialised by the companies DTM Corporation, Austin, USA (first system – DTM Sinterstation 125 – was introduced in 1992) and EOS GmbH Electro Optical Systems, Munich, Germany (first system – EOSINT P – was introduced in 1994). DTM had a worldwide, exclusive, licence to commercialise the SLS Technology from the University of Texas. Since 2002, 3D Systems Inc., USA, acquired the rights to produce and commercialised this technology worldwide.

Presently, SLS is the second more used process of RP in the USA and Europe, after SL and before FDM [Wohlers, 2003].

As previously referred to, the SLS process creates three-dimensional objects, layer by layer, from powdered materials, with the heat generated by a CO_2 laser within the apparatus.

The main aspects to be considered in the SLS technology are the properties of the powders used in the process and the fabrication parameters. The fabrication parameters depend strongly on the materials used, and have significant influence in the mechanical properties (e.g. tensile strength, surface hardness and density), dimensional accuracy and surface quality (i.e. roughness) of the parts produced.

Default fabrication parameters recommended by DTM for commercial powders cannot satisfy the demands of all the applications, as new materials are being added to the SLS range of materials.

With SLS it is not possible to achieve the best quality in appearance with the best mechanical properties. Maximum density of the parts is only achieved with parameters that result in excess powder sticking to the surface and low speed of construction (which is time-consuming). Using the correct combination of coating and finishing in post-processing, the mechanical properties and surface of SLS parts can be improved.

The main SLS process variables are:

Part bed temperature, laser power, scan size, scan spacing, slice thickness, and part position and orientation.

A detailed description of the SLS technology (principle of the process, sinter bonding mechanisms, fabrication parameters, SLS fabrication of metals and ceramics, medical applications of SLS) can be found elsewhere [Cruz, 2004].

4 EXPERIMENTAL WORK

The ultimate goal of this research work was to produce, by selective laser sintering, customised parts from biocompatible and bioactive materials, which have sufficient mechanical and biological properties to allow for these components to be used in bone replacement applications.

In this section an overall view of the experimental work is presented.

The final goal of the research reported in this paper was to produce models from HA / PLLA's biomaterial composites, in which the biopolymer acts as a binder, using the Selective Laser Sintering Technology. This step of the experimental work was carried out in order to check the availability of this route to produce synthetic porous models for bone implants.

Based on previous screening trials, a choice of 40% ratio of PLLA by weight has been made. This choice was based on the fact that adding lower than 30% wt. of PLLA to HA will not add significantly for the biodegradation or to enhance the ductility properties, and higher than 50% wt. the implant could be too plastic and therefore not behave as bone tissue [Cruz, 2004].

4.1 Materials

The materials used in this work (in powder form) were as follows:

- HA: Capital® 120 grade – sintering powder, supplied by Plasma Biotal Limited (UK), with mean particle size (D50) of $111 \pm 5\,\mu m$, melting point of $1250 \, (-0 + 50)°C$ and $1.30\,g/cm^3$ of density ($\sim£300/Kg$);
- PLLA: Purasorb® L, supplied by PURAC (The Netherlands), with mean particle size (D50) of $163 \pm 5\,\mu m$ (after sieving), melting range of $182.4–192.3°C$ and $0.47\,g/cm^3$ of density (after sieving) ($\sim£1700\,/Kg$).

4.2 Equipment

The experiments were performed on a SLS Sinterstation 2000 apparatus installed at CRDM. This apparatus was modified in order to allow operating with non-standard materials.

4.3 Methods

The aim of the first phase of the experimental work was to identify those factors (process parameters) that have large effects on the response variables (physical properties of the parts produced). In this phase a screening experimental design and analysis, based in the fractional factorial design technique, was applied to determine the most influential factors of interest on fabrication of HA/PLLA models by SLS. A DoE (Design of Experiments methodology) was used [Montgomery, 1997].

The data was analysed using PARETO charts of effects and the ANOVA tables. For that purpose the software STATISTICA was used.

The results obtained in the tests were focused on three (3) response variables (density, geometric accuracy and surface quality of the parts produced).

In the screening design and analysis procedure, eleven (11) different factors (independent variables) were identified as being important in the laser sintering of HA/PLLA composites. Each factor was evaluated at two (2) different values/characteristics (levels) of the factor, namely the lowest and the highest values of the range within, were considered, each factor could work. Specialists in the SLS operation at CRDM assessed these values.

In Table 1, the list of all factors and respective levels considered in the screening analysis procedure is presented [Cruz, 2004].

The results obtained show that the three factors with most influence on the physical properties of the parts produced by SLS were laser power, scan speed and scan space, i.e., the Applied Energy Density (AED). In fact, these three factors together constitute the

Table 1. Factors and levels used in the screening process.

Factors	Levels	
	Low	High
Laser power (W)	5	32.5
Laser scan speed (mm/s)	100	500
Laser scan spacing (mm)	0.10	0.15
Part orientation	Horizontal (XX direction)	Vertical (YY direction)
Roller speed (mm/s)	100	127
Part bed temperature (°C)	160	172
Outline laser power (W)	2.5	12.5
Outline scan speed (mm/s)	100	500
Slice thickness (mm)	0.10	0.15
Warm-up thickness (mm)	2	6
Cool-down thickness (mm)	0.5	1

Table 2. Data for the 2^3 factorial design.

Factors	Levels	
	Low	High
Coding	−1	1
Laser power (W)	5	7.5
Laser scan speed (mm/s)	200	300
Laser scan spacing (mm)	0.10	0.15
Applied energy density (cal/cm²)	2.66	8.96

Andrew's Equation [Cruz, 2004], corresponding to the energy density applied to the powder bed, which determines the scanning strategy during the SLS operation.

After this preliminary stage of the experimental work, the next step was focused on the measurement of the ultimate compressive strength and the elastic modulus in compression of the parts produced, by means of a 2^k full factorial design and analysis, in order to achieve more detailed information about the most influent factors selected in the previous phase

The results obtained in the previous section indicated that physical characteristics of the parts produced are fundamentally affected by laser power, laser scan speed and laser scan spacing (i.e., by the AED – Applied Energy Density in the part bed).

In order to investigate the influence of those parameters on the mechanical properties of the parts, a 2^3 full factorial design was applied [Montgomery, 1997]. The objective was to obtain a mathematical equation to express the effects of the AED in the ultimate compressing strength, bending and density of HA/PLLA specimens.

A design consisting of sixteen ($16 = 8 + 8$) experiments (i.e., a 2^3 experimental design: 3 independent variables at 2 levels each, with one (1) genuine replicate added to the design) was conducted in a randomised sequence.

Table 3. Comparison of properties between HA_PLLA and human bone.

	Compression		Bending		
	σc	Ec	σc	Ec	Density (g/cm³)
Cortical Bone (hard bone)	200 ± 36 MPa	23 ± 4.8 GPa	35–283 MPa	5–23 GPa	1.7–2.1
Cancellous Bone (soft bone)	1.5–38 MPa	10–1570 MPa	–	<1 GPa	0.14–1.1
HA/PLLA (Laser sintered)	2.4–4.6 MPa	17.2–40.8 MPa	1.6–4 MPa	140–257 MPa	0.78–1.1

The experimental design was conducted according with the data presented in Table 2 [Cruz, 2004].

In order to investigate the degradation behaviour of the HA/PLLA parts produced by SLS, namely its PLLA component, in vitro tests were performed.

To study the in vitro degradation mechanism of the HA/PLLA specimens, an aqueous media (physiological fluid) – saline phosphate buffer (SPB, pH = 7.4) was used. This media has been taken as a model of biological fluids [Taddei et al., 2002].

HA/PLLA samples as SLS produced, were weighed and immersed in 10 ml of a saline phosphate buffer media (SPB) at 37°C. The samples were immersed into glass jars, filled with SPB (one sample per jar), and placed into an incubator set at 37°C. The study was conducted on their physical properties, namely their mass loss along the time (six months).

4.4 Results

In the following Table 3 and Figures 1 and 2, the main results obtained are summarized [Cruz, 2004].

4.5 Discussion

The main goal of the experimental work was to demonstrate the feasibility of producing HA/PLLA parts by selective laser sintering. The results were successful in proving that feasibility.

In fact, the main results of the trials performed, have shown that:

– It is possible to produce bioceramic parts (HA based) using Rapid Prototyping Techniques, namely the Selective Laser Sintering (SLS) process, using a biocompatible/biodegradable polymer as binder;
– SLS seems to be a process well suited to the production of bone replacement components;
– For that purpose it is necessary to use a polymeric binder (40% wt.) to promote the agglomeration of the ceramic particles during the laser operation (sintering). The component, in this phase, is the so-called "green body";

Figure 1. SEM image of a HA/PLLA part with measurement of internal porosity (Example in the Z direction – magnification 100×).

Figure 2. HA/PLLA bones has SLS produced.

– In the case of using a biocompatible binder (e.g. PLLA), the blend HA/PLLA produces a part that can be used directly in the body without removal of the binder (green body). This technique is a direct route to fabricate laser sintered HA based ceramic parts for medical purposes;
– Thus, this route allows the production of parts directly, without the need of post-process operations, therefore reducing the time and cost of production;

- In such a case the mechanical properties of the parts produced constitutes a major limitation, allowing their application only in non-load bearing situations or supported load bearing uses;
- The internal porosity of the HA/PLLA parts produced (mean value ⊘150⊘m) is suitable to promote the growth of new tissues in the implant. In fact, this condition is essential to allow extensive blood supply of the new bodies, providing a rich supply of cells, growth factors and other nutrients needed to make bone grow. Moreover, these small holes allow access of bone marrow elements from the host tissues to the bone graft, providing nutrients for bone healing;
- The density of the HA/PLLA sintered parts (mean value = 0.883 g/cm^3) is within the density range values for the cancellous bone (0.14 to 1.10 g/cm^3) as previously referred. The green density in nearly the same of the powder bed. This result is important because, in principle, the higher the density of bone implant, the larger its strength.

The results obtained by applying the full factorial design have shown that the mechanical properties of the HA/PLLA parts as SLS produced are relatively poor when compared with other biomaterials (σ_c between 2.44 and 4.57 MPa / E_c between 17.19 and 40.80 MPa), but, even so, the Ultimate Compressive Strength and the Elastic Modulus lie in the lower limits of reported values for cancellous bone, as presented in Table 3 (1.5–38 MPa and 10–1570 MPa, respectively).

According to the analysis made, the main reason for this behaviour is caused by the internal porosity (average 200 µm, measured in the SEM analysis), which on one hand is positive because it allows the ingrowths of new tissues as previously referred, but on the other hand reduces dramatically the mechanical properties of the parts. This compromise is not easily solved.

Results have shown that at lower levels of AED (2.66 cal/cm^2), increases in AED raise the binder temperature and decrease binder viscosity, allowing the binder to fuse to itself more effectively. Beyond about 6 cal/cm^2, further increases in AED evaporate the binder and decrease the strength of the green part. In fact, there exists a competition between binder fusion and binder decomposition to gas as the AED increases. As a result of the experiments, it was possible to determine that the best values of strength and modulus of the HA/PPLA green parts produced by SLS were obtained for values of AED in the range 4–6 cal/cm^2.

The mechanical properties of the HA/PLLA sintered parts by SLS are expected to rise with the optimisation of the various parameters involved, particularly the energy density applied and the grain dimension of the starting powder.

To complement the results obtained in the compressive tests, new parts were produced and submitted to three-point bending tests. The reason for this study is that the human bones are mainly submitted to compressive and/or bending loads rather than tensile solicitations.

The results obtained show that the elastic modulus ranges between a minimum of 140.47 MPa and a maximum of 257.27 MPa (average = 181.92 MPa, Standard deviation = 0.02), whereas the bending strength ranges between 1.57 MPA and 4.05 MPA (average = 2.35 MPa, Standard deviation = 0.025). These values, like the ones obtained in the compression testing, are relatively low but situated within the range of reported bending strength and modulus for cancellous bone.

These results confirm the low mechanical strength of the HA/PLLA parts, as SLS produced, thus constituting a limitation for the applications in load-bearing situations.

5 CONCLUSIONS

The major contribution of this research is the establishment of data and formulation of guidelines for the rapid manufacture of hydroxyapatite based components, by means of the SLS technology, to be used for bone tissue implantation, to replace and/or repair bone defects, due to traumatised, damaged or lost bone.

The overall feasibility of producing hydroxyapatite based bone shapes, from SLS, has been established.

The research has shown that the Rapid Prototyping of Biomaterials is possible with a reasonable accuracy, strength and geometry of the models produced.

In summary, the research has proved the potential to produce customised components for niche applications.

REFERENCES

Barlow, J. W.; Lee, G. H.; Snyder, T. M.; Vail, N. K.; Swain, C. D.; Fox, W. C.; Aufdlemorte, T. B. 2000, Preparation of Calcium Phosphate Implants, *University of Texas Web Site,* June 2000, Austin, Texas, www.utexas.edu, USA.

Chelule, K. L. 2002, *Characterisation of Hydroxyapatite for Direct Manufacture of Bone Implants using CAD/CAM Techniques*, Ph.D. Dissertation, Staffordshire University, UK.

Cruz, F. 2004, *Direct Manufacture of Hydroxyapatite Based Bone Implants Using Selective Laser Sintering*, Doctoral Dissertation, Brunel University/West London, UK.

Hench, L. L. 1991, Medical and Scientific Product, In ASM International (eds), *Ceramics and Glasses, of Engineered Materials Handbook*, Section 13, Vol. 4, pp. 1007–1013, USA.

Kesenci, K.; Fambri, L.; Mibliaresi, C.; Piskin, E. 2000, Preparation and properties of poly (L-lactide)/

hydroxyapatite composites, *J. Biomaterials Society, Polymer Edition*, Vol.11, No.6, pp. 617–632.

Lee, G.; Barlow, J. W., 1994, Selective Laser Sintering of Calcium Phosphate Powders, *Solid Freeform Fabrication Symposium Proceedings – 1994,* September 1994, The University of Texas, Austin, USA, pp. 191–197.

Montgomery, D. C. 1997, *Design and Analysis of Experiments – Fourth Edition*, John Wiley & Sons, Inc., USA.

Stevens, E. S. 2000, *Green Plastics: An Introduction to the New Science of Biodegradable Plastics*, Princeton University Press, USA.

Taddei, P.; Monti, P.; Simoni, R. 2002, Vibrational and thermal study on the in vitro and in vivo degradation of a poly (lactic acid) – based bioabsorbable periodontal membrane, *Journal of Materials Science: Materials in Medicine*, Kluwer Academic Publishers, Vol. 13, No. 5, pp. 469–475.

Wohlers, T. 2003. Wohlers Report 2003: Executive Summary, *TCT (Time-Compression Technologies)*, Vol. 11, No. 6, pp. 26–29, UK.

Wohlers, T. 2003, Wohlers Report 2003: Executive Summary, UK.

Virtual modeling and rapid manufacturing – Bártolo (eds)
© *2005 Taylor & Francis Group, London, ISBN 0 415 39062 1*

Layer manufacturing of polymer/bioceramic implants for bone replacement and tissue growth

L. Hao, M.M. Savalani & R.A. Harris

Rapid Manufacturing Research Group, Wolfson School of Mechanical and Manufacturing Engineering, Loughborough University, Leicestershire, United Kingdom

ABSTRACT: This research aims to develop a rapid layer manufacturing technique to provide tailor made bone implants and tissue scaffolds with greater functionality than those currently available. Selective laser sintering (SLS), which is a particular additive layer manufacturing technique and capable to produce the required product directly and automatically form a 3D computer model representation, is selected to build implant and scaffold structures using composite material consisting of a polymer and a bioactive ceramic. Hydroxyapatite (HA), a bioceramic that can bond with natural bone, can be combined with high density polyethylene (HDPE), a biocompatible polymer, to form a material with appropriate stiffness, toughness and bioactivity for use in the body. A miniature SLS system was built to minimize the usage of the expensive experimental material. The results revealed that the SLS is a suitable technique to fabricate clinical grade HA-HDPE composite.

1 INSTRUCTION

1.1 Layer manufacturing

Layer manufacturing, also known as solid free form fabrication, rapid manufacturing or rapid prototyping, is an advanced mouldless manufacturing technique which allows physical parts to be created immediately, directly and automatically from a three dimensional (3D) computer-aided-designed [CAD] model. Unlike conventional machining processes which involve the removal of materials from a stock, layer manufacturing builds parts by selectively adding materials, layer by layer with each layer representing the shape of the cross-section of 3D model. This additive approach permits very complex shapes to be produced and enables the economically viable production of customised product.

1.2 Selective laser sintering

Selective laser sintering, a particular additive layer manufacturing technique, employs a laser beam to selectively sinter powder to form layers by a high precision laser scanning system. Subsequent layers are built directly on top of previously sintered layers with new layers of powder being deposited by a roller. SLS process deals with materials such as polymer powder, ceramic powder, metallic powder and their composite powder. SLS process has properties like, a wide range of materials to be used, building process without the need of supports, which enables its use in a vast range of applications.

1.3 Bio-composite

Because metals are too stiff to prevent stress shielding, polymers tend to be too flexible and weak to meet the mechanical demands for an internal fixation device, and bioceramic are too brittle and have unfavourable mechanical properties where weight bearing is concerned, the emphasis on the development of biomaterials has shifted from monoliths to composites in recent years. By controlling the volume fraction and distribution of the second phase in the composite, the properties of composites can be tailored to meet mechanical and physiological requirements as an implant. Recently attention has been paid to the application of hydroxyaptite (HA) in combination with polymer. HA has been successfully combined with different thermoplastic (polyethylene, polypropylene) to produce composites. For example, HA-reinforced high-density polyethylene (HDPE) composites (HAPEX™) (Bonfield 1988) have been developed successfully for clinical application such as orbital floor implants for patients suffering from post-enucleation socket syndrome and in middle ear implants.

1.4 HA/Polymer implants

The production methods for available HA/polymer implants are currently restricted. The HA/polymer matrix material is compounded in powder form and subsequently moulded to form the bulk material. The matrix material can be processed into several formats, including extruded solids, pellets and powders, but consequently is slow to produce as implant devices. Presently, HA/polymer implants are shaped by conventional techniques such as moulding or cutting. Such techniques have several disadvantages when utilised in implant production. For example, these indirect techniques require more process steps, they are less automated, they have high material waste (cutting techniques), and they require individual tooling for one-off jobs (moulding techniques). These conventional techniques also impose geometrical restrictions on the shapes that may be produced. In addition, the conventional techniques impose a long lead-time for implant production and present a stumbling block that dilutes many of the advantages beyond even those of unlimited shape complexity and increased speed and efficient of treatment.

The layer manufacturing of the HA/polymer implant has advantages beyond even those of unlimited shape complexity and increased speed and efficiency of treatment. This technique is therefore potential to provide customised implants which offer potential to provide improved bone defect repair modalities for individual patients and enhance the longevity of the implants by providing a securer fit (Mercuri et al. 1995). There also exist further additional unique and exciting characteristics that may be realised by utilising the proposed technique. These are controlled compositional and geometrical structure, which would be possible in these implants due to the nature of the processing technique.

2 THE COMPOSITE GRAINS FOR SLS SINTERING

According to the preparing approach, the composite grains for SLS sintering can be divided into three groups, namely separate grains, coated grains and composite grains. Separate grains are prepared by the physical blending or mixture of structure particles and binder particles. However, it is important ensure a homogenized particle mixture due to the differences in density and size of the particle. If the powder particles are not well mixed, segregation of different elements can happen after melting. Segregation of different elements can also take place during the deposition of the powder onto the powder bed. Coated grains are combined with a structural material coated by a binder material. Composite powder grains contain both the binder and structural material within each individual grains. Segregation of the different elements can be avoided.

In this study, composites grains consisting HA and HDPE are prepared for SLS process. The HDPE

(Rigidex HM4560XP, BP Chemicals Ltd) and HA particles (P218R, Plasma Biotal Ltd., UK) were provided to produce HA/HDPE composites. The HA particles have a median size ($d_{0.5}$) of 3.80 μm (Zhang and Tanner 2003). The processing route of HA/HDPE composite materials consists of blending, compounding and powderising. HA and HDPE powders with 40% HA by volume ratio were mixed using a high speed commercial blender (Kenwood Mixer). The mixture of HA and HDPE was then fed into the extruder hopper of a twin screw extruder (Betol BTS40L) for the composite compounding. Then, the pelletized HA-HDPE composites were powderized in an ultra centrifugal mill (Retsch powderizer) using sieve of 0.250 mm. Liquid nitrogen was used as a coolant to cool and thus embrittle the composite during powderizing.

A differential scanning calorimetry (DSC) was used to analyse the melting point of HA/HDPE. The melting temperature of the HA/HDPE composites is 136.47°C as shown in Figure 1.

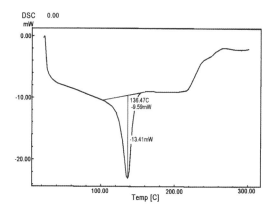

Figure 1. DSC analysis of HA/HDPE powder with HA ratio of 40%.

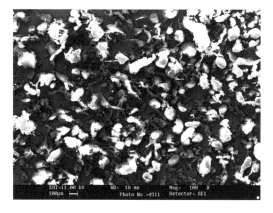

Figure 2. SEM micrograph of HA/HDPE composite powder with 40% HA ratio.

The scanning electron microscope (SEM) was used to investigate the morphology of composites powders following sputter coating with gold. The HA/HDPE powders are in irregular particle shape and most of the particles have particle size from 100 to 150 μm as shown in Figure 2.

3 SLS SYSTEMS DEVELOPMENT

A miniature SLS system was built to minimize the usage of the expensive experimental material. The system includes an "experimental powder bed chamber" and CO_2 laser system and has the function like the commercial SLS system.

3.1 An experimental powder bed chamber

An "experimental powder bed chamber" with a rig and transparent plastic box simulates the commercial SLS processing platform. The rig as shown in Figure 3 has two 103 mm diameter cylinders serving as the operation and feed powder beds. A roller is used to spread and layer the powder. Three control units are drive motors for the build and feed pistons and roller. A zinc celandine glass window is located on the upper cover of the chamber for transmission of the applied laser radiation.

An inert atmosphere is a requirement of the basic system to ensure minimum to no oxidation. By filling the nitrogen gas into the closed chamber, an inert atmosphere was created with oxygen level being maintained below 5.5%, the level which the 3D SLS system has to reach before the SLS building. An oxygen sensor (Pico Technology) is integrated to the chamber to measure the oxygen level.

Two heater bands and one infrared lamp were installed to preheat the powders for reducing laser power and the distortion in the sintering process.

Figure 3.   Photograph of experimental chamber.

Heater bands with PID temperature controller surrounded the middle area of cylinder to heat up the powder by convection. The infrared lamp (1 KW) in rectangle shape was situated at the upper area of side plastic wall to heat up the surface of the powers by radiation. Since the lamp does not vertical radiate to the surface, the powders at different areas were not evenly heated up due to the varied radiation distances. In order to minimise this problem, the experiments conducted on the relatively narrow region. In the later stage of the optimization of the experimental system, a ring shape heater element will replace the lamp to vertically heat up the surface of powders and allow the laser beam to go through. The thermal image analysis will be used to characterise the surface temperature distribution and help to achieve more temperature distribution. The current system allows us to investigate the potential of the SLS to build up the HA-HDPE part and effect of laser process parameters on the sintering.

3.2 Laser system

A Synrad 48-1-28 carbon dioxide (CO_2) laser has a wavelength of 10.6 μm and an output power of 0.1–10 W in continuous mode. The scanning speed of the laser beam varies from 0.2 to 10,000 mm/second. For the sintering, the laser beam was focused to a spot size of 193 μm at a focal length of 421 mm.

Scan Spacing is defined as the distance between two adjacent scan vectors. The laser is coupled with a marking software namely Winmark which doesn't provide this option. However, a resolution option is available which is measured in dots-per-inch (dpi). Hence, the higher the number of dots per inch the smaller the scan spacing vector. However, a means of quantification of this is needed. The system has limits between 0 to a 1000 dpi. To provide a theoretical quantification of these values, geometrical calculations from the values used on the 3D sinterstation have been used.

To equate the same amount of overlap on both the systems:

$$\frac{Exp_{SPOTSIZE}}{Exp_{SCANSPACING}} = \frac{3D_{SPOTSIZE}}{3D_{SCANSPACING}} \tag{1}$$

Where spot size (3D sinterstation) = 454 μm; scan spacing (3D sinterstation) = 150 μm; spot size (experimental system) = 193 μm. Hence, scan spacing for the experimental system is 63.33 μm.

4 SLS LASER SINTERING OF HA/HDPE

Laser power and scanning speed determine the energy imparted to the layer in the SLS process. The

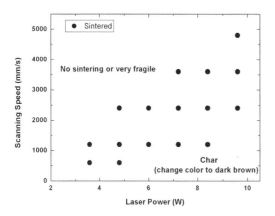

Figure 4.   Schematic of the operating window for the laser sintering of HA/HDPE composite.

Figure 5.   A 5-layer HA/HDPE block (40% HA) with 0.4 mm layer thickness fabricated by this SLS experimental system (Laser power of 7.2 W and scanning speed of 1200 mm/s).

energy imparted is termed the energy density and is assigned a dimension known as the Andrew number (A_N) that measures the effective energy exposure of the lay built by SLS (Williams and Deckard 1998). The relationship is given by

$$A_N = \frac{P}{VHS} \qquad (2)$$

Where P = laser power; V = scanning speed; HS = scan spacing.

The laser power imparts energy to the part by directly bombarding radiation on to the part bed during the scanning of a part section profile. As the amount of energy absorbed is determined by the duration of radiation on a unit area, the faster the laser beam travels, the less energy is absorbed. Thus, the amount of energy imparted, which will determine the effects of the sintering, can be controlled by laser power and scanning speed.

Since the energy is a function of both laser power and scanning speed (that is, if either one is independently altered then such a change can be compensated for through an independent alternation of the other), it is necessary to determine the exact operating window for the laser sintering of HA/HDPE by means of experimentation. Under the fixed scanning space applied, the scanning speed was varied from 300 mm/s to 4800 mm/s and the laser power was varied from 1.2 W to 9.6 W for the sintering experiments. The HA/HDPE layers were sintered at the parameters shown in Figure 4. When the power was too low or scanning speed was too high, the layers were in general not sintered or very fragile. On the other hand, the layers typically became dark brown when the power was too high or scanning speed was too low. The colour change of the layer should be caused by the char in the SLS process. Figure 5 shows a 5-layer HA/HDPE block fabricated at 7.2 W and 1200 mm/s.

(a)

(b)

Figure 6.   SEM micrograph of surface morphology of sintered layer's surface with 40% HA ratio composite and laser power of (a) 3.6 W and (b) 7.2 W at the scanning speed of 1200 mm/s.

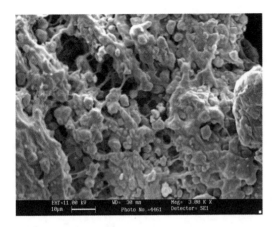

Figure 7. The SEM image of sintered HA/HDPE particles.

The different process parameters resulted in the different effects on the sintered layers. The layer fabricated at 3.6 W and 1200 mm/s presents apparently porous morphology with individual particle still being identified as shown in Figure 6(a). In contract, the layer fabricated at 7.2 W and 1200 mm/s exhibits relative dense morphology owing to the fusion of the particles as shown in Figure 6(b).

Figure 7 shows the morphology of a HA/HDPE matrix particle in the layer fabricated at 4.8 W and 1200 mm/s. Small round particles with 3 to 10 μm size are HA particles as in the matrix. The connections, which are appeared as the bridges between the gaps existing in the matrix, would be formed by the melted HDPE in the laser sintering. Thus, the liquid-phase sintering would have been occurred in SLS processing. The main advantage of liquid-phase sintering is the very fast initial binding. This binding is based on capillary forces, which can be very high. This transformation is faster than physical diffusion in solid state sintering. In addition, the matrix particle exhibits no segregation between the HA particle and HDPE, indicating the composite grains are suitable for laser sintering.

5 CONCLUSION AND FUTURE WORK

An experimental SLS system has been developed by integrating an "experimental powder bed chamber" and a CO_2 laser system. The successful building of HA/HDPE block attested that the SLS is potential to fabricate HA/HDPE products and will provide the advantages of fast leading time and customised geometry over the conventional methods.

Future work includes optimising experimental chamber for even temperature distribution, controlling the porosity and pore interconnectivity by means of process parameters and using HA/HDPE powders with different particle sizes and investigating the mechanical properties and biocompatibility of SLS fabricated parts.

REFERENCES

Bonfield, W. 1988. Hydroxyapatite-reinforced polyethylene as an analogous material for bone replacement. Annual New York Academic Science, 523: 173–177.

Boyan, B. D., Hummert, T. W., Dean, D. D. & Schwartz, Z. 1996. Role of material surfaces in regulating bone and cartilage cell response. Biomaterials, 17(2): 137–146.

Leong, K. F., Cheah, C. M. & Chua, C. K. 2003. Solid freeform fabrication of three-dimensional scaffolds for engineering replacement tissues and organs. Biomaterials, 24(13): 2363–2378.

Leong, K. F., Phua, K. K. S., Chua, C. K., Du, Z. H. & Teo, K. O. M. 2001. Fabrication of porous polymeric matrix drug delivery devices using the selective laser sintering technique. Proceedings of the Institution of Mechanical Engineers, Part H: Journal of Engineering in Medicine, 215(2): 191–201.

Mercuri, L. G., Wolford, L. M., Sanders, B., White, D., Hurder, A. & Herderson, W. 1995. Custom CAD/CAM total emporomandibular joint reconstruction system: preliminary multicenter report. Journal of Oral Maxillofacial Surgery, 53: 106–115.

Mikos, A. G., Sarakinos, G., Lyman, M. D., Ingber, D. E., Vacanti, J. P. & Langer, R. 1993. Prevascularization of porous biodegradable polymers. Biotechnology and Bioengineering, 42(6): 716–723.

Tachibana, A., Kaneko, S., Tanabe, T. & Yamauchi, K. 2005. Rapid fabrication of keratin-hydroxyapatite hybrid sponges toward osteoblast cultivation and differentiation. Biomaterials, 26(3): 297–302.

Tan, K. H., Chua, C. K., Leong, K. F., Cheah, C. M., Cheang, P., Abu Bakar, M. S. & Cha, S. W. 2003. Scaffold development using selective laser sintering of polyetheretherketone-hydroxyapatite biocomposite blends. Biomaterials, 24(18): 3115–3123.

Wang, M., Porter, D. & Bonfield, W. 1994. Processing, characterization, and evaluation of hydroxyapatite reinforced polyethylene composites. British Ceramic Transactions, 93(3): 91–95.

Williams, J. D. & Deckard, C. R. 1998. Advances in modeling the effects of selected parameters on the SLS process. Rapid Prototyping Journal, 4(2): 90–100.

Zhang, Y. & Tanner, K. E. 2003. Impact behavior of hydroxyapatite reinforced polyethylene composites. Journal of Materials Science: Materials in Medicine, 14(1): 63–68.

Fabrication of bone substitute material by rapid prototyping

A. Ott, R. Pelzer & F. Irlinger
Lehrstuhl für Feingerätebau und Mikrotechnik, Technische Universität München

ABSTRACT: A key requirement in the field of bone tissue is the development of scaffold structures, on which cells adhere. This can be done by fabricating scaffolds by direct procedures like 3D-printing or by indirect procedures like casting. With the 3D-printing process different structures were build up by using hydroxyapatite powder (HA) and a special binder material. Afterwards the printed 3D structures were sintered. For the casting process molds have been made of different resins by stereolithography and other processes using polymers and waxes. These structures were filled by a suspension of HA. Compared to the 3D printing a better resolution can be obtained here. But there are restrictions for the level of porosity.

1 INTRODUCTION

In the field of biomedical implants attempts to design suitable systems have been made from different directions. One current research field focuses on tissue engineering [Chu 2002]. The research group Fortepro is developing a process for fabricating bone substitute material by using rapid prototyping techniques. These implants are designed to replace bone defects at the head and musculoskeletal system. They are made of hydroxyapatite (HA) and will be replaced by endogenous bone material after implantation by building new tissue. Figure 1 shows the complete process sequence. In a first step data from the defect is collected

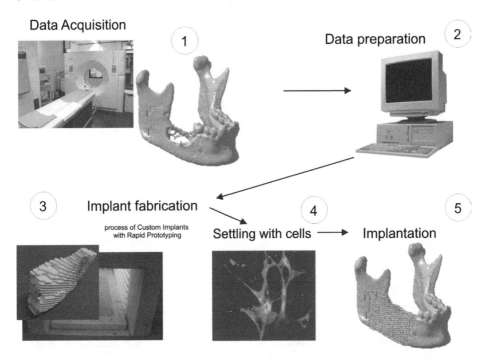

Figure 1. Principle process sequence.

with computer tomography (CT). Then the three-dimensional geometry of the implant is designed with a special software and converted in a standard triangulated language (STL) file [Sun 2002]. With different RP processes the implant geometry is produced and later prepared with cells. Afterwards the prepared scaffold can be implanted. A key requirement is the development of scaffold structures on which cells can adhere. The aim of the institute "Feingerätebau und Mikrotechnik" is to fabricate these scaffolds with controlled internal porosity.

2 REQUIREMENTS ON THE IMPLANTS

This study aims at the creation of a porous artificial extracellular matrix or scaffold to accommodate cells and guide their growth and tissue regeneration in three-dimensions (3D). These scaffolds should be created individually for each patient, with interconnected and controlled porosity and pore distribution [Hollister 2002]. Furthermore the mechanical strength not only depends on the 3D structure of these scaffolds but also on the used process parameters for example the temperature profile during the sintering process [Shoufeng et al. 2002].

2.1 Part accuracy

The process allows to generate patient individual implants. There is no need of post processing the implant by the medicines at the following surgery. Moreover, it is necessary for cell growth and flow transport of nutrients and metabolic waste that the defined inner structure and porosity can be produced consistent. The desired parts should have interconnecting pores with a size of about 400 to 600 μm for cell proliferation and the porosity should be around 50% [Taboas 2003].

2.2 Mechanical requirements

Desired is an elastic modulus, breaking strength and fatigue strength with valves pores similar to those of natural bones. Our research group wants to develop an implant which can be handled by the medicines without exhaustive care and will serve for implantation in regions without high loading eg. the head region.

2.3 Medical requirements

Biocompatibility is necessary for the implantation. Furthermore the scaffold structures should allow the infiltration of stem cells. This is necessary to replace the biodegradable artificial bone substitute material by autogenous material [Koch 2000]. Literature shows that the hydoxyapatite ceramics have been well established because of their high biocompatibility qualities. The degradation of the implant also depends on the used material and is so far not clarified completely.

2.4 Process requirements

For the different processes the used materials have big influence on the results. For the 3D printing process the particle size and a spherical particle form are very important. Furthermore it is necessary to prevent the particles from agglomeration. A lot of printing experiments were done to investigate the influence of different HA and binder combinations. On the other hand the quality of the filling of the negative molds depends strongly on the prepared slurry. Different proportion of HA ceramic, deflocculant, solvents and water have big influence on the results. Moreover the slurry and the temperature profile can be matched to the different materials of the molds.

3 PROCESSES 3D-PRINTING

This process creates parts by a layered printing process with adhesive bonding, using HA powder as a base material. After loading the STL-file, which defines the geometry and the interconnecting pores, the software of the Rapid Prototyping machine slices the three-dimensional data into two-dimensional pictures according to the cross-sectional area of the object. Each layer of powder is selectively joined where the scaffold is to be formed by ink-jet printing. This process is repeated layer by layer until the green body of the scaffold is complete. Afterwards the powder which is not joined together is removed and the green body is sintered. This printing process requires a special powder. The resolution depends on the form and particle size and the grain size distribution. The major problem is to choose an appropriate binder for micro dosing. The used hydroxyapatite powder is spray dried with particle sizes between 100–200 μm diameter. The binder, which joins the powder particles, was also specially developed for this process. Each powder layer is about 200 μm thick and finished with a counter rotating roll, which sleeks the HA. The micro dosing of the binder is realized with a piezoelectric drop on demand system or an electromagnetic valve. Therewith it became possible to dose drops with a diameter of 60–80 μm. The temperature profile for sintering only depends on the ceramic. Starting at room temperature the temperature is increased to 1250°C at 100°C/hour. This temperature is hold for about 1 hour min and then the ceramic is cooled down to room temperature with about 100°C/hour.

4 RESULTS 3D-PRINTING

The first printed bodies have been fabricated without internal structure. These tests proved, that the green bodies can be sintered and led to an overview of the quality of the printing results by using different powders and binders. On scanning electron microscope pictures, (Fig. 2) the sinter-necks between the agglomerates can be observed as well as the sintered powder-particles in the agglomerate. These results prove the feasibility of the process [Pelzer 2003]. The next step was the production of scaffolds with defined inner structures in three dimensions. Figure 3 shows a jaw bone which contains a individual interconnected porosity and is prepared patient individual.

The strength of the printed parts (green bodies) is low but after sintering at 1250°C they are solid enough for handling. The breaking strength depends on the porosity. Cylindrical test items shows a breaking strength of 39 MPa, the porous parts reaches only MPa. The powder-binder interaction influences the strength of the part and the behaviour of the wetted powder.

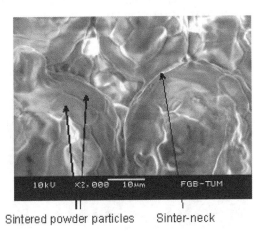

Figure 2.   Sintered necks between the spray dried particles.

The strength of the printed scaffolds is about 2,5 MPa. Scanning electron microscope pictures of cross sections from a printed and sintered scaffold shows that the surface remains very rough and spherical ceramic particles are visible. At the fractured surface of the ceramic micro porosity is remaining.

5   PROCESSES CASTING

After presenting the results of the 3D printing process now the second RP process is described. Figure 4 shows the flow chart of the implant manufacturing process by casting. On the left side the process of the hydroxy-apatite suspension preparation can be seen. This is done by the Friedrich-Baur-Institute which is highly experienced in ceramic materials.

In this chapter the designing of molds from negative images of implants and the fabrication of the molds by RP processes is described. At the beginning of the project we focused on the stereolithography as the highest accuracy for getting the desired internal structure of the molds were expected. But also, the 3D Wax printing process is under investigation. In the next step the prepared slurry is filled into the mold (Impregnation of the mold). Followed by a thermal process to remove the binder components from the suspension and to combust the mold. By rising the temperature up to 1250° C the HA ceramic is sintered. In this cage the sinter profile has to be adjusted not only to the ceramic but also to the material of the mold. The heating for combusting the molds has to be done very slowly with holding the temperature at a special point. After the combustion of the mold is completed, the heating process for sintering is comparable with the temperature profile described for the 3D printing process [Ott 2003,

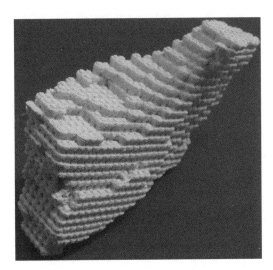

Figure 3.   3D-printed jaw bone implant with internal porosity.

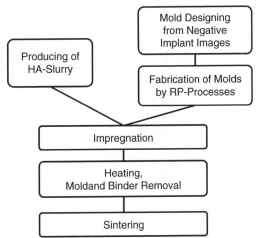

Figure 4.   Flow chart of the manufacturing process.

Ott et al. 2004]. The critical points are the correct impregnation (filling of the molds) and the prevention of cracks in the ceramic during the first heating process, when the mold causes stress because of its thermal expansion. Figure 5 demonstrates the feasibility of this process. The mold was fabricated by stereolithography and filled by a standard procedure of slip casting. The geometry of the parts is similar to the geometry used in the 3D printing process. The rough surface due to the layer by layer fabrication on the mold (left) can also be observed at the sintered scaffold (right). But the cracks in the structure led to a failure of the whole scaffold. The interconnecting pores size for cell proliferation and the maximum wall thickness for degradation are limiting the design of the implants. To avoid cracks due to thermal stress, the ratio between resin and ceramic had to be decreased. This was realized by inserting hollow structures in the mold which are not filled by the ceramic slurry. This way, the stress on the ceramic during combusting the molds can be reduced. But still, there was a high reject due to defects. Therefore a new design of the parts and a better filling process were established.

5.1 Filling process

Figure 6 shows a micro computed tomography (μ-CT) of an impregnated mold prepared at the beginning of the project. The structure in dark-grey is the shot of the mold. The white color presents the ceramic. The origin of the cracks during the drying of the ceramic can be traced very well. The insufficient filling of the mold made it necessary to improve this process. By adding different tensides to the slurry and by investigating different filling methods the characteristics of the slurry have been changed. The vacuum-, the centrifuge- and the pressure/vacuum methods have been investigated. The pressure/vacuum method (Fig. 7) solved most of the problems during the impregnation of the molds. A small chamber is filled with the HA slurry and the mold. By exceeding pressure on one side and low pressure on the other side of the chamber the small cavities could be reasonable filled without any air bubbles. By using low pressure it became possible to dry the slurry during the filling process. This way the swelling of the resin molds by the water-based slurry can be reduced. The used materials for the molds are standard resins for stereolithography and standard wax of 3D printing respectively. Therefore one major question is the influence of the toxic vapor on the ceramic parts during combustion of the mold material. This is done by technical analysis like EDX or by analyzing the cell proliferation on the material. EDX analysis of sintered ceramics produced with different stereolithography materials and waxes have been curried out. No toxic residua could be found at the surface of the ceramic parts. First experiments with cell proliferation on the surface of the sintered parts show good results.

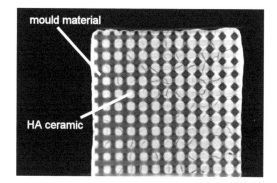

Figure 6. Micro CT of an impregnated mold.

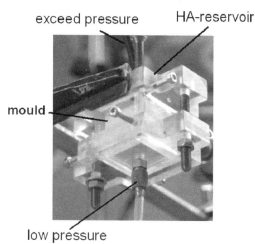

Figure 7. The exceed pressure & vacuum filling station.

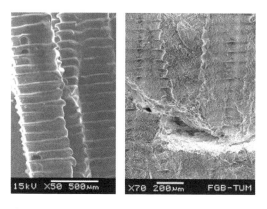

Figure 5. ESEM micrographs of the RP mold and the resulting scaffold.

Next investigations with stem cells have to prove if the chosen wall and pore size are adequate for cell proliferation on the sintered implants.

6 RESULTS OF THE CASTING PROCESS

The desired channel and wall size of less than 500 μm can easily be achieved by using the SLS technology. But due to the thermal expansion and the exited stress it is necessary to keep a ratio of polymer to ceramic of less than 50%. Using a honeycomb structure with hollow blocks, allows the reduction of the polymer ratio dramatically. Therefore the filled mold is more suitable for the thermal process. Moreover sharp edges can be reduced which is expected to be better for cell

proliferation and the filling process. In contrast with the rough surface of the 3D printing scaffolds the results by casting are very smooth.

7 RESULTS OF THE RESORPTION PROCESS

First tests with regard to biocompatibility were done at the Heinz Nixdorf-Lehrstuhl für Medizinische Elektronik at the Technische Universität München. In Figure 9 you see fibroblast cells coating the implant. The investigation of the resorption was made by the "Friedrich-Baur-Forschungsinstitut für Biomaterialien" in Bayreuth. In Figure 10 you can see a rat osteoclast cell cultured on HA and in Figure 11 the lacuna is shown.

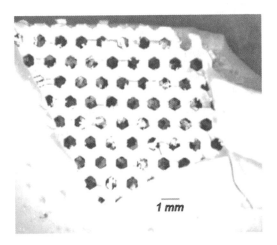

Figure 8.   Ceramic parts with internal porosity.

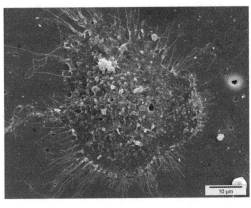

Figure 10.   SEM image of a rat osteoclast cell cultured on HA.

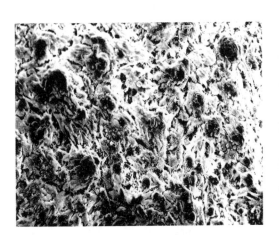

Figure 9.   SEM image (Area 1 × 1 mm). Cells on Hydroxylapatit (HA). [Wolf 2005].

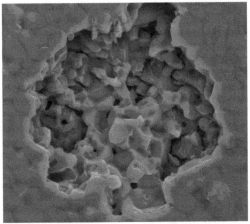

Figure 11.   SEM image of the surface after osteoclast cultivation for 48 hours [Detsch 2005].

8 CONCLUSION

Both RP processes are suitable for producing scaffolds for tissue engineering. The 3D-printing process is the winner of the golden EuroMold AWARD 2004. First investigations proofed the possibility of cell proliferation on the surface of the structures and the resorption of the material. Further steps will include the habitation of the scaffold structures with stem cells and bone morphogenic protein induced osteogenesis.

ACKNOWLEDGEMENTS

We thank the "Bayrische Forschungsstiftung" for founding the project Fortepro. Furthermore our industrial partners Alphaform Ag, Forschungszentrum Caesar, KL Technik and Siemens and the involved institutes: Chirurgische Klinik und Poliklinik, München; Forwiss, Passau; Friedrich-Baur-Forschungsinstitut für Biomaterialien, Bayreuth; Hightech Forschungszentrum, München; Institut für Umformtechnik, München; Zentralinstitut für Medizintechnik, München.

REFERENCES

Chu, G. 2002. Mechanical and in vivo performance of hydroxyapatite implants with controlled architectures. Biomaterials, Vol. 23, No. 5 , pp 1283–1293.

Detsch, R., Mochau, F., Mayr, H. & Lefevre, A. 2005. Osteoclast Generated Resorption of Synthetic Bone Substitute Materials. Faenza Proceeding,

Hollister, S.J. 2002. Optimal design and fabrication of scaffolds to mimic tissue properties and satisfy biological constraints. Biomaterials (23), No. 20, pp 4095–4103.

Koch, K.U. 2000. Verfahren zur Herstellung von individuellen Implantaten auf Basis resorbierbarer Werkstoffe. Thesis, Stuttgart, Germany.

Ott, A. 2003. Fabrication of scaffold structures by casting. In Proceedings ESEM 2003. Halle Germany, pp 169.

Ott, A., Heinzl, J., Janitza, D. & Pelzer, R. 2004. Fabrication of bone substitute material by Rapid Prototyping. In Solid Freeform Fabrication Symposium, Austin, Texas, 2004, p 784–793.

Pelzer, R. 2003. The 3D-printing process with new biocompatible materials. In Proceedings ESEM 2003. Halle Germany, pp 171.

Shoufeng Yang, et al. 2002. The Design of Scaffolds for Use in Tissue Engineering. Tissue Engineering, Vol. 8, No.1, pp 1–11.

Sun, W. 2002. Recent development on computer aided tissue engineering – a review. Computer Methods and Programs in Biomedicine, Vol. 6, No. 2, pp 85–103.

Taboas, J.M. 2003. Indirect solid free form fabrication of local and global porous biomimetic and composite 3D polymer-ceramic scaffolds. Biomaterials, Vol. 24, No.1, pp 181–194.

Wolf, B. & Grothe, H. 2005. Heinz Nixdorf-Lehrstuhl für Medizinische Elektronik, Technische Universität München. E-mail conversation Grothe / Irlinger.

Virtual modeling and rapid manufacturing – Bártolo (eds)
© *2005 Taylor & Francis Group, London, ISBN 0 415 39062 1*

Digital manufacturing of biocompatible metal frameworks for complex dental prostheses by means of SLS/SLM

J.-P. Kruth, B. Vandenbroucke, J. Van Vaerenbergh
Division PMA, Department of Mechanical Engineering, Katholieke Universiteit Leuven, Belgium

I. Naert
School for Dentistry, Oral Pathology and Maxillofacial Surgery, Department of Prosthetic Dentistry, Katholieke Universiteit Leuven, Belgium

ABSTRACT: In recent years, digitizing and automation have gained an important place in the manufacturing of medical products. However, many dental parts are still being produced by manual and inefficient conventional methods. This paper presents a fully digital and fast procedure for the design and manufacturing of implant-supported frameworks for complex dental prostheses by means of Selective Laser Sintering (SLS) or Selective Laser Melting (SLM). Offering a digital solution to the dental profession implies a real challenge because patient and dentist set high requirements on quality, material and precision. During this study, frameworks were manufactured by SLS/SLM of stainless steel, Ti6Al4V and a dental CoCr-alloy. The accuracy of the final framework has been analyzed by studying the error accumulation of the successive steps in the proposed procedure. SLS/SLM allows an efficient and customized production of the complex biocompatible frameworks without lengthy manual pre- or post-processing and with a medically acceptable accuracy corresponding to the precision of a conventionally fabricated framework.

1   INTRODUCTION

Selective Laser Sintering (SLS) and Selective Laser Melting (SLM) are layer-wise material addition techniques that allow generating complex 3D parts by selectively consolidating successive layers of powder material on top of each other, using thermal energy supplied by a focused and computer controlled laser beam [6, 8, 10]. Different binding mechanisms can be responsible for the consolidation of the powder: Solid State Sintering, Liquid Phase Sintering, Partial Melting or Full Melting [3]. The competitive advantages of SLS/SLM are geometrical freedom and material flexibility.

Over the last decade SLS/SLM processes have gained a wide acceptance as Rapid Prototyping (RP) techniques. Recently, a shift to Rapid Manufacturing (RM) has come up because of technical improvements of layer manufacturing processes [2, 6]. SLS/SLM techniques are no longer exclusively used for prototyping and the possibility to process all kind of metals yields opportunities to manufacture real functional parts. Medical and dental applications could take advantage of this evolution by using SLS/SLM not only for plastic devices like visual anatomical models

or one-time surgical guides, but also for functional implants or prostheses with long-term consistency made from a biocompatible metal. Dental applications are very suitable for processing by means of RM due to their complex geometry, low volume and strong individualization. The manufacturing of multiple unique parts in a single production run enables mass customization. Moreover, computer controlled production corresponds to the global trend of digitizing the fabrication of medical and dental parts.

This paper presents a fully digital solution for the design and manufacturing of implant-supported biometal frameworks for complex dental prostheses (Fig. 1) [9]. The intended framework (2) is the metal base structure of the prosthesis (1). It is supported by oral implants (4) placed in the jawbone (5) or by remaining teeth. Such framework is fixed to the jaw by screws retaining it on the implants or by luting it on the remaining teeth. The framework supports the artificial teeth (3) which are attached on the support surfaces on top of the framework.

The framework is made from a titanium or CoCr (cobalt chromium)-alloy because these metals combine good mechanical and biocompatible properties. The framework is patient specific and has to meet

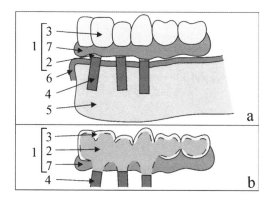

Figure 1. Scheme of implant-supported prosthesis: (a) drawing of lower jaw with implant-supported prosthesis (b) transparent view of prosthesis; 1. prosthesis, 2. framework, 3. teeth, 4. oral implants, 5. jawbone, 6. soft tissue, 7. artificial gums.

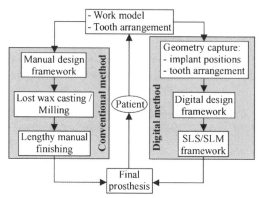

Figure 2. Global scheme of the conventional and digital method to manufacture a dental framework.

strict requirements of accuracy to minimize the risk of mechanical or biological failures of the prosthetic system. To distribute forces evenly and to avoid high stresses in the jawbone causing the oral implants to loose and to diminish the risk for colonization of bacteria resulting in infection and eventually bone loss, a good passive fit at the prosthesis-implant or prosthesis-tooth junction and severe fit criteria below 40 μm are necessary [1, 12].

The proposed digital procedure for the design and manufacturing of dental frameworks replaces the conventional system based on a manual design by 'clay' modeling and a production by lost wax casting or milling. Although these methods lead to high precision, they are time consuming and inefficient. The lost wax method is a lengthy and labor-intensive process and comprises many manual steps: fabricating, embedding and burning out the wax pattern, metal casting and post-processing. By digitizing the manually designed 'clay' model and using CAM-instructions, the frameworks can be produced by CNC-milling. However, this milling process is also time consuming because of the complex tool path calculations and because of the manual finishing needed to obtain the required shape of the framework. Moreover, most of the expensive material is wasted and spatial restrictions limit the production of complex shapes.

Figure 2 shows the different steps of the conventional and digital method for producing a framework. After installing the oral implants into the jawbone of the patient, a plaster work model with implant replicas is made representing the position of the implants. Upon this work model a tooth arrangement is shaped from which the patient validates the aesthetics. This tooth arrangement will look the same as the final prosthesis but the internal metal framework is still absent. The developed digital procedure consists of

three main steps: the digital geometry capture of implant positions and tooth arrangement, the digital design of the framework and the computer-driven production of the framework by means of SLS or SLM. The complex shape of the framework can be fabricated without the need for manual intervention and with less pre- or post-processing. SLS/SLM allows an efficient and customized production of the complex framework for different materials. Remaining unprocessed powder can be reused. The prosthesis is finished by gluing or pressing the teeth upon the support surfaces of the framework and finally installed in the mouth of the patient.

The following presents the fully digital procedure to design and manufacture dental frameworks and investigates if this procedure, including SLS/SLM, fulfills the strict requirements of quality, accuracy and material.

2 PROCEDURE

The possibility and repeatability of manufacturing dental frameworks by means of SLS/SLM have been proved in a preliminary research where neither the complex design of the framework, nor the production in a biocompatible metal have been dealt with [5]. The following procedure explains the different steps to digitally design and manufacture complex dental frameworks with high precision in a biocompatible metal.

2.1 *Geometry capture*

The first step of the digital procedure is the measurement of the twofold input geometry: firstly the position of the implant replicas in the plaster work model, corresponding with the position of the implants in the jawbone of the patient, and secondly the tooth arrangement, approved by the patient.

Figure 3. Measurement of position of implants: (a) registration elements mounted upon implant replicas of work model, (b) point cloud of scanned elements, (c) fit of top plane and cylinder surface on identified points of a registration element.

The position and inclination of each implant vary depending on the presence of qualitative and quantitative surrounding bone and the anatomical location of nerves and blood vessels. The measurement of the position of the implants has to be very accurate with regard to the final fit between framework and implants. Due to the complexity of the upper part of the implants, mechanical or optical techniques have difficulties to measure directly the position of the implants. Therefore the use of registration elements is preferred. Figure 3a shows cylindrical registration elements mounted upon the implant replicas of the work model. An optical scan of the mounted registration elements leads to an accurate point cloud (Fig. 3b).

Using numerical algorithms the position and the inclination of each registration element are computed by fitting a top plane and cylinder surface on the identified points of each element (Fig. 3c). Based on these data and the well-known dimensions of the registration elements, the position and inclination of each implant can be calculated, providing sufficient information to design the lower part of the framework which fits on the implants.

The precision of digitizing the tooth arrangement is not very critical, in contrast to the position of the implants, because no major effect is stated with regard to possible biological or mechanical failures. Teeth used for the tooth arrangement have a standard shape and thus the measured point cloud (Fig. 4a) only has to indicate the relative positions of the teeth. Because the data of the implant positions will be combined with the tooth arrangement data, both measurements are mathematically matched with the work model as reference object.

2.2 Digital design of the framework

The second step of the procedure is the digital and automated design of the framework based on the computed position of the implants and the captured geometry of the tooth arrangement. This digital design replaces the conventional manual design, requiring a lot of experience of the dental technician who builds up the framework in a wax material. Based on the many manual actions of the technician, several design

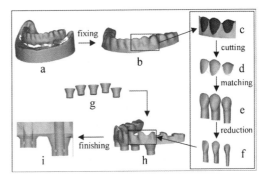

Figure 4. Digital design strategy: (a) scan of tooth arrangement, (b) 3D digital teeth model, (c) identified tooth surfaces, (d) incomplete tooth surfaces, (e) completed tooth surfaces, (f) support surfaces of framework, (g) fitting structures, (h) connection of support surfaces and fitting structures, (i) finished framework.

rules are defined and put in a certain sequence to determine the design strategy. This strategy is implemented in a software module by translating the design rules to computer tools. An important issue during the design step is the manipulation of the complex dataflow. The different and large data files of the complex medical shape have to be filtered and combined to one geometrical model in a fast and convenient way without any mistakes.

The design process starts with the modeling of the tooth arrangement. Based on the scan (Fig. 4a) a digital 3D teeth model (Fig. 4b) is designed by fixing measurement errors like gaps, scatter and inverted normals of the surfaces. Each artificial tooth of the final prosthesis needs a support surface on top of the framework, obtained by an offset of the real tooth surface (Fig. 1b). The different teeth can be identified by means of a curvature analysis (Fig. 4c). Each separate tooth surface, obtained by cutting the 3D teeth model, is incomplete because the side surfaces are missing (Fig. 4d). These side surfaces are needed to compute the offset of the total tooth surfaces. Therefore each incomplete tooth surface is matched with a full point cloud of the corresponding standard tooth (Fig. 4e). By scanning all standard teeth once, a digital library of standard tooth surfaces is available. The support surfaces of the framework are then computed by a defined offset of the completed tooth surfaces (Fig. 4f), depending on the tooth material (acrylic, composite or porcelain).

Using the mathematical match calculated in the previous step, the position and inclination of the implants are transferred to the design environment. Conical fitting structures are designed according to these data (Fig. 4g) and connected to the support surfaces of the teeth (Fig. 4h). Finally, the digital design

141

Table 1. Process parameters of SLS/SLM, divided into material, laser, scan and environmental parameters.

Material	Laser	Scan	Environm.
composition	mode*	scan speed*	preheating*
powder density	wave length	hatching space*	pressure*
morphology	power*	layer thickness*	gas type*
diameter of grains	frequency*	scan strategy*	O₂ level*
distribution	pulse width*	scan sectors*	
thermal properties	offset*	pulse distance*	
flow properties	spot size	scaling factors*	

* Studied parameters to reach high precision (material parameters and some laser parameters are not varied due to machine dependency).

of the framework is finished by adding detail features like screw holes and fillets (Fig. 4i).

2.3 Production by means of SLS/SLM

The third step of the digital procedure is the production of the framework by means of an SLS or SLM technique. These Rapid Manufacturing processes use a computer controlled laser beam for scanning successive layers of powder material to create the 3D framework. Based on the slicing of the digital design from the previous step, scanning patterns of each layer are computed automatically.

2.3.1 Binding mechanisms
For the consolidation of the powder, two different binding mechanisms are used depending on the material. Other mechanisms are possible, but not considered in this study [3].

The first mechanism is a kind of Liquid Phase Sintering where the polymer coating of steel powder grains is liquefied by the laser beam and acts as a binder for the structural stainless steel grains. This technology needs an additional furnace cycle, in which the polymer is burnt out and the green part is further sintered and infiltrated with e.g. bronze to reach high density. The second used mechanism is Full Melting of a biocompatible metal like Ti6Al4V or a dental CoCr-alloy. Near full density is reached within one step by melting the metal powder completely by the laser beam, thus avoiding lengthy post-processing steps.

2.3.2 Process parameters
SLS/SLM is a complex thermo-physical process and the determination of parameters is very important to reach high precision. Table 1 divides the many process

parameters into four groups: material, laser, scan and environmental parameters. The optimal parameter setting can be found by a combination of empirical research and numerical simulations.

Due to the contraction of molten material that solidifies and cools down and due to high thermal gradients during SLS/SLM processes, distortions like curling or delamination can appear [7,11]. An optimal scan strategy and appropriate energy density can avoid these harmful effects. The energy density is an absolute process parameter for a certain material powder. This parameter represents the energy supplied by the laser beam to a volumetric unit of powder material and combines some important laser and scan parameters:

$$E_{density} = \frac{P_{laser}}{v_{scan} \cdot s_{hatching} \cdot t_{layer}} \tag{1}$$

where $E_{density}$ = energy density, P_{laser} = laser power, v_{scan} = scan speed, $s_{hatching}$ = hatching space, t_{layer} = layer thickness. A few loops of benchmark tests can help to reach optimal parameters. For example, offset and scaling values, used to compensate for dimensional changes due to the laser beam spot size, are optimized iteratively based on dimensional analyses of produced benchmark parts [4].

2.3.3 Production strategy
Before the production of the framework can be started, some geometrical decisions have to be taken with regard to the position of the framework within the build volume of the SLS/SLM machine. Firstly, the framework is positioned upside down to guarantee well finished fitting planes and secondly, the framework is tilted to reduce the stair effect and the volume of eventual support structures.

Top surfaces of a part manufactured by SLS/SLM have a relatively low roughness and high accuracy. Bottom or overhanging surfaces are not finished well since the laser beam penetrates deeply into the powder bed. The fitting planes of the framework which make the connection with the implants need to be very accurate. Therefore the framework is positioned upside down to guarantee an accurate finish of the fitting planes (Fig. 8d).

Due to the layer-wise production of SLS/SLM a stair effect appears on the inclined fitting planes of the framework (Fig. 5a). Gaps arise between the framework and the implants, leading to local inaccuracies (Fig. 5b). Since the sloping angles of the various fitting planes differ, only one plane can be positioned horizontally and thus the stair effect can not be avoided completely. Yet the gap size can be reduced by decreasing the layer thickness or by increasing the sloping angle.

The layer thickness is difficult to change because its value depends on the powder grain size. A possible,

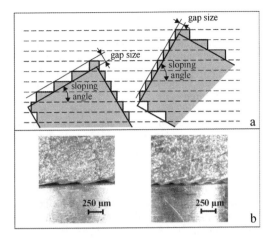

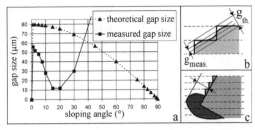

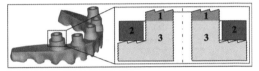

Figure 6. (a) Theoretical and measured gap sizes related to sloping angle (b) Rounded edges of the stairs reduce the gap size (c) Inaccurately built bottom surfaces increase the gap size.

Figure 5. The stair effect due to the layer-wise production of SLS/SLM: (a) Small sloping angle leads to few stairs with large gap size (left), large sloping angle leads to many stairs with small gap size (right); (b) Gaps between fitting plane of framework and implant: sloping angle of 7.5° leads to few large gaps of 40 μm (left), sloping angle of 15° leads to many small gaps of 12 μm (right).

Figure 7. During a simple post-process, the extra material (1) is removed by a tool to avoid the stair effect on the fitting structure (3). A ring (2) is the reference object and its top surface defines the border condition for the tool path.

but in this study not tested solution to decrease the layer thickness is a combined process: firstly, the layer is scanned with the usual layer thickness and secondly, this layer is partially taken away by laser erosion.

Theoretically, the gap size due to the stair effect decreases proportionally with the cosine of the sloping angle. When the fitting plane is steeper, more stairs appear but the gaps are smaller (Fig. 5). This property is experimentally studied by producing stainless steel cylinders with a different sloping angle and by measuring the gap size. Figure 6a shows that the measured gap size decreases with increasing sloping angle up to 15–20°. The gaps are smaller than the theoretical size due to the rounded edges of the stairs (Fig. 6b: $g_{meas.} < g_{th.}$). For higher sloping angles the total gap size increases, in spite of the decreasing stair effect, because appearing bottom surfaces are not finished accurately (Fig. 6c). Consequently, an optimal sloping angle, depending on process and material, minimizes the gap size.

Because the inclinations of the different implants do not vary more than 10 degrees for an average patient, all fitting planes can be positioned around the optimal sloping angle by tilting the framework in the build volume. Full melting processes (SLM) need support structures for overhanging surfaces. When calculating the tilt angle of the framework to minimize the gap size of all fitting planes, the need for support structures has to be taken into account. When the inclination of one implant differs much from the others,

the tilt operation doesn't succeed in reducing the gap size between the framework and that implant. Therefore, some extra material is designed upon the fitting plane of the framework and during a simple post-process, this extra material is removed by a tool (Fig. 7). A well known reference object defines the border condition of this removal process so that the final fitting plane is on the right position and without any stairs.

3 RESULTS AND DISCUSSION

The proposed digital procedure has been applied to fabricate frameworks in stainless steel, Ti6Al4V and a dental CoCr-alloy (Fig. 8). The energy density, introduced by equation 1, amounts to 1 J/mm³ for liquid phase sintering of polymer coated stainless steel and to 100 J/mm³ for full melting of Ti or CoCr-alloys. Good geometrical and mechanical properties are reached with densities over 97%.

In dental literature, there is a growing consensus that the final gap size between framework and implant should not exceed 40 μm to avoid failures of the prosthetic system. Therefore, the accuracy of a stainless steel framework is analyzed by studying the gap size accumulation of the different steps of the procedure. The impression of the jaw of the patient, needed to fabricate the work model, implies a small deviation on the position of the implants, but is neglected in this

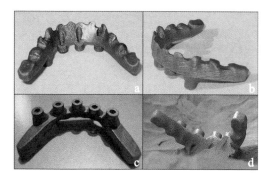

Figure 8. Produced frameworks by SLS/SLM from stainless steel (a, b) and from Ti6Al4V (c, d). Figure d shows the framework emerging from the powder.

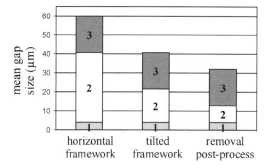

Figure 9. Mean gap size accumulation: 1. gap size due to measurement system and SLS/SLM process accuracy, 2. stair effect, 3. infiltration deformation.

study. A first error is caused by the geometry capturing system, including the precision and mounting of the registration elements and the accuracy and resolution of the measurement equipment. The production of the framework by SLS leads to the most important errors. The stair effect is inherent to the layer-wise manufacturing but can be reduced by tilting the framework in the build volume. The SLS process accuracy, limited by deformations due to thermal stresses and dimensional errors, can be increased by using optimal parameters. The infiltration process, needed for Liquid Phase Sintering, is responsible for a last deformation error because the complex framework is difficult to support during the furnace cycle.

The total gap size between framework and implant is divided into three errors which can be quantified by different experiments: firstly the combined gap due to the measurement system and the SLS process accuracy, secondly the stair effect and finally the infiltration deformation. This error accumulation is studied for three differently produced frameworks.

Framework 1 is positioned horizontally in the build volume, framework 2 is tilted to position all fitting planes around the optimal sloping angle and framework 3 is built horizontally and subjected to the removal post-process. Figure 9 shows the gap size accumulation for these frameworks, averaged for the different implant positions.

The first error amounts to 4 μm and proves that the measurement method is very accurate and that the frameworks are produced with optimal process parameters. The stair effect of the first framework is very high (37 μm) due to the low sloping angles of the fitting planes. This gap size is reduced with 50% (to 18 μm) by tilting the second framework in the build volume. The third framework has no stairs but the error is not completely removed (9 μm) because the post-process was applied with a manual tool. The infiltration deformation (19 μm) can not be reduced due to lack of process control during the furnace cycle, but is absent when Full Melting is used as binding mechanism (SLM). The total mean gap size of framework 2 amounts to 41 μm and for framework 3 to 32 μm and would be reduced to about 22 and 13 μm for SLM. The analysis of the gap size accumulation proves that the developed procedure, including the SLS or SLM process, allows an accurate production of the framework.

4 CONCLUSIONS

The proposed procedure provides an efficient and fast method to digitally design and manufacture biocompatible metal frameworks for complex dental prostheses. Based on the computed position of the implants and the captured geometry of the tooth arrangement, the framework is designed by implementing specific design rules. Frameworks are produced with good mechanical and geometrical properties by means of SLS/SLM of stainless steel, Ti6Al4V and a dental CoCr-alloy. SLS/SLM allows an efficient and customized production of the complex framework without lengthy manual pre- or post-processing. Optimal process parameters and an appropriate production strategy guarantee an accurate fit between the framework and the implants, needed to avoid mechanical or biological failures of the prosthetic system. Analysis of the fitting gaps demonstrates a medically acceptable accuracy.

ACKNOWLEDGEMENTS

This research is supported by a grant of 'Fonds voor Wetenschappelijk Onderzoek Vlaanderen (FWO-V)' and a patent covering the presented content has been filed.

REFERENCES

[1] Jemt, J. & Lekholm, U. 1998. Measurements of Bone and Framework Deformations Induced by Misfit of Implant Superstructures, a Pilot Study. *Clinical Oral Implants Research 1998* Vol. 9 (Issue no. 4): 272–280.

[2] Kruth, J.-P., Mercelis, P., Van Vaerenbergh, J., Froyen, L. & Rombouts, M. 2003. Advances in Selective Laser Sintering, Invited keynote paper. *Proc. of the 1st Int. Conf. on Advanced Research in Virtual and Rapid Prototyping (VRAP2003), Leiria, 1–4 October 2003*: 59–70.

[3] Kruth, J.-P., Mercelis, P., Van Vaerenbergh, J., Froyen, L. & Rombouts, M. 2005. Binding Mechanisms in Selective Laser Sintering and Selective Laser Melting. *Rapid Prototyping Journal, January 2005* Vol. 11 (Issue no. 1): 26–36.

[4] Kruth, J.-P., Vandenbroucke, B., Van Vaerenbergh, J. & Mercelis, P. 2005. Benchmarking of Different SLS/SLM Processes as Rapid Manufacturing Techniques. *Proc. of 1st Int. Conf. of Polymers and Moulds Innovations (PMI2005), Gent, 20–23 April 2005* (accepted).

[5] Kruth, J.-P., Van Vaerenbergh, J., Mercelis, P., Lauwers, B. & Naert, I. 2004. Dental Prostheses by means of Selective Laser Sintering. *Les 10èmes Assises Européennes du Prototypage Rapide, Paris, 14– 15 September 2004* S5-4.

[6] Levy, G.N., Schindel, R. & Kruth, J.-P. 2003. Rapid Manufacturing and Rapid Tooling with Layer Manufacturing (LM) Technologies, State of the Art and Future Perspectives. *CIRP Annals 2003* Vol. 52/2.

[7] Matsumoto, M., Shiomi, M., Osakada, K. & Abe, F. 2002. Finite Element Analysis of Single Layer Forming on Metallic Powder Bed in Rapid Prototyping by Selective Laser Processing. *Int. Journal of Machine Tools and Manufacture 42 2002*: 61–67.

[8] McAlea, K., Forderhase, P., Hejmadi U. & Nelson, C. 1997. Materials and Applications for the Selective Laser Sintering Process. *Proc. of the 7th Int. Conf. on Rapid Prototyping, San Francisco 1997*: 23–33.

[9] Ortrop, A. & Jemt, T. 2000. Clinical Experiences of CNC-milled Titanium Frameworks Supported by Implants in the Edentulous Jaw: a One Year Prospective Study. *Clinical Implant Dentistry and Related Research 2000* Vol. 2: 2–9.

[10] Over, C., Meiners, W., Wissenbach, K., Lindemann, M. & Hutfless, J. 2002. Selective Laser Melting a New Approach for the Direct Manufacturing of Metal Parts and Tools. *Proc. of SME conf. on Rapid Prototyping and Manufacturing, Cincinnati, May2002.*

[11] Pohl, H., Simchi, A., Issa, M. & Dias, H.C. 2001. Thermal Stresses in Direct Metal Laser Sintering. *Proc. of the Solid Freeform Fabrication Symp 2001*: 366–372.

[12] Riedy, S., Lang, R.B. & Lang, E.B. 1997. Fit of Implant Frameworks Fabricated by Different Techniques. *The Journal of Prosthetic Dentistry 1997* Vol. 78: 596–604.

Virtual modeling and rapid manufacturing – Bártolo (eds)
© 2005 Taylor & Francis Group, London, ISBN 0 415 39062 1

Customisation of bio-ceramic implants using SLS

T. Coole
Faculty of Technology/Buckinghamshire Chilterns University College – England

F. Cruz & J. Simões
Escola Superior de Tecnologia de Setúbal/Instituto Politécnico de Setúbal – Portugal

C. Bocking
CRDM/Buckinghamshire Chilterns University College – England

ABSTRACT: Rapid Prototyping (RP) is an excellent manufacturing process to produce medical models. These models are designed and optimised using anatomical scan information obtained from internal structures of the patient's body (e.g. Computed Tomography (CT), Magnetic Resonance Imaging (MRI) and Ultrasound) and the 3D shape is manufactured straight from 3D medical image data, from an appropriate material exhibiting desired proprieties. This route is a reverse engineering (RE) process.

Several of the existing RP processes are able to produce medical models in different materials and are being used for different applications. However, those models cannot be used directly as implants due to its non-biocompatibility.

This paper presents research work which investigates the industrial production of biocompatible ceramic medical models to be used directly as implants in human body, using the hydroxyapatite (HA) ceramic material and polymers. This osseous-inductive calcium phosphate bioceramic has a tissue similar to human bones and is classed as a biological active material, which encourages bone growth on its surface. The objective is to produce implants using the SLS (Selective Laser Sintering) process, from poly (L – lactide)/hydroxyapatite composites (PLLA/HA). Results of initial findings show that this process can be used as an accurate method of bone reconstruction and will enable the direct implant of the prosthesis made as a bespoke component for the patient. From the results tit can bee seen that there are some issues with the tolerance accuracy of the reconstructed part these will be presented and discussed. Some of the results on time and cost reduction to produce the implants for specific bones, as well as its long-term biocompatibility are highlight.

1 INTRODUCTION

This paper reviews the work and experiments conducted over the last 4 years on the application of RP to the reconstruction of porous bone tissue. This work is part of an on going programme conducted through the University College and with the cooperation of CRDM. This paper describes the construction process and the results of the production of a bone profile looking at the experimental issues and problems associated with using this type of process for bone tissue construction and reconstruction.

The average adult skeleton contains 206 bones, but the actual number varies from person to person and decreases with the age as some bones become fused. The average density of human bones (which constitutes about 18% of the mass of the body) is $1,9\,g/cm^3$ [Enderle *et al.*, 2000].

2 TYPICAL BONE MATRIX COMPOSITION [WEINANS, 2002]

- Mineral ≈ 60% (Hydroxyapatite – HA)
- Organic ≈ 30% (Collagen protein of type I, i.e. collagen specific of bone – there are other types of collagen in human tissue)
- Water ≈ 10%

The mechanical bone quality is measured by bone *stiffness* (resistance against deformation) and *strength* (resistance against fracture). Both parameters can be defined and evaluated by the *compression test* as shown in Figure 1.

The skeleton has an important load-bearing function, therefore numerous studies have quantified strength, stiffness and failure load of cortical (the external part of bone) and trabecular (the internal porous part, also

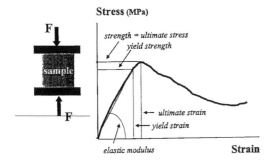

Stress (MPa)

strength = ultimate stress
yield strength

← ultimate strain
← yield strain

elastic modulus   **Strain**

Figure 1.   Bone test to failure in compression [Huiskes and Kaastad, 2002].

Mechanical Characterization of Bone

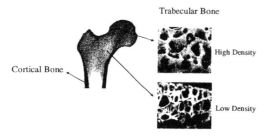

Trabecular Bone

High Density

Cortical Bone

Low Density

Figure 2.   Laser technology for medical applications.

known by cancellous) bone. Around 70% of bone structure is cancellous. In Figure 2 the mechanical characterization of bone is illustrated.

As a heat source the lasers normally used are CO_2, Nd-YAG, Nd-Glass and Excimer.

In recent years new applications of the laser technology occurred, namely in two main areas:

- **Medicine**. The laser technology is nowadays extensively used in medical surgery, in such different applications as ophthalmology, dermatology, urology, otorrinolaringology, physical and rehabilitation medicine, gynaecologic, and cosmetic surgery. The types of lasers used include CO_2, Nd-YAG, Ruby, Argon and Excimer, depending on the goal.
- **Rapid Prototyping technologies**. Among the different RP methods available today (around 25) 70% are based in the application of a laser. The main rapid prototyping techniques using lasers are *Stereolithography (SL)*, *Selective Laser Sintering (SLS)* and *Laminated Object Manufacturing (LOM)*.
  - *Stereolithography* uses an ultra-violet light of wavelength 325 nm from a low power He-Cd laser to polymerise a plastic monomer. The laser shines into a vessel of liquid monomer causing it to selectively polymerise and set.

- *Selective Laser Sintering* produce parts by selectively fusing (sintering) thermoplastics, ceramic or metallic powders with the localised heat from an infra red laser, usually a CO_2 laser of 50 W.
- *Laminated Object Manufacture* uses also a CO_2 laser to cut the required shape of each layer of the part consisted by a sheet of paper, polyester film or other laminates. The laser also cuts the unwanted material into small squares (tiles) for later removal.

3   THE SLS PROCESS

3.1   *Part accuracy*

As previously referred to, the SLS process creates three-dimensional objects, layer by layer, from powdered materials, with the heat generated by a CO_2 laser within the apparatus.

First, three-dimensional CAD data must be output in the industry-standard .STL (Standard Triangulation Language) format.

The process comprises 4 phases [DTM website, 2001]:

1. As the selective laser sintering process begins, a thin layer of the heat-fusible powder is deposited into the part build chamber;
2. An initial cross-section of the object under fabrication is selectively "drawn" (or scanned) on the layer of powder by a heat-generating CO_2 laser. The interaction of the laser beam with the powder elevates the temperature to the point of melting, fusing powder particles and forming a solid mass, i.e., sinters the powder particles (heats and bonds selected portions of each layer). The intensity of the laser beam is modulated to melt the powder only in areas defined by the object's design geometry;
3. An additional layer of powder is deposited, via a roller mechanism, on top of the previously scanned layer;
4. The process is repeated, with each layer fusing to the layer below it. Successive layers of powder are deposited and the process is repeated until the part is complete.

After the building the part is removed from the build chamber and the loose powder falls away. Parts may then require some post-processing, such as sanding, depending upon the intended application.

There is no need to create support structures with the CAD design prior to or during processing and, therefore, no support removal is needed when the part is complete.

The software components of the Sinterstation Systems include a Unix operating system and proprietary application software.

In Figure 3 a schematic view of the SLS Sinterstation 2000 Process Chamber is presented.

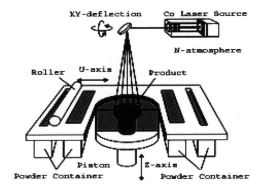

XY-deflection   Co Laser Source

N-atmosphere

Roller   U-axis   Product

Piston   Z-axis
Powder Container   Powder Container

Figure 3. Process chamber of SLS Sinterstation 2000 [Adapted from www.cs.hut.fi, 2001].

4 ADVANTAGES AND DISADVANTAGES OF THE SLS PROCESS

The most important *advantages* of the SLS process are:

a. The variety of materials usable, both currently and in the near future (polycarbonate, investment casting wax, PVC, ABS – plastic, nylon, sand, ceramic, and metals);
b. The speed of construction, in comparison with SL and FDM. Parts are produced at a rate of up to one inch per hour (5 to 25.4 mm per hour);
c. Parts present good mechanical (strength and toughness) and thermal properties;
d. Parts do not require additional support during the building process or subsequent post-curing (as SL, for example, does).

The main *disadvantages* are:

a. The parts do not have full density, due to the porosity inherent to the process, which do not allow the production of parts for structural applications;
b. The parts present rough surfaces;
c. Dimensional accuracy of the models lower than the other RP processes (SL, LOM);
d. Necessity to provide the process chamber continuously with nitrogen, to ensure safe material sintering operation, leading to additional expenditure. The oxygen content has to be limited to a value below 5.5% (default machine parameter value).

5 THE DIRECT FABRICATION OF HIGH-DENSITY CERAMIC STRUCTURES USING THE SLS

The main approach developed for the purpose of producing direct ceramic parts by SLS is the two-phase powder approach. Similar to liquid-phase sintering, the powder bed consists of at least two different powders of significantly disparate melting points. The laser beam heats the powder bed locally, inducing melting only of the low-melting-point species. The liquid wets and binds the high-melting-point solid.

This two-phase powder approach has been used successfully in SLS processing of metals and ceramics. Examples ceramic systems include Alumina–ammonium dihydrogen phosphate, SiC-Si, WC-Co, and alumina – boron oxide [Beaman et al., 1997].

The HA/PLLA parts produced in the scope of this research work are included in this group of methods, thus allowing the direct production of biomaterial composites for medical applications using the SLS technology, without any subsequent post-operative processing. The advantages of this direct route are the substantial reduction of time and cost to produce the customised parts, compared with the previous indirect and transfer methods. According with the classification described in this case there is a Liquid Phase Sintering, with Partial Melting of the binder (PLLA), without melting of the structural grains (HA).

6 HA/PLLA BONES MANUFACTURE

In this preliminary stage hydroxyapatite [HA]/poly (L-lactide) [PLLA] composites of different ratios (by weight) were produced being the powders submitted to die compaction to obtain dense samples of these composite biomaterials. An evaluation of some physical properties (density, hardness, microstructure) was performed. The aim of this preliminary study was to establish a baseline for further studies using non-conventional methods.

Hydroxyapatite (HA) is a calcium phosphate bioceramic used for several years as a substitute of natural bone. This material offers a large potential to be used as an implant material due to its biocompatibility, non-toxicity and unique osteoinductivity properties [Enderle et al., 2000]. This material allows bone to grow and attach onto its surface as well as promotes the ingrowths of bone tissue into the interior of the implant. However, the mechanical characteristics of HA limit the use on load-bearing applications due to its brittleness, low toughness and hardness [Hench, 1991]. Therefore, studies have been carried out to improve the mechanical properties of HA by using biodegradable polymers, e.g. poly (L-lactide) – PLLA-as binders for particulate bioceramics, to produce HA/PLLA composites with improved mechanical properties, allowing their use for both low and hard tissue repair [Kesenci et al., 2000].

This preliminary phase of the experimental work was performed at ESTSetúbal/Portugal (Mechanical Engineering Department) and University of Aveiro/Portugal (Ceramic and Glass Engineering Department).

Table 1. Process parameters used in SLS.

Process parameters	Adopted values
Temperature in the part bed (°C)	**160**
Roller speed (mm/s)	**127**
Outline laser power (W)	**2.5**
Outline laser speed (mm/s)	**200**
Laser power (W)	**5**
Laser scan speed (mm/s)	**300**
Laser scan space (mm)	**0.10**
Applied energy density (cal/cm^2)	**3.98**
Slice thickness (mm)	**0.10**
Warm-up thickness (mm)	**5**
Cool-down thickness (mm)	**2**
Binder ratio (%wt.)	**40**

Table 2. Dimensional accuracy of parts.

	Average deviation from the CAD model (%)		
Specimen	XX direction (width)	YY direction (length)	ZZ direction (height)
Bone 1	+5.1	+10.6	+16.1
Bone 10	+8.2	+10.1	+15.3
Bone 100	+9.7	+12.4	+14.8

Table 3. Dimensional accuracy of the implant manufactured by SLS.

	Average deviation from the CAD model (%)		
Specimen	XX direction (length)	YY Direction (width)	ZZ direction (height)
Bone 60	−0.15	+2.5	+5.0

One of the aims of the experimental work was to study the behaviour of powder based biocomposites during the laser sintering process in order to evaluate their responses for different sets of parameters. Therefore cost and resources spent were not the priority.

After defining the range of ideal operational values, based in the DoE performed, the final step has been to produce intricate human bones using SLS.

Four different types of human bones (referred respectively as bone1, bone10, bone 60, and bone100) from a HA/PLLA composite were successfully produced in the Sinterstation 2000 (two specimens of each type).

The bones produced were initially obtained from data collected from a human cadaver's foot, by means of MRI. The data generated was then prepared using the MIMICS software (from Materialise, Belgium) and the correspondent .STL files created at Leeds University. These files were sent to CRDM by e-mail and then manipulated and corrected using the MAGICS 6.32 software (also from Materialise, Belgium). The resulting .STL files were finally sent to the Sinterstation 2000 trough out internal network.

The models were scaled down in all the axes (X, Y, Z) to allow them to fit the centre of the modified part platform (∅ 130 mm).

The set of SLS parameters used in the Sinterstation 2000 are presented in Table 1.

In order to check the dimensional accuracy of the parts generated with different geometries, an analysis was made to evaluate the standard deviation of the length, thickness and height in relation with the correspondent dimension of the CAD geometries. An average of five measurements in each direction was taken. An electronic digital caliper (Mitutoyo, 0–150 mm) was used for those measurements. The results are summarised in Table 2.

As can be seen the geometric accuracy of the bones produced is poor, namely in the ZZ direction (direction of construction). No offsets were applied.

It was mentioned previously, the standard dimensional deviation of the SLS process is +/−0.25 mm.

The main explanation for this is that there is an area affected by the laser heat around the part during the build due to the low transition temperature of the polymer, even considering that the outline laser power and scan speed are dramatically reduced to minimise this effect. This means that there is an agglomeration of the ceramic particles all around the parts by the effect of the binder action, which creates an over-dimension very difficult to remove as it adheres to the part walls. Therefore, and as often happens in the standard operation of the Sinterstation 2000, the application of dimensional offsets to compensate the contraction/expansion of the polymeric parts is necessary. Based on the results previously achieved in this phase a new model (an intermediate component of a knee joint implant) was manufacture at the Sinterstation, in the same conditions of the previous bones, but with the use of offsets in all directions. The result, in terms of dimensional accuracy, is presented in Table 3.

7 CONCLUSIONS

From the work of the experimental work HA/PLLA human foot bones were produced by SLS, from MRI medical scan data. The geometric and dimensional accuracy of the parts produced as well as the capability of the SLS technology to produce such intricate shapes were evaluated. Images of the manufactured parts are presented in the figures below, which depict the bones in both the CAD image as recreated from the MRI image and the SLS models. Because the accuracy was not a major issue (unlike the overall feasibility to

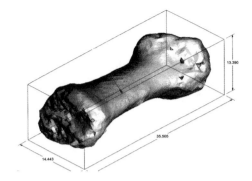

Figure 4.   CAD model of the metatarsal.

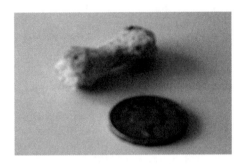

Figure 5.   The metatarsal bone used in the first trial (£1 coin showing size).

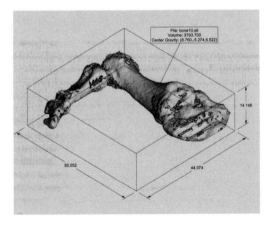

Figure 6.   CAD model of foot bone.

Figure 7.   SLS output.

produce HA/PLLA bones by SLS), the dimensional results obtained were considered satisfactory.

In the figures below are the CAD drawings and the resultant synthetic bone developed on the SLS machine.

REFERENCES

Beaman, J. J., Barlow, J. W., Bourel, D. L., Lee, G., Vail, N. K., Marcus, J. J., 1997. Solid Freeform Fabrication: A New Direction in Manufacturing, Kluwer Academic Publishers, USA. ISBN 0-7923-9834-3.

Enderle, J., Blanchard, S., Bronzino, J., 2000. Introduction to Biomedical Engineering, Academic Press, USA. ISBN 0-12-238660-4.

Hench, L. L., 1991. "Medical and Scientific Products", In Ceramics and Glasses, of Engineered Materials Handbook, 1991, ASM International, USA, Section 13, Vol. 4, pp. 1007–1013. ISBN 0-87170-282-7.

Huiskes, R., Kaastad, T. S., 2002. "Biomechanics, Bone Quality and Strength – Lecture Notes", *Advanced Course and Workshop on Bone Mechanics – Mathematical and Mechanical Models for Analysis and Synthesis*, June 2002, Instituto Superior Técnico (IST), Lisbon, Portugal.

Kesenci, K., Fambri, L., Mibliaresi, C., Piskin, E., 2000. "Preparation and properties of poly (L-lactide)/hydroxy-apatite composites", *J. Biomaterials Society*, 2000, *Polymer Edition*, Vol.11, No.6, pp. 617–632.

Weinans, H., 2002. "Bone Remodelling Models – Lecture Notes", Advanced Course and Workshop on Bone Mechanics – Mathematical and Mechanical Models for Analysis and Synthesis, June 2002, Instituto Superior Técnico (IST), Lisbon, Portugal.

CAD and 3D data acquisition technologies

Virtual modeling and rapid manufacturing – Bártolo (eds)
© 2005 Taylor & Francis Group, London, ISBN 0 415 39062 1

Computer aided reverse engineering of human tissues

Kevin D. Creehan
Virginia Tech, Center for High Performance Manufacturing

Bopaya Bidanda
University of Pittsburgh

ABSTRACT: The objective of this study is to extend the applications of reverse engineering technology from manufacturing industries to the biomedical industry. By obtaining nearly exact geometric data of human tissues, such as bones, tendons, and ligaments, in a high-speed and inexpensive manner, potentially groundbreaking research becomes possible for applications in injury rehabilitation, injury prevention, and strengthening. Previous applications of reverse engineering technologies in the biomedical community have dealt largely with prosthetic design and plastic surgery. This study expands this research to include muscular and skeletal applications.

1 INTRODUCTION

Reverse engineering, the process by which an existing item is identically reproduced, becomes necessary when a physical prototype exists but accurate geometric data of the part does not. In order to recreate the existing part, a computerized (CAD) model of the part must be drawn or otherwise acquired. This file provides the coordinates of multiple points on the product surface, which is then used to develop the drawing of the product for redesign or manufacturing. The resulting data may be analyzed within the CAD software, used to create a three-dimensional reconstruction of the part, or exported to a machine capable of rebuilding the new design, such as rapid prototyping equipment.

Collecting accurate geometric data of human tissues is complicated by the flexibility of the items and their numerous curves and odd shapes. Inexact data can be obtained using simple calipers, but the time required to gather the data becomes extensive and the results become susceptible to inevitable human error.

Digitizing, or scanning, is the term used to describe the process of gathering information about an undefined three-dimensional surface. It is used in several fields of study, wherever there is a need to reproduce a complex freeform shape. Scanning technologies provide extremely precise three-dimensional geometric information of nearly any item. Of these technologies, laser scanning has become increasingly popular in several manufacturing industries due to its speed, accuracy, and low cost.

The objective of this research is to extend the applications of the laser scanning technology from manufacturing industries to the biomedical industry. By obtaining high precision geometric data of human tissues, such as bones, tendons, and ligaments, in a high-speed and inexpensive manner, potentially groundbreaking research becomes possible for applications in injury rehabilitation, injury prevention, and strengthening.

1.1 Contribution and importance of this research

The primary advantage provided by scanning technologies would be improved quality of the data. Laser scanning technologies are known to obtain repeatable results accurate to within less than 0.001 inch. Secondly, the time investment for the researcher and clinical practitioner would be greatly reduced. Using scanning technologies requires very little user intervention until the data analysis stage. Thirdly, the use of a scanner would reduce the current need for three-dimensional reconstruction software as reconstruction algorithms are already built into the scanning software programs. And finally, the use of a scanning technology will usually provide a less expensive alternative to the medical imaging counterparts.

From an engineering perspective, the primary contribution of this research details an application of reverse engineering by part family, rather than through the more typical job shop approach that requires developing application-specific methods using very general instructions and substantial operator influence. The factors presented in this methodology include part mounting and orientation, geometric considerations such as concavity, expected scan duration, and optimal parameter settings such as the scanning range and horizontal and vertical point spacing. Because these technologies are not widely used among medical researchers, their future research burden will be reduced once more research applications of this technology are realized.

Additionally, this research provides guidelines, or a structured methodology, for all potential users of these scanners. At present, no structured methodology based on part families exists on any level for geometric data acquisition scanning equipment. All users of this equipment, such as those in the manufacturing community, could examine these methodologies and apply them to their specific need. For example, at present, each user relies on experience to determine scanning parameters that may or may not be optimal. This research provides a framework for decision making for all researchers that limits the human intervention in the scanning process that is considered essential today.

1.2 *Example of research importance*

Shoulder injuries in athletics, especially throwing sports, are a rather common occurrence. The assessment and evaluation of shoulder problems requires a systematic approach that must be both comprehensive and efficient. The evaluation of shoulder injuries in athletes is a complex process that relies on accurate diagnosis before proper treatment can be successful. It is one of the most difficult areas of the body to assess given its complicated composition and structure and the considerable demands placed on it in overhand athletics (Stone, 2001).

The complex joints are able to provide an extraordinary range of movements, but all at a substantially increased risk for injury. For example, a National Collegiate Athletic Association study of injuries suffered in intercollegiate competition showed that shoulder injuries accounted for the highest number of baseball injuries between 1985 and 1999 (National Collegiate Athletic Association, 2000). Thus, increased diagnostic ability and rehabilitation knowledge could assist in limiting the rehabilitation time required for these numerous injuries, and would impact a large segment of the athletic population.

Researchers at the University of Pittsburgh, Department of Orthopaedics, study joint motion, including the effects of the static and dynamic restraints at the glenohumeral joint. They utilize a shoulder testing apparatus, which has been developed over the past several years to examine joint motion. In its use, "simulated muscle forces are used to produce joint motion through hydraulic cylinders while tendon excursions and six degree-of-freedom joint motions are recorded

using a magnetic tracking device" (Musculoskeletal Research Center, 2001). Other ongoing work in the laboratories includes the examination of the structural properties of the superior glenohumeral and coracohumeral ligaments, mechanical properties of the long head of the biceps tendon, and the examination of the length and orientation changes of the ligament at the glenohumeral joint. Further, the research has grown to include measurement and evaluation of strain in the glenohumeral ligaments during passive and active motion and the determination of the different structures to joint stability during active motion using the shoulder testing apparatus.

These researchers have used reverse engineering technology to assist their efforts, such as the examination of the structural properties of the shoulder ligaments, and the examination of the length and orientation changes of the ligament at the glenohumeral joint. Further implementation of the results of this research will substantially benefit medical research in these areas, which in turn will benefit those rehabilitating shoulder or other musculoskeletal injuries. The availability of extremely precise geometric data would allow researchers to develop highly accurate virtual models of joint structures and their motions. With these accurate models, researchers could perform joint-testing procedures virtually, enabling them to demonstrate the results of applying forces to the virtual model that would be unsafe or impossible to simulate on living specimens.

1.3 *Research overview*

The goal of the research herein is to develop guidelines and methodologies for scanning human tissues that will assist medical researchers and reverse engineers alike. In order to do so completely and thoroughly, the research was conducted in the following manner.

The first challenge of this study was to develop a thorough taxonomy of human tissues and structures based on traits such as shape, texture, tissue type, and curvature, which may affect various scanning parameters. Although two human structures may be functionally very different, they may have similar characteristics with respect to scanning parameters and thus should be reverse engineered in very similar ways. As such, the guidelines for a medical researcher to follow need not be customized to each structure, but rather to each classification, or part family, with only the occasional subtle deviation for specific members of the category.

Once the taxonomy was complete, the research focused on developing a strict methodology for reverse engineering each category. These methodologies detail the best manner in which the parts may be scanned, including the optimal scanning parameters and part configuration, with respect to scan quality and speed, while also detailing potential trouble areas that may lead to poor scan quality or speed.

2 LASER SCANNING METHODOLOGIES

The human tissues on which this methodology will concentrate will be taken from the human musculoskeletal system. The focus of this project will be to examine bones (both axial and appendicular), tendons, and ligaments. The functional effectiveness of the system is important because the tasks of everyday life require integration of activity within the entire system (Martini, 1998).

The geometric similarities and the resulting orientation and mounting methods that these geometric similarities require determine the classifications. The following are the structural and geometric characteristics that were evaluated: size, shape, texture, tissue type, flexibility, fragility, curvature, and the presence of holes or concavities.

Classification of the tissues is based primarily on part orientation and mounting requirements, and secondarily on similarities needed in the scanning parameters. As such, due to the similarities of their geometric qualities, tendons and ligaments are grouped together and classified in the "Soft Tissue" category. The bones were grouped into seven categories, or part families, based upon the required part orientation, similarities in part mounting, and common geometric characteristics.

2.1 Scanning methodologies

Upon completion of the taxonomy, the research turned to the development of strict methodologies for each anatomical classification developed by the taxonomy. These methodologies detail the best manner in which the parts may be scanned, including the optimal scanning parameters and part configuration, with respect to the scan quality and speed, while also detailing potential trouble areas that may lead to poor scan quality or speed. Once the experiments were completed, the data was evaluated to determine the appropriate scanning parameters. For each classification, a fractional factorial experimental design was planned using several levels within the following factors: horizontal and vertical point spacing, coating material, edge detection method, and part orientation. From these conclusions, the methodology for each class of parts was derived.

2.2 Methodology development

For each classification, five factors were expected to contribute to the measurable variables, scan quality and duration. These factors were selected based on information provided by experts in the use of laser scanning and other reverse engineering equipment, and are widely considered to be the most important variables in obtaining high-quality scan data.

- *Coating* – the type of material used to coat the part for maximum reflectivity.

- *Horizontal Point Spacing* – the horizontal distance between measured points, defining the density of the scanned point cloud.
- *Vertical Point Spacing* – the vertical distance between measured planes.
- *Edge Detection Method* – manner in which a surface edge is detected by the laser scanner.
- *Part Orientation and Mounting* – the position of the piece during the scanning procedure and the manner in which it is held in this position.

The quality of the scan was measured as the dependent variable. As defined for this research, the quality measure is simply the ratio of horizontal levels that do not require additional manual editing at the completion of the scan to the total number of cross-sectional levels in the scan.

2.3 Experimentation

Because the availability of the scanning equipment was somewhat limited, and the biological tissues in question are time-sensitive due to their propensity for losing their natural moisture (thus causing a change in their shape and surface characteristics), the manner in which the experiments were constructed was exceedingly important so that each variable and variable setting would be appropriately considered in the available time frame.

A one-quarter fractional factorial design was implemented because high-order interactions are negligible and there was limited experimental capacity. The experimental design, as derived by Montgomery

Table 1. Experimental design.

Exp.	Vert.	Coat.	Orient.	Horiz.	Edge.
1	−	−	−	+	+
2	+	−	−	−	−
3	−	+	−	−	+
4	+	+	−	+	−
5	−	−	+	+	−
6	+	−	+	−	+
7	−	+	+	−	−
8	+	+	+	+	+

Table 2. Experimental design factor levels for Flat Bone category.

Factor	−	+
Vert. Pt. Spacing	0.75 mm	1.0 mm
Coating	Powder	Paint
Orientation	Horizontal	Vertical
Horiz. Pt. Spacing	0.75 mm	1.0 mm
Edge Detection	Auto	Delta

Table 3. Data results for Flat Bone category.

Horizontal Point Spacing	Vertical Point Spacing	Orienta- tion	Coating type	Edge Alg.	Scan quality
1.0 mm	0.75 mm	Vertical	Powder	Auto	0.475
0.75	1.0	Vertical	Powder	Auto	0.536
0.75	0.75	Vertical	Paint	Delta	0.624
1.0	1.0	Vertical	Paint	Delta	0.568
1.0	0.75	Horiz.	Powder	Delta	0.435
0.75	1.0	Horiz.	Powder	Delta	0.478
0.75	0.75	Horiz.	Paint	Auto	0.577
1.0	1.0	Horiz.	Paint	Auto	0.535

Table 4. Analysis of variance results for the Flat Bone category.

Source	DF	SS	MS	F	P
VPS (A)	1	1.800E-05	1.800E-05	0.28	0.6513
Coat (B)	1	0.07220	0.07220	1110.77	0.0009
Orient (C)	1	0.01584	0.01584	243.72	0.0041
HPS (D)	1	0.02040	0.02040	313.88	0.0032
EDGE (E)	1	1.620E-04	1.620E-04	2.49	0.2552
A*B*C*D*E	2	1.300E-04	1.620E-04		
TOTAL	7	0.10875			

(Montgomery, 1997), that was utilized in this research is below.

The Flat Bone category was the first considered, and the sample bone studied was the scapula. The following table outlines the values of the factor levels for this experiment.

Each of the factor levels given above has been previously explained in detail with the exception of the orientation factor. The terms that describe the manner by which the scapula was mounted refer to the direction of the scapula's major axis during the scanning procedure. Thus, the "Horizontal" orientation depicts the scapula lying on its side edge, with its longest axis nearly parallel to the surface of the scanning plate and the "Vertical" orientation describes the scapula standing upright on the scanning plate. While there are infinite possible orientations, these were selected based on their substantial difference and the previous knowledge that such orientations were conducive to acceptable quality.

The results of the experimental design are shown below.

2.4 Analysis of variance

Outlined below are the results obtained through the use of a statistical software package for the "Flat" category design given above.

This analysis of variance shows three of the five variables having a significantly low p-value, which leads one to reject the standard null hypothesis that the factor level means are equal. These results, along with those from the other classifications, demonstrate credible evidence that the variability witnessed in the scan quality was largely the result of the change in coating type, Horizontal Point Spacing, and Part Orientation. Further, the change in values of the Vertical Point Spacing and Edge Detection method does not significantly affect the quality of the resulting scan data.

Multiple replications of scans of a single structure, and multiple scans of many of the same kinds of structures, have enabled specific conclusions to be developed for each independent variable and each structure.

In each case, substantial evidence exists to suggest that the three aforementioned factors contribute most to the change in quality.

2.5 Scanning results and general conclusions

Within these three factors deemed most important to the success of the scan, it is clear that a more dense point cloud, or smaller horizontal point spacing, produces a more accurate scan with fewer imperfections. However, the increase in density of the point cloud often leads to additional minor editing of the data, all of which are easily solved using repair algorithms developed in conjunction with this research effort. Never does the increase in point cloud density lead to algorithmically unsolvable data problems that would not have existed with a less dense data set. The converse is often true, however. Occasionally a low-density data set will exhibit data imperfections, such as a hole of missing points, which a high density set will eliminate through its more defined search for data.

However, the variable that most affects the quality of a scan is the part mounting and orientation. Any gains in quality seen as a result of optimizing the other variable settings can quickly be lost when improperly orienting the specimen. Furthermore, a subtle alteration of a part's orientation can lead to a substantial increase or decrease in scan quality. Often, in cases of oddly-shaped parts, a small number of iterations of orientations can lead to a considerable improvement in quality. Below are observations and general guidelines for improving scan quality through part mounting and orientation.

- The part should be mounted so that as little surface area as possible is parallel to the laser source.
- Concavities and holes should never be oriented perpendicular to the source unless is has been determined that obtaining their geometric data is unnecessary. If the geometry of a deep concavity is a region of interest, the hole should be oriented directly facing the laser. This will give the laser a line of sight to gather any information it can from the deep concavity.

- Often there is a specific region of interest along the geometry of the part. In these cases, the orientation should be such that the laser has an unobstructed sight line to all faces of this region of interest.
- If the geometry of a hole is a region of interest, the hole should be oriented facing the laser source at an angle between 30° and 45° from vertical. This enables the laser to have a direct line of sight to most interior faces of the hole and will minimize the loss of points. However, in any case, the internal geometries of deep holes are very difficult to obtain because the line of sight or the reflective beam is often blocked by another edge of the hole.
- The ideal Vertical Point Spacing is at most 0.75 mm. The intricate irregularities in the shapes require a very dense cloud.
- The Horizontal Point Spacing should generally be set between 0.5 and 0.75 mm. It is recommended to set the minimum and maximum values unequal to each other to allow the scanning algorithm to select a point within a range rather than at an exact location.

Thus, for each of the bone categories, the following settings are recommended:

- *Coating* – a dull, matt, white paint is the best coating substance for bony structures.
- *Edge Detection Protocol* – this protocol had little or no effect on the quality of the scan data.
- *Vertical Point Spacing* – a change in this variable did not affect the output quality of the scan.
- *Horizontal Point Spacing* – in much the same way as the Vertical Point Spacing setting, this setting may continue to be altered by classification. Good results were observed for settings as low as 0.4 mm, but for larger, less detailed items a value of between 0.75 mm and 1.0 mm produced acceptable results.
- *Part Mounting and Orientation* – the most volatile of variables, this setting often requires a small number of iterations before excellent results are obtained.

2.5.1 *Flat bones specific conclusions*
The orientation of the part should be such that its longest axis sits nearly vertically on the work volume, with a long, thin rod of any sort balancing the upper portion along an easily interpolated, or unimportant face of the bone.

2.5.2 *Long bones specific conclusions*
The first method of orientation is a very flat, nearly horizontal orientation. This method is only effective, and should only be used when the geometry of one end is the only important geometry. The scan radius should only include the area of the important head. This minimizes the lost data at the top of the scan to a small surface area. The top face of the geometry comes nearly to a single point. The draw back to such a method is a

significant loss of data along one face of the shaft and at the opposite end of the bone.

The next method of mounting a long bone rests the shaft of the bone on two long posts at an angle of approximately 30° from vertical and the bottom of the bone rests on nails that have been hammered through the wood base. The need for this orientation arises when the geometry of both ends are important. It minimizes the total lost data at the top and bottom faces.

The final method of mounting long bones is to drill a hole at one end and stand the bone upright on a mounted post. This method produced the highest quality of data along the shaft and at the top end. At the bottom, the data quality is good with the exception of the area of drilling. However, if this area is small, this orientation tends to perform the best of any of these possibilities.

2.5.3 *Short bones specific conclusions*
To mount the short bones, a 1/16" drill bit may be used to drill a small hole in a region of the bone where data loss is minimized, such as at a pointed end. The bone can then be mounted by placing the end of a toothpick into this hole and into a stabilizing base.

2.5.4 *Concave bones specific conclusions*
Bones of this class should be scanned using the same parameter settings as the Flat category, with the notable exception of insisting that the concave surface face the laser directly. In addition, it is important to make sure that the top or bottom faces of the concave surface do not cover any portion of the interior of the concavity, thus blocking the laser's line of sight.

The orientation of the part should be such that its longest axis sits nearly vertically on the rotating plate, with a long, thin rod of any sort balancing the upper portion along an easily interpolated, or unimportant face of the bone.

2.5.5 *Curved bones specific conclusions*
Bones of this class should be scanned using the same parameter settings as the Long category, with the notable exception of insisting that the curvature face the laser directly, so that while sitting on the plate the bones look as close to the letter "C" as possible. In addition, it is important to make sure that the top or bottom heads of the bone do not cover any portion of the interior of the curvature, thus blocking the line of sight from the laser.

2.5.6 *Vertebral bones specific conclusions*
The orientation of Vertebral bones should be such that the transverse axis is mounted vertically on the scanning plate. As always when negotiating a hole in the part, the bone should be slightly angled so that the laser may have lines of sight into the interior of the hole. However, in these cases, be sure that the angle of lean

is very small, i.e. very close to vertical, due to the great depth of many of these holes.

2.5.7 *Soft tissue specific conclusions*

Initially perceived as a poor application to the laser scanning technology as well, the data acquisition of the soft tissue category can perform moderately well. However, there are several limitations to the use of this method for a soft tissue application. First, the nature of soft tissue is such that if the tissue is not kept sufficiently moist, the loss of moisture will cause a significant change in geometry in a very short period of time, a matter of only minutes. As a result, scanning could only take place in a period of three to five minutes, which drastically limits the scan area.

Because the item must be kept so moist, it is impossible to use a coating substance to maximize the reflective beam. Thus, unless the tissue is completely white, as it is in many cases, there will be some loss of data. Further, the translucent nature of the soft tissue often results in a poor reflective beam, which in turn results in lost data.

3 SUMMARY

The objectives of this research were to standardize and automate the reverse engineering process in the medical and orthopaedic research arena. To achieve this objective, a detailed, structured methodology for obtaining accurate geometric data has been developed for all classifications of the human musculoskeletal system. In so doing, this research has led to the further conclusion that the reverse engineering process can be facilitated through the use of group technology, by utilizing part families.

In general, the development of the scanning methodologies was successful. Clear evidence was obtained that demonstrates that substantial improvement in scan quality can be obtained by utilizing appropriate parameter settings, and more significantly, by orienting the item optimally. The traditional mentality that suggests that scanning methods, and specifically orientations, should be determined on a case-by-case basis is inaccurate and unnecessary. Although some parts have notable obstacles that may require modifications to a scanning methodology, there is no question that most structures can be classified into part families that have similar, if not identical requirements for reverse engineering.

Most importantly, this research provides a framework to expand the orthopaedic research of reverse engineering technologies. For example, orthopaedic researchers may desire the ability to distinguish between rough and smooth texture of a tissue simply by viewing a data file of a part that they have never seen or touched. To determine the methodology and scanning protocol required to obtain such precise data would facilitate the distribution of data throughout the research community.

REFERENCES

Martini, Frederic 1998. *Fundamentals of Anatomy and Physiology*. Upper Saddle River: Prentice Hall, Inc.

Montgomery, Douglas 1997. *Design and Analysis of Experiments, 4th Ed*. New York: John Wiley & Sons.

Musculoskeletal Research Center, University of Pittsburgh, Department of Orthopaedics 2001. http://www.pitt.edu/~msrc

National Collegiate Athletic Association, Injury Surveillance System 2000. Spring study shows many baseball, softball injuries occur on base path. *The NCAA News*.

Stone, Kevin 2001. The Stone Foundation for Sports Medicine and Arthritis Research, http://www.stoneclinic.com

Virtual modeling and rapid manufacturing – Bártolo (eds)
© 2005 Taylor & Francis Group, London, ISBN 0 415 39062 1

Medical application of single point incremental forming: Cranial plate manufacturing

J.R. Duflou, B. Lauwers, J. Verbert, F. Gelaude & Y. Tunckol
Department of Mechanical Engineering, Katholieke Universiteit Leuven, Heverlee, Belgium

ABSTRACT: Single Point Incremental Forming (SPIF) is a new sheet metal forming technique that, unlike other forming processes, does not require a dedicated tool set. Because of the absence of a die, SPIF is ideally suited for small batch or tailored production of sheet metal parts. Medical applications typically fit within these categories. The case presented in this paper deals with the production of a cranial plate used in reconstructive skull surgery. The SPIF process is compared to the conventional manufacturing methods.

1 INTRODUCTION

Single Point Incremental Forming (SPIF) is an emerging sheet metal part production technique. In this process a sheet metal part is formed in a stepwise fashion by a CNC controlled spherical tool without the need for any supporting (partial) die. This technique allows a relatively fast and cheap production of customized or small series of sheet metal parts.

This paper compares the SPIF process to the more conventional manufacturing methods for creating a cranial plate used in reconstructive skull surgery. Both processes are compared in terms of time and cost effectiveness.

2 CRANIAL MEMBRANE

A cranial membrane is a metal plate that is used for cranioplasty (see Figure 1). This could be, for example, a permanent reconstruction of a large vulnerable

opening in the human cranium. These openings can originate from trauma or tumor. Other applications are maxillo-facial corrections, where the surgeon wants to correct a deformation of the patient's head.

The cranial plate is usually made in medical grade Titanium with a thickness between 0.2 mm and 0.5 mm. Titanium is chosen because of its inertness to body fluids, its limited weight and its favourable strength properties.

2.1 Design of the cranial membrane

Every cranial plate is custom made for a particular patient. If a large part of the skull is still intact, the cranial plate can be modelled based upon a CT scan (computerized tomography) of the patient's head. Otherwise the design of the plate is done based on similar, more or less standardized skull shapes. A CT scan produces a series of cross-sectional images of the patient's head. Computer programs can combine these images to create a three dimensional CAD model of the skull of the patient.

The reconstruction techniques that are available to model a cranial plate can either be manual or (semi) automated in a CAD approach. The most commonly used reconstruction technique is modelling by clay and spatulas in a lab. The data obtained by the CT scan of the patient's head is sent to a rapid prototyping firm, which produces a replica of the skull with the defect using stereo-lithography. On that replica of the skull, the desired reconstruction is sculpted (see Figure 2).

Mirroring is a second approach. The CT data is loaded in a computer program and a bone piece from the healthy side is cut out and mirrored to the side of the skull with the defect. Extensive manual fine-tuning is

Figure 1. Front (left image) and top view (right image) of a cranial membrane (Courtesy: Uni-dent N.V., Leuven, Belgium).

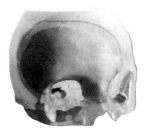

Figure 2. Clay reconstruction of a defect (Courtesy: Unident N.V., Leuven, Belgium).

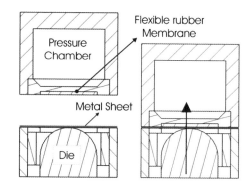

Figure 3. Sheet hydroforming process.

still required to obtain a good fit in the overlap region between skull and implant.

A third approach is a layer-by-layer reconstruction in a specialised software environment. On every CT slice, the operator identifies the defect region, and outlines a correction. Once all slices are processed, a CAD model of the cranial plate is made using those curves. This approach has reached some degree of automation, but highly relies on manual input [V. Pattijn et al. 2002].

Current research at the Department of Mechanical Engineering (KULeuven) aims at automating the design process for custom-made surface based implants, such as cranial plates. This design results in a STL representation and accompanying outline of the plate, thus offering a direct link to CAD based production techniques.

2.2 Requirements for the cranial membrane

The faster the skull of a patient can be reconstructed after a trauma, the more chances the patient has for survival. Since conventional sheet metal forming methods require at least one die to operate and since every cranial plate is unique, these processes lack the flexibility needed to optimally support emergency surgery.

The accuracy of the final part is important as well. When specifying accuracy requirements distinction can be made between two geometrical regions, namely the edge of the cranial plate and the central reconstruction area. The accuracy requirements at the edge of the plate are more strict than in the centre of the plate, since the edge has to provide a good fitting overlap with the remaining skull of the patient. The accuracy in the centre of the plate is more of importance for aesthetic reasons, to insure the symmetry of the skull of the patient after the procedure.

3 CONVENTIONAL MANUFACTURING METHODS

3.1 Conventional methods

Manufacturing the cranial plates with a conventional stamping press would be economically unfeasible.

Since every cranial plate is custom made, stamping would require a custom made positive and negative die for every plate to be manufactured.

The cranial membrane could also be manufactured by milling, starting from a square block of Titanium. This approach is too expensive as well due to the tooling cost and waste of Titanium.

Another approach is casting of the cranial plate. The drawbacks of this approach are that currently only thicker plates (>0.6 mm) can be obtained and that for each cranial plate, a custom made mould has to be manufactured.

3.2 Hydroforming

The operational principle of sheet hydroforming consists of pressing a metal sheet on a die by means of hydrostatic pressure (see Figure 3). A blank metal sheet is placed between the positive die and the pressure chamber. The pressure chamber is lowered until contact is made with the metal sheet. The pressure in the pressure chamber is increased to a predetermined setting and then the die is raised step by step by the machine operator, to produce the final cranial plate. Because this sheet hydroforming process only requires a positive die, it is currently the most commonly used manufacturing method for cranial plates.

3.2.1 Procedural steps for manufacturing a cranial plate by means of hydroforming

The procedural steps for manufacturing cranial plates by means of hydroforming are shown in Figure 4.

The first three steps, needed to design the cranial plate, were already outlined in the previous section. The cranial plates are modelled manually using clay and spatulas. This modelling process takes about two days because the surgeon needs to give his input on the final shape of the plate. The total time required for the design process is typically four days.

After the design, the manufacturing process can start. Since hydroforming machines are quite expensive

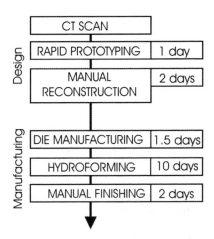

Design
- CT SCAN
- RAPID PROTOTYPING — 1 day
- MANUAL RECONSTRUCTION — 2 days

Manufacturing
- DIE MANUFACTURING — 1.5 days
- HYDROFORMING — 10 days
- MANUAL FINISHING — 2 days

Figure 4. Manufacturing pipeline for hydroforming.

Figure 5. Top view of a cranial membrane made by hydro-forming, before trimming (Courtesy: Uni-dent N.V., Leuven, Belgium).

and complicated to operate, most companies do not own a dedicated machine. Generally, the actual production of the die and cranial plate is subcontracted. The clay model of the cranial plate is used to make a negative die in silicone. From this negative die, the final positive die is made in a plastic and metal powder composite. Since the final die has to harden for about 24 hours in an oven, this last step takes at least one and a half day. Only then can the actual hydroforming start. The time needed for hydroforming varies, with a lead time average of about ten days.

The last step is manual finishing of the cranial plate. This finishing includes hole drilling, trimming of the plate (see Figure 5), final adjustments to the shape of the plate, cleaning, etching and sterilisation. The etching step is to ensure an optimal biocompatibility of the Titanium plate. The sterilisation step is performed just before implantation of the plate.

The average throughput time varies from one implant to another. Generally it takes about two weeks for a cranial plate to be completed.

3.2.2 Process features

Some of the advantages of the sheet hydroforming process are [Zhang, S.H et al. 2004]: minimal and uniform thinning of the material, less spring back than

in conventional stamping, the large drawing ratio that can be obtained and generally a good surface quality.

However sheet hydroforming has some drawbacks as well. As can be seen in the schematic overview, many intermediate steps are required to create the die used in the hydroforming process. Each of these steps induces a penalty on the process, both in manufacturing time and in achievable accuracy of the final part.

The subcontracting of the actual hydroforming typically induces a significant time delay as well.

Another important drawback is the fact that a lot of know-how of the sheet hydroforming process still needs to be explored [Lang, L.H et al]. That is the reason why cranial plates are usually manufactured by trial and error. This results in a material consuming production process, as one generally uses up to four times the theoretically required material quantity for a cranial plate.

4 SINGLE POINT INCREMENTAL FORMING

Single Point Incremental Forming (SPIF) is an emerging sheet metal forming technique that uses a spherical CNC controlled tool. The process starts from a flat metal sheet. To form a part, the tool follows a pre-programmed contour, just like in a conventional milling operation. For the manufacturing of a standard cone, the start position of the tool for the first and the last contour of the tool path are given in Figure 6.

SPIF is a unique sheet metal forming process in the sense that it does not require a specific die to operate. As a result, the lead time and cost of manufacturing a die do not apply for SPIF.

SPIF does need a backing plate for some applications. This backing plate supports the part at the edges where the metal sheet should not deform. In Figure 7, the difference in final shape of a standard cone formed with and without a backing plate is schematically shown. For the manufacturing of a cranial plate, the backing plate can be omitted by carefully selecting the location and orientation of the cranial plate.

4.1 Machining requirements

SPIF does not require an expensive and complicated machine, but can be preformed using a standard 3-axis milling machine equipped with a spherical tool. The maximum forces recorded in previous tests when forming parts using aluminium 3003-O sheets with a thickness of 1.2 mm, were about 700 N [Duflou, J.R. et al. 2005]. In other tests, peak forces of around 1500 N were recorded when using aluminium 3003-O sheets of 2 mm thickness. Because higher forming forces could be expected when forming Titanium, these forces should be taken into account when selecting an incremental forming machine.

The feed rate and tool rotation of the machine should be set to the value that creates a rolling effect of the tool

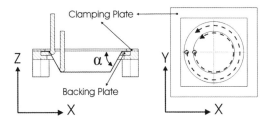

Figure 6. SPIF process: section view (left) and top view (right).

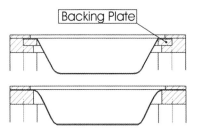

Figure 7. Side section view of a standard cone with backing plate (top) and without backing plate (bottom).

on the sheet. This ensures optimal forming conditions for the metal sheets and increases tool life by minimising friction between tool and sheet.

In order to further limit the tool wear, the tool material should be harder than the plate material to be deformed. Lubrication was found to have a profound impact on tool life as well. Lack of lubrication causes tools to fail within a few contour steps of the tool path, even when forming a softer material like Aluminium 3003-O.

4.2 *Accuracy*

The accuracy of the SPIF process for manufacturing the cranial plate using Aluminium 3003-O was examined [Duflou J.R. & Lauwers B. et al. 2005]. Several tool paths were tested and their dimensional accuracy was verified by means of a laser scanner with an accuracy of ±15 micron.

The first tested strategy was a standard single pass contouring tool path. This is an ordinary milling tool path with a constant step-down, used for finishing operations in milling. The second strategy was a double pass tool path consisting of the single pass contouring tool path, followed by a downward spiral finishing tool path. The last tool path was made by using the spring back information obtained in previous tests. The aim was to make a single tool path, which takes into account the expected spring back of the material [Bambach, M. et al. 2004]. Table 1 shows the error deviations from the CAD model for each strategy.

Table 1. Accuracy of the tested tool paths (given in mm).

(mm)	Strategy		
	1	2	3
Average absolute deviation	0.293	0.091	0.109
Avg. neg. dev.	−0.302	−0.345	−0.413
Avg. pos. dev.	0.103	0.058	0.082
Minimum	−1.39	−0.909	−0.906
Maximum	0.379	0.293	0.523
Sigma (σ)	0.199	0.138	0.171

These tests showed that good results can be obtained using adjusted tool paths. A nice feature of the manufactured cranial plate is that the largest errors are concentrated in the centre of the plate. This is acceptable when looking at the accuracy requirements outlined in the second paragraph.

It should be noted that, to be able to accurately measure the deviations, the plates were not removed from their clamping before measurement. Residual stresses inside the cranial plates are expected to deform the plates slightly after unclamping and trimming of the plates. However in practice, the dimensional accuracy would also not be verified in absolute terms (with a fixed coordinate system attached to the clamping device) but rather as a "best-fit" matching. This last analysis method more closely resembles the prosthesis fitting as performed by the surgeon.

4.3 *Wall thickness*

The final wall thickness of the cranial plate was investigated as well. A cranial plate was manufactured in a sheet of Aluminium 3003-O, with a thickness of 1.2 mm. As can be seen in Figure 8, the wall thickness in the centre of the cranial plate is between 1.2 mm and 1.0 mm. The wall thickness at the edge of the plate is smaller than in the centre, with a minimum of 0.7 mm at the very edge.

For less complex shapes, such as the simple cone, the wall thickness can be derived from the so-called Sine Law (see Equation 1). This law states that the final wall thickness of the part is linked to the wall angle of the part in that area (see Figure 6).

$$T_f = T_i \times \sin(90 - \alpha) \tag{1}$$

where
T_f = final wall thickness
T_i = initial wall thickness
α = wall angle

With different tool path strategies [Young, D. & Jeswiet, J. 2004] or a different orientation of the cranial plate, the wall thickness variation in a part can be changed.

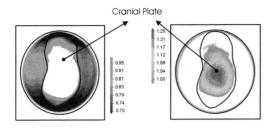

Cranial Plate

Figure 8. Top view of the wall thickness variation of the cranial plate made starting from a sheet of Aluminium 3003-O of 1.2 mm thickness. On the left image, the part of the cranial plate with a thickness of 0.95 mm to 0.70 mm is given. On the right image the part with a thickness between 1.25 mm and 0.95 mm.

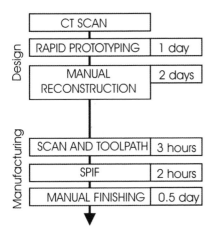

Figure 9. Manufacturing pipeline for SPIF.

4.4 Procedural steps for manufacturing a cranial plate by means of SPIF

The manufacturing pipeline for the SPIF process is given in Figure 9. As can be seen, SPIF greatly simplifies the production process. The intermediate steps to create the die for hydroforming are eliminated. This significantly reduces the total throughput time and cost of the production process.

After the design of the cranial plate, the clay plate was reverse engineered by means of a laser scanner to obtain the CAD model of the plate (see Figure 10). This CAD model was then used to create the tool path needed in the SPIF process. Since the hole drilling and the trimming of the plate can be performed on the incremental forming machine, the manual finishing step can be simplified as well.

In fact, with SPIF, the manufacturing takes less time than the current manual design process. Together with an optimisation of the design, the lead-time can be further reduced. Developed algorithms start from a

Figure 10. CAD model of the manufactured cranial plate.

CT scan of the skull and finally derive a CAD model of the cranial plate, making rapid prototyping, manual reconstruction and scanning superfluous. The semi-automatic design of middle-sized skull defects takes about 30 minutes; reconstructions of larger defects, as presented above, demand a model based approach and are currently under investigation. Linking this design methodology with the swift and flexible SPIF process should result in a next day delivery, which significantly lowers the threshold for application in trauma.

4.5 Manufacturing the cranial plate in Titanium

The above test results were obtained while forming the skull part in Aluminium 3003-O. The improvements in time and money, compared to the hydroforming approach will still hold when forming Titanium.

Forming Titanium does pose a challenge. The formability of the thinner sheets of Titanium is not known. Also the forces required to form Titanium are unknown. Further research has to be performed to determine the optimal forming conditions for forming cranial plates in Titanium.

5 CONCLUSION

The SPIF process seems well suited for the manufacturing of cranial plates. The use of the SPIF process for medical applications is not limited to cranial plates: many implants that are nowadays manufactured using hydroforming could also be manufactured using SPIF. For example plates used in hip reconstructions, which are much more common than cranial reconstructions, could also be made making optimal use of the flexibility of the single point incremental forming process.

REFERENCES

Pattijn, V. et al. 2002. Medical image based, preformed Titanium membranes for bone reconstructions: Design study and first clinical evaluation. *Proceedings of the Institution of Mechanical Engineers*, Part H: Journal of Engineering in Medicine, v 216, n 1, 2002, p 13–21

Zhang, S.H. et al. 2004. Recent developments in sheet hydroforming technology. *Journal of Materials Processing Technology*, v 151, n 1–3 SPEC. ISS, p 237–241

Lang, L.H. et al. 2004. Analysis of key parameters in sheet hydroforming combined with stretching forming and deep drawing. *Proceedings of the Institution of Mechanical Engineers*, Part B: Journal of Engineering Manufacture, v 218, n 8, August, 2004, p 845–856

Duflou, J.R. et al. 2005. Force Measurements for Single Point Incremental Forming: An Experimental Study. To be published in the proceedings of the 11th International Conference on Sheet Metal. Erlangen 2005

Duflou, J.R. & Lauwers, B. et al. 2005. Achievable Accuracy in Single Point Incremental Forming: Case Studies. To be published in the proceedings of the 8th Esaform Conference

Young, D. & Jeswiet, J. 2004. Wall thickness variations in single-point incremental forming. *Proceedings of the Institution of Mechanical Engineers*, Part B: Journal of Engineering Manufacture, v 218, n 11, 2004, p 1453–1459

Bambach, M. et al. 2004. Modeling of optimisation strategies in the incremental CNC sheet metal forming process. *NUMIFORM 2004*, 1969–1974

A critical review on acquisition and manipulation of CT images of the maxillofacial area for rapid prototyping

M.I. Meurer & L.F.S. Nobre
Federal University of Santa Catarina (UFSC), Florianópolis, Brazil

E. Meurer
University of the South of Santa Catarina (UNISUL), Tubarão, Brazil

J.V.L. Silva & A. Santa Bárbara
Renato Archer Research Center (CenPRA), Campinas, Brazil

M.G. Oliveira, D.N. Silva & A.M.B. Santos
Pontifical Catholic University of Rio Grande do Sul (PUCRS), Porto Alegre, Brazil

ABSTRACT: Regarding the development of rapid prototyping (RP) and image diagnosis technologies, the process of manufacturing biomedical prototypes has been evaluated by merging these two technologies. However, this process is complex. In order to obtain good results, a number of aspects requires special attention along the image acquisition by computed tomography (CT) and the further manipulation by dedicated software. This demands an interaction between (1) the biomedical sciences and (2) the computer science and engineering. The present article aims at discussing an experience of this multidisciplinary group with the acquisition and manipulation of CT images of the maxillofacial complex as well as the construction of biomedical prototypes for surgical purposes.

1 INTRODUCTION

Reaching accuracy in diagnosis and treatment procedures has been a challenge to dentomaxillofacial surgery groups. Especially in the most complex cases – namely facial deformations – the technological advancement has contributed to the improvement of diagnostic techniques and, in the last instance, to enhance the planning of therapeutic quality.

Facial deformations are relatively frequent and their treatment is time-consuming, complex and onerous. Besides the congenital causes of severe facial alterations, accidents associated to labour, traffic and social violence also pose as important acquired causes, especially in countries under development. According to data from the Brazilian Ministry of Health (Ministério da Saúde do Brasil, 2005), in the technical area of accidents and violence, hospital admission related to this causes are alarming.

Patients with severe facial alterations often present a low level of social integration. The difficulty of inserting these people into society is generally greater than the one found by people with alterations in members or other parts of the body. In parallel, resources invested in functional, aesthetic and social rehabilitation of people with facial deformations are exorbitant (Peckitt, 1999, Sanguera et al., 2001). These expenses are due to diverse factors, such as hospitalisation period, expenses with social providence during the beginning of the patient's recovery until her/his return to work, number of reconstructive surgeries, available time of surgical teams and auxiliary staff involved, medicine used, among others (ATLS, 1997, Sanguera et al., 2001).

Therefore, there is a need for the adoption of a series of measures to tackle this problem in a consistent way. One possible measure to ease the effects of this situation's results is to improve the services provided by health systems, by using technological resources and research applied to the improvement of life of citizens that, by chance, have facial deformations.

The integration of different technologies and multidisciplinary teams has led to results considerably favorable within the many scientific and technological fields, due to their complimentary resources and knowledge. An example of this integration is found between (1) medical images acquisition/manipulation technologies, and (2) CAD (Computer-Aided Design) and Rapid Prototyping (RP) Systems. Analogically to manufacture, in which prototypes are used to enhance products' quality, which allows for error detection in the first stages of the development cycle (Silva et al., 1999a), these technologies' integration permits predicting the complications and problems that might occur during and after complex surgeries of facial reconstruction. It also signals the reduction of costs and the enhancement of quality in surgeries' outcomes.

The concepts of prototypes' production, when applied to the biomedical field, involve the Biomedical Sciences, Informatics, and Engineering. Here, it is presented a critical review of the key-points needed for the construction of biomedical prototypes – as a result of the integration of a team composed by researchers linked to the Pontifical Catholic University of Rio Grande do Sul (PUCRS) Graduate Program in the field of Dentomaxillofacial Surgery, by researchers from the Product Development Division of Renato Archer Research Center (CenPRA), and by radiologists from the Federal University of Santa Catarina (UFSC).

2 THE RELEVANCE OF BIOMODELS IN BUCOMAXILLOFACIAL SURGERIES

Craniofacial surgeries involve a set of surgical techniques that aim at correcting congenital or acquired deformations resulting from traumas, resection of tumours, infections and other illnesses. Surgical treatments of such cases require quantitative and qualitative knowledge on anatomy. This pre-operation information is generally obtained through the acquisition of images of the human body's internal structures. Among the techniques of diagnosis through imaging, computed tomography (CT) is in evidence. This technique captures two-dimensional (2D) images of the human body region of interest, through slices that are sequential and parallel among themselves.

Although the data provided by 2D images are essential, additional information can be obtained from visualization through a three-dimensional (3D) reconstruction of the structures involved in a surgery. The computer screen is usually the means used to visualize, where the images generated by specific software can be analysed and eventually manipulated by a radiologist and/or surgeon. 3D visualization enables a privileged access to structures of interest and puts in evidence some aspects that would only be available to access through the mental reconstruction of slices.

Despite the fact that virtual images provide data which are indisputably elucidative, there remains a certain distance between the model visualized on the screen and the handling of "real" anatomic structures during a surgery. CT of 2D images, after properly treated, generates files that can be recognized by rapid prototyping equipment, which makes possible to physically build models that reproduce the human anatomy.

Biomodels – as they have been called – offer a direct visualisation and manipulation of areas that need surgery, which allows for previous measurement of structures and simulation of osteotomies and resection techniques. This aspect tends to reduce the room time of surgical procedure and, consequently, the anaesthesia period and risks of infection, thus improving final outcomes and lowering treatment global costs (Kermer et al., 1998, Peckitt, 1999, Meurer, 2002). Moreover, biomodels improve the communication between professional and patient, and may still serve as a basis for the manufacture of custom prosthesis implants (D'Urso et al., 1998; James et al., 1998; Haex and Poukens, 1999; Petzold et al., 1999).

3 STEPS IN THE PRODUCTION OF BIOMODELS

For manufacturing biomedical prototypes, the following steps must be correctly interwoven: (1) selection of patient; (2) acquisition of biomedical images; (3) transfer of image files; (4) manipulation of images in specific software; (5) conversion of files to a rapid prototype file format; and (6) construction of biomodels.

3.1 Selection of patient

The use of biomodels for diagnostic-surgical purposes is correctly indicated only when a cost-benefit relation is favorable, thus justifying the financial investment in biomodeling. Hence, opting for the production of biomodels, instead of less onerous techniques, must be only applied to cases in which there exist real benefits to the patient, disregarding fashion and mercantilism (Perry et al., 1998; Kermer, 1999; Meurer, 2002).

Biomodels use is of greater value to those surgical procedures in which there is the absense of widely used surgical techniques or when these techniques need to be modified and/or improved. With a physical model at hand, a surgeon may, during the planning phase, elaborate a technique, evaluate details, optimise a procedure, predict and, mainly, solve difficulties. The experience in using biomodels held by a team of surgeons at PUCRS has demonstrated that, for the cases in which difficulties during the simulation of a surgical procedure in a biomodel are observed, these difficulties are considerably similar to the ones faced in the moment of real surgical procedure (in the patient). This gives strength to the great validity of these prototypes during planning.

3.2 Acquisition of images

In the scan of craniofacial images for the purposes of prototyping, the traditional CT technique is not altered, thus it is kept the protocol of thin slices, given the anatomic complexity and the presence of thin structures peculiar to the facial bones. Nevertheless, some caution steps should be taken during the scan of images, in order to optimise them for further manipulation:

- The scan in a helicoidal mode provides better reconstructions due to the fact that the acquisition is volumetric. However, scans in a circular mode also allow for the construction of prototypes, provided that protocols of thin slices be used. It is important to underscore that, in the helicoidal mode, the scan time is shorter, thus reducing the possibility of patient's motion.
- In theory, the thickness of a slice must be as small as possible in order to achieve an adequate 3D reconstruction (thickness of 1 mm is a good trade-off). Nonetheless, if the area to be scanned is too big, it is often not technically possible to obtain the data with the less thick slice allowed by the equipment, due to, for instance, the heating up of the X-ray tube. In case the area of interest is the face, a change from axial to coronal in the scan planning may reduce the number of slices substantially.
- In case the acquisition is in a helicoidal mode, the increase of the pitch may allow for the scan of a greater volume, thus keeping slices thin. This solution is better than increasing the thickness of the slices. If increasing the thickness of slices is inevitable, one must choose the volume reconstruction as thin as possible.
- Patient exposure to radiation is a limiting factor and there exists a commitment solution to obtain a number of interest volume layers as high as possible with the least of patient exposure to radiation. Radiologists are responsible for choosing the most adequate protocol to acquire tomographic images, from those technical possibilities offered by a given machine, and to use common sense in relation to the model quality and radiation exposure level.
- The field of view (FOV) chosen must encompass the entire area of interest. To the face, even though it is possible to work with even smaller FOVs, a FOV of about 250 mm is enough, depending on the intended area to be represented in the prototype. Theoretically, the smaller the FOV used, the greater the image quality, for this way the matrix available is applied to a smaller area, thus reducing the pixel size and, as a result, the partial volume effect.
- The machine's gantry must not be inclined because some manipulation software still do not allow for the compensation of this inclination, thus producing prototypes with two-dimensional alterations (Kragskov et al., 1996).
- The use of filters during the acquisition is a controversial issue. Some studies have reported on a greater quantity of image artefact formation with the use of filters for bones during acquisition (Choi et al., 2002; Silva, 2004). This issue needs further careful research.
- The presence of metallic structures, such as dental restorations, leads to the production of artefacts in CT images, which add noise to the image. Afterwards, these artefacts need to be removed through graphics computing, thus making it a time-consuming and boring process, which frequently interferes with the prototype's final result. Aiming at minimising the effects of production of metallic artefacts, the patient must be positioned with the occlusal plan (line of dental occlusion) in parallel to the axial slice plan. This restricts the range to the dental crowns and thus reduces the number of slices to be manually edited.
- Whenever the aim is at reproducing the bone tissue, the reduction of mA during the acquisition allows for lessening radiological effects without the loss of diagnostic information (Bongartz et al., 2002). However, very reduced parameters of mA may produce artefacts of technique due to the low quality of radiation that reaches the detectors, thus compromising the quality of image (Romans, 1995).
- An essential condition to the process is that the CT machine probably exports the images captured (via optical disk, CD, DAT cassettes).

3.3 Storage and transference of image files

One of the problems in acquiring data from medical equipment is perhaps the lack of standardization of image formats. International organizations (National Electrical Manufacturers Association – NEMA/ American College of Radiologists – ACR) have invested in a standardized format of medical images data, and created a protocol named DICOM (Digital Imaging and Communications in Medicine). Today, the DICOM represents a standard to the medical equipment industry for the transfer of radiological images, as well as other kinds of medical information among computers, diagnosis and therapeutic equipment, and systems from various manufacturers (DICOM, 2004). Since the protocol has become a standard which is accepted worldwide, it guarantees higher interoperability between computing systems and medical equipment. Such interoperability is important to the cost-benefit relation in heath systems. The medical images analysis and treatment systems can import data and convert them into a more appropriate format for processing.

However, some tomography machines cannot export files in the DICOM format, particularly the older ones. For this reason, it is important that the diagnosis clinic previously ensure that the prototype center images be compatible with the software to be further used for the manipulation.

The files can be stored in any available media capable of storing large amounts of data: Usually CD's, DAT cassette, and optical disks. RW CDs (the ones which can be rewritten) should be avoided, since they may not be recognized by some manipulation software (Kernan & Wimsatt, 2000).

During the image file transfer, the amount of data may represent a problem. Every image in the DICOM format, using an acquisition matrix of 512×512 pixels, generates a file of approximately 512 Kbytes per slice. A complete tomography of a cranium, with adequate spacing to obtain 3D models, is around 100 Mbytes (or 500 Kbytes per slice). Compacting tools can be useful during the transfer. Other options are broadband network, FTP transfer (File Transfer Protocol) directly to the prototype center, availability of files in sites where access is restricted to the prototype center, or, still, posting the images in a CD.

3.4 *Image manipulation*

The manipulation of images for biomodeling is done with specific software. At this stage, a straight interaction between the biomedical specialization and the engineering is essential (Silva et al., 2003). These special pieces of software allow for 3D images manipulation, edition visualization until the ideal 3D model is obtained.

The manipulation stage aims to segment the image in order to isolate the data of interest from the set of information made available by the CT, which is an essential step for the success of the process. In case of prototypes for oral and maxillofacial surgeons, whose object of study is the bone, the segmentation aims to separate the bone portion from other tissues, so that only the bone tissue 3D model is generated (Figures 1, 2). If the segmentation parameters are not adequately defined, the prototype will present variations which may lead to the suppression of important structures and even the reproduction of undesirable ones.

In some cases, it is necessary to edit the images manually. Also, the presence of a professional with knowledge in sectional anatomy is fundamental during this stage. The manual edition of the images, by using tools like cut, delete, select, among others, is specially useful in areas with image artefacts proceeding from dental prosthesis, which aims to determine the appropriate contour of the region.

Among the available tools for the image segmentation, the threshold is the simplest and widely used. This tool is based on the definition of density intervals that can express, for example, only the voxels of the image corresponding to bone tissue. If the interval is not correctly determined, an effect named *dumb-bell* occurs, which may result in, for example, the suppression or thickening of structures during the process (Choi et al., 2002).

The CenPRA, by means of the PROMED project (Prototipagem Rápida em Medicina – Rapid Prototyping in Medicine), developed the software InVesalius, one of the pioneers in Brazil for medical images processing. Its Beta version, which has been used by PUCRS, is already available to professionals and health institutions within the biomedical area, following the free software policy. InVesalius implements algorithms which made resources available for 2D and 3D visualization, segmentation and reformatting. It allows for detailed analysis and access to information about the human body internal anatomic structures. Moreover, the software still offers an additional function: The conversion process, which allows to export the images in a format that can be recognized by the rapid prototyping equipment.

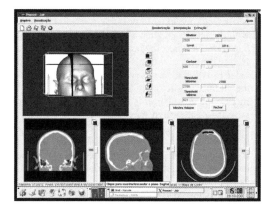

Figure 1. 3D image view and the coronal, sagital and axial planes before the segmentation process. Note the presence of soft tissues. Software InVesalius, CenPRA.

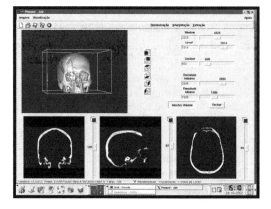

Figure 2. 3D image view and the coronal, sagital and axial planes after the segmentation process. Note the absence of soft tissues, and that only bone tissue correspondent structures remain included inside the cross-section interval. Software InVesalius. CenPRA.

3.5 Conversion of images

The images obtained from the tomography machine cannot be directly processed by the prototyping equipment for two main reasons: First, the prototyping equipment does not recognize the file format sent by the tomography machine; second, the thickness of tomographic slices usually varies from 1 to 5 mm (considerably thicker slices than those used in RP, which measure around 0.1 mm). Furthermore, RP processes use data from CAD_3D systems moulded through surfaces or solids, whereas tomographic images are represented by voxels. Thus, 2D tomographic images require 3D reformatting and conversion to an image format used in the RP processes (Souza et al., 2001, Souza et al., 2003). The most usual format is the STL.

In order to represent the human cranium complex structure properly, the model is defined by hundreds of millions triangular facets like in the Figure 3. Since the model is a simple representation and the anatomic structures are extremely complex, such as foramen and bone details, files of an excessive size may be generated, which can make the prototyping process unworkable, due to the processing required for reslicing the model. For this reason, an specific prototyping software is required to handle the STL file, which will be used to correct eventual surface inconsistencies, to close triangles, to optimize the number of triangles in the grid, as well as to choose the most appropriate construction orientation to the geometry of the piece in question – phase known as verification (Souza et al., 2003).

After the verification, the virtual STL model is resliced, in parallel layers, and then sent to the RP equipment for prototype reconstruction.

3.6 Construction of prototype

For the construction, the STL files from the workstation or from the PC must to be transmitted to the prototyping station computer (usually, the workstation is not at the same place as the prototyping system computer). The file transmission in formats like the IGES and the STL can be made by CD, ZIP drive floppy disk, e-mail, FTP, a local computer network, or any other means of supporting the transmission of these data.

During the transfer of large files, compacting tools may again be useful. After the transfer, the model construction in most processes is automatic. The construction process may take several hours, depending on the number of layers and on the height of the model.

Some systems need a post-processing stage after the prototype construction. On this stage, some operations are manual, thus, when post-processing is necessary, the possibility of introducing errors and reducing the final model precision must be considered.

3.7 Construction of moulds for customized prosthesis

Applications of CAD systems integrated with medical systems facilitate to a great extent manipulation and modeling of objects through its embedded resources, thus allowing virtual images of structures to be manipulated and simulated within this environment. Moreover, CAD systems are appropriate to define mirror procedures by using a collateral symmetry of the face for diagnosis and for specific prosthesis creation. The study of specific prosthesis can also be performed through the simulation of the prosthesis assembly within the structure. Figure 4 illustrates the creation of moulds for modeling customized prosthesis as well as the virtual prosthesis assembly within the 3D model.

In order to obtain a custom prosthesis it is necessary to mould the structure that will substitute the lesion

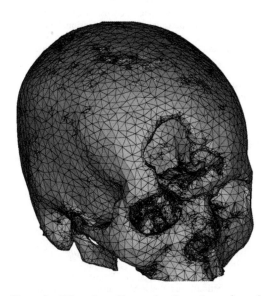

Figure 3. Triangular grids covering the entire surface of the image.

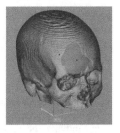

Figure 4. Obtaining a custom prosthesis. On the left, a virtual mould. On the right, a virtually positioned mould in the area of bone tissue absence.

171

area. The operations in CAD systems generate a 3D model from the mould, which will be sent to manufacturing in Rapid Prototyping. The mould obtained through RP is then used to shape the material to be implanted. It is important, whenever necessary, to consider typical contractions of these materials while manufacturing the moulds.

4 CONCLUSIONS

Many reports on the fields of both Medicine and Odontology have revealed the excellent potential of biomodels to improve and definitely modify many current treatments. The use of biomodels in Brazil is still restricted, particularly due to the costs involved in the production of prototypes and to a lack of equipment available. The lengthy nature of the process (from obtaining the image to the prototype production), associated with the high cost, represents obstacles to the application in routine surgery procedures, even when there is indication. It is likely, though, that such limitations melt in short time, be it attributable to technological advancements or due to the interdisciplinary use of the RP technology, thus increasing biomedical prototypes accessibility and benefiting a growing number of patients.

The experience acquired by this group when obtaining biomodels has offered for researchers from CenPRA, PUCRS and UFSC an exchange of information that is fundamental to the development of each knowledge area.

The CenPRA, an Institute of the Science and Technology Ministry in Brazil, faced the opportunity to have its software, the InVesalius, used by a group of specialized professionals, who represent, in the last instance, the software intended users. The idea is that the InVesalius is to be used not only in radiology clinics workstations, but mainly in surgeons' and students' personal computers to process biomedical images and to allow not static but dynamic interpretations, according to the diagnosis needs, even during surgical procedures. In addition, the CenPRA develops, in association with its partners from the biomedical area, methodologies, surgical planning and the construction of biomodels.

The PUCRS Post-Graduation Program in Oral and Maxillofacial Surgery has used biomodels to deal with several surgical cases, especially in patients with facial deformities. The use of biomodels has granted both patients in PUCRS institution and developing professionals with access to modern technologies.

At UFSC, a group of radiologist researchers have been studying the relation between parameters for obtaining tomographic images and the quality of biomedical prototypes, which is of great importance for the construction of qualitatively more precise models.

This whole undertaking has a primary aim: To dispose technology at offering better treatments, respecting the dignity of those who need these methods to better reestablish both form and function.

REFERENCES

ATLS – Advanced Trauma Life Support – Student Manual. 1997. 6. Ed. Chicago: American College of Surgeons.

Bongartz, G., Kalender, W., Golding, S.J, Gordon, L., Murakami, T., Shrimpton, P. 2000. Managing Patient Dose in CT. In: www.if.ufrgs.br/ast/med/images/ctdose.pdf. access: July 2002.

Choi, J.Y., Choi, J.H., Kim, N.K., Kim, Y., Lee, J.K., Kim, M.K., Lee, J.H., Kim, M.J. 2002. Analysis of errors in medical rapid prototyping models. *Int J Oral Maxillofac Surg.* 31(1):23–32.

DICOM Standards Committee. DICOM Home Page. In: http://medical.nema.org. Acess: Mar. 2005.

D'Urso, P.S. et al. Stereolithographic (SL) biomodelling in craniofacial surgery. *British Journal of Plastic Surgery,* Edinburgh, v. 51, n. 7, p. 522–530, Oct. 1998.

Haex, J.K.T., Poukens, J.M.N. 1999 Preoperative planning with the use of stereolithographic model. *Phidias Newsletter,* n. 3. In: www.phidias.org. Acess: Feb. 2005.

James, W.J. et al. 1998. Correction of congenital malar hypoplasia using stereolithography for presurgical planning. *J Oral Maxillofac Surg.,* 56: 512–517.

Kernan, B.T., Wimsatt, J.A. 2000. Use of a stereolithography model for accurate, preoperative adaptation of a reconstruction plate. *J Oral Maxillofac Surg.* 58(3):349–51.

Kermer, C., Lindner, A., Friede, I., Wagner, A., Millesi, W. 1998. Preoperative stereolithographic model planning for primary reconstruction in craniomaxillofacial trauma surgery. *J Craniomaxillofac Surg.* 26(3):136–9.

Kragskov, J., Sindet-Pedersen, S., Gyldensted, C., Jensen, K.L. 1996. A comparison of three-dimensional computed tomography scans and stereolithographic models for evaluation of craniofacial anomalies. *J Oral Maxillofac Surg.* 54(4):402–11.

Meurer, E. 2002. *As tecnologias CAD-CAM em Cirurgia e Traumatologia Bucomaxilofacial.* Tese de Doutorado, Pontifícia Universidade Católica do Rio Grande do Sul, Porto Alegre.

Ministério da Saúde do Brasil. In: http://www.ms.gov.br. Acess: Feb. 2005.

Peckitt, N.S. 1999. Stereoscopic lithography: customized titanium implants in orofacial reconstruction. *British Journal of Oral and Maxillofacial Surg.* 37, 353–369.

Petzold, R., Zeilhofer, H.F., Kalender, W.A. 1999. Rapid protyping technology in medicine–basics and applications. *Comput Med Imaging Graph.* 23(5):277–84.

Romans, L.E. 1995. *Introduction to computed tomography.* Philadelphia: Williams & Wilkins.

Sanghera, B. et al. 2001, Preliminary study of rapid prototype medical models. *Rapid Prototyping Journal,* 7, 275–284.

Silva, J.V.L. et al. Rapid Prototyping: Concepts, Applications, and Potential Utilization in Brazil. 15th International Conference in CAD/CAM Robotics and Factories of the Future, 1999.

Silva, J.V.L., Meurer, E., Zavaglia, C.A.C., Santa Barbara, A., Gouvêia, M.F. 2003. Rapid Prototyping Applications in the Treatment of Craniomaxillofacial Deformities – Utilization of Bioceramics. *Key Engineering Materials,* v. 254–256, p. 687–690, Trans Tech Publications, Switzerland.

Silva, D.N. 2004. *Análise do erro dimensional dos biomodelos de sinterização seletiva a laser (SLS) e de impressão tridimensional (3DP), a partir de imagens de tomografia computadorizada, na reprodução da anatomia craniomaxilar: estudo in vitro.* Tese de Doutorado, Pontifícia Universidade Católica do Rio Grande do Sul, Porto Alegre.

Souza, M.A., Centeno, T.M., Pedrini. H. Integrando recon- strução 3D de imagens tomográficas e prototipagem rápida para a fabricação de modelos médicos. *Rev. Bras. Eng. Biomédica.* [S.l.], v. 19, n. 2, p. 103–115, ago., 2003.

Souza, M.A., Ricetti, F., Centeno, T.M., Pedrini, H., Erthal, J.L., Mehl. A. 2001. Reconstrução de imagens tomográficas aplicada à fabricação de próteses por prototipagem rápida usando técnicas de triangulação. *Memorias II Congresso Latinoamericano de Ingeniería Biomédica,* La Habana, Cuba. In: www.hab2001.sld.cu/arrepdf/00256.pdf. Acess: Mar. 2005.

173

Virtual modeling and rapid manufacturing – Bártolo (eds)
© *2005 Taylor & Francis Group, London, ISBN 0 415 39062 1*

Costs reducing of maxillofacial RP biomodels

M.S. Saddy
Universidade Nove de Julho – UNINOVE, São Paulo, SP, Brazil

J.V.L. Silva, A. Santa Bárbara & I.A. Maia
Renato Archer Research Center – CenPRA, Campinas, SP, Brazil

ABSTRACT: Dimensional characterization was performed on a real mandible and their derivatives virtual and physical biomodels to compare dimensional accuracy regarding the combination of the available techniques involved in building these biomodels by Rapid Prototyping. The data acquisition techniques were Multislice Computerized Tomography (CT) and Cone Beam Computerized Tomography (CBCT). The medical image processing software were the commercial Analyze (Mayo Clinic, USA) and the freeware InVesalius (CenPRA, Brazil). Finally the RP techniques were the Three Dimensional Printing (3DP) and Selective Laser Sintering (SLS). Statistical treatment shows that the maximum dimensional error found was lower than 5%, and the average error was lower than 2.5%. These results indicate that cost effective exams as CBCT, associated to a free available software for medical image processing, running in affordable computers and integrated with low-end rapid prototyping systems, can deliver good solutions in terms of reducing costs without considerable dimensional losses. In being so these results represent a major progress towards widespread the RP medical applications in poor countries where the current costs of biomodels make them prohibitive for many people that necessitate maxillofacial reconstruction.

1 INTRODUCTION

1.1 *Motivation*

Costs reducing is a crucial measure to widespread the medical applications of Rapid Prototyping in poor countries. The applications of Rapid Prototyping (RP) in medicine consists of building the so called biomodels which are 3D physical models of the injured region of the body. Having the biomodels, the surgeon can planning and training the surgery and also construct prosthesis with biocompatible materials for implant purposes. The biomodel construction requires the integration of three major technologies, namely, acquisition of medical images with tomography equipments, software for medical image processing and, finally, rapid prototyping techniques (Chilvarquer et al. 2004; Mazzonetto et al. 2002; Silva et al. 1999; Van Lierde et al. 2002). A brief explanation of these technologies is given as follows:

1.2 *X-ray data acquisition*

The most common data acquisition tomography technique is the Multislice Computerized Tomography which is commonly known as CT (Chilvarquer et al.

2004). Its advent made possible the reproduction of the anatomical structures with high accuracy and good spatial resolution of hard and soft tissues. Images from slices thinner than 0.5 mm can be obtained (Cavezian et al. 2001; Chilvarquer et al. 2004; Rothman 1998).

An alternative to CT is called the Cone Beam Computerized Tomography (CBCT) which was specially developed for hard tissues in maxillofacial area. It utilizes the principle of cone beam. The volume of the whole bone is acquired in a 360° rotation angle of the X-ray tube around the anatomical structures. Mozzo et al. (1998) observed a geometric distortion of 1% in horizontal plane and 2.2% in vertical plane. In the same year, Bianchi & Lojacono (1998), reported the possibility to make two and three dimensional reconstruction with a low dose of radiation, as maximum as 15 mA. The main interest for using CBTC images is its low cost and simplicity. Nowadays, this kind of X-ray scanner is becoming very popular.

1.3 *Image processing*

The dataset generated from the both tomography equipments mentioned above have DICOM file format (ISO Standard for Digital Imaging and Communication in

Medicine). The DICOM 2D and 3D images are processed using specialized medical image processing software such as the commercial Analyze (Mayo Clinic, USA) and the free InVesalius (CenPRA, Brazil). The image process involves the following procedures: (1) import the data obtained from the tomography equipment, (2) segmentation which consists of removing material others than bones of the digital images (3) rendering which consists of construction of the digital 3D model, (4) transformation of the DICOM file format into STL file format that is that read by the RP machine of (5) export the STL file to the rapid prototyping machine.

The CenPRA, by means of the PROMED project (Prototipagem Rápida em Medicina – Rapid Prototyping in Medicine), developed the software InVesalius, one of the pioneers in Brazil for medical images processing. Its Beta version is already available to professionals and health institutions within the biomedical area, following the free software policy. InVesalius runs in affordable PCs instead of complex and expensive workstations.

1.4 *Rapid prototyping techniques*

The Rapid Prototyping technology was, originally, created for the industries to assure higher quality products. This technology aims at the early detection of errors during the stages of product development that reduces dramatically the costs and further problems in production with tangible benefits in efficiency and quality (Chilvarquer et al. 2004; Silva et al. 1999).

The SLS process is based on additive deposition of a polyamide powder in layers that are sintered selectively by the incidence of a servo controlled CO_2 low power laser. This process takes place in a pre-heated and inert chamber whose set points are controlled automatically. The prototype (in that case a biomodel) grows layer-by-layer, continuously, until the end of the process. The areas of the powder bed where the laser traces out melts down and the surface becomes harder after the material in process is cooled down. In the sequence the biomodel can be taken out from the powder and sand blasted or flushed being ready for the use. In the 3DP process a print head deposits a liquid binder, layer-by-layer, over a powder bed provoking a chemical reaction. The powder material that is subjected to the chemical reaction becomes harder and generating a green part. This green part shall be gently manipulated and flushed for cleaning the excess of powder. After it has to be infiltrated, manually, with special resins or wax to get strengthener.

1.5 *The present studies*

The present studies look for a combination of the techniques mentioned above to build affordable and good quality maxillofacial biomodels.

2 MATERIALS AND METHODS

A human dry mandible, called here gold standard, were used with eight small metal references (orthodontic braces) glued with wax and distributed on the mandible surface as references for dimensional measurements. These eight references are disposed on the mandible, in order to assess x, y and z axis accuracy's. They are localized as follows: two in retro molar region, two in the cortical area of the bicuspids, one in the middle lane and the remaining three in the low border. These references were combined two by two creating 28 measurement groups.

The mandible was submitted to a CT exam (Picker-Elscint, U.S.A.), setting up 1 mm slice thickness and parameter values of 120 KV e 60 mA. The mandible was also submitted to CBCT with 80 KV e 5 mA parameter values which were automatically adjusted by the machine. In order to perform the automatic parameter adjustment, the mandible was immersed in a plastic box with water to allow the scanner recognize a low density structure. After this procedure, the data were reformatted in axial slices of 1 mm thickness. The both DICOM files were stored in CD ROM media. Digital dimensional measurements were obtained from the 3D images (Figure 1) using Analyze and InVesalius software.

The Figure 2 shows the combination scheme to produce the biomodels from the two source of data (CBTC and CT) manipulated by the two medical image processing software (InVesalius and Analyze) and prototyped by two RP technologies (3DP and SLS) as shown in Figure 3. The SLS biomodels were built in polyamide and those 3DP in plaster.

The dimensional measurements on the gold standard, and physical biomodels were carried out utilizing a digital caliper ruler (0.001 mm precision). The nine groups of measurements were classified as:

I   – Mandible (gold standard);
II   – Analyze with CBCT;
III   – InVesalius with CBCT;
IV   – SLS biomodel with CBCT;
V   – 3DP biomodel with CBCT;
VI   – Analyze with CT;
VII – InVesalius with CT;
VIII – SLS with biomodel CT and;
IX   – 3DP biomodel with CT.

The measurement groups were calculated, compared and the errors statistically analyzed. As the sample has a normal and homogeneous distribution were applied the ANOVA (Analysis of Variance between groups) test (Sincich 1999) with a unique variation factor and, after that, applied the Tukey's method

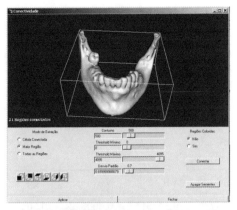

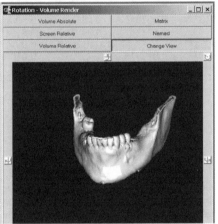

Figure 1. Three-dimensional reconstruction using InVesalius (upper) and Analyze (lower).

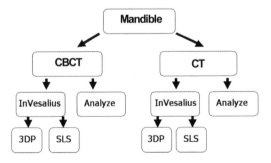

Figure 2. Scheme of the biomodels construction.

(Vieira 1991) to verify which groups have statistical significance.

Technical incompatibility, because of bad image quality, makes unable the Group II (Analyze with CBCT) to be used in this research.

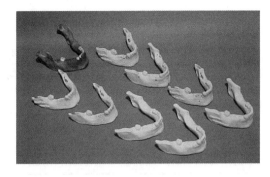

Figure 3. Dry skull mandible (dark) and SLS and 3DP prototypes.

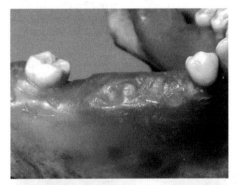

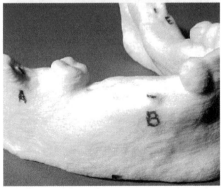

Figure 4. Metallic references used as measurements points: In the mandible (upper) and reproduced in the bio-model (lower).

3 RESULTS

Using ANOVA test, the value of critical F (2.146813) is lower than the calculated F (2.645276) indicating that the samples are different (Table 1).

In order to verify which samples have statistically significant differences between them, the Tukey's method was applied resulting 0.0053 value. If the

177

Table 1. ANOVA test with a unique factor of variation between groups and within measurement groups.

Variation	SQ	d.f.	MQ	F	P value	Critical F
Between groups	0.002895	6	0.000482	2.645276	0.017364	2.146813
Within groups	0.034473	189	0.000182	–	–	–
Total	0.037368	195	–	–	–	–

Table 2. Tukey's method with the averages between the measurement groups.

Groups	III	IV	V	VI	VII	VIII	IX
III	–	0.0029 n.s.	0.0025 n.s.	0.0087**	0.1470**	0.0006 n.s.	0.0019 n.s.
IV	0.0029 n.s.	–	0.0055**	0.0057**	0.0023 n.s.	0.0036 n.s.	0.0031 n.s.
V	0.0025 n.s.	0.0055**	–	0.0112**	0.0078**	0.0018 n.s.	0.0006 n.s.
VI	0.0087**	0.0057**	0.0112**	–	0.0033**	0.0093**	0.0106**
VII	0.1470**	0.0023 n.s.	0.0078**	0.0033**	–	0.0120**	0.0060**
VIII	0.0006 n.s.	0.0036 n.s.	0.0018 n.s.	0.0093**	0.0120**	–	0.012 n.s.
IX	0.0019 n.s.	0.0031 n.s.	0.0006 n.s.	0.0106**	0.0060**	0.012 n.s.	–

Where: n.s. stands for "non significance" and ** means "significance".

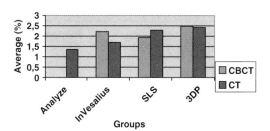

Figure 5. Averages (%) between the measurements groups and the gold standard, in CBCT and CT.

differences between the group averages are greater than the Tukey's value, then these averages can be considered different (p = 0.05) as shown in Table 2.

As a result, the averages of alteration of the measurement groups compared to the gold standard and the type of tomography technique (CBCT or CT) is shown in the Figure 5.

With the averages of alteration, the samples were classified from the more accurate to the less accurate:

1. Group VI – 1.35%
2. Group VII – 1.69%
3. Group IV – 1.93%
4. Group III – 2.22%
5. Group VIII – 2.29%
6. Group IX – 2.42%
7. Group V – 2.48%

4 DISCUSSION

This study produced interesting results as follows: 1) the linear measurements point by point when compared with the gold standard do not get higher than the 5% of the distortion, and 2) the highest level of average distortion found is 2.48%, in group V, using 3DP technology and CBCT.

Analyzing Graphic 1 it is possible to verify that the higher difference between the measurement groups is obtained in those groups analyzed with the InVesalius, and the lower difference is found in the group of the 3DP biomodels, where the average is almost zero.

All the measurement groups have differences when compared to the gold standard. However these differences ranges from 1.93% to 2.48%. These low differences values indicate that any combination of tomography technique, image processing software and RP technique give quite accurate biomodels.

The CBCT images dataset needed more treatment in the medical image software processing because of the lower quality of image. These excessive manipulations can degraded the 3D model in the computer, promoting distortion in the final results. Figure 6 shows details of two 3DP biomodel. The upper picture, originated from CBCT dataset, reveals bone defects, like fake alveolus. These defects are created artificially due to the incompatibility between its dataset and image processing software. The incompatibility arises mainly from the attempt to make the references visible by

178

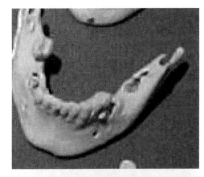

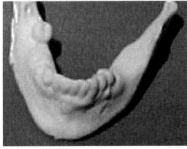

Figure 6. Detail of the reproduction of the anatomical structures in 3DP biomodels, from CBCT (upper), and from CT (lower).

image processing. The Figure 6, on the bottom, shows a biomodel produced with CT dataset where even the wax surface, used to fix metallic references can be seen.

5 CONCLUSIONS AND FURTHER WORK

Biomodel based on dataset from CBCT and CT is an available technology that, for clinical applications, reproduces the structures of interest with acceptable accuracy, independently of the technology used. The main interest for using CBCT images is its low cost and simplicity. Nowadays, this kind of scanner is becoming very popular. Another interesting resource is the availability of free software, like InVesalius, for medical image processing running in affordable PC's instead of complex and expensive workstations. Integrating these two initial solutions with inexpensive RP machines available today can deliver an excellent trade-off for underdeveloped countries where it is more applicable

due to health problems bring about by the uneven social conditions.

Finally, present evaluations are in progress to encompass more systems and technologies for medical image processing and rapid prototyping. The 3D models and biomodels are being evaluated using a laser scanner for reverse engineering and a special software to compare the whole surface of the models and not only group of measurements.

REFERENCES

Asaumi, J. 2001. Comparison of the three-dimensional computed tomography with rapid prototype models in the management of coronoid hiperplasia. *Dentomaxillofacial Radiology*, 30 (6) 330–335.

Bianchi, S. D., Lojacono, A. 1998. *2D and 3D images generated by Cone Beam Computerized Tomography (CBCT) for Dentomaxillofacial investigations.* http:// www. qrverona.it/htm/Articles.htm,

Cavezian, R., Pasquet, G., Bel, G., Baller, G. 2001. *Imagerie dento-maxillaire – Approche radio-clinique.* (2.ed.) Paris: Masson.

Chilvarquer, I., Chilvarquer, L. W., Hayek, J. E., Saddy, M. S. 2004. A prototipagem na Odontologia do novo milênio. In: Querido, M. R. M., Fan, Y. L.(1.ed.) *Implantes Osseointegrados – Inovando Soluções:* 317–328. São Paulo: Artes Médicas.

DICOM 2005. Standards Committee. DICOM Home Page. Disponível em: http://medical.nema.org. Access: Mar. 2005.

Mazzoneto, R. 2002. Uso de modelos estereolitográficos em cirurgia buco-maxilo-facial. *Rev Assoc Paul Cir Dent*, 56 (2): 7–13.

Mozzo, P. 1998. *A new volumetric CT machine for dental imaging based on cone-beam technique: preliminary results.* Eur Radiol, Spring 8(9) 1558–1564.

Rothman, S. L. G. 1998. *Dental Applications of Computerized Tomography – Surgical Planning for implants placement.* Chicago: Quintessence.

Silva, J. V. L. 1999. Rapid Prototyping: concepts, applications and potential utilization in Brazil. *15th International Conference in CAD/CAM Robotics and Factories for Future.*

Silva, J. V. L. 2004. Rapid prototyping applications in the treatment of craniomaxillofacial deformities – utilization of bioceramics. *Key Eng Materials* 254 (256) 687–690.

Sincich, T., Levine D. M., Sthephan D. 1999. *Practical Statistics by Example Using Microsoft Excel*, Upper Saddle River, New Jersey: Prentice Hall.

Van Lierde, C. 2002. Curvature accuracy of RP skull models. *Phidias*, 8, 1–14.

Vieira S. 1991. *Introdução à Bioestatística*, (1.ed.) Rio de Janeiro: Campus.

Virtual modeling and rapid manufacturing – Bártolo (eds)
© 2005 Taylor & Francis Group, London, ISBN 0 415 39062 1

Digital parametric models and anthropometric information of human orbits

H. Klein & K. Broekel

Institute of Engineering Design, Construction Technology/Computer Aided Design, University of Rostock

ABSTRACT: In the last couple of years Computer Aided Engineering (CAE) became more important in medicine. One of the reasons for this development is digitalization of bony structures based on x-ray tomography. But this process applied on human craniums brings new problems. High personal expenditure and low resolution of x-ray tomography make a detailed cranium model impossible. Therefore a parameterized 3d model from anthropometric data should produce better results. First experiences and problems are presented in this contribution.

1 INTRODUCTION

Goal of this contribution is to discuss the geometric-topologic oriented modeling procedures and their applicability on human orbits (bony part of the eye socket). Thereby the following finite element analysis for simulation of complicated fractures of the central face should be considered. Procedures for digitalization of bones and tissues, used today, are not applicable due to the geometrical complexity of human craniums. Scanning methods with an adequate resolution generate a great number of points, which are difficult to manage in software solutions. Also the separation of bones and tissue particularly inside the cranium is time-consuming and sometimes impossible. The further use of Computer Aided Design (CAD) models for finite element analysis additionally demand topological correctness and geometrical accuracy.

Digitalization of orbits with x-ray tomography is unsuitable due to their small resolution for detailed modeling. Because of this one or more generic and parameterized CAD models should be developed. These can be flexibly adapted to an individual case, in order to produce fast geometrical models for characterization of the mechanical behavior of the model. This process should be verified with rapid prototyping models. In this way it should be possible to generate finite element analysis suited 3d models and analyze fracture types of the central face and their causes.

This should be a first step for developing a complete virtual cranium model, in order to simulate complex, the entire cranium referring fractures and their mechanisms. For this purpose finite element simulations and experimental investigations should draw conclusions from forces which results cranium fractures.

2 MEDICAL BACKGROUND

Central face fractures are divided into fractures of the nasal bone, the upper jaw and the zygomatic bone massif (Schwenzer 1967). Beside this there are also isolated and combined fractures of orbital walls, which require special interest. Isolated fracture of the orbit bottom became established in Converse & Smith (1957) as "blow out fracture".

Injuries within the human face range, particularly within the range of the results from various causes. Principal reasons for example are traffic accidents, recreational accidents, sport accidents, domestic accidents and facial injuries which resort to violence. These causes of injuries within the facial range are evaluated in numerous German and Anglo American statistics. In Schröder et al. (1982) causes of facial injuries from 1956–89 are represented. Here traffic accidents were determined as main cause with 44%. In Germany sport and recreational accidents and particularly facial injuries, which resort to violence, are strongly increased (Weerda 1992). This tendency is also confirmed in numerous other studies (Scherer et al. 1989, Prokop & Boeckler 1990, Muaoka et al. 1995, Müller 1969). In the area of Rostock in 1985 most of the blow out fractures appeared to people who are between 18 and 22 years of age.

Today goal of medical care of all facial injuries is the complete aesthetic and functional rehabilitation of patients with re-establishment of appearance, facial expression, sensitivity, as well as chew, breath and vision function. That's why medical care changed to a more operational therapy in the last couple of years. In this way fractured structures become stabilized and reconstructed. Thus sagittal, transversals and vertical

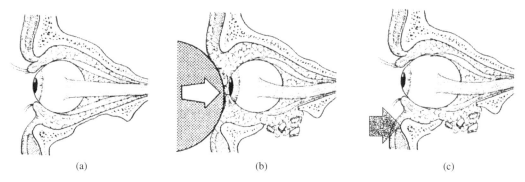

Figure 1. Mechanism of blow out fracture (left to right: orbit, hydraulic pressure theory and buckling force theory).

dimensions of the face can be restored and the full functional rehabilitation can be ensured.

In order to lower the risks of serious facial injuries in accidents and to clear up reasons of injuries which resort to violence, more basic investigations of fracture mechanisms are necessary.

Blow out fractures are a special type of central face fractures. The exposed orbital bones keep unhurt. Blow out fractures affect only the lower orbital walls. They occur isolated or combined. More exact investigations are for example in Edwards & Ridley 1968, Dodick et al. 1971, Curtin et al. 1982, Raflo 1984. As reasons for fracture pattern of blow out fractures two different theories are postulated. These are the hydraulic pressure theory (Fig. 1b) and the buckling force theory (Fig. 1c). The hydraulic pressure theory (Pfeiffer 1943) proceeds from a power transmission by the incompressible tissue volume of the orbit. A blunt trauma on the eye produces an increasing, all-side spreading pressure, which leads to a fracture of the thin orbital bottom. Critics of this theory miss the absence of serious eye injuries and placed the buckling force theory (King 1944). After a blunt trauma at the exposed orbital parts it comes through direct force transmission via the orbital walls to typical bending fractures at the orbital bottom. Both theories were re-enacted in newer experiments (Ahmad et al. 2003) at death craniums (1–7 days post mortem). Both, the hydraulic pressure theory as well as the buckling force theory, produced fractures of the orbital bottom, but had different patterns of injuries. Furthermore forces of the buckling force theory, which cause blow out fractures, are higher then forces of the hydraulic pressure theory. However these attempts might be unjustifiably because of medico legal and ethical reasons today. The causes and mechanisms of blow out fractures are well examined and documented. The available results and case examples make it possible to undertake a comparison between virtual blow out fractures with a parameter supported virtual orbit model and the reality.

In the past finite element investigation at human craniums were made from Bandak et al. 1995, Hartmann 1999, Power et al. 2001, Takizawa et al. 1988 and Voo et al. 1996. But these studies are insufficient for simulations of complex, the entire cranium referring, fractures and blow out fractures. Furthermore their models are neither detailed enough nor individual.

3 AUTOMATIC AND MANUAL MODELING OF SCANNING DATA

3.1 Mathematical algorithms

This attempt includes direct and indirect procedures. The source is a point cloud of x-ray tomography data record. Single points of this cloud are called voxel (be composed of volume and pixel). The grey tone of the voxel is measured in hounsfield units. The grey tone corresponds with the density of the irradiated material and permits conclusions whether it is bone, cartilage or tissue. The resolution of the x-ray tomography used in this investigation was 1.2 mm. With application of direct procedures, voxel were converted to knots of a finite element mesh. The necessary rework (smoothing of the surfaces) was very time intensive and in some areas impossible due to complexity of geometry. Results have been found useless for further computational investigation. With application of indirect procedures (Fig. 2) voxel were converted with different mathematical algorithms to three dimensional surface or volume models. These models should be used as basis for finite element analyses. Also the indirect procedures did not work very well. Reason for this is the very complex geometry of the cranium (small bones, cavities, thin bones). Another obstacle represents the temporal expenditure.

3.2 Commercial software MIMICS®

Further Attempts of automatic generation of three dimensional models from x-ray tomography data with

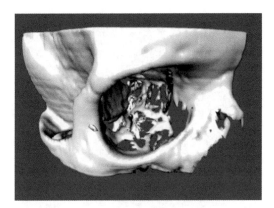

Figure 2. Orbit model.

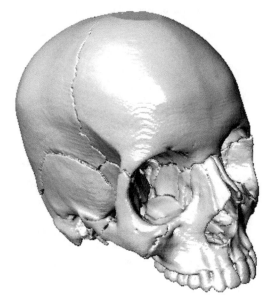

Figure 3. Cranium model with MIMICS®.

the commercial software MIMICS® (Materialise's Interactive Medical image control system) supplied promising results (Fig. 3) against past investigations. Here a cranium copy made from plastic was scanned with a technical x-ray tomograph with a resolution of 0.5 mm. Also an advantage was the very sharp boundary edge, because only a model of the bony part became scanned, so that no additional filtering of voxel for separating bones and tissue had to be made. Main problems, which make all methods useless till here are the "partial volume" effect and the small resolution of the computer tomograph. The "partial volume" effect comes out by the fact that the grey tone of a voxel represents an average value of the radiant absorption of

the material lying in it. In boundary regions this leads to blurred edges, which can only be reduced, where same material extends over wide areas, which are larger than the resolution of the computer tomograph. The small resolution causes another problem. There are bone structures with a thickness of 0.5 mm within ranges of the orbit. These can only be displayed sufficiently, if the resolution of the x-ray tomography is clearly higher.

3.3 Shell model for finite element investigation

Independently to the proceeding above a none patient-specific finite element model from shell elements was generated (Martin et al. 2004). Thereby the cranium was developed with data of an individual head with the help of a coordinate measuring table.

Thickness of the sections was piecewise set constant. Several load cases were examined with static boundary conditions, which represented different variants of load (buckling force theory and hydraulic pressure theory). The cranium was held at the foramen magnum. On basis of the distribution of stress it was possible to made conclusions on the validity of the different theories (hydraulic pressure theory and buckling force theory). Load cases applied on the exposed orbital part resulted stress peaks within the range of the orbit bottom, which corresponds to that, where blow out fractures are to be expected. More investigations are necessary for quantitative comparison. In order of this the eye and the associated tissue (muscle/fat) must be modeled with. In this way pressure distribution can be identified.

4 PARAMETRIC MODELS

4.1 Parametric with x-ray tomography

Depending on the generic parameter height, width and depth a small number of points from x-ray is used for reconstructing the surface of the orbit with splines. The coordinates x_i, y_i, and z_i of these points are all related to the generic parameter s_{xi}, s_{yi} and s_{zi} and the scanned coordinates of the points x_s, y_s, and z_s These relation are shown in Formula 1 or as matrices notation in Formula 2.

$$
\begin{bmatrix} x_i \\ y_i \\ z_i \\ 1 \end{bmatrix} = \begin{bmatrix} s_{xi} & 0 & 0 & 0 \\ 0 & s_{yi} & 0 & 0 \\ 0 & 0 & s_{zi} & 0 \\ 0 & 0 & 0 & 1 \end{bmatrix} \cdot \begin{bmatrix} x_s \\ y_s \\ z_s \\ 1 \end{bmatrix}
\tag{1}
$$

$$
\mathbf{p}_i = \mathbf{S}_i \cdot \mathbf{p}_s
\tag{2}
$$

Thereby the placement of these points to each other is defined through x-ray data. In this way the topology of the orbit surface is always the same. But the generic model can be scaled and stretched in all three directions. In this manner it is possible to transform the virtual orbit in making it smaller or bigger. This is the first step to a parameterized model.

Disadvantage of this procedure is the use of voxel from x-ray scans, because this is not accurate enough. Especially in range of the orbital walls more resolution is needed. Beside this you need more freedom in setting points for defining the topology. Without this freedom it is impossible to create smooth splines and consequently surfaces too.

Furthermore you need more parameter to adapt the model in more than only through scaling in three directions.

4.2 Parametric with defined grid scanning

In order to create a more powerful parameterized model than from x-ray data, another scanning method is necessary. This should be more accurate and should have no restrictions in defining points. A tactile scanning method was used. Disadvantage of this procedure is the need of macerated craniums or plastic models. It is impossible to apply this scanning method on a living individual.

The procedure is similar to 4.1. After scanning a plastic cranium model, three anthropometric points and three anthropometric measurement values are used. In Fig. 4 the anthropometric points and values for the direction x and y are described. The anthropometric point in x direction is *point dakryon* with its distance *iow* (intra orbital width) from the y-z-plane. The anthropometric value in x direction is *ow* (orbital width). There is the same interrelationship in y and z direction. The anthropometric point in y-direction is *point fissure* with its distance *ioh* (intra orbital height) to the x-z-plane and the anthropometric value *oh* (orbital height). In z-direction the distance of the point to the x-y-plane is *iod* (intra orbital depth) and the anthropometric value *od* (orbital depth).

The relation between the points of the virtual model $p_i(x_i, y_i, z_i)$ and the scanned points $p_s(x_s, y_s, z_s)$ is defined as in Formula 3–5.

$$x_i = \left(x_s - iow_s\right) \cdot \frac{ow_i}{ow_s} + iow_i \qquad (3)$$

$$y_i = \left(y_s - ioh_s\right) \cdot \frac{oh_i}{oh_s} + ioh_i \qquad (4)$$

$$z_i = \left(z_s - iod_s\right) \cdot \frac{od_i}{od_s} + iod_i \qquad (5)$$

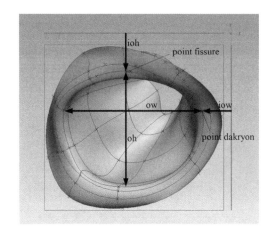

Figure 4. Orbit with defined anthropometric points.

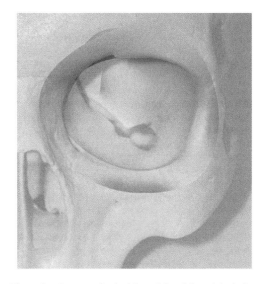

Figure 5. Parameterized orbit model and the original plastic cranium model.

Formula 3–5 converted into a matrix operation is shown in Formula 6 and as matrices notation in Formula 7.

$$\begin{bmatrix} x_i \\ y_i \\ z_i \\ 1 \end{bmatrix} = \begin{bmatrix} \dfrac{ow_i}{ow_s} & 0 & 0 & iow_i - iow_s \cdot \dfrac{ow_i}{ow_s} \\ 0 & \dfrac{oh_i}{oh_s} & 0 & ioh_i - ioh_s \cdot \dfrac{oh_i}{oh_s} \\ 0 & 0 & \dfrac{od_i}{od_s} & iod_i - iod_s \cdot \dfrac{od_i}{od_s} \\ 0 & 0 & 0 & 1 \end{bmatrix} \cdot \begin{bmatrix} x_s \\ y_s \\ z_s \\ 1 \end{bmatrix}$$

$$(6)$$

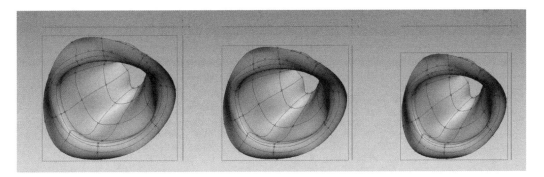

Figure 6.   Surface model of the orbit (left to right: maximum, average and minimum orbit from a 20 years old person after Lang & Oehmann 1976.

$$\mathbf{p}_i = \mathbf{M}_i \cdot \mathbf{p}_s \qquad (7)$$

Advantage of this method is, that you can span a special designed grid over the orbit. Dependent on the gradient of the area it is possible to set more or less points which are defining the topology. In this way it is possible to generate a high quality model with a small set of data. The correlation between plastic and virtual model is shows Fig. 5, where both are printed super imposingly.

The results of this method are shown in Fig. 6. There are three orbit surfaces presented. They show a big, an average and a small orbit. Used parameter and values are collected in Lang & Oehmann (1976).

5   FURTHER INTENTIONS

To implement more anthropometric points is the next step in developing a parameterized orbit model. Therefore a new conversion matrix and a different scan point vector will be needed. If this procedure works, two methods for generation of individual craniums should be established.

First method is to use anthropometric investigation, where used anthropometric points and values are listed in dependency on sex and age of the individual, like in Lang & Oehmann (1976). For a more realistic model anthropometric points and values should be measured at the individual itself. This can be done with x-ray for example.

After successful parameterized modeling of the orbit, this procedure should be applied on the middle face region and result volume models of human craniums. These can be used for first finite element investigations (Fig. 7).

6   CONCLUSION

The procedures for the digitization which are used today seem inapplicable for human craniums. Particularly

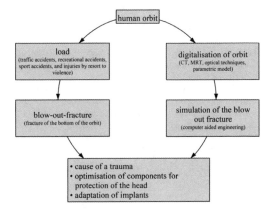

Figure 7.   Systematic for investigations of blow out fractures.

regarding the following finite element analysis, the quality of the surface through triangulation is insufficient. Furthermore the density of the point cloud controls the minimum number of elements when automatic meshing is used. Due to this a generic parameterized model should be developed, which can be adapted flexibly to the individual form of the orbit. Within a short time a complete geometrical and material description of individual parts of the head can be generated by the variation of the describing parameters. The verification of this procedure takes place through comparison of original geometry with rapid prototyping models.

Further rapid prototyping models provide a basis for impact tests. In this manner boundary conditions for finite element analysis can be examined. With typical examples of blow out fractures and results of impact tests finite element simulation can be verified. Subsequently, special blow out fractures with different loads can be simulated. Results compared with original injury admit conclusions on the cause of fracture.

185

REFERENCES

Ahmad, F., Kirkpatrick, W. N., Lyne, J., Urdang, M., Garey, L. J., Waterhouse, N. 2003. Strain gauge biomechanical evaluation of forces in orbital floor fractures. *Br J Past Surg.* 56: 3–9.

Bandak, F. A., Vander Vorst, J., Stuhmiller, L. M., Mlakar, P. F., Chilton, W. E., Stuhmiller, J. H. 1995. An Imaging-Based Computational and Experimental Study of Skull Fracture: Finite Element Model Development. *Journal of Neurotrauma* 12(4): 679–688.

Converse, J. M., Smith, B. 1957. Enophthalmos and Diplopia in Fractures of the Orbital Floor, *Britisch Journal of Plastic Surgery 9:* 265–274.

Curtin, H. D., Wolfe, P., Schramm, V. 1982. Orbital roof blow-out fractures. *Am J Roentgenol, 139:* 969–72.

Dodick, J. M., Galin, M. A., Littleton, J. T., Sod L. M. 1971. Concomitant medial wall fracture and blow-out fracture of the orbit. *Arch Opthalmol, 85:* 273–6.

Edwards, W. C., Ridley, R. W. 1968. Blow-out fracture of the medial orbital wall. *Am J Ophthamol, 65:* 248–9.

Hartmann, U. 1999. Ein mechanisches Finite-Element-Modell des meschlichen Kopfes (Dissertation). *MPI Series in Cognitive Neuroscience 7.*

King, E. F. 1944. VI. Diseases of the Orbit and Sphenoidal Sinus. *1. Fractures of the Orbit Transactions of the Ophthalmic. Society of the United Kingdom 64:* 134–139.

Lang, J., Oehmann, G. 1976. Formentwicklung des Canalis opticus, seine Maße und Einstellung zu den Schädelebenen. *Verh. Anat. Ges. 70:* 567–574.

Muaoka, M., Nakai, Y., Nakai, Y., Nakagawa, K., Yoshioks N., Nakaki Y., Yabe T., Hyodo T., Kamo R., Wakami S. 1995. *Fifteen-year statistics and observation of facial bone fracture.*

Müller, W. 1969. Häufigkeit, Sitz und Ursachen der Gesichtsschädelfrakturen. Leipzig.

Pfeiffer, R. L. 1943. *Traumatic Endophthalmus Archives of Ophthalmology.*

Power, E. D., Stitzel, J. D., West, R. L., Herring, I. P., Duma, S. M. 2001. A nonlinear finite element model of the human eye for large deformation loading. *25th Annual Meeting of the American Society of Biomechanics.*

Prokop, D., Boeckler, H. H. 1968. *Morbidity analysis of 1484 facial bone fractures between 1968 and 1987.*

Raflo, G. T. 1984. Blow-in and blow-out fractures of the orbit: clinical correlation and proposed mechanisms. *Ophthalmic Surg. 14:* 114–9.

Schwenzer, N. 1967. Zur Röntgendiagnostik als Voraussetzung für die operative Behandlung von Mittelgesichtsfrakturen. *Dtsch. Zahnärztin 22:* 221.

Schröder, H.-G., Glanz, H., Kleinsasser, O. 1982. Klassifikation und "Grading" von Gesichtsschädelfrakturen. *HNO 30:* 174–179.

Scherer, M., Sullivan, W.-G., Smith, D.-J. Jr, Phillips, L.-G., Robson, M.-C. 1989. *An analysis of 1,423 facial fractures in 788 patients at an urban trauma center.*

Takizawa, H., Sugiura, K., Baba, M., Tachisawa, T., Kadoyama, S., Kabayama, T., Ohno, H., Fuseya, Y. 1988. Structural Machenics of the Blowout Fracture: Numerical Computer Simulation of Orbital Deformation by the Finite Element Method. *Neurosurgery Vol.22, No.6, Part 1.*

Voo, L., Kumaresan, S., Pintar, F. A., Yoganandan, N., Sances, A. 1996. Finite-element models of the human head (Review). *Medical & Biological Engineering & Computing 34:* 375–381.

Weerda, H. R. 1992. Verletzungen der Nase, der Nasennebenhöhlen und des Gesichtsschädels. *Oto-Rhino-Laryngologie in Klinik und Praxis Band 2.*

Dipl.-Ing. Hendrik Klein
Institute of Engineering Design, Construction Technology/ Computer Aided Design
University of Rostock
Albert-Einstein-Str. 2
18055 Rostock, Germany
hendrik.klein@uni-rostock.de

Prof. Dr.-Ing. habil. Klaus Broekel;
Institute of Engineering Design, Construction Technology/ Computer Aided Design
University of Rostock
Albert-Einstein-Str. 2
18055 Rostock, Germany
klaus.broekel@uni-rostock.de

Virtual modeling and rapid manufacturing – Bártolo (eds)
© *2005 Taylor & Francis Group, London, ISBN 0 415 39062 1*

Digital denture manufacturing – Application of abrasive computer tomography, CNC machining, and rapid prototyping technologies

Yuh-Si Liu & S. Wang
Department of Mechanical Engineering, Kun-Shan University of Technology, Tainan, Taiwan

C.C. Chang
Department of System Engineering, National University of Tainan, Tainan, Taiwan

ABSTRACT: The fixed partial denture produced by conventional method is greatly relied on the skill and experience of dental technician. Its quality and accuracy depends mostly on the technician's subjective judgment. In addition, the manual process involves many procedures which require a long time to complete. Most important, it does not preserve any quantitative information for future retrieval.

In this article, a new device for scanning denture images and reconstructing 3D digital information of dental model by abrasive computer tomography (ACT) was designed and proposed. The fixed partial denture was then suggested to be produced by the rapid prototyping (RP) and computer numerical control (CNC) machining methods based on the digital information. A force feedback sculptor (FreeForm system, Sensible Technologies Inc., USA), using 3D Touch technology was applied to modify the morphology or design of denture. In this article, the comparison of conventional manual operation, digital manufacture using RP and CNC machining technologies for denture production was presented and the digital denture manufacturing protocol using proposed computer abrasive teeth profile scanning, computer-aided denture design, 3D touchable feature modification and numerical denture manufacturing were reported. These proposed methods provide a solid evidence that digital design and manufacturing technologies may become a new avenue for custom-made denture design, analysis, and production in 21st century.

1 INTRODUCTION

According to the American Dental Association, approximately 113 million American adults are missing at least one tooth, and 19 million have no teeth at all. Many of these people use removable dentures or fixed partial dentures to restore their missing teeth. But removable dentures can be uncomfortable and troublesome, and they often provide a patient with only a fraction of the chewing force of natural dentition. Fixed prostheses are more effective than removable dentures but require grinding healthy adjacent teeth to stubs before fabrication new prostheses. The current manufacturing of removable denture, crowns, or bridges is a labor-intensive work that undergoes substantial processing, with the result that a lot of information is lost and the process takes a long time to complete (Schmitt, 2001).

Digital denture manufacturing (DDM), a combination of digital imaging, Computer-Aided-Design/Computer-aided-Manufacturing (CAD/CAM), and Rapid Prototyping (RP), can alleviate these problems. The DDM appears to be a highly promising production process for dental prostheses such as crowns and

bridges in the future. By using technologies of CAD/CAM and RP, DDM is able to make complicated, customer-specific product immediately, without involving time-consuming intermediate stages such as the manufacture of dental molds (Chang & Chiang, 2002).

Therefore, a new devise which is a more convenient and economic approach to obtain all information of structures in a complicated object and rebuild the 3D CAD model is proposed (Chang, 2003). The device uses the abrasive method to remove the inlaid object layer by layer and capture the cross-sectional image of each layer with a CCD camera. We named the device as abrasive computer tomography (ACT) apparatus. Therefore, digital denture manufacturing is proposed here by using the integrated technologies of abrasive computer tomography, CNC machining and rapid prototyping.

2 MATERIAL AND METHODS

In this article, an integrated digital denture manufacturing process was proposed. Various techniques such as ACT teeth profile image capture, 3D computer-aided denture design and modification with 3D touch technology, and digital manufacturing (i.e. CNC machining and RP casting) were utilized in the study. First, an ACT apparatus was designed to capture the image of the sliced plaster dental model and to reconstruct 3D CAD model from image data. Once the 3D CAD denture model was build, two different computerized digital manufacturing techniques (i.e. CNC machining and RP wax prototyping) can be used for denture production. The 3D touch technology was integrated to modify denture model.

3 IMAGE CAPTURE BY ACT

Figure 1 show the design configuration and actual picture of in-house made ACT apparatus. The concept of ACT apparatus is similar to the X-ray CT scanner. The only difference is that ACT apparatus is to remove the inlaid physical dental plaster model layer by layer abrasively and capture the cross-sectional image of each layer by a CCD camera (1.67 million pixels, Pixera Corporation, USA). The images are then transmitted to a computer for image processing and 3D reconstruction. The physical dental plaster model was embedded in silicone or gypsum material and sliced layer by layer through the rolling tool (Figure 1). The

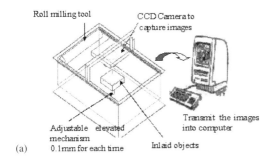

(a)

(b)

Figure 1. (a) Concept design of Abrasive Computer Tomography, (b) Slicing teeth plaster model embedded in silicon or gypsum.

working dimension of ACT apparatus is 80×80 mm. Its dimensional precision is 0.063 mm. The precision can be increased, if the higher quality of CCD camera is used.

4 BINARY SEGMENTATION PROCESS

The binary segmentation of image processing is to threshold the gray gradient of the image under the control of the threshold value. Mathematic algorithms to threshold the gray gradient of images are described as follows:

For a second order of threshold function:

$$f(x,y) = \begin{cases} 1 & f(x,y) \geq T \\ 0 & f(x,y) < T \end{cases} \quad (1)$$

For n-order of threshold function:

$$f(x,y) = \begin{cases} N_1 & f(x,y) \geq T_1 \\ N_2 & T_2 \leq f(x,y) < T_1 \\ \vdots & \vdots \\ N_n & T_n \leq f(x,y) < T_{n-1} \end{cases} \quad (2)$$

Where f(x,y) is the value of gray gradient of pixel and T is threshold value.

In the second order of binary segmentation, the pixel in the image will be kept when the value of gray gradient is larger than the threshold valve, otherwise

Figure 2a. Images of the sliced plaster teeth model by ACT apparatus.

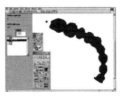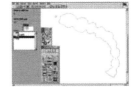

Figure 2b. A cross-sectional teeth image after binary segmentation and a cross-sectional boundary of teeth model.

it will be treated as a void. For n-order of binary segmentation, the pixel will be kept when the value of gray gradient is in the range of threshold value.

Figure 2a shows images of the sliced plaster dental model captured by ACT apparatus. After the binary segmentation of image processing in the LabView (National Instruments, USA) a cross-sectional dental image and the boundaries of images for each slicing layer are shown in Figure 2b.

5 RECONSTRUCTION OF 3D CAD MODEL

There are several methods to rebuild 3D CAD from the image processing, such as stack region growing, quad tree split etc. (Chang & Chiang, 2002). In this paper, a different approach to reconstruct 3D CAD model was proposed and described as follows:

1. To stack point clouds of all boundaries.
2. To generate Stereolithography (STL) format of point clouds.
3. To rebuild 3D CAD Model of teeth model.

We use the software of LabView, CopyCAT (Delcam International, Birmingham, UK), and PowerSHAPE (Delcam International, Birmingham, UK) to assist finishing above steps.

The complicated object that includes the undercut or inner structure, is difficult to use traditional CMMs to obtain data and reconstruct 3D CAD model through the technique of reverse engineering. Most of medical applications are classified to those kind cases, such as the artificial teeth, bone prosthesis, auricular prosthesis, cranial implants. . . . Basically, CTM or MRI is adopted to scan objects in the aspect. Those facilities are harmful to human because of X-ray. CTM and MRI are only available in research hospitals or medical center. Also, it needs a well-trained technician to operate those expensive instruments. Thus, we apply ACT to obtained CT images of physical models. The format of images is Bitmap. The boundaries of each layer are generated through the ways of binary segmentation and numerical scheme. Then, 3D CAD model can be reconstructed in the procedure of reverse engineering as Figure 3.

6 3D MODEL MODIFICATION USING FREEFORM SYSTEM

FreeForm System (Sensible Technologies, Inc.) is a 3D touch technology that allows users to directly interact with digital objects exactly as they do in the real world. Modelers sculpt as naturally in digital form as they do in clay and surgeons perfect their crafts with virtual patients instead of actual people. The technology makes working in three dimensions more natural, efficient and intuitive by allowing users to have continuous, two-way interaction with their work. It provides the ability to directly manipulate models and data, which means less time is spent setting parameters and more is spent working. Direct manipulation metaphors can also considerably reduce the learning curve required for professional 3D applications.

The digital denture data (STL format) can be easily modified by FreeForm as shown in Figure 4. Dentists can directly access and modify without relying on CAD/CAM technician or laboratory technician. The flow chart is shown in Figure 5. Three elements together deliver 3D Touch technology. The PHANTOM hardware (Sensible Technologies Inc., USA) provides precision positioning input and high fidelity, force- feedback output. GHOST software (Sensible Technologies Inc., USA) works as the engine, handling complex computations and allowing developers to deal with simple, high-level objects and physical properties such as location, mass, friction and stiffness.

7 DISCUSSION AND RESULTS

The integrated technologies of CAD/CAM and RP are initially applied in automotive and aerospace industries. Now, many dentists begin to use these technologies to design and fabricate dental restorations. There are some CAD/CAM systems adopted by dentists, such

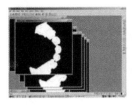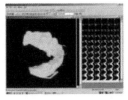

Figure 3. Segmentation of teeth model composed by all layers and STL format of teeth model.

Figure 4. The digital denture data (STL format) modified by FreeForm.

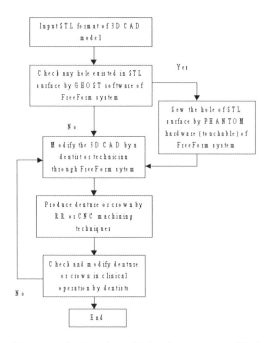

Figure 5.   The flow chart of digital denture data modified by 3D touch technology of FreeForm system.

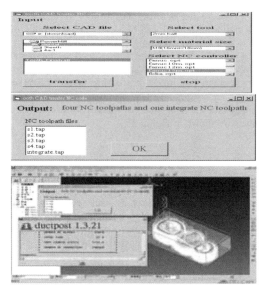

Figure 6.   Input dialog box of interface, output dialog box of interface, and generate simulation tool path of interface.

as Duret, DentiCAD, CEREC, etc. Duret et al. (1988) obtained dental data by laser scan and produce the prosthesis through a milling machine. DentiCAD was developed by Rekow in 1989. A micro mechanical probe is used to obtain the teeth profile information, and the abutment of crown is intolerance of 40–60 μm. The CEREC system consisted of image capture, prosthesis design and milling processes, is developed by SIEMENS Ltd. The image is captured by a CCD camera, and prosthesis is produced by a 3-axis milling machine (Calamia, 2000; Tinschert, Zwez, Marx & Anusavice, 2000).

7.1   *Physical model generation by CNC machining*

When the 3D CAD model is reconstructed and modified, the tool-path can be generated and transmitted into 4-axis CNC milling machine to produce the prosthesis. There are two different ways to generating the tool-path, one is for the above three-unit denture and another is for single-unit prostheses. Four tool-paths are produced to mill prosthesis above three units since the bending force is too large to stand cutting, while one integrated tool-path is directly applied in machining a crown.

The digital interface of CNC machine is developed in-house using visual basic (VB) language in PowerMILL (Delcam International, Birmingham, UK)

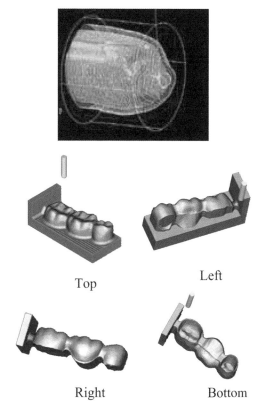

Figure 7.   One integrated tool path of one tooth, four different tool paths of three teeth.

Figure 8.   Bridge machining by 4-axis CNC milling machine.

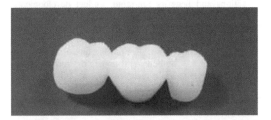

Figure 9.   Rapid prototyping of teeth model by Actua 2100 (3D System Inc.).

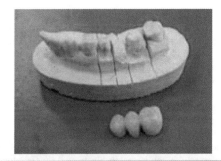

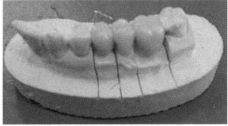

Figure 10.   Artificial teeth fabricated by means of ACT techniques.

environment. The macro commands are specifically assembled for machining denture. Input dialog box of interface, included the type of tool, material size and brand of controller, are indicated in Figure 6.

There are four different tool-paths to mill the three-unit prosthesis as shown in Figure 7. The real machining condition is shown in Figure 8.

7.2 *Generation of rapid prototyping physical model*

Once the digital denture data was created into a stereolithography (STL) format, the data were then read by a rapid prototyping (RP) machine to produce stereolithographic denture models. The RP is a new layer by layer material adding manufacturing process. It can rapidly produce physical products without any mold. The different kinds of rapid prototyping machine (RPM) were developed by using different mechanisms or materials, such as stereolithography apparatus (SLA), laminated object manufacturing (LOM), selective laser sintering (SLS), fused deposition modeling (FDM), and Multi-jet modeling (MJM) (Onuh & Yusuf, 1999) Rapid prototyping casting patterns for implant crowns are used as patterns to cast the parts in gold or dental alloy (Schmitt, 2001).

Figure 9 shows the denture physical model fabricated by rapid prototyping machine. After that, the wax pattern of teeth is produced using Actua 2100 system (3D system Inc., USA), and then used for casting metal prosthesis. The final custom denture was produced as shown in Figure 10.

8   CONCLUSIONS

In this article, abrasive computer tomography method using in-house made ACT device is proposed to quickly capture that geometric characteristics of teeth. The advantages of this approach are:

a. ACT apparatus is a more convenient and economic way to obtain the images of CT for complicated objects, existed under-cut or inner structure.
b. All the geometric data is saved in the form of digital format, which can be easily manipulated during design and shape modification process. In this case, when subject lost his/her false teeth (denture), the digital data file can be quickly retrieved and a replacement can be produced without go through tedious conventional denture fabrication procedures.

c. The images from ACT can be quickly converted to STL format for fabricating rapid prototyping, or reconstructed to 3D CAD model for CNC machining through the scheme proposed here.
d. In medical applications, ACT is another approach to provide 3D CAD models for surgical rehearsal, and produce all kinds of prosthesis through the techniques of RP/RT.
e. Currently, fingerprints and eyeball characteristics are used for personal identification in case of accident. However, the finger and eyeball are soft tissues which can be easily destroyed by fire during accident. Therefore, the digital teeth information can be used for comparison for the forensics purpose of personal identification.

REFERENCES

Apholt, W., Bindl, A., Luthy, H. and Mormann, W. H., "Flexural strength of Cerec 2 machined and jointed InCeram-Alumina and InCeram-Zirconia bars", Dent Mater, 17(3), pp. 260–267, 2001.

Chang, C. C. and Chiang, H. W., "Reconstruction the CAD model of complex object by abrasive computed tomography", 2002 IEEE/ASME International Conference on Advanced Manufacturing Technologies and Education in the 21st Century, Taiwan, 2002.

Chang, C. C. and Chiang, H. W., "Three-Dimensional Image Reconstruction of Complex Objects by Abrasive Computed Tomography Apparatus", accepted by International Journal of Advanced Manufacturing Technology, 2002.

Chang, C. C., "Three-Dimensional Image Reconstruction of Complex Objects by Abrasive Computed Tomography Apparatus", Taiwan Patient NO. 164809.

Chua Chee Kai, Lim Chu-Sing and Leong Kah Fai, "Rapid Prototyping Principles and Applications in Manufacturing," World Scientific Pub. Co., November 2002.

Calamia, J. R., "Advances in computer-aided design and computer-aided manufacture technology", Current Opinion in Cosmetic Dentistry.

Pham, D. T. and Gault, R. S., "A comparison of rapid prototyping technologies," International Journal of Machine Tools & Manufacture, 38 pp. 1257–1287, 1998.

Duret, F., Blouin, J. L. and Duret, B., "CAD/CAM in dentistry", JAM Dent Assoc., 117, pp. 715–720, 1988.

Duret, F. et al., "Method of Making a Prosthesis, Especially a Dental Prosthesis", Technical report, US Patent Nr. 4742464, 1988.

Onuh, O, Y. Yusuf Y., "Rapid prototyping technology: applications and benefits for rapid product development," Journal of Intelligent Manufacturing, 10, pp. 301–311, 1999.

Rekow, D. E., "The Minnesota CAD/CAM System Denti-CAD", Technical report, U. of Minnesota, 1989.

Rekow, D. E., "Method and apparatus for modeling a dental prosthesis", Technical report, US patent Nr. 5273429, 1993.

Schmitt, S., "Rapid prototyping custom dental implants", Rapid Prototyping Report, February 2001.

Tinschert, J., Zwez, D., Marx, R. and Anusavice, K. J, "Structural reliability of alumina-, feldspar-, leucite-, mica- and zirconia-based ceramics", J Dent, 28(7), pp. 529–535, 2000.

Varady, T., Martin, R. R. and Cox, J., "Reverse engineering of geometric models-an introduction", Computer-Aided Design, 29(4), pp. 255–268, 1997.

Virtual modeling and rapid manufacturing – Bártolo (eds)
© *2005 Taylor & Francis Group, London, ISBN 0 415 39062 1*

The use of a rapid prototyping biomodel with vascular anatomy in the treatment of temporomandibular joint ankylosis – A case report

Eduardo Meurer[1], Maria Inês Meurer[2], Grasiela Paiano[1], Aira Bonfim Santos[3],
Marília Gerhardt de Oliveira[4], Jorge Vicente Lopes da Silva[5] & Ailton Santa Bárbara[5]

[1]*University of South of Santa Catarina – UNISUL, Tubarão, Brazil*
[2]*Federal University of Santa Catarina – Florianópolis, Brazil*
[3]*Florianópolis Hospital – Florianópolis, Brazil*
[4]*Pontific Catholic University of Rio Grande do Sul – Department of Dentistry, Porto Alegre, Brazil*
[5]*Renato Archer Research Center – CenPRA Campinas, Brazil*

ABSTRACT: The treatment of patients with temporomandibular joint ankylosis is a challenge to the surgical team considering its high complexity and poor success rate. Regarding the recent developments of the rapid prototyping and image diagnosis technologies it was built a biomodel representing osseous and vascular anatomy to help the patient's treatment. To this end an exam generated by computerized tomography with an intravenous contrast agent was obtained from an eight years old girl with temporomandibular joint ankylosis. The images were dealt by a biomedical CAD-CAM software (*InVesalius* – developed at CenPRA). Based on the STL files that were generated, one biomodel was manufactured using the SLS (Selective Laser Sintering) technology. This model was used for planning the treatment of this patient. In this particular case, where complex and customized surgical procedures are necessary, the biomodel with vascular anatomy facilitated the surgery and improved the quality of results with a decrease of the vascular risks concerning to the maxillary artery. New studies have to be carried out in order to evaluate the use of these biomodels in specific surgical procedures like surgeries for temporomandibular joint ankylosis. Also it is mandatory a study to evaluate its potential reduction in costs for the health system.

1   INTRODUCTION

Ankylosis is the development of significant or complete limitation of movement of the temporomandibular joint by bone or fibrous tissue. Temporomandibular joint (TMJ) ankylosis in children disturbs not only mandibular growth, but also facial skeletal development. The resultant deformity will be severe if the ankylosis occurs before 5 years of age. The mandibular ramus and body on the ankylosed side are underdeveloped (Williams, 1994, Singh, Bartlett, 2005). It may be unilateral or bilateral. Ankylosis in a young child produces a failure of mandibular development with deficiency in ramus height leading to a downward inclination of the mandible. This alteration is associated with an anterior open bite, sometimes an incompetent lip action, obligatory tongue thrusting in order to achieve an anterior oral seal and proclination of the incisor teeth from unopposed tongue pressure. Many times this combination of deformities produces a typical "bird-face" deformity (Williams,

1994, Singh, Bartlett, 2005). Oral hygiene will be a variable feast with evidence in some neglected cases of gross caries and periodontal disease with the restriction of the mouth opening by ankylosis (Williams, 1994).

Diagnosis and treatment planning for patients presenting with symptoms of TMJ ankylosis requires clinical examinations and image exams. A surgical treatment shall be considered. Treatment to ankylosis should be instituted as early as possible with the objective of establish joint movement and jaw function, to prevent relapse and, if it is possible, to achieve normal growth and occlusion. Interceptive surgery during childhood is of great importance and it can be countered by freeing the ankylosis early in combination with reconstruction to minimize or prevent secondary deformities (Puricelli, 1997, Singh, Bartlett, 2005).

The surgical management of the TMJ ankylosis is undoubtedly a challenge for the oral and maxillofacial surgeon, being the procedure that involves the greatest number and gravity of complications in this specialty.

Surgical access is difficult because there is a risk of damage to branches of the facial nerve. Others difficulties and risks of this surgery are resulting of the altered anatomy of the region and its proximity of noble craniofacial structures with potential chance of complications (Schmelzeisen *et al.*, 2002). The nobel structures of this area includes the mandibular nerve, the internal jugular vein, the additive conduit, the internal carotid artery and, mainly, the maxillary artery and its ramifications: medium meningeal artery, posterior tympanic artery and the deep temporal artery (Meurer *et al.*, 2003).

The maxillary artery finds-itself below the curvature of the neck of the head of the mandible, normally is located in the level of the sigmoid notch or in a more superior plan and should be protected during the procedures that present a high risk of arterial wound. To hemorrhage of this artery represents an important complication and in some cases is difficult to control, being necessary carry out the ligature of the carotid artery (Schmelzeisen *et al.*, 2002).

The better way to avoid these complications is an adequate knowledge of the anatomy of this region, which finds-itself altered due to the ankylotic process.

The surgical planning with aid of the RP biomodels is very important for an adequate anatomical comprehension. With the RP biomodels the osteotomies can be planned and quantified, being transferred by a guide in the moment of the surgery, avoiding injuries of the noble anatomical structures and increasing the security of the procedure. The possibility of this kind of planning has a great potential for reduce the surgical risk, the time of surgery and of anesthesia, the risk of infection, as well as relapses because of providing better results. The diminution of the global cost of the treatment may be improved (Meurer *et al.*, 2003).

The concept of using a three-dimensional model in planning the operation has amazed maxillofacial surgeons since its first application in the 90s. It was followed by applications in the field of facial reconstructive surgery. To assist in diagnosis of facial fractures, joint ankylosis and even impacted teeth. The surgery can be simulated prior the operation of complex craniofacial syndromes, facial asymmetry, ankylosis and distraction osteogenesis (Cheung, Wong, Wong, 2002).

The rapid prototyping system, e.g. stereolithography, selective laser sintering and fused deposition model, is a new technology that allows three-dimensional (3D) imaging data to be used in the manufacture of accurate replicas of anatomical structures to construct a biomodel. The data of a specific part of human body can be manipulated as three-dimensional data. It is of enormous value for diagnosis, operative planning, surgical simulation, instruction for less experienced surgeons, educational teaching and as a patient information tool when obtaining the consent for surgery.

This data set can be obtained from any medical imaging modality (Silva *et al.*, 2003).

We report a case of an eight years old girl with temporomandibular joint ankylosis where biomodel representing osseous and vascular anatomy was used to help in the surgery treatment.

2 CASE REPORT

An 8-years old girl was referred to the Florianópolis Hospital (Florianópolis/SC Brazil) in dez/2002 with history of diminution of the mandibular mobility increasing along the time and presenting at this moment a significant limitation of the temporomandibular joint movements, having a mandibular hipoplasia, the restriction of the mouth opening by ankylosis, just 9 mm, impossibility of having a normal life and normal meal and consequently a bad oral hygiene. A history of trauma was reported at age of 2 years. Past medical history was unremarkable.

As surgery was really mandatory, with the developments of the rapid prototyping and image diagnosis technologies it was decided to build a biomodel representing osseous and vascular anatomy to help in the treatment of this patients.

Acquisition of the axial images was done using a spiral CT scanner (GE HighSpeed, General Eletric) with intravenous contrast agent and no gantry/tilt. The dataset acquisitions followed a basic protocol considering the necessary quality of the images for RP purpose and the dose of radiation. The slice thickness used was 2 mm, the FOV – Field of View – with resolution of 512×512 and 23 cm yield a good dataset. High contrast and none filters were used.

The image processing was accomplished by the *InVesalius*, a software developed at CenPRA, to import and process data from medical scanners in DICOM. The segmentation threshold used was 110 – 450. Based on this CAD study an STL file was created and reproduced in a physical model using a Sinterstation 2000 (DTM) [Figs 1 and 2].

The surgical planning in the biomodel consisted mainly in the evaluation of the specific bone and vascular anatomy of the region. The surgery was simulated on the biomodel to determine the angulations, the extension and depth. of osteotomies.

The biomodel was not sterilised but the information's (data base) were transferred for the surgical time with an acrylic template that was built-based on the data base of the SLS informations which permitted a rapid and safe gap osteotomy (Fig. 3).

After the bone ankylosis was complete released the articulation was reconstruct with a biconvex arthroplastia (Puricelli, 1997).

The six month follow-up showed a satisfactory function and a better mandibular growing (Fig. 4).

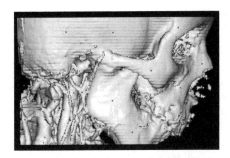

Figure 1. View of the 3D model of the patient skull rendered with the *InVesalius* software.

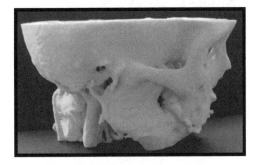

Figure 2. View of the SLS model represent bone and vessels anatomy.

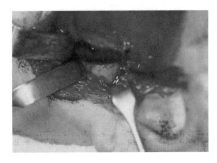

Figure 3. The osteotomy angulations, extension and depth determined in the biomodel being done in a safe and fast surgical maneuver.

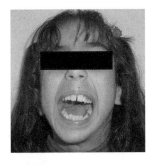

Figure 4. Post-op six months.

3 DISCUSSION

There is fairly general agreement that early release of an ankylosed joint is desirable. However, not all workers agree that transplantation of growth centers such as costochondral grafts is necessary. Mandibular growth rate increases after surgical release of the ankylosis, however it is not possible to make up the total growth deficit (Williams, 1994).

It is considered that computerized tomography scanning gives a first class image of the disorganized joint and with the advent of three-dimensional reconstruction both medial and lateral aspects are clearly shown. A number of operations have been carried out in countries where radiographic assessment has been limited to plain films (Williams, 1994).

Biomodels replicated the anatomical structures and corresponded to the intraoperative findings. Advances in imaging resolution and post processing methods helped overcome the initial limitations of the image threshold. The major advantage of this technology is that the surgeon can closely study complex vascular anatomy from any perspective by using the biomodel, which can be held, allowing simulation of intraoperative situations and anticipation of surgical maneuvers (Meurer *et al.*, 2003). In this particular case, where complex and customized surgical procedure are necessary, the biomodel with vascular anatomy facilitated the surgery and improved the quality of results with a decrease of the vascular risks concerning the maxillary artery.

One drawback of SLS biomodeling is the time it takes for the model to be manufactured and delivered. This biomodels demonstrates the feasibility and clinical utility of this new visualization method for ankylosis surgery. This helps the intuitive imagination of the surgeon and can be effectively added to conventional imaging techniques (Wurm *et al.*, 2004).

In order to enhance the surgical planning the virtual and physical biomodel are very useful. They can improve the communication between surgical team members, radiologist, doctors and the patient (Meurer *et al.*, 2003).

This biomodels can be used as an accurate anatomic guide, increasing safety, indicating the osteotomy sites for the surgery, minimizing blood loss, help to fitting miniplates, increase the understanding of the procedure and promote better results. All these advantages permit reduce patient's discomfort, the number of reconstructive surgeries, improving the speed of patient recovery and by the end save costs for the national health system (Meurer *et al.*, 2003).

REFERENCES

Cheung, L.K., Wong, M.C., Wong, L.L., 2002. Refinement of facial reconstructive surgery by stereo-model planning, *Ann Australas Coll Dent Surg*. Oct; 16: 129–32.

Meurer, E. *et al.*, 2003. Rapid Prototyping Biomodels in Oral and Maxillofacial Surgery. *Rev. Brasileira Cirurgia e Periodontia,* v.1, n.3, 172–80.

Perry, M. *et al.*, 1998. The use of computer-generated three-dimensional models in orbital reconstruction, *British Journal of Oral and Maxillofacial Surgery*, Edinburgh, Aug 36 (4), 275–284.

Puricelli, E., 1997. Artroplastia biconvexa para tratamento da anquilose da articulação têmporo-mandibular. *Revista da Faculdade de Odontologia* UFRGS, Porto Alegre, v. 38, n. 1, p. 23–27.

Schmelzeisen, R., Gellrich, N., Schramm, A., Otten, J., 2002. Navigation-Guided resection of temporomandibular joint ankylosis promotes safety in skull base surgery. *Journal of Oral and Maxilofacial Surgeons*, 60, 1275–1283.

Silva, J. V. L., Gouvêia, M. F., Santa Bárbara, A., Meurer, E., Zavaglia, C.A.C., 2003. Rapid Prototyping Applications in the Treatment of Craniomaxillofacial Deformities – Utilization of Bioceramics, *Key Engineering Materials.*, v.245–256, pp 687–690. Trans Tech Publications. Switzerland.

Silva, J.V.L. *et al.*, Rapid Prototyping: Concepts, Applications and Potential Utilization in Brazil, 15th International Conference in CAD/CAM Robotics and Factories of the Future, Águas de Lindóia, Sp, Brazil, 1999.

Singh, D.J., Bartlett, S.P., 2005. Congenital mandibular hypoplasia: analysis and classification. *J Craniofac Surg.*, Mar; 16(2): 291–300.

Williams, J.Ll. *Rowe and William's Maxilofacial Injuries* v.1, p 387–455, Churchill livingstone, NY, 1994.

Wurm *et al.*, 2004. Cerebrovascular stereolithographic bio-modeling for aneurysm surgery, *Neurosurg.*, Jan; 100(1): 139–45.

Virtual modeling and rapid manufacturing – Bártolo (eds)
© 2005 Taylor & Francis Group, London, ISBN 0 415 39062 1

Development of a finite element model to predict stress and strain fields in real bones

C. Pereira
Dep.º Eng.ª Mecânica, I.S.E.C., Coimbra, Portugal

F. Ventura
Dep.º Eng.ª Mecânica, F.C.T.U.C., Coimbra, Portugal

A. Mateus
Dep.º Eng.ª Mecânica, E.S.T.G./I.P.L., Leiria, Portugal

M.C. Gaspar
Dep.º Eng.ª Industrial, E.S.T./I.P.C.B., Castelo Branco, Portugal

ABSTRACT: Finite element models have been widely employed in an effort to quantify the stress and strain distribution around implanted prostheses and to explore the influence of these distributions on their long-term stability. In order to provide meaningful predictions, such models must contain an appropriate reflection of mechanical properties. The aim of this paper is to develop a finite element model to predict stress and strain fields in bones. Considering the different properties that can be found in the literature, a sensitivity study is fundamental. Isotropic and heterogeneous properties were considered in bone modeling. A conventional flatbed scanner was used to obtain the model's geometry by the 2-D digitalization of sections of real bone. The resulting nurbs were used in CAD and finite element software to generate surfaces of the studied model. To validate the resulting numerical model, an experimental setup was defined. Bones were loaded with different weights and the displacements were measured. The hardness of bones was measured locally at different cross-sections and the density of each bone slice was evaluated. The experimentally acquired data are important due to the variability of the material properties of real bones, and can be used to improve the numerical model.

1 INTRODUCTION

Some of the most important of bone's various functions in the body are structural in nature: protecting vulnerable body parts, supporting the body, and providing muscle anchorages. Knowledge of the mechanical and adaptive properties of bone is useful in the design and use of prostheses which replace a bone or a portion of a bone. Mechanical properties of bone are also of interest in trauma biomechanics and in efforts to prevent injury to the body (Lakes R. S. 1988).

As bones have a complicated geometry, and are composed of nonhomogeneous material, one of the main difficulties is the precise modeling of this geometry, simulating the bone's mechanical properties (Muller-Karger et al. 2004). Nonetheless, accurate mathematical or finite element models can be used to predict mechanical failure when designing customized orthopaedic implants (Knopf et al. 2001) or

their pre-clinical testing (Stolk et al. 2002). Many pathologies (Viceconti et al. 2004, Rietbergen et al. 2002 and Kopperdahl et al. 2002) or simply ageing (Testi et al. 1999) can also make use of these models.

Bones are constituted by cancellous regions at the extremities and a cortical region in their centre, with quite distinct properties. Therefore, it is necessary for finite element modeling to distinguish the two types of material areas. Different parts must also be considered in the computed model of bone to allow an accurate modeling of its global mechanical behavior. In fact, it is accepted that bone has an anisotropic behavior (Dempster and Liddicoat 1952; Bargren et al. 1974; Martens et al. 1983; Goulet et al. 1994), and so any attempt at modelling with isotropic constitutive properties is an approximation to the complexities of the real material (Clayton et al. 2003). The numerical problems of anisotropic simulation of bone modeling (Beaupre et al. 1990; Jacobs et al. 1997) arise from the lack of a comprehensive data bank incorporating the material properties of bone as a function of the orthotropic load directions (Dieter *et al.* 2000).

In this study, a fresh animal femur was cut in parallel slices to assess localised mechanical properties and to perform the scanning of its cross-sections. The resulting nurbs were used in CAD and finite element software to generate surfaces both of the cut bone slices and for the global model. To validate the resulting numerical model, an experimental setup was defined and bone was loaded with different weights to measure the corresponding displacements. Hardness and density were assessed in the different bone slices.

2 BONE LOADING

2.1 *Bone characterization*

In this work, a fresh animal femur taken from a pig's leg (Fig. 1) was studied. Soft tissues were removed and the bone's mechanical properties were assessed according to the following procedures, only a few hours after death.

The bone was 194 mm long, with a maximum transversal dimension of approximately 67 mm at each extremity, and weighed about 304 g.

2.2 *Bone loading*

In order to validate the numerical predictions an experimental setup (Fig. 2) was defined to measure the transversal displacements resulting from several vertical loadings.

The experimental setup was used to perform 3-point bending tests on the studied bone. Although this is not a usual physiological loading type, it has been found to be suitable for easy numerical reproduction.

A hydraulic press was used to constrain the vertical displacement of one end of the bone, while different weights were applied to the other end to promote its bending over a cylindrical bar, positioned near its centre. The vertical displacements were measured using an analog measuring device.

3 BONE CUTTING

3.1 *Bone cutting apparatus*

A hand-cutting apparatus was used in order to section the bone in parallel slices, all 5 mm thick. A major concern in selecting this hand-cutting process was to minimize contamination and local heating of the bone section surface during cutting. Figure 3 shows the bone sections obtained.

3.2 *Bone slices scanning*

A conventional flatbed scanner was used to obtain the 2-D digitalization of the bone's cut sections. The 2-D image files were obtained with the same angular orientation and a fixed reference, to allow their subsequent 3-D alignment. The planes are perpendicular to the bone's main axis, as illustrated in Figure 4. Each 2-D image file was converted into nurbs to permit 3-D modelling of the bone. One of the main advantages of this 2-D image scanning process is that it provides views of local inner-structure regions, like cancellous or cortical regions, as well as the bone's inner geometry.

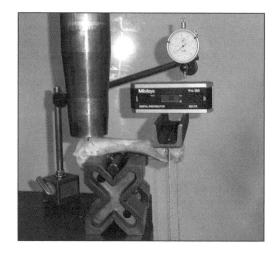

Figure 2.   Bone loading experimental setup.

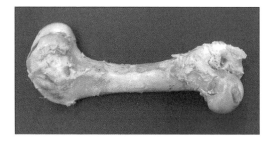

Figure 1.   Studied animal femur.

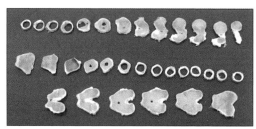

Figure 3.   Bone cut sections.

4 3-D BONE MODELING

4.1 *Part modeling*

Based on the previously obtained nurbs, the bone was built in conventionally available 3-D modeling software. Surfaces were created based on the equally spaced nurbs and subsequently converted into solid 3-D bone slices. Each thickness removed during the cutting process was included in the global model, shown in Figure 5.

4.2 *Assembly modeling*

To carry out a sensitivity study, considering heterogeneous properties along the bone, the global 3-D model (Fig. 5) was obtained by assembling each of the previously defined bone slices. Therefore, each bone slice can have a separate set of mechanical properties to improve the accuracy of the numerical model.

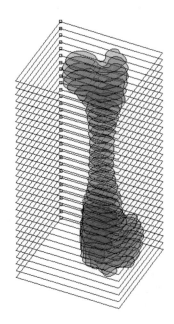

Figure 4. Spatial positioning of 2-D bone sections.

Figure 5. Complete assembly of the bone's digital model.

5 ACQUISITION OF MECHANICAL PROPERTIES

5.1 *Surface finishing*

The surfaces of the specimens were prepared according to the following procedure. Grinding was performed using 500, 800 and 1000 grit water-cooled silicon carbide papers to remove the scars produced by cutting and from each previous grinding step. After grinding, specimens were rough polished using 6 and 3 μm diamond paste. Wheel speeds of approximately 150 rpm at a moderate pressure were used.

5.2 *Vibration*

A preliminary study was developed to determine resonant vibration modes and resonant frequencies. The material was assumed to be isotropic with: E = 18 GPa, $\nu = 0.3$, $\rho = 1.5\,\text{g/cm}^3$.

Table 1 presents the resonant frequencies for free vibration conditions. In general the values are quite high, mainly due to the reduced dimensions of each bone slice.

Considering only a half slice, the resonant frequencies increase and the first 3 modes are distinct. Reducing radial thickness produces a significant reduction of resonant frequencies. Figure 6 presents the first two resonant vibration modes of a cortical half slice.

Table 1. Resonant vibration frequencies.

Mode	Frequency [Hz]
1	13600
2	14530
3	17970
($\rho = 1.5\,\text{g/cm}^3$)	
1	14550
2	21700
3	31380
($\rho = 1.5\,\text{g/cm}^3$)	
1	12930
2	19280
3	27880
($\rho = 1.9\,\text{g/cm}^3$)	
1	6857
2	14360
3	17020
($\rho = 1.9\,\text{g/cm}^3$)	

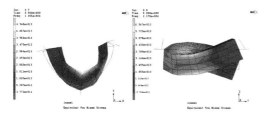

Figure 6.   Resonant vibration modes.

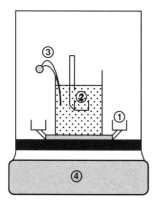

Figure 7.   Schematic representation of a *Mettler Toledo AG*.

The analysis of a slice instead of the whole body is interesting since the density and material properties vary along the bone. Numerical prediction of natural frequencies for the whole body is a complex task and is not recommended, as it induces significant errors.

5.3 *Density*

The density of each bone section (mass per unit volume of a material) was measured using a *Mettler Toledo AG 204* device, schematically presented in Figure 7. The bone specimens were first washed in deionized water and warm-air dried. They were then covered with a synthetic layer to keep them impermeable and prevent porous water absorption.

Specimens were weighed in air on the *Mettler* Balance's plate (1) and their mass value − m_1 (4) – was registered. These specimens were weighed immersed in distilled water at 24°C using a sinker and a wire to keep the specimens completely submerged (2). The apparent mass value − m_2 (4) – was obtained and registered. The temperature of the distilled water was monitored during the process (3). Based on the water temperature, its corresponding density value ρ_{H_2O} – is given by the ASTM D792 (ISO 1183) test norm. The density of each bone slice was determined using following expression:

$$\rho_{bone} = \left(\frac{m_1}{m_1 - m_2} \right) \times \rho_{H_2O} \qquad (1)$$

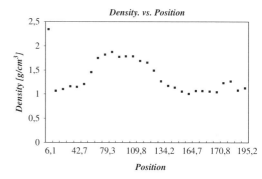

Figure 8.   Density variation along the bone length.

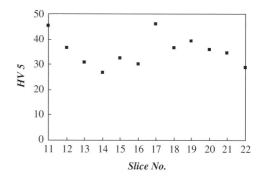

Figure 9.   Hardness variation along the cortical bone.

Figure 8 shows the results obtained for the bone slices. For the cancellous bone region the density value is approximately 1 g/cm³ (6–67 mm). For the cortical bone region, the mean value is 1.8 g/cm³. These values are in keeping with literature reference values (Carter *et al.* 1976). The first value corresponds to a slice that presents predominantly cartilaginous tissues; hence it has a higher density. The same reason explains the values for the 176.9 and 183 mm sections.

5.4 *Hardness*

Vickers hardness values were measured on different slices from the cortical region alone. The objective of these tests was the qualitative assessment of the variation in the material properties. Figure 9 presents the hardness results obtained.

Significant oscillation can be observed, which can indicate important variations in mechanical behavior.

However, further work is required for a better understanding of hardness variation. With respect to cancellous bone, a value of 19 HV 5 was obtained.

6 FINITE ELEMENT MODELING

6.1 CAD based meshing

A numerical model was developed from the CAD representation of the bone. The geometry was automatically divided into linear tetrahedral elements, which are suitable for approximating highly non-regular objects like bones. The total number of elements and nodes was 17690 and 27975, respectively. Figure 10 presents the finite element mesh and the loading, which is meant to reproduce the real loading that can be seen in Figure 2.

Different slices, with distinct elastic properties, were considered, as described above. In each slice the material was assumed to be homogeneous, isotropic with linear elastic behavior. The elastic properties values considered were based on (Wirtz et al. 2000): $E = 18\,GPa$, $\nu = 0.3$ for cortical bone and $E = 0.5\,GPa$, $\nu = 0.12$ for cancellous bone. In a second run, the Young's modulus of cortical bone was assumed to be 18 GPa only for the two central layers, and a reduction of 0.2 GPa was considered from slice to slice up to the extremities.

Figure 10 shows the deformed shape and the displacement field for a load of 500 N considering a magnification of 15.8.

Figure 11 is a plot of force versus displacement. A linear variation is observed, as expected.

The highest rigidity was obtained for the CAD based model and the difference relative to the experimental results is explained by inaccurate loading condition positioning in the finite element analysis.

In the experimental results, the zero load displacement is different from zero and is positive. This is explained by the local deformation at the central constraint, which is a cylinder with relatively small radius. Noting that the displacement for zero load is 2.4 mm, the bone deformation at the central constraint is expected to have a depth of 1.4 mm. The numerical models do not consider this deformation.

6.2 Parametric meshing

A second approach involved building a numerical model with commercial finite element software directly from digitalized sections. Points were defined from the 2-D bone slice scans, which were used to define curves, surfaces and volumes. Meshing and boundary conditions were defined parametrically and therefore can easily be changed.

Figure 12-right shows one of the sections, comprising 56 points, 24 curves and 17 surfaces. This parametric meshing procedure facilitates the definition of distinct material properties in different bone regions. Brick isoparametric elements with 8 nodes were considered for meshing. The total number of

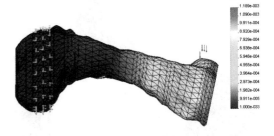

Figure 10. Boundary conditions and displacement field.

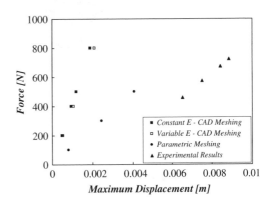

Figure 11. Numerical predictions of force versus displacement.

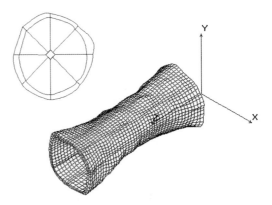

Figure 12. Geometry of each section (left). Cortical bone (right).

elements and nodes considered was 27909 and 34253, respectively. Figure 12–left illustrates the geometry and mesh of cortical bone.

Figure 11 shows a plot of force versus displacement for the obtained results. A linear variation is also observed in this case and the corresponding slope is

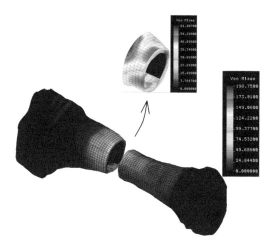

Figure 13.   Stress field.

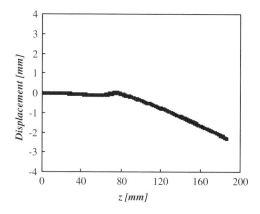

Figure 14.   Deformation along bone axis.

very similar to that given by the experimental data. Figure 13 gives the stress field. Bone was submitted to three-point bending, and therefore maximum stresses occur at the center of the bone and a long way from the neutral line. The detail shows that cancellous regions inside cortical bone have negligible stresses due to their lower rigidity. Therefore, this type of bone does not have a structural function. Finally, Figure 14 shows the displacement at different points along the bone axis.

The deformation between the two constrained points is quite small, indicating that they work as a rigid constraint.

7   FUTURE WORK

7.1   Cutting solutions

In order to improve the slices' parallelism, an alternative cutting procedure could be considered. Having different bone slice thicknesses would permit a more detailed study in specific bone regions and also a better geometry definition in the resulting model.

7.2   Loading solutions

To improve the accuracy of experimental results, a different fixing mechanism is required, since the setup used exhibited some instability during the tests.

7.3   Material properties

Hardness will be extensively used to identify cortical and cancellous regions, and to identify material changes within each region.

In future work, the elastic properties of cortical and cancellous bone will be determined with nanoindentation. Therefore, in each bone section, localized elastic modulus assessment will be possible. Although this technique requires perfectly parallel surfaces, it has been found to be suitable to determine rigidity in each bone slice. Results obtained with the resonant technique can be used for comparison with nanoindentation, although major improvements are still required.

8   CONCLUSIONS

Two different finite element models were developed to predict stress and strain fields in bones. In a first approach, a 3-D CAD based model was used to perform the analysis and, although a linear variation was observed, the results were significantly different from the experimental data due to some inaccurate positioning of constraints and loading conditions.

A second approach consisted of building a numerical model with commercial finite element software, directly from the scanned sections. As meshing and boundary conditions were parametrically defined, they were easily changed. The results obtained with this approach were found to agree better with the experimental data.

To validate the resulting numerical model, an experimental setup was defined. Bones were loaded with different weights and the displacements were measured.

The geometry of the bone model used in both finite elements approaches was obtained from a fresh animal femur. The bone was cut into slices of equal thickness and subsequently digitized in a flatbed scanner. These bone slices were used to obtain some localized mechanical properties, such as hardness and density. The slices' resonant vibration frequencies and modes were also evaluated.

REFERENCES

Bargren, J.H., Andrew, C., Bassett, L., Gjelsvik, A., 1974. Mechanical properties of hydrated cortical bone, *Journal of Biomechanics* 7: 239–245.

Beaupre, G.S., Orr, T.E., Carter, D.R., 1990. An approach for timedependent bone modeling and remodeling – theoretical development, Journal of Orthopaedic Research 8: 651–661.

Carter, D.R., Hayes, W.C., 1976. Bone compressive strength: the influence of density and strain rate, *Science* 194: 1174–1176.

Clayton Adam, Mark Pearcy, Peter McCombe, 2003. Stress analysis of interbody fusion-finite element modelling of intervertebral implant and vertebral body, *Clinical Biomechanics* 18: 265–272.

Dempster, W.T., Liddicoat, R.T., 1952. Compact bone as a non-isotropic material, *American Journal of Anatomy* 91: 331–362.

Dieter Christian Wirtz, Norbert Schilers, Thomas Pandorf, Klaus Radermacher, Dieter Weichert, Raimund Forst, 2000. Critical evaluation of known bone material properties to realize anisotropic FE-simulation of the proximal femur, *Journal of Biomechanics* 33: 1325–1330.

Goulet, R.W., Goldstein, S.A., Ciarelli, M.J., Kuhn, J.L., Brown, M.B., Feldkamp, L.A., 1994. The relationship between the structural and orthogonal compressive properties of trabecular bone, *Journal of Biomechanics* 27: 375–389.

Jacobs, C.R., Simo, J.C., BeaupreH, G.S., Carter, D.S., 1997. Adaptive bone remodeling incorporating simultaneous density and anisotropy considerations, *Journal of Biomechanics* 30: 603–613.

Knopf, G.K., Al-Naji, R., 2001. Adaptive reconstruction of bone geometry from serial cross-sections, *Artificial Intelligence in Engineering* 15: 227–239.

Kopperdahl, D.L., Morgan, E.F., Keaveny, T.M., 2002. Quantitative computed tomography estimates of the mechanical properties of human vertebral trabecular bone, *Journal of Orthopaedic Research* 20: 801–805.

Lakes, R.S., 1988. Properties of bone and teeth. *Encyclopedia of Medical Devices and Instrumentation*, ed. J.G. Webster, J. Wiley, NY, 501–512.

Martens, R., Van Audekercke, R., Delport, P., DeMeester, P., Mulier, J.C., 1983. The mechanical characteristics of cancellous bone at the upper femoral region, *Journal of Biomechanics* 16: 971–983.

Muller-Karger, C.M., Rank, E., Cerrolaza, M., 2004. p-version of the finite-element method for highly heterogeneous simulation of human bone, *Finite Elements in Analysis and Design* 40: 757–770.

Rietbergen, B. van, Majumdar, S., Newitt, D., MacDonald B., 2002. High-resolution MRI and micro-FE for the evaluation of changes in bone mechanical properties during longitudinal clinical trials: application to calcaneal bone in postmenopausal women after one year of idoxifene treatment, *Clinical Biomechanics* 17: 81–88.

Stolk, R., Verdonschot, N., Cristofelini, L., Toni, A., Huiskes, R., 2002. Finite Element and Experimental Models of Cemented Hip Joint Reconstructions Can Produce Similar Bone and Cement Strains in Pre-Clinical Tests, *Journal of Biomechanics* 35: 499–510.

Testi, D., Viceconti, M., Baruffaldi, F., Cappello, A., 1999. Risk of fracture in elderly patients: a new predictive index based on bone mineral density and finite element analysis, *Computer Methods and Programs in Biomedicine* 60: 23–33.

Viceconti, M., Davinelli, M., Taddei, F., Cappello, A., 2004. Automatic generation of accurate subject-specific bone finite element models to be used in clinical studies, *Journal of Biomechanics* 37(10): 1597–1605.

Wirtz, D.C., Schiffers, N., Pandorf, T., Radermacher, K., Weichert, D., Forst, R., 2000. Critical evaluation of known bone material properties to realize anisotropic FE- simulation of the proximal femur, *Journal of Biomechanics* 33: 1325–1330.

Virtual modeling and rapid manufacturing – Bártolo (eds)
© *2005 Taylor & Francis Group, London, ISBN 0 415 39062 1*

An approach of modeling and representation of heterogeneous objects for rapid prototyping

M.Y. Zhou

Department of Computer Science, Guilin University of Electronic Technology, Guangxi, P. R. China

ABSTRACT: An approach of modeling and representation of heterogeneous objects based on STEP is presented in this paper. To model a heterogeneous object, Boundary representation is used for geometry representation and a feature-based representation is used to represent material distribution. Using this modeling method, functions of the current CAD software can be employed for geometry modeling of heterogeneous objects and the material distribution is modeled in the heterogeneous solid modeling system. STEP standard format is adopted to exchange geometry information between the CAD system and the heterogeneous solid modeling system. For material modeling, reference features and material distribution functions were incorporated together by attaching material distribution functions to reference entities as their properties.

1 INTRODUCTION

In more and more engineering applications, there can be special and conflicting functional material property requirements on an object. For example, the thermal deformation of satellite's paraboloid antenna (10 m in diameter) should be controlled within 0.2 mm in order to work well under the environment with large variations in temperature (-180 to 120 □). To fulfill it, its thermal expansion coefficient should be very small as to close to zero. Another example is that Poisson's ratios of sensors should be negative in order to increase their sensitivities to hydrostatic pressures. When Poisson's ratio of a sensor is changed from an ordinary value of 0.3 to -1, its sensitivity will be increased by almost 10 times. The third example is the cylinders of pressure vessels, which are subjected to a high temperature/ pressure on the inside while the outside is subjected to ambient conditions. It is desirable to have ceramic on the inner surface due to its good high temperature properties while have metal on the outer surface for its good mechanical properties. Often homogeneous objects cannot meet these requirements satisfactorily. In such cases, a multi-material object is suitable where the object is made of different materials, each material catering to a different property requirement. These objects usually have a sharp interface between dissimilar materials. Stress concentration may develop at these interfaces which eventually separate the materials creating a crack in between. Therefore, it is desirable to eliminate the sharp interfaces and obtain a smooth variation among different materials.

Functionally graded/gradient material [1], also called heterogeneous material, has certain gradation in material constituent composition from one side to another side and can join two different materials on two sides without stress concentration at their interface to help reducing thermal stress, preventing peeling off of coated layer, preventing micro-crack propagation etc. Functionally graded material has become a subject of research in the material science, composites, ceramic engineering and metallurgy communities. Functionally graded material objects, also called heterogeneous objects, can exhibit continuously varying composition and/or microstructure thus producing gradation in their properties [2, 3].

Layered Manufacturing (LM), also called Rapid Prototyping (RP), automatically generates physical objects directly from 3D CAD data by depositing material layer by layer [4]. Due to its additive nature, LM is well suited to the fabrication of heterogeneous objects. It can deposit several materials in varying compositions within a layer and with variations between adjacent layers. For the purpose of design, representation, analysis and fabrication of heterogeneous objects, a CAD model of the object is required which contains not only the geometry information but also the information on material, property, etc. at each point of the object. As mentioned above, Material compositions vary continuously throughout the geometric domain of a heterogeneous object to satisfy the conflicting functional property requirements of an object. Therefore, besides geometry, the material variations also need to be designed so that the object can satisfy all the functional requirements. However, currently available commercial CAD systems do not have the capability to design heterogeneous objects. The design of heterogeneous objects involves representing the material variations inside the object and obtaining heterogeneous solid model. It should represent the continuous material variation with relation to the object's geometric features.

In this paper we propose a new approach based on STEP (ISO10303) for modeling and representation of heterogeneous objects for layered manufacturing. Heterogeneous objects with complex geometry shape (e.g. NURBS surfaces) can also be modeled conveniently using this method because the geometry of the object is modeled in commercial CAD systems by using its powerful geometry modeling ability.

The rest of the paper is organized as follows. Section 2 reviews the literature. Section 3 presents the representation of heterogeneous objects. Section 4 focuses on the modeling of heterogeneous objects and the data structure of heterogeneous solid model. Validation of the proposed approach for modeling of heterogeneous objects and discussion are presented in Section 5. The paper concludes with some observations and issues for future research in Section 6.

2   LITERATURE REVIEW

In the context of heterogeneous objects, modeling is the process of establishing and storing the geometry and material composition information in a computer. In the literature, several modeling techniques have been proposed.

Kumar and Dutta [5] introduced an approach for modeling heterogeneous objects by using r-sets extended to include composition r_m-sets with accompanying Boolean operators. An r_m-object is defined as a finite collection of these r_m-sets with each r_m-set being a material domain with an analytic material function. Shin and Dutta [6] extended the r_m-object approach into constructive methods of heterogeneous object representation. Jackson et al. [7] exploited an approach to produce heterogeneous objects based on sub-dividing a model into sub-regions (tetrahedrons). For each region, an analytic composition blending function is assigned to define the material composition variation. Pegna and Safi [8] proposed a model by representing multi-material models as point sets including Cartesian coordinates plus material composition. Decomposition modeling, however, is cumbersome during design because it does not maintain topological information about the model and is not as memory efficient as the more commonly used B-rep methods. Patil et al. [9, 10] proposed an information model to represent heterogeneous objects using the information modeling methodology developed for STEP (ISO10303). Siu and Tan [11] proposed a 'source-based' heterogeneous solid

modeling scheme and defined extended operations (e.g. insertion, merge, immersion) in addition to the CSG type Boolean operations to model the grading sources. Qian and Dutta [12] have presented a heterogeneous object design method that uses B-spline volumes to model the object geometry. The material variation is represented by a physical process called diffusion. In the paper, the material properties are specified at the object control points, assuming that the material variation would always be represented as B-spline functions of geometry. Chen and Feng [13] provide an approach based on axiomatic design principles. The method creates a 3D variational geometric model of the component before it is divided into several 'regions' using commercial FEA software. Material composition in each region is assumed to be constant. Optimal material constituent composition are selected using optimization techniques such as sensitivity analysis and steepest descend method. Since the material attributes are not expressed as a function of geometry, the material composition must be stored explicitly in every region. Liu et al. [14] has proposed a parametric feature-based methodology for the design of solids with LCC (Local Composition Control). Material composition features are related to the geometry of the designed object which allows changing the geometry and material composition simultaneously until a satisfactory result is obtained. The Euclidean digital distance transform and Boundary Element Method are used to calculate the material composition. Apart from developing the modeling schemes, some researchers also studied and developed a number of LM processes for the fabrication of heterogeneous objects, such as 3DP [7], shape deposition manufacturing (SDM) [15], multi material selective laser sintering (M2SLS) [16], laser engineered net shaping (LENS) [17]. However, the current LM processes for manufacturing heterogeneous objects are still in research and new commercial LM processes have not been emerged yet.

The above heterogeneous object representation schemes focus on a modeling combination of material information and geometry information of heterogeneous objects while the proposed method described in this paper emphasize on the memory efficiency of the heterogeneous solid model and the ease of modification of the material grading distribution inside the heterogeneous objects. Only the material distribution functions, which represent the material distribution information, are stored in memory. Therefore, much memory space can be saved for storage of material information.

3   EXTENDED B-REP SOLID MODEL FOR HETEROGENEOUS OBJECTS

A B-Rep solid model of an object is composed of geometry entities, such as points, curves and surfaces, which represent the boundary of the object. For a homogeneous object, the model of the object is intact as soon as its boundary (geometric shape) is defined. For a heterogeneous object, however, its model is half-baked if only its boundary (geometric shape) is defined. The material composition at each point inside the heterogeneous object must be also defined. In our approach, some material related entities, which make up an extended B-Rep solid model of heterogeneous objects in company with geometry entities, are created to represent material information.

In current solid modeling schemes, a bounded region in E^3 is defined as denoting the geometry of an object. Traditional CAD modeling systems only describe geometry information of the object. Heterogeneous solid modeling must also describe material composition information apart from geometry information.

A heterogeneous object is composed of two or more than two kind of elementary material compositions where the volume fraction of the materials varies continuously. Assuming that the number of the elementary material compositions is n, which is finite. The material value at a point p inside a heterogeneous object S is a blend of these elementary material compositions and can be described as the volume fraction of the elementary materials. The n dimensional space $\mathbf{M^n}$ is used to model the material information of heterogeneous objects, where each dimension represents one elementary material composition. The material composition at point p inside a heterogeneous object can be represented by a point in material space $\mathbf{M^n}$. Considering that the sum of volume fractions of all the elementary material compositions, the set V composed of the material points of a heterogeneous solid model is defined as a subset of $\mathbf{M^n}$, where v_i represents the volume fraction of ith elementary material composition.

$$V = \left\{ v \in M^n \mid \|v\|_1 = \sum_{i=1}^{n} v_i = 1 \quad and \quad v_i \geq 0 \right\} \tag{1}$$

For a point v in material space, it represents a significative material point on condition that $v \in V \subset \mathbf{M^n}$.

The space to model heterogeneous objects is a product space $\mathbf{T} = \mathbf{E^3} \times \mathbf{M^n}$ where $\mathbf{E^3}$ is 3 dimensional geometry space and $\mathbf{M^n}$ is n dimensional material space. Each point inside the heterogeneous object could be modeled as a point $(x\square\mathbf{E^3}, v\square V)$ in $\mathbf{T}$, where x and v represent the geometric and material points respectively.

By applying the concept of material space, material composition information of heterogeneous objects can be represented and incorporated with geometry information through some kind of mapping between geometry space and material space. For each point p inside object S, the relationship between material point and

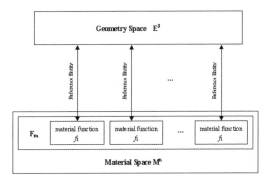

Figure 1. Mapping of the material information from material space to geometry space.

geometry point is described by a mapping from geometry space $\mathbf{E^3}$ to material space $\mathbf{M^n}$.

$$F : E^3 \rightarrow V \left\| \left\| F(x \in E^3) \right\|_1 = 1 \right.$$
$$F(x) = \{f_i(\underline{x})\} \equiv \underline{v}(\underline{x}) = \{v_i\} \tag{2}$$

Mapping F is also called material function which is a vector function. The material distribution function F can be used to define the material gradings inside the object S by putting appropriate restrictions. Figure 1 shows the mapping of the material information from material space to geometry space.

The material grading information is represented by material distribution functions. The material distribution function is defined in the entire Euclidean space E^3, but it is only valid within the boundary of the object. The material function is represented in terms of a distance function $f(d)$ where d represents the distance between query point inside the object and the reference feature. The distance function can be a linear or nonlinear function in real field. The form of material distribution function is determined by the requirement of graded properties of heterogeneous objects. Determining the form and parameters of material function is problem in design domain, which can be obtained designers using optimization designing method. The material distribution function is a vector function. The fraction $f_i(d)$ represents the material distribution function of the ith elementary material. The value of material function must be a nonnegative real number which is less than 1 and all the fraction functions $f_i(d)$ must sum to 1.

$$\sum_{i=1}^{n} f_i(d) = 1 \quad 0 \leq f_i(d) \leq 1 \tag{3}$$

where n is the number of elementary materials and $f_i(d)$ represents the material distribution function of the ith elementary material.

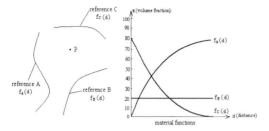

Figure 2. Different material gradings inside a heterogeneous object.

A material grading inside a heterogeneous object always relatives to a certain reference geometry feature of the object. A geometry feature can be a feature surface, curve or point. Different kinds of primary material can relative to different reference features. When there are more than one material grading, as shown in Figure 2, the material value at point p inside the object S can be evaluated by union operation:

$$m(\underline{p}) = f_1(d) \cup f_2(d) \cdots \cup f_n(d) = \frac{k_1 f_1(d) + k_2 f_2(d) \cdots + k_n f_n(d)}{k_1 + k_2 \cdots + k_n}$$

$$\sum_{i=1}^{n} k_i = 1$$

$$(4)$$

where $m(p)$ represents the material value at point p, n is the number of elementary materials and k_i represents the weight of ith material grading. The weights of material gradings are defined by designer during the designing process of heterogeneous objects according to the materials property requirements.

4 MODELING OF HETEROGENEOUS OBJECTS

Current implementations in the CAD/CAM industry do not easily permit manufacturing heterogeneous parts because the solid model does not represent any material information of the object. With the possibility of fabricating graded compositions through RP, methods to design and represent models need to be reconsidered and extended. An approach of modeling functionally graded material based on continuous geometry model is proposed in this paper. In this approach, geometry modeling of heterogeneous objects is done in commercial CAD systems and the STEP standard format is used to exchange and transfer geometric and topologic data between CAD systems and material modeling system.

4.1 Geometry modeling and STEP solid model

The first task of heterogeneous solid modeling is geometry modeling. This step generates the geometric data

that define the physical boundary of the model. In our approach, it is implemented in a commercial CAD system, such as Unigraphics, Pro/Engineer etc. The geometric data is exported from the CAD system to the heterogeneous solid modeling system as a standard exchange format such as STEP for the following reasons. Stereolithography (STL) file format, which has become the de facto standard in the RP community, has many disadvantages, such as loss of topologic information and large file size besides approximation of the boundary of the object [18]. Furthermore, STL file format cannot be used to represent material information. ISO10303 or STEP (Standard for the Exchange of Product model data) is an international standard that describes the product data completely during the life cycle of a product from the design to its manufacture [12]. Compared to STL file format, STEP provides complete and accurate description of the geometric and topologic information of a product using NURBS patches. STEP also can be used to describe material information of a product. Given the limitations of the STL file format, the logical choice for the data exchange heterogeneous solid models is STEP.

The basis of solid modeling within STEP is Part 42 of the integrated resources: geometric and topological representation. Part 203 is used for exchanging 3D design information of mechanical parts and assemblies. A CAD solid model is represented in manifold solid B-Rep scheme as a STEP file, which containing a precise mathematical description of all of the curves and surfaces as well as the connectivity between all of the topological entities. The STEP description provides a complete and accurate description of complex product. The detailed information in a neutral format allows the exchange of the product data to another system without loss of information.

After geometry modeling, the geometry data is exported from CAD system to material modeling system by the format of STEP file. The geometry information in STEP file must be extracted to construct geometry model before material modeling. EXPRESS language, which is a kind of object-oriented language, is used to describe product data in STEP. C++ language, which is also a kind of object-oriented language, is used to instance the entities in geometry schema of STEP. The process of extracting geometry information is composed of three steps: scanning the STEP file, checking file's structure and extracting geometry data inside the STEP file to reconstruct geometry model in computer.

4.2 Modeling of heterogeneous material

For material modeling, the incorporation of geometry information and material information is done through assigning the material composition grading information to a reference geometry feature. The reference feature could be an arbitrary geometry entity of the model, such as a point, curve, surface or the entire outer boundary of the heterogeneous object. The material composition grading is represented by a distance function $f(d)$. The material composition at any point inside the heterogeneous object can be evaluated through this distance function $f(d)$. The distance between the reference geometry feature and query point is used as a parameter d in the material distribution function. When the reference feature is the entire outer boundary of the object, the distance used as a parameter in the material distribution function refers to the minimum of all the distances between query point and all the surfaces of the outer boundary. By selecting a geometry entity as the reference and specifying a material distribution function $f(d)$, the designer accomplishes the design of material grading distribution of a primary material made of the heterogeneous object. The material composition distribution information can be incorporated with geometry information through storing the distribution function $f(d)$ as an attribute of the reference entity. Different compositions can be assigned to different references and can have different material distribution functions. By reselecting a reference geometry feature, the material distribution can be easily modified without changing the geometry of the object.

Because the material volume fraction should be positive and not greater than 1, the material distribution $f(d)$ should be truncated to 0 or 1 when it is less than 0 or greater than 1. That is, the material distribution $f(d)$ turns to the following forms:

$$f_i(d) = \begin{cases} f_i(d) & 0 \le f_i(d) \le 1 \\ 0 & f_i(d) < 0 \\ 1 & f_i(d) > 1 \end{cases} \quad (5)$$

where i = material index.

Figure 3 shows the process of material design. The heterogeneous object in Figure 3 is composed of two kinds of materials: A and B. The material A is on the outer surface and the metal B is inside the object away from the outer surface. The material of the object varies continuously from A to B. The volume fraction of material A turns from 1 to 0 along the composition grading direction from the outer surface to the center and the fraction of material B turns from 0 to 1. Followed are their distribution functions.

$$\begin{cases} f_A(d) = (1-d)^2 \\ f_B(d) = 1 - f_A(d) \end{cases} \quad (6)$$

where d represents the ratio of distance between query point to the reference.

For more than one material grading distribution, multi-references can be selected. Different materials

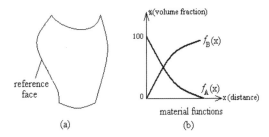

Figure 3. Material design of heterogeneous object.

can be assigned to the same reference or different references. If the same reference is selected, different materials can have different distribution functions so that they have different material grading distributions just like in Figure 3. When different materials are assigned to different references, they can have the same distribution function or different distribution functions. Whether single reference or multi-references are selected for material modeling, the formula (3) should be submitted.

When a primary material is assigned to more than one reference, it has different material distribution functions and it will have different material composition values at the same point inside the heterogeneous object. To deal with this kind of conflict, minimum distance principle is adopted. That is the reference which has minimum distance from the query point is used to evaluate the material composition at that point by its material distribution function.

4.3 Representation of heterogeneous solid model in STEP

The most commonly used representation schemes for solid model are Constructive Solid Geometry (CSG) and Boundary Representation (B-Rep). In STEP, solid model is represented in the manifold solid B-Rep scheme. The basis of the representation of heterogeneous solid model is also a manifold solid B-Rep scheme. Some material related entities are created to describe the material information of heterogeneous objects. The entire data structure representing a heterogeneous solid model is shown in Figure 4. The data structure shown on the left represents geometry information of heterogeneous objects. A material related data structure, which is shown on the right, is included to the original data structure to represent material information.

The entity MBODY represents a heterogeneous object. The entity SHELL represents the entire boundary of the object including an outer boundary and all inner boundaries. A SHELL entity is composed of one or more ADVANCED_FACE entities and an ADVANCED_FACE entity is represented by a surface

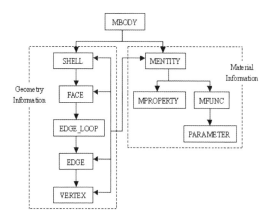

Figure 4. Computer representation of heterogeneous solid model.

entity that composes it and the boundaries (an outer boundary and/or one or more inner boundaries) of the surface, which are represented by EDGE_LOOP entities. EDGE_LOOP is a closed loop composed of one or more edges. An EDGE entity is represented by a curve and the two end points of the curve. End points and origin is represented by VERTEX entities. The entity MENTITY, which has material property and a material distribution function, represents material composition and distribution information inside the object. The MENTITY entity is always associated with a geometric reference entity. The geometric reference entity can be a SHELL, FACE, EDGE or VERTEX entity. When the reference feature entity is SHELL, it is assumed that the reference refers to the entire outer boundary of the heterogeneous object. The MFUNC entity, which represents the material distribution function, is a distance function. The material distribution function is used to evaluate the material composition at all points inside the object. The material distribution function can be all kinds of mathematical functions such as polynomial functions and exponential functions. The distance parameter in material function refers to the minimum distance form a query point to the reference entity. The entity MPROPERTY represents the type and properties of primary material composed of the heterogeneous object. The PARAMETER entity, which is a real number, stores the coefficient of the material distribution function.

As a developing international standard, STEP does not establish application protocols (APs) for some developing and immature application fields, such as LM. In our approach, some material related entities are created to represent material composition and grading distribution of heterogeneous solid model on the base of STEP Part 42 (geometry and topology representation) and Part 45 (material). The EXPRESS representation of some material related entities and representation

of a heterogeneous object in STEP are presented as following.

```
ENTITY MEntity;
    Reference: geometric_representation_item;
    Local_coordinate: axis2_placement_3D;
    Properties: LIST [1:?] OF MProperty;
    Material_functions:    LIST    [1:?]    OF
MFunction;
    END ENTITY;
    ENTITY MProperty;
    Name: STRING;
    Value: REAL;
END ENTITY;
    ENTITY MFunction;
    Function_type: STRING;
    Coefficients: LIST [1:?] OF Parameter;
    Exponents: LIST [1:?] OF Parameter;
END ENTITY;
    ENTITY Parameter;
    Value: REAL;
END ENTITY;
```

5 CASE STUDY

The example here is a cutting tool made of diamond/SiC/Polymer functionally graded materials, with S-pattern material distribution functions which can result in smooth thermal stress inside the object. The material on the edge axis of cutting tool is diamond which has high hardness and good cutting performance. The material on the body of the cutting tool is SiC/Polymer which has fine mechanical intensity. The materials of the cutting tool alternate from diamond to SiC/Polymer continuously from the edge to the body. The edge axis of cutting tool is selected as the reference geometry feature of materials distribution and materials composition variation is defined as functions of distance between the query points and the reference. The material distribution functions are assigned as:

$$\begin{cases} f_A(d) = \begin{cases} 2d^2 & 0 \le d \le 0.5 \\ 1 - 2(1-d)^2 & 0.5 < d \le 1 \end{cases} \\ f_B(d) = 1 - f_A(d) & 0 \le d \le 1 \end{cases} \quad (7)$$

where d is the ratio of distance between query point and the reference and material A, B respectively represent SiC/Polymer and diamond. Figure 5 is the material distribution functions and mechanical performance of the cutting tool.

The process of heterogeneous solid modeling is divided into three steps as follows. First, the geometry of the part is modeled in a commercial CAD system such as Unigraphics. Then the CAD model is exported

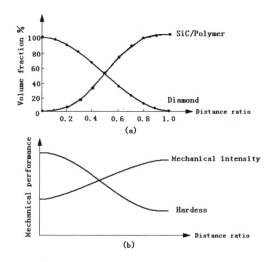

(a)

(b)

Figure 5. Material distribution functions and mechanical performance of the cutting tool. (a) material distribution functions; (b) mechanical performance.

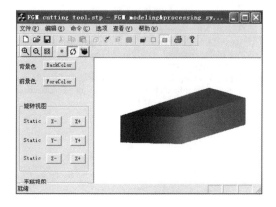

Figure 6. Heterogeneous solid model of the cutting tool.

```
#10=SHAPE_REPRESENTATION_RELATIONSHIP('none',
'relationship between FGM_1 and FGM_1',#94,#11);
#11=EXTENDED_ADVANCED_BREP_SHAPE_REPRESE
NTATION(' FGM_1',(#95),(#125);#5680);
...
#95=MANIFOLD_SOLID_BREP('',#96);
#96=CLOSED_SHELL('',(#214,#215,#216,#217));
...
#125=MATERIAL_REPRESENTATION('',(#1001,#1002));
...
#1001=MENTITY(#300, #3015, (#1250, #1252),#2115);
#1002=MENTITY(#301,#3015,(#1251,#1252), #2116);
...
#2115=MFUNCTION('POLYNOMIAL', (#3750,
#3751),(#4115,#4116));
#2116=MFUNCTION('POLYNOMIAL',
(#3752,#3753,#3754),(#4117,#4118,#4119));
```

Figure 7. Representation of the cutting tool in STEP (ISO10303).

is generated manually. Here the representation does not provide the details of geometry representation of the part. The software for heterogeneous solid modeling is developed using Visual C++6.0.

6 CONCLUDING REMARKS

An approach of modeling and representation of heterogeneous objects based on STEP for rapid prototyping is presented in this paper. In this approach, material information is represented in terms of a group of distance functions, which are accompanied with one reference or multi-references geometry features. The material distribution information is incorporated with geometry information by attaching material distribution functions to reference geometry entities as their properties. The material distribution information is only affected by the reference and material distribution function. The modeling of material grading does not affect the geometry of the object. Therefore, the modification of the material grading such as shifting, deletion and re-assignment can be done without modifying the geometry of the object. The geometry information is exchanged by STEP standard format between CAD system and our heterogeneous solid modeling system.

The design and fabrication of heterogeneous objects is very complicated in comparison with that of homogeneous objects for the representation and processing of the functionally graded material information. Apart

by STEP standard format. Second, after the STEP file is loaded into our heterogeneous object modeling system, the correctness of schema and contents of the STEP file is checked. Then the geometry data of the STEP file is extracted and an inner geometry model is constructed in the computer. The inner geometry model is displayed on the screen using OpenGL for being browsed by designer. Last, the user selects reference geometry feature and models the material gradings. The system incorporates the geometry information and material distribution information and generates an entire heterogeneous solid model. The heterogeneous solid model can be saved as a binary file outside the modeling system. Figure 6 shows the heterogeneous solid model of cutting tool. The representation of this part in STEP (ISO10303) is shown in Figure 7, which from the modeling and representation of heterogeneous objects, analysis, process planning and manufacturing of heterogeneous objects should also be studied in greater detail in future work considering the material distribution information.

REFERENCES

1. Holt B, Koizumi M, Hirai T, Munir Z. Functionally gradient materials. Ceramic Trans 1993:34.
2. Ilschner B. Processing-microstructure-property relationships in graded materials, Journal of Mechanics and Physics of Solids, 1996, 44(5):647–656.
3. Markworth A, Ramesh K, Parks Jr W. Modeling studies applied to functionally graded materials, Journal of Materials Science, 1995, 30:2183–2193.
4. Yan X, Gu P. A review of rapid prototyping technologies and systems, Computer Aided Design, 1996, 28(4): 307–318.
5. Kumar V, Dutta D. An Approach to Modeling & Representation of Heterogeneous Objects, Journal of Mechanical Design, 1998, 120(12):659–667.
6. Shin KH, Dutta D. Constructive representation of heterogeneous objects, Journal of computing and information science in engineering, 2001, 1(3):205–217.
7. Jackson TR, Liu H, Partikalakis NM. et al. Modeling and designing functionally graded material components for fabrication with local composition control. Materials & Design, 1999, 20:63–75.
8. Pegna J, Sali A. CAD modeling of multi-modal structures for free-form fabrication. Solid Freeform Fabrication Symposium, Austin, Texas, 1998.
9. Lalit Patil, Debasish Dutta, Bhatt AD, Jurrens K. Lyons K, Pratt MJ, Sriram RD. Representation of Heterogeneous Objects in ISO 10303 (STEP), Proceedings of the ASME Conference, Orlando, FL, November 2000.
10. Lalit Patil, Debasish Dutta, Bhatt AD. et al. A proposed standards-based approach for representing heterogeneous objects for layered manufacturing, Rapid prototyping journal, 2002, 8(3):134–146.
11. Siu YK, Tan ST. 'Source-based' heterogeneous solid modeling, Computer-Aided Design, 2002, 34(1):41–55.
12. Qian X, Duta D. Physics based B-spline heterogeneous object modeling. Published In: Proceedings of ASME DETC and Computers and Information in Engineering Conference, Pittsburgh; September 2001.
13. Chen K, Feng X. Computer-aided design method for the components made of heterogeneous materials. Computer-Aided Design, 2003, 35(5):453–66.
14. Liu H, Maekawa T, Patrikalakis NM, Sachs EM, Cho W. Methods for feature-based design of heterogeneous solids. Computer-Aided Design, 2004, 36:1141–1159.
15. Fessler J, Nickel A, Link G. et al. Functional gradient metallic prototypes through shape deposition manufacturing. Solid Freeform Fabrication Proceedings, Austin, 1997:521–528.
16. Jepson L, Beaman J, Bourell D. et al. SLS processing of functionally gradient materials. Solid Freeform Fabrication Proceedings, Austin, 1997:67–80.
17. Griffith ML, Harwell LD, Romero JT, Schlienger E, Atwood CL, Smugeresky JE. Multi-material processing by LENSe Proceedings of Solid Free-form Fabrication, Austin; August 1997.
18. Kumar V, Dutta D. An assessment of data formats for layered manufacturing. Advances in Engineering Software, 1997, 28(3):151–164.

Virtual modeling and rapid manufacturing – Bártolo (eds)
© *2005 Taylor & Francis Group, London, ISBN 0 415 39062 1*

Modelling of perception

S. Roth-Koch

Fraunhofer Institute for Manufacturing Engineering and Automation IPA, Stuttgart, Germany

ABSTRACT: Within the concurrent design and manufacture a new trend is rapidly emerging: a shift towards the virtual reality. Thus, the virtual product model in its different stages is an essential prerequisite for the virtual modelling of processes and their virtual simulation. Product design starts with conceptual design including the concepts for shape and function. The conceptual designers work on fuzzy product data applied to perception modelling and shape creating. Fuzziness means different aspects: the shape of the product is interpreted in scribbles which are partly coloured, handmade patterns, verbal descriptions, photographs and montages etc, which define the appearance and the function of the product. In the paper we will suppose a method to create virtual models out of scribbles which construe the perception of the subject. The mathematical basis is given by Interval Fuzzy-Splines. The method combines an image-based modelling system with usual CAD modelling strategies.

1 INTRODUCTION

1.1 *Integrating the early stages of conceptual design in computer aided design*

Product design starts with conceptual design including the concepts for shape and function. The transfers between the basic steps of planning, concept, outline and elaboration are smooth and the creative process is characterized by a high extent of individuality and is not stringent and difficult to analyse. These first design steps work with unfocused perception, where the initial design is represented as a fuzzy, inaccurate sketch or scribble.

'Fuzziness' means different aspects: the shape of the future product is interpreted in a couple of pencilled scribbles which are partly coloured, handmade patterns, verbal descriptions, photographs and montages etc, which define the appearance and the function of the product. Thus, the product firstly is abstracted as a 'subject' which will be translated intuitively into a multifaceted 'vocabulary' of curves and shapes. It can be used for background information to develop and to build-up the three-dimensional shape model in CAD systems. This means that new 'computer' sketches arise which have to be derived from the paper scribbles, demand a particular contour description (eg Spline curves) and represent the result of conception (the form and styling content is fixed more or less) more than the conceptual steps themselves. CAD technology provides but insufficient support for this intuitive work.

The conceptual designer uses design modeller systems (3D-Studio, AliasWavefront) to create computerized, virtual models as well as sheet sketches and handmade models to stimulate his imagination while forming curves and shapes and verifying the haptics, function and ergonomics of these conceptual models. Thus even the early conceptual stages comprise different coexisting model types – virtual and physical models. In a certain way, these models are different and independent of one another because they cannot, or can only with great difficulty be derived from one another or transformed into each other in the absence of related information processing and transferring techniques, eg of related digitising equipment with software tools for geometrical shape reconstruction. This means that the concept modelling has to be carried out separately in each stage and cannot be executed iteratively.

This is the crucial question for future strategies of product development. Considerable research effort has been done in this area investigating the CAD support for styling, to support sketching recognition and to develop sketching systems. Here, freehand sketches are considered to be fragmentary and imprecise but are regarded as a mapping of proper objects. The conceptual scribbles depict more the perception of an object, however.

1.2 *Modelling of perception*

The application described in this paper focuses on the creation of virtual sketches out of paper scribbles obeying their perceptual character. The method enables the conceptual designer to scan paper scribbles and afterwards convert them into a virtual sketch model which includes all the 'fuzziness' of the paper scribble and the therein expressed conceptual steps but is automatically represented in a sequence of real B-Spline curves.

The most important factor in this regard is the enhancement of these virtual tools with fuzziness and spontaneity (the design engineer may say: 'inaccuracy'). It shall be possible to transfer an arbitrary sketch, scribbled on a sheet and representing more a 'frozen movement' than an object itself, into a design modeller system. Individual design curves and shape characteristics which form the basis of typical CAD wireframe models shall be extracted and visualized without losing the fuzziness of their appearance. The conceptual designer can work on the scribble-like resulting two-dimensional geometrical representation and further develop it into the final concept and into the three-dimensional CAD surface model.

This paper presents a semi-automatic method of transferring conceptual sketches (pencilled sheets) into fuzzy virtual sketches as a basis for CAD modelling. It is organised as follows: section 2 describes the basic functions and techniques of conceptual design and the transfer into virtual conceptual design, section 3 comprises the background of the solution while section 4 depicts the exemplified single steps.

2 THE BASICS

2.1 What conceptual design means

The traditional design process comprises a requirement profile for the future product which defines the scope of the search for ideas. The conceptual design of the product first of all begins with a plan (a creative search for ideas) as well as a list of requirements as the basis of a goal-oriented conception (a transformation of ideas into outlines). These phases entail the use of traditional-real aids which are supposed to rouse the imagination and provide inspiration. The sketches are often supplemented with texts or suitable photographs and are explained so as to breathe as much life as possible into the underlying idea and to verify the 'physical' perception of this idea. Eventually, outlines are produced which, as a final step, are elaborated into three-dimensional, physical models – the shape models.

This ultimately means that the extent of detail in the sketches continually increases and that special details can be specified in individual sketches while the rest of the sketch may remain vague. Inaccuracies are therefore accepted. The following steps summarize the main functions of such sketches:

– representation of shapes,
– representation of space,
– representation of colour concepts,
– interpretation of outline alternatives,
– clarification of proportions,
– delineation of technical functions and processes,
– comparison of the relations of shapes and surfaces and
– illustration of ideas with additional texts.

Given this diversity of functions, various sketching techniques are employed. Space and three-dimensional objects are depicted using perspective, for example. Perspectives, on the other hand, may be created with the help of several sketching methods, but they can never really be drawn true to scale. One can furthermore distinguish between line and surface sketches or combinations of both. Often individual styles and/or preferences for certain drawing instruments are also important.

The manually prepared initial models (shape models) therefore allow the only truly three-dimensional assessment of the outline and the laying down of the conceptual statement in its entirety. Initial models may also be line models (wireframe models) or surface models.

The transfer of the sketches or the initial models into virtual models for the purposes of design development and verification is not carried out in practice. One either draws sketches directly using the design modelling system – thereby reducing the shaping opportunities described above, since the modelling strategies demand that an analytically constructive approach be adopted for further model construction immediately after the freehand input of curves – or one directly proceeds to construct a CAD model. The virtual sketch models thus generated are based on the same geometrical elements as the CAD models, eg the free-form curves are always defined as B-Spline curves or as NURBS curves and therefore embody a comparatively 'tangible' shape representation. Their 'scribble' nature is thus lost.

2.2 Required steps for computer aided design

The further development of virtual conceptual design models into CAD models demands that several conditions be met. Surface-oriented 3-d CAD systems and design modeller systems utilize the same basic geometrical functions for the description of regular and free-form geometry. Their incorporation into a system-specific data model is, however, carried out in different ways. In particular, the incorporation of the generated curve features or surface features into topological models (ie their attachment to the object surface) takes different forms. This variety definitely makes sense since the computed models all represent different stages of outlining and development.

In order to incorporate the outlining phase into the virtual development of the product, these differing models must be integrated and transferred into one another. Either sketches or the conceptual designer's handmade initial model are the starting point. Both kinds of product outlines still exhibit certain inaccuracies, however. Therefore a transfer into a computed, virtual model cannot result in a 1:1 copy which accurately preserves the established shape, but is merely supposed, for example, to precisely depict a few essential forming lines while schematically visualizing the 'remaining geometry' to amplify the human perception.

If the conceptual designer's 2D sketches are used as pattern, then two-dimensional computerized contours have to be extracted from the fuzzy description of forming lines (eg multiple pencilled contours) after the sketches have been scanned. From this a 3-d wireframe model must then be constructed. Since the only available perspective views are indistinct and therefore not to be automatically evaluated, the conceptual designer must visually monitor and manually influence the process of contour specification. The wireframe model initially interprets the product's form as consisting of edges (wires) which are either forming lines, the object's genuine edges or surface borders with adjoining conditions, depending on the extent of detailing.

3 THE IDEA AND THE TECHNICAL BACKGROUND

Methods for integrating the early stages of product design into the CAD/CAM process must take into account the fact that current CAD developments tend to become process-oriented systems which manage a specialized, object-oriented data structure. The methodical basis described in this paper uses the methodical and geometrical kernel of the development toolkit OpenCASCADE which deals with a (STEP compatible) non-manifold geometric description. The data exchange can be executed by standardized interface structures or via the transfer of brep models.

Normal flatbed scanners are used to digitise sketches rapidly. The result is a grey level image or colour image in a discrete dpi-resolution (dots per inch) where neither geometrical properties nor features (curves, edges, profiles, forming lines etc) can be read directly. These are only interpreted as visual mappings.

Further problems to be considered are the fuzziness and 'incompleteness' of the sketches. The fuzziness of the sketch means different amounts of thickness of lines, multiple lines, shadings or open contours (Fig. 1). That is why only selected forming/shaping elements or features (shaping curves, light reflection curves etc) are to be transferred.

The definite identification of such forming/shaping elements in grey level images as a prerequisite to detecting geometrical data is generally the task of

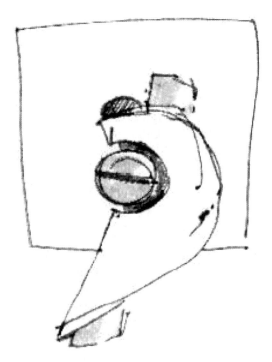

Figure 1.   Conceptual sketch.

image processing techniques. Conventional image processing strategies (clustering, regions of interest, texture analysis) normally use specifiable and qualifiable information from given geometric shapes. This kind of information cannot be obtained from sketches at first glance. The 'fuzzy' elements of a sketch must be identified by iterative analysis and/or user interaction.

First of all we had to investigate search methods driven by user targets for regions and contours of fuzzy or ambiguous objects. We have found a rich field of solutions in the geosciences. The implemented procedures can be expanded to other applications dealing with ambiguous objects (eg reflecting objects).

The elaborated strategies allow the definition, partly by automatic extraction, of the relevant regions of geometrical basis elements in scanned sketches. Furthermore, the conceptual designer can define relevant regions where he imagines the object's contour by interaction.

Using this information, a virtual model is built which fits the sketch visually (with a few omissions and simplifications) and is modelled in a CAD style (with B-Spline curves). This makes the conversion into the initial CAD wireframe model of selected model curves possible. This wireframe model and some additional surface points are normally the informational basis for CAD modelling.

3.1 What is the gist of sketches?

The pictured perspective plays no essential role from the point of view of geometrical evaluation in preparing the exact geometrical model (CAD model) because it is not possible to extract three-dimensional information automatically.

But it is important for the further development of the virtual model. Furthermore, conceptual sketches comprise elements which have no relevance to the geometrical shape, like coloured markings or background shadings. These would falsify the evaluation result and have to be removed. Since they are mostly typical expressions of the designer's personal style, they must to be removed by the designer himself before starting the feature extraction. This can be done with the usual image handling software (Photoshop, CorelDraw etc) and is accepted by the designer.

3.2 What is the gist of the virtual model?

The preserved but still fuzzy contours, which may possibly be in warped perspective, should be transferred into a design modelling system to develop them into a real three-dimensional model.

The character of the sketch, ie the visualization of fuzzy contours should be preserved in the initial virtual mapping. On the other hand, the visualized curves should be defined as geometrical elements for using usual modelling procedures. Our method of resolution fulfils these demands by implementing an array of curves which represents the fuzzy contour of the object. The individual curves in this array can be manipulated with parameters derived from the manner of representation in the sketch – for example: a kind of 'importance' may be assigned to contour areas near thick lines or a cluster of tightly packed lines.

4   APPLICATION

The implementation method is based on the development toolkit OpenCASCADE. Additional algorithms (for image processing) are implemented and their results converted into an OpenCASCADE data model after the evaluation is complete for the purpose of enabling the use of OpenCASCADE functions.

The configuration of the functions and the user interaction will be described in the following:

1. User handling of the sketches:
   The conceptual designer scans the sketch in a suitable resolution and adjusts it with image handling software by removing elements which do not belong to the objects (markings etc).
2. Image pre-processing/Image decoding:
   The scanned TIFF-formatted sketches are decoded into a bitmap to make the pixels with their grey

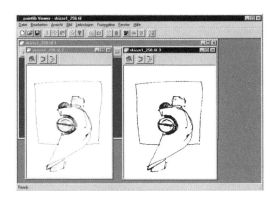

Figure 2.   Bitmap of the scanned conceptual sketch.

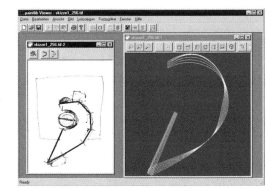

Figure 3.   Guiding polygon defining the Fuzzy-Spline area.

code and their x,y-values available. The bitmap is displayed in a separate window (Fig. 2).
3. Contour and edge analysis/Vectorising:
   For getting information on the run of drafted lines out of the bitmap, the user primarily has to determine the regions where lines have to be picked up for the CAD drawing. This is done by specifying 'guiding polygons'. The picked lines then depict the area of probable locality of the object's contours (Fig. 3). Two possibilities of determining regions are: Indication of points located on/nearby one line in the sketch to be transferred, they form a (guiding) polygon following the line.

   For the other option, further image processing step must be carried out: Transforming the bitmap into a binary image. This binary image is skeletonised automatically (ie detecting one particular line of one pixel width) and the skeleton lines are vectorised for guiding polygons. Afterwards, all 'sufficiently joined' skeleton lines or individually selected skeleton lines can be transferred into the design modeller (Fig. 4).

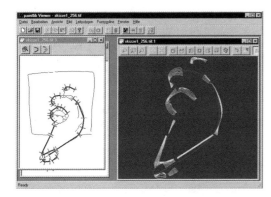

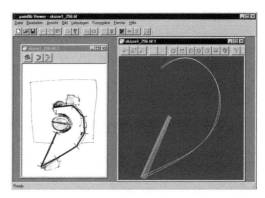

Figure 4.   Skeleton lines are not sufficient in this example.

Figure 5.   Modified Fuzzy-Spline array describing the object's contour.

4. Determination of the fuzziness of the resulting B-Spline curves:

A suitable parameterisation of the B-Spline curve to be generated is to be computed from the guiding polygon. The fuzzy points are to be computed over a grey level adjustment along the guiding polygon. Firstly, extremal values of grey level distribution are determined orthogonal to the lines of the guiding polygon and within of a tube of flexible width. The computation of the fuzzy member function then adds all distances of the extremal values among each other. These fuzzy points reflect the point fuzziness of the curve which is to be constructed and in so doing they specify the most likely location for this curve.

5. Generation of the fuzzy array of B-Spline curves as a basis for wireframes in design modeller systems:

Individual B-Spline curves are to be fixed using these points. They interpret the fuzziness of the original sketch in such a way that the user does not have to opt for a single 'sharp' curve even at the beginning. The conceptual designer can vary the location and shape of the B-Splines by (Fig. 5):
   – modifying one 'weight' parameter displaying one discrete curve computed out of weighted sketching lines,
   – deleting and redrawing of 'guiding polygons',
   – changing the width of the considered region around the 'guiding polygon',
   – displaying of individual levels of the Fuzzy-Splines (two 'max-min' curves, one 'middle' curve and two inner curves respectively) which can be modified with a slider.

In this manner different shapes for the final curve may be inspected. The selected curves may further be processed in design modeller systems or in CAD systems. A further development of the geometrical shape to be created in the sketch is thus possible (Fig. 6).

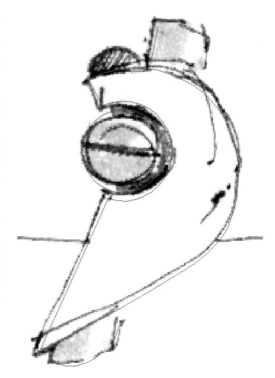

Figure 6.   The developed object contour.

5   CONCLUSION

The methods described for the transfer of outline sketches drawn on paper into virtual models are the first step towards integrating the traditional working methods of perception modelling and shape creation with the virtual product development process. The conceptual designer can scan his pencilled scribble

217

and 'transfer' the included fuzzy shape description (approximate contours, shades etc) into the Fuzzy-Spline curves of a virtual model. At a very early stage the conceptual designer is thus presented with entirely new tools for developing and verifying his outline without relinquishing the shape outlines (eg pencil sketches or wooden initial models) that are so important for creative work. At the same time, virtual product development is supplemented with the outlining stage, for which iterative model construction leading to the complete product data model is now also possible. The styling of the products is also granted more leeway – this furthermore involves an increasing influence on the development and the characteristics of innovative products.

These are, of course, only a few preliminary steps. The fuzziness of the sketch is preserved in the virtual sketch model and the conceptual designer is able to digitise his first stylistic idea without having to specify individual geometrical features. The next crucial steps shall consist of methods for developing the two-dimensional sketch model, which may possibly contain perspective, into a three-dimensional model using virtual tools, a suitable means for conceptual designers. The experience gained by conceptual designers working with virtual sketches is to supply the basis for the drafting and implementation of such methods. Ways to take advantage of shading in order to specify a spatial direction or to 'slip' the sketch over a pre-modelled basic structure are currently under consideration – also obeying related research work in this area.

Moreover, the research work has shown that there is a strong requirement for new conceptual-design-related modelling strategies, eg a semi-automatic proceeding which is combined with intuitive tools.

These efforts were sponsored by the German 'Stiftung Industrieforschung' (Foundation for the Advancement of Research in Trade and Industry) based in Cologne. The graduate theses of Effenberger (Effenberger 2000) and Stotz (Stotz 2000) were essential contributions to the research project.

REFERENCES

Effenberger, I. 2000. Aufbau eines CAS.CADE-basierten Wireframe-Modellierers. *Graduate Thesis*. Stuttgart: Faculty of Computer Science University Stuttgart
Stotz, M. 2000. Interpretation von Designerskizzen in drei-dimensionalen CAD-Modellen. *Graduate Thesis*. Stuttgart: Faculty of Mathematics University Stuttgart

Virtual modeling and rapid manufacturing – Bártolo (eds)
© *2005 Taylor & Francis Group, London, ISBN 0 415 39062 1*

A new surface reconstruction method based on volume warping for RE & RP applications

Sergei Azernikov & Anath Fischer
Laboratory for CAD & LCE, Mechanical Engineering Department, Technion – Israel Institute of Technology, Haifa, Israel

ABSTRACT: Volumetric models of 3D objects are recently introduced into the reverse engineering process. The grid-based methods are considered as the major technique for reconstructing surfaces from these volumetric models. This is mainly due the efficiency and simplicity of these methods. However, these methods suffer from a number of inherent drawbacks, resulted from the fact that the imposed Cartesian grid in general is not well adapted to the surface, nor in size neither in orientation. In order to overcome the above obstacles a new iso-surface extraction method is proposed for volumetric objects. The main idea is first to construct a geometrical field which is induced by the object's shape. This geometrical field represents the natural directions and a grid cell size for each point in the domain. Then, the imposed volumetric grid is deformed by the produced geometrical field toward the object's shape. The iso-surface meshes can be extracted from the resulting adaptive grid by any conventional grid-based contouring technique. These meshes provide better approximation of the unknown surface and exhibit anisotropy, which is present in this surface. Since the produced meshes are all-quad, Catmull-Clark subdivision surfaces can be directly constructed from these meshes. Moreover, accurate physical models of the reconstructed objects can be produced using RP technology.

1 INTRODUCTION

Typical reverse engineering pipeline consists of four main phases (Varady et al. 1997): (a) laser scanning of a 3D object, (b) registration of the data into a single point cloud, (c) meshing of the cloud, and (d) fitting of the NURBS patches. This approach is suitable for objects with simple topology. However, meshing the cloud scanned from a complex objects is ought to be difficult. Moreover, the point clouds are often incomplete and noisy, which makes the meshing process to be very unstable.

In order to deal with these obstacles, volumetric approaches were introduced. With these approaches, the implicit volumetric model is reconstructed from the point cloud (Carr et al. 2001; Ohtake et al. 2003). For the downstream applications however, i.e. analysis and manufacturing, explicit representation is required. Therefore, an implicit representation is converted into an explicit one by iso-surface mesh extraction or *contouring*.

Segmentation of the resulted polygonal mesh and fitting of the NURBS patches are usually done manually by a skilled designer. This is a time consuming, and therefore, expensive process. Almost three decades ago, Catmull and Clark (Catmull and Clark 1978) have proposed the *subdivision* scheme for construction of tensor product B-Spline surfaces directly from quadrilateral meshes. Recently, subdivision surfaces have gain a popularity in the computer graphics community as a more flexible alternative to NURBS. These surfaces can be *automatically* generated from the polygonal mesh and efficiently fitted to the implicit models reconstructed from the initial cloud of points (Litke et al. 2001).

2 THE APPROACH

The current research, proposes new approach to RE based on *volumetric* representation of the object's shape. For the point cloud data, the proposed approach consists of four main phases as shown in Figure 1: (a) reconstruction of volumetric implicit model from the scanned cloud of points, (b) geometric tensor field propagation, (c) anisotropic surface meshing, and (d) subdivision surface fitting. The proposed approach can be also applied on the discrete volumetric data.

The process begins by reconstructing an implicit model from the scattered points. In the current work, the *multi-level partition of unity (MPU)* method is used (Ohtake et al. 2003). With this method, a set of

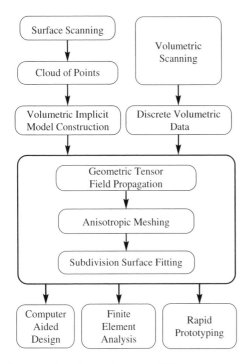

Figure 1.   Flow chart of the proposed RE approach.

overlapping quadric patches is fitted to the points. Then, these patches are blended in order to produce a piecewise-smooth implicit representation of the surface.

In the next phase, the *geometric tensor field* is evaluated from the implicit model. This geometric field is based on tensors of curvature of the implicit model and induced by the shape of the desired object. As a result, the problem domain becomes *anisotropic*. This is analogously to the influence of a gravitational field in physics as it was proposed by Einstein (Einstein 1953): "*...the gravitational field influences and even determines the metrical laws of the space-time continuum. If the laws of configuration of ideal rigid bodies are to be expressed geometrically, then in the presence of a gravitational field the geometry is not Euclidean.*"

In order to extract an *explicit* representation of the object, the implicit model is meshed. Since the meshing is performed in the anisotropic space, the resulted polygonal mesh is adaptive and exhibits anisotropic properties presented in the object's shape. This mesh provides much better approximation of the surface, then mesh extracted from the uniform grid of the same size.

The RE process is resumed by fitting a subdivision surface to the implicit model, while the adaptive mesh is used as a control polyhedron. The fitting is performed iteratively using the *quasi-interpolation* method (Litke et al. 2001). With this method, the control polyhedron

is subdivided and projected onto the implicit surface until convergence is achieved. The proposed approach has several important advantages:

- Implicit model of complex shapes can be robustly reconstructed from noisy or incomplete clouds of points with non-uniform density.
- Anisotropic meshing produces high quality, quad meshes. These meshes are very suitable for modeling and analysis applications.
- Subdivision surfaces can be automatically generated from the anisotropic polygonal meshes and be fitted to complex objects, without any user intervention.
- The reconstructed subdivision surface accurately follows the surface of the scanned object and can be directly used for high-precision rapid prototyping.

Since the geometric tensor field is the core of the proposed method, we start from a detailed description of this subject.

3   GEOMETRIC TENSOR FIELD

3.1   *Riemannian metric*

We formally define the geometric field as a *Riemannian metric* induced by the shape of the surface S on the domain Ω. For each point $\mathbf{x} \in \Omega$ this metric defines a tensor $\mathcal{M} = \mathcal{R}\Lambda\mathcal{R}^{-1}$, where $\mathcal{R}$ is an orthonormal rotation matrix and Λ is a diagonal scaling matrix. This tensor defines the mapping of the desired hexahedral voxel to a unit axis-aligned cube for $\forall \mathbf{x} \in \Omega$ (Frey and George 2000). As in (Tchon et al. 2003), we associate the lower 2×2 minor of $\mathcal{M}$ with the tensor of curvature $\mathcal{C}$ of the surface. But for the normal direction we give a different interpretation, which is more appropriate in our context. Tchon (Tchon et al. 2003) deduce the target size of an element at $\mathbf{x}$ in the normal direction from the local *thickness* of the solid enclosed by S. In terms of the normal field $\hat{\mathbf{N}}$ produced by S, this is twice the distance from $\mathbf{x}$ to the closest singular point of $\hat{\mathbf{N}}$. This thickness is approximated from the digital skeleton of the solid bounded by S. In our case however, S does *not* necessarily bound a solid, and anyway we are interested only in a *narrow band* surrounding S and not in the bounded volume. In our implementation, the size of this band is set to twice the surface-voxel intersection evaluation precision ϵ. The geometric tensor components can be now rewritten as follows:

$$\mathcal{R} = \begin{pmatrix} \hat{\mathbf{n}} & \hat{\mathbf{t}}_1 & \hat{\mathbf{t}}_2 \end{pmatrix}^T \tag{1}$$

$$\Lambda = \begin{pmatrix} 1/4\epsilon^{-2} & 0 & 0 \\ 0 & \alpha\kappa_1^2 & 0 \\ 0 & 0 & \alpha\kappa_2^2 \end{pmatrix} \tag{2}$$

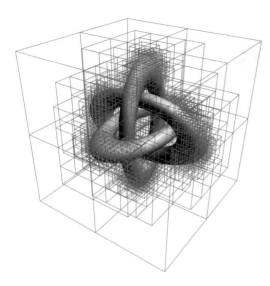

Figure 2. The constructed octree for the rings model (3032 nodes), with maximal depth 6.

where $\hat{\mathbf{t}}_1$, $\hat{\mathbf{t}}_2$ are the principal directions which span the tangent space of the surface, $\hat{\mathbf{n}}$ is the normal vector and α is a coefficient which allows control of the introduced anisotropy level.

3.2 Background octree

For the geometric field to be represented faithfully, it should be appropriately sampled on the background mesh. Sampling it on a dense uniform grid is possible, but such an approach is extremely memory-intensive. A better solution is to sample this field *adaptively*. Octrees have been used for adaptive sampling of scalar distance fields by Frisken (Frisken et al. 2000), and geometric tensor fields by Tchon et al. (Tchon et al. 2003). Octrees are very suitable for adaptive field representation since they provide local refinement and efficient point localization. In our implementation, the octree is constructed by the *top-down* approach. With this approach, the process starts from the root voxel, which occupies the whole domain of support Ω. This voxel v is then subdivided recursively until the minimal allowed size is not reached and one of the following conditions hold:

1. the voxel v contains more than one disconnected component of S,
2. the size of v is bigger than the target size at v.

These conditions occur in regions where the gradient field *divergence div*($\mathbf{N}$) is high.

Figure 2 presents the octree constructed for the rings model. After the octree is constructed, the geometric tensors are set in the leaf voxels which contain a portion

of the iso-surface S. In the remaining voxels, these tensors are initialized to a diagonal matrix,

$$\mathcal{M} = \begin{pmatrix} 1/R^2 & 0 & 0 \\ 0 & 1/R^2 & 0 \\ 0 & 0 & 1/R^2 \end{pmatrix} \qquad (3)$$

where R is the radius of the domain Ω circumsphere. Then, the geometric field is propagated from the iso-surface to the neighboring voxels by iterative Laplacian smoothing, as in (Tchon et al. 2003),

$$\mathcal{M}_i^{n+1} = \mathcal{M}_i^n + \frac{\sum(\mathcal{M}_i^n - \mathcal{M}_j^n)/l_{ij}}{\sum 1/l_{ij}} \qquad (4)$$

where summation is performed on the face neighboring voxels j, and l_{ij} is the distance between centers of voxels i and j. Since we are interested in propagating the field into a *narrow band* surrounding S, we apply only several iterations of the above operator.

4 ANISOTROPIC MESHING

4.1 Anisotropic grid adaptation

The grid adaptation approach was introduced to implicit surface contouring by Moore and Warren (Moore and Warren 1991). In this work, the imposed grid vertices are perturbed in order to reduce the number of triangles produced by the Marching Cubes algorithm. However, this perturbation is indifferent to the surface geometry. Balmelli (Balmelli et al. 2002) proposed to warp the grid according to an *importance map*. This map makes imposed grid vertices attract to the regions of interest chosen by the user. As a result, the extracted iso-surface mesh is denser in these regions and sparser elsewhere. The importance map constitutes a *scalar* metric field, which guides the adaptation of the grid. Since the field is scalar, the adaptation is *isotropic*. Recently, it was shown by Tchon (Tchon et al. 2003) that employing an *anisotropic* geometric field for grid adaptation is very natural and effective. The adapted grid tends to follow the shape of the surface and reflects clearly anisotropic properties of this shape. These results suggest that a geometric field is a very natural and general representation of shape for applications, where its anisotropic properties are important. Indeed, these properties must be considered during high quality surface meshing (Frey and George 2000). Therefore, we employ the anisotropic grid adaptation approach for adaptive polygonization of implicit surfaces.

The grid adaptation is performed by relaxation or iterative Laplacian smoothing in the metric field. Grid boundaries must be constrained in order to prevent

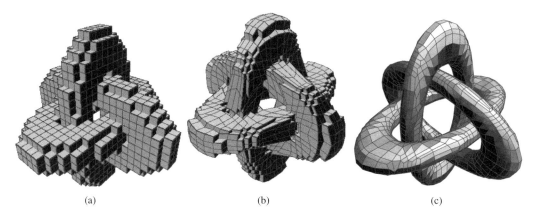

(a)   (b)   (c)

Figure 3.   Synthetic example with complex topology: (a) initial grid $25 \times 25 \times 25$ (shown 1752 voxels which contain a portion of the iso-surface), (b) after 100 relaxation iterations (2450 voxels) and (c) mesh extracted from the adaptive grid (2374 faces).

shrinkage. In our implementation, these boundaries are projected onto a *circumsphere* of the domain Ω. This technique provides maximum flexibility, since the grid boundary vertices can travel freely on the sphere and are not constrained to the specific cube sides, corners or edges, as in (Tchon et al. 2003) and (Balmelli et al. 2002). Then, for each grid vertex $\mathbf{v}_i$ the optimal location is calculated by the standard iterative scheme:

$$\mathbf{v}_i^{n+1} = \mathbf{v}_i^n + w \frac{\sum l_{ij}^{\mathcal{M}}/l_{ij}(\mathbf{v}_j^n - \mathbf{v}_i^n)}{\sum l_{ij}^{\mathcal{M}}/l_{ij}} \qquad (5)$$

where summation is applied on the neighboring vertices $\mathbf{v}_j$, w is a relaxation factor, l_{ij} is a Euclidean distance between the vertices $\mathbf{v}_i$ and $\mathbf{v}_j$ and $l^{\mathcal{M}}_{ij}$ is the *metric* distance between these vertices. This metric distance is calculated by integrating the metric dot product along the edge $\{\mathbf{v}_i, \mathbf{v}_j\}$,

$$l_{ij}^{\mathcal{M}} = \int_0^1 \sqrt{(\mathbf{v}_j - \mathbf{v}_i)^T \mathcal{M}(\mathbf{x}(t))(\mathbf{v}_j - \mathbf{v}_i)}dt \qquad (6)$$

where $\mathbf{x}(t) = \mathbf{v}_i + (\mathbf{v}_j - \mathbf{v}_i)t$ and $\mathcal{M}(\mathbf{x}(t))$ is the metric tensor at $\mathbf{x}$. This tensor is set to the tensor associated with the background octree voxel which contains $\mathbf{x}$. Figure 3 shows the grid adaptation for the synthetic example with complex topology.

4.2   *Iso-surface mesh extraction*

The iso-surface mesh is extracted from the produced adaptive grid. Since the grid is adapted with an anisotropic metric, the extracted mesh should have anisotropic properties. Traditional contouring techniques (Lorensen and Cline 1987; Bloomenthal 1994) produce triangular or mixed meshes. However, for exploiting the anisotropy of a surface, employing quads

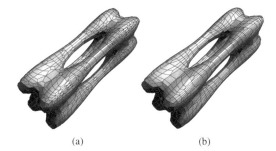

(a)   (b)

Figure 4.   Tangle cube surface: (a) iso-surface extracted from the adaptive grid with dual contouring (2248 faces) and (b) dual contour after decimation (1950 faces).

is more appropriate (Alliez et al. 2003). Recently, *dual contouring* techniques have been proposed (Ju et al. 2002; Azernikov et al. 2003). Applying these techniques on adapted grids produces anisotropic, all-quad meshes as can be seen in Figure 3. These meshes often contain topologically redundant faces, which can be collapsed in order to improve the overall mesh quality and to reduce the number of faces (see Figure 4).

5   SUBDIVISION SURFACE FITTING

The adaptive quadrilateral mesh produced on the previous phase, provides piecewise-linear approximation of the object's surface. This approximation is used as a control polyhedron for a subdivision surface generation. This surface is iteratively fitted to the implicit model. Each iteration of the fitting process consists of two steps: (a) applying the Catmull-Clark subdivision scheme (Catmull and Clark 1978), (b) projecting the produced vertices onto the implicit surface using the Newton-Raphson method. Figure 5 (f) presents the subdivision surface fitted to the screwdriver model.

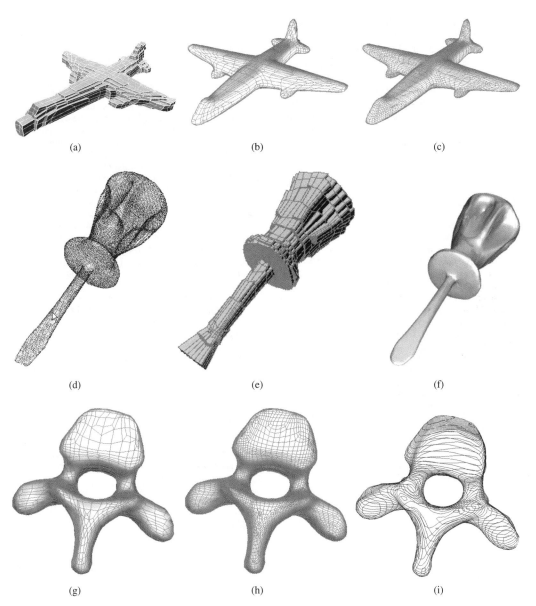

Figure 5. Example applications of the proposed approach: (a) adapted grid for the airplane model, (b) anisotropic mesh extracted from the adapted grid, (c) anisotropic mesh subdivided with Catmull-Clark scheme; (d) point cloud scanned from a screwdriver, (e) adapted grid for the screwdriver model, (f) fitted subdivision surface; (g) anisotropic mesh of a spinal-cord model, (h) anisotropic mesh subdivided with Catmull-Clark scheme; (i) slices for layered manufacturing extracted from the fitted subdivision surface.

6 CONCLUSIONS AND FUTURE WORK

In this paper, the volumetric approach to RE of 3D objects is described. The proposed approach is based on representing shape of the solid object as a geometric tensor field, which defines the metrical properties of the problem domain. As a result, the surface recon- struction process becomes sensitive to the shape of the object. The reconstructed quadrilateral meshes are adaptive and exhibit anisotropic properties of the object's shape. Fitting of subdivision surfaces to these meshes produces high-quality, piecewise smooth approximations of the desired objects.

Accurate physical models of the reconstructed objects can be produced using RP technology. Figure 5(i) shows slices for layered manufacturing extracted from the fitted subdivision surface.

Currently, the proposed method has been applied on modeled algebraic surfaces and partition of unity implicits. In the future, we plan to apply the proposed method on discrete volumetric data sets, such as medical images.

REFERENCES

Alliez, P., D. Cohen-Steiner, D. Devillers, B. Lévy, and M. Desbrun (2003). Anisotropic polygonal remeshing. In *Proceedings of ACM SIGGRAPH 2003*, Volume 22(3) of *ACM Transactions on Graphics*, pp. 485–493. ACM Press.

Azernikov, S., A. Miropolsky, and A. Fischer (2003, June). Surface reconstruction of freeform objects based on multiresolution volumetric method. In *8th ACM Symposium on Solid Modeling and Applications*, Seattle, WA, USA, pp. 115–126.

Balmelli, L., C. J. Morris, G. Taubin, and F. Bernardini (2002, 27–). Volume warping for adaptive isosurface extraction. In *Proceedings of the 13th IEEE Visualization 2002 Conference (VIS-02)*, Piscataway, NJ, pp. 467–474. IEEE Computer Society.

Bloomenthal, J. (1994). An implicit surface polygonizer. In *Graphics Gems IV*, pp. 324–349. Boston: Academic Press.

Carr, J. C., R. K. Beatson, C. Cherrie, T. J. Mitchell, W. R. Fright, B. C. McCallum, and T. R. Evans (2001). Reconstruction and representation of 3D objects with radial basis functions. In *SIGGRAPH 2001, Computer Graphics Proceedings*, Annual Conference Series, pp. 67–76. ACM Press/ACM SIGGRAPH.

Catmull, E. and J. Clark (1978, spline). Recursively generated B-spline surfaces on arbitrary topological meshes. *Computer-Aided Design 10*, 350–355.

Einstein, A. (1953). *The Meaning of Relativity*. Princeton University Press.

Frey, P. J. and P.-L. George (2000). *Mesh Generation Application to Finite Elements*. HERMES Science Publishing.

Frisken, S. F., R. N. Perry, A. P. Rockwood, and T. R. Jones (2000). Adaptively sampled distance fields: A general representation of shape for computer graphics. In *Siggraph 2000, Computer Graphics Proceedings*, Annual Conference Series, pp. 249–254. ACM Press/ACM SIGGRAPH/ Addison Wesley Longman.

Ju, T., F. Losasso, S. Schaefer, and J. Warren (2002, 3). Dual contouring of hermite data. *ACM Transactions on Graphics 21*, 339–346.

Litke, N., A. Levin, and P. Schröder (2001). Fitting subdivision surfaces. In T. Ertl, K. Joy, and A. Varshney (Eds.), *Proceedings of the Conference on Visualization 2001*, Piscataway, NJ, pp. 319–324. IEEE Computer Society.

Lorensen, W. E. and H. E. Cline (1987). Marching cubes: A high resolution 3D surface construction algorithm. *Computer graphics 21*(4), 163–168.

Moore, D. and J. Warren (1991, September). Mesh displacement: An improved contouring methods for trivariate data. Technical Report COMP TR91-166, Department of Computer Science, Rice University, P.Ø. Box 1892, Houston, TX 77251-1892.

Ohtake, Y., A. Belyaev, M. Alexa, G. Turk, and H.-P. Seidel (2003). Multi-level partition of unity implicits. In *Proceedings of ACM SIGGRAPH 2003*, Volume 22(3) of *ACM Transactions on Graphics*, pp. 463–470. ACM Press.

Tchon, K.-F., M. Khachan, F. Guibault, and R. Camarero (2003, September). Constructing anisotropic geometric metrics using octrees and skeletons. In *12th International Meshing Rountable*, pp. 293–304.

Varady, T., R. Martin, and J. Cox (1997). Reverse engineering of geometric models – an introduction. *Computer Aided Design 29*(4), 255–268.

Virtual modeling and rapid manufacturing – Bártolo (eds)
© 2005 Taylor & Francis Group, London, ISBN 0 415 39062 1

A novel haptic interface and its application in CAD

Zhan Gao & Ian Gibson
University of Hong Kong, Hong Kong, China

ABSTRACT: In this paper, we propose a new general haptic interface, which is able to haptically render point clouds, NURBS curves and surfaces, and rigid heterogeneous material objects. A feature of this approach is the haptic cursor, which has a shape and size, instead of just a point. We also present two CAD applications based on that haptic interface; an experimental haptic sculpting system for B-spline surfaces, and a haptic based method to draw a NURBS curve upon a point based model. By taking advantage of our haptic rendering techniques, various haptic sculpting tools are developed to facilitate sculpting B-spline surfaces, resulting in more intuitive virtual sculpting paradigms. To draw a NURBS curve upon a point-based model is a difficult problem. However, this problem can be solved with relative ease using our haptic interface.

1   INTRODUCTION

Applications of haptic-based CAD/CAM systems require intuitive approaches for haptic sense of curves, surfaces, volumetric models or even point clouds. This paper presents a new general haptic rendering approach for several current popular CAD model representations.

The key issue of the new haptic rendering technique is the implicit surface to point clouds haptic rendering algorithm. Recently, point-based modeling has received more and more attention. The reason is convincing: for modern computer graphics, 'in complex models the triangle size is decreasing to pixel resolution', therefore, using non-organized point clouds to represent primary surface model is reasonable (Azariadis et al. 2005). Also, in reverse engineering, modern acquisition technology can generate highly dense point clouds. The point clouds can be so dense that triangulation becomes meaningless (Azariadis & Sapidis 2005).

The implicit surface is the representation of the haptic interface object, or haptic cursor, which we call probe or tool in this paper. Users manipulate the stylus of PHANToM robotic haptic feed back device to drive the probe of implicit surface (SensAble). The system checks any collision between the probe and the point clouds. If any collision is detected, the magnitude and direction of force is calculated. This algorithm can be extended to other popular geometric model representations if the model can be discretized into point clouds. Since NURBS is the de facto standard for surface representation in CAD/CAM systems, the most valuable extension of our algorithm is a new haptic rendering interface for NURBS surfaces which features a surface-surface haptic interaction

style. Similarly, the prototypic haptic rendering method can also have good applications in haptic rendering of NURBS curves, rigid heterogeneous material objects and point-based models.

Based on our new haptic interface, two applications, an experimental haptic sculpting system for B-spline surfaces and a haptic based method to draw a NURBS curve upon a point based model, are presented in this paper to verify the validity of this novel haptic rendering techniques.

2   RELATED WORK

Haptic rendering is the process of applying forces in order to give the operator a sense of touch and interaction with physical objects (Zilles et al. 1994). Earlier researches mainly dealt with haptically rendering objects with a point interaction paradigm. A point-object based haptic interface generates only 3 degrees of freedom (DOFs) of force feedback because a point cursor, or a Haptic Interface Point (HIP), cannot simulate torques. To provide 6 DOFs force feedback, the haptic cursor must have a shape and a size. A ray-object haptic interface uses a ray as the haptic cursor to provide both torques and forces (Ho et al. 1997). The surface-object haptic interface uses surfaces for geometric representation of the haptic cursor. This method can introduce a much more complex haptic cursor into the haptic simulation, thus improving the degree of realism.

Gregory et al presented an algorithm for haptic display of moderately complex polygonal models with a polygonal haptic cursor by making use of incremental algorithms for contact determination between convex primitives (Gregory et al. 2000). McNeely et al put forward a simple, fast, and approximate voxel-based approach (McNeely et al. 1999). This approach enables the manipulation of a modestly complex haptic cursor within an arbitrarily complex environment of static rigid objects. Nelson derived a novel velocity formulation for use in a parametric surface-surface tracing paradigm and integrated it into a three step tracking process to compute reaction force between two NURBS surfaces (Nelson et al. 1999).

Because implicit surface is convenient for collision detection, some researchers have managed to present point-surface haptic rendering techniques with implicit surface. Kim introduced a haptic algorithm which is mainly based on an implicit surface representation and represents the surface with potential values in a 3D regular grid (Kim et al. 2002). Kim's method also allows the user to feel a smooth surface without force discontinuity.

3 HAPTIC RENDERING OF POINT CLOUDS

3.1 *Fundamental mathematical formulations of implicit surface*

The implicit representation of the surface S is described by the following implicit equation

$$S = \{(x,y,z) \in R3 \mid f(x,y,z) = 0\} \qquad (1)$$

where f is the implicit function, and (x, y, z) is the coordinate of a point in 3D space.

Here $f(x, y, z)$ could be polynomials, discreet grids of points or some black box functions. When $f(x, y, z)$ is polynomial, it yields an implicit algebraic surface. When $f(x, y, z)$ is linear, it describes a plane. When $f(x, y, z)$ is quadratic, it describes a quadric surface, such as an ellipsoid, a sphere or a cylinder.

If the potential value of $f(x, y, z)$ is 0, then the point (x, y, z) is on the surface. The set of points for which the potential value is 0 defines the implicit surface. If the potential value is positive, then the point (x, y, z) is outside the surface. If $f(x, y, z) < 0$, then the point (x, y, z) is inside. We use a closed quadric surface to represent our probe; more specifically, an ellipsoid. We choose an ellipsoid surface as our probe/tool because it is mathematically simple enough to evaluate thousands of points for collision detection in 1 kHz frequency, but also can be shown to represent a tool with both sharp and blunt region.

3.2 *Collision detection*

The implicit surface of the probe, driven by the PHANToM device, moves as the user manipulates the PHANToM stylus. The definition function of the probe is transformed by the PHANToM transformation matrix at every time step of the haptic servo loop. Model surfaces are presented as a layer of uniformly distributed sampling points. Now, the inside/outside property of the implicit function makes the collision detection between the discrete sampling points on model surfaces and the implicit surface of the probe very easy. Each point's coordinates are input into the implicit function to evaluate, and then we know if the point is inside or outside the probe by judging the sign of the potential value of $f(x, y, z)$. If a point is inside, a collision is detected.

3.3 *Force magnitude*

According to classical mechanics, the probe should be viewed as a rigid body instead of a single mass point, therefore forces applied on the surface of the probe not only forms a force vector but also a torque. However, the PHANToM device of our setup provides only 3 DOFs force feedback but no torque output. To simplify the problem, we only consider force at this stage. However, note that our probe is a rigid body in nature, so it would be quite easy to adapt the algorithm to 6 DOFs haptic rendering.

Penalty and constraint haptic rendering methods determine force magnitude by Hooke's law:

$$F = k \cdot s$$

where k is stiffness of a spring and s is the displacement of mass point connecting to the spring. To make the system more stable, a damping force is added, hence:

$$F = k \cdot s - d \cdot \dot{s}$$

where d is the damping factor and $\dot{s}$ is the velocity of the point. In our method we need an array of distributed damping springs instead of only one spring. Figure 1 shows a probe penetrating a planar point cloud of a model. White dots denote the points inside the probe and black ones are sampling points outside. The depths of colliding points are the distances between those

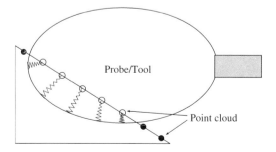

Figure 1.   Distributed springs of probe.

inside points to the correspondingly closest points on the implicit surface of the probe/tool.

Note the potential value from $f(x, y, z)$ is not Euclidean distance from a point to the implicit surface of the probe. Although Lagrange multiplier method can be used to find the minimum distance to the algebraic surface, it requires several iterations to find the numerical roots for the equation. We find the approximate answer by shooting a normal vector from the point P_0 and finding out the intersection between the vector and the implicit surface of probe. The normal vector of point $P_0(X_0, Y_0, Z_0)$ can be found as $N_0(N_x, N_y, N_z)$. Since points are inside the implicit surface instead of on the surface, the normal vector N_0 can be viewed as a normal vector of a smaller surface offset from the original one. If the implicit function is quadratic, in this case an ellipsoid, an analytic solution for the intersection points can be found easily. The distance between P_0 and the closer point is the depth.

Finally, we define the magnitude of the force as:

$$|F| = \sum_{i=0}^{n} (k \cdot s_i - d \cdot \dot{s}_i) \qquad (2)$$

Figure 2 shows what happens when a spherical probe, with a diameter of 800, passes through a uniform planar point cloud with continuous force magnitude. The output force reaches the maximum when the intrusion depth is 400. At that moment the center of the probe touches the plane, the probe has the largest contact area with a surface. After this peak the force magnitude decreases.

3.4 Impact of size of probe on force magnitude

According to Equation (2), force magnitude is proportional to the number of points colliding with a probe. Therefore, even at the same intrusion depth, a bigger probe generates bigger force because it has more points in collision with the model than a smaller probe has. Figure 3 shows two spherical probes, at the same depth, the bigger probe has more sampling points in collision.

Figure 4 shows force magnitudes recorded when two spherical probes pass through a planar point cloud. When the smaller probe reaches the force peak at depth 200, the force magnitude is less than 20 units, while the force feedback of the larger probe is over 40 units at the same depth. This is an undesirable effect that makes the haptic interaction unstable and changeable. In practice, when user increases the size of probe/tool, he may find the force feedback increases drastically and the surface of model becomes increasingly harder. The force feedback level may even exceed the safe threshold of the PHANToM device and result in hardware error.

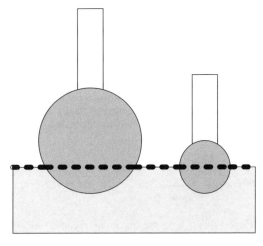

Figure 3. Two probes of different size generate different force feedbacks.

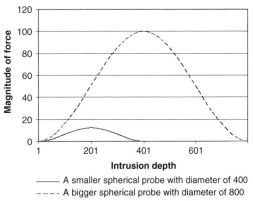

Intrusion depth

——— A smaller spherical probe with diameter of 400
– – – A bigger spherical probe with diameter of 800

Figure 4. Intrusion depth and force magnitude of two probes of different size.

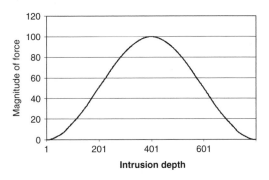

Intrusion depth

Figure 2. Intrusion depth and force magnitude.

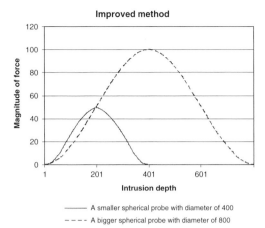

Improved method

—— A smaller spherical probe with diameter of 400
---- A bigger spherical probe with diameter of 800

Figure 5. Intrusion depth and force magnitude of two probes with new force feedback formula.

To resolve this effect, the equation (2) has to be modified to make the force output less related to the size of probe/tool. Since the intrusion depth and force magnitude is not linearly related, to totally eliminate the impact of size is not realistic. A simple way is to modified the equation as:

$$|F| = \frac{\sum_{i=0}^{n}(k \cdot s_i - d \cdot \dot{s}_i)}{Probe_{MaxSec}}$$ (3)

where $Probe_{MaxSec}$ is the maximum sectional area of probe.

Figure 5 shows nearly the same situation of the experiment of Figure 4 except the equation (3) is used in this example. We found at the force output peak of the smaller probe, the levels of force of both probes are nearly the same. Before the depth of 200 units, the smaller probe generates a slightly bigger force than the bigger one does. According to the author's test, the difference is nearly indiscernible.

3.5 Force direction

Penalty-based approaches and constraint-based approaches generally compute force vectors according to the normal vector of contacted surface (Zilles & Salisbury 1994). However, an unorganized point cloud does not have any surface normal information.

Our approach is called the *probe normal* method, where force vectors are not derived from the surface of the model but from the implicit surface of the probe instead. If there is no friction between the probe and model surface, the pressing force and reaction force are normal to the surface of contact area. Therefore, if

we can obtain the approximate normal vector of contact area on the implicit surface of the probe, we can use the sum of those normal vectors to determine the reaction force direction.

We therefore formulate the direction of force as:

$$\vec{F} = \frac{\sum_{i=0}^{n} \vec{N}_i \cdot d_i}{n}$$ (4)

where n is the number of surface sampling points inside probe, N_i as the normal vector of surface sampling point and d_i as the depth. The direction of feedback force is an average of the normal vectors weighted by the depth. This makes sense because the point having bigger depth plays a bigger role in determining the direction of force.

4 EXTENSIONS OF THE POINT CLOUDS HAPTIC RENDERING METHOD

4.1 Haptic rendering of NURBS surface patches

A two-stage haptic rendering approach has been implemented to haptically render B-rep models composed of NURBS surface patches. Each NURBS surface patch is firstly discretized into a dense point cloud, in which sampling points are evenly positioned. A bounding box is then generated for each NURBS patch to contain the corresponding point cloud.

In the first stage of haptic rendering, the system checks if the bounding box of probe/tool intersects with any bounding box of point clouds. In the second stage, point clouds corresponding to the collided bounding boxes found in the first stage are checked against the probe/tool implicit surface for collision detection. The two-stage strategy makes collision detection more efficient. Figure 6 shows the haptic rendering of a B-rep model.

4.2 Haptic rendering of NURBS curves

A NURBS curve with zero thickness is impossible to be touched with a zero-size point. Ye offsets NURBS curves to NURBS general cylinders and tessellates them to generate triangular meshes so that people can touch general cylinders with point-based 3D haptic cursor (Ye 2005). However, we found that it is not easy to access a specific curve with this method and tracing action is also not available. Tracing action permits movements of a cursor upon a surface or along a curve and helps the designer to quickly locate a position.

To haptically render NURBS curves, we evaluate the curve at a certain parametric density and get a chain of 3D sampling points. Then we can detect collision between the chain of 3D sampling points and our implicit surface probe and hence find out the feedback

228

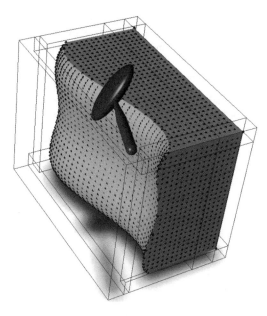

Figure 6.   Haptic rendering of NURBS surface models.

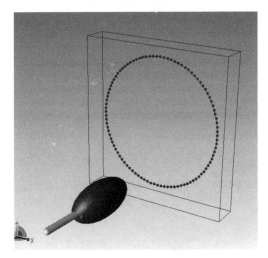

Figure 7.   Haptic rendering of NURBS curve.

force. Force feedback direction can be determined by the probe normal method. Figure 7 shows haptic rendering of a NURBS circle.

4.3   *Haptic rendering of rigid heterogeneous material models*

Traditionally, haptic rendering strategies are only developed for homogeneous material objects. However, due to the development of heterogeneous material

modelling, it is helpful to let the user touch heterogeneous material object. For example, in haptic evaluation of heterogeneous product has been explored.

Two haptic properties can be explored for rigid heterogeneous material models. One is friction, the other is stiffness. Because in our strategy surfaces are discretized into evenly-spaced point clouds, local friction factors and stiffness factors can be easily associated with points. Heterogeneous material patterns of models can be pre-defined before the haptic exploration or specified by the user interactively. Precision of a heterogeneous material pattern is proportional to the density of point clouds of a model surface. Different stiffness of different surface areas can be easily realized by introducing a local constant stiffness coefficient.

Haptic model of friction is:

$$\vec{f} = -\mu_f \cdot \vec{v} \tag{5}$$

where $\vec{f}$ is friction force vector, μ_f is a constant friction force coefficient and $\vec{v}$ is the velocity vector of probe/tool.

Note friction force is proportional to the velocity, therefore when there is no relative movement between model surface and the probe/tool, no friction force is outputted. This feature makes the system stable and less likely to vibrate. Also, since the friction force is related to the coefficient, different coefficients can be assigned to different areas to represent different materials and therefore results in different sense of smoothness. The friction force is then summed up with the normal force vector to yield the final force feedback.

5   CAD APPLICATIONS

In this section, we present two CAD applications based on our haptic interface. One of them is an experimental haptic sculpting system for B-spline surfaces, and another is a haptic based method to draw a NURBS curve upon a point based model.

5.1   *Haptic sculpting of B-spline surface*

Haptic sculpturing is a modelling technique based on the notion of sculpting a solid material while providing haptic feedback at the same time. The sense of touch, in combination with our kinaesthetic sense, adds a new modality to virtual sculpture, especially in presenting complex geometry and material properties.

The experimental haptic sculpting system features our new haptic rendering technique and shaped probe/tool (Gao et al. 2005). Because the probe/tool is an implicit surface, the probe/tool has a certain shape and size. The shape and size of sculpting tool plays an important role because the final shape of the model not only depends on the material property of the model,

229

the moving path of the tool, but also on the shape and size of the tool. The shapes of tools in our experimental sculpting system also serve as visual hints that correlate the desired deformation on models with the shape of selected tool. Users hold the stylus of the haptic device to drive the probe/tool to touch the B-spline surface model. A physics-based B-spline surface sculpting method has been developed by taking advantage of the new tool-model haptic interaction technique. A B-spline surface, which is constrained to a mass-spring system (MSS) mesh, deforms dynamically according to the interaction between the tool and the mass-spring mesh.

When the user sculpts the surface with a tool, the surface reacts in a physics-based manner and deforms according to the manipulation of the user. The system runs at 1 kHz updating rate, which is mainly because of the principle of haptic rendering. Euler integration for MSS mesh deformation also requires very small time steps to maintain stability. At each time step, the system samples the 6 DOF position of the PHANToM device and updates the position of the tool. Forces due to interaction between tool and surface are also computed in every time step. When haptic stylus clicks are detected, sculpting forces are applied to the MSS. Then, the mass points of MSS evolve to new positions. To visually display the deformation of a B-spline surface under sculpting, the control points of the B-spline surface must be updated from the transformation of MSS.

Besides the moving path of tool, the shape and size of tool also play roles in forming the final shape of surface. In Figure 8 and Figure 9, the deformations on surfaces copy the shape of sculpting tool. The final shape of the surface not only depends on the shape of the tool, but also on the stiffness of the model. Changing the overall mass of the surface not only changes the extent of deformation, allowing the surface to react either wildly or mildly to sculpting forces, but also changes the stiffness and material behaviour of B-spline surface. In Figure 8 and Figure 10, two B-spline surfaces, both at 30 × 30 grids but with different overall mass, are sculpted with the same tool. They demonstrate different behaviour during deformation. Figure 10 shows a lighter MSS having a bigger deformation extent and generating a feeling of softer material. In Figure 8, the mass of that MSS is 100 times heavier than the one in Figure 10. The result shows that the deformation is much more local and a sharp edge along the bulge appears.

Figure 11 and 12 demonstrate two more complex models. In particular, one should note the organic form of the model, which is the intended effect, as well as the continuity of the model, which is a direct consequence of using B-spline surface representation.

5.2 Drawing B-spline curves on point based models

Point based models are becoming more and more popular in both the computer graphics and CAD field.

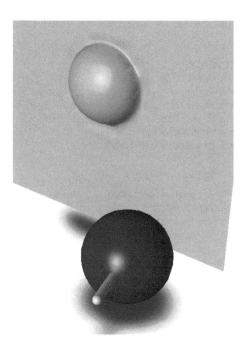

Figure 8. Sculpting with a spherical tool.

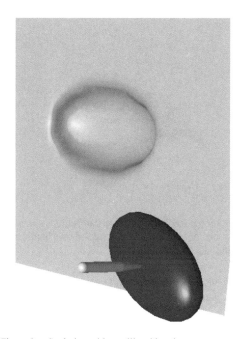

Figure 9. Sculpting with an ellipsoid tool.

One problem of point based modelling is drawing curves onto unorganized point clouds. Drawing curves on point clouds has applications not only on reverse engineering, but also on point based modeling. Azariadis et al. present a set of design tools to solve this problem (Azariadis & Sapidis 2005). Their

primary tool is a point projection algorithm. There are 3 steps to draw a curve onto a point cloud. In the first step, a user-defined poly-line is projected onto the point cloud by taking advantage of the point projection algorithm. The second step smooths the projected polyline on the point clouds. Then, the polyline is interpolated to a B-spline curve. Because the polyline projection of the first step often introduces wrinkling, the smoothing process in the second step is very important. A parameter has to be specified by the user to smooth the polyline and the choosing of the parameter is of trial-and-error style.

We propose a haptic based method which is more intuitive and free of parameter adjustment. Our approach also includes three components: polyline sketching, polyline checking and modification, B-spline curve interpolation.

In the first step, the user manipulates the PHANToM stylus to pick points on the point clouds and hence forms the polyline. The point picking is the primary tool in our approach. The user can explore the point-based model with PHANToM device freely. When he wants to pick a point in contact with the probe, he can click the PHANToM stylus button click to signal the system. According to the algorithm described in chapter 3, a number of points have been found in collision with the probe/tool by the collision detection unit of haptic rendering component. Therefore, a point can be easily found by averaging the points in collision with the probe/tool.

Figure 10.   Sculpting on a softer and lighter material.

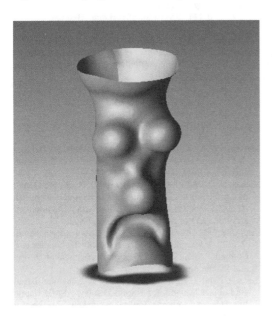

Figure 11.   A vase of funny face.

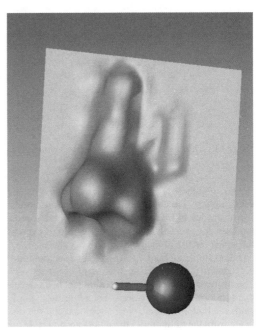

Figure 12.   A nose.

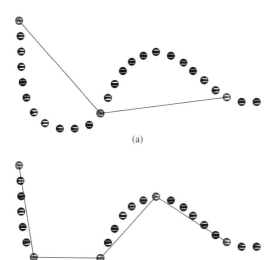

(a)

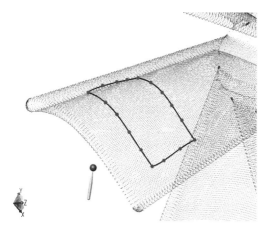

(b)

Figure 13.   Polyline examination and modification.

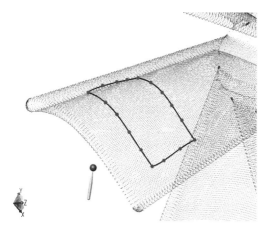

Figure 14.   Drawing NURBS curve on a point cloud.

Our approach is highly interactive and it is the user's duty to sketch a polyline on the point clouds. However, due to the complex shape of point clouds, the user defined polyline may not be able to express the shape feature of the point clouds very well.

For example, in Figure 13 (a) two line segments are not enough to describe such a highly curved area. Therefore, an examination algorithm has been developed to locate these situations. The algorithm simply finds the midpoint of every line segment and the distance from the midpoint to the points around it. If the distances are all bigger than a tolerance value, the line segment is then labelled as problematic segment and highlighted. Then the user can insert more points in the highly curved area to improve the polyline in Figure 13 (b). This process can be repeated until all of the line segments are close enough to the point cloud.

The last component of our approach is simply the B-spline curve local interpolation algorithm. This part interpolates the polyline into a B-spline curve. Figure 14 shows a curve drawn on a point cloud of a fan.

6   CONCLUSION

A new technique of haptic rendering has been implemented. This technique leads to a more intuitive and versatile haptic interface for the CAD/CAM system. Our haptic rendering technique can deal with multiple CAD model representations, such as point clouds, NURBS curves and surfaces, and rigid heterogeneous material objects.

Two CAD applications based on that haptic interface, including an experimental haptic sculpting system for B-spline surfaces and a haptic based method to draw a NURBS curve upon a point based model, are demonstrated to verify the proposed haptic rendering technique.

REFERENCES

Azariadis, P. and N. Sapidis (2005). "Drawing Curves onto a Cloud of Points for Point-Based Modelling." *Computer-Aided Design* **37**(1): 109–122.

Gao, Z. and I. Gibson (2005). "Haptic B-spline Surface Sculpting with a Shaped Tool of Implicit Surface." *Computer-Aided Design and Applications* **2**(1–4).

Gregory, A., A. Mascarenhas, et al. (2000). Six Degree-of-Freedom Haptic Display of Polygonal Models. Proceedings Visualization 2000.

Ho, C. H., C. Basdogam, et al. (1997). Haptic rendering: point- and ray-based interactions. The second PHANToM Users Group Workshop.

Kim, L., A. Kyrikou, et al. (2002). An Implicit-based Haptic Rendering Technique. IEEE IROS 2002, Switzerland.

McNeely, W. A., K. D. Puterbaugh, et al. (1999). Six Degree-of-Freedom Haptic Rendering Using Voxel Sampling. SIGGRAPH 99, Los Angeles, CA USA.

Nelson, D. D., D. E. Johnson, et al. (1999). Haptic Rendering of Surface-to-Surface Sculpted Model Interaction. 8th Annual Symp. Haptic Interfaces for Virtual Environment and Teleoperator Systems, Nashville, TN.

SensAble. "PHANTOM™." from http://www.sensable.com.

Ye, J. (2005). Integration of Virtual Reality Techniques into Computer Aided Product Design. Department of Design and Technology, Loughborough University.

Zilles, C. and J. K. Salisbury (1994). A Constraint-based God-object Method For Haptic Display. ASME Haptic Interfaces for Virtual Environment and Teleoperator Systems.

Virtual modeling and rapid manufacturing – Bártolo (eds)
© *2005 Taylor & Francis Group, London, ISBN 0 415 39062 1*

Product customization in a virtual environment

Y.H. Chen, L.L. Lian & Z.Y. Yang
Department of Mechanical Engineering, The University of Hong Kong, Hong Kong, China

ABSTRACT: In today's competitive market place, product designers need not only to elaborate their design with respect to forms and functions but also physical properties that make product safer and more comfortable to use. Current computer aided (CAD) systems mainly support product geometric properties and visualization. Physical properties such as stiffness are difficult or even impossible to be perceived in existing CAD systems. In this paper, a haptic system is developed to customize product physical properties such as product surface textures and stiffness. In the proposed system, a designer can perceive, for example, the surface roughness of a handle, the stiffness of a toothbrush, and exam the trigger force of a pushbutton, or even feel the vibration while an electrical razor is powered on as if the designer is operating a real product. If any of the above physical properties is not desirable by the designer (or customer), the designer can easily make change in terms of geometries, materials, or combinations until customized properties are perceived. Since these physical properties of a product, traditionally evaluated based on physical prototypes, can now be perceived in a virtual environment, the product development cycle can be shortened and the cost in making prototypes can be reduced.

1 INSTRUCTIONS

In the modern society, sophisticated consumers desire that products match their own feelings of design, function, and price (Nagamachi 1995). Product developers have to cater to the consumers' personalized requirements. This has lead to the wide spread use of Mass Customization (MC), producing customized goods for a mass market, in today's product development (MacCarthy 2003). There are two key issues in mass customization (Duray 2000). The first issue concerns the customer involvement in determining the degree of customization. The second issue is related to modularity. Customer involvement provides the customization while the modularity provides the basis for repetitiveness in production or the "mass" in MC.

In hand tools or hand hold product development, the comfort of product operation relies a lot on the condition of interaction between hands and the product. McGorry proposed a system for the measurement of grip forces and applied moments during hand tool use (McGorry 2001). The magnitude and direction of the grip force vector and the simultaneous determination of applied moments produced in a task simulation is estimated. The result could be used as guidelines in ergonomic design of single handled tools. Smith *et al.* investigated the musculoskeletal discomfort or disorder experienced by hygienists during their careers and alternative methods for viewing teeth while performing

simulated dental procedures were presented (Smith *et al.* 2002). The methods allowed participants to assume postures requiring less neck flexion than the standard direct view. More computer-aided ergonomics softwares have been developed to evaluate ergonomic properties in a computer such as 3DSSPP (Feten *et al.* 2000). The use of ergonomics analysis software in conjunction with CAD has distinct advantages: a reduction in overall project time, the ability to apply ergonomics information early in the design process, improved communication of both ergonomic concerns and design alternatives, and cost effectiveness.

External pressure at the hand during object handling and work with tools was measured in Hall's work (Hall 1997). Pressure-pain threshold of 15 subjects with variant hand size and hand strength was recorded. Obviously, this information is useful to design personalized tools which are to be used safely and comfortably. Kawai *et al.* analyzed the manipulative force control in grasping an augmented object (Kawa *et al.* 2002). Virtual reality allows changes to be made to the visual perception of object size even while the tangible components remain completely unaltered.

However, current virtual systems are only applicable to product dimensional factors. Other ergonomic factors are seldom provided as an option in mass customized product design. One of the reasons is that not all the ergonomic characteristics can be easily presented to the end users via existing techniques. Some

Table 1. Ergonomic properties and related virtual reality techniques.

Sense	Human sensors	Virtual techniques	Ergonomic properties
Sight	Eyes	Computer graphics	Visual comfort Targeting effectiveness . . .
Sound	Ears	Digital audio	Noise level, tune, . . .
Touch	Hands, Skin	Computer haptics	Stiffness, textures, force, CDTs . . .
Smell	Nose	Virtual smell	?
Taste	Tongue	Virtual Taste	?

Figure 1. The system set-up.

ergonomic properties and their corresponding simulation techniques are listed in Table 1. Not all of the five channels through which human beings collect information from the environment can be simulated faithfully via virtual reality techniques till now. We have reached the point where virtual sound is almost indistinguishable from real sound, and virtual sight is acceptable in many cases of simulation. Therefore, aural and visual ergonomics properties are easier to be simulated and presented to designers or consumers in a virtual system. In contrast, haptic ergonomic properties, such as weight, pressure points, contact stress, cumulative trauma Disorders (CTDs) are difficult to be presented if no real product is available. Haptic modeling has been used in industrial and conceptual design since 1990s (Srinivasan and Basdogan 1997). Because of the high fidelity in today's haptic devices, some ergonomics factors, such as tactile, force and pressure can be modeled and felt by users. Haptic aided ergonomics design is therefore proposed in this paper, within the context of mass customization in the early stage of product design. In this paper, we limit our work to virtual real time stiffness evaluation of hand tools, such as a toothbrush.

2 HAPTIC RENDERING

In this research, the haptic device is a Phantom system from SensAble Technologies®. The overall system is developed using C++ together with the device interface GHOST® API (Sensibe 2005). As in Figure 1, the haptic arm has both position encoders and servo actuators. A positioning cursor can be shaped as a finger or any other shapes shown on a display panel. In Figure 2, a hand shape is intuitively used as a positioning cursor for modeling the activation of a pushbutton. When the pointer figure comes into contact with an object such as a pushbutton button in Figure 2, a feedback force is activated and felt by the user. The magnitude and direction of the force are mainly dependent on the mechanics model of the actual application. In this paper, a tooth brush is simulated as an example.

Figure 2. Pushbutton simulation.

As can be seen in Figure 1, holding the haptic device is like holding a toothbrush. For best visual effect, a tooth model will be scanned and used as the environment in the system implementation, and the haptic cursor will be modeled as a toothbrush.

When simulating a toothbrush operation, the applied force F and the resulting displacement x relationship must be established. In this research, the load is simplified as a cantilever beam under a bending force F shown in Figure 3. The bending force will result in a corresponding displacement x. Due to the geometric and material change of toothbrush cross sections, the deduction of a equivalent spring constant k that relates F and x is nontrivial. As the recent development of

234

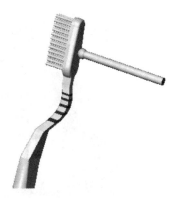

Figure 3.　The simplified toothbrush operation.

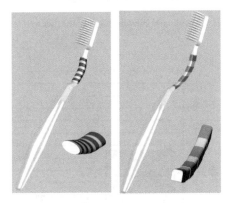

Figure 4.　Toothbrush with different neck design.

soft touch material molding technologies, more and more toothbrush designs use soft materials in the neck regions so that the toothbrush is more elastic and good looking which are thought to be good features that attract customers.

Figure 4 shows two examples of toothbrush design with different neck shapes and color patterns. The different colors may also represent different materials. Customer preference of the color patterns can be easily assessed by the rendered images. However, it is difficult to assess the elasticity of the toothbrush. A change of toothbrush neck shape and material composition results in a change of elasticity. Elasticity is an important factor that has a large impact to customer satisfaction. Based on a two material composition, a force and displacement relationship $F = kx$ has been deduced in a previous research paper (Yang and Chen 2004). In that paper, the spring constant k is found to be dependent on the materials and size of the neck. With the availability of the spring constant, a real time haptic simulation can be easily done. Figure 5(a) shows a toothbrush design under no force. When a force is

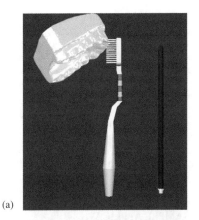

(a)

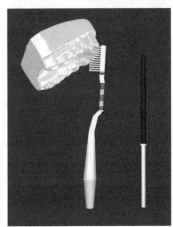

(b)

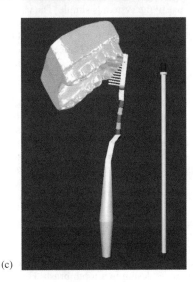

(c)

Figure 5.　A virtual toothbrush in operation. (a) Toothbrush and tooth model in contact, (b) A moderate force is applied, (c) A large force is applied.

applied, the toothbrush deflects. An indicator showing the relative force magnitude is given beside the toothbrush. Figure 5(b) shows a moderate force and the corresponding deflection. Figure 5(c) indicates an obvious deflection when a much larger force is applied. While applying the force, the user can observe the deflection and feel the force. It is desirable that a change of material composition and geometric shape of the toothbrush neck should result in a comfortable deflection and force relationship, that is the spring constant k, a factor that is normally assessed based on physical prototypes, can now be subjectively assessed in the proposed haptic virtual system.

3 A SAMPLE APPLICATION

A prototype system is developed to customize product design based on age groups, gender groups, physical body weight and size groups. It is believed that data collected from these kinds of virtual experiments could help the mass customization design of a product. Figure 6 shows a sample implementation of a graphical user interface. It allows the input of user particulars. By adjusting the stiffness scale from 0–10, the user can evaluate the virtual toothbrush to see if the toothbrush stiffness is appropriate. In Figure 6, the current stiffness is set to 4.5. If the stiffness value is not desirable by the user, the user can easily reset the stiffness value on the scale bar. By doing trial and error test on a haptic system, the desired stiffness value of a toothbrush can be found. This value together with the user's particulars can be saved for further use.

The use of the system is simple. The user can hold the haptic device Phantom as circled in Figure 1. This gesture is quite similar to the way we hold a toothbrush while we actually brush our teeth. Since the toothbrush is controlled by the motion of the haptic device, a force feedback is felt by the user when the toothbrush is brought into contact with the tooth model. When the user group is large, statistic data such as those shown in Figure 7 can be got and plotted. The chart in Figure 7 clearly indicates that the stiffness of toothbrush design should be different for women and men groups. When this visible difference in stiffness is incorporated in a product design, a customized feature is added to the product which is believed to be more appealing to customers. The design process will be more valuable if a material data base is imbedded so that the impact of different material can also be evaluated in the haptic system.

In fact, the simple graphical interface in Figure 6 provides users with a lot of opportunities for customized stiffness design in a tooth brush. For examples, the stiffness can be designed based on age groups, or body weight groups, or body height groups, or combinations such as gender-age groups, gender-weight

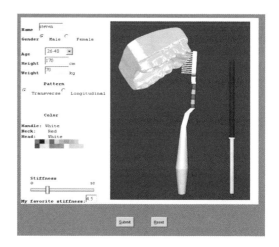

Figure 6.   A sample interface for stiffness perception.

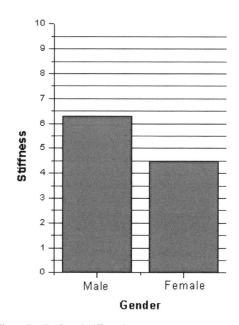

Figure 7.   Preferred stiffness by user groups

groups, etc. This kind of customization provides not only customized properties, but also product choices.

4 DISCUSSIONS AND CONCLUSIONS

The life cycle of products is becoming shorter while mass customization is inevitable, resulting more intense requirement on design techniques, manufacturing technologies, and equipments than ever. With the aid of virtual reality techniques, the need of making physical prototypes is reduced. When the proposed system is extended to a web-based haptic evaluation platform, it can even reach more people with different culture and background at a very short time. Since some properties of a product design can be evaluated in a virtual system, the product design cycle is reduced, and the cost in making physical prototypes is also reduce. This paper reported a sample study only on a simple part, more studies on part with complex geometry and load patterns will benefit product designers in a broader industry.

ACKNOWLEDGEMENTS

This research is supported by a grant from Hong Kong Research Grants Council under the code HKU 7073/02E.

REFERENCES

Duray, R., Ward, P., Milligan, G., and Berry, W. 2000. Approaches to mass customization: configurations and empirical validation. *Journal of Operations Management*. Vol. 2, No. 2: 605–625.

Elofson, G., and Robinson, W. N. 1998. Creating a custom mass-production channel on the Internet. *COMMUNICATIONS OF THE ACM*. Vol. 41, No(3): 56–62.

Feyen, R., Liu, Y. L., Chaffin, D., Jimmerson, G., and Joseph, B. 2000. Computer-aided ergonomics: a case study of incorporating ergonomics analyses into workplace design. *Applied Ergonomics*. Vol(31): 291–300.

Hall, C. 1997. External pressure at the hand during object handling and work with tools. *International Journal of Industrial Ergonomics*. Vol(20): 191–206. http://www.sensable.com

Kawai, S., Summers, V. A., Mackenzie, C. L., Ivens, C. J., and Yamamoto, T. 2002. Grasping an augmented object to analyse manipulative force control. *Ergonomics*, Vol. 45, No(15): 1091–1102.

MacCarthy, B., Brabazon, P. G., and Bramham, J. 2003. Fundamental modes of operation for mass customization. *International Journal of Production Economics*. Vol. 85: 289–304.

McGorry, R. W. 2001. A system for the measurement of grip forces and applied moments during hand tool use. Applied Ergonomics Vol(32): 271–279.

Nagamachi, M. 1995. Kansei Engineering: a new ergonomic consumer-oriented technology for product development. *International Journal of Industrial Ergonomics*. Vol. 15: 3–11.

Smith, C. A., Sommerich, C. M., Mirka, G. A., and George, M. C. 2002. An investigation of ergonomic interventions in dental hygiene work. *Applied Ergonomics*. Vol(33): 175–184.

Srinivasan, M. A., and Basdogan, C. 1997. Haptics in virtual environments: taxonomy, research status, and challenges. *Computers & Graphics*, Vol. 21, No(4): 393–404.

Wu, S. P., and Hsieh, C. S. 2002. Ergonomics study on the handle length and lift angle for the culinary spatula. *Applied Ergonomics*. Vol(33): 493–501.

Yong, Z. Y., and Chen Y. H., 2004. Haptic-based function analysis and shape modeling of multi-material product: A case study. DETC2004-57782, *ASME 2004 Design Engineering Conferences and Computers and Information in Engineering Conference*, October, 2004, Salt Lake City, USA.

Virtual modeling and rapid manufacturing – Bártolo (eds)
© *2005 Taylor & Francis Group, London, ISBN 0 415 39062 1*

3D laser scanner technology: building data integration system

P. Gamito & B. Caetano
Universidade Lusofona de Humanidades e Tecnologias, Campo Grande, Lisbon, Portugal

Y. Arayici & A. Hamilton
School of Construction and Property Management, University of Salford, Greater Manchester

ABSTRACT: Advanced digital mapping tools and technologies are enablers for effective e-planning, consultation and communication of users' views during the planning, design, construction and lifecycle process of built and human environments. The regeneration and transformation of cities from the industrial age (unsustainable) to the knowledge age (sustainable) is essentially a 'whole life cycle' process consisting of; planning, development, operation, reuse and renewal. In order to enhance the implementation of build and human environment solutions during the regeneration and transformation of cities, advanced digital applications can have a significant impact. 3D laser scanners offer a fast and reliable procedure to capture data from terrains and buildings. This technology is beginning to be used to acquire 3D point clouds from buildings to be refurbished that are lacking of detailed and up-to-date architectural plans. The fast development of the technology presently allows a rapid production of 2D and 3D CAD models from the point cloud. This paper presents a methodology for producing CAD models from point clouds and discusses the integration of this technology on a vast platform of IT systems, such as, workbenches, 3D printers, nD modeling, GPS and VR projection system. A building data integration system for a visual decision making supported system is also proposed.

1 INTRODUCTION

The ever-increasing development on both software and hardware technologies is stimulating the speeding of processes and the creation of different working methodologies. Such is the case of 3D laser scanner technology. This recent technology enables an accurate and fast way of acquiring 3D data from objects or terrain areas. This data forms a 3D point cloud that reproduces on a PC the entire object or portion of landscape with a degree of accuracy that can go up to 5 mm.

For each acquired point, a distance is measured on a known direction: X, Y, and Z coordinates of a point can be computed for each recorded distance direction. Laser scanners allow millions of points to be recorded in a few minutes. Because of their practicality and versatility, these kinds of instruments are beginning to be widely used in the field of architectural, archaeological and environmental surveying. From these point clouds, 3D models can be generated with sufficient accuracy for built documentation, restoration plans, virtual environment generations, and interactive manipulation.

Intervention within areas of historical and architectural importance it is not confined to the analysis of an accurate 3D model of a building. Other dimensions are needed to be accounted for. A multidisciplinary knowledge base, created from the analysis of a myriad of aspects such as: historical, architectural, archaeological, environmental, social, economic is compulsory. Only with comprehensive information on the complete lifecycle of an historical landscape enables the development of policies and the later decision-making on the part of the owners and managers of the buildings to be refurbished and of the areas to be reused. This "clear picture" can be provided when 3D laser scanner technology is associated to other IT systems and platforms.

2 3D LASER SCANNER TECHNOLOGY

The 3D laser scanner is targeted to the physical objects to be scanned and the laser beam is directed over the object in a closely spaced grid of points. By measuring the time of laser flight, which is the time of travel of the laser from the scanner to the physical objects and back to the scanner, the position in three-dimensional space of each scanned point on the object is established. The result is a "cloud of points" which consists in thousands of points in 3-dimensional space that are

Table 1. Advantages and disadvantages of 3D laser scanning technology.

3D laser scanning advantages	Disadvantages
Applicable to all 2D and 3D surfaces.	Some systems do not work in sun or rain.
Rapid 3D data collection-near real time-requiring substantially less site time.	Large, high resolution 3D data sets require post-processing to produce a useable output.
Very effective due to large volumes of data collected at a predictable precision.	Difficulty in extracting the edges examples from indistinct data clouds.
Ideal for all 3D modeling and visualisation purposes.	Output requires manipulation to achieve acceptable recording quality.
Both 3D position and surface reflectance generated which, when processed, can be viewed as an image.	No common data exchange format-such as DXF-currently in use to allow ease of processing by third parts.
Rapidly developing survey technology.	Difficult to stay up-to-date with developments.
Extensive world-wide research and development currently undertaken on both the hardware and software tools.	With hardware still expensive and sophisticated software required to process data, cost is prohibitive for many projects.

Figure 1. 3D CAD models extracted from the raw scan data Jactin house. Manchester, UK.

a dimensionally accurate representation of the existing object (Schofield, 2001). This information can then be converted in a 3D CAD model that can be manipulated using CAD software, and to which the design of new equipment can be added. Terrestrial laser scanning offers fast 3D terrain and building data acquisition. It has advantages over current survey techniques including EDM, GPS and photogrammetric applications obtaining high density point data without the need for a reflector system. Merged data clouds have sufficient points to negate the need for DEM interpolation techniques potentially providing the optimum representation of any scanned surface. The advantages of speed and high data point density must be viewed against the data point accuracy, which may reduce instrument performance below that achievable using EDM techniques (such techniques are, however, much more time consuming). Any improvement to measurement range, resolution, field of view and error/accuracy would further fill the research gap relating to spatial and temporal measurement of space and change in the natural environment and resolve the accuracy issue with regard to EDM techniques.

3D laser scanning technology advantages and disadvantages are summarised in Table 1.

3 PRODUCING 3D CAD AND PHYSICAL MODELS FROM THE POINT CLOUD

Due to its characteristics, one main advantage of using a 3D laser scanner for refurbishment purposes resides on the possibility of CAD lines extraction from the 3D mesh model. This process only requires a simple command operation on Polyworks software and the lines are ready to be exported to a CAD system. This means that CAD plans and DEM can be obtained within a reduce frame of time. The benefits of having 'ready-made' accurate CAD lines from an old building can be mammoth in such cases where original plans are missing or are simply not up to date. Besides, this technology has the potential to accurately record inaccessible and potentially hazardous areas such as pitched rooftops. Consequently, it facilitates "virtual refurbishment" of the buildings and allows the existing structure and proposed new services to be seen in an effective manner (Arayici et al, 2003). Figure 1 below shows an example of 2D and 3D CAD models produced from the building data captured by the 3D laser scanner.

Regarding architecture and construction, architects frequently need to determine the "as built" plans of building or other structures such as industrial plants, bridges, and tunnels. A 3D model of a site can be used to verify that it was constructed according to specifications. When preparing for a renovation or a plant upgrade, the original architectural plans might

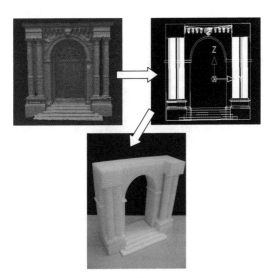

Figure 2. The process of creating physical models from digital laser scanned data.

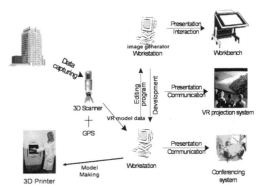

Figure 3. Integration of spatial data with 3D printer and the virtual environments.

be inaccurate, if they exist at all. A 3D model allows architect to plan a renovation and to test out various construction options.

Within the built environment, the use of the 3D laser scanner enables digital documentation of buildings, sites and physical objects for reconstruction and restoration. Furthermore, the use of the 3D scanner in combination with the 3D printer provides the transformation of digital data from the captured CAD model back to a physical model at an appropriate scale – reverse prototyping (Figure 2). The use of these technologies is key enablers to the creation of new approaches to the "Whole Life Cycle" process within the built and human environment for the 21st century.

Overall, this advanced digital mapping tools and technologies are enablers for effective e-planning, consultation and communication of users' views during the planning, design, construction and lifecycle process of built and human environments. The regeneration and transformation of cities from the industrial age (unsustainable) to the knowledge age (sustainable) is essentially a "Whole Life Cycle" process consisting of planning, development, operation, reuse and renewal. In order to enhance the implementation of build and human environment solutions during the regeneration and transformation of cities, advanced digital applications can have a significant impact.

4 CONCEPTUAL SYSTEM OF BUILDING DATA INTEGRATION

The benefit of the 3D laser scanner technology is not confined to the production of CAD models. When

used along other IT platforms can be a powerful tool in enhancing decision making in the area of cultural heritage. As illustrated in below figure (Figure 3), laser scanner technology can be integrated with various systems such as CAD systems for 2D and 3D CAD plans, with the GPS system for linking the OS data, with the virtual environment tools such as workbench, VR projection system, Video conferencing system for visualisation and communication, with the 3D printing system for hard modelling and with the nD modeling repository for storing the information produced with the laser scanning system in a database that embraces information in various formats for future use during the refurbishment process.

5 INTEGRATION WITH GIS THROUGH ND MODELING DATABASE FOR A VISUAL DECISION MAKING SUPPORT SYSTEM

The building data captured by the laser scanner is processed to generate information, which will be stored in the nD modeling database in the appropriate formats including IFC. For example, the pattern matching approach is elaborated for how to define building objects from scan data. The information stored in the database will be available for decision making process and reuse for the refurbishment process in the future.

The aim of this section is to describe a framework which includes the nD modeling database, the 3D laser scanner and a series of analytical tools that will enable various stakeholders in the building design in the regeneration process to make informed decisions relating to building maintenance works (Ahmed et al, 2004).

The geo-spatial nD modeling database will describe the geometries of both the building frame and its components. Simple open geometric descriptions will be used, but each entry will also be associated with data

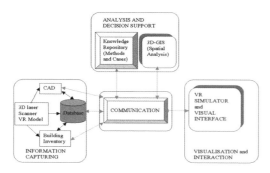

Figure 4. Structure of a visual decision making support system.

on inventory information such as name, supplier, date installed/replaced, number of previous replacements, etc. (Ahmed et al, 2004). The system will enable the capture of all geo-spatial data of the using laser scanner. This data will be processed and building frame and components will be determined in CAD environment. The building frame and components will also be stored in the nD modeling database in IFC (Industry Foundation Classes) data model. Inventory information relating to each frame and component will also be captured within the relational structure of the database. Such information will be accessible in real-time with some of the attributes (e.g. component supplier information) hyperlinked to the World Wide Web. This framework is depicted in Figure 4.

Following the development of the nD modeling database, information captured will be linked to a knowledge repository developed purely for rule base and/or case-based interpretation of possible building maintenance schedules. This component of the system will facilitate the generation of alternatives based on user-specified queries (Mahdjoubi & Ahmed, 2004).

GIS software will be used to generate and analyse thematic developments relating to the building properties and associated maintenance management strategies. The GIS software will retrieve the building information from the nD modelling database through the communication layer, which provides the users situated in various locations with access to the nD modeling database.

The VR environment described in section 4 and illustrated in figure 4 will be developed using the laser scanner technology which also provides data models in different formats including the Virtual Reality Modeling Language (VRML). This approach is complimentary to previous work done on the information infrastructure developed through the OSCON, VIRCON and HyCon projects and the ongoing research for nD modeling project (Lee et al, 2003). Therefore, the results of the spatial analysis obtained within the 3-D GIS environment can be evaluated in real-time with the

options of viewing building maintenance alternatives developed from querying the knowledge repository.

The research is of potential benefits and practical applications to the construction industry and professions. It will provide a better support for evaluation and visualisation of building maintenance works so that informed policies can be effectively targeted. It will benefit construction companies, facility and estate managers, and all those concerned with building maintenance issues (Ahmed et al, 2004).

The ultimate beneficiaries of this work will be professionals and stakeholders of the construction industry involved with the:

– building maintenance,
– improved predictability of building maintenance requirements,
– reduced maintenance planning and execution time,
– increased safety,
– increased productivity.

To make the information flow as the arrows illustrate in the figure 4 several problems need to be overcome. One basic problem is the nature of scanned 3D data. Whereas 2D data captured by a camera confirms to accepted data standards (JPEG, TIFF, etc.) current 3D data output varies in format from scanner to scanner. This data conversion issue is discussed in the next section. It can be seen that to make progress in the area of modeling, conversion, issues of 3D scanned data need to be solved.

6 FINAL CONSIDERATIONS

Among others, there is one question that remains unanswered: can different objects be automatically recognized within the above proposed visual decision supported system? i.e. is it possible to develop a pattern matching interface so that elements of the "virtual" building could be automatically detected? Despite being on its early stages, it is already possible to trace a direction. 3D CAD models using semi-automated and fully automated techniques are extracted from the polygonal mesh model developed in the laser scanner system. The pattern matching interface being designed will invoke the 3D CAD model to its display screen. Highlighting any building frame in the model will enable the pattern matcher to define geometric features as criteria for matching process such as shape, sides, width, height, thickness, line type, line thickness, line colour and so on. Beside the automatic feature recognition, users can also input further criteria for matching process.

Microstation triforma employs a single building model concept. All information about a building (or at least as much as possible) is recorded in a 3 dimensional model. Traditionally a given door in a building would be drawn in at least three or four places (plan, building elevation, building section, interior elevation, etc). In the single building model, it is constructed once and these various drawings are later extracted automatically. The single building model requires building objects, which are defined, edited and stored in the triforma library.

REFERENCES

Ahmed, V., Arayici, Y., Hamilton, A., Aouad, G., 2004. Virtual Building Maintenance: Enhancing Building Maintenance Using 3D-GIS and 3D Laser Scanner (VR) Technology. In *Proceedings of The European Conference on Product and Process Modelling in the Building and Construction Industry (ECPPM)*, Istanbul.

Arayici, Y., Hamilton, A., Hunter, G., 2003. Reverse Engineering in Construction. In *Proceedings of The Conference of World of Geomatics 200*3: Measuring, Mapping, and Managing", Telford, UK, 2003.

Lee, A., Marshall_Ponting, A., Aouad, G., Song, W., Fu, C., Cooper, R., Betts, M., Kagioglou and Fischer, M., 2003. Developing a Vision of nD-Enabled Construction.

Mahdjoubi, L., and Ahmed, V., 2004. Virtual Building Maintenance: Enhancing Building Maintenance using 3D GIS and Virtual Reality (VR) Technology. In *Proceedings of The Conference of Designing, Managing, and Supporting Construction Projects through Innovation and IT solutions (INCITE2004)*, February 2004, Langkawi, Malaysia.

Schofield, W., 2001. Engineering Surveying 5th Edition: Theory and Examination Problems for Students. Butterworth-Heinemann.

Virtual modeling and rapid manufacturing – Bártolo (eds)
© *2005 Taylor & Francis Group, London, ISBN 0 415 39062 1*

3D-digitising and reverse engineering, aspects of production engineering

D. Fichtner, C. Schoene & S. Schreiber
University of Technology Dresden, Germany

ABSTRACT: The user of Reverse Engineering methods is confronted with a variety of different 3D digitising systems and data recording methods for further processing. However, the powerful digitising equipment can only be used successfully in connection with qualified and customised data conditioning and application in the follow-up computer-aided strategy. In recent 10 years, 3D digitising and Reverse Engineering have been established in mock-up and diemaking of mechanical engineering. Apart from these ranges we know a lot of other tasks to be solved successfully with Reverse Engineering. Supporting for these processes is one research and transfer focus of the chair of Production Automation and Cutting Technology. Strategies to extend the teeth shape variety at Heraeus Kulzer, Reverse Engineering in art, culture, as well as in the automotive industry are represented.

1 INTRODUCTION

In the strictest sense, Reverse Engineering describes the procedure of 3D digitising of workpieces with sculptured surfaces, conditioning the 3D point data and converting them to CAD models (Beyer 2002). As we have learnt from our own experience in using powerful 3D digitising equipment, positive results will only be obtained by 3D data recording in conjunction with qualified and problem-oriented data conditioning and application in the follow-up computer-aided strategy (Schöne et al. 2004, Schöne & Paul 2001, Schöne et al. 2001, Schöne & Hoffmann 2003). Today workpieces with sculptured surfaces are still very demanding and dependant on the expert's knowledge as well as the software and hardware components in the areas of data recording, CAD modelling, production planning and manufacturing.

Reverse Engineering is to be considered in the wider sense, if the digitised data is converted into CAD models thereby not infringing the marginal conditions demanded, on the one hand. Another equally important fact is that the CAD representation successfully withstands production planning, manufacturing and quality inspection at the corresponding user's firm (Figure 1).

At present, physical prototypes are frequently made from processed points that are cross linked and represent triangulated data.

This process chain variant bypasses Reverse Engineering and guarantees the reproduction of initial prototypes commensurably quickly. This approach can also be regarded as the method of choice in some cases of application. This paradigm is predominantly recommended for organic objects. For VR techniques, such as for instance 3D projection, the 3D objects are subjected to discretisation. For this reason, immediate use of the 3D model based on the triangulated data is indicated even in this case.

For objects that have to fulfil styling requirements, Reverse Engineering is frequently necessary to smooth and obtain CAD models of continuous curvature.

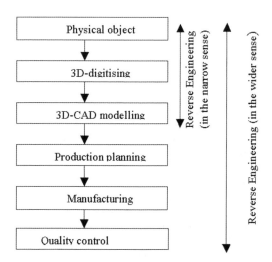

Figure 1. Reverse Engineering elements in the wider and narrow sense.

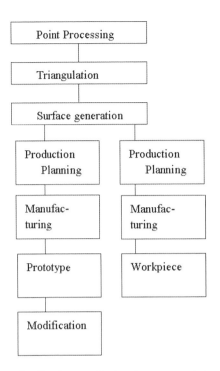

Figure 3.   Process steps to reduce the lateral teeth (lower jaw) by keeping the occlusal dimension.

In computer-aided processes, the geometry of the individual teeth is determined with 3D scanners developed especially for the dental industry. If we receive digital 3D data for these 3D objects after scanning, e.g. in the form of point clouds, this data may be processed in a follow-up process chain.

The point cloud is triangulated for visual evaluation of the digitising results. Powerful algorithms are available for triangulation. The point cloud is subsequently filtered to reduce white noise. Then, the following local changes of the individual teeth's geometries are carried out based on the triangulation data to provide the desired data. The objective includes abstracting the new tooth shapes from available digital data so that 3D digitising and data conditioning becomes inapplicable. Now, the product portfolio may be extended from existing 3D geometries by intentional scaling along the 3 main axes. The corresponding teeth are generated by mirroring around each main axis. This procedure to extend the tooth shapes for front teeth cannot always be used for lateral teeth as scaling results in changing the occlusal dimension. The quantity and depth of fissures on the masticatory surface should be kept even after changing the model. For this objective, we tested a sequence coping with these requirements (Figure 3).

The 3D CAD models have to be subsequently prepared for manufacturing the production teeth moulds. First, the geometry has to be aligned in a way that enables mould parting. In the next step, the parting line separating the upper and lower dies is defined. Afterwards the parting surface, which has the functions of the sealing surface between the mould halves and intermediate moulds, is designed according to the fixed parting line (Figure 4).

A tooth surface representation via Spline surfaces with continuous curvature transitions between the individual surfaces has to be created from STL data in a few minutes. Consequently, this route was followed in parallel to the STL representation. The Spline surface description of the tooth's outer shape was analysed in order to compare it with the STL representation. Mean deviations between both representation types are -0.004 to $+0.006$ mm. Values from -0.04 to $+0.08$ mm were calculated as maximal deviations. Let us mill the moulds for 14 teeth in model material. At first, for each tooth mould, the combined representation of the tooth geometry – described by STL data – and

Figure 2.   The Reverse Engineering process chain, (a) Based on triangulation (left), (b) based on surface representation (right).

These models frequently provide an essential precondition for NC programming and multi-axis milling (Figure 2).

2   EXTENDING THE WORLD OF SHAPES AT HERAEUS KULZER

The application of new technologies for mould making at the Heraeus Kulzer denture production location Wasserburg (D) is to be implemented by the deliberate extension of an existing spectrum of the PREMIUM teeth product line. This action is aimed at optimising the development of new denture lines in terms of time, cost and quality. The shape and size of the new PREMIUM teeth geometries are determined in accordance with market wishes and manufactured using STL data already available. The intermediate shapes are also converted from available data. This series foresees:

- a greater upper jaw (OK) front tooth set
- a more narrow and longer lower jaw (UK) front tooth set
- an OK/UK lateral tooth set as short variant (short shape)

keeping the occlusal dimensions and morphology.

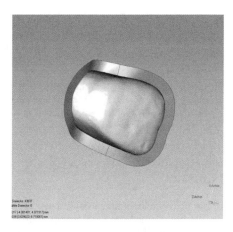

Figure 4. Mould halve with designed parting surface.

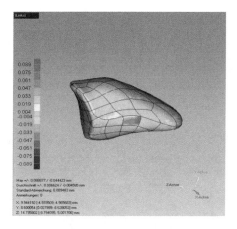

Figure 5. Deviation between spline surfaces and the STL-representation.

the sealing surface – identified by a Spline surface – were maintained (Figure 5). The roughing and finishing paths are calculated. Milling is carried out on a 3 axis milling machine of the type MICROMAT HSC 4V fitted with an ANDRONIC 2000.

One mould was selected for comparing evaluations. This mould was programmed and milled based on the Spline surfaces generated. Within the NC programming system, it is much easier to handle the Spline surfaces of the tooth outer shapes in comparison with surfaces basing on the STL data description. In the next step, the final steel moulds are programmed (CNC programming). The test moulds milled from model material are evaluated by the tool maker at Heraeus Kulzer. For final evaluation, the following criteria are taken into account:

- Roughness of the negative shape due to milling path distance
- The quantity of details performed at the surface
- Quality and fit of the sealing surface.

Visible roughness was eliminated by reducing the milling path distances during further manufacturing. As can be seen, it's apparent that the test shape programmed and milled based on the Spline surface representation has a very smooth surface, but a surface like this is not desired for these teeth shapes. The test mould milled in model material were used for manufacturing model teeth. These model teeth were repeatedly estimated by the head of product development, the senior sales manager and other Heraeus Kulzer employees. Repeated alterations of the tooth geometries of the front teeth resulted from the evaluation of the real 3D objects. Here it can obviously seen that it is only possible to state that the Reverse Engineering procedure was successful if the objects to be made correspond to the customer wishes.

3 REVERSE ENGINEERING IN ART AND CULTURE

For restoration, it is of interest to build up physical models. To achieve a higher acceptance of applying Reverse Engineering in art and culture, the authors mainly considered a tremendous reduction of costs and time, as well as to handle the process chain elements to less effort. Furthermore, the Reverse Engineering – CAM – process chain was analysed with its requirements. Summarising this knowledge, a guideline was made available to prospective customers. Various problems in art, culture and restoration were successfully solved. The guideline not only summarises the used equipment, methods and occurring problems, but also the resulting costs and times to produce replacement parts, replicas and doubles.

In summer 2005, the golden statue on the top of the Dresden City Hall was removed for restoration. This sculpture was created in 1905. It was embossed from copper sheet and afterwards gold-plated. In a sense, the Rathausmann is a symbol of the city of Dresden and for the town's population.

This figure is called the Dresden City Hall Man, the "Dresden Rathausmann". The figure's height is almost 8 m and consequently it can be seen atop the City Hall for miles around. The last time the Rathausmann was newly gold plated was in the 60s. A new gold plating is planned.

The Chair of Production Automation and Cutting Technology of the Dresden University of Technology made an offer to the town of Dresden to perform 3D data scanning of the Rathausmann free of charge. Scanning was started at the head, since it can be removed. Since the object is golden in some areas, i.e. shiny, and black on others, problems in data recording were expected at first.

However, optical recording proved free of problems. An optical scanner, which is based on strip projection, was used for data recording. For measurement, the equipment was set up to the maximum available measuring field of 600 × 800 mm. The head was fitted with 150 checking marks that are required to match the single records. It was possible to completely record the head with 45 individual scans in 1.5 hours (Figure 6).

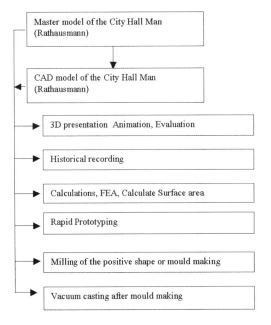

Figure 6. Head of "Dresden Rathausmann", 900.000 polygons, 1.5 h scanning time, 6 h point processing.

```
┌─────────────────────────────────────────┐
│ Master model of the City Hall Man        │
│ (Rathausmann)                            │
└─────────────────────────────────────────┘
                    │
                    ▼
┌─────────────────────────────────────────┐
│ CAD model of the City Hall Man           │
│ (Rathausmann)                            │
└─────────────────────────────────────────┘

    ┌──────────────────────────────────────┐
    │ 3D presentation Animation, Evaluation │
    └──────────────────────────────────────┘

    ┌──────────────────────────────────────┐
    │ Historical recording                 │
    └──────────────────────────────────────┘

    ┌──────────────────────────────────────┐
    │ Calculations, FEA, Calculate Surface area │
    └──────────────────────────────────────┘

    ┌──────────────────────────────────────┐
    │ Rapid Prototyping                    │
    └──────────────────────────────────────┘

    ┌──────────────────────────────────────┐
    │ Milling of the positive shape or mould making │
    └──────────────────────────────────────┘

    ┌──────────────────────────────────────┐
    │ Vacuum casting after mould making    │
    └──────────────────────────────────────┘
```

Figure 7. Application potential for 3D scanned data in Reverse Engineering.

It took a few hours to process the model into a CAD data record. There are many ways to use digital information. Figure 7 illustrates some opportunities for use.

A Rapid Prototyping model of the head was produced with Thermojet 3D in 4 hours.

What is to be particularly emphasized is that after 3D scanning of the whole figure, complete detailed 3D object data are available, with which it is much easier to create a duplicate.

4 REVERSE ENGINEERING IN CAR INDUSTRY

Even now, it takes a lot of time to style a new car body shape. The free-formed geometry is still drafted in a number of steps. Thereby, manual rework at the model is still to be taken into account.

Figure 8 shows the car body, added by checking marks for data recording. The checking marks are foreseen for matching the single records of the scanner.

After scanning, the digitised data are processed to STL data. In this procedure, data are smoothed,

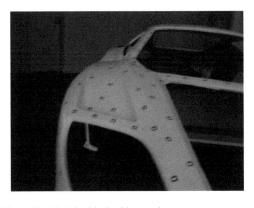

Figure 8. Model with checking marks.

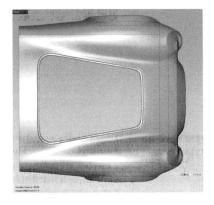

Figure 9. Front panel segmented for NC programming.

248

customised, surface areas are segmented for NC programming.

5 CONCLUSIONS AND OUTLOOK

In parallel to investigate research topics aimed at 3D digitising and Reverse Engineering, the following problems are studied:

- Tryout customised process chain
- Create special 3D software tools
- Calculate Reverse Engineering problems
- Support in the introduction of Reverse Engineering

As a result of solving tasks from several business lines and research directions, multiplier effects for going on tasks are expected to rise.

REFERENCES

Beyer (2002) Nutzung der 3D-Digitalisierung bei der Entwicklung von Produkten, Dissertation, TU Magdeburg, Fakultät Maschinenwesen

Schöne et al. (2004) Digitizing and Reverse Engineering, 33. International Conference on Computers and Industrial Engineering, Jeu, Korea, 25.-27.03.2004

Schöne & Paul (2001) Leitfaden mit prototypischer Erprobung einer Reverse Engineering – CAM – Prozesskette für den Kunst- und Kulturbereich, Leitfaden zum AIF-Forschungsvorhaben 8ZBR, TU Dresden, Institut für Produktionstechnik, Gesellschaft zur Förderung angewandter Informatik Berlin

Schöne et al. (2001) Erarbeitung einer Reverse Engineering – CAM – Prozesskette für den Bereich der Konstruktion und Fertigung zahnärztlich-prothetischen Restaurationen, Vortrag zur Rapromed

2001, Rapid Prototyping in Medizin und Medizintechnik, Beckmann-Institut für Technologieentwicklung e.V. Lichtenstein, 24.10.2001

Schöne & Hoffmann (2003) Reverse Engineering und Rapid Prototyping für medizinische Anwendungen, Internationales Kolloquium TU Dresden, TU Liberec, TU Dresden, September 2003

Virtual modeling and rapid manufacturing – Bártolo (eds)
© *2005 Taylor & Francis Group, London, ISBN 0 415 39062 1*

Virtual and rapid prototyping by means of 3D optical acquisition and CAD modeling: application to cultural heritage and to the automotive domain

G. Sansoni, F. Docchio & M. Trebeschi
Laboratory of Optoelectronics, University of Brescia, Italy

S. Filippi & B. Motyl
DIEGM Dept., University of Udine, Udine

ABSTRACT: In this paper, the activity aimed at the contact less acquisition, the generation of polygonal descriptive models, the generation of NURBS-based models and the reproduction by means of rapid prototyping tools of pieces of interest in both the heritage and the automotive domains are presented. The work gives an insight of how three-dimensional range cameras and suitable data modeling can be helpful in the reverse engineering and in the virtual representation of complex shapes. In addition, the research represents a valuable example of Computer Support Collaboration Work (CSCW) for the restoration and the rapid prototyping of the objects.

1   INTRODUCTION

In the last years, there has been an increasing demand of 3D optical sensors for fast acquisition of complex, free-form shapes (Blais 2004). The advantages with respect to contact probes are non-invasiveness, higher speed of measurement and, in many cases, lower cost. In addition, the availability of powerful software tools for the editing of the raw 3D data, and the modeling of the shapes opened the door to a considerable number of applications. Typical processes that can take a great benefit from the combined use of optical 3D sensors and suitable elaboration chains are the reverse engineering (RE) and the rapid prototyping (RP) of physical objects. In the manufacturing industry, RE and RP are key elements in design and production, to fulfil today's needs of reducing time to market and overall costs, and of allowing the product restyling (Boulanger et al. 1994).

In the field of cultural heritage, the availability of 3D descriptive models of monuments, buildings, statues, and findings of cultural value, opens the door to (i) the remote study, by the scientists and/or their students, of the pieces of interest, from a common and reliable data base, (ii) the timely monitoring of the degradation of the pieces, (iii) the restoration of lost parts, and (iv) the reproduction of the pieces, both virtual, to fully exploit the concept of "virtual museums", and physical, to obtain high accuracy scaled replicas of the original pieces (Levoy 1999).

In the automotive industry, they represent a valid aid to the stylist's creativity. The typical process involves the fast, contact-less acquisition of the 1:1 scale model of the car body and its transformation into a three-dimensional model. This model can be further manipulated by using specialized Computer-aided Styling tools, validated by using scaled replicas, and further refined, in the framework of collaborative design (Varady et al. 1997).

All the activities expected in these domains could gather a great help by using methods and tools of the Computer Supported Collaborative Work – CSCW. The different competencies involved in the processes (Museums, Computer Graphic labs, RP centres, etc.) are often far each other, so the experts working there could exploit videoconferencing, application sharing, etc., to be more productive (Bandera et al. 2002).

In this paper, valuable examples of the application of full optical RE and RP processes to the cultural heritage and to the automotive industry are presented. The target objects, shown in Figure 1, are the 'Winged Victory' (Fig. 1a) a 2 meters high bronze statue located at the Civici Musei in Brescia, one fragment of a precious majolica of the XVIII Century, found in the lagoon of Venice, and located at the Civic Museums of Udine (Fig. 1b), and the 1953 Ferrari 250 MM, an historical, unique racing car (Fig. 1c).

The goal of the work in the case of the 'Winged Victory' was the achievement of descriptive models based on triangle meshes of the whole statue, in order to study the overall proportions and to allow the archaeologist, by means of an inductive approach, to determine the archetype from which these proportions have been generated.

In the case of the fragments of majolica, the objective was the physical reconstruction of the lost parts, in view of the real full restoration of the original object.

(a)

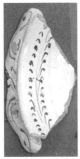

(b)

(c)

Figure 1. (a) The statue of the 'Winged Victory', (b) the XVIII century majolica fragment (15 cm × 5 cm) and (c) the '1953 Ferrari 250 MM'.

In the case of the 'Ferrari', the work was aimed at proving the applicability of non-contact gauging to perform the RE of the car body instead of using contact probes. This objective was thought to be crucial, considering that, in the automotive domain, contact probes require a considerable amount of time, especially when free-form shapes are to be modeled. In addition, they result in an overwhelming accuracy when the RE of full-size cars, and the production of polygon or CAD models are required for restyling applications.

In all the considered cases, the acquisition of the shapes has been performed by using a three-dimensional vision system based on the projection of suitable patterns of non-coherent light. This system, called OPL-3D, represents one of the research products of the Laboratory of Optoelectronics, located in Brescia (Sansoni et al. 2003). OPL-3D has been used to perform the multi-view acquisition of all the shapes. After their alignment, the views were fused into a single one; then the triangle meshes were created and edited. These models were then sent via the Internet to the RP Laboratory located in Udine. Here, scaled replicas of the statue head and of the car were obtained, by using a RP machine. Special elaboration was applied to the majolica fragment: it was used as the initial knowledge to derive the shapes of the missing parts. In this way, all the pieces could be reconstructed, and subsequently reproduced.

The following sections detail the hardware and the software facilities used to implement the whole process, and shall give a comprehensive description of the results.

2 OPL-3D: THE OPTICAL DIGITIZER

OPL 3D is shown in Figure 2. The optical head is composed of an LCD projector (ABW LCD320) and a CCD video camera (Hitachi K3P-M3) mounted on a rigid bar, fixed on a robust tripod.

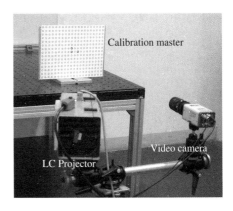

Figure 2. The optical digitizer OPL-3D.

The projector projects on the surface under measurement a sequence of eleven patterns of incoherent light, formed by black and white stripes of different spatial period. The patterns belong to the well-known Gray Code-Phase Shift (GCPS) method. The video camera synchronously frames each pattern: the stripes appear to be deformed by the object shape, due to the fact that the acquisition direction is angled with respect to the projection direction. The 3D shape of the object is retrieved from this deformation, and by exploiting the optical triangulation principle.

The in-field optical measurement is carried out in three steps. In the first step, a menu-driven interface sets the triangulation geometry of OPL-3D according to the measurement range and to the required resolution. In addition, it suggests the values of the optical parameters of the projector-camera pair, in order to adapt to the environmental light and to the surface appearance of the object.

In the second step, the operator mounts the optical devices onto the rigid bar in the suggested configuration: the mount is fully configurable, to optimize the flexibility of OPL-3D to a wide range of experimental configurations. Then, a suitably designed calibration module estimates the pose and the orientation of the projector-camera pair with respect to the reference system, in which the measurement is expressed, and compensates for radial and tangential distortion of the lenses (Weng et al. 1992).

In the last step, OPL-3D performs the measurement by projecting the GCPS sequence of fringe patterns on the surface. The projection-acquisition step is performed in 2 sec, and the elaboration is completed in 4 sec, (data storage included). A very dense point cloud is produced (typically up to 70% of the number of pixels of the video camera). Typical values of the Type A uncertainty of the measurement are about 1/1500 of the depth range.

3 ACQUISITION OF THE POINT CLOUDS

The digitization of the 'Winged Victory' showed up to be critical under a number of aspects. These are (i) the high variability of color of the surface, and the low levels of brightness and contrast, (ii) the high variability of the local curvature of the shapes, and (iii) the overall extension of the surfaces to be measured. The critical aspects in the digitization of the majolica fragment were the very high reflectance and the very small surface details, especially in correspondence with the borders. The problems in the case of the Ferrari 250 MM were represented by (i) the uniform glossy red paint, (ii) the high regularity of the surfaces without edges to be used as 'anchoring' points to align the different views, and (iii) the large overall dimensions.

The speed of the calibration procedure and the high flexibility of the instrument, turned out to be very helpful to overcome these problems.

As an example, in the case of the statue, it was possible to perform the calibration by choosing different colors for the master (light and dark blue were used instead of white and black). In the case of the car, it was possible to vary the integration time of the video camera: it was set to 1/250 sec during the calibration, and increased to 1/50 sec during the measurement. As a result, in both cases, the system was calibrated in experimental conditions very similar to those typical of the measurement. Hence, the quality of the images acquired in correspondence with the target surfaces was optimal both in the contrast and in the brightness.

In addition, different set-ups of OPL-3D were used, depending (i) on the degree of definition required for the surface details and (ii) on the use of suitable acquisition strategies aimed at keeping the alignment errors under control.

In the former case, the objective was to find an optimal trade-off between the adherence of the models to the original shapes and the file dimension.

In the latter, the goal was to guarantee a good accuracy during the alignment of the views. It is well known, in fact, that the higher the number of the views, the higher the probability that the alignment errors accumulate in an unpredictable way, and decrease the quality of the surface representation.

Table 1 shows the different set-ups used for the acquisition of the objects. In the case of the statue, OPL-3D was firstly configured in the so-called skeleton set-up: a shell of the whole statue (head excluded),

Table 1. Summary of the geometrical set ups used for the measurement of the targets. FW: width of Field of View, FH: height of Field of View, Z-Range: measurement interval; R_z: measurement resolution along the Z-Range; U_A: Type A uncertainty of the measurement.

Geometrical set-ups	FW [mm]	FH [mm]	Z-Range [mm]	R_z [mm]	U_A [mm]
Set-up used for the skeleton of the statue	450	348	150	0,3	0,12
High resolution (Statue head)	160	123	40	0,1	0,07
Medium resolution (Statue body and wings)	300	232	100	0,2	0,09
Medium resolution (Ferrari)	550	423	120	0,3	0,13
High resolution (Ferrrai)	370	284	60	0,2	0,1
High resolution (majolica fragments)	60	50	20	0,08	0,05

253

was obtained by acquiring and merging together 110 views at low resolution (Guidi et al. 2003). Then it was configured at high resolution to acquire the head (41 views). After that, the set-up at medium resolution was used to acquire 480 views. These were overlapped to the shell, and aligned together by using it as the reference. The shell was then removed. In this way it was possible both to adjust the resolution of the measurement to the surface details and to minimize the influence of the alignment errors.

The acquisition of the 'Ferrari' was performed, in the first step, by acquiring 280 views, at medium resolution. Then, OPL-3D was configured at high resolution to acquire the car details, such as the handles and the windows borders. However, to compensate for the absence of 3D features on the car body, it was necessary to place physical, tape markers on the surface, to guarantee sufficient accuracy of the alignment.

In the case of the majolica fragment, the system was configured at very high resolution, to allow a precise acquisition of the borders, in view of obtaining an optimal fitting of the parts to be added. However, the reduced dimension of this object allowed us to limit the number of the acquired views to 10.

To acquire the shape of large, complex objects such as the 'Winged Victory' and the 'Ferrari 250 MM', OPL-3D was mounted on wheels. It was moved around with its tripod and oriented with respect to the surfaces to have optimal focus condition, and minimal disturbance of reflectivity. In the case of the fragment, the object was moved with respect to the optical sensor.

The alignment of the views and their fusion into a single point cloud was performed, in each experiment, by using the IMAlign module (PolyWorks, from InnovMetric Inc., Canada), a very robust, well-engineered market available software was used. Figure 3 shows the point clouds corresponding to the statue (Fig. 3a), to the majolica fragment (Fig. 3b) and to the car (Fig. 3c).

In the case of the 'Winged Victory', an overall average measurement error from 90 μm to 400 μm was observed. The maximum error is 1.5 mm in correspondence to the dress folding and the hand-made junctions of the arms and wings (Sansoni et al. 2005).

In the case of the majolica fragments, the setup allowed a fast acquisition of all the surfaces, with an accuracy of 60 microns.

The quality of the 3D raw data of the 'Ferrari' was evaluated by means of suitable color error maps calculated by IMAlign: they highlighted that the mean value of the alignment error was equal to 0,5 mm. In the figure, the black areas of the border of the markers, and the areas that, due to under cuts and spurious local surface reflections, were not be acquired by OPL-3D, are well visible. They could be filled up and smoothed using suitable tools within both the IMAlign and the IMEdit modules. In order to optimize the time

(a)

(b)

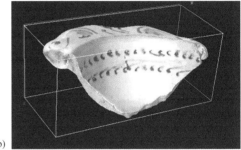

(c)

Figure 3. The point clouds of the statue (3.a), of the fragment (3.b) and of the car (3.c).

efficiency of the process, the editing was postponed to the creation of the triangle mesh, and performed by the IMEdit tool (Sansoni et al. 2004).

4 MODELING OF THE POINT CLOUDS

Following their alignment, and fusion, suitable triangle models were created and edited. The IMMerge and the IMEdit modules of the Polyworks suite of programs were used to perform this task. The editing session had the purpose of (i) filling the residual gaps between points, due to shadowed regions, undercuts, and holes in the surface, (ii) reconstructing the surfaces which were not visible, and (iii) controlling the overall topology of the triangles.

4.1 The work on the 'Winged Victory' and on the 'Ferrari'

Two triangle meshes, respectively of 16 millions triangles for the 'Victory' and of 1,3 million triangles for the 'Ferrari' were obtained. Further compression resulted into the generation of suitable smaller models, representing good trade-offs between the adherence of the meshes to the original shapes and the file dimension.

The triangle representation of the statue perfectly responded to the archaeologist's demand. In response to the selection of any triangle pair in the model, the system provided the distance between them in real time. Since the model was very accurate, the measurement of a distance between the selected triangle pair was very reliable (Morandini 2002). This model was not only ideal for the archaeologist's work, but also for a number of other issues. A first application was the rendering of the statue for multimedia and virtual reality. As an example, Figure 4 shows the appearance of the 'Victory' before (Fig. 4a) and after (Fig. 4b) the virtual removal of the wings (the surface in correspondence to the body-wing connection has been reconstructed).

The work carried out on the 'Ferrari' resulted into the creation of four CAD models with 28, 24, 12 and 8 control points respectively. The Geomagic Studio 4.1 software (Raindrop Geomagic Inc, US) carried out the elaboration, starting from the original triangle mesh. Figure 5 shows the rendered view of the 8 control points CAD model.

4.2 The work on the majolica fragments

For what concerns the majolica fragment, the triangle mesh was sent through the Internet to the Computer Graphic Laboratory of the University of Udine. Then it was imported both into a geometrical-based CAD (Rihnoceros), a solid-based CAD (SolidEdge), and into MagicsRP, a software package dedicated to the STL files manipulation. As shown in Figure 6a, it was used to reconstruct the bowl. In addition, it was used as the initial knowledge to model the handle of the majolica (Fig. 6b).

4.3 RP activities

One of the major advantages of the RE process is to obtain, from either the triangle model or the CAD

(a)   (b)

Figure 4.   Virtual removal of the wings. a: original 16 million triangle model; b: the edited model after suitable compression.

Figure 5.   CAD model of the 'Ferrari'.

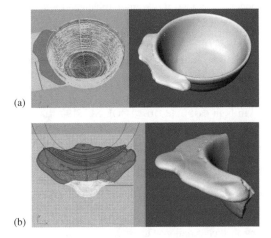

(a)

(b)

Figure 6.   Generating the CAD model of the majolica, starting from the fragment acquired by RE. a: generation of the bowl; b: generation of the handle.

Figure 7.   1:8 scaled copy of the 'Victory' head.

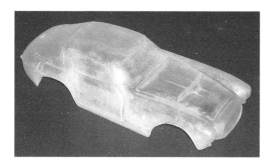

Figure 8.   Scaled replica of the 'Ferrari'.

model, a topological description that can be fed to a RP machine for the creation of replicas. In order to test the feasibility of the prototyping step, the mesh of the head statue and the mesh of the Ferrari were 1:8 and 1:10 respectively scaled. Then, a 4 mm thickness was added. The models were topologically controlled and saved in the STL format. Then, they were sent to the Laboratory located in Udine.

In both cases the CIBATOOL SL 5190 has been used as the material. Figures 7–8 show the two copies. The overall dimension of the statue head is 140 mm × 10 mm × 133 mm, whilst the replica of the Ferrari is 370 mm × 150 mm × 90 mm. The time required to obtain the copies was 0.20 hours for the elaboration of the data, plus 15 and 12 hours respectively for the machining.

The CAD model of the majolica was used without any further variation: Figure 9a shows the 1:1 scale replica resulted from the RP process. Figure 9b shows the fully reconstructed piece, after the assembling of original fragment and of the machined parts.

(a)

(b)

Figure 9.   Reconstructed part by means of RP activities. a: machined part; b: assembled majolica.

5   CONCLUSIONS

In the fields of RE and RP, this paper has described the activities aimed at the contact less acquisition, the generation of polygonal descriptive models, the generation of NURBS-based models and the reproduction by means of rapid prototyping tools of pieces of interest in both the heritage and the automotive domains. The results have been already tested and validated positively by the experts of the two fields. The same experts have encouraged the prosecution of the researches. Some new experiences are going to start, both in the same fields and in others, i.e. the medical/surgical and the mechanical ones.

REFERENCES

Bandera, C. Filippi, S. & Motyl, B. 2002. Computer-Supported-Cooperative-Work Strategies in Cultural Heritage Preservation. *Proc. of Eurographics '02*, Milano (Italia).

Blais, F. 2004. A review of 20 years of range sensors development. *Journal of Electronic Imaging* 13(1):231–240.

Boulanger, P. Roth, G. & Godin, G. 1994. Application of 3D active vision to rapid prototype development. *Proc. Intelligent Manufacturing System (IMS), International Conference on Rapid Product Development,* 147–160.

Guidi, G. Beraldin, J.A. Ciofi, S. Cioci, A. & Atzeni, C. 2003. 3D acquisition of Donatello's Maddalena: protocols, good practices and benchmarking. *Proc. Electronic Imaging & the Visual Arts: EVA*: 174–178.

Levoy, M. 1999. The Digital Michelangelo Project. *Proc. of 3DIM99, Second International Conference on 3-D Digital Imaging and Modelling:* 2–11. Varady, T. Martin, R.R. & Cox, J. 1997. Reverse engineering of geometric models – an introduction. *Computer Aided Design* 29(4):255–68.

Morandini, F. 2002. Rilievo tridimensionale della Vittoria-tavole delle misure. *In Nuove ricerche sul Capitolium di Brescia: scavi, studi e restauri*: 165–173.

Sansoni, G. Patrioli, A. & Docchio, F. 2003. OPL-3D: a novel, portable optical digitizer for fast acquisition of free-form surfaces. *Rev. Sc. Instrum.* 74(4): 2593–2603.

Sansoni, G. & Docchio, F. 2004. Three-dimensional optical measurements and reverse engineering for automotive applications. *Robotics and Computer-Integrated manufacturing* 20: 359–367.

Sansoni, G. & Docchio, F. 2005. 3-D optical Measurements in the Field of Cultural heritage: The Case of the Vittoria Alata of Brescia. *IEEE Trans. Instr. Meas.* 54(1): 359–368.

Weng, J. Cohen, P. Herniou, M. 1992. Camera calibration with distortion models and accuracy evaluation. *IEEE Trans. Patt. Anal. Mach. Int.* 14: 965–980.

Virtual modeling and rapid manufacturing – Bártolo (eds)
© 2005 Taylor & Francis Group, London, ISBN 0 415 39062 1

Photogrammetry techniques applied to reverse engineering

S. Aranda, D. Cruzado, T. de la Fuente & M. San Juan

Área de Ingeniería Mecánica, E.T.S. Ingenieros Industriales, University of Valladolid, Spain

ABSTRACT: The process of capturing object form through surface data sampling and generating a CAD model of the part is termed reverse engineering because the process is the opposite of the normal design and manufacturing sequence. This paper presents a new development of the photogrammetry and its applications in reverse engineering systems. In the context of manufacturing, reverse engineering is an important process for instances where a product initially exists as a designer's model. In this case, a design system based in this technology will be set up and validated by means of two experimental works. We will try to consider how the industrial engineering can be influenced by this new technology.

1 INTRODUCTION

1.1 *Reverse engineering*

Manufacturing systems based on reverse engineering are beginning their implementation in the industry as a complement to fulfil some requirements that traditional engineering is not able to sort out. The trouble appears when engineers do not have CAD data of the geometry of a pre-existing real part of a device, which is desired to be manufactured. In these cases it is necessary to obtain a digital model of that object, in order to get the indispensable CAD model to machine it. Until today, usual digitization methods were slow or very expensive, or presented disadvantages that reduced their industrial application. However, recent advances achieved in photogrammetry allow engineers to use it as an innovative digitisation technology fully applicable to goods manufacturing.

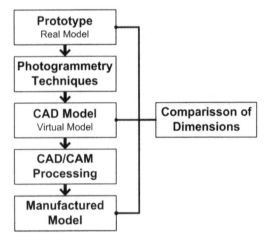

Figure 1. Reverse engineering stages.

1.2 *Photogrammetry*

Photogrammetry is the art, science, and technology of obtaining reliable information about physical objects and the environment through the processes of recording, measuring and interpreting photographic images and patterns of electromagnetic radiant energy and other phenomena.

The most frequent industrial application of photogrammetry is as a digitisation technique in reverse systems engineering. Furthermore, it is useful in production inspection systems and to locate and measure devices in machines or installations.

Photogrammetry has also a role to play in fields not directly related to industry, like archaeology and anthropology; topography and aerial cartography, to create maps and spatial images; in architecture and preservations; in medicine to the reconstruction of prostheses and bones and in forensics and accidents reconstruction.

1.3 *Objectives and methodology*

The objective is to validate this kind of design systems in reverse engineering by means of photogrammetric techniques, and get an evaluative study of the whole manufacturing process and of each stage independently as well.

We can distinguish three different stages in the design system: 'Manufacturing', 'Digitisation and CAD improvement' and finally 'Comparison'.

2 TOOLS AND WORKING ENVIRONMENT

We have used a Nikon Copix 4500 digital camera. The main characteristics of this camera are the following: CCD of 4,13 million of pixels; focal length 38–155 mm equivalent to 35 mm; f number between f/2,6 to f/5,1; shutter speed between 1 to 1/2.300 seconds; approach distance over 2 cm.

A big amount of pixels is helpful to obtain better accuracy, but the most decisive factor is that the lens has the minimum distortion, or at least a known distortion. In the calibration process of a camera, parameters that characterize Radial Distortion and Decentring Distortion of the group of lenses are estimated as well as other properties, like Focal Length, Format Size Width and Principal Point.

In addition to estimating correctly optical parameters of the camera, is very important to be careful preparing an appropriate photo scenery. We must consider some guidelines: to take photographs with good angular separation (close to 90°); do not take photographs from similar points of view; surround the object of interest and get above and below it; maximize the number of photographs that each point is marked on; this points cover a greater percentage of the photograph area to get an accurate orientation; use targets and sub-pixel marking algorithms and introduce in the scene a test object with known dimensions to scale the model (Hanke 2000).

Once photographs are already made, we continue the process marking, referencing and processing, using the commercial software package PhotoModeler Pro 5.2.2, of EOS Systems Inc. *Marking* is the process of creating and positioning an element on a photograph: points, straight lines, curves and surfaces drawn up over the photograph. Is in this stage where resolution has influence over the final accuracy. *Referencing*: the process relating marks on two or more different photographs that represent the same physical element in space. *Processing*: to create a 3D model from the Calibration and Photograph marking information, using a numerical algorithm. This algorithm uses advanced mathematical techniques to adjust the input data, to create the 3D point data and to minimize errors in order to maximize accuracy. Processing (Sánchez 2000) is an iterative process. It repeats a sequence of steps as many times as necessary to determine the location of each point in three dimensions and to minimize the total error.

Finally, we can export the complete set of 3D model data from PhotoModeler to use in external CAD, graphics, animation and rendering packages.

3 EXPERIMENTAL WORK

3.1 *Part of revolution geometry*

In order to validate this type of theories, we tried to set up a system of manufacture of pieces from a model or prototype.

We choose a body of revolution, whose complex geometry made difficult to digitize it, allowing us to evaluate the design system in the worst conditions.

The first stage consists on digitizing the model by means of photogrammetry techniques using the software PhotoModeler 5.2. In order to make the test field we introduced two calibration grids and we marked a generatrix in the prototype (Figure 2).

Once the model was oriented, we created the generatrix in the software marking it on the photos (Figure 3).

When the generatrix is constructed, we can recreate the model easily, revolving it around its axis (Figure 4).

When the generatrix is constructed, we can recreate the model easily, revolving it around its axis.

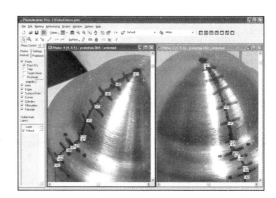

Figure 2.   Test field made on the prototype.

Figure 3.   Generating elements with PhotoModeler.

Once the piece is digitised in PhotoModeler, the model can be completed by means of some software.

The following step, with the geometry of the piece already collected in a file CAD, would be the processing CAM that allows the manufacture of pieces automatically with a numerical control lathe Danumeric NI-650.

After mechanizing the manufactured piece from the numerical control program generated, the last step is to compare geometries between prototype, digitized CAD, and manufactured piece. For this process, a coordinate measuring machine DEA Mistral is used, that works with the software PC-DMIS 3.5, Figure 5.

3.2 3D model of a milling chip

One of the most important utilities of numerical simulation in machining is the possibility of evaluating some thermomechanical variables involved in cutting-chip processes. These variables are difficult to estimate through experimental tests.

Nevertheless, the knowledge of the behaviour equations of materials is one the biggest drawbacks to obtain numerical models (Arrazola 2003). In order to overcome this difficulty have emerged reverse methods based in experimentation to identify those behavioural laws.

These techniques are based on minimizing on the existent discrepancies between results from simulations and results from experiments (cutting forces, contact length, plastic deformations, etc.) by iteration of parameters of law of material, like Johnson-Cook equation. As input variables in the minimization process we may consider those whose experimental study could be done with adequate accuracy (San Juan 2005), e.g. the geometry of a chip.

The experimental characterization of the process has been focused on conditions of orthogonal milling; that allows us to fit more accurately the numerical models, including in the analysis cutting forces, contact length, and geometry of the chip. Therefore, it is very important the modelling of chip geometry. Is commonly studied the profile of a mounted chip, but by means of photogrammetry we can generate a solid model from a set of photographs (Figure 7).

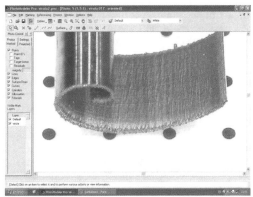

Figure 6.   Chip snapshot.

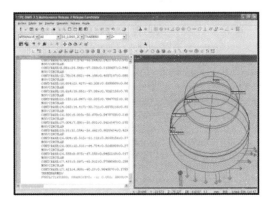

Figure 4.   Model built with photogrammetry techniques.

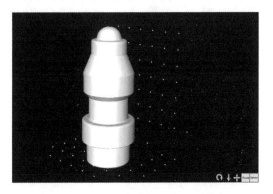

Figure 5.   Comparison of geometries with feelers.

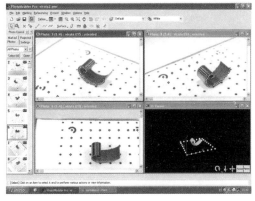

Figure 7.   3D Model Construction.

261

To achieve the best precision when creating the model, operator should follow the guidelines described in Section 2. The calibration process must be carefully completed. Angles between cameras should be around 90°. About 30 points will be marked over each one of the 12 snapshots using the subpixel technique, making sure that orientation process will be reliable.

4 RESULTS AND DISCUSSION

The results obtained when comparing the dimensions of the prototype with the mechanized piece, are acceptable, but they would not be valid if we needed high accuracy. In any case, it is necessary to stand out that the geometry of the chosen piece is very unfavourable to digitise with PhotoModeler, because the program models worse curves and nurbs, that edges and corners.

Obtained models allow measures over the geometry of the chip in a more accurate and continuous way along its section. Also is possible to determine the volume of the chip, and to correlate this data with those obtained from the study of cutting conditions. The integration of such models in reverse simulation opens new possibilities, because they provide information not only about profile but also about chip curvature, as illustrated in Figure 8.

The photogrammetry has some advantages over other digitisation techniques, especially for inaccessible and very small objects, or when we do not previously know which measures are going to be taken. It is a versatile technique, useful for objects of any size and superficial aspect, and does not deform the model because there is no contact.

The biggest disadvantage of photogrammetry is the low precision offered compared to other digitisation systems when is the user who has to mark the elements manually on the photos. This contrasts with the high accuracy obtained by PhotoModeler when it detects automatically some geometry (control points of the calibration grid). Some studies (Pappa 2000), (Fedak 2000), relate a precision about 0,020 mm when PhotoModeler detects automatically these control points, using a powerful sub-pixel marker algorithm.

At the moment, we could reject the photogrammetry for objects with restrictive tolerance dimensions. However, we can use in those cases the hybrid systems of digitisation: the model is reconstructed by photogrammetry easily, and we assure those restrictive dimensions in the definitive CAD, by means of some other system measurement.

The last progresses developed in reverse engineering and photogrammetry, offer many possibilities for design and improvement of the productivity, that is not known yet.

ACKNOWLEDGEMENTS

This study is being developed under the Project DPI2003-09676-C02-02 of the National Program of Industrial Design and Manufacturing financed by the Ministry of Science and Technology of Spain and FEDER.

REFERENCES

(Arrazola, 2003) Arrazola, P.J.: "Modélisation numérique de la coupe: étude de sensibilité des paramètres d'entrée et identification du frottement entre outil-copeau." Ph.D. Thesis. E.C. Nantes. France. 2003
(Fedak, 2000) Fedak, M.: "3D Measurement Accuracy of a Consumer-Grade Digital Camera and Retro-Reflective Survey Targets". InSpec Engineering Services. 2000
(Hanke, 2000) Hanke, K.: "Accuracy Study Project of Eos Systems´ PhotoModeler". InSpec Engineering Services. 2000
(Pappa, 2000) Pappa R.S.; Giersch, L.R. & Quagliaroli, J.M.: "Photogrammetry of a 5 m Inflatable Space Antenna With Consumers Digital Cameras." NASA Langley Research Centre. 2000
(San Juan, 2005) San Juan, M.; Arrazola, P.; Mostaza, R. & Montoya, J.: "Le contact outil-copeau dans le cas du fraisage orthogonal." 17eme Congrès Française de Mécanique. Troyes. 2005
(Sánchez, 2000) Sánchez, F.M.: "Reconstrucción tridimensional de escenas con iluminación láser: Aplicaciones a la fotogrametría industrial". Ph.D. Thesis. Universidad Politécnica de Madrid. Spain. 2000

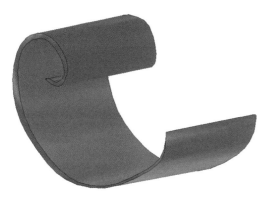

Figure 8. Colour map of chip curvature.

Best practices for 3D CAD interoperability

D. Prawel

Longview Advisors Inc., Loveland, Colorado, USA

ABSTRACT: In the tireless pursuit of innovation and cost management in design and engineering, faster cycle times and greater process efficiencies have become core requirements for competitiveness in global product development. Few issues present a more formidable impediment to rapid product development than poor product data exchange. Important progress continues to be made, however. This session will discuss 3D interoperability best practices, infrastructures and tools, along with the latest trends in standardization, feature translation, and healing.

1   BACKGROUND

Manufacturers face relentless pressure to develop new products and bring them to market faster than ever before. Twenty to thirty percent of all sales by industrial companies come from products that have been on the market for less than 5 years. Competitive pressures drive innovation and demand quality, while increasingly complex products have led most companies to depend on global business partners for component parts.

To win, companies must integrate their customers and partners directly into their product development operations. Success depends on how well business partners share information and react to changes in design and process, regardless of their geographic location. Much of this success relies on fast, efficient sharing of product data.

The challenges and difficulties of CAD data exchange are well known and publicized. Studies have suggested that the cost of problems in sharing engineering data to the automobile industry alone could be $10 Billion per year (Tassey, 1999). Few single issues present a more formidable impediment to accelerating product time-to-market. Large companies have implemented significant procedures, tools and infrastructures to try to alleviate the huge impact of these problems, but generally push the problem to their suppliers. Suppliers, generally unable to afford sophisticated tools, have little more than brute force methods to address these problems, as pressure builds to further accelerate product development, improve quality and cut cost.

Interoperability best practices are poorly understood. This paper presents the results of research that investigated interoperability best practices at 50 small to medium discrete manufacturers, to better understand what works and what doesn't work for manufacturers who build products on a local or global level.

Our highly qualitative research discovered interoperability best practices from a broad range of discrete manufacturing industries and company sizes, and representing a variety of CAD platforms.

Our study confirmed that the costs and frustration resulting from poor product data interoperability, and the negative impact on manufacturing productivity are indeed very, very large. Although the most costly and visible problem derives from poor product data quality resulting from different CAD systems and formats, significant costs are also associated with lack of data communication tools to make the processes easier and smoother, and lack of management support to address and manage these problems.

2   DATA EXCHANGE STRATEGY

2.1   *Technical considerations*

The first and most important consideration in building an effective interoperability strategy is the intended use of the data. Doug Cheney, senior consultant at ITI TranscenData (www.transcendata.com), agrees that "a clear understanding of which business functions will receive the CAD data and how will it be used is the most important element of a strong interoperability strategy."

Product designers will be sharing data with others as they iterate toward their product release. In this case they generally need a "robust" translation that provides as much of the CAD model as possible, including features, history and other critical design elements. For this class of translation, powerful software tools from experienced vendors mentioned in this article can achieve a high percentage of good quality translations. If the transactions are internal to the company, robust translations are usually best. If models are sent outside the company, confidentiality requirements may outweigh robustness requirements, so this also needs to be considered in a good interoperability strategy. Designers may be sharing CAD data with analysis or simulation teams in engineering. These transactions usually require less robustness, for example, perhaps only the geometry, topology and some physical properties of a model. Manufacturing, rapid prototyping and other downstream processes usually require the least robust forms of CAD data, often only the outer surfaces of a model or its geometry. These processes also usually require relatively low precision. These "lean" translations can often be successfully accomplished through industry standards like STEP, IGES or STL. A more robust translation may be required if a designer is collaborating on a project with a supplier who is developing a quote than if the collaboration is for feedback on the visual aspects of a design. JT format from the JT Open Consortium (www.jtopen.com) has become quite popular for visual collaboration, but is not robust enough for most cooperative CAD design work.

If you are changing CAD systems or moving a lot of CAD data to a different system for a project, you will need as much of the CAD data as possible. Any data that doesn't translate generally must be re-created in the new system, at great cost. These CAD migration scenarios are well suited to tools that provide feature-based interoperability, for example from Proficiency and other vendors.

2.2 Complexity

The complexity of the CAD data is also a very important consideration when developing an interoperability strategy. If the data is 2D, then IGES will suffice. If the model contains 2fi dimensional surfaces intended for rapid prototyping or some machining applications, then STL will probably suffice. If the CAD models are fully featured 3D models, then the proprietary CAD formats are probably your best solution, although STEP may also work, depending on considerations mentioned earlier.

If transactions require physical properties, such as mass and surface area, or other information such as metadata, GD&T, annotations, coordinate systems or ERP data (e.g. part ID, pricing), custom services will probably be required. The cost for these services would be evaluated based on the amount of data that has to be translated and standard assessments of return on investment.

2.3 Data quality requirements

The third most important technical consideration when building an effective interoperability strategy is the requirement for quality in the end result. In many applications, high quality is essential. In others it's not. There

are two common interpretations of quality *vis a vis* interoperability: robustness and accuracy. As a general rule, quality can be assessed in very simple terms by considering the intended use of the data, as discussed above.

Wolfgang Winstel, managing director at ProCAEss, agrees that data quality is a very important consideration, because rework is so very expensive. In addition, Mr. Winstel suggests, "An unappreciated aspect of quality is methodology. Not only must model accuracy and precision meet requirements, but the model must also be designed correctly, using the customer's design methodology." This is becoming another important aspect of quality as suppliers are more often required to deliver CAD data in accordance with customer design standards. You can expect to see model quality checking tools start to accommodate this trend by checking design methodologies.

Accuracy is another very important consideration. CAD modelers work at different levels of precision. Sometimes translations take place between two models created at the same precision, and sometimes between different levels of precision. It is important to consider both. Decide in advance what the requirements are and make sure the tools chosen are tested adequately in this regard. Healing software is available from numerous vendors to help repair poor quality (primarily accuracy) translations, for example, CAD-healer from Theorem Solutions (www.theorem.co.uk) and CADdoctor from Elysium (www.elysium.com). Another choice is CADfix from ITI TranscenData, which can repair problems whether or not there are features or parameters.

In many cases, for example in many manufacturing applications, the CAD models provide far more accuracy than the application requires. Higher accuracy may slow performance of the software and cause larger model files than necessary. When building an effective interoperability strategy, keep in mind that high levels of accuracy are not always important.

The question of how to assess the "quality" of translations is too large a topic to cover in this article. Suffice it to say that it is essential to use quality checking tools to assess the robustness and accuracy of translations. Todd Reade, president at Transmagic (www.transmagic.com), suggests that when exchanging CAD files, as a minimum, always "Round-trip, then ZIP", to help reduce errors that occur when saving files in a non-native CAD file format such as IGES or STEP. After you save the file, re-open it in the originating CAD system to verify and validate that the file is usable as expected.

2.4 Business considerations

In today's competitive landscape, it is impossible (and unwise) to be good at everything. A key component of any strategy is the tools you use for product development. It's best to try to standardize on a core set of tools and processes, and maintain the skills and relationships you need to use and support these tools.

The best approach to building an effective interoperability strategy is to determine your core strategic strengths and form partnerships with experts who understand your products and business processes. Use data exchange automation tools where they work well, and outsource to experts who have the knowledge, experience, and contacts to cost-effectively solve the special cases. Many good tools are available to help with CAD interoperability, but even relatively simple models almost always exhibit small differences that prevent flawless automation. In addition, requirements and products change constantly. So, whichever tools you choose, additional expertise will be needed.

Vinay Wagle, vice president of sales and marketing at CADCAM-e (www.cadcam-e.com), refers to this approach as the "core chore" strategy. When considering interoperability strategy Mr. Wagle notes that "a company's core strength is its product development and its people. They know the core business, and to develop sufficient expertise in interoperability can be counter-strategic." The goal in the end is to deliver maximum customer value. Consider the experience of your team in CAD data exchange, understand limitations and decide if you want to compensate with training or outside help.

Security is another important factor to consider from a strategic perspective. The technical aspects of data security are generally well understood by most good IT people and will be left to discuss in another article. The simple truth is, many projects live or die by email, a relatively insecure, "brute force" mode of data exchange. Especially smaller companies with smaller IT budgets depend on email to send and receive CAD files. Companies who can afford to purchase and implement a PDM system usually count on their PDM system for security. Data is usually encrypted and password protected. This method works as long as all suppliers also have the same PDM system–an expensive proposition. Furthermore, if the destination application is only view and markup, this is clearly overkill.

The lifecycle of your product will also play a role is determining your best strategy. If a product is in the later stages of its lifecycle, it may not be worth the cost of translating its data. It is always important to consider how or if your interoperability strategy fits into your PIM/PDM/PLM/ERP initiative(s). One of the most important keys to efficient interoperability is how users interact with data management systems to get and put data, get approvals to send data, find correct contact information, and many others key details of a typical transaction.

3 DATA TRANSLATION PROBLEMS

Problems that occur in the data translation process are the most publicized and the most costly of interoperability problems. Poor quality data translation results from a myriad of issues–geometry, topology, model precision, features, and countless more. The most loudly voiced issues in data translation have to do with the inability of current systems to effectively and consistently translate the feature, parametric and history information in a CAD model. Although this is indeed a significant problem if you truly need this information, we found that feature translation is actually required less than 25 percent of the time, and in these cases almost exclusively for internal use. In fact, most people we interviewed remove this information from a model before sending it to anyone, to protect proprietary information or to make the translations work better.

Common causes of poor data quality also cited in our research include poor CAD model creation, resulting from a lack of training or sloppiness or both, and mathematical inconsistencies naturally inherent in CAD systems that use different modeling techniques and algorithms. For example, edges and faces that have small gaps between them, "sliver" faces (very thin aspect ratio and generally very hard to detect), and duplicate vertices and edges, and faces with inverse orientation. These errors often result in an engineer re-creating the offending part of the model, resulting in the largest single direct cost factor in data exchange.

Further magnifying the cost of poor quality data translation is the fact that many error conditions go undetected until later in the design process, where they create additional, potentially much more expensive problems. For example, an error in a CAD file may not appear to CAD users, but may appear during finite element analysis, or when engineers are creating physical prototypes, or even further downstream (and more costly!) when manufacturers are developing tooling. Different applications put model data to different uses, each potentially exposing different effects of poor data quality, and all causing model re-work and project delay. The further downstream data quality problems are detected, the more costly they are, both directly and indirectly. It is not uncommon for companies to report that downstream functions, such as rapid prototyping, finite element analysis, or CNC programming, spent as much as 50 percent of their time working with CAD data files that exhibited quality problems of some sort that impeded their efficient use in downstream applications.

A survey reported in *Interoperability Cost Analysis in the Automobile Industry*, published by the Research Triangle Institute in March, 1999, (Tassey, 1999) revealed that in about 51 percent of the jobs, the CAD data had to be repaired. The job shop had to completely recreate CAD data in an additional 25 percent of the cases. In about 15 percent of all cases, these errors were not discovered until after the part tooling had already been cut, resulting in significant cost and delay because the company had to scrap and re-cut the parts.

The largest manufacturers are accustomed to dictating the CAD systems used by their suppliers and have tried for many years to enforce this requirement. This may appear to be the simplest way to minimize interoperability problems, but unfortunately it's not that easy. In fact, this may be the most expensive method when all direct and indirect costs are considered. Supplier costs have to be absorbed either in quality, time or price. It is common for suppliers to be forced to have one of each CAD system, and fully trained staff to use it. At the end of the day, this cost comes back to the OEM indirectly as higher cost to their suppliers. Other factors also contribute to an apparent trend away from this approach to interoperability. No single CAD system can be the best for all, so sacrifices have to be made. These sacrifices translate into indirect, often intangible, cost, such as in lost opportunity for innovation or time to market. A very large, global manufacturer we interviewed reported no cost advantage to having everyone using the same CAD system, so they have ceased this practice, making these decisions on a program basis.

Among the most popular neutral formats, such as STEP, VDA-FS and IGES, IGES is by far the most popular industry standard neutral file format among the professionals we interviewed. We found many users quite satisfied with the results from IGES file transfer in many 3D situations. One of the disadvantages of neutral format translation is that interoperability involves many organizations, such as engineering, purchasing and marketing, and product data includes a very wide range of data types, not just CAD data. IGES is very limited in its ability to deal with different data types and applications. STEP was designed for this reason – to be extensible into many more functional and business requirements.

STEP did not appear in our research to be widely used, except among the largest aerospace companies and in some specific applications where only the simplest STEP files are required (e.g. with no features or history). The main reasons expressed for the lack of STEP usage were high cost of the translators from the CAD vendors, and the fact that the standards bodies generally move too slowly, so the formats never seem to evolve quickly enough. High cost and huge scope create challenges to STEP's broader adoption.

Most data exchange discussions center on CAD data. However, a huge amount of non-CAD data is also shared on a daily basis. We found that the most preferred method for sharing these data is to email Adobe PDF files.

Direct translators are available for a wide variety of CAD formats. These can be purchased from the vendors of the CAD systems or from independent suppliers. In general, companies we interviewed noted that after several years of testing and use, they felt that translators from the independent suppliers were better and less expensive than those from the CAD vendors. Commercial translators are considered expensive, but a thorough analysis often reveals acceptable ROI. We found that some popular commercial translation software worked quite well, especially for common problems with CATIA file translation. For example, Theorem Solutions (www.theorem.co.uk) was often mentioned, and there are many other good vendors. There is no substitute for good testing, including a wide range of data typical of your business, and "round trip" testing, to and from another format, producing the same results as verified by good quality checking tools, such as CADIQ from ITI TranscenData.

4   DATA TRANSMISSION

But translation problems are not the only challenges to effective interoperability. Significant hidden costs are incurred due to a lack of support for transmitting data to another party. Huge amounts of product data are sent and received daily in manufacturing. The Research Triangle Institute's *Interoperability Cost Analysis in the Automobile Industry*, reported that one OEM estimates that as many as 453,000 product data exchanges occur each year within their company and among their company and their suppliers. Another OEM estimates that electronic exchange of computer-aided design (CAD) data alone occurs at least 7,000 times per month; that quantity rises as high as 16,000 transfers per month during peaks. This last estimate does not include transfers that take place using physical media such as tape and CD-ROM; nor does it include transfers of data besides CAD/computer-aided manufacturing (CAM) data. These numbers match the magnitude we discovered in our research.

A tremendous amount of engineering resource is spent performing manual tasks associated with sending and receiving data, confirming receipt, tracking contact information, etc.–tasks that could easily be automated. A large aircraft manufacturer we interviewed reported that up to 25 percent of an engineer's time can be spent solving data exchange problems, in larger projects. This time is mostly spent transmitting and retransmitting data and re-creating faulty models. At their quoted rate of thousands of models exchanged in a typical month, this adds a huge cost to the program. So it's easy to see why the cost of interoperability is so high. If problems exist in only a small percentage of all product data files exchanged, and these problems cause just an hour of rework or lost progress. This represents a huge detriment to productivity.

Email is by far the most common form of communication between parties. Email is used throughout the business process, internally and externally. Most communications between engineers and purchasing or suppliers is done through email. "It's not efficient, and all the back and forth costs a lot of time, but it works with relatively little hassle", was often heard in our research.

The second most common form of communication is to put a CD in the mail–low tech but it gets the data there, usually. When files reach a size of about 10 Mb (compressed), they are generally put on an ftp site, or put on CD and sent through the mail.

A significant and virtually unaddressed aspect of interoperability is the underlying infrastructure to support effective data transmission. Currently, it is up to the engineers to know who should receive file(s) at each partner company, email addresses and contact information, required formats and versions, log and archive transmission events, notify receipt of data, know who's authorized to get data and not, etc. We have found this to be the second largest source of cost and wasted time associated with interoperability problems. A large worldwide manufacturer we interviewed felt that this is the best area to invest in for the biggest ROI– minimizing and streamlining cycles between relatively high paid professionals. Some companies have developed powerful tools to help manage interoperability methodology by helping maintain information like CAD system and version for transaction parties, authorized contact information, confirmations, security and the like. For example, Fusion-DX from Datranet (www.datranet.com) and DDX from ProCAEss (www.procaess.de) are popular tools for this purpose. These tools can save a lot of time and cost by helping automate a methodology so engineers can spend more time engineering rather than resending CAD data because it went to the wrong email address, or any of the other seemingly countless pitfalls that waste time in data exchange.

4.1 *Security*

One surprising discovery is that security is generally *not* a concern. In fact, management fears aside, the common perception among the interviewees is that security fears are highly overstated and those security provisions that were in place were generally considered a waste of resource and an impediment to getting the job done. Among those companies who are concerned about security, practices vary by industry. For example, automotive companies generally use a secure exchange, such as ANX or ENX. Secure ID cards are generally viewed as a burden, but they are very secure, if that's the requirement. Web-based, secure ftp is much easier to implement and use, and sufficiently safe for most applications.

5 MANAGEMENT & BUSINESS ISSUES

Interoperability is typically performed from within the engineering or IT departments, as might be expected, sometimes both. In some cases, it is the responsibility of procurement, since they are the keepers of contact information, authorizations and supplier relationships. In larger companies, engineering or IT management is assigned responsibility for interoperability throughout the extended enterprise. But this is rare. Most often, an engineer takes on the job simply by doing it. He or she gains experience in data exchange in their operation, understands the importance of efficient data exchange, and simply takes responsibility. This person usually owns this task alone–a very thankless job. Engineers we spoke to lamented that they received very little support or recognition from their colleagues or from management.

The more support your interoperability champions get, the more effective they can be. Management visibility and recognition of the magnitude of the problem can help support the interoperability person or team, at the highest possible level, and help reduce your interoperability costs.

Consider implementing an interoperability team–a cross-functional team that should be managed as a corporate function. Empower them to develop and implement policies and processes for interoperability. Good methodology can prove much more valuable than the latest technology. Capturing and communicating the insights of your most experienced people, perhaps through a "Portal" or message board, will help reduce interoperability problems and costs, since success rates appear to improve dramatically between experienced data exchangers.

Some companies understand the cost of poor interoperability business practices, but we found no one who attempts to measure it. A few companies appear to recognize the significant opportunity cost associated with poor interoperability practices, and the productivity value of efficient supplier interaction and communication, but for the most part, these costs are passed on to the program or project.

Perfect interoperability should never be the goal for all transactions–just sufficient to do the job, and scalable enough to adapt to new projects where the requirements may be more rigorous.

CAD training is a extremely valuable. We found that there is no substitute for investments in training for CAD users. Interoperability training and consulting is valuable, especially when new problems arise or new projects involve new types of data or other requirements. Engineering managers we interviewed told us that training their CAD users can reduce by 15 percent to 25 percent the interoperability problems they experience.

Your business process analysis should include consideration of who will take responsibility for conversions and the final digital product delivered. In general practice we found the supplier will be expected to deliver 100 percent correct, fully inspected data, on time. CAD quality checking tools can help in this area, but these tools they are not sufficient, but they have not yet been certified by standards bodies like SASIG or AIAG. So be careful assuming too much liability protection from quality checking tools.

Data exchange is a non-value-add activity. Thus a dedicated data transfer person is a wasted engineering resource. Invest in good translation and communication tools and develop win-win relationships with their vendors. Let your engineers do engineering and the purchasers do purchasing. While it is dangerous to generalize too much, we recommend that if your engineering team transmits more than about 100 files in a typical month to and from a few dozen suppliers, you may be able to justify the cost of a communication solution with the time your engineers will save in manual file sharing tasks.

6 CONCLUSION

Competitiveness in the global manufacturing industries has amplified the importance of interoperability solutions and the urgency of competence in sharing huge amounts of product data. As trade export imbalances make front-page news, global manufacturers have looked at many forms of business practice optimization, such as lean manufacturing, six-sigma, concurrent engineering, and the like. Interoperability problems are highly costly, but often underestimated and generally not noticed by management, as they secretly undermine productivity and time to market. Problems in interoperability affect business between companies and between different groups in the same company. It should be respected as core business process in any manufacturing operation, and tantamount to success and competitiveness for any manufacturing company.

REFERENCES

Tassey Ph.D. G., Brunnermeier, S., & Martin, S., Interoperability Cost Analysis of the U.S. Automotive Supply Chain, *Research Triangle Institute Pub. 7007-03, March 1999.*

Virtual modeling and rapid manufacturing – Bártolo (eds)
© *2005 Taylor & Francis Group, London, ISBN 0 415 39062 1*

Fitting B-splines with draft angle to scanned data

K. Schreve
University of Stellenbosch, Stellenbosch, South Africa

ABSTRACT: Ensuring that geometric models have the required draft angle is very important for many manufacturing applications. In this paper an algorithm is presented that ensures that 2D B-splines fitted to point data will have a sufficient draft angle. First, sufficient conditions are derived that will be used as constraints in the least squares fitting of B-splines to the point data. The algorithm uses an Augmented Lagrange Multiplier method to solve the system of equations with inequality constraints. The implementation of this method is described. The paper concludes with a case study.

1 INTRODUCTION

Reverse engineering parts for moulding processes (e.g. sand casting, die casting, injection moulding, etc.) requires that certain manufacturing constraints must be satisfied. One of these is that the part must have a positive draft angle. (A positive draft angle means that there are no undercuts. Figure 1 shows a cross section illustrating the sign convection for the draft angle used in this paper.) The draft angle is dependent on the manufacturing process and the materials used. Often, it may also vary from one manufacturer to another. It is therefore important that the designer must be able to control this parameter.

Current reverse engineering practice is that the designer will fit surfaces to the point cloud, create a complete B-rep model and then impose all the manufacturing constraints. The last step can be called a beautification step (Langbein et al. 2004). This is difficult if the fitted surfaces are not simple planes and

cylinders. Adding draft angles to general free form surfaces is not a trivial process.

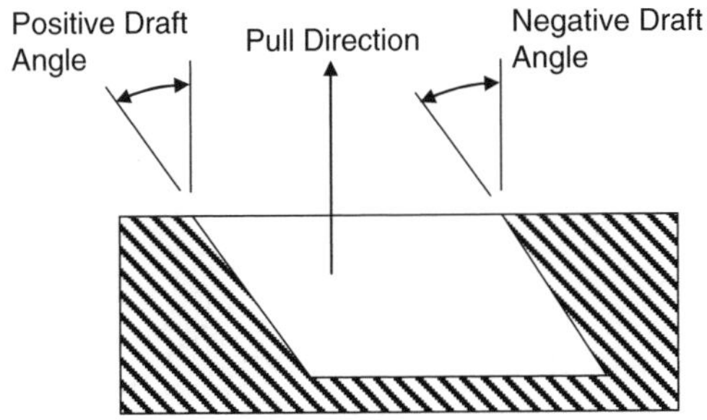

A better approach would be if the draft angles could be incorporated at the surface construction stage. This paper proposes such an approach. A draft angle constraint will be imposed when approximating the cross sectional scan lines with B-spline curves, thus aiding the surface construction process. This can eliminate, or at least simplify, the beautification step.

In this paper sufficient conditions to ensure a minimum positive draft angle on a curve is derived using the convex hull property of B-splines as well as a subdivision scheme. It will be demonstrated that by constraining the control polygon, the draft angle of the curve will be satisfied. These conditions result in linear inequality constraints.

An Augmented Lagrange Multiplier (Rao 1996) method is used to solve the system of equations. The implementation of this algorithm will be discussed.

Finally the method will be demonstrated with a case study. It will be shown how the algorithm can be used to find the best curve that approximates point data on an object while satisfying a draft angle constraint. It is not necessary that the draft angle exists on the object itself.

This algorithm will be useful for doing reverse engineering of dies and moulds as well as parts that will be manufactured by such methods. It not only saves time by eliminating some of the modelling work, but it also aids in ensuring that the final model will be manufacturable.

In the next section related work is discussed. Next, sufficient conditions for approximating curves with draft angle constraints are derived. Then the optimisation method used will be discussed. Finally a case study is presented.

Figure 1. Draft angle sign convention used in this paper.

2 RELATED WORK

In a previous paper by the author and colleagues (Dimitrov et al. 2004), it was shown that in many instances the final reverse engineered model will not be an exact replication of the original scanned part. Various aesthetic and manufacturing constraints mean that the model will deviate from the original part, sometimes by significantly more than the general surface fitting tolerance. From an extensive literature review, the importance of applying engineering knowledge to reverse engineering problems was demonstrated. However, the tools for doing this efficiently are not always available.

Available tools include algorithms for fitting surfaces with boundary conditions. These conditions are positional, tangential and curvature continuity (Kruth & Kerstens 1998, Ma & Zhao 2002, Krause et al. 2003). Such tools are well implemented in most good reverse engineering packages. Tools for controlling surface smoothness (Renner & Weiss 2001) and constraining the control points to be co-linear or co-planar (Rogers & Fog 1989) are also available. However, there are still some outstanding problems, such as ensuring curvature continuity at surface corners (Langbein et al. 2004).

A number of researchers are attempting to solve multiple constraint reverse engineering problems (Werghi et al. 1999, Fischer 2004, Langbein et al. 2004, Karniel et al. 2005). According to Fischer (2004) "frozen-in" errors often result from the normal surface approximation approach to reverse engineering, e.g. holes that do not line up, surfaces that are not parallel, etc. He also agrees that these problems can be solved more efficiently by incorporating knowledge. He does not advocate a fully automatic reverse engineering process, saying that humans are more efficient at providing intelligence.

The current work on multiple constraint reverse engineering problems normally only addresses constraints between planes, cylinders, quadric surfaces, lines, conical and toroidal surfaces. The above mentioned researchers only consider equality constraints. Some (Werghi et al. 1999, Karniel et al. 2005) try to minimise the distance between the surfaces and the point cloud while maintaining the constraints. Langbein et al. (2004) fit the surfaces ignoring the constraints and then apply a beautification step. The first approach can be very computationally intensive since a typical mechanical component can have hundreds of surfaces each with a few constraints between them.

The above authors recognised an important need in reverse engineering. If the quality of the reverse engineered CAD model must be comparable to standard CAD models, then the design intent must be captured. Finding quick ways of incorporating the design intent will most likely also speed up the reverse engineering process. It may take longer to create the initial model, but rework may be reduced. This may shorten the reverse engineering process.

It is towards this work that this paper makes a contribution. It appears that fitting B-spline curves with draft angle constraints can be useful, but was not done before.

3 SUFFICIENT CONDITIONS TO ENSURE POSITIVE DRAFT ANGLE

If the angle between the pull direction and all the line segments of the control polygon of a 2D B-spline satisfies the draft angle constraint, then the B-spline will also satisfy the same. If the line segments are orientated as in Figure 2, this means that the scalar product of the line segment $(\mathbf{P}_2 - \mathbf{P}_1)$ and the pull direction $(\mathbf{Y})$ must be smaller than or equal to zero. This statement can be proved by subdividing the control polygon. A formal theorem is proposed below and the proof is given in Appendix A.

3.1 *Theorem*

For any planar B-spline, if $(\mathbf{P}_{i+1} - \mathbf{P}_i) \cdot \mathbf{Y} \leqslant 0$ and $(\mathbf{P}_{i+1} - \mathbf{P}_i) \cdot \mathbf{M} \leqslant 0$ are satisfied for all i (or if $(\mathbf{P}_{i+1} - \mathbf{P}_i) \cdot \mathbf{Y} \geqslant 0$ and $(\mathbf{P}_{i+1} - \mathbf{P}_i) \cdot \mathbf{M} \geqslant 0$ are satisfied for all i), then the B-spline will not have undercuts where:

- $\mathbf{Y}$ is the pull direction.
- $\mathbf{M}$ is a vector orthogonal to $\mathbf{Y}$, indicating the material side.
- $\mathbf{Y}$, $\mathbf{M}$ and the B-spline are co-planer.
- $\{\mathbf{P}_i\}$ is the set of $n + 1$ control points of the B-spline.
- The degree of the spline is p.
- The knot vector, $\mathbf{U}$, is defined as follows:
  $\mathbf{U} = \{u_i\}$ with
  $u_i = 0$ for $i = 0. . .p + 1$
  $u_i = 1$ for $i = m - p. . .m + 1$ ($m + 1$ is the number of knots)
  $0 < u_i < u_{i+1} < 1$ otherwise.
- The dot, or scalar, product is indicated by "$\cdot$".

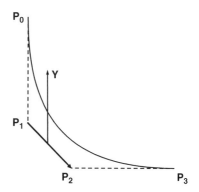

Figure 2. B-spline with control polygon.

This is a simple way of ensuring that the draft angle requirement will be satisfied everywhere on the B-spline. It results in a set of $2n$ inequality constraints. The constraints are simple to evaluate for any draft-angle and involves only a scalar product.

4 ALGORITHM

The Augmented Lagrange Multiplier (ALM) method (Rao 1996) is used to solve the minimisation problem with inequality constraints. The equations used in the algorithm are first derived and then the algorithm is presented.

4.1 The augmented Lagrange multiplier method

The minimum of the following function must be found (Rao 1996):

$$A(\mathbf{P}, ,r_h) = f(\mathbf{P}) + \sum_{j=0}^{2n-1} \lambda_j \alpha_j \beta_j + r_h \sum_{j=0}^{2n-1} \beta_j \alpha_j^2 \quad (1a)$$

where $f(\mathbf{P})$ is the least squares objective function:

$$f(\mathbf{P}) = \sum_{i=0}^{N} (\mathbf{Q}_i - \mathbf{C}(u_i)) \cdot (\mathbf{Q}_i - \mathbf{C}(u_i)) \quad (1b)$$

$\mathbf{Q}_i$ are the $N + 1$ data points. $\mathbf{P}$ are the control points of the curve $\mathbf{C}(u)$.

β_j $(0 \leq j \leq 2n - 1)$ is 1 if the jth constraint is active. β is introduced to simplify the differentiation in equations 8 and 9. If all the constraints are inactive, equation (1a) reduces to the normal least squares problem. β_j is defined as follows:

$$\beta_j = 1 \text{ if } g_h > \frac{-\lambda_h}{2r_h}, 0 \text{ otherwise} \quad (1c)$$

The constraint function for the hth iteration is g_h. This ensures that only the active constraints participate in the calculation of $\mathbf{P}$.

Using the notation of Piegl & Tiller (1997), the B-spline, $C(u)$, is:

$$C(u) = \sum_{j=0}^{n} N_j(u) \mathbf{P}_j \quad (1d)$$

where $N_j(u)$ are the B-spline basis functions.

The constraints are included in the definition of α_j, given in equation 4. The constraint functions are:

$$g_j(\mathbf{P}) = (\mathbf{P}_{j+1} - \mathbf{P}_j) \cdot \mathbf{Y} \leq 0 \text{ for } 0 \leq j \leq n-1 \quad (2)$$

and

$$g_j(\mathbf{P}) = (\mathbf{P}_{i+1} - \mathbf{P}_i) \cdot \mathbf{M} \leq 0 \text{ for } n \leq j \leq 2n-1, i=j-n \quad (3)$$

Equations (2) and (3) are defined for the case where $(\mathbf{P}_{j+1} - \mathbf{P}_j) \cdot \mathbf{Y} \leq 0$ and $(\mathbf{P}_{j+1} - \mathbf{P}_j) \cdot \mathbf{M} \leq 0$. The equations are easily changed for $(\mathbf{P}_{j+1} - \mathbf{P}_j) \cdot \mathbf{Y} \geq 0$ and $(\mathbf{P}_{j+1} - \mathbf{P}_j) \cdot \mathbf{M} \geq 0$.

According to (Rao 1996), α_j is then defined as:

$$\alpha_j = \max\left\{ g(\mathbf{P}), -\frac{\lambda_j}{2r_h} \right\} \text{ for } 0 \leq j \leq 2n-1 \quad (4)$$

The Lagrange multipliers, λ_j, are updated after each iteration as follows (Rao 1996):

$$\lambda_j^{(h+1)} = \lambda_j^{(h)} + 2r_h \alpha_j^{(h)} \text{ for } 0 \leq j \leq 2n-1 \quad (5)$$

The penalty parameter, r_h, is updated after each iteration as follows (Rao 1996):

$$r_{h+1} = c r_h, c>1. \quad (6)$$

In this implementation of the ALM method $c = 1.5$.

4.2 Finding the solution vector

The solution vector can be found explicitly if the parameterisation is known. In this algorithm a chord length parameterisation is used and it remains unchanged through all the iterations. $\mathbf{P}$ can now be found by taking each term on the right hand side of equation (1a) and differentiating with respect to the unknowns. If $\mathbf{P}_k \in \mathbf{P}$, $\mathbf{P}_k = (\mathrm{P}x_k, \mathrm{P}y_k, \mathrm{P}z_k)$, $\mathbf{Q}_i = (\mathrm{Q}x_i, \mathrm{Q}y_i, \mathrm{Q}z_i)$ and $0 \leq k \leq n$, then the following equations can be derived.

The first term in equation (1a) can be differentiated with respect to $\mathrm{P}x_k$ as follows:

$$\frac{\partial}{\partial \mathrm{P}x_k} f(\mathbf{P}) = -2 \sum_{i=0}^{N} \mathrm{Q}x_i N(u_i)$$

$$+ 2 \sum_{i=0}^{N} N_k(u_i) \sum_{j=0}^{n} N_j(u_i) \mathrm{P}x_j \quad (7)$$

For the above and all the following equations, similar equations can be derived for differentiation with respect to $\mathrm{P}y_k$ and $\mathrm{P}z_k$.

When a constant is inactive, α_j is not a function of $\mathbf{P}$ therefore its derivative will be zero. This is why β was introduced. It simplifies the writing of the equations below. If $\mathbf{Y} = (\mathrm{Y}x, \mathrm{Y}y, \mathrm{Y}z)$ and $\mathbf{M} = (\mathrm{M}x, \mathrm{M}y, \mathrm{M}z)$ the second term in equation (1a) is differentiated as follows:

$$\frac{\partial}{\partial \mathrm{P}x_0} \left[\sum_{j=0}^{2n-1} \lambda_j \alpha_j \beta_j \right] = -\mathrm{Y}x(\lambda_0 \beta_0) - \mathrm{M}x(\lambda_n \beta_n) \quad (8a)$$

$$\frac{\partial}{\partial Px_k}\left[\sum_{j=0}^{2n-1}\lambda_j\alpha_j\beta_j\right]=Yx\left(\lambda_{k-1}\beta_{k-1}-\lambda_k\beta_k\right)+$$
$$Mx\left(\lambda_{k-1+n}\beta_{k-1+n}-\lambda_{k+n}\beta_{k+n}\right) \qquad (8b)$$

with $0 < k < n$

$$\frac{\partial}{\partial Px_n}\left[\sum_{j=0}^{2n-1}\lambda_j\alpha_j\beta_j\right]=Yx\left(\lambda_{n-1}\beta_{n-1}\right)+Mx\left(\lambda_{2n-1}\beta_{2n-1}\right)$$
$$(8c)$$

The third term in equation (1a) is differentiated as follows:

$$\frac{\partial}{\partial Px_0}\left[r_h\sum_{j=0}^{2n-1}\alpha_j^2\beta_j\right]=-2r_h Yx\beta_0 Y\cdot\left(P_1-P_0\right)$$
$$-2r_h Mx\beta_n Y\cdot\left(P_1-P_0\right) \qquad (9a)$$

$$\frac{\partial}{\partial Px_k}\left[r_h\sum_{j=0}^{2n-1}\alpha_j^2\beta_j\right]=$$
$$2r_h Yx(Y)\cdot\left[-P_{k+1}\beta_k+\left(\beta_{k-1}+\beta_k\right)P_k-P_{k-1}\beta_{k-1}\right]+$$
$$2r_h Mx(M)\cdot\left[-P_{k+1}\beta_{k+n}+\left(\beta_{k-1+n}+\beta_{k+n}\right)P_k-P_{k-1}\beta_{k-1+n}\right]$$
$$(9b)$$

with $0 < k < n$

$$\frac{\partial}{\partial Px_n}\left[r_h\sum_{j=0}^{2n-1}\alpha_j^2\beta_j\right]=2r_h Yx\beta_{n-1}Y\cdot\left(P_n-P_{n-1}\right)$$
$$+2r_h Mx\beta_{2n-1}M\cdot\left(P_n-P_{n-1}\right) \qquad (9c)$$

Equations (7) to (9) can now be combined in $2(n + 1)$ linear equations to solve for **P** by setting the derivative equal to zero to find the minimum. Note that usually the least squares problem for unconstrained B-spline fitting results in $(n + 1)$ equations. However, the nature of the constraints applied here, means that Px_k will depend on Py_k say. Therefore there are twice as many equations for 2D curves. Still, it is useful that **P** can be solved explicitly in each iteration.

4.3 *Selection of initial values*

The data points are parameterised using the chord length method. The parameterisation is not changed again. The knot vector is calculated using Piegl and Tiller's method (Piegl & Tiller 1997 pp. 412). This is also not changed again. These assumptions have the significant advantage that the only unknowns that must be determined during each iteration are the control points. They can therefore be solved from the resulting set of linear equations as described above.

Since the parameterisation is not updated, this algorithm does not truly minimise the distance between the curve and point data. However, experience has shown that in most engineering cases these assumptions are good enough.

The control points calculated from an unconstrained curve fit are used as the starting values for the ALM algorithm.

Initially all the Lagrange multipliers are 1. The initial value of the penalty parameter is the sum of the distances between the data points and the unconstrained curve fitted through them.

4.4 *Algorithm*

The algorithm starts by selecting the initial values as described in section 4.3. The next step is to minimise equation 1a using the equations 7 to 9 to find the control points and using a given set of Lagrange multipliers and penalty parameters. Then the convergence of the control points and Lagrange multipliers are tested. If the values have not converged, a new set of Lagrange multipliers and penalty parameters are calculated using equations (2) to (6). This is illustrated in Figure 3.

4.5 *Results*

In this section some of the test cases are reported. The first case is a simple B-spline with degree 1 and only two control points. (Effectively a line.) The result as shown in Figure 4 is as may be expected. There is a 5° angle between the pull direction (**Y**) and the curve. The curve goes through the centroid of the data.

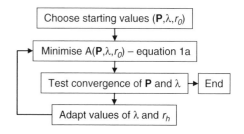

Figure 3. Algorithm procedure.

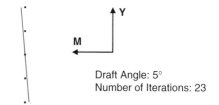

Figure 4. Test result No. 1.

272

The second test case (Figure 5) shows the ability of the algorithm to approximate data with different draft angles using a cubic B-spline with 4 control points. Note the undercut formed in the upper section of the unconstrained curve. This is resolved in the constrained curves.

The third case (Figure 6) is similar to the second case, except that the B-spline now has 10 control points. It took 26 iterations to converge, significantly more than the previous case.

Note that in Figure 6 the upper section of the constrained B-spline is a straight line. The constraints are active on all the segments of the control polygon in this segment.

Although the initial results (Table 1) are encouraging, the algorithm is not stable under all conditions. In some tests, the algorithm did not converge. The reason for this instability is not yet understood. It can also not yet be predicted when the algorithm will converge. This is part of ongoing research.

5 CASE STUDY

The algorithm was used to generate the curves needed to construct the surface of a boat hull. B-splines were fitted to cross section point data and a constraint was added to change the draft angle to 5°. A surface was then interpolated through the curves. B-splines fitted without constraints are shown in Figure 7 (left). B-splines fitted with the draft angle constraint are shown in Figure 7 (right) and the surface is shown in Figure 8.

This algorithm minimises the distance between the B-spline and the point data while maintaining the draft angle constraint. It is therefore not surprising that the constrained B-splines deviate from the point data at the radius at the bottom left corner of the point data. Note also that the curves form a straight line in the "vertical" section. The segments of the control polygon in this part of the curve form a straight line because the constraints are active here.

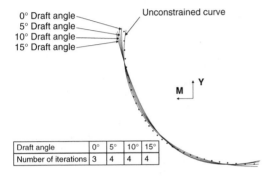

Draft angle	0°	5°	10°	15°
Number of iterations	3	4	4	4

Figure 5. Test result No. 2.

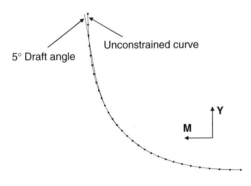

Figure 6. Test result No. 3.

Table 1. Summary of results.

B-spline parameters

Degree	# Control points	# Sampled points	Draft angle [°]	Iterations
1	2	5	5	23
3	4	30	0	3
3	4	30	5	4
3	4	30	10	4
3	4	30	15	4
3	10	30	5	26

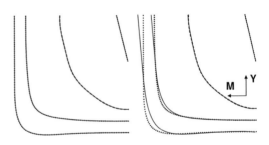

Figure 7. Unconstrained (left) and constrained (right) B-splines fitted to hull contour point data.

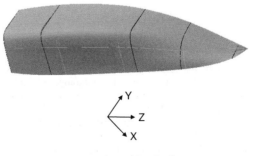

Figure 8. Boat hull surface with point data.

273

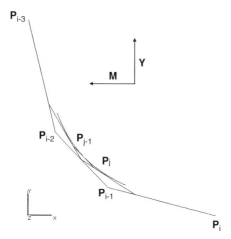

Figure 9.   Subdivision of a segment of a cubic B-spline.

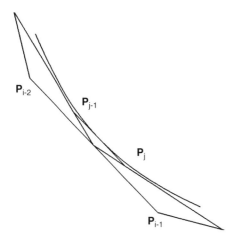

Figure 10.   Curve subdivision – enlargement.

6   CONCLUSIONS

A theorem was presented that gives sufficient conditions for a B-spline to satisfy a draft angle constraint. This is the main contribution of the paper. This theorem can also be used to add a draft angle to an existing B-spline. The constraints result in a set of linear inequality equations. Currently an Augmented Lagrange Multiplier method is used to approximate the B-spline with the inequality constraints. This was used successfully in several test cases and a case study. However, the algorithm does not converge for all cases. This algorithm will be useful in many practical reverse engineering problems, especially in the moulding industry.

APPENDIX A

Theorem proof

Any B-spline is contained in the convex hull of its control polygon. Specifically, if $u \in [u_i, u_{i+1}], p \leq i < m-p-1$, then $\mathbf{C}(u)$ is in the convex hull of $\mathbf{P}_{i-p}, \ldots, \mathbf{P}_i$. $\mathbf{C}(u)$ is a point on the spline and u_i are the knots (Piegl & Tiller 1997 pp. 83).

Points on the spline and tangents can be calculated by subdividing the control polygon (Piegl & Tiller 1997 eq. 3.7). If the tangent $\mathbf{C}'(u)$ must be calculated, then the curve must be subdivided by inserting $p-t-1$ new knots at u where t is the original multiplicity of u, in this case either 0 (if $u \notin \mathbf{U}$) or 1 (if $u \in \mathbf{U}$). This is shown for a segment of a cubic B-spline in Figure 9 and enlargement in Figure 10. The original control points, $\mathbf{P}_i$, are shown. The last segmentation, parallel to the tangent line, is the line $\mathbf{P}_{j-1}$ to $\mathbf{P}_j$. This segment will intersect the curve at u, at which it is tangent to the curve. The figure also shows the pull direction, $\mathbf{Y}$, and material side, $\mathbf{M}$.

Since $(\mathbf{P}_{i+1} - \mathbf{P}_i) \cdot \mathbf{Y} \leq 0$ and $(\mathbf{P}_{i+1} - \mathbf{P}_i) \cdot \mathbf{M} \leq 0$ for all i, it follows from simple geometric reasoning that $(\mathbf{P}_{j+1} - \mathbf{P}_j) \cdot \mathbf{Y} \leq 0$ and $(\mathbf{P}_{j+1} - \mathbf{P}_j) \cdot \mathbf{M} \leq 0$.

A similar argument can be done for $(\mathbf{P}_{i+1} - \mathbf{P}_i) \cdot \mathbf{Y} \geq 0$ and $(\mathbf{P}_{i+1} - \mathbf{P}_i) \cdot \mathbf{M} \geq 0$. This concludes the proof.

ACKNOWLEDGEMENTS

The author wishes to acknowledge the advice of Mr. DNJ Els and Dr. JD Buys regarding the optimisation methods. Prof. AH Basson made his graphics library available for testing the algorithm and generously assisted in integrating it with the rest of the C++ code.

REFERENCES

Dimitrov, D., Schreve, K. & Bartel, K. 2004. Analysis of Current Reverse Engineering Practice and Possibilities for Improvements to the Process, *COMA '04 International Conference on Competitive Manufacturing, 4–6 February 2004*: 131–136. Stellenbosch, South Africa.

Fischer, R.B. 2004. Applying Knowledge to Reverse Engineering Problems. *Computer-Aided Design* 36(6): 501–510.

Karniel, A., Belsky, Y. & Reich, Y. 2005. Decomposing the Problem of Constrained Surface Fitting in Reverse Engineering. *Computer-Aided Design* 37(4): 399–417.

Krause, F.-L., Fischer, A., Gross, N. & Barhak, J. 2003. Reconstruction of Freeform Objects with Arbitrary Topology Using Neural Networks and Subdivision Techniques. *Annals of the CIRP* 52(1): 125–128.

Kruth, J.-P. & Kerstens, A. 1998. Reverse Engineering Modelling of Free-form Surfaces from Point Clouds Subject to Boundary Conditions. *Journal of Materials Processing Technology* 76(1–3): 120–127.

Langbein, F.C., Marshall, A.D. & Martin, R.R. 2004. Choosing Consistent Constraints for Beautification of Reverse Engineered Geometric Models. *Computer-Aided Design* 36(3): 261–278.

Ma, W. & Zhao, N. 2002. Exact G1 Continuity Conditions for B-spline Surfaces with Applications for Multiple Surface Fitting. *Proceedings of the Intl. Conf. On Manufacturing Automation, Hong Kong, 10–12 December 2002*: 47–56, Hong Kong, China.

Piegl, L. & Tiller, W. 1997. *The NURBS Book*, 2nd Edition, Berlin: Springer.

Rao, S. 1996. *Engineering Optimization: Theory and Practice*. 3rd Edition, New York: Wiley-Interscience.

Renner, G. & Weiss, V. 2001. Reconstruction of Complex Freeform Objects with Automatic Smoothing. *Intl. CIRP Design Seminar, KTH Stockholm, 6–8 June 2001*: 375–379 Stockholm, Sweden.

Rogers, D.F. & Fog, N.G. 1989. Constrained B-spline Curve and Surface Fitting. *Computer-Aided Design* 21(10): 641–648.

Werghi, N., Fischer, R., Robertson, C. & Ashbrook, A. 1999. Object Reconstruction by Incorporating Geometric Constraints in Reverse Engineering. *Computer-Aided Design* 31(6): 363–399.

Virtual modeling and rapid manufacturing – Bártolo (eds)
© *2005 Taylor & Francis Group, London, ISBN 0 415 39062 1*

The development of an augmented reality system for the design and retail of ceramic tableware

R. Omar, D.G. Cheshire & W. Wu
Faculty of Computing, Engineering and Technology, Staffordshire University, UK

ABSTRACT: Augmented Reality (AR) is a technique whereby real world data is overlaid with computer generated data in real-time. This duality of data types opens the possibility of a new methodology to product design and retail. This paper describes work being undertaken at Staffordshire University to focus and apply these methodologies and develop practical applications to assist the product design industry. Using AR technology product designers can combine real-time video and virtual models to view a new product in real world surroundings without having to create actual physical prototypes. AR can also be used to involve consumers in the design process and to aid sales and marketing. Although these techniques are suitable for any product design process this paper describes the work undertaken in conjunction with the UK tableware industry based in Stoke-On-Trent. The initial aim is to develop systems to support the early stages of the design process.

1 INTRODUCTION

1.1 *Computer aided prototyping*

Computers have changed the way we live. The diverse application of computers in our daily lives can be attributed to various factors such as the decreasing price of computers and the ever increasing processing ability. Other equally important contributing factors are the advances and the multitude of software developed to meet the diverse demand of users. The advances in computer applications like CAD/CAM, CAID software have resulted in the gradual replacement of the manual modeling process with 3D computer-simulation models for product design, mode and prototyping process (Papamichael et al. 1999). These 3D prototypes aid the visualization of problems and their solutions allowing the designer to refine the products at the earliest possible stage thereby lowering the design cost of the product and producing higher quality products (Chen & Cheng 1999).

The use of computer generated 3D prototypes has given rise to two new prototyping approaches in the past decade, the rapid prototyping and the virtual prototyping process. Rapid Prototyping (RP) refers to the process of rapidly constructing physical models by adding a layer-by-layer deposition of material based on computer data in the form of 3D digital models (Chelule et al. 2000). Computers equipped with advanced Graphics Processing Unit (GPU) together with the advancements made in the Virtual Reality (VR) research domain, have spurred the concept of Virtual Prototyping (VP). VP can be defined as the application of VR for prototyping physical mock-ups (PMUs) using product and process data (Gomes & Zachmann 1999). The aim of VP is to build a full virtual artifact which can be used in a cooperative work environment to foresee and discuss probable design and manufacturing flaws which might arise (Gomes et al. 1998). VP uses virtual prototypes – computer-based simulations possessing a certain degree of functional realism – instead of physical prototypes for these evaluation purposes. According to Kalawsky, one of the most important breakthroughs of VP practice in future is expected to be the integration of real and virtual environments, i.e. Augmented Reality (Kalawsky 1999).

1.2 *Integration of real and virtual environments*

Augmented Reality (AR) can be defined as enhancing the users' perception of and interaction with the real world by overlaying 3D virtual objects in the real world so that the real and virtual object appear to coexist in the same space (Azuma et al. 2001). In contrast to Virtual Reality (VR) technology, where users are totally immersed in a wholly computer generated world (virtual world); an AR system needs to maintain the feeling of presence in the real world (Vallino 1998). Milgram and Kisihino have described the relationship and taxonomy of AR technology along the "Virtuality Continuum" (VC) in a Mixed Reality (MR) environment; see Figure 1 (Milgram & Kisihino 1994).

Figure 1. Simplified representation of a VC Continuum (Milgram & Kisihino 1994).

Although Augmented Reality can be considered relatively new compared to other areas of research within the realm of the computer graphics' domain such as photorealistic rendering and Virtual Reality, remarkable growth has been made in the last few years (Azuma et al. 2001). These significant improvements in AR technology are due to research done in various key areas, such as tracking and registration and real-time rendering.

Malik S. has stated that a typical augmented reality system consists of, a capture device such as a camera, a display device and for certain applications user interface devices for interacting with the computer generated synthetic objects (Malik 2002). There are several classification made for AR systems. One significant classification method for AR systems is based on their display technology. Display technology is considered to be one of the enabling technologies for AR environments (Azuma et al. 2001). Generally display technologies employed in an AR system are either based on optical or video technologies. The four major classes of AR display system are (Isdale 2000):

1. Optical See-Through AR – transparent Head Mounted Display (HMD) directly overlays real world with virtual environment (VE).
2. Projector Based AR – real world objects are used as projection spaces for VE.
3. Video See-Through AR – an opaque HMD is used for display of merged video streams of VE and views from cameras mounted on the HMD.
4. Monitor Based AR – uses merged video streams of AR on desktop monitors or hand held displays.

1.3 *Applications of AR*

The idea of enhancing perception of reality can be dated as early as the 13th century when Roger Bacon commented on the use of optical lenses (Sairo 2001). Recently there have been substantial numbers of reports on the use of AR in industrial design related processes, largely in automotive and building industries (Poh et al. 2005). Two car manufacturers BMW and Volkswagen AG have tested AR technology in the design process of their cars (Fruend et al. 2002) (Klinker et al. 2002). AR has been used for real-time augmentation of broadcast video (Azuma et al. 2001). Other areas of application where AR technology is used include

military training, entertainment, medical and surgery robotics, engineering design and manufacturing.

Verlinden et al. have introduced the concept of Augmented Prototyping where digital images are projected on objects manufactured by rapid prototyping techniques (Verlinden et al. 2003). They have developed a system called Workbench for Augmented Rapid Prototyping (WARP). WARP system, utilizes the concept of Projector Based AR (see section 1.2). Digital images displaying a variety of materials and lighting conditions are projected onto models which have been manufactured using rapid prototyping methods. Nam et al. have proposed the use of AR to integrate hardware and software of products in the early phase of the design process (Nam & Lee 2003). This integration is achieved by overlaying a virtual display onto (limited) functional hardware prototypes using two AR display technologies namely, the video see through HMD and the projector based AR display system.

2 CERAMIC PRODUCT DESIGN PROCESS

2.1 *The evolution process of ceramic ware design*

Traditionally ceramic product manufacturing processes occur sequentially and are time consuming (Bin et al. 2001).

Ten serial steps can be identified from Figure 2. If the porcelain samples do not pass the evaluation stage, the manufacturer will have to return to the preceding steps and redo the earlier processes, resulting in wastage of time and resources. This traditional ceramic tableware design and development techniques which have remained unchanged for many decades means the lead time for new products development is too long, hence uncompetitive for today's competitive market (Wormald et al. 1994). The introduction of CAD/CAM and RP technologies has helped to reduce the lead time in ceramic tableware design process. CAD/CAM applications have been employed by Wedgewood, Royal Doulton and others to help in the design process of their ceramic tableware.

2.2 *Using VR for ceramic tableware prototypes*

In 2000 researchers from Staffordshire University together with The HotHouse, Wedgwood and Virtalis developed a VR system for the design review process and as an aid to marketing (Danneels & Cheshire 2001). A 3D virtual dining room was modeled to showcase different ranges of ceramic tableware, see Figure 3. This Virtual Environment (VE) was created using Studio Max and then exported to Lightscape for radiosity computation to generate photorealistic rendering of the virtual dining room.

To display the photorealistic virtual dining room in real-time with the highly detailed 3D models of the

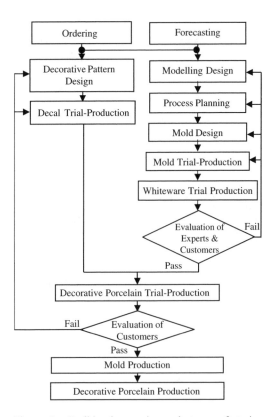

Figure 2. Traditional ceramic product manufacturing processes (Bin et al. 2001).

Figure 3. The virtual dining room (Danneels & Cheshire 2001).

ceramic tableware, a four 250 MHz R10000 processors Onyx2 with a single pipe Infinite Reality2 Graphic Board and 2 GB of RAM memory was used as this was one of the few machines available at the time with sufficient graphics processing power.

Figure 4. Various decal designs for ceramic plates (Danneels & Cheshire 2001).

Various different models of tableware were designed with the ceramic industries standard surface modeler CAD system called DeskArtes. To achieve low latency (real-time) drawing, a Level of Detail (LOD) algorithm was applied where three versions of each tableware models were created with different numbers of polygons. These 3D models of the ceramic tableware are then imported in the form of STL files into 3D Studio Max where textures, material properties and layers are added to each model. To achieve high quality rendering of 3D models a layering technique is used. Three different layers are defined; material, texture and environment layer; which are render separately and then composited together (see Figure 4).

Users can navigate within the virtual dining room using a hand controller tracked by Polhemus Fastrack. Users can examine selected 3D items by picking up, rotating or replacing the selected items with various other items from different ranges of product. Movement within the virtual environment can be either in walk through or orbit mode.

2.2.1 Advantages of VR technology

Using VR technology in the ceramic product prototyping process has several advantages over the traditional approach. Using VR, users can visualize, be involved in and interact with realistic 3D representation of ceramic products. VR reduces or eliminates the need for expensive physical mock-ups (Danneels & Cheshire 2001). Virtual ceramic products make it possible for customers to view the final show of products and give suggestions to designers before production (Bin et al. 2001). Virtual ceramic products are very useful in electronic warehouses or even in shops (electronic showrooms), where using point of sale (POS) computers allow customers to click on products which are of interest to them and view these products in virtual environments (Bin et al. 2001) (Danneels & Cheshire

279

2001). The advancement in the internet and the increase in bandwidth capacity (with broadband connection to home users) opens the possibility of using virtual ceramic products via the internet.

As mentioned in section 1.1, the ability to merge virtual and real environment in one single environment is the next leap forward in VP (Kalawsky 1999). This merger gave rise to a domain called Augmented Reality, which opens new avenues for using 3D computer generated models in the ceramic tableware design process.

2.3 Using AR technology for ceramic tableware design process

Realizing the immense potential of using AR technology in the ceramic tableware design and retail processes has prompted researchers at Staffordshire University to develop new prototype systems using AR technology as an alternative to the earlier developed VR based system, (see section 2.2).

The AR system being developed is based on the monitor based video displays or Window-on-the-world (WoW) display setup (Milgram & Kisihino 1994), see Figure 5. There are several reasons for choosing this type of display setup:

1. Simplicity of the AR system design architecture; according to Vallino this design of AR system is the simplest (Vallino 1998).
2. Compared to video based AR, optical see-through AR HMD displays typically requires more complicated camera calibration and registration requirements (Kato et al. 2001).
3. Due to the fact that real world images are digitised to the video based display, it is easier to match the brightness and the common illumination effect between real and virtual objects in synchronous space and time. Hence the appearance of virtual objects can be enhanced to look as realistic as possible, by applying appropriate real-time rendering techniques.

The merging (see "Video Merging" in Fig. 5) of the real scene video images with virtual objects is achieved by aligning the computer graphics virtual camera and the real camera. Once the virtual camera is aligned with the real world camera, virtual objects can be overlaid correctly in the real world video. For the development of these AR prototype systems, the alignment between virtual and real objects is achieved using ARToolKit. This software development toolkit is primarily being developed by Dr. Hirokazu Kato of Osaka University, Japan.

ARToolKit uses computer vision techniques for calculating the real camera viewpoint relative to real world 2D fiducial markers (Kato et al. 2001). First the live video image is captured (Fig. 6a), and then the

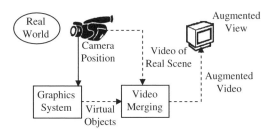

Figure 5.   Monitor based AR system (Vallino 1998).

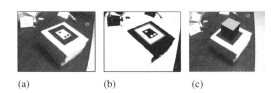

(a)   (b)   (c)

Figure 6.   Augmentation of video streams (Kato et al. 2001). (a) Input video, (b) Binary image video with threshold value. (c) Final result, augmented video.

captured live video image is transformed to binary (black and white) image (Fig. 6b). ARToolKit generates 6DOF (degree-of-freedom) matrices for each of the detected 2D fiducial markers, which is used to position the virtual camera. Since the real world and the virtual camera coordinates are the same; computer generated objects can be rendered precisely over the real marker in the final scene using the camera calibration frustum (Fig. 6c).

2.4 The development of an AR based system for the ceramic tableware design and retail process

The AR system is developed on the Windows® platform, using MS-Visual Studio 6.0. To be able to offer end-users with easier ways to access various additional application-specific functionalities and to create a more user-friendly environment means compatible Graphical User Interface (GUI) libraries have to be incorporated on top of the native standard ARToolKit tracking library.

Window-Icon-Mouse-Pull-down menus (WIMP) style interaction is chosen due to the nature of the of the AR prototype application developed, a monitor based AR system. Several libraries have been investigated and as the result several applications were developed incorporating these GUIs' widgets. One of these GUI based prototype system was developed using Picoscopic User Interface (PUI) GUI library, see Figure 7.

The previous GUI libraries tested are lacking in terms of the MS-Windows GUI "look-and-feel", for

280

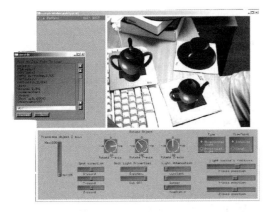

Figure 7. An AR application, with PUI's widgets integrated. The smaller window is a pop-up window which only appears when users select the file open sub-menu from the pull-down menu.

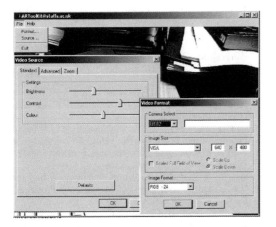

Figure 8. An SDI implementation of MS-Win GUI based AR application, showing two pop-up windows for video format and source controls.

this reason the next prototype was built using the Windows's GUI library, see Figure 8. Using MS-Win3 GUI library gives users the typical MS-indows's user interface look-and-feel and characteristics. Standard MS-Win file selector and loading pop-up windows appear when users want to augment the real world video sequences with new 3D computer generated objects or when users wants to change the texture of a certain selected 3D object.

A new functionality, which is provided to users, is the ability to control and change the video resolution and other typical video image control functions such as the brightness and contrast levels, in real-time during

Figure 9. A Windows' MDI version of the AR application, there are various tools which users can use to evaluate the ceramic tableware. Apart from the selection of various models (e.g. trying various designs for the handle or the spout of a teapot) and decals, users can also change the scale of these 3D meshes, to compare the models with real world objects. This data can then be used by the modelling package to remodel the prototypes.

the execution of the AR application. Users can change the video resolution size without affecting the tracking and alignment of virtual objects during the execution of the application.

The second improved version of the MS-Windows based GUI is developed using the Windows' Multiple-Document Interface (MDI) implementation, see Figure 9. Apart from "inheriting" the same basic functionalities of the earlier version, the MDI version offers some additional enhanced functions, such as being able to generate BMP images of the real world overlaid with various virtual ceramic tableware models. The interfaces are also much more user-friendly and intuitive. Users have easier accessibility, to functionalities such as the selection of various ceramic tableware items, ranges and various different pattern designs.

The 3D virtual objects of ceramic tableware used, are in the format of Wavefront OBJ files and the rendering processes are done using OpenGL. These OBJ files are modelled and textured using WinSurf, a Windows based surface modeler developed for industrial designers to design products based on freeform shapes developed by David Cheshire of Staffordshire University. Obtaining digital data of decals used for texturing of the OBJ models is easily done due to the fact that Ceramic Industry has been using digital tools such as Adobe Illustrator for many years to create tableware decorations (Danneels & Cheshire 2001). A wide range of texture file formats can be loaded to decorate the virtual objects; this flexibility is achieved using an image loading library.

281

3 CONCLUSIONS

Computer generated 3D prototype used in RP and VP have helped to reduce the cost of developing new high quality products. Too much time is taken in the development of new ceramic tableware products in the traditional serial sequences design process and this will affect the ability of manufacturers to compete in today's highly competitive market. Therefore realizing this need for quick and effective product design and evaluation processes, researchers at Staffordshire University are looking into new technologies to evaluate and deliver high quality 3D computer data to be used for the ceramic tableware production.

Two new technologies have been identified; using VR and AR technologies. VR enables CAD data to be displayed in real-time with very high degree of realism without the need for any prototyping of physical mock-ups. Although VR has proven itself to be an effective tool in the design and evaluation processes of new products, for the most realistic experience in VR, expensive viewing setups and large projection walls are needed, and in spite of these, a recent observational study shows that the presence of physical objects during the design process helps in enhancing design thinking and communication (Poh et al. 2005). Another major drawback of using VP is the lack of a sense of scale, therefore depriving the experience of the relationship between the product's scale and its physical context (Verlinden et al. 2003).

A new prototype AR system (which possesses the same advantages of a VR system, see section 2.2.1) has been developed to overcome the drawbacks of a VR based system. From its early conception the prototype system has been developed with a cost-effective solution in mind. The system setup is very simple. Video images of real world scenes are captured using a (low cost) USB webcam. Although the prototype is developed on a computer equipped with an NVidia Geforce 4 graphics card, tests done on computers with lower specifications have given similar results, without much loss of quality. The simple system setup and requirement means that this AR system can be used on most computers utilizing the already available hardware. This will make an ideal tool for consumer marketing, as prospective buyers can download the software and then view the virtual ceramic tableware within the real surrounding of their own home. Being able to view the virtual objects within the augmented real environment also helps designers to evaluate the product without depriving the sense of immersion in real world.

4 FUTURE WORKS

Currently the prototype system uses a simple environment mapping technique to simulate reflections of a ceramic's surface; future enhancements include more advanced rendering techniques such as bump mapping. For increase in visual and spatial cues, the current system uses a simple shadowing technique to simulate the shadow of the virtual objects. More realistic shadows are planned for future systems. We are also planning to migrate the current system to portable devices such as PDA's and develop a web based application.

REFERENCES

Azuma, R. Baillot, Y. Behringer, R. Feiner, S. Julier, S. & MacIntyre, B. Recent Advances in Augmented Reality. *IEEE Computer Graphics and Applications, Nov–Dec 2001:* 33–47.

Bin, D. Fei, L. & Le, C. Computer-Integrated Ceramics Manufacturing. *American Ceramic Society Bulletin.* Vol. 80, No. 3. March 2001.

Chelule, K.L. Coole, T. & Cheshire, D.G. Fabrication of Medical Models from Scan Data via Rapid Prototyping Techniques. *Time Compression Technologies 2000 Conference.* Cardiff International Arena, UK.

Chen, D. & Cheng, F. 1999. Integration of Product and Process Development Using Rapid Prototyping and Workcell Simulation Technologies. *Journal of Industrial Technology.* Volume 16, Number 1 (Nov. 1999 – Jan 2000).

Danneels, S. & Cheshire, D.G. Virtual Reality for Tableware Prototypes. *Virtual Reality International Conference (VRIC).* 16–18 May, 2001. Laval.

Fruend, J. Matyszok, C. & Radkowski, R. AR-based Product Design in Automobile Industry, *Proceedings of First IEEE International Augmented Reality Toolkit Workshop* 2002. Darmstadt, Germany.

Gomes, A. & Zachmann, G. Virtual Reality as a Tool for Verification of Assembly and Maintenance Processes. *Computers and Graphics,* 23(3), 1999: 389–403.

Gomes, P.R. Feijó, B. Cerqueira, R. & Ierusalimschy, R. Reactivity and Pro-Activeness in Virtual Prototyping. *2nd International Symposium on Tools and Methods for Concurrent Engineering.* Manchester, UK, 1998.

Isdale, J. Augmented Reality. URL: http://vr.isdale.com/vrTechReviews/AugmentedReality_Nov 2000.html#types (accessed 21st May 2004).

Kalawsky, R.S. Keynote Address: The Future of Virtual Reality and Prototyping. *Actes du colloque scientifique international. Réalité virtuelle et prototypage, 3–4 June 1999.* Laval.

Kato, H. Billinghurst, M. & Popyrev, I. ARToolKit Manual 2.33. 2001.

Klinker, G. Dutoit, A.H. Bauer, M. Bayer, J. Novak, V. & Matzke, D. Fata Morgana – A Presentation System for Product Design. *International Symposium on Augmented and Mixed Reality (ISMAR 2002)* 2002. Darmstadt.

Malik, S. Robust Registration of Virtual Objects for Real-Time Augmented Reality. *Master of Science Thesis.* The Ottawa-Carleton Institute for Computer Science, School of Computer Science, Carleton University, 2002.

Milgram, P. & Kisihino, F. A Taxonomy of Mixed Reality Visual Displays. *IEICE Trans. Information Systems,* vol. E77-D, 1994: 1321–1329.

Nam, T.J. & Lee, W. Integrating Hardware and Software: Augmented Reality Based Prototyping Method for Digital Products. CHI 2003. 5–10 April, 2003. Florida, USA.

Papamichael, K. Chauvet, H. LaPorta, J. & Dandridge, R. 1999. Product modeling for computer-aided decision-making. *Automation in Construction 8:* 339–350. The Netherlands. Elsevier Science B.V. 1999.

Poh, Y.L. Nee, A.Y.C. Youcef-Toumi, K. & Ong, S.K. Facilitating Mechanical Design with Augmented Reality. *Singapore-MIT Alliance Symposium Symposium,* 19–20 January 2005. Singapore.

Sairo, M. Augmented Reality. *Presentation Topic for the Research Seminar on Digital Media pervasive Computing Tik-111.590 (2–3 CR) fall 2001.* Telecommunication, Software and Multimedia Laboratory, Helsinki University of Technology.

Vallino, J.R. Interactive Augmented Reality. *PhD Thesis.* Department of Computer Science, University of Rochester, Rochester, New York, 1998.

Verlinden, J.C. de Smit, A. Peeters, A.W.J. van Gelderen, M.H. Development of a flexible augmented prototyping system. *Journal of WSCG,* Vol.11. WSCG'2003, 2003, Czech Republic.

Wormald, P.W. Harrison, D.K. & Cheshire, D.G. Novel Techniques for Tableware development. *3rd European Conference on Rapid Prototyping and Manufacturing,* University of Nottingham, July 1994.

Virtual modeling and rapid manufacturing – Bártolo (eds)
© 2005 Taylor & Francis Group, London, ISBN 0 415 39062 1

Real-time simulation environment for machine-tools

S. Röck, H. Rüdele, G. Pritschow
Institute for Control Engineering of Machine Tools and Manufacturing Units, Stuttgart, Germany

ABSTRACT: By taking advantage of the ever increasing capabilities of information systems, simulation technology can nowadays be applied to very complex systems. This paper outlines the development of an integrated environment that allows simulation of machine-tools in real-time. Both the kinematic and dynamic behaviour of the simulated machine-tools can be analysed. The numerical algorithms used for solving the corresponding differential equations were optimised for their real-time capabilities, hence the simulation environment can be utilised in hardware-in-the-loop configurations. Thus it is possible to operate the real motion control system in combination with the simulation environment. The paper discusses the software architecture of the simulation environment, its numerical nucleus and its well-defined interfaces. Finally specific usage scenarios for the simulation environment are described and analysed in regard to the benefit they provide for the stake holders.

1 INTRODUCTION

Simulation methods have been employed in many branches of industry and science for quite some time. By utilising simulation technology data can be gathered in a fast, reproducible and cost-efficient way. In contrast, on real systems the acquisition of this data would require more effort. If the behaviour of mixed systems consisting of both real and virtual elements is to be investigated, the simulation system has to meet strict temporal requirements. The requirements for these types of configurations – so called hardware-in-the-loop-simulation (HILS) – can only be fulfilled by systems that exhibit time deterministic behaviour. The capacity to meet temporal requirements is the chief distinction between simulation systems that are real-time capable and non-real-time systems. These restraints have to be taken into account in all development phases of projects that intend to realise hardware-in-the-loop-simulation. System properties influenced by these constraints are:

- the selection of numerical algorithms needed for the solution of the fundamental mathematical problems
- the models' level of detail and granularity
- the means of communication between the system's single components
- the choice of operating system and hardware environment on which the whole project will be based

Until recently only specialised computing hardware could meet the requirements for hardware-in-the-loop-simulation. The hardware was either very costly or rather primitive and hence could only simulate restricted systems (e.g. logic systems). The automotive industry played a significant role in the adaption of HILS as it acted as a pioneer and utilised real-time simulation very early (Schäuffele and Zurawka 2003). Due to the evolution of semiconductor technology described by Moore's Law, standard personal computers (PCs) nowadays offer ample computing power to simulate even complex models.

Additionally low-level enhancements have been developed by third-party companies that provide Microsoft Windows with the capability to assign small slices of computing time (sub-milliseconds) and other resources to several tasks in a time deterministic manner. Hence tasks can be executed in compliance with real-time requirements (Burmberger 2002). These developments established the basis for HILS on standard PCs that incorporate the advantages of small expenses, simple extensibility and well-known user interfaces.

The increasing market pressure forces manufacturers of machine-tools and motion control systems for machine-tools to explore innovative ways to reduce time-to-market as well as costs and simultaneously improve the quality of their products (Hinkel 2004). The evolution in the area of information technology as described above offers such a possibility. HILS can now be executed on inexpensive hardware that can be applied to a wide range of scenarios during development, commissioning and deployment of machine-tools. At the Institute for Control Engineering of Machine Tools and Manufacturing Units (ISW) at the

University of Stuttgart, Germany a simulation environment was conceived in cooperation with ISG (Industrielle Steuerungstechnik Stuttgart), a commercial partner. This environment was designed as a platform for the analysis of potential applications of HILS in the machine-tooling industry. A further design goal of the environment is to provide a platform for the development of solutions based on this analysis.

2 REQUIREMENTS

The main objective of the simulation environment is to simulate the kinematic and dynamic behaviour of machine-tools in real-time. The environment has to be designed in a way as to allow the simulation of a wide range of different types of machines. Additionally, the novel nature of this approach may create new and possibly unexpected areas of application in the future. Hence importance should be attached on the extensibility of the platform. In the following sections these general requirements will be discussed in more detail.

2.1 Requirements on the system's overall structure

The general requirement on the simulation platform that it must be able to simulate multiple types of machines leads to the conclusion that a parametric model based approach should be used. Although the other alternative – a non-parametric solution – would offer the advantage that no a priori knowledge about the machine-tools' structure would be necessary, its disadvantages are considerable. The modelled systems (i.e. machine-tools) are dependent on many state variables that can additionally be coupled via non-linearities. Hence non-parametric modelling would require a significantly larger computational effort for system identification and simulation. As the environment has to adhere to strict temporal restraints, this approach cannot be pursued. Due to this assessment the simulation environment needs to provide means for the straightforward creation of parametric models. These models should be displayed and handled in an intuitive way, so models and their parameters can be generated in an efficient manner.

One of the main advantages of HILS lies in the fact that parts of the real system can be combined with other, simulated components. The non-simulated parts of the system work agnostically with respect to the nature of the other components. In the specific application of simulated machine-tools this leads to the conclusion that the actual motion control system should behave identically regardless of whether it is connected to a simulated or a real machine. As the motion control system communicates with the drives of the machine-tool via a fast communication bus (IEC 61491) the

simulated machine must offer the corresponding drive interfaces to the communication bus (Kreusch 2001). This yields the additional advantage of independence of the motion control system. If the simulation platform implements the established communication interfaces no further knowledge about the specific type of motion control system is required.

The temporal requirements to the simulation platforms are defined by the non-simulated motion control system. The simulation has to produce valid data in the time frame of the motion control system's clock-cycle. This hard real-time constraint demands a real-time capable foundation (hardware and operating system) for the simulation system.

It is very likely that the simulation platform or parts thereof will be used in different applications and new configurations. To achieve easier portability and extensibility a modular design of the system architecture is required.

2.2 Interfaces

The aforementioned modular system design leads to the need for interfaces between the system's components. Depending on the properties of the involved components the interfaces have to comply to different requirements. Parts of the simulation platform have to run in accordance with real-time conditions; other parts of the system may run with less strict temporal constraints. The interfaces between the different components of the simulation environment have to allow for these different demands. To avoid loss of data between the real-time components and the non real-time part of the environment, a mechanism for the buffering of data between the environment's components has to be provided. The interfaces to external components (e.g. motion control systems) have to comply with the expected behaviour of non-simulated components.

2.3 Numerical considerations

The simulation has to adapt to the interpolation cycle of the motion control system. Current cycle-periods are in the range of two milliseconds and less. Hence high demands on computational efficiency and determinism are placed on the utilised algorithms and models.

Each simulation cycle has to terminate before a predefined deadline. If this deadline was exceeded, the real-time constraints would be violated. The duration of each computation must not be dependent on the current state of the simulation (Pritschow and Röck 2004, 2002). However the numerical solution's stability and the accuracy of nonlinear models are dependent on the current state of the model.

Conventional simulation systems vary the increment of computation in order to guarantee the accuracy and stability of the solution. As this approach obviously cannot result in time-deterministic behaviour another solution has to be found. This solution has to adhere to the constraints that the time increment can be varied only within certain boundaries because the maximum number of computations must not be exceeded and a fixed point of time (the point of time at which the calculated data must be transmitted) has to be hit.

Additionally numerical robustness has to be considered. The simulation platform has to supply valid data at the end of each simulation step. This results in high demands on the stability of the numerical algorithms. In general the numerical solution will differ from the correct analytical solution of the simulation equations. In order to minimise this deviation the selection of algorithms and the calculation's increment must be optimised. Obviously the demand for high numerical robustness and accuracy conflicts with the demand for efficiency of the calculation. Hence a compromise between accuracy and computational efficiency has to be found while also taking into account that additional errors may be introduced by erroneous parameters and the quality of the model in general (Pritschow and Röck 2004).

3 SYSTEM DESCRIPTION

The extensive array of requirements listed in the previous section form the base on which the implementation of the simulation platform is erected. The following sections describe the current status of the simulation platform and point out the interrelationship to the underlying requirements.

3.1 System architecture

In accordance with the requirements of extensibility and portability the system architecture conforms to a modular design philosophy. Figure 1 illustrates the modules and their interconnections. The real motion control system generates command position values for each of the machine-tool's separate drives according to its input data (i.e. NC programme). The command data are read by an interface driver that substitutes all of the actual drives. It then sends these values via a specialised interface to the numerical nucleus. The nucleus simulates the behaviour of the machine-tool and transmits the computed actual values to the interface driver. The data is subsequently written onto the communication bus by the interface driver and hence can be accessed by the motion control system. All of these actions have to happen during one cycle of the motion control system (i.e. in real-time).

The drives' command and actual values (and all of the simulation's internal data) can also be accessed by the modelling and monitoring module. As the monitoring does not necessarily have to take place in

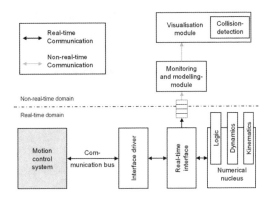

Figure 1. System architecture.

accordance to real-time demands, the data are transferred into the non-real-time domain of the simulation platform. The monitoring module also offers the possibility to transmit the geometrical position of each machine component based on the drives' actual values to a three-dimensional visualisation module.

The requirement for deterministic behaviour leads to the demand for support of real-time capabilities in the underlying layers (e.g. operating system) (Pritschow et al. 2003). Unfortunately the usage of these capabilities conflicts with the demand for easy portability of the software. This stems from the fact that the different operating system enhancements that provide real-time support implement different programming interfaces which are generally not compatible to each other. The platform's real-time capabilities must not be compromised under any circumstances. Hence the requirement for portability is treated secondarily to the real-time-requirements.

The numerical nucleus of the platform and the whole area of inter-modular communications will be depicted in separate sections. Here the system's other modules are described in a more perfunctory way.

The intended purpose of the simulation environment allows for a simple implementation of the interface to the communication bus as only the standard (closed) operation of the motion control system is to be covered. Nevertheless the protocols needed for the initialisation of communication and signalling during operations still have to be implemented. Otherwise the communication will not be conducted in a realistic manner.

The so-called modelling and monitoring module (see Figure 2) is used for two tasks in the course of a simulation project. First it is used to create the model of the machine-tool. This modelling phase is supported by a graphical user interface that is implemented in the style of standard modelling tools. The end user can use a wide variety of predefined elements that are available in a library. The library contains elements that realise mathematical functions,

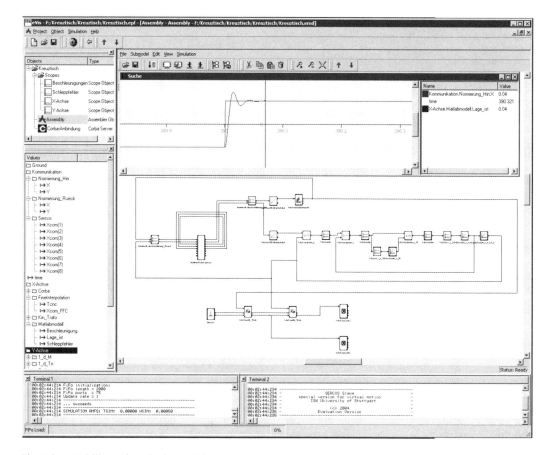

Figure 2. Modelling and monitoring module.

dynamic and kinematic machine properties, drive models, nonlinearities and I/O components. Additionally existing linear models that were created in Matlab/Simulink can be imported and executed immediately in the environment's real-time domain without the need for explicit code-generation. This offers the chance to import linearised model structures up to reduced finite element models (FEM). The second application of the module lies in the control of the simulation and its monitoring. The module offers mechanisms to display and plot the current values of the actual and command values as well as of all internal values that are computed during the simulation. Additionally data can be recorded for later analysis or transmitted to the visualisation module.

The visualisation module (see Figure 3) displays the motion of the machine-tool's components by using three-dimensional models. The machine-tool's geometric models can either be constructed out of primitives in the module itself or the output from external CAD tools can be used by importing geometry data from

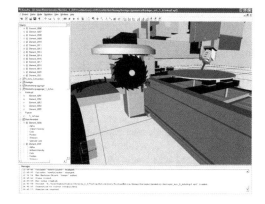

Figure 3. Visualisation module.

VRML files. The geometric objects are combined into machine objects. Each of these objects can be moved independently according to the simulated spatial position of each machine component.

3.2 Numerical solver

Generally speaking, two types of simulation seem possible. The behaviour of the machine-tools could either be simulated in the frequency domain or in the time domain. As the external components expect the simulation results in synchronicity to their behaviour the approach of a time based simulation suggests itself. The computations necessary for the simulation can generally be reduced to the numerical solution of differential equations. Due to the real-time constraints only methods that will work satisfactory with fixed step sizes can be considered. To allow for greater flexibility, several integration methods have been implemented that can be selected by the user of the monitoring and modelling module. It is also possible to combine these methods in order to gain their combined benefits (e.g. accuracy and robustness). Less advanced users can avoid the selection and utilise a preselected default method. During the simulation's run time the current numerical properties can be shown in order to detect possible numerical problems.

3.3 Communications/real-rime-interface

The different modes of interaction among the modules demand different mechanisms for the means of communication. All of the modules that reside in the real-time domain of the environment have to interact in a time deterministic manner. Transmission delays or lost information are not tolerable. On the other hand restrictions for information that passes into the non-real-time domain are more relaxed. To accommodate for all these different needs, a general purpose interface entity was implemented. The deterministic data transfer between modules in the real-time domain is achieved by utilising underlying mechanisms of the operating system's real-time enhancements. The data is stored in shared memory areas that are controlled by the real-time environment. The nondeterministic parts of the operating system cannot access or influence these memory regions and hence the access to it is guaranteed to be of a deterministic nature. To avoid inconsistencies, access to all relevant memory areas is synchronised by utilising semaphore variables. As already mentioned above, mechanisms for the transmission of data into the non-real-time domain also have to be implemented. To avoid delays (and the resulting loss of real-time behaviour) a buffering mechanism was realised. Data from the real-time domain is written into a FIFO queue whose data can be accessed by modules in the non-real-time domain. Hence even if the clients from the non-real time domain experience nondeterministic delays, these delays will not affect the modules in the real-time domain. All of these mechanisms are bundled together into one entity – the so-called I/O-interface. It also offers methods for running and

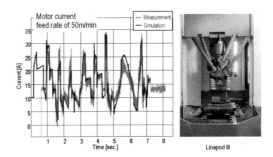

Figure 4. Validation.

stopping the environment's numerical nucleus. Therefore this interface (and its mechanisms) can be employed by external programmes as an API for the numerical nucleus.

The transport mechanisms needed for the transfer of the machine objects' simulated spatial positions from the modelling and monitoring module to the visualisation module are implemented using the CORBA protocol. CORBA is scalable across different hosts by using network protocols (e.g. TCP/IP). This allows for the distribution of modules on multiple hosts. As three-dimensional visualisation can demanding significant computing power, this separation seems beneficial. A further advantage of the separation lies in the possibility for the visualisation module to receive data from several servers. This offers the chance to simulate kinematically decoupled parts of the same machine-tool on different hosts and still visualise them conjointly.

Furthermore the models that were created in the monitoring and modelling module have to be transferred to the numerical nucleus. As this transfer has to happen only once during the course of a simulation project, this data is stored in a conventional file.

3.4 Validation

The plausibility of the simulated data has been validated in several public projects and in co-operations with industrial partners. If all the relevant parameters of the model can be ascertained accurately the consistency between simulated and real machine-tools is very high. Figure 4 shows the simulated and actual characteristics of a hexapod's motor current. The hexapod machine-tool (*LINAPOD III*) was developed at ISW in the project *DYNAMIL*. Therefore exact knowledge of the machine's parameters already existed and facilitated the modelling of this machine-tool (Pritschow and Röck 2004, Pritschow et al. 2003).

4 APPLICATIONS

The simulation environment leads to possible cost reductions for three parties: the manufacturers of the motion control system as well as the machine-tool and the end user. Advantages for the motion control system's manufacturer lie in the possibility to test new functionality of the motion control system without the need for a physical test bench (Vogt and Schröder 2004). Additionally it is possible to test the reaction of the motion control system to changes in the machine parameters' values (e.g. friction) during simulation time.

The manufacturer of the machine-tool can use the simulation environment to gain knowledge about the behaviour of the machine before it is physically manufactured. The simulation of the machine-tool's dynamic behaviour enables the manufacturer to identify processing times much more reliably than by conventional estimation. This knowledge can be used to optimise the machine-tool for maximum efficiency even before its physical assembly. Furthermore parts of the machine-tool's commissioning can already be carried out during the development or construction phases. Especially for manufactures of special purpose machines this can drastically reduce the time-to-market. The end user of the machine-tool benefits from the simulation platform by simulating manufacturing processes in advance. By analysing the simulated behaviour he can detect unused potential of the machine-tool or machining faults. These deficits can subsequently be corrected in the corresponding NC programmes. Additional usage scenarios lie in the fields of training and marketing.

5 CONCLUSION AND FURTHER RESEARCH

The simulation environment presented in this paper fulfils the requirements necessary for hardware-in-the-loop simulation of machine-tools. The environment supports the work flow for a complete simulation project from the creation of the models until the analysis of the acquired data. The simulation platform will be constantly enhanced for new applications and is commercially available.

The novel character of the simulation platform yields many opportunities for further research. One of the most interesting is to operate the real machine-tool in parallel to the simulation platform instead of substituting the machine-tool. As the properties of the simulated machine will never change after its initial modelling, any deviation between the actual and simulated values will stem from changes in the real machine-tool or in the manufacturing processes and could be used as an indicator for errors or inaccuracies. Another possibility is to integrate the simulation nucleus into the motion control system itself. This offers new possibilities for adaptive algorithms in the motion control system.

ACKNOWLEDGEMENTS

We would like to thank the German Research Foundation (DFG) and the Federal Ministry of Education and Research (BMBF) for the funding of projects related to this work. The authors would also like to thank the industrial partner ISG for their valuable support.

REFERENCES

Burmberger, G. 2002. *PC-basierte Systemarchitekturen für zeitkritische technische Prozesse.* Ph. D. thesis, TU München.

Hinkel, P. 2004. Digitale Fabrik: Starker Start. *LINK – about realities and visions 01/2004.*

IEC 61491. Electrical equipment of industrial machines – Serial data link for real-time communication between controls and drives.

Kreusch, K. 2001. *Verifikation numerischer Steuerungen an virtuellen Werkzeugmaschinen.* Ph. D. thesis, BTU Cottbus.

Pritschow, G. and Röck, S. 2002. Echtzeitfähige Maschinendynamik-Modelle in der Steuerungstechnik. In *Datenmodelle in der Produktion.* Düsseldorf: W. Adam, G. Pritschow, E. Uhlmann, M. Weck.

Pritschow, G. and Röck, S. 2004. Hardware in the Loop Simulation of Machine Tools. *Annals of the CIRP 53.*

Pritschow, G., Rogers, G., Bauer, G. and Kremer, M. 2003. Open Controller Enabled by an Advanced Real-Time Network. In *CIRP 2nd International Conference on Reconfigurable Manufacturing*, Ann Arbor, MI, USA.

Pritschow, G., Rogers, G. and Röck, S. 2002. Echtzeitfähige Maschinenmodelle – Abbildung von Antriebsstöorkräften bei Parallelkinematiken. *wt Werkstattstechnik online 92.*

Schäuffele, J. and Zurawka, T. 2003. *Automotive Software Engineering. Grundlagen, Prozesse, Methoden und Werkzeuge.* Wiesbaden: Vieweg Verlag.

Vogt, R. and Schröder, T. 2004. Den Systemtest automatisiert. *Computer & Automation 8-2004.*

Virtual modeling and rapid manufacturing – Bártolo (eds)
© *2005 Taylor & Francis Group, London, ISBN 0 415 39062 1*

Rapid development of composite prototypes with rapid tooling in soft materials

J.C. Ferreira, L. Simões, J. Borges & R. Beira
Technical University of Lisbon, Portugal

ABSTRACT: The manufacturing process of composite parts can be greatly improved through the integrated use of advanced computer assisted technologies. This paper proposes a new strategy for the rapid product development and optimisation of the manufacturing process of composite prototypes, integrating conceptual design, virtual prototyping, rapid prototyping, rapid tooling and reverse engineering to perform metrology control of the assembled rapid tooling parts, as well rapid manufacturing to produce functional and optimised composite prototypes. This strategy enables to reduce the lead-time of composite prototypes designs and optimise the manufacturing process for functional composite prototypes. An intake manifold for a competition motor of a racing car was chosen as a case study to test this new approach.

1   INTRODUCTION

Today, manufacturers experience immense pressure to provide a greater variety of complex products in shorter product development cycles. The evolution of the market requires the time-to-market reduction for two main reasons: the product life cycle is shorter and also due to the importance to produce more rapidly, integrating initial conceptions or "ideas" with faster material processing technologies. New approaches in rapid product development (RPD), from the design point of view, are needed, (Bernard & Fischer 2002). The late discovery of product requirements is fairly common, and the use of certain methods and product representations may reduce this problem. The identification and implementation of designer requirements in the early stages of RPD are key issues for successful product development reducing time (Engelbrektsson & Soderman 2004). These needs, coupled with the request to further reduce costs and improve quality have urged manufacturers to turn their focus towards integration of product design with advanced processing technologies. One possible methodology to achieve this goal is the use of rapid prototyping (RP) technologies, giving the ability to quickly transit from concept to manufacturing, improving product quality. The evolution of RP allows obtaining some parts as final products within a relative short time. The initial laboratory RP demonstration occurred in 1984, although the first operational RP systems to manufacture conceptual prototypes has been commercially available since 1988 (Jacobs 1992).

In a RPD context, the main advantages of virtual prototyping (VP) consist of the ability to change the shape at the designer's will or to comply with assembling/disassembling, directly in a CAD format. The design visualization or virtual prototyping can also be carried out handling stereolithography tessellated models (STL), the actual standard for layer manufacturing technology (LMT) (Jee & Campbell 2003). VP allows virtual simulation techniques with intermediate geometric models, prior to fabricating the corresponding physical models. Careful examination of the virtual simulated model, before the actual fabrication, can help minimizing unwanted design iterations (Jee & Sachs 2000). Solid models can be manufactured, from the improved VP model, by materials processed with a rapid prototyping apparatus within a relative short time (or mask time overnight), to evaluate the set up corrections.

The basis of rapid tooling (RT) technology allows the direct manufacturing of tools, or tools manufactured with indirect processing technologies based on a model built by RP (Pham & Dimov 2001). The rapid production of a tool allows the manufacturer to have a better overall control of the new product development process, not only of the product itself – visual aids for engineering and fit – but also of the material processing technology, by having a prototype tool at an early stage of the manufacturing process. RT responds better to the growing interest of the industry in reducing the time to market of new products and respective costs.

The RT in soft thermopolymers (synthetic wax material) can be obtained directly by RP multi-jet modelling (MJM) techniques. The surface finish and dimensional precision of parts obtained through this RP material processing technology is often highly important, especially in those occurrences where the manufactured prototypes are used as functional lost patterns, like in the case of investment casting (Perez & Calvet 2002). RT thermopolymer pattern dimensions are determined by both synthetic wax's thermophysical and thermomechanical properties and material processing parameters (Sabau & Viswanathan 2003).

The main bridge between RP/RT and reverse engineering (RE) is based on the indirect use of STL geometry data files. These tessellated representations of products have several acceptable drawbacks linked to the shape and topology of the triangular mesh (Galantucci et al. 2003). The reverse engineering tasks for RP/RT metrology control is somehow time consuming but allows comparing the generated STL model with the initial 3D-CAD, using adequate software to determine and analyse geometry deviations (Ferreira & Alves 1999, Ferreira et al. 2001).

The main contribution of this work is to propose an original methodology to integrate advanced computer assisted technologies with advanced materials processing technologies for rapid product development. The new methodology provides meaningful information for industry integrating these advanced technologies best suited to their production requirements, in order to reduce the lead-time of composite prototypes designs and optimise the manufacturing process for functional composite prototypes.

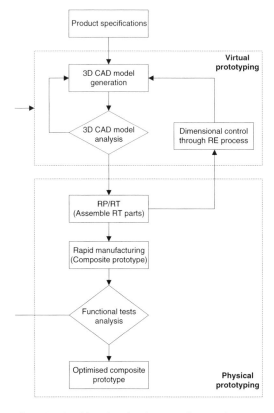

Figure 1. Rapid product development of composite prototypes.

2 METHODOLOGY

The methodology developed for the rapid product development of composite prototypes is illustrated in Figure 1. It comprises two main phases: the virtual prototyping (VP) and physical prototyping (PP).

2.1 Virtual prototyping

In the virtual prototyping phase, two subtasks must be considered. The first one includes the three-dimensional computer model (3D CAD) generation, based on specifications from designers, and a subsequent 3D CAD preliminary analysis regarding the functionality and assembling/disassembling of new products. A second subtask involves the dimensional control of RP/RT parts based on reverse engineering (RE) processes to assure the geometrical quality of the parts, at an early stage of the RPD process.

Computer representations of an object/product play a significant role in exploring design ideas and solving design problems. Product development is based on the exploration and search processes for optimal solutions, comprising a great variety of knowledge, to enable an

effective generation of a good concept for the required product. Thus, virtual prototyping is a crucial phase within the RPD process providing a pre-optimised model for physical prototyping in a shorter time.

2.2 Physical prototyping

The physical prototyping phase comprises four subtasks. The first one includes the rapid prototyping of conceptual models, used to easily identify problems related to aesthetics, functional and manufacturing issues. Consequently, modifications of a part or tool should be made earlier in the design process, reducing costs and lead-time. Additionally, the use of reverse engineering techniques for the prototype control dimension enables the evaluation of any geometrical deviations, which can be corrected.

The second subtask involves the direct RT from RP. Rapid manufacturing of composite prototypes by composite processing technology, and a subsequent surface finishing characterises the third subtask. Functional tests with the physical composite prototypes to analyse functionality and performance are carried out in the fourth subtask. After some iterations to improve the product performance and enhance the processing technology within a short lead-time, the product and the rapid manufacturing process are adequately developed for real life applications.

3 CASE STUDY

The methodology developed for the rapid product development of composite prototypes was utilised to rapidly enhance the intake manifold of a competition motor mounted in a formula student-racing car. The intake manifold constructed in two parts (venturi's and funnel) presents some requirement, in terms of low weight and geometric shape, related to computational fluid dynamics' simulation for optimising the non-stationary fluid motions (not presented in this work).

3.1 *Virtual prototyping*

The design process for the intake manifold device can understandable in the context of developing the motor assembly. In doing so, the specifications for the device, derived from specific needs and constraints held by the motor, namely those of the venturi's base to separate the air flow that intake in four cylinders and the superior funnel for airflow intake through a cylinder with a throttle. The air intake manifold shape between the top and the base was designed to allow dynamic laminar airflow inside the funnel.

Figure 2 shows the CAD model of the defined intake manifold device. Virtual models were used to optimise the design of the part and its functionality in terms of assembly/disassembly.

3.2 *Physical prototyping*

3.2.1 *RT manufactured by 3D printing*

In this R&D work, following the RPD methodology to manufacture the rapid soft tooling from the tool design, it was utilised a ThermoJet™ solid object printer from 3D Systems. This RP machine uses the multi-jet modelling (MJM) technique to process the polymeric materials (Chua et al. 2003).

To manufacture the rapid tooling together with the rapid prototyping apparatus involves the conversion of STL files of the design tools into the machine's software.

As the machine selected is a concept modeller with limits to the dimension of the parts that could be printed, there was a need to divide the two parts of the intake manifold geometry into two halves each: two symmetrical halves for both the venturi's base and the intake funnel. An extended height of about 20 mm on the four cylindrical extremities was considered from the developed intake manifold venturi's base model to allow the composite processing technology.

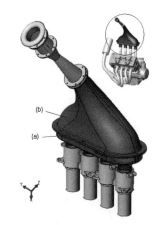

Figure 2. Virtual prototyping via CAD models for assembling.

Figure 3. Rapid tool of the lower base design in STL format.

The rapid tooling was designed in 3D-CAD, as shown in Figure 3.

The 3D printing to manufacture the soft tool is carried out layer by layer over the apparatus metallic platform processing a Thermojet 88 thermopolymer material from 3D Systems, which principal properties are listed in Table 1.

It took some time for the rapid tool (venturi's) processing with the ThermoJet™, nearly 780 minutes for the two halves built together at the machine platform area.

In this case study, the half base tools are printed upside-down with the ribs sitting on the inside, consequently the direct soft rapid tool present an external smooth surface finishing adequate for the direct application of carbon composite fiber to build up the functional prototypes. The RT for the funnel intake manifold was similarly made as the venturi's RT. The two halves of the RT venturi's base manufactured upside-down with inside support ribs are assembled as shown in Figure 4.

Table 1. Properties of thermopolymer material.

Property		Thermojet 88″
Melt temperature	[K]	358–368″
Softening temperature	[K]	343″
Density @ 413 K	[g/cm³]	0.846″
Density @ 403 K	[g/cm³]	0.846″
Density @ 296 K	[g/cm³]	0.975″
Volumetric shrinkage from C413 K to room temperature [%]		12.9″

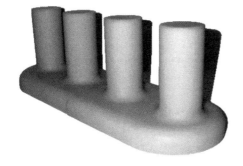

Figure 4. Rapid tool manufactured processing RP thermopolymer.

3.2.2 Dimensional control of the RP/RT

Dimensional control of the RP/RT parts was performed based on a RE process to assure the geometrical quality of the parts.

A 3D scanning technique with a FARO™ arm was used that allowed the digitising of parts by taking coordinates on the surfaces at selected points, as shown in Figure 5. A RE technology was utilised to control the geometrical accuracy of the rapid tooling halves, which were joined together.

The data interface between the 3D scanner and the digital model was done through Rhino™ software. The digitised array of data ("points cloud") allows redesigning RT real surfaces by an RE process, i.e., a new CAD model is rebuilt from the several points digitised on the rapid tool surface. The point array matrix, set up on Rhino from the RT surface transformed into a STL format, was compared with the original 3D CAD model, imported in IGES format, allowing the dimensional control, as shown in Figure 6.

The dimensional control from RE data was performed in order to analyse the accuracy of the direct RT. The RT using a ThermoJet thermopolymer has been measured using 3D scanning, and the coordinate data was compared with the initial 3D CAD. A total of 128 evenly distributed array of points over the entire outer RT surface have been measured for the RT

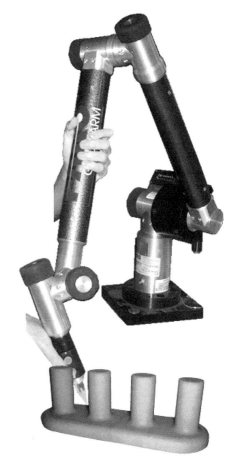

Figure 5. 3D digitising with FARO™ arm for dimensional control.

dimensional control. The measured points were directly compared with similar ones on the CAD model, allowing to check the error distribution function (EDF), as represented in Figure 7.

It seems, based on the analysis of the EDF histogram, that the xyz dimensions peaks within the tolerance 0.0 to −0.3 mm with a frequency of 79.4% for the RT. It can be noticed a negative bias of the distribution in the thermopolymer material for the RT, meaning, that the tool material can present contraction throughout RP processing. This problem can be solved by changing the CAD scale parameters. The CAD scale factors should be specified individually for the x-, y-, and z-axis. However, there is a need to firstly necessary to understand all the causes that influence the processing and the contraction of the fused Thermojet 88 thermopolymer material, before setting new scaling factors (or functions), because the contraction is inferior to the expected one.

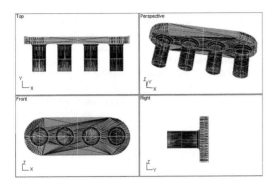

Figure 6. RE from RT and CAD data for dimensional control.

TOOLING MEASURING DEVIATION vs NOMINAL DIMENSIONS
RE - Metrology (X , Y , Z)

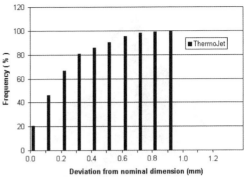

Figure 7. EDF histogram from RT compared with CAD data.

The cumulative error distribution (CED) for the RP-model measured, i.e., the normalised integral of the EDF from zero error to infinity, have also been calculated for a graphical bar representation. A CED plot of the percentage of measurements, positioned within a given error versus the magnitude of the error (steps = 0.1 mm), is represented in Figure 8. It is possible to choose practical tolerance levels from the CED, as well as realistic accuracy expectations to stay within the design (3D CAD) specifications. Confidence limits for a given specification can be obtained by the value $\in (90)$, i.e., the value within which 90% of all dimensions measured fall. The $\in (90)$ tolerance level for the thermopolymer RT lie within 0.0 to -0.5mm.

3.2.3 Rapid manufacturing of composite functional prototype

Rapid manufacturing polyester resin layers were successively deposited on the RT surfaces. The first polyester resin thin layers are applied on the RT by

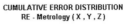

CUMULATIVE ERROR DISTRIBUTION
RE - Metrology (X , Y , Z)

Figure 8. CED histogram from RT compared with CAD data.

spray to allow a high-quality surface smoothness, which is an important asset for the final result of the intake manifold inside. Each resin layer was applied, and then left to cure during a short time, ensuring that the under layers were ready to be covered with polyester resin, to acquire some mechanical compression resistance. At this stage, the coated rapid tools are ready to start the composite manufacturing process.

To manufacture the intake manifold it was selected a carbon fibre reinforced polymer, since these materials are extremely solid and light for functional prototypes. The process of applying the successive carbon fibre cloths over the surface, and consecutively apply the polymeric resin to impregnate fibres, had to be promptly done due to the short curing time of the resin that starts to cure and solidify very quickly, once the catalyser is soaked to the carbon fibre cloths.

The process allowed the application of 3 to 4 layers of carbon fibre on each intake manifold part, which ensured a considerable resistance and rigidity to the functional prototype, once it was completely cured.

Finally, there was an excess of composite material on the prototype's border that was applied on RT moulds in order to ensure that the whole surface was covered. This excess of carbon fibre reinforced with polymeric resin was cut out to comply with the intake manifold design. An array of peripheral holes was also drilled on the functional prototypes, for fixation with screws and nuts.

The finished prototypes had a good quality level, after the application of all rapid technologies to the manufacture of the intake manifold, taking into account the objective for which the product was conceived, to produce a functional prototype, as visualised in Figure 9, that allows carrying out practical tests with the motor running.

With the RPD methodology developed the whole process involves lead-time and manufacturing in a

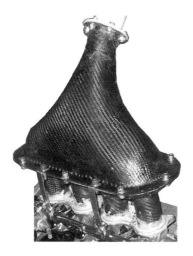

Figure 9.   Functional prototype in carbon composite fibre.

total average time of 4 working days considering RP/RT automatic-manufacturing processes executed by night. As for materials, the raw materials used were: thermopolymer RP; de-moulding release agent; carbon fibre cloths; polymeric epoxy resin; catalyser and accelerator. This methodology relied also on CAE facilities and equipments, like the CAD software and the computers to design the parts; the ThermoJet™ 3D printer with proprietary software to make physical direct soft tools out of the CAD model; the FARO™ harm with the Rhinoceros™ software to measure and control the geometries; and some manual tools to manufactured and finish the functional prototypes.

4   CONCLUSIONS

An integrated virtual and physical prototyping methodology was proposed for rapid manufacturing of composite functional prototypes and a case study was presented.
   The findings of this research study can be described as follows:

- The virtual prototyping simulation and analysis in the RPD methodology context has provided the ability to alter earlier the intake-manifold shape to fulfil the requirements of assembling/disassembling, prior to fabricating the corresponding physical direct soft tools.
- The simultaneous RT and RE dimensional control of the rapid soft tooling allows predicting the thermopolymer material contraction throughout RP processing. This contraction is inside the tolerance 0.0 to −0.3 mm with a frequency around 80%. A practical tolerance level for 3D CAD utilising the Thermojet 88 thermopolymer for RT should be within 0.0 to −0.5 mm. As well, these dimensional corrections could be integrated in design phase.

- The RPD methodology includes RM to quickly fabricate a functional prototype as the composite device for a racing-car motor has demonstrate that is really possible to design and rapidly obtain direct rapid tools to manufacture finished functional prototypes, ready to be tested and analysed.
- For manufacturing, the RPD methodology integrating the VP and PP could become standard for specific materials processing technologies, consequently the lead-time can be shortened and hidden cost of product and manufacturing process can be reduced.

REFERENCES

3D Systems, 2005 [www.3dsytems.com].

Bernard, A. & Fischer, A. 2002. New trends in rapid product development. CIRP Annals-Manufacturing Technology, vol. 51 (2), pp. 635–652.

Chua, C.K., Leong, K.F. and Lim, C.S. 2003. Rapid Prototyping – Principles and Applications. 2nd Edition, World Scientific Publishing Co. Pte. Ltd., Singapore.

Engelbrektsson, P. & Soderman, M. 2004. The use and perception of methods and product representations in product development: a survey of Swedish industry. Journal of Engineering Design, vol. 15 (2), pp. 141–154.

Ferreira, J.C. & Alves, N.F. 1999. Image Processing after 3D-scanning for Quality Control of Rapid Prototyping Pieces. Proceedings IWK'99 Congress, edn. TU Ilmenau, Germany, vol.1, pp. 445–450.

Ferreira, J.C., Alves, N.F. & Mateus, A.S. 2001. 3D Digitising for Reverse Engineering to Manufacture Rapid Tooling. Proceedings of EURO RP'2001, Paris/France, pp. 1–7.

Galantucci, L.M., Percoco, G. & Spina, R. 2003. Evaluation of rapid prototypes obtained from reverse engineering. Proceedings of The Institution of Mechanical Engineers Part B-Journal of Engineering Manufacture, vol. 217 (11), pp. 1543–1552.

Jacobs, P.F. 1992. Rapid Prototyping & Manufacturing – Fundamentals of StereoLithography. Book edited by Society of Manufacturing Engineers, Dearborn.

Jee, H.J. & Campbell, R.I. 2003. Internet-based design visualization for layered manufacturing. Concurrent Engineering Research and Applications, vol. 11 (2), pp. 151–158.

Jee, H.J. & Sachs, E. 2000. A visual simulation technique for 3D printing. Advances in Engineering Software, vol. 31 (2), pp. 97–106.

Perez, C.J.L. & Calvet, J.V. 2002. Uncertainty analysis of multijet modelling processes for rapid prototyping of parts. Proceedings of the Institution of Mechanical Engineers Part B-Journal of Engineering Manufacture, vol. 216 (5), pp. 743–752.

Phan, D.T. & Dimov, S.S. 2001. Rapid Manufacturing – The Technologies and Applications of Rapid Prototyping and Rapid Tooling. Springer-Verlag Edition.

Sabau, A.S. & Viswanathan, S. 2003. Material properties for predicting thermopolymer pattern dimensions in investment casting. Materials Science and Engineering a-Structural Materials Properties Microstructure and Processing, vol. 362 (1–2), pp. 125–134.

Materials

Virtual modeling and rapid manufacturing – Bártolo (eds)
© 2005 Taylor & Francis Group, London, ISBN 0 415 39062 1

Dimensional accuracy of CastForm Polystyrene patterns produced by Selective Laser Sintering

S.S. Dimov, D.T. Pham, K.D. Dotchev, & A.I. Ivanov
Manufacturing Engineering Centre, Cardiff University, Cardiff, United Kingdom

ABSTRACT: Selective Laser Sintering, also known as Laser Sintering (LS), is a rapid prototyping technique extensively used to produce not only concept models and functional prototypes, but also patterns for investment casting and cores and cavities for different types of moulding tools. A variety of powdered materials are used with this technology to build parts for different applications. LS with the CastForm™ Polystyrene (CF) material is one of the fastest and most cost effective techniques for rapid fabrication of small quantities of wax-like patterns for shell or flask investment casting. The material properties and the process parameters affect the dimensional accuracy of the patterns and ultimately the final accuracy of the metal castings produced from them. This paper analyses the dimensional accuracy of CF patterns and also the technological factors affecting the process accuracy and proposes ways to improve it.

1 INTRODUCTION

Selective Laser Sintering or Laser Sintering (LS) is one of the most popular Rapid Prototyping (RP) technologies. A variety of thermoplastic, metal and ceramic composite powders can be processed with this technology to build prototypes for different applications (EOS 2005, Pham & Dimov 2002c, 3D Systems 2005).

CastForm™ (CF) is a polystyrene material system introduced by DTM Corporation (DTM 1999) to replace the material called TrueForm and used in LS for rapid fabrication of patterns for shell or flask investment casting (DTM 1999a, b, 3D Systems 2005).

The CF fabrication process involves laser sintering of a "green" part and wax infiltration. The main idea behind this two-stage process is to fabricate a pattern with properties very close to those of conventional wax patterns and therefore to be compatible with standard foundry practices. The drawback is obvious, an additional processing stage – wax infiltration is required that increases the processing time and also introduces additional errors affecting the accuracy of the CF pattern and therefore the final metal casting accuracy.

The research published on technological capabilities of the LS process is mostly focused on process accuracy and reports mainly case studies (Dickens et al. 1995, Pham et al. 1999a).

Investigations of the accuracy of different RP technologies (Childs & Juster 1994, Ippolito et al. 1995, Kruth 1991) were also reported but so far there was no study providing a systematic analysis of dimensional accuracy of the LS process.

The dimensional accuracy data released by vendors is very limited and applies either only to dimensions in X-Y plane or the specified accuracy could be achieved in ideal conditions. It is not difficult to conclude that the accuracy is material and also machine dependent.

There are some publications dealing with the accuracy of the LS process in the X-Y plane only (Childs & Juster 1994, Dickens et al. 1995, Shellabear 1999) however for other polymer materials than CF and different applications.

This paper investigates the capabilities of the LS process to produce CF patterns for investment casting. It provides also an analysis of the main factors affecting the dimensional accuracy of the patterns and proposes ways and ideas how to improve it.

2 CF PROCESS ANALYSIS

2.1 *CF process and material properties*

The CF fabrication process normally involves input data preparation, LS of "green" parts, cleaning and wax infiltration. The input data, an STL or a slice file is loaded into the machine software to prepare a build. During the LS process the part bed temperature is raised to 80°C below the CF glass transition temperature of 89°C. The laser scanning system draws each layer of the part sequentially thus applying a sufficient amount of energy to fuse the powder particles together. Then, the part is cleaned and infiltrated with liquefied casting wax. The tensile strength of the CF patterns at

20°C (DTM 1999a) is sufficient for them to withstand further processing although it reduces with the increase of the temperature. The patterns are in their weakest state at 79–80°C during the wax infiltration stage.

2.2 *Input data preparation*

The part geometry is normally transferred to the RP machine as an STL file that approximates CAD model surfaces using triangular facets. The STL files are then sliced in order to create the 2D data of all layers. The accuracy of the tessellation process, the slicing algorithm, the part orientation and also the layer thickness are the main generic sources of geometrical errors. They are very small compared to others and could be reduced but not completely eliminated. If the CAD model contains not properly connected surface patches the produced STL file do not define closed volume and may contain errors like bad edges, bad contours and missing or double surfaces. This could disturb the slicing process and the laser beam movements and ultimately result in poor part strength and possible part disintegration during cleaning or wax infiltration.

During the whole fabrication process, the pattern shrinks along X, Y and Z directions. In order to compensate this shrinkage the input part geometry is modified applying scaling factors in the X, Y and Z directions correspondingly. It is important to note that they are products of two sets of scaling factors. The first set compensates the shrinkage of the "green" part while the second one the shrinkage resulting from the wax infiltration. The part should be scaled further to reflect the shrinkage during the metal casting process. The shrinkage depends very much on part geometry and its orientation during the LS and wax infiltration processes. Sometimes the selected orientation during wax infiltration differs from that during LS. Thus, the scaling factors should be calculated taking into account the specific part orientations in the whole fabrication process.

The inside and outside surfaces are offset to compensate for the size of the sintering spot of the laser beam (beam offset). The right selection of the scaling factors and beam offset determines the extent to which the systematic factors due to input data could be compensated.

2.3 *The LS machine and process parameters*

The mechanical inaccuracy of the re-coating system, the roller and the part piston; the inaccuracy of the laser beam delivery system, the scanner and the control system; the friction between powder particles and the re-coating system result in systematic and random errors. These errors cannot be fully eliminated but their impact on machine accuracy could be reduced by proper servicing and maintenance of the equipment.

It should be noted that the "green" part cools down while being built. It shrinks and distorts during the LS process (in build shrinkage) and this continues during the cooling down of the part bed to room temperature (post build shrinkage). The uneven temperature distribution and heat dissipation in the build chamber causes non-linear shrinkage rates across the part bed. Thus, parts positioned in the middle of the part bed would shrink slower than those in its periphery. At the same time the part bed piston moves down with constant increments (layer thickness) at the start of each new layer that compensates the shrinkage of the previous layers. As a result of this parts placed in the middle of the part bed experience less in-build shrinkage and continue to shrink after the end of the process. In contrast, parts placed around the periphery of the part bed have more in-build and less post build shrinkage. This phenomenon is the cause of the observed non-linear shrinkage in the vertical or Z direction. In addition, the weight of subsequent layers affects the shrinkage in Z. Therefore parts positioned on the bottom of a tall build would shrink more during the LS process then those placed on the top.

The Z shrinkage is also dependent on build time. The reason for this could be the different number of parts contained in each build and the processing speed of the PC controller. This results in different sintering times for the builds and thus more or less in-build shrinkage.

The LS process parameters, laser power, laser beam scanning speed, step size, scanning strategy (sorted fill enable/disable, skin and core methods) and the powder feed and part bed temperatures determine the total amount of energy delivered to the top surface of the part bed. If this energy is higher than required the part would experience a growth. In particular, all external dimensions would be bigger and on the contrary all internal would become smaller than their nominal dimensions. Also, this would result in higher density and strength of the "green" part and subsequently less wax would be infiltrated in the part. On the contrary if the sintering energy is not sufficient then the part would be weak and some features could erode or break during the cleaning or wax infiltration procedures.

Most of the errors during the LS process are systematic and therefore could be compensated. However, there are some random errors that are geometry and build dependent which could not be compensated easily. A possible solution would be to build the same part twice on the same machine using the same processing parameters, if the aim is to achieve higher pattern accuracy. The first build is used to calculate the correct scaling factors for each particular part and then the build is repeated.

2.4 Cleaning and wax infiltration

After removing them from the machine, parts are further cleaned from unsintered powder using only a low-pressure air and fine brushes. During the wax infiltration the "green" parts are placed onto a dipping tray and preheated up to 79–80°C in an oven. Then, they are immersed into a vat with molten wax. The time for preheating and infiltration depends on the part height, volume and wall thickness. The parts are in their most fragile state during this stage. It is very likely that some areas of the part would deform or even collapse under its own weight. Specific rules, careful processing and experience are required for a successful cleaning and infiltration.

3 DIMENSIONAL ACCURACY

3.1 Test parts

To assess the combined effect of all factors on the final process accuracy two test parts, a "pyramid" and a "staircase", were designed (Fig. 1). The "pyramid" is used to evaluate the errors along the X and Y directions while the "staircase" is designed to evaluate the accuracy in Z direction.

The test parts were built several times with DTM Sinterstation 2500CI applying the following processing parameters: fill laser power 13 W, part bed temperature 82°C, left/right feed cartridges temperature 48°C, scan spacing 0.15 mm, outline laser power nil, wax infiltration oven temperature 79–80°C, preheating time 10 min. and infiltration time 10 min.

In order to analyze the impact of the two main processing stages on the final accuracy all parts were measured twice: at the "green" stage after the LS process and then after the wax infiltration.

3.2 Accuracy in X and Y directions

The results of measurements are shown in Figures 2–3 which depict the error of the actual dimensions of the three test parts in X (**X1, X2, X3** data sets) and

Y (**Y1, Y2, Y3** data sets) directions against their nominal values before and after the wax infiltration stage.

Each data set contains systematic and random errors. The systematic error depends mainly on the material shrinkage, applied laser power, part-bed temperature and on the quality of the carried out laser calibration.

The errors for each data set at "green" stage and after infiltration show a linear dependence with the nominal dimensions. To a large extent these errors could be compensated by applying constant scaling factors.

The shrinkage is a percentage change of the nominal dimension. In particular, it is calculated applying a regression analysis and implying a linear dependence:

$$E = \alpha \times N + \beta \qquad (1)$$

where E = error; N = nominal dimension (nominal); α = shrinkage; β = spot size sintered by the laser beam.

The scaling factor S is then calculated as:

$$S = \frac{1}{1-\alpha} \qquad . \qquad (2)$$

Using the experimental data the average shrinkages α_x, α_y and scaling factors S_x, S_y in X and Y directions respectively were estimated (Table 1).

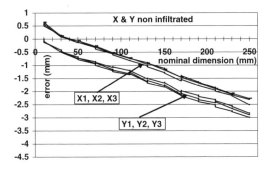

Figure 2. The error in X and Y directions at "green" stage.

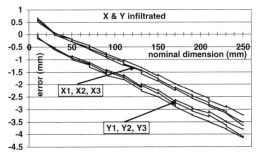

Figure 3. The error in X and Y directions after wax infiltration.

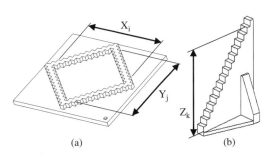

(a)   (b)

Figure 1. The test parts: (a) "pyramid"; (b) "staircase".

Table 1. The shrinkage and scaling factors in X and Y directions.

	"Green" part	Infiltration only	After infiltration
Scaling factor S_x	1.0119	1.0044	1.0163
Scaling factor S_y	1.0115	1.0043	1.0159
Shrinkage α_x, mm	0.0117	0.0044	0.0161
Shrinkage α_y, mm	0.0114	0.0043	0.0157

Table 2. Random errors in X and Y directions (6σ).

	1	2	3
X direction, mm	0.404	0.417	0.408
Y direction, mm	0.466	0.441	0.415

Table 3. The laser beam sintered spot size variation in X and Y directions.

Sintered spot size, mm ($i = 1,2,3$)	βx_i	βy_i
"Green" CF part	0.5312	−0.067
	0.5228	−0.0822
	0.575	−0.1047
The normal deviation of the spot size, 6σ	0.16817	0.1138
After wax infiltration	0.6249	−0.0339
	0.6138	−0.0597
	0.652	−0.0584
The normal deviation of the spot size, 6σ	0.1179	0.087

Where $\beta x_i, \beta y_i$ is the laser spot size in X and Y directions.

The random component of these errors is calculated for each data set and is given in Table 2.

The X data sets are offset from those in Y direction and the reason is the elliptical shape of the spot sintered by the laser beam. The size of this spot depends mostly on the specific machine calibration, the sintering parameters (laser power, scanning speed and part bed temperature) and material properties. The variation of the spot size is given in Table 3 together with the computed error when a normal distribution, 6σ, is assumed with a probability of 99.73%.

In order to estimate the CF process capability in X and Y the standard deviations, 6σ, are calculated and shown in Figure 4. Comparing the results in Figure 4 against the standard IT tables (BS 1993), the closest accuracy class achievable in X and Y directions is IT13. This could be used as a guideline for assigning process tolerances to CF patterns.

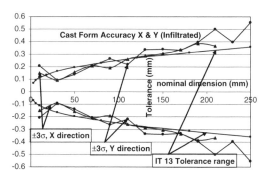

Figure 4. Process capability in X and Y directions after wax infiltration.

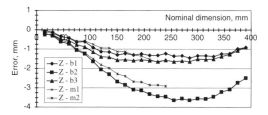

Figure 5. The error in Z direction at "green" stage.

3.3 Accuracy in Z direction

Five "staircase" test parts were fabricated simultaneously in one build. Three of them, **b1, b2** and **b3** have overall height 390 mm while the height of the other two, **m1** and **m2**, is 250 mm. All parts were positioned at the same height in the building chamber but at different X-Y locations in the part bed.

All five staircases were measured and then the data processed in the same way as in Section 3.2.

Figure 5 depicts the error (deviation) of the actual dimensions in Z direction from their nominal values at the "green" stage.

Figure 6 shows the total error after the wax infiltration. It could be seen in these figures that the part shrinkage has a clear non-linear character.

The shrinkage depends not only on parts' nominal dimensions in Z but also on their location in the X-Y plane. Part **b2** and **m2** positioned in the middle of the part bed exhibit a shrinkage that is significantly higher than that of the parts positioned along its periphery, **b1, b3** and **m1**.

Because of the varying part geometry and build arrangements it is practically impossible to predict precisely the non-linear shrinkage that would result in each particular case. This systematic error cannot be estimated accurately and thus cannot be compensated off-line. The only option is to fabricate each part twice. The first time the part is built without applying

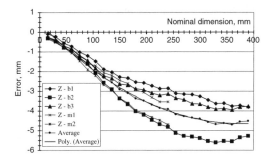

Figure 6.   The error in Z direction after the wax infiltration.

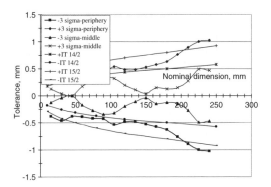

Figure 7.   The process accuracy in Z direction after wax infiltration.

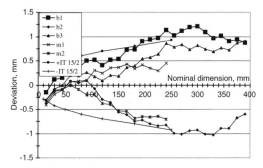

Figure 8.   The error that is not compensated in Z direction after wax infiltration.

any scaling factors in order to calculate its shrinkage and Z scaling function that depend on the part Z height and X-Y placement. Then, the build is repeated by applying the scaling factors that are specific for that part. This improves the part accuracy; however the total fabrication cost and delivery time would double for the production of a single pattern.

The errors at "green" stage (Fig. 5) were deducted from the total error after wax infiltration (Fig. 6). The results represent only the errors of the wax infiltration stage. They also indicate a linear dependence between the errors and nominal dimensions. Hence, the non-linear character of the Z shrinkage shown in Figure 5 and Figure 6 is caused mainly by the LS fabrication process.

In order to estimate the process accuracy in Z direction the data from the five sets were re-grouped into two sets taking into account the part positions. The first set contains the data of the parts placed around the periphery and the second one those placed in the middle of the build. Figure 7 provides an indication about the process capabilities after the wax infiltration stage assuming a normal distribution. Comparing these results (Fig. 7) against the IT tables (BS 1993) it

could be judged that the achievable accuracy in Z direction is IT14 for parts placed in the middle of the part bed and IT15 for those around the periphery.

Usually in the RP business the delivery time and cost have the highest priority and often there is no time to build the parts twice for a more precise process calibration. In that case, a non-linear function depending only on the Z height can be applied to scale the parts. This function is calculated from the average error shown in Figure 6 and compensates only partially the systematic error.

It takes into account the nominal Z height but not the part placement in the X-Y plane. Thus, the final dimensional accuracy would be worse especially for very tall parts.

Figure 8 provides an indication about the error that will not be compensated if the above approach is applied. For relatively short parts up to 200 mm the deviations that are not compensated are within IT15 however the randomness of the process could increase the error. For dimensions above 200 mm the error could be much bigger although it starts decreasing for nominal dimensions exceeding 320 mm due to the shrinkage compensation effect discussed in section 2.3.

4   CONCLUSIONS

This research analyses the factors influencing the accuracy of the CF patterns. The carried out experimental study helps not only to determine the influence of different factors on the final pattern accuracy but also to identify ways for improving the process.

The paper demonstrates how the influence of different systematic factors throughout the whole process could be estimated and then compensated.

Part shrinkage and deformation during the LS and infiltration processes are the main sources of systematic and random errors. The thermal expansion and contraction that the parts undergo during these processes lead to geometrical errors, which despite their generally systematic behavior also introduce random errors.

With regard to the wax infiltration stage, the part rigidity at temperature of around 80°C, the material and wax properties combined with part specific geometry and placement in the infiltration plate are the main factors influencing the final accuracy of the pattern. It has been found that the shrinkage from wax infiltration has a linear dependence from the nominal in X, Y and Z directions.

The shrinkage of the CF patterns in X and Y directions is linear during the whole CF process. The accuracy along these axes is consistent and within IT13. Unfortunately, the shrinkage along the Z axis during the LS stage is non-linear. It depends mostly on specific part geometry; part orientation, X-Y placement into the build envelope and total build height. The main reasons for this behavior are that the parts experience a non-uniform cooling in the machine building chamber during the LS process and that the in-build shrinkage is partially compensated by the recoating system.

It should be noted that if the scaling factor in Z-direction is computed as a function of X, Y and Z co-ordinates of the part in the build, it is possible to calibrate the process more accurately. However, this requires the part to be built one more time just for calibration purposes. If this technique is applied the best achievable accuracy in Z direction is the range of IT14–IT15.

In case a general scaling function depending only on Z height is used then the error is within IT15 for parts up to 200 mm height. For taller parts the error could be even higher. More investigations are necessary to develop new techniques for improving the accuracy of the CF patterns.

ACKNOWLEDGEMENTS

The authors would like to thank the European Commission, the Welsh Assembly Government and the UK Engineering and Physical Sciences Research Council for funding this research under the ERDF Programme 52718 "Support Innovative Product Engineering and Responsive Manufacture" and the EPSRC Programme "The Cardiff Innovative Manufacture Research Centre".

Also, this work was carried out within the framework of the EC Networks of Excellence "Innovative Production Machines and Systems (I*PROMS)" and "Multi-Material Micro Manufacture: Technologies and Applications (4M)".

REFERENCES

BS EN 20286-1:1993, ISO system of limits and fits, 1993.
Childs T.H.C. & Juster N.P. 1994, Linear accuracies from layer manufacturing, CIRP Annals, Vol. 43-1, pp 163–167.
Dickens, P.M., Stangroom, R., Greul, M., Holmer, B., Hon, K.K.B., Hovtun, R., Neumann, R., Noeken, S. & Wimpenny, D. 1995, Conversion of RP models to investment castings, Rapid Prototyping Journal, Volume 1, Number 4, p. 4–11.
DTM Corporation 1999a, The Synterstation systems. Guide to materials: CastForm PS, April 1999.
DTM Corporation 1999b, SLS® Selective laser sintering technology for investment, DTM white paper.
DTM Corporation 2001c, SLS – Selective laser sintering, DTM Corporation, 1611 Headway Circle, Building 2, Austin, Texas 78754.
EOS Web page 2005, EOS GmbH – Electro Optical Systems, Headquarters – Munich, Robert-Stirling-Ring 1, D-82152, Kreilling/Munich, Germany, http://www.eosint.com.
Ippolito R., Luliano L., & Gatto A. 1995, Benchmarking of rapid prototyping techniques in terms of dimentional accuracy and surface finish, CIRP Annals, 44(1), pp. 157–160.
Kruth J.P. 1991, Material ingress manufacturing by rapid prototyping technologies, CIRP Annals, Vol. 40-2, pp 603–614.
Pham D.T., Dimov S.S. & Lacan F. 1999a, Selective Laser Sintering: Applications and Technological Capabilities, Proc. IMechE, Part B: Journal of Engineering Manufacture, Vol. 213, pp 435–449.
Pham D.T. & Dimov S.S. 2002b, Rapid Manufacturing, Springer-Verlag, London Berlin Heidelberg.
Pham D.T. & Dimov S.S. 2003c, Rapid prototyping and rapid tooling – the key enablers for rapid manufacturing, Proceedings of the ImechE, Vol. 217, Part C, pp 1–23.
Shellabear M. 1999, Benchmark study of accuracy and surface quality in RP models, Task 4.2 Report 2, BRITE/EURAM Project BE-2051, Process chains for Rapid Technical Prototypes (RAPTEC).
Shen, J., Steinberger, J., Gopfert, J., Gerner, R., Daiber, F., Manetsberger, K. & Fetstl, S. 2000, Inhomogenous shrinkage of polymer materials in Selective Laser Sintering, Solid Freeform Fabrication Symposium Proceedings, Austin, USA, pp 298–305.
3D Systems Web page 2005, 3D Systems – Valencia, CA – Headquarters, 26081 Avenue Hall, Valencia, California, USA, http://www.3dsystems.com.

Virtual modeling and rapid manufacturing – Bártolo (eds)
© *2005 Taylor & Francis Group, London, ISBN 0 415 39062 1*

Materials issues in rapid prototyping

D.L. Bourell & J.J. Beaman, Jr.
Laboratory for Freeform Fabrication, Mechanical Engineering Department,
The University of Texas, Austin, USA

ABSTRACT: The evolution of rapid prototyping towards rapid manufacturing requires acquisition and utilization of knowledge of materials. This paper reviews material behavior important to most rapid manufacturing. Included are viscosity, sintering and infiltration. Assessment of properties of rapid manufactured parts is described with emphasis on the role of residual porosity.

1 INTRODUCTION

Materials have always played a key role in rapid prototyping. Their importance is even more significant for rapid manufacturing, where part performance is critical to success.

It is the present purpose to describe important materials properties and behavior as they apply to various rapid manufacturing processes. While stereolithography is a mainstay of the industry, its fundamental chemical processes focus on photocuring and are treated in great depth elsewhere (Beaman et al. 1997). Materials considerations here are limited to selective laser sintering (SLS), fused deposition processing (FDM), 3D printing (3DP) and fused metal deposition (FMP).

2 VISCOUS FLOW

Liquid motion is the key feature in much rapid manufacturing. For example, polymers melt and flow in FDM and SLS. Metals are molten in FMP and in direct SLS. Post-process infiltration of porous RM preforms also involves liquid flow. For all these, a critical feature of materials flow is the viscosity η:

$$\tau = \eta\dot{\gamma} \tag{1}$$

where $\dot{\gamma}$ is the shear strain rate at an applied shear stress τ. Viscosity is temperature dependent, decreasing with increasing temperature. For both polymers and liquid metals, a general relationship is

$$\eta = \eta_o \exp\left(\frac{+Q}{RT}\right) \tag{2}$$

where η_o is a constant, Q is the activation energy for flow, R is the universal gas constant and T is absolute temperature. Low viscosity is generally desirable for rapid manufacturing, since material flow is easily facilitated.

Viscosity is lowered significantly by increasing the temperature according to Equation 2. This explains in part why most of the RM processes operate at elevated temperature, maintaining low viscosity. The viscosity of polymers varies significantly with temperature. The melt viscosity of metals ranges between about $0.2\,\text{mPa} \cdot \text{s}$ for alkali metals to as high as $5\,\text{mPa} \cdot \text{s}$ for d-transition metals.

3 SINTERING

Sintering is the time-dependent consolidation of a porous medium. Viscous sintering, applicable to polymers, is described by neck growth in a two-particle system (Frenkel 1945, Scherer 1986a, b, Sun et al. 1991), Figure 1. The rate of growth of the neck $\dot{y}$ is given by Frenkel as

$$\dot{y} = \frac{2}{3}\left(\frac{\Gamma}{\eta}\right)\frac{R}{y} \tag{3}$$

where Γ is the surface tension, η is the viscosity, R is the particle size and y is the neck radius.

Scherer (1986a, b) described the densification rate for viscous sintering using a free strain term e that may be related to the porosity ε. Based on a cubic array of cylinders, the change in free strain e with time is written as

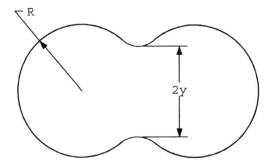

Figure 1.   Schematic of two particle sintering.

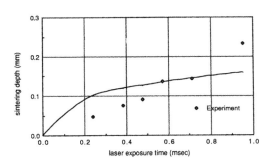

Figure 2.   Powder densification as a function of laser exposure time. ABS powder, heat input = $8.88 \times 10^7\,\text{W/m}^2$, bed
temperature = 20°C. From Sun et al. 1990.

$$\frac{\partial e}{\partial t} = -\frac{M}{\eta}\frac{(3\pi)^{1/3}}{6}\frac{2-3cx}{\sqrt[3]{x(1-cx)^2}} \qquad \Delta \leq 0.94$$

$$\frac{\partial e}{\partial t} = -\frac{M}{2\eta}\left(\frac{4\pi}{3}\right)^{1/3}\left(\frac{1}{\Delta}-1\right)^{2/3} \qquad \Delta > 0.94 \tag{4}$$

where $M = (\Gamma/R)(3/4\pi)^{1/3}$, $c = 8\sqrt{2}/3\pi$, η is the viscosity, Δ is the relative density defined as the ratio of
the part density to the material theoretical density, Γ
is the surface tension, R is the particle initial radius
and x is defined as the ratio of the cylinder structure
to the length of the cylinder. The porosity $\varepsilon = (1 - \Delta)$
is related to the free strain e by

$$\varepsilon = 1-(1-\varepsilon_o)\exp(-3e) \qquad \Delta \leq 0.94$$

$$\varepsilon = 1- 3\pi x^2 +8\sqrt{2}x^3 \qquad \Delta > 0.94 \tag{5}$$

where ε_o is the initial porosity when t = 0.

Figure 2 shows the results of these relationships
output as the sintering depth as a function of laser
exposure time on an ABS powder bed.

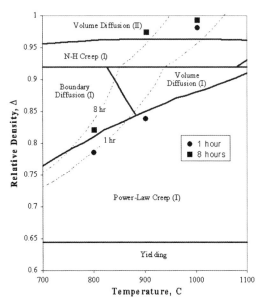

Figure 3.   Densification map for Nickel alloy 625 powder
after SLS. The particle diameter was 25 μm. From Bourell
et al. 2000.

Sintering of metals and ceramics is limited to pre-
and post-processing since the times required to
accomplish significant densification are long. Ashby
1990 converted the surface area component of the driving force to an effective stress and summed it with a net
applied HIP gas pressure. The resulting driving pressure
was written for open and closed porosity as a function
of the initial applied pressure, the starting and current
relative density, the particle radius and powder surface energy. In all stages of sintering, several densification mechanisms act simultaneously. By combining
the driving force and densification mechanisms, it is
possible to construct a sintering map, a plot of the relative density as a function of sintering temperature
and time. An example is shown in Figure 3 for Nickel
Alloy 625 powder SLSed to $\Delta = 0.6$ and HIPped for
1 hr or 8 hr at temperatures between 800°C and 1000°C
(Bourell et al. 2000). The computed isochronal lines
show excellent agreement with experimental results.
Also shown are the dominant sintering mechanisms.

Sintering maps can also be used to aid in powder pre-
processing. An example is pre-processing of nanocrystalline zirconia prior to SLS (Bourell et al. 2000).
Nanocrystalline (D = 24 nm) zirconia was available
in polymer-bound, agglomerated form (50 μm). This
was unacceptable for SLS processing because the
binder would melt and flow in an unpredictable fashion
during SLS, and the coarse particles would not hold
together. The need then was to determine a thermal

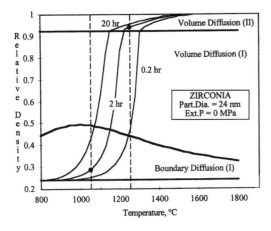

Figure 4. Pressureless densification map for 24 nm zirconia powder. Densification occurs between 1050°C and 1250°C. From Bourell et al. 2000.

(a)

(b)

(c)

(d)

Figure 5. SEM Micrographs of Agglomerated Zirconia. (a) Held in air at 1050°C for 2 hr. The agglomerated 50 μm powder holds together even though the binder has been burned off. (b) Held in air at 1050°C for 2 hr. Fine, low-density nanocrystalline structure present within 50 μm particles. (c) Held in air at 1250°C for 2 hr. The agglomerated structure remains. (d) Held in air at 1250°C for 2 hr. Within 50 μm particles, the nanocrystalline particulate has fully densified. From Bourell et al. 2000.

excursion by which the nanocrystalline particulate within agglomerates might be debound and sintered together to high density, but for which the 50 μm agglomerates would not sinter.

A pressureless temperature densification map for zirconia with a particle size of 24 nm appears as Figure 4.

It is observed that zirconia fine particles sinter lightly in 2 hr at 1050°C and almost completely densify in 2 hr at 1250°C. At 1250°C, the dominant densification mechanism is boundary diffusion initially which transitions to volume (bulk) diffusion at about 45% relative density. Coarse grained 50 μm zirconia is quite resistant to pressureless densification to at least 2000°C.

Figures 5a and 5b show low and high magnification SEM micrographs of yttria-stabilized zirconia powder held 2 hr in air at 1050°C. In Figure 5b, the fine particulate is visible and has lightly sintered. Figures 5c and 5d are SEM micrographs of powder held 2 hr at 1250°C. The coarse particles are still unsintered, but the fine-particle interiors are completely densified. This is consistent with the densification map observations.

4 INFILTRATION

Infiltration is a technique for increasing the end-use properties of RM parts. It is viable when the part has open porosity which is porosity that forms a continuous tunnel-like network throughout the part.

The wetting angle θ is related to the various surface energies involved according to Young's Equation (Heady & Cahn 1970):

$$\cos\theta = \frac{\gamma_{sv} - \gamma_{ls}}{\gamma_{lv}} \quad (6)$$

where γ_{sv} is the surface energy associated with the solid surface being wet, γ_{ls} is the surface energy associated with the liquid-solid interface and γ_{lv} is the surface energy associated with the liquid infiltrant. Low wetting angles promote wetting.

For metals, the liquid surface energy γ_{lv} varies from about 300 mN/m for alkali metals and semi-metals to as high as 1800 mN/m for d-transition metals (Handbook of Chemistry and Physics 1991). Solid metal surface energies γ_{sv} are theoretically 9% higher than the metal liquid surface energy at the melting point, although actual values range between 4% and 33% higher (Brophy et al. 1964). Liquid–solid surface energies γ_{ls} vary widely depending on the materials involved. It may be measured experimentally using the wetting balance test (Vianco et al. 1990).

A threshold pressure P* for spontaneous infiltration into particulate media is given as (Mortensen and Cornie 1987, Oh et al. 1989):

$$P* = \frac{6\lambda(-\gamma_{lv}\cos\theta)\Delta}{(1-\Delta)D} \quad (7)$$

where D is the particle size and λ is the ratio of the particle actual surface area to the surface area of a

307

Figure 6. A silicon-infiltrated silicon carbide preform produced using SLS. (a) The part is 5 cm by 6 cm in cross section. (b) A micrograph of a cross section showing dark SiC and light Si infiltrant. From Bourell et al. 2004.

sphere of the same volume. When the wetting angle for Equation 7 is less than 90 degrees, wetting occurs and some infiltration will take place. The extent of infiltration depends on the magnitude of the effective pressure which is impacted not only by the degree of wetting but also other factors in the equation.

If infiltration works against gravity, there exists an equilibrium height associated with the balance between the wetting and gravitational forces (Gern 1995):

$$h_{max} = \frac{4\gamma_{lv}\cos\theta}{dg\rho} \qquad (8)$$

where g is the gravitational acceleration and ρ is the liquid density.

Figure 6 is a photograph of a silicon-infiltrated silicon carbide part produced by indirect SLS of silicon carbide followed by post-process pressureless infiltration with silicon. The threshold pressure for the onset of infiltration according to Equation 7 is -83 kPa (Wang 1999). This negative pressure is conceptually a vacuum on the molten metal offsetting capillarity

forces for infiltration into the pore structure. The maximum vertical height for self-infiltration in this system was calculated using Equation 8 to be 2.5 m, which compares well with other research, in which a maximum height of about two meters was calculated (Gern 1995).

5 MECHANICAL PROPERTIES OF RM PARTS

The in-use properties of RM parts define the scope of applicability for the technology. Parts that are fully dense and homogeneous manifest properties associated with their microstructure. For a number of direct processes, the parts are a solidified structure of polymer or metal. The associated cast properties are generally obtained. For example, Ti-6Al-4V was direct SLSed to create a nominally fully dense part (Das et al. 2004). Table 1 lists the properties and structure information for wrought, cast and selective laser sintered material as well as information on the impurities in PREP and Ar atomized powder, precursors for the SLS processing. It is seen that the mechanical properties and impurity concentrations are comparable to cast and annealed wrought material.

The authors are indebted to Neil Hopkinson of Loughborough University for his input on polymer mechanical properties. Polymer mechanical properties of layer manufactured parts are often inferior to injection molded counterparts due to issues such as porosity and the absence of high pressures and consolidation during manufacturing. Table 2 compares tensile properties of laser sintered nylon with the range of tensile properties found in cast and molded nylons. While the stiffness of SLSed parts is comparable with other processes, the strength and ductility fall short. ABS parts made by fused deposition modeling also fail to match the mechanical properties of molded parts, especially in the build direction.

Residual porosity negatively impacts all the mechanical properties, in order of decreasing severity: fracture, fatigue, strength, ductility, modulus. The effects of relative porosity ε on each is described.

For strength, the general dependence takes the form (German 1984):

$$\sigma = K\sigma_0(1-\varepsilon)^m = K\sigma_0(\Delta)^m \qquad (9)$$

where σ is the strength, K and m are constants, and σ_0 is the wrought strength of the same alloy. K is geometry and processing dependent. The relationship is generally applied to yield strength, tensile strength and three- or four-point bend strength. Figure 7 is a logarithmic plot of the tensile strength of SLS processed bronze-nickel powder. The linear relationship for as-SLS processed and post-process liquid-phase sintered parts confirms the effect of porosity on strength.

Table 1. Properties of Ti-6Al-4V in cast, wrought, SLS and powder form.

Processing	Hardness HRC	Tensile strength MPa	Elongation %	Oxygen %	Nitrogen %
Cast grade C-5 (ASTM B367 1993)	39	895	6	0.25	0.05
Cast grade C-6 (ASTM B367 1993)	36	795	8	0.20	0.05
Ann. wrought grade 5 (ASTM B348 1995)	–	895	10	0.20	0.05
PM HIP (Das et al. 1998)	34–36	–	–	–	–
SLS direct; Ar atom. (Das et al. 1998)	36	1120	5	0.23	0.037
PREP powder (Das et al. 1998)	–	–	–	0.19	0.01
Ar atomized (Das et al. 1998)	–	–	–	0.196	0.02

Table 2. Tensile properties of conventionally processed and selective laser sintered nylon.

Material/Process	UTS (MPa)	E (GPa)	Elongation at break (%)
Nylons/Cast/molded Kalpakjian and Schmid 2001	55–83	1.4–2.8	60–200
Nylon 12/Selective laser sintered Zarringhalam and Hopkinson 2003	46	1.8	12

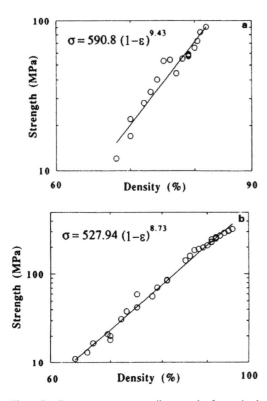

Figure 7. Room-temperature tensile strength of pre-mixed SLS (90Cu-10Sn) bronze and commercially pure nickel powder as a function of relative density $\Delta = 1 - \varepsilon$. (a) As SLS Processed, (b) SLS processed and sintered at 900–1100°C for 1 to 10 hr. From Agarwala et al. 1993.

Ductility of parts with residual porosity may be predicted using the relationship (Haynes 1977):

$$Z = \frac{(1-\varepsilon)^{3/2}}{\left(1+C\varepsilon^2\right)^{1/2}} \tag{10}$$

where Z is the ratio of the porous material ductility to equivalently processed full density ductility, and C is an empirical constant related to the sensitivity of the ductility to the presence of pores. Large values of C correspond to highly sensitive ductility. For example, a residual porosity of 10% causes a 20% reduction in ductility if C equals 10, but it causes a 97% reduction if C equals 100,000. Generally, C varies between about 100 and 100,000.

Fatigue and fracture behavior of RM parts are strongly influenced by residual porosity (ASM Handbook Volume 7 1998). Not only is the volume fraction of pores important, but also are the spacing between pores, the average pore size and the morphology of the pores, particularly ones that reside on the surface. Pores generally increase the threshold stress intensity for crack initiation but lower the resistance to crack propagation.

For many materials with low residual porosity, the fatigue endurance limit is about 35% of the tensile strength (ASM Handbook Volume 7 1998), compared to about 50% for bulk materials. The stress field around an isolated spherical pore is about twice the far-field stress level.

The plane strain fracture toughness K_{IC} has been shown to be a strong function of porosity (ASM

309

Handbook Volume 7 1998). For quenched and tempered steels, the fracture toughness decreases by about 100 MPa $\sqrt{m}$ for each percent of porosity.

6 SUMMARY

Important properties of materials for rapid manufacturing (RM) have been presented for non-photocuring processes. Viscosity often drives material flow and densification. Polymer sintering has been described by Frenkel and further developed by others. Ashby sintering maps capture the effects of time, temperature and pressure on the densification of metals and ceramics. Infiltration of polymers and metals into porous preforms is an effective method to improve performance. Finally, service properties of RM parts are strongly dependent on residual porosity. The effect is described for strength, ductility, fatigue and fracture.

ACKNOWLEDGMENTS

This project was funded by 2001 Texas Technology Development/Transfer Grant #003658 and the US Office of Naval Research Grant N00014-00-1-0334.

REFERENCES

Agarwala, M.K., D.L. Bourell, B. Wu, J.J. Beaman, 1993, "An Evaluation of the Mechanical Behavior of Bronze-Ni Composites Produced by Selective Laser Sintering", SFF Symposium Proceedings, H.L. Marcus, J.J. Beaman, J.W. Barlow, D.L. Bourell and R.H. Crawford, eds., Austin TX, 193–203.

American Society for Testing of Materials, ASTM B367-93, Standard Specification for Titanium and Titanium Castings, 1993.

American Society for Testing of Materials, ASTM B348-95a, Standard Specification for Titanium and Titanium Alloy Bars and Billets, 1995.

ASM Handbook, Volume 7, "Powder Metal Technologies and Applications", 10th ed., 1998, 957–964.

Ashby, M.F., 1990, "Sintering and Isostatic Pressing Diagrams", Published by Author, Department of Engineering Cambridge, England.

Beaman, J.J., J.W. Barlow, D.L. Bourell, R.H. Crawford, H.L. Marcus, K.P. McAlea, 1997, "Solid Freeform Fabrication: A New Direction in Manufacturing", Kluwer Academic, Norwell MA, 104.

Bourell, D.L., M. Wohlert and N. Harlan, 2000, "Powder Densification Maps and Applications in Selective Laser Sintering", in 'Deformation, Processing and Properties of Structural Materials – A Symposium Honoring Oleg D. Sherby', E.M. Taleff, C.K. Syn and D.R. Lesuer, eds., TMS, pp. 219–230.

Bourell, D.L., R.S. Evans, S-W. Chen, S.L. Barrow, 2004, "Rapid Manufacturing of Functional Parts Using Infiltration Post-Processing", Proceedings of Laser Assisted Net Shape Engineering 4, Erlangen, Germany, M. Geiger, A. Otto, eds., Meisenbach-Verlag, Bamberg, 2004, pp. 93–104.

Brophy, J.H., R.M. Rose, J. Wulff, 1964, "The Structure and Properties of Materials: Volume II – Thermodynamics of Structure", John Wiley and Sons, New York, 49–52.

Das S.M. Wohlert, J.J. Beaman, D.L. Bourell, 1998, Processing of Titanium Net Shapes by SLS/HIP, SFF Symposium Proceedings, D.L. Bourell, J.J. Beaman, R.H. Crawford, H.L. Marcus and J.W. Barlow, eds., Austin TX, 469–477.

Frenkel, J., 1945, "Viscous Flow of Crystalline Bodies Under the Action of Surface Tension", J. Phys. (USSR), 9, 385.

German, R.M., 1984, "Powder Metallurgy Science", Metal Powder Industries Federation, Princeton NJ.

Gern, F.H., 1995, "Interaction Between Capillary Flow and Macroscopic Silicon Concentration in Liquid Siliconized Carbon/Carbon", Ceramic Trans, 58, 149.

Handbook of Chemistry and Physics 72nd ed., CRC Press, 1991 4.136.

Heady, R.B., J.W. Cahn, 1970, "An Analysis of the Capillary Forces in Liquid-Phase Sintering of Spherical Particles", Met Trans, 1#1, 185–189.

Haynes, R., 1977, "A Study of the Effect of Porosity Content on the Ductility of Sintered Metals", Powder Met., 20, 17–20.

Kalpakjian, S and Schmid, S., 2001, Manufacturing Engineering and Technology, 4th edition, Prentice Hall, ISBN 0-201-36131-0.

Mortensen, A. and Cornie, J.S., "On the Infiltration of Metal Matrix Composites", Met Trans A, 1987, 18A, 1160–1163.

Oh, S.-Y., J.A. Cornie, K.C. Russell, 1989, "Wetting of Ceramic Particulates with Liquid Aluminum Alloys: Part I. Experimental Techniques", Met Trans A, 20A, 527–532.

Scherer, G.W., 1986a, "Viscous Flow under a Uniaxial Load", J Am Cer Soc, 69 #9, 206–7.

Scherer, G.W., 1986b, "Sintering of Low Density Glasses: I-Theory", J Am Cer Soc, 60#5–6, 236–9.

Sun, M-s. M, J.J. Beaman, J.W. Barlow, 1990, "Parametric Analysis of the Selective Laser Sintering Process", SFF Symposium Proceedings, J.J. Beaman, H.L. Marcus, D.L. Bourell and J.W. Barlow, eds., Austin TX, 146–154.

Sun, M-s. M, J.C. Nelson, J.J. Beaman, J.W. Barlow, 1991,"A Model for Partial Viscous Sintering", SFF Symposium Proceedings, H.L. Marcus, J.J. Beaman, J.W. Barlow, D.L. Bourell and R.H. Crawford, eds., Austin TX, 46–55.

Vianco, P.T., F.M. Hosking, J.A. Rejent, 1990, "Solderability Testing of Kovar with 60Sn-40Pb Solder and Organic Fluxes", Welding Journal, 6, 230s–240s.

Wang, H., 1999, "Advanced Processing Methods for Microelectronics Industry Silicon Wafer Handling Components", Ph.D. Dissertation, The University of Texas at Austin, Austin TX, 37–44.

Zarringhalam, H. and Hopkinson, N., 2003, Post-Processing of Duraform™ Parts for Rapid Manufacture, Proceedings from the 14th SFF Symposium, Austin, Texas, pp. 596–606, ISSN 1053–2153.

Virtual modeling and rapid manufacturing – Bártolo (eds)
© *2005 Taylor & Francis Group, London, ISBN 0 415 39062 1*

Processing characteristics and mechanical properties of a novel stereolithographic resin system for engineering and biomedicine

C. Tille, A. Bens & H. Seitz
Caesar research center, Bonn, Germany

ABSTRACT: Stereolithography, as a well-known rapid prototyping process, has been used in a wide field of technical and also medical applications. Due to the stereolithography principle – the curing of a liquid photopolymer by a UV laser – the number of commercially available reaction mechanisms and related material classes is very limited. On the other hand, new applications lead to a growing need for resins with advanced material characteristics. Our paper presents a novel class of polyether(meth)acrylate based resin formulations with outstanding flexible material characteristics. In contrast to the mostly rigid commercial materials in the engineering world, these polymeric formulations are able to meet the demand for very soft to even stiffer manufacturing materials. Depending on the individual formulation, the cured resins can show a Young's modulus from 10 MPa up to 2000 MPa. We give an overview over basic formulations and processing characteristics for this material class. Process parameters were studied in a commercial Viper Si^2 system (3D Systems); mechanical properties of different formulations were tested using standard tensile testing methods.

1 INTRODUCTION – MATERIALS IN STEREOLITHOGRAPHY

Rapid prototyping describes a group of technologies that allow the manufacturing of three-dimensional prototype models directly from 3D surface data. In recent years, this technology is being used in a broad field of technical as well as medical applications.

The best-known rapid prototyping process, stereolithography, uses a liquid photopolymer that is locally cured by a UV laser (Kodama 1981). In the last years, there has been a growing need from customers for resins with novel material characteristics. Nowadays, standard stereolithographic resins consist mainly of epoxy based oligomer components. Those materials are optimized to mimic stiff technical polymers like PP, PE or ABS. On the other side, softer materials cannot be represented. Furthermore, epoxy based materials are not suitable for medical applications due to known cytotoxic effects on human cells.

To overcome the described problems, a new polyether(meth)acrylate based resin material class was developed (Bens et al. 2005). The resin class is called FlexSL because of its outstanding material properties and its adjustable characteristics for stereolithography and other photolithographic applications (Figure 1). The new polymeric formulation class is able to fulfill

Figure 1. Sample part demonstrating the high flexibility of FlexSL materials.

the extended need for a more flexible resin material with a broad range of adjustable hardness and strength.

Depending on the concrete application, the FlexSL photopolymeric formulations can contain various components like basic linear monomers, cross-linking polyfunctional monomers (e.g. triacrylates or tetraacrylates), a photoinitiator and other compounds, e.g. antioxidants und UV stabilizers. For special applications, dyes and pigments can also be added.

Figure 2. Chemical structure of one basic component of all tested FlexSL resins.

2 MATERIALS AND METHODS

2.1 Compounds of the polyether(meth)acrylate resin class

The basic resin class consists of mainly two components:

1 A very flexible oligomeric, polyether(meth)acrylate monomer with high molecular weight

and

2 A short and therefore hard-segment polyfunctional cross linking polyether(meth)acrylate monomer

or

3 a polyfunctional tri-acrylate or tetraacrylate with lower molecular weight.

There is an option to add different filler materials that can significantly alter the material properties. Figure 2 shows the chemical structure of a typical component 1, in this case a bisphenole A ethoxylated dimethacrylate.

Different resin compounds were tested in order to find optimized resins for standard and highresolution stereolithography. The main influencing factor on the resolution of a resin is found in the photoinitiator itself and its concentration. Table 1 gives an overview of the two investigated compound classes. While Flex-AH-01 to Flex-AH-07 uses Irgacure 369 as photoinitiator at different concentrations, Flex-AH-11 to Flex-AH-16 contains Irgacure 907, both from Ciba Specialty Chemicals Inc.

All samples were mixed for 24 hours under darkness at room temperature using a laboratory dissolver.

Table 1. Compounds of tested resin samples.

Compound	Core component	Photoinitiator concentration (w.-%)
Flex-AH-01	Polyether based resin AH*	0,20% Irgacure 369**
Flex-AH-02	Polyether based resin AH*	0,30% Irgacure 369**
Flex-AH-03	Polyether based resin AH*	0,50% Irgacure 369**
Flex-AH-04	Polyether based resin AH*	0,75% Irgacure 369**
Flex-AH-05	Polyether based resin AH*	1,00% Irgacure 369**
Flex-AH-06	Polyether based resin AH*	1,25% Irgacure 369**
Flex-AH-07	Polyether based resin AH*	1,50% Irgacure 369**
Flex-AH-11	Polyether based resin AH*	0,75% Irgacure 907**
Flex-AH-12	Polyether based resin AH*	1,00% Irgacure 907**
Flex-AH-13	Polyether based resin AH*	1,25% Irgacure 907**
Flex-AH-14	Polyether based resin AH*	1,50% Irgacure 907**
Flex-AH-15	Polyether based resin AH*	1,75% Irgacure 907**
Flex-AH-16	Polyether based resin AH*	2,00 % Irgacure 907**

* Polyether based resin AH: 100% Bishenol-A-ethoxylated Polyetherdi(meth)acrylates/Tri- and Tetracrylates and Stabilizers (e.g. 4-Methoxy-phenol).
** Irgacure: Registered Trademark of Ciba Specialty Chemicals Inc.

Figure 3. UV curing and silicone mould of tensile test specimens.

2.2 Material processing

For mechanical characterization, all resin samples were cast into a silicone mould designed for tensile probes. The liquid resin is cured under an Hg highpressure lamp (UVA plus UVB) with 30 mW/cm^2 and an energy dose of 1.8 J/cm^2 (Figure 3). The specimens were acetone-cleaned after complete curing.

After initial mechanical tests, optimized resin formulations were selected and used for manufacturing complex parts. A Viper Si^2 (3DSystems, Valencia) stereolithography apparatus was used for manufacturing parts with standard resolution. For testing as well as small batch production on the Viper Si², a customized vat with very low resin volume of 3 liters was developed.

Furthermore, a high-resolution stereolithography apparatus with 20 μm spot size and 20 μm layer thickness was used for processing tests of very small parts.

2.3 Mechanical characterization

Before the properties of the two selected resins were optimized, it was necessary to carry out further tests on cured material probes based on the basic formulations FlexAH-01 to Flex-AH16.

Therefore, a universal testing machine was used for mechanical tensile tests according to DIN EN ISO 527-1 of all cast specimens. Young's modulus, elongation at break and tensile strength were measured using 5 specimens of each resin compound.

2.4 Processing characteristics

To describe a resin's behavior, the well-known Windowpane technique is widely used to capture the working curve of an unknown material. In this method, the resin surface is exposed with a pattern of laser light using different energy doses (Jacobs 1993). Each exposed area shows an individual thickness of the cured resin. A linear regression of the logarithmized relative energy dose in the working curve equation

$$C_D = D_P \, \ln\left(\frac{E_{max}}{E_c}\right) \qquad (1)$$

leads to the characteristic resin values E_C (polymerization energy dose [mJ/cm²]) and D_P (penetration depth [mm]) of a stereolithography resin (Jakubiak & Rabek 2000).

Because of the free-floating geometry that is exposed by the laser a high distortion and thus a high error have to be accepted.

First tests showed, that an improvement of this standard method was necessary for an exact analysis of the influence of different compounds on the curing behavior even for thin layers: The developed protocol uses a quartz-glass window with optical quality as a reference plane. In a first step, the quartz-glass window is fixed in a box with a distance of 10 mm to its ground. A small liquid resin sample that has to be tested is poured in such a way that no air bubbles remain beneath the window. The box is placed in the building chamber of a stereolithography apparatus and a pattern is exposed with an increasing energy dose. Remaining resin is allowed to drip off for 20 minutes. In the next step, the cured structure is gently rinsed with TPnB. After drying on a clean tissue for 5 times, each

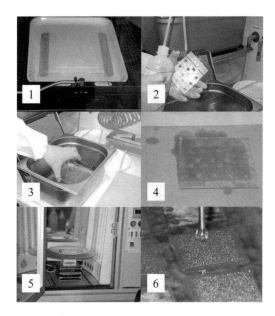

Figure 4. Overview of all steps of the quartz-glass window method (1 – UV laser exposure; 2, 3 – cleaning of a quartz-glass window; 4 – drying of the cured sample; 5 – post-curing in a UV-oven; 6 – thickness measurement using height measuring instrument).

30 seconds, the window is post-cured for 10 minutes in a UV-oven.

With the help of a height measuring instrument Z-CAL 150/300 (Sylvac, Crissier) with a predefined small contact force of 1 N, the thickness of each cured area within the pattern was measured against the quartz-glass surface. This method allows a significantly higher precision (approx. ±5 μm) in comparison to the standard Windowpane method. Figure 4 depicts the described steps of the quartz-glass window method.

3 RESULTS AND DISCUSSION

3.1 Mechanical properties

Figure 5 displays the results of the described tensile tests. It shows that the Young's modulus can be considerably lower in comparison to most commercial stereolithography resins.

To show that the mechanical parameters are customizable in a very broad range, two further compounds with strongly differing chain lengths, called *FlexSL Hard* and *FlexSL Soft*, are displayed in Figure 5.

The characteristics of the new optimized resin family can be compared to standard stereolithography resins using Table 2.

Due to known difficulties with high-viscosity resins during the recoating process, it was necessary to limit

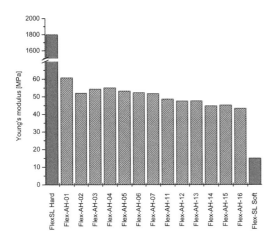

Figure 5.   Mechanical properties (Tensile Young's modulus [MPa]) for different flexible FlexSL specimens.

Table 2.   Typical properties of the FlexSL material class.

	FlexSL family	Typical SL resin
Young's modulus	15–1800 MPa	1400–2800 MPa
Tensile strength	2–50 MPa	30–110 MPa
Elongation at break	2%–15%	5%–10%
Shore **A/D** hardness	Shore **A** 30–95	Shore **D** 75–95

the target value for the dynamic viscosity to approx. 1500 mPas (Tille 2004b). All resins can meet this criterion. Typical values for the dynamic viscosity are in the range of 700 mPas.

3.2   Windowpane results

Using the described quartz-glass window protocol, both resin formulations were characterized in order to identify the process parameters for stereolithography (Figure 6).

For the Irgacure 369 resins, the values for D_P can be adjusted from 23 to 173 µm by changing the photoinitiator concentration. In this case, E_C ranges from 0.5 to 6 mJ/cm^2.

Using Irgacure 907 as a photoinitiator, D_P is significantly larger 175 from 386 µm depending on its concentration. E_C lies now in the range from 9.3 to 20.7 mJ/cm^2.

From this measurement it can now be concluded, that the resin process parameters of this material class cannot only be optimized for mechanical properties but also for either high building speed or high resolution.

For example, a high-speed resin would be achieved by using 2% Irgacure 907 (E_C = 9.3 mJ/cm^2, D_P = 175 µm). A high-resolution resin can be obtained with 1% Irgacure 369, resulting in E_C = 1.2 mJ/cm^2,

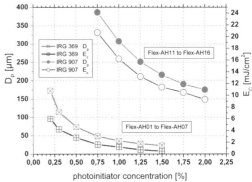

Figure 6.   Process parameters E_C [mJ/cm^2] and D_P [µm] of resin formulations Flex-AH-01 to Flex-AH-16.

D_P = 35 µm. Of course, further optimization (esp. addition of UV blockers) is necessary to find a suitable combination for all customer's needs.

4   CONCLUSIONS

We showed the basic formulations of a new stereolithography resin class. Important mechanical parameters were investigated showing a high potential for adjustment to customers' needs, especially to mimic rubberlike materials.

Its was also shown, that by changing the photoinitiator resp. its concentration, this material class is adjustable for high resolution stereolithography as well as the standard stereolithography process.

After the described initial tests, optimized resin formulations based on the described basic compounds were developed. Numerous different stereolithography models were manufactured using either a 3D Systems Viper Si^2 stereolithography apparatus or a microstereolithography apparatus.

A further advantage of the material is the good biocompatibility in comparison to other stereolithography materials. Results from biocompatibility tests will be published elsewhere in the next months.

In the near future, a group of three different soft resin materials for technical applications will undergo applications tests. Those resin products will cover typical demands on very low Shore values (skin-like), mid-Shore values (rubber-like) and harder properties (hard rubber-like). Sample parts out of the new FlexSL materials will shortly be available from caesar[1]. A further distribution strategy for all three resins is currently under instigation.

---

[1] info@3mat.de

REFERENCES

Bens, A.T., Tille, C., Bermes, G., Emons, M., Seitz, H.: Novel Biocompatible Polyether(meth)acrylate-based Formulations for Stereolithography – A New Flexible Material Class for Three-Dimensional Applications. e-Polymers 2005.

Bens, A.T., Seitz, H., Tille, C., Patent Application (2004).

Jacobs, P.F.: Rapid Prototyping & Manufacturing: Fundamentals of Stereolithography. McGraw-Hill, 1993.

Jakubiak, J., Rabek, J.F.: Three-Dimensional (3D) Photopolimerization in Stereolithography. Part I. Fundamentals of 3D Photopolimerization, Polymeri 45 (11–12) 759–770.

Jakubiak, J.; Rabek, J.F.: Three-Dimensional (3D) Photopolimerization in Stereolithography. Part II. Technologies of 3D Photopolimerization. Polymeri 46 (3) 164–171.

Kodama, H.: Automatic method for fabricating a threedimensional plastic-model with photo-hardening polymer. Review of Scientific Instruments 52 (11) 1770–1773.

Kovacs, L., Zimmermann, A., Wawrzyn H., Schwenzer, K.; Seitz, H.; Tille, C.; Papadopulos, N.A., Sader, R.; Zeilhofer, H.F., Biemer, E.: Computer aided surgical reconstruction after complex facial burn injuries–opportunities and limitations. Burns 31 (1) 85–91 (2005).

Lindén, L.-A., Jakubiak, J.: Three-Dimensional (3D) Photopolimerization in Stereolithography, Part III. Medical Applications of Laser-Induced Photopolymerization and 3D Stereolithography. Polymeri 46 (4) 227–292.

Seitz, H., Tille, C., Irsen, S., Bermes, G., Sader, R., Zeilhofer, H.-F.: Rapid Prototyping models for surgical planning with hard and soft tissue representation. CARS2004: Computer Assisted Radiology and Surgery, Chicago, June 23–26, 2004, International Congress Series 1268 (2004), 567–572.

Tille, C., Seitz, C., Schaefer, D.J., Irsen, S., Ritter, L., Zeilhofer, H.F.: New approach to modeling 3D-shaped ceramic bone substitutes for craniofacial surgery via rapid prototyping – a case study. CMBBE2004: 6th Symp. on Computer Methods in Biomechanics and Biomedical Engineering, Madrid, February 25–28, 2004.

Tille, C.: Process errors and aspects for higher resolution in conventional stereolithography. SFF2004: 15th Solid Freefrom Fabrication Symposium, Austin, August 2–4, 2004.

Yan, X.; Gu, P.: A review of Rapid Prototyping technologies and systems. Computer-Aided-Design 28 (4) 307–318.

Virtual modeling and rapid manufacturing – Bártolo (eds)
© 2005 Taylor & Francis Group, London, ISBN 0 415 39062 1

New developments in steel formulations for Direct Laser Sintering: Characterization and evaluation of joining mechanisms

E. Bassoli & A. Gatto
University of Modena and Reggio Emilia, Dept. of Mechanical and Civil Engineering, Modena, Italy

L. Iuliano
Polytechnic of Turin, Department of Manufacturing Systems and Economics, Turin, Italy

ABSTRACT: Rapid Prototyping and Tooling are playing a more and more important role in the achievement of compressed time-to-market solutions. Even so, the spread of these techniques is hardly supported by scientific knowledge about the micro-mechanisms ruling the macroscopic performances of the part. In the present research, parts produced by Direct Laser Sintering technique have been studied, using DirectSteel 20 and the innovative DirectSteel H20. The research aimed at investigating the sintering mechanisms and their influence on the failure modes and mechanical performances. Tensile specimens have been produced, with different orientations in regard to laser path, to study the effect of laser sintering strategy on the anisotropy of parts performances. The tensile tests results have been correlated to the SEM observation of the specimens rupture surfaces. The results proved that no anisotropy can be noticed in the mechanical response for loads applied in the different directions within the plane of powder deposition.

1 INTRODUCTION

Among the techniques for layer-by-layer construction, Selective Laser Sintering (SLS) is largely spread thanks the wide variety of available materials (Morgan et al. 2002, Kruth et al. 2003). In particular, SLS of metal powders is suitable not only for Rapid Prototyping, but also for Rapid Tooling applications, where tools for small series are obtained directly from digital data, ensuring wide benefits in terms of compressed time-to-market (Gatto & Iuliano 1998, 2000, 2001; Iuliano et al. 2003).

Many authors tried a classification of the different versions of the process (Kathuria 1999, Khaing et al. 2001, Morgan et al. 2002). The present research regards the technique allowing to obtain metal parts directly through laser sintering without any post-treatment, known in literature as Direct Laser Sintering (DLS). Binary powder mixtures are processed and parts consolidation is ensured by the low melting phase, which acts as a matrix.

As to the considered process, Tolochko et al. 2000 outline that the powder bed absorbs the incident energy through much more complex modalities than a bulk material. The effective thermal supply depends strongly on the powder morphology and on the porosity, but also on the presence of gases such as oxygen.

Regarding the consolidation phenomena, Agarwala et al. 1995 state that the leading effect is particles rearrangement due to capillary and hydrodynamic forces in the liquid phase. The described phenomenon is dependent on a good wettability between the melted phase and the solid particles (Anestiev & Froyen 1999); otherwise antagonist mechanisms prevail and poor consolidation is obtained (Niu & Chang 1998, 1999a, b; Tolochko et al. 2003). Simchi et al. 2003 outline that an accurate development of the powders composition is crucial to obtain satisfying mechanical properties.

Previous studies were performed by the authors (Bassoli et al. 2004) on DLS of a bronze- and a steel-based powder: DirectMetal and DirectSteel. Both exist in two versions, with different mean grain dimension (20 and 50 μm): the experimental characterisation regarded the finest version of the two materials. The results confirmed that for both the materials a high densification is obtained thanks to good wettability and capillary forces, which lead to an isotropic structure within the plane orthogonal to the growth direction.

In the present research, a comparison has been carried out between DirectSteel 20 and the innovative formulation DirectSteel H20. The mechanical properties of the two materials are reported in Table 1, as stated by the producer. The recently developed DirectSteel H20 appears to have very interesting mechanical properties,

Table 1. Mechanical characteristics of DirectSteel 20 and DirectSteel H20 (as stated by the producer).

	DirectSteel 20	DirectSteel H20
Minimum remaining porosity (%)	2	0.5
Tensile strength (MPa, MPIF 10)	600	1100
Hardness (HV)	225	400
Surface roughness after shot-peening (μm)	$R_a = 4$	$R_a < 6$

comparable to austempered steels. Thanks to its full density and very high hardness and strength, combined with a higher ductility with respect to the previous material, it is addressed to the production of tooling as well as of heavy-duty parts. For both the materials, data based on an accurate experimental characterization is lacking in literature. The research objective was the investigation of the sintering mechanisms and their influence on the failure modes and mechanical performances of parts produced by DLS with the two materials formulations.

2 EXPERIMENTAL PLAN

Tensile specimens have been produced with DirectSteel 20 and DirectSteelH20, following the specifications of standard ASTM E8M, for flat unmachined specimens obtained by powder metallurgy (Fig. 1). The specimens main dimensions are reported in Table 2.

Standard processing conditions have been adopted for the samples construction, using a 200 W CO_2 laser, moving in the X direction. Layer thickness was of 20 μm. Specimens different orientations in regard to laser path have been adopted, to investigate the effect of laser sintering strategy on the anisotropy of parts performances. In particular, samples have been constructed with the axis parallel to the X- and Y-axis of the working area, as well as inclined of 45° (XY samples). After the construction, the specimens were shot-peened to improve the surface quality.

Tensile tests were carried out with a free-running crosshead speed of 5 mm/min, using self-centring equipment. At least five samples have been tested for each combination of material and orientation.

Thanks to a software tool for statistical analysis, the t-test has been carried out on the data obtained from the tensile tests, to evaluate the presence of significant differences in the mechanical properties of the specimens produced with the various orientations.

Macroscopic characterisation has been constantly combined with the observation of joining phenomena and failure mechanisms at a microscopic level. For this purpose, rupture surfaces after the tensile tests have been observed through the SEM, using secondary and

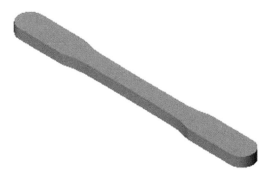

Figure 1. CAD model of a tensile specimen following ASTM E8M.

Table 2. Main dimensions of a tensile specimen following ASTM E8).

Length of reduced section (mm)	31.8
Overall length (mm)	89.6
Width at center (mm)	5.7
Width at the end of reduced sec. (mm)	5.9
Thickness (mm)	3.6

Table 3. Mean values and std. deviations of the mechanical characteristics measured for DirectSteel 20 and DirectSteel H20.

	UTS MPa		ε_b %	
	mean	std. dev.	mean	std. dev
DirectSteel 20				
X direction	535	24	1.12	0.20
XY direction	513	37	0.99	0.44
Y direction	512	31	0.92	0.22
DirectSteel H20				
X direction	1272	43	2.62	0.15
XY direction	1280	58	2.53	0.05
Y direction	1382	68	2.94	0.16

back-scattered electrons, with the aid of semi-quantitative EDS micro-analysis.

3 RESULTS AND DISCUSSION

3.1 *Tensile tests results*

Table 3 shows the values of tensile strength at break (UTS) and percent elongation at break (ε_b) measured on DirectSteel 20 and DirectSteel H20 specimens, for the different orientations in the XY plane. The low values of standard deviation prove a good repeatability of the results.

Thanks to a software tool for statistical analysis, the t-test has been carried out on the data, to evaluate the presence of significant differences in the mechanical

Table 4. P-values resulting from the t-test for the variables UTS and ε_b among the groups X, Y and XY; for DirectSteel 20 and DirectSteel H20.

	UTS	ε_b
DirectSteel 20		
X vs Y	0.653	0.469
X vs XY	0.985	0.555
Y vs XY	0.584	0.152
DirectSteel H20		
X vs Y	0.869	0.381
X vs XY	0.078	0.071
Y vs XY	0.118	0.014

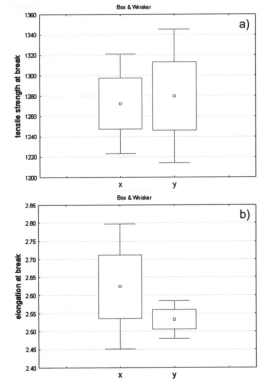

Figure 2. Box & whiskers diagrams resulting from the t-test comparing tensile strength (a) and elongation at break (b) between X and Y specimens, for DirectSteel H20.

properties of the specimens produced with the various orientations. Table 4 reports the p-values calculated for the variables UTS and ε_b between the groups X, Y and XY.

Assuming a level of significance of 0.01, the statistical analysis does not allow to reject the hypothesis that no significant differences exist in the mechanical characteristics among the various groups. For an easier interpretation of the results, in Figure 2 the box &

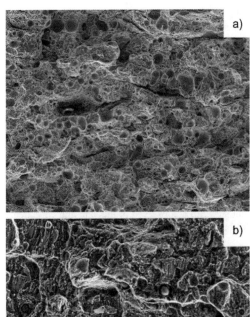

Figure 3. Rupture surface of an X direction specimen of: a) DirectSteel 20 and b) DirectSteel H20.

whiskers diagrams are shown, comparing tensile strength and elongation at break between X and Y directions for DirectSteel H20.

Thus, the mechanical response appears isotropic within the plane of powder deposition XY, for both the investigated materials.

3.2 Failure mechanisms observation

In Figure 3 the rupture surfaces of X direction specimens produced with the two materials can be compared.

The rupture morphologies appear considerably different. For DirectSteel 20 (Fig. 3a) failure occurs mainly by micro-voids coalescence in the matrix, with evident cavities preexistent or due to particle detachment. In DirectSteel H20 (Fig. 3b) a ductile failure mode can still be observed, but cavities are almost absent. This result is consistent with the much higher density referred to the most recent material formulation.

Figure 4 shows in detail the rupture surface of a DirectSteel 20 specimen, where particle disconnection and failure of the dendritic matrix can be appreciated.

Previous studies (Bassoli et al. 2004) revealed, for DirectSteel 20, that steel particles are immersed in a low-melting phase with a strongly dendritic structure, with a good wettability ensured by the presence of copper, nichelius and phosphorus (Fig. 5a). This evidence supported the hypothesis that the consolidation modalities, based on the particles' rearrangement and densification under the effect of hydrodynamic and capillary forces in the liquid phase, smooth the anisotropy due to laser tool-path.

As to DirectSteel H20, EDS semi-quantitative analysis on the matrix, shown in Figure 5b, reveals only the presence of Iron, Nichelius and Chromium.

Regarding the failure mechanisms of the most recent material, at a higher level of detail two distinct morphologies can be observed in the skin and core areas, the former having a thickness of about 0.2 mm. Figure 6

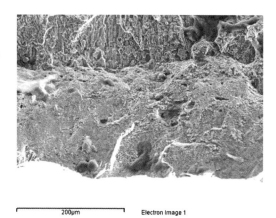

Figure 6.   Rupture surface of an X direction specimen of DirectSteel H20: transition zone between the skin and the core area.

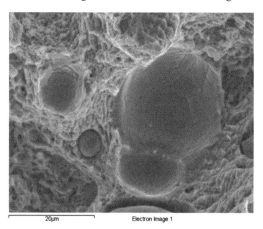

Figure 4.   Rupture surface of an X direction specimen of DirectSteel 20.

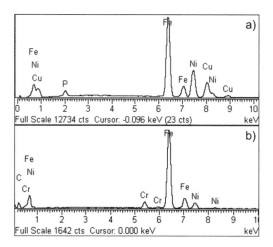

Figure 5.   Spectra obtained by EDS semi-quantitative analysis on the matrix of: a) DirectSteel 20 and b) DirectSteel H20.

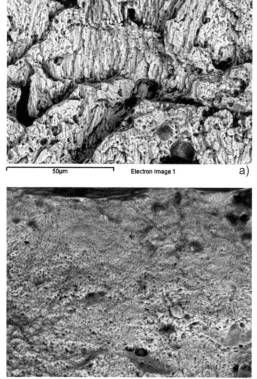

Figure 7.   Rupture surface of an X direction specimen of DirectSteel H20: a) core and b) skin.

allows to appreciate the transition zone between the two zones, which are shown separately in Figure 7 a and b. The outer layer evidences a much finer structure than the inner area, with failure mechanisms that are considerably different.

The effect of the two distinct areas in the sections of DirectSteel H20 specimens on the macroscopic mechanical characteristic still needs to be investigated, as well the consolidation mechanisms leading to the observed structure.

4   CONCLUSIONS

An experimental investigation has been carried out on metal parts produced by the liquid-phase selective laser sintering process. Mechanical performances have been evaluated for the steel-based material DirectSteel 20 and for the innovative formulation DirectSteel H20. SEM observation of the rupture surfaces allowed to combine the macroscopic characterization with the investigation of joining phenomena and failure mechanisms at a microscopic level.

The proposed experimental approach allowed to investigate eventual anisotropies in the mechanical response due to the construction orientation. The statistical analysis of the results proved that no considerable differences can be noticed in the mechanical performances between samples produced in the X, Y and XY direction within the plane of powder deposition.

For DirectSteel 20 material, the results are consistent with an explanatory model of the consolidation phenomena, based on the particles' rearrangement and densification under the effect of hydrodynamic and capillary forces generated in the liquid phase. Thus, the anisotropy due to laser tool-path can be smoothed.

For the newest material formulation, a ductile failure mode has been observed, with different morphologies in the skin and core areas. A very high level of densification can be noticed, where cavities are almost absent. This result is consistent with the data reported by the producer. For DirectSteel H20, the mechanisms of densification and there influence on the mechanical performances and failure means still need to be investigated, but yet no significant anisotropy can be noticed in the mechanical response.

Thus, it can be stated that the considered process produces parts with isotropic characteristics within the XY plane, for any of the two investigated materials. The relevance of the exposed result can be easily understood, in terms of definition of the construction strategy. There is indeed good evidence for the considered technology to benefit from an advanced know-how regarding materials and process development.

On the whole, the mechanical performances were found to be very close to the values stated by the producer, with an excellent repeatability of the results.

Interesting future developments will regard a better understanding of the phenomena occurring during the consolidation process, in particular for the newest metal mixture.

REFERENCES

Agarwala, M. et al. 1995. Direct selective laser sintering of metals. *Rapid Prototyping J.* 1 (1): 26–36.

Anestiev, L.A. & Froyen, L. 1999. Model of the primary rearrangement processes at liquid phase sintering and selective laser sintering due to biparticle interactions. *J. of Applied Physics* 86 (7): 4008–4017.

Bassoli, E. et al. 2004. Direct Laser Sintering of Metal Parts: Characterisation and Evaluation of Joining Mechanisms. *Proc. of 2004 MRS Fall Symposium, vol. 860E, Boston, November 29–December 3, 2004.*

Gatto, A. & Iuliano, L. 1998. Comparison of rapid tooling techniques for moulds and dies fabrication. *Proc. of Time Compression Technology, Nottingham, 13–14 October 1998.*

Gatto, A. & Iuliano, L. 2000. Micro joining mechanisms between metal particles in the SLS technique. *Proc. 9th European Conference On Rapid Prototyping & Manufacturing, Athens, 17–19 July 2000.*

Gatto, A. & Iuliano, L. 2001. Evaluation of Inserts for Modular Thermoplastic Injection Moulds Produced by Spin Casting. *J. of Materials Proc. Tech* 118 (1–3): 411–416.

Iuliano, L. et al. 2003. Selective Laser Sintering Of Metal Parts: Comparison Of Two Material Systems. *Proc. of 6th A.I.Te.M. Conference, Gaeta, 8–10 September 2003.*

Kathuria, Y.P. 1999. Microstructuring by selective laser sintering of metallic powder. *Surface & Coatings Technology* 116–119: 643–647.

Khaing, M.W. et al. 2001. Direct Metal Laser Sintering for rapid tooling: processing and characterisation of EOS parts. *J. of Materials Proc. Technology* 113: 269–272.

Kruth, J.P. et al. 2003. Lasers and materials in selective laser sintering. *Assembly Automation*, 23 (4): 357–371.

Morgan, R.H. et al. 2002. High density net shape components by direct laser re-melting of single-phase powders. *J. of Mat. Science* 37: 3093–3100.

Niu, H.J. & Chang, I.T.H. 1998. Liquid Phase Sintering of M3/2 High Speed Steel by Selective Laser Sintering. *Scripta Materialia* 39 (1): 67–72.

Niu, H.J. & Chang, I.T.H. 1999a. Selective Laser Sintering of Gas And Water Atomized High Speed Steel Powders. *Scripta Materialia* 41 (1): 25–30.

Niu, H.J. & Chang, I.T.H. 1999b. Instability Of Scan Tracks Of Selective Laser Sintering Of High Speed Steel Powder. *Scripta Materialia* 41 (11): 1229–1234.

Simchi, F. et al. 2003. On the development of direct metal laser sintering for rapid tooling. *J. of Materials Proc. Technology* 141: 319–328.

Standard Test Methods for Tension Testing of Metallic Materials – Metric, ASTM E8M-01e1, vol. 03.01.

Tolochko, N. et al. 2000. Absorptance of powder materials suitable for laser sintering. *Rapid Prototyping J.* 6 (3): 155–160.

Tolochko, N. et al. 2003. Selective laser sintering of single- and two-component metal powders. *Rapid Prototyping J.* 9 (2): 68–78.

Virtual modeling and rapid manufacturing – Bártolo (eds)
© 2005 Taylor & Francis Group, London, ISBN 0 415 39062 1

Laser deposition of pure Ti powder onto a cast CoCrMo substrate using Laser Engineered Net Shaping (LENS®)

A.R. Smith & B.E. Stucker
Utah State University, Logan, Utah

ABSTRACT: An experimental study to determine the feasibility of depositing pure Ti powder onto a cast CoCrMo substrate using Laser Engineered Net Shaping was performed. Excess mixing of Ti and CoCrMo at high temperatures can cause interlayer cracking and delamination upon cooling of the Ti from the surface of the CoCrMo substrate. This is most likely due to brittle phase formation and resultant failure due to residual stresses. The LENS process was proposed as a viable solution to this problem, as it produces a minimal heat affected zone compared to other deposition processes, resulting in less melting of the substrate and, therefore, little mixing of the elements involved. Under standard operating conditions one deposited layer of Ti approximately 0.010–0.015 inch thick showed cracking and delamination. LENS parameters were varied to attempt to alleviate residual stresses and mixing. Location of the laser focus with respect to the substrate, traverse speed of the laser, flow rate of the Ti powder, and deposition method were all varied. With the addition of a second layer of titanium and an appropriate use of process parameters, resulting in a total deposit approximately 0.025–0.030 inch thick, interlayer cracking was minimized and delamination did not occur. A possible explanation for success is stress relief of the transition region due to the deposition of the second layer. Further work will be undertaken to investigate the microstructure of the transition region between the Ti and the CoCrMo to help determine how cracks were reduced with 2 layer depositions.

1 INTRODUCTION

1.1 Project description

The overall goal of this project is to successfully join Ti to medical grade CoCrMo. The Laser Engineered Net Shaping (LENS) process was chosen as a likely successful candidate process. The initial objective was to deposit an approximately 0.030 inch thick layer of Ti onto a CoCrMo substrate with minimal cracks and porosity in the Ti deposit, the transition region between the Ti and CoCrMo, and the CoCrMo mushroom substrate (see Figure 1). The least crack-prone deposition settings are to be used in future tensile tests to determine the strength of the bond between the Ti and CoCrMo.

1.2 LENS technology

LENS is an excellent process for the laser deposition of metal powders onto a substrate due to its design. The Optomec LENS machine at Utah State University is a 3-axis deposition system equipped with a 500 W Nd:YAG laser and a powder feeder that allows the powder to be fed directly into the laser focus area. The deposition apparatus is enclosed in an airtight, argon environment that can be purged down to below 5 ppm of oxygen. This low oxygen content is necessary for

Figure 1. CoCrMo tensile mushroom.

the deposition of Ti since titanium oxidizes readily at elevated temperatures.

The LENS machine is controlled by means of software that converts a 3D STL file into a tool path to deposit the metal powder onto the substrate. The tool paths are composed of a series of lines that are scanned by the laser as the powder is fed into the melt pool created at the substrate surface due to the laser. The melted metal from each deposit slightly overlaps, allowing for a relatively uniform thickness of deposit.

1.3 Prior research

Research on laser welding and cladding is abundant. Although a wide range of materials have been studied using LENS by many groups of investigators, including CoCrMo (Stucker, Esplin and Justin, 2004) and Ti (Kobryn and Semiatin, 2001), little is known about the effects of laser processing on resultant microstructure and properties when laser depositing Ti on CoCrMo. Therefore, this study became necessary to ascertain the feasibility of joining Ti to CoCrMo.

An important issue regarding the strength of the bond between Ti and CoCrMo is the presence of residual stresses after laser processing. Two articles of particular interest were identified. Rangaswami et al. (2003) used electron diffraction to map residual stresses in a rectangular sample deposited by LENS. Stresses were found to be tensile and maximum at the edges and compressive at the center of the deposit. Dai & Shaw (2002) used finite element modeling to determine residual stresses in deposited samples. Stresses were also found to be greatest at the edges of rectangular samples, but scanning pattern had a large effect on the magnitude of the residual stress. These findings are important to the deposition of Ti onto CoCrMo because the nature of the mixing of the two materials leads to brittle phase formation, which could crack under high residual stress.

1.4 Mixing of Ti and CoCrMo

The primary concern when laser depositing Ti onto CoCrMo is that when the cast CoCrMo is heated under laser energy and comes in contact with molten Ti there are many undesirable microstructural combinations that can occur in the transition region between the two materials.

Phase diagrams for the combination of Co and Cr with Ti are shown in Figure 2. As can be seen from the figure, there are several different intermetallic phases which can form based on the weight fractions present. Most of these are brittle ceramic compounds. The presence of residual stresses with brittle phases may lead to premature failure in the presence of stress concentrations (cracks). Under these circumstances delamination of the deposited Ti from the CoCrMo will occur, thus rendering the deposit useless.

2 METHODOLOGY

2.1 Experimental variables

A number of experiments were conducted to attempt to minimize thermal stresses and inhibit mixing of Ti

Co-Ti

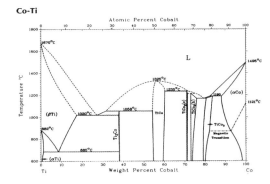

Cr-Ti

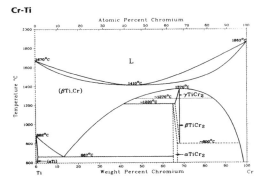

Figure 2. Top: phase diagram of cobalt and titanium. Bottom: phase diagram of chromium and titanium (ASM International 1992).

and CoCrMo. The variables that were considered for this purpose were:

1) Laser focus height
2) Scan direction and scan pattern
3) Number of layers of material deposited
4) Powder feed rate (PFR)
5) Laser Traverse Speed (LTS)
6) Laser Power

The laser focus height is defined as the height above the substrate at which the laser beam is focused to its maximum power density. Generally, depositions are conducted with the laser focus "buried" into the substrate about 0.050 in. This causes melting of the substrate as well as the powder material being deposited and will typically result in a stronger weld. Since the mixing of the powder and the substrate in this case is undesirable, this parameter was considered a variable and z height was varied from 0.05 in above the substrate to 0.10 in below the surface of the substrate.

Different scan types were used for this series of depositions. In three of them the scan direction is the variable; these are vertical, horizontal, and 45 degree. The other types are referred to as spiral and raster. Spiral uses a combination of horizontal and vertical

324

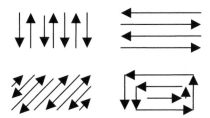

Figure 3. Top Left: vertical scanning pattern. Top Right: horizontal scanning pattern. Bottom Left: 45 degree scanning pattern. Bottom Right: spiral scanning pattern.

scanning. Raster follows either the horizontal or vertical, except every other line is skipped in the first pass and are filled in by the subsequent pass. All scan types are illustrated in Figure 3, where the lines represent the location of the laser focus and the arrow indicates the direction of the laser scan.

The number of layers deposited has an effect on the thermal stresses that build up in the deposit. Thermal stresses due to rapid cooling of the deposits can be large; however, subsequent layers of deposit may have the effect of annealing the previous layers, relieving these stresses to some degree.

The powder feed rate, laser traverse speed, and laser power are all interrelated. For an increase in feed rate, the number of powder particles the laser must melt increases, therefore reducing the laser power available to form a melt pool on the substrate. Similarly, as the scan speed increases, the laser energy dwells over a location for less time and as laser power is decreased there is less total energy input, thus a smaller melt pool is created. Since these parameters are interrelated, the laser power and the powder feed rate are generally held constant while the powder feed rate varies.

2.2 Composition and details

The composition of the cast CoCrMo mushrooms is, by mass percent, 59% Co, 30% Cr, and a balance of Mo and other elements. The diameter of the deposition and the deposition surface is 0.875 in. The powder used was a $-100/+325$ mesh size, gas atomized, commercially pure Ti.

3 RESULTS

Delamination, or significant cracking, occurred in all samples where the deposit was only one layer thick, resulting in approximately 0.011–0.015 in of Ti. This meant that the deposit could be removed with only a small amount of force. Large cracks could be seen in the mixing region closest to the CoCrMo mushroom, see Figure 4. These cracks were visible throughout

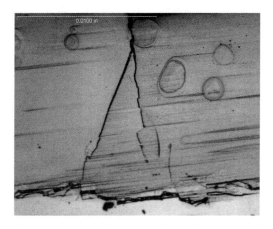

Figure 4. Large cracks in the mixing region between Ti deposit and CoCrMo mushroom.

the entire cross-section of the deposit. The reason for the cracking is assumed to be due to the residual stresses caused by the transient thermal effects of the melted material as it rapidly solidifies.

For a deposit of a single layer, some parameters performed better than others. Burying the laser deep into the substrate, 0.01 in below the surface, caused significant mixing of the Ti and CoCrMo as expected and delamination readily occurred, regardless of the other variables. However, focusing the laser at or above the surface of the substrate in the hopes of creating little mixing did not fully alleviate this problem. Relatively little mixing of the materials was noted, but delamination continued to occur. Better bonding (less cracks) appeared to be obtained with the laser buried into the substrate 0.05 in. This is consistent with prior experience with materials that bond well, but was not necessarily expected with Ti and CoCrMo. This effect may be due to the mass percentages of Ti and CoCrMo that mix during laser processing, creating a more ductile alloy, but the weight fractions of the constituents in the transition region have not yet been investigated and this is simply conjecture.

Scanning direction did not seem to have any positive affect on a deposition of a single layer. Complete delamination only occurred with the raster pattern, which proved to be the worst deposition pattern. Significant cracking in the mixing region was present in all other depositions, including the spiral pattern, which was expected to be the best pattern. In a single layer, no visible difference was noted in the depositions using horizontal, vertical, 45 degree, or spiral patterns.

The effects visible in a deposition of a single layer were somewhat alleviated as subsequent layers were added. For a deposit with two layers totaling 0.020–0.025 in of Ti, cracks can still be seen, but are

325

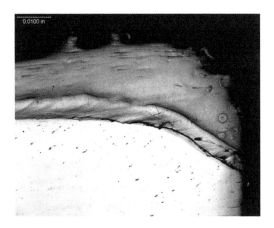

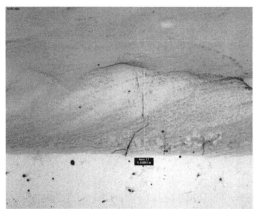

Figure 5. Localized cracking only at edges under 5X magnification with two layers of Ti deposited.

Figure 6. Three layer deposit under 20X magnification. Only microcracks can be seen.

much more localized, see Figure 5. This suggests some alleviation of the thermal stresses.

The only scanning patterns used for depositions above one layer were the horizontal and vertical patterns, due to time constraints in the research. The first layer was deposited in a vertical pattern, the second was horizontal, and in three layer depositions, the third layer was again vertical. It is possible that the spiral pattern could be used to alleviate stresses in multi-layer deposits. If the strength of the bonds is not adequate in the scheduled tensile tests, the spiral pattern may help to alleviate thermal stresses and allow for greater strength.

A problem that arose aside from cracking in multi-layer deposits is porosity between the layers. Bonding of the layers was visibly good, except for this porosity. This was remedied by lowering the powder feed rate between layers. This lowering of powder feed rate did not significantly affect the thickness of the second layer.

The best deposit parameters, based on optical microscopy observations of minimal of cracking and porosity, were used subsequently to make the depositions for tensile testing, see Figure 6. The depositions were three layers thick, resulting in 0.025–0.030 in of Ti. The parameters for the final depositions are shown in Table 1. For these deposits, the feed rate of Ti powder was reduced for each subsequent layer, resulting in an amount small enough on the third layer that probably the most prominent effect of adding the third layer was additional annealing of the mixing region. Due to the low powder flow rate, much of the powder entrapped in the molten pool during the third layer of depositions was powder left clinging to the surface of the previous deposit, rather than newly delivered powder. The laser traverse speed and laser power remained constant from the earlier. Twenty-five

Table 1. LENS parameters for the tensile test deposition.

| Parameter | Value | | |
	Layer 1	Layer 2	Layer 3
z height (in)	−0.050	−0.050	−0.050
Scan type	Vertical	Horizontal	Vertical
PFR (g/min)	2.63	1.74	0.77
LTS (in/min)	20	20	20
Power (W)	173	173	173

mushrooms were finished with the parameters listed in Table 1 and were sent off for tensile testing.

4 CONCLUSIONS

The laser deposition of Ti onto CoCrMo mushrooms proved to be a difficult task based on the residual stresses from the laser process and the undesirable mixing characteristics of the materials involved. Residual stresses in a single layer deposit proved to be too large and delamination occurred. Upon deposition of subsequent layers, the thermal stresses appeared to be relieved to some degree, limiting cracking to localized areas and strengthening the samples. Tensile testing of the samples is pending to determine values for the ultimate tensile strength of the laser deposited Ti onto CoCrMo.

REFERENCES

ASM International 1992. ASM handbook: Alloy phase diagrams.

Dai, K. and Shaw, L. 2002. Distortion minimization of laser-processes components through control of laser scanning patterns. *Rapid Prototyping Journal.* 8(5): 270–276.

Kobryn, P. A., and Semiatin, S.L. 2001. Mechanical properties of laser-deposited Ti-6Al-4V. *Solid Freeform Fabrication Proceedings*, edited by D.L. Bourell, J.J. Beaman, R.H. Crawford, H.L. Marcus, J.W. Barlow, and K.L. Wood. August, 2001, Austin, TX.

Rangaswami, P. et al. 2003. Residual stresses in components formed by the laser engineered net shaping (LENS) process. *Journal of Strain Analysis.* 38(6): 519–527.

Stucker, B.E., Esplin, C.A., and Justin, D.J. 2004. An Investigation of LENS-Deposited Medical-Grade CoCrMo Alloys. *Solid Freeform Fabrication Proceedings*, edited by D.L. Bourell, J.J. Beaman, R.H. Crawford, H.L Marcus, J.W. Barlow, and K.L. Wood. August, 2004, Austin, TX.

Virtual modeling and rapid manufacturing – Bártolo (eds)
© 2005 Taylor & Francis Group, London, ISBN 0 415 39062 1

Roughness after laser melting of iron based powders

M. Rombouts & L. Froyen
Department of Metallurgy and Materials Engineering, Katholieke Universiteit Leuven, Leuven, Belgium

D. Bourell
Texas Materials Institute, Laboratory for Freeform Fabrication, The University of Texas, Austin, Texas, USA

J.P. Kruth
Department of Mechanical Engineering, Katholieke Universiteit Leuven, Leuven, Belgium

ABSTRACT: This paper studies the influence of process and material parameters on the behavior of the melt pool during laser melting of iron based powders. The investigated process parameters are the laser power, scan spacing, scan speed and preheating temperature. Design of experiments reveals that all the parameters, except the laser power, have a significant effect on the roughness. The first order interaction effects scan spacing – scan speed and scan spacing – preheating temperature are also significant. These results combined with visual inspection of the solidified tracks provide an understanding in the flow behavior of the melt during processing. The effect of material parameters is investigated by using different iron based powders containing carbon, copper, silicon and phosphorous. These elements alter physical properties including surface tension, surface tension gradient, viscosity, melting point, solidification range,... These changes are thought to be the main cause for the different morphology of the scan tracks.

1 INTRODUCTION

Selective laser melting is a solid freeform fabrication process whereby a three dimensional part is built layer wise by laser scanning a powder bed. Full melting of the powder can lead to parts with a high density (>95% relative density) at optimized parameters. Study of the effect of process parameters on the density of iron parts has been performed in Simchi (2004), Simchi & Pohl (2003) and Dingal et al. (2004). The roughness is, along with the density, a part property that needs to be well controlled. The present study aims at obtaining a better understanding in the factors influencing the roughness after laser melting of powder beds. Firstly, Design of Experiments (DOE) was performed to investigate the effect of the laser power, scan spacing, scan speed and powder bed preheating temperature on the roughness of iron layers. Secondly, the effect of alloying elements such as copper, phosphorus, silicon and carbon on the roughness of single layers of iron is studied. Several studies have been performed that indicate a significant effect of carbon on the quality of iron parts (Simchi & Pohl (2004), Xie et al. (2003), Rombouts et al. (2004)).

2 EXPERIMENTAL

The first series of tests aimed at investigating the effect of process parameters on the roughness of iron layers. These tests were performed using a research machine built at the University of Texas at Austin. It uses a Nd:YAG laser having a maximum output power of 250 W in continuous wave, a wavelength of 1.064 μm and a beam diameter of 0.5 mm. The tests were performed in a $N_2 + 4\%H_2$ environment and at an absolute pressure of 0.65 atmosphere. Molybdenum heating wires were located around the building platform that enable preheating of the powder bed up to 800°C. Rectangular single layers 10 mm wide and 20 mm long were produced on a 15 mm thick powder bed. The laser scanning direction was parallel to the width of the layer. Spherical Fe powder (99.5%) with a grain size smaller than 10 μm was used. A full factorial design of experiments with 3 replicates and 4 factors having 2 levels was performed. Table 1 represents the factors and the corresponding values at low and high levels. The velocity range was limited to 200 mm/s since higher velocities resulted in very fragile parts, making roughness testing impossible. Velocities lower

Table 1. Factors with their values at low and high level used in the design of experiments.

Factors	Low level (−1)	High level (+1)
Laser power (A)	100 W	250 W
Scan spacing (B)	0.25 mm	0.5 mm
Scan speed (C)	50 mm/s	200 mm/s
Bed temperature (D)	25°C	400°C

than 50 mm/s and scan spacings lower than 0.25 mm were not considered because they led to excessive build times for commercial applications. At preheating temperatures higher than 400°C, significant caking of the powder bed due to solid state sintering was observed. The influence of the factors was evaluated by measuring the roughness of the parts perpendicular to the scanning direction. The measurements were performed on the last portion of the layer since the beginning showed a significant higher roughness due to the 'first line scan balling' instability (Das (2001)). This region captured a large part of the surface at high scan spacings. The roughness was measured by a Dektak 3 surface profile measuring system having a diamond stylus with a tip radius of 12.5 μm. A total evaluation length of 40 mm and a cut off length of 8 mm was used. The Center Line Average (R_a) was used. The significance of the effects was evaluated using analysis of variance (ANOVA).

The second series of tests aimed at investigating the effect of alloying elements on the roughness. These tests were performed using a research machine built at the Katholieke Universiteit Leuven. The laser is the same as in the first series of tests except for the maximal output power which is only 100 W and the beam spot which is about 0.7 mm at 100 W. All tests were performed at 100 W and at velocities of 10, 50 and 200 mm/s. The scan spacings were 0.25 mm and 0.5 mm. The atmosphere was N_2 and tests were performed at an absolute pressure of 1 atmosphere. The main constituent of the powder mixtures was gas atomized iron powder. This powder has been mixed with other gas atomized powders: pure copper, prealloyed Fe-2.5 wt% C and prealloyed $Fe_{3+x}P$. All the gas atomized powders were sieved below 63 μm. Silicon powder with a grain size smaller than 8 μm has also been mixed with the iron powder.

3 RESULTS AND DISCUSSION

3.1 *Effect of process parameters on Fe layer*

Analysis of variance on the effects obtained by DOE was performed (Table 2). The calculated F-values were compared with the critical F-value at a confidence interval of 90%. From this could be concluded

Table 2. ANOVA table for the effect of factors and first order interaction effects using Fe powder. The bold numbers correspond to significant effects at a confidence interval of 90% (critical F-value = $F_{1,32}$ (90% C.I.) = 2.87). The last column contains the value of the effects. SS = Sum of squares; DOF = degrees of freedom; MS = mean square; A = laser power; B = scan spacing; C = scan speed; D = preheating temperature.

Source	SS	DOF	MS	F-value	Effect
Mean (μm)					35.2
A	1.9	1	1.9	0.04	−0.39
B	**578.4**	1	**578.4**	**10.97**	**6.94**
AB	2.2	1	2.2	0.04	−4.30
C	**221.8**	1	**221.8**	**4.21**	**−5.27**
AC	101.0	1	101.0	1.92	0.43
BC	**171.1**	1	**171.1**	**3.25**	**2.90**
D	**333.1**	1	**333.1**	**6.32**	**−3.78**
AD	4.9	1	4.9	0.09	0.64
BD	**203.7**	1	**203.7**	**3.86**	**−4.12**
CD	36.5	1	36.5	0.69	1.74
error	1686.4	32	52.7		

that a low scan spacing (B), high scan speed (C) and high preheating temperature (D) led to a significant decrease in roughness. The first order interaction effects scan spacing-scan speed (BC) and scan spacing-bed temperature (BD) were also significant. The laser power was the only factor that did not have any significant influence. The last column of Table II contains the values of the effects. Visual inspection of the surfaces was performed in order to explain these results in terms of the behavior of the melt pool as a function of the process parameters.

From DOE it followed that a reduction in scan spacing led to the largest decrease in roughness. The beneficial effect of a small scan spacing has also been observed by others (Das (2001), Song (1997)). The effect was larger at low scan speeds as indicated by the positive first order interaction effect scan spacing – scan speed BC (Table 2). Figure 1 shows that without preheating the powder, the surfaces of the layers have different characteristics depending on the scan spacing and speed. The surfaces scanned at small scan spacings have a waviness with a period equal to twice the scan spacing in a direction perpendicular to the scan direction (Fig. 1). This waviness was caused by the high thermal interaction between consecutive scan tracks. The first passage of the laser created a molten pool that attracted powder particles lying next to it. The molten pool was enlarged by the second passage of the laser, which overlapped largely with the molten pool created by the first passage. More neighboring particles were attracted to the melt pool and a zone deficient in powder was created at the position of the third scan track. Consequently, only a small molten pool was formed during the third scan track

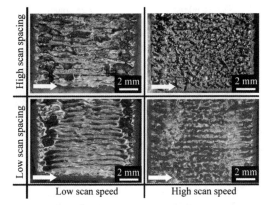

Figure 1. Top surface of iron layers processed at a power of 250 W, without preheating and high/low scan spacing and low/high scan speed.

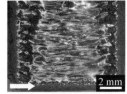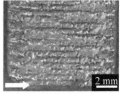

Low scan speed   High scan speed

Figure 2. Surface after scanning at 250 W, with preheating at 400°C, high scan spacing and low/high scan speed.

resulting in a depression after solidification. Then everything started over again. This mechanism was independent of the scan speed but caused deeper depressions at low velocities due to the larger melt pool formed at higher energy input.

It is obvious that the behavior of the molten pool is different when there is only a small thermal interaction between consecutive scan lines as when using a high scan spacing and no preheating. The behavior is similar as when scanning a single line in a loose powder bed. When scanning a single line balling takes place at high scan velocities while at low velocities an elongated track is formed. The physical phenomenon controlling the instability of the molten track is dependent on the material properties and process parameters (Niu & Chang (1999)). The instability is mainly controlled by Marangoni convection when processing a material with positive surface tension gradient $\delta\gamma/\delta T$, as for instance iron with high oxygen content, and when a large thermal gradient is present (Niu & Chang (1999)). In such materials the direction of Marangoni flow is from low to high temperature and thus from the edge to the middle of the scan track (Mills et al. (1998)). The magnitude of the force exerted by this flow on the liquid is larger at higher values of the surface tension gradient and thermal gradient. The latter implies that instability is more likely to occur at lower preheating temperatures and higher scan velocities. In iron with low oxygen contents the Marangoni flow is radially outward and this kind of instability does not occur. There are no data available concerning the effect of N and H, present in the atmosphere, on the surface tension gradient of liquid iron.

Another instability observed in liquid cylinders is the Rayleigh instability: the cylinder breaks up in order to lower its surface tension if the wavelength λ of a sinusoidal perturbation exceeds the circumference πD of

the cylinder (Coriell et al. (1977), Chandrasekhar (1970)). This might lead to instability of the molten pool at certain process parameters due to the varying shape of the molten pool. Especially higher scan speeds result in a rapid decrease in D and thus less stable melt pools (Niu & Chang (1998)).

In the single layers scanned at low speed and high scan spacing, tracks extending over a large distance along the scanning direction could be distinguished. This is in correlation with the situation when scanning a line. These tracks were interrupted at some places, especially near the edges of the layer. This led in conjunction with the deep depressions between neighboring tracks to a high roughness. At high speed balling occurred and no elongated tracks were formed as in the situation when scanning a single line. Balling did not have a large impact on the measured roughness since the roughness was not measured along but perpendicular to the scanning direction. This was together with the larger amount of melting at low velocities the main cause for the higher roughness at lower velocities. At small scan spacings elongated tracks were formed at high speed since the molten pool was then bounded on one side by a solid of its own material, which resulted in a reduced driving force for breaking up.

Another factor that had a significant influence on the surface roughness and morphology is the preheating temperature. The sign of the effects in Table 2 indicate that a high preheating temperature reduced the roughness, especially at high scan spacings. Figure 2 shows the surfaces processed with preheating and high scan spacing. There is a large difference with the surfaces obtained at the same process parameters but without preheating (Figs. 1, 2). No discontinuous molten tracks were formed at high speed. This can be caused on the one hand by the smaller thermal gradient decreasing Marangoni convection in the melt. On the other hand, preheating leads to a broader melt pool suppressing Rayleigh instability. Due to the broader scan tracks an interconnection between neighboring scan lines was formed at low speed and high scan spacing resulting in a smaller surface roughness than without preheating. The surfaces processed with preheating and high scan spacing were similar to the ones produced without preheating and small scan spacing. This is in correlation with the observation that small scan spacings have the effect of preheating the powder in front of the laser beam (Das (2001)). This interaction was also revealed by the results of DOE where a negative inter-action effect preheating temperature – scan spacing was found.

Important to notice is that there was no interaction effect between the laser power and preheating tem-perature in the investigated process window. The input of a homogeneously distributed energy had thus a different effect on the roughness than the input of a localized one. This is caused by the dissimilar effect of these inputs on the solidification time, melt pool dimensions and thermal gradient.

3.2 *Effect of alloying elements*

In this section the effect of alloying elements on the surface morphology of a single layer is investigated.

Following powders were tested: pure Fe, Fe + 10 wt% Cu, Fe + 30 wt% Cu, Fe + prealloyed Fe-C (total wt% C: 2), Fe + prealloyed $Fe_{3+x}P$ (total wt% P: 0.8, 4), Fe + 1 wt% Si, Fe + 10 wt% Si. The mix-tures containing 2 wt% C, 10 wt% Si and 4 wt% P have about the same liquidus temperature (1400°C).

Single lines were scanned at 100 W and velocities ranging from 10 to 200 mm/s to obtain insight in the stability of the melt pools (Fig. 3). The Fe powder showed no balling at low speed while it did at high speed, as observed in section 3.1. The tracks were continuous in the entire velocity range for the powders containing Cu, C and 4 wt% P. The molten track broke up in elongated cylinders at velocities equal to 25 mm/s when only 0.8 wt% P was added. At the other velocities continuous tracks were formed for the Fe – 0.8 wt% P powder. The Fe – Si powders showed balling at all velocities.

Single layers were scanned at the same parameters and with a scan spacing of 0.25 and 0.5 mm. Compari-son of the layers processed at a speed of 200 mm/s was difficult since most of the layers were very fragile. Figure 4 shows the surfaces scanned at a speed of 50 mm/s and a scan spacing of 0.25 mm. The Fe lay-ers contained at intermediate energy irregular ripples in the direction perpendicular to the scan direction (Fig. 4 (a)). The Fe – Cu powders produced the smoothest layers. Continuous tracks at a distance equal to the scan spacing from each other could be distinguished on all layers. The Fe – 2 wt% C powder formed at higher energy a waved surface with holes at the positions where the melt had penetrated the pow-der bed underneath. Individual scan tracks could not be distinguished at high energy. The layers made of Fe – 4 wt% P powder were similar to the ones produced with Fe – 2 wt% C powder. The surface was less smooth for the Fe – 0.8 wt% P powder. The scan lines

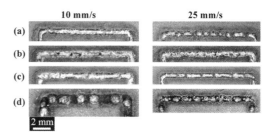

Figure 3. Single lines scanned at 100 W and at a scan speed of 10/25 mm/s of (a) Fe, (b) Fe – 10 wt% Cu, (c) Fe – 0.8 wt% P and (d) Fe – 1 wt% Si.

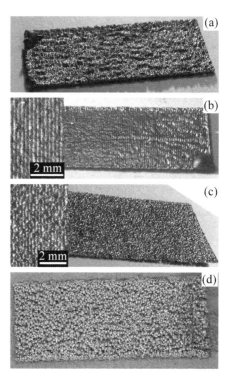

Figure 4. Single layers scanned at 100 W, 50 mm/s and a scan spacing of 0.25 mm using (a) Fe, (b) Fe – 10 wt% Cu, (c) Fe – 0.8 wt% P and (d) Fe – 1% Si. The laser scans vertically.

could be distinguished but they were not continuous throughout the whole length of the scan, as observed when scanning a single line (Figs. 3, 4). When using the Fe – Si powders individual balls without intercon-nection were formed at all parameters. The balls had a larger size at higher energy input and reached a diameter of 16 times the scan spacing at the lowest speed and smallest scan spacing.

The correlation between these results and the mate-rial properties was investigated. Table 3 summarizes

Table 3. Effect of alloying elements on properties of pure Fe % O = % oxygen dissolved in molten iron; γ = surface tension; δγ/δT = surface tension gradient; η = viscosity; L + S = liquidus + solidus region.

Addition	% O	γ	δγ/δT	η	L + S
Oxygen	/	↑↑	Positive	↑	/
10% Cu	≈	↓↓	Unknown	↓	~50°C
30% Cu	≈	↓↓	Unknown	↓	~350°C
2% C	↓	≈	Positive for	↓	~270°C
			T < 1600°C		
0.8% P	↓	↓	Unknown	↓	~60°C
4% P	↓	↓	Unknown	↓	~350°C
1% Si	↓↓	↓	Negative	↓	~0°C
10% Si	↓↓	↓↓	Negative	↓	~50°C

the effect of the alloying elements on the properties that affect the stability and spreading behavior of the iron melt.

3.2.1 Indirect effect due to change in oxygen content

The amount of oxygen dissolved in molten iron is influenced by alloying elements. The main sources for oxygen are the atmosphere, gases adsorbed on the powder surface or contamination of the base material itself. All the investigated alloying elements except copper, reduce the content of dissolved oxygen. Lower amounts of dissolved oxygen lead to lower surface tension gradients. At oxygen contents lower than 0.002 wt% the surface tension gradient becomes negative (Keene (1988)). This leads to a change of the direction of Marangoni convection, as discussed in section 3.1. Lower oxygen contents lead therefore to more stable melt pools. However, a decrease in oxygen content also results in a large increase in surface tension which is about 7500 mN/m per at% O at low oxygen concentrations (Keene (1988)). This favors breaking up of the scan track due to Rayleigh instability (Coriell et al. (1977), Chandrasekhar (1970)). The kinetics of breaking up is also affected by the viscosity (Chandrasekhar (1970)). There is however no unanimity in literature concerning the effect of oxygen on the viscosity of molten iron (Bergquist (1999)). Some do not detect an influence of oxygen while others detect a large increase in viscosity. In this case, the increase did not occur in iron melts containing a small amount of carbon.

3.2.2 Direct effect on Rayleigh instability

Alloying elements can also have a direct effect on the melt pool behavior. Their influence on the stability of the molten pool with regards to Rayleigh instability and on the kinetics of Rayleigh instability will first be considered.

The former is a consequence of a change in dimensions of the molten pool induced by a change in the

heating and cooling behavior. This behavior is controlled by material parameters as laser coupling to the powder bed, thermal conductivity, specific heat, melting point,... From the investigated additions, particularly copper is known to have a large effect on the thermal behavior. This is mainly due to its lower absorption coefficient (Olsen & Flemming (1989)) and higher thermal conductivity in comparison with Fe. This results in a smaller, more equiaxed and consequently more stable molten pool.

On the other hand, the kinetics of breaking up of the melt depends in the first place on the surface tension. All the investigated alloying elements decrease the surface tension of liquid iron, which is at the melting temperature equal to 1865 + 50 mN/m (Keene (1988)). The effect of phosphorus and especially carbon is however very small at the concentrations considered here. The effect of 10 wt% Si is a decrease of about 230 mN/m. The addition of 10 wt% Cu leads to about the same decrease. However these are only small reductions in comparison with the effect of oxygen. Nitrogen, which is present in the atmosphere, results also in a large decrease being 1400 mN/m per at% N at low concentrations.

The second factor affecting the kinetics of Rayleigh instability is the viscosity of the entire melt pool. Carbon decreases the viscosity of molten iron at constant temperature (Barfield & Kitchener (1955)). The viscosity at the bottom of the melt pool is however increased because the viscosity at the melting temperature increases with increasing carbon content (Barfield & Kitchener (1955)). A second factor contributing to a higher viscosity at the bottom is the mushy zone being between the solidus and liquidus temperature. This zone extends over ~270°C at 2 wt% C in case of perfect homogenization upon melting. The viscosity of liquid Fe – P will probably be lower than that of pure Fe (Bunnell et al. (1995)). The addition of P leads also to an increased solidification range being approximately 70°C and 350°C at resp. 0.8 and 4 wt% P. Si reduces the viscosity of molten iron (Secco (1995)). No mushy zone is formed at 1 wt% Si, but at 10 wt% Si the zone extends over approximately 70°C. No data on the viscosity of binary Fe – Cu alloys is found. However, from comparison of the viscosities of the pure elements (Chhabra & Seth (1990)) it is expected that the viscosity of a Fe – Cu alloy will be lower than that of pure Fe. On the other hand a large solidification range, being about 350°C, is formed at weight percentages larger than about 10 wt% Cu. The apparent viscosity of this region increases with decreasing copper content.

3.2.3 Direct effect on Marangoni convection

There is not much information available in literature concerning the surface tension gradient of binary Fe alloys (Keene (1988)). For alloys containing Cu, N and P no reliable data are available. For the Fe – C alloys positive gradients for temperatures up to ~1600°C are reported. The Fe – Si alloys have negative gradients over the entire temperature range at the concentrations used in this study. Marangoni convection also determines the connection between neighboring scan tracks, which is better at negative surface tension gradients (Niu & Chang (1999)).

3.2.4 Effect on spreading behavior

The surface morphology of a layer is affected by the spreading behavior of the melt as well as by the presence of instabilities. The spreading behavior is improved at lower surface tensions and lower viscosities. The overlap between neighboring scan tracks depends also on the thermal behavior (melting temperature, thermal conductivity, laser absorption).

3.2.5 Correlation between results and effect of alloying elements

The irregular ripples present in the Fe layer (Fig. 4) are probably caused by instability due to Marangoni convection. These ripples are not present for the Si, Cu and P containing powders, mainly due to their deoxidizing effect and increased solidification range. The connection between neighboring scan tracks at high scan spacings is worst for the Fe powder. This is caused by the higher instability and higher melting temperature. The decreased tendency for balling in the presence of copper has also been observed in Simchi et al. (2003) and is in correlation with the presence of a mushy zone, lower surface tension and increased thermal conductivity (Table 3). From Table 3, it follows that carbon and phosphorous both have a similar effect on the properties of iron. The increased stability of the melt is in correlation with the presence of a mushy zone at the bottom of the melt and the slight decrease in surface tension, as also noted by Simchi et al. (2001). The latter is based on the assumption that the oxygen content is very low, because otherwise the surface tension would be increased due to the deoxidizing effect of P. At high oxygen contents the decrease in surface tension gradient would be the main cause for the suppression of balling. The same remark applies for the Fe – C powders. The effect of carbon has been investigated in Simchi & Pohl (2004) and Xie et al. (2003). However these results were obtained from 3D parts and can not be compared with the situation of a single layer. The addition of silicon led in the present study to a less stable melt pool (Fig. 3) while in Xie et al. (2003) a minor reduction in roughness of 3D parts was observed upon additions of Si up to 2 wt%. It is not completely clear what causes this decrease in stability. It might be due to a strong deoxidizing effect, which results in an increase in surface tension. This favors in conjunction with the decrease in viscosity the kinetics of Rayleigh instability. Another possibility is

the different thermal behavior caused by exothermic reactions and the different powder size.

4 CONCLUSIONS

1. A smaller scan spacing, higher scan speed and higher preheating temperature led in the investigated process window to a significant lower roughness for layers scanned in a loose powder bed.
2. Cu, P and C led, in contrast to Si, to a more stable melt pool in the investigated concentration range.
3. Since Cu led to an increased stability of the melt pool, the dissolved oxygen content is not thought to be the only factor having a large impact on the melt stability.

REFERENCES

Barfield R.N. & Kitchener J.A. 1955. The viscosity of liquid iron and iron-carbon alloys. *J. of the Iron and steel Institute* 180: 324–329.
Bergquist B. 1999. New insights into the influence of variables of water atomization of iron. *Powder Metall.* 42(4): 331–343.
Bunnell et al., D.E. 1995. Fundamentals of liquid phase sintering during selective laser sintering. *Proc. Solid Freeform Fabrication Symp. 1995*, Texas, 7–9 August 1995: 440–447.
Chandrasekhar S. 1970. *Hydrodynamic and hydromagnetic stability*. Oxford: Clarendon Press.
Chhabra R.P. & Sheth D.K. 1990. Viscosity of molten metals and its temperature dependence. *Zeitschrift fur Metallkunde*. 81(4): 264–271.
Coriell at al., S.R. 1977. Stability of liquid zones. *Journal of Colloid and Interface Science* 60(1): 126–136.
Das S. 2001. On some physical aspects of process control in direct selective laser sintering of metal parts – Part III. *Proc. Solid Freeform Fabrication Symp. 2001*, Texas, 6–8 August 2001: 102–109.
Dingal et al., S. 2004. Experimental investigation of selective laser sintering of iron powder by application of Taguchi method. *Proc. LANE 2004*, Erlangen, 21–24 Sept. 2004: 445–456.
Keene B.J. 1988. Review of the data for the surface tension of iron and its binary alloys, *Int. Mat. Rev.* 33(1): 1–36.
Mills et al., K.C. 1998. Marangoni effects in welding. In Hondros E.D. (ed.), *Marangoni and interfacial phenomena in materials processing*. London: IOM communications London.
Niu H.J. & Chang I.T.H. 1998. Liquid phase sintering of M3/2 high speed steel by selective laser sintering. *Scripta Materialia* 39(1): 67–72.
Niu H.J. & Chang I.T.H. 1999. Instability of scan tracks of selective laser sintering of high speed steel powder. *Scripta Materialia* 41(11): 1229–1234.
Olsen F. & Flemming O. 1989. Theoretical investigations in the fundamental mechanims of high intensity laser light refelectivity. *Proc. of SPIE* 1020: 114–122.

Rombouts et al., M. 2004. Production and properties of dense iron based parts produced by laser melting with plasma formation. *Proc. Euro PM2004* 5, Vienna, 17–21 October 2004: 115–121.

Secco R.A. 1995. Viscosity of the outer core. In Ahrens T. J. (ed.), *Mineral Physics and Crystallography: A Handbook of Physical Constants*. Washington: Amer. Geophys. Union.

Simchi et al., A. 2001. Direct metal laser sintering: material considerations and mechanics of particle bonding. *International Journal of Powder Metallurgy* 37(2): 49–61.

Simchi et al., A. 2003. On the development of direct metal laser sintering for rapid tooling. *Journal of Materials Processing Technology* 141: 319–328.

Simchi A. & Pohl H. 2003. Effects of laser sintering processing parameters on the microstructure and densification of iron powder. *Mat. Sc. and Eng. A* 359: 119–128.

Simchi A. 2004. The role of particle size on the laser sintering of iron powder. *Met. and Mat. Trans. B* 35(5): 937–948.

Simchi A. & Pohl H. 2004. Direct laser sintering of iron-graphite powder mixture. *Mat. Sc. and Eng. A* 383: 191–200.

Song Y.A. 1997. Experimental study of the basic process mechanism for direct selective laser sintering of low-melting metallic powder. *Annals of CIRP* 46(1): 127–130.

Xie et al., J.W. 2003. Direct laser re-melting of tool steels. *Proc. EuroPM2003* 3, Valencia, 20–22 Oct. 2003: 473–478.

Virtual modeling and rapid manufacturing – Bártolo (eds)
© *2005 Taylor & Francis Group, London, ISBN 0 415 39062 1*

SLS new elastomer powder material sintaflex expanding applications perspectives in layer manufacturing

Gideon N. Levy[2], Paul Boehler[1], Raffaele Martinoni[1], Ralf Schindel[2], Peter Schleiss[2]

[1] *The Valspar Corporation AG, Switzerland*
[2] *Institute for Rapid Product Development, University of Applied Sciences, St. Gallen, Switzerland*

ABSTRACT: A new powder material for the SLS (Selective Laser Sintering) process was recently released. The material is a result of fruitful research programs involving industry and university. The well known and widely used DuraFom™ (PA12) and theCastForm™ (PS) SLS-materials were developed by the same team.

In the search for new powder materials many properties of the candidate polymer, e.g. the pulverization, the laser absorption and sintering parameters have to be tuned carefully. Previous Elastomer options materials were poor in strength, detailing, and long-term use. The new product overcomes most of the known deficits. It open completely new practices in many branches like: automotive, house appliances, office equipment, foot ware, medical, and many more. The Sintaflex has a Shore hardness variability 45–75 A and Elongation around 150–300%. The attainable yield strength range is 1.3–4.2 MPa. The resolution on the SLS is up to 0.6 mm. It is positioned in good agreement compared with other commonly used injection plastics.

The paper describes the basic material properties. Further the main sintering parameters are describes and indications on machine settings are given. RP (Rapid Prototyping) applications and the recent practical experience are illustrated. Some conclusions are stated.

1 INTRODUCTION

The SLS processes, which generate parts from powder in a layered way, have more than 13 years of history. This process is not exclusively used for prototyping any longer. New opportunities and applications in appropriate manufacturing tasks open up, even though the economical impact is still modest.

The main appeal is towards novel possibilities and applications offered by this technology. However the attractive almost absolutely geometry freedom is mainly blocked-up by the limited materials that can be processed. New processable materials means gradual enlargement of new opportunities spectrum and technological innovation enhancement. The development of new materials depends tightly on the particular physical process feature which has to be rigorously respected whereas the achieved process performance (production speed, product strength for the given material, accuracy etc.) have to be optimised.

The search for SLS materials is summarizes in Figure 1. It is a balance between user needs, process technology and the selected candidate material properties. The relation between the basic material properties and the finally obtained properties after the SLS

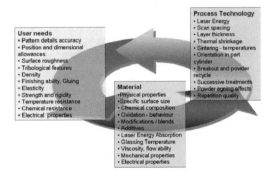

Figure 1. Integral development of SLS powder materials.

processing are different and cannot be predicted. An intensive experimental way is the only viable way [6].

2 THE SINTAFLEX IN RELATION TO COMMONLY USED ELASTOMERS

The target was to meet user needs. We wanted to find an elastomer covering the traditional uses. The typically

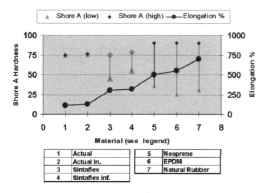

1	Actual	5	Neoprene
2	Actual In.	6	EPDM
3	Sintaflex	7	Natural Rubber
4	Sintaflex inf.		

Figure 2. Sintaflex in relation to commonly used Elastomers.

Table 1. Typical build parameter setting for the SLS Sintaflex processing.

Build-Parameter 2500plus	
Layer thickness [mm]	0.1
Part bed temperature [°C]	116
I/O Ratio [%]-	75
L/R Feed distance [mm]	0.285
L/R Feed temperature [°C]	93
Piston heater temp. [°C]	105
Cylinder heater temp. [°C]	100
Roller speed [mm/min]	177

Part Parameter			
	LP	OLP	SP
Parts	11	4	0.15

used final part materials are numerous; starting with natural rubber (NR), butadine- caotchouc (BR) Nitric caotchouc (NBR) and others, a widely used material in injection molding is the Ethylene Propylene Terpolymer (EPDM). Previous attempts to mimic flexible rubber-like materials were also considered [1] [3]. The previous proposed solution failed due to unacceptable final part properties disintegrating and processing difficulties [8]. Nevertheless user evidenced the need and benefits the problem remained latent [2].

This project targeted to provide valid alternatives. As reference properties Neoprem, EPDM, and natural rubber in respect to elongation at break and shore hardness, both the most significant properties in thermoplastic elastomer material selection, were considered. Anticipating the final findings as described later, Figure 2 gives the comparison and positioning of Sintaflex. It demonstrates the ability to fulfill Shore A spec as well as the elongation requirements [8].

3 THE MECHANICAL PROPERTIES RANGE OF SLS SINTERED SINTAFLEX PARTS

The development process is extensive and involves numerous iterations and modifications of the final product. We intended to present here the final optimized properties; and the influence of a single parameter the laser power, since it opens up new opportunities [4].

The experiments were carried out on a 3D systems Sinter-station 2500$^+$ in a second step prior to commercialization the optimized parameters settings for all actual Sinter-station platforms were elaborated.

Typical parameters are shown in table 1.

It was found that the SP scan space has a moderate influence on the achieved mechanical properties mainly the strength. The most significant properties elongation at break and shore A hardness were found to be well correlated to the laser beam power (LP) during hatching.

In order to seal the parts an infiltration follow-up process was developed. The infiltration is explained later however from here on we have to consider natural parts as coming of the Sinter-station and Infiltrated part in parallel. Diagrams will show the two corresponding states.

In Figures 3 and 4 we can see the shore hardness in function of the elongation with the laser beam power as parameter, Fig. 3 relates to not infiltrated and Fig. 4 to the infiltrated state. In both case we see a clear correlation. It becomes also apparent that we have a properties range. The Sintaflex has a Shore hardness variability 45–75 A. For the hardness measurement, in accordance with DIN EN ISO 868, we use a Hildebrand a manual Durometer Model HD3000 with a Drag pointer or a similar instrument [5, 9].

It is very important at for repeatability and proper results to calibrate the sinter system used. The laser power in an open loop parameter, the value on the screen does not correspond to the actual values. The time dependent descent in laser power output may have significant uncontrolled influence on the hardness and the final part's mechanical properties.

The stress stain tests were done on a Zwick testing machine according the ISO 527–2 standard.

The mechanical properties E module [MPa] and the tensile strength at rupture [MPa] are quasi linear in function of the laser power (Figures 5, 6). However the influence of the infiltration is very small. Actually from the practical point of view the differences are negligible. The seal infiltrate has a dedicated influence on seal and some influence on the hardness.

In summary the E module varies in the range 3 to 11.5 [MPa] and the tensile strength has the range 1.3 to 4.2 [MPa].

On the other hand as shown in Figure 7 and said before the influence of laser power and infiltration are

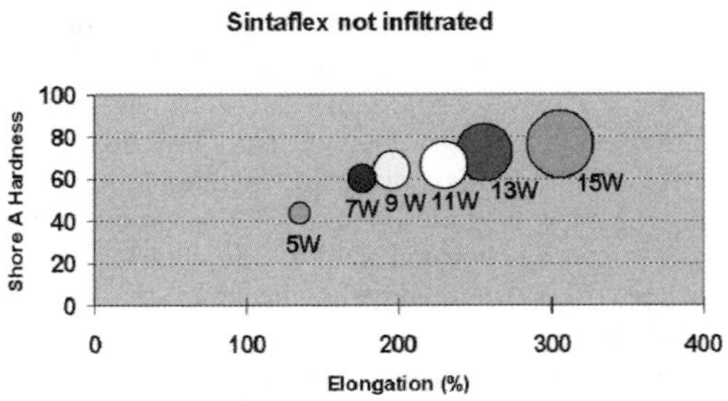

Sintaflex not infiltrated

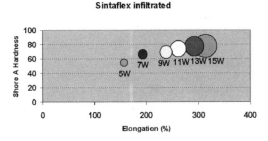

Figure 3. Laser power influence on shore A hardness vs. elongation top for non-infiltrated Sintaflex.

Sintaflex infiltrated

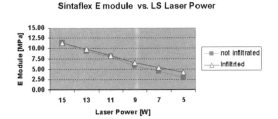

Figure 4. Laser power influence on shore A hardness vs. elongation top for infiltrated Sintaflex parts.

Sintaflex E module vs. LS Laser Power

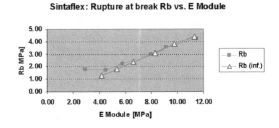

Figure 5. The relation E module vs. laser power for infiltrated and non infiltrated parts.

Sintaflex: Rupture at break Rb vs. E Module

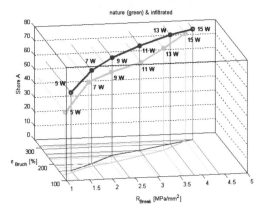

Figure 6. The relation E module vs. Rupture at break Rb for infiltrated and non infiltrated parts.

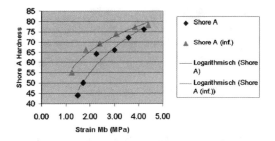

Figure 7. Sintaflex: Strain Mb vs. shore A hardness range.

nature (green) & infiltrated

Figure 8. Sintaflex material properties: elongation, durability and tensile strength in function of the laser power.

interesting. We end up with a properties range in dependence of the laser power. The first commercialized material with variable major properties the local shore hardness. Unfortunately the existing system control soft ware doesn't allow the straight forward use of this ability for practical application.

In conclusion by controlling the locally the laser power we are able to influence the local material properties in the range of [shore A] hardness 45 to 75; and the elongation around 150 to 300 [%]. The corresponding tensile yield strength range is 1.3 to 4.2 [MPa]. (Fig. 8).

4 THE SINTAFLUID INFILTRATION WITH FINISHING AND GLUING

The infiltration is done after the SLS process in a following process. We use a propriety water based infiltrate. The non infiltrated parts are in ivory color and useable in this condition, the decision whether to infiltrate or not depend on the application and required properties. The natural parts are porous and able to absorb liquids, a spongy effect. With the use of the

Figure 9.   The Sintafluid infiltration process.

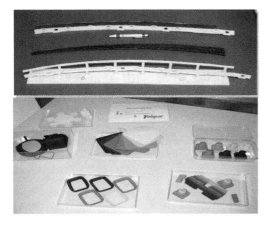

Figure 10.   Colored sealing (bottom) and assembly gluing for large parts also in combination with Duraform (PA 12) and other materials.

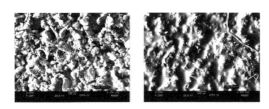

Figure 11.   Sintafluid wetting SEM of a non infiltrated on the left hand side and an infiltrated surface on the right side.

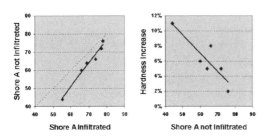

Figure 12.   The influence of the infiltration of natural parts on the shore A hardness.

infiltrate Sintafluid we can obtain stain parts, colored parts. The treatment implies a change in the Shore A hardness. The water based infiltrate has a low surface tension resulting in very good wetting ability, Figure 11.

The infiltration is a manually operation, after dipping the parts they are dried at 80°C in an oven, the process can be repeated until a complete sealing is obtained (Fig. 9) colored sealing infiltrate are available Figure 10.

No changes in the geometrical size were observed, this resulted in a great practical advantage confronting to other earlier alternative. The changes in durability are in the range of up to 10% as seen in Figure 12.

Due to the limitation in build sizes for the available sinter station particular glue was retrieved. The glue is compatible with the well known PA12 Duraform material and allows interesting combinations enhancing product innovation ideas.

5   THE APPLICATIONS

5.1   Branches, geometries

During the development the applications and possible uses were identified numerous practical applications were carried out proving the "fit for industry" proof of the new material. A non complete list in table 2.

5.2   Accuracy

The attainable parts accuracy and resolution were further important criteria in the validation. We know that this parameters imply restriction on applications for the SLS process. In this case we have to do with a forgiving material that can be stretched or adopted to its counterparts. In Figure 13 we present a measurements data evaluation the dimensional deviations in percentage

Table 2.   Overview of application branches and technical geometries.

Geometries	Application fields
Gaskets and seals	Aerospace
Hose tubes	Automotive
Isolation	Aviation
Springs and damping	Bed wear
Grips and knobs	Footwear
	House appliances
	Machinery
	Medical
	Office equipment
	Protective masks
	Toys
	Power tools
	Furniture
	Sports free time

Sintaflex Deviations in % of Numinal Values(>100 measurments)

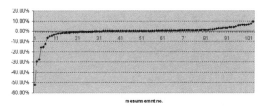

Figure 13. Sintaflex deviations in % of the nominal target dimension.

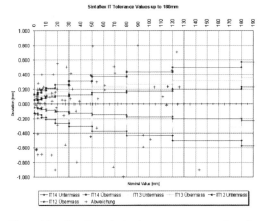

Figure 14. Sintaflex obtained accuracy compared to the ISO IT classes (IT 12, IT 13, IT 14).

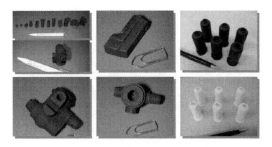

Figure 15. Some technical small Sintaflex parts.

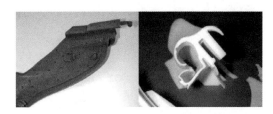

Figure 16. Typical car seals with details up to 0.6 mm.

of the nominal value. More the 100 measurements were considered. Some errors in the lower size range are explainable by measurements difficulties using ordinary calibers and not non-contact probes. We can confirm a deviation of only several percents of the nominal value. Typical resolution is shown in Figure 16.

In a different presentation we compare the deviation to the ISO tolerance standards IT 12–14 (Fig. 12) the attainable results are comparable to the SLS praxis and to the injection molded parts excluding the high divergence. This non repeatable stochastic character of the SLS process is a research due for the coming years.

5.3 Application examples

See Figure 15.

6 SUMMARY

- A new Elastomer Powder SINTAFLEX for the SLS (Selective Laser Sintering) was presented.
- The process parameters and its technology and post processing were developed.
- The Powder material SINTAFLEX and Infiltrate SINTAFLUID are ready for use.
- The new material by itself or in combination with the well known DuraForm opens completely new opportunities.
- The material allows, in function of the sintering parameters and infiltration, to mimic a range of hardness properties.
- It is well positioned in the traditionally used elastomers range for part injections.
- The main material characterizes were measured.
- Further practical applications in many branches are shown.
- Medium term use were successfully tested RM (Rapid Manufacturing) uses have to be further explored.

7 CONCLUSIONS

A new Elastomer for the SLS (Selective Laser Sintering) process, its technology and post processing were developed. The Powder material SINTAFLEX and Infiltrate SINTAFLUID are ready for use. The new material by itself or in combination with the well known DuraForm opens completely new applications.

The material allows in function of the sintering parameters and infiltration to mimic a range of hardness properties. It is well positioned in the traditionally used elastomers for part injections. The main material characterizes were measured. Further practical applications in many branches are shown. Medium term use were successfully tested the RM (Rapid Manufacturing) uses have to be further explored.

Valspar chose 3D Systems to commercialize the Sintaflex and Sintafluid products under 3D's Dura-Form® brand. The products will be known as: DuraForm® Flex plastic LS Material, and DuraForm® FlexSeal (Black, Blue, Yellow, Red, and Neutral) infiltration fluids.

REFERENCES

[1] Anonym Somos 201 Product data Sheet , DSM
[2] Anonym Somos 201 Helps Reebok Stay On Course, DSM application note, June, 2000
[3] Clausen et al., laser sintering thermoplastic powder US 6.110.411 Patent
[4] Childs T.H., Tontowi, 2001; Selective laser sintering of a crystalline and a glass-filled crystalline polymer: experiments and simulations; Proceedings of the Institution of Mechanical Engineering, Journal of Engineering Manufacture; Vol 215 B11 ISSN 0954–4054; pp 1481–1495
[5] DIN EN ISO 868, Kunststoffe und Gummi; Bestimmung der Eindruckhärte mit einem durometer (Shore-Härte)
[6] Levy G.N., 1997, Systematische Werkstoffentwicklung erweitert den Nutzen beim Einsatz von Selektive LASERsintern SLS; Proceedings WZMO Lausanne October 1997
[7] Levy G.N., Schindel, R., Kruth J.P., 2003, Rapid Manufacturing and Rapid Tooling with Layer Manufacturing (LM) Technologies, State of the Art and Future Perspectives, CIRP Annals 2003 Volume 52/2/2003, pp 589–60
[8] Boehler P., Levy G.N., Martinoni R., Schindel R., Schleiss P., 2005, Sintaflex a new Elastomer powder material for the SLS Process, Rapid Prototyping & Manufacturing, May 9–12, 205 Dearborn MI
[9] http://www.3trpd.co.uk/index.htm We run the largest sintering capability in the UK
[10] http://www.hildebrand-gmbh.com/eng/index.htm, Durometer

Virtual modeling and rapid manufacturing – Bártolo (eds)
© *2005 Taylor & Francis Group, London, ISBN 0 415 39062 1*

Comparison between Laser Sintering of PEEK and PA using design of experiment methods

T. Rechtenwald, G. Esser & M. Schmidt
Bavarian Lasercenter, Erlangen, Germany

D. Pohle
Friedrich-Alexander University, Erlangen-Nürnberg, Germany

ABSTRACT: PEEK (Polyetheretherketone) as well as Polyamide – the standard laser sintering material – are both semi-crystalline thermoplastics. Moreover, PEEK offers outstanding temperature resistance, mechanical strength and stiffness. PEEK possesses an outstanding chemical resistance, which leads to an excellent bio-compatibility making the material a good choice especially for the manufacturing of medical instruments or even implants. Up to now these parts typically are produced by conventional manufacturing methods like injection moulding. But especially the production of individually shaped parts like implants would benefit a lot from a more flexible manufacturing technique. Laser Sintering (LS) can offer the required flexibility. Moreover, it is the most promising Rapid Prototyping Technology for the production of thermoplastic parts with a good geometrical accuracy on one hand and satisfying mechanical properties on the other. LS consequently is the ideal production technology for individual parts. It enables the direct manufacturing of products with complex geometries, including undercuts and defined porosity. Due to its high melting temperature of about 343°C laser sintering PEEK is a challenge compared to the standard polymer polyamide with a moderate melting point of about 172°C. This paper describes results on selective laser sintering of PEEK in comparison to laser sintering of the standard polymer, polyamide, using paired comparison and a full factorial design. The investigated influence coefficients on the process include the operating parameters of the laser system, the properties of the applied PEEK powder and the preheating temperature of the powder bed. Their effects on the sintering results are described and discussed in detail.

1 INTRODUCTION

Polyetheretherketone (PEEK) is a high temperature resistant, semi-crystalline thermoplastic polymer. PEEK combines a very good strength and stiffness with an excellent thermal and chemical resistance – e.g. against oils and acids. Its mechanical properties remain stable up to temperatures of about 200°C for prolonged periods of time. Depending on the specific sort of PEEK used, long-term operating temperatures reach up to 260°C.

Due to its excellent biocompatibility PEEK is also a good choice for the manufacturing of medical implants (Jahur-Grodzinski, 1999). Up to know these parts typically are produced by conventional manufacturing methods like injection moulding. But especially the production of individually shaped implants would benefit a lot from a more flexible manufacturing technique. Layer based Rapid Manufacturing techniques can offer the required flexibility.

The idea of manufacturing three-dimensional objects by stacking planar elements (or layers) on top of each other without requiring part-specific tooling already has been discussed in the early 1970's. But only in the late 1980's new additive processes (Kruth, 1991) that automatically build up a three-dimensional object layer by layer have been introduced to the markets (Levy, Schindel, Kruth, 2003). Selective laser sintering (Figure 1) is one of these additive processes (Hon, Grill 2003). Of special interest for a lot of applications is its high potential for the direct manufacturing of functional parts with good mechanical properties.

Like in all modern layer based manufacturing processes also in the SLS process the part is directly build up based upon digital computer data. The laser selectively cures the powder material within the top layer of the building cavity according to the cross section data of the CAD model. After finishing the curing of one layer and lowering the building platform a new layer is applied on the top of the previous one and

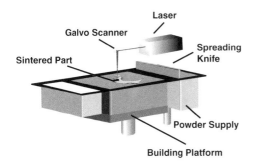

Figure 1. Caption of a typical figure. Photographs will be scanned by the printer. Always supply original photographs.

the curing process starts again. A big advantage – when compared with stereolithography – is the fact that the part is built into the powder. As the powder itself supports the part, less extra support structures are needed.

With respect to the direct production of individually shaped implants laser sintering can be seen as the ideal production technology. It enables the direct manufacturing of products with complex geometries, including undercuts and defined porosity. Although its productivity can not compete with mass production technologies it is still sufficiently high for the direct fabrication of single parts such as implants. Taking into account the good flexibility of laser sintering and the high quality of laser sintered parts made from commercially available polymer precursors such as polyamide it seems very promising to investigate the laser sintering PEEK in more detail.

2 EXPERIMENTAL

2.1 *Precursor materials*

The experiments shown in this paper were carried out using four different precursor materials. As a reference material a commercially available polyamide – trademark PA 2200, a standard powder for the laser sintering process distributed by EOS, Krailling – has been used. Additionally three different PEEK powders with varying average particle sizes and melt viscosities, marketed under the brand names PEEK 450PF, PEEK 150PF and PEEK 150XF by Victrex Plc, Lancashire, UK were examined. To our best knowledge all investigated PEEK materials have not been qualified for the laser sintering process so far. Investigation of the powders by SEM has shown that in contrast to the polyamide powder which is characterized by a nearly spherical particle shape, the PEEK grains exhibit a very irregular geometry. This is a disadvantage with respect to the laser sintering process as the achievable homogeneity of a coated powder layer suffers from

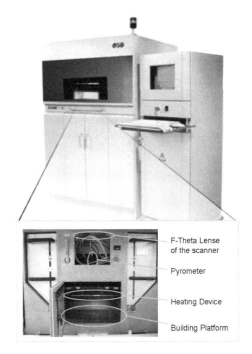

Figure 2. Laser sintering machine used for experiments.

the irregular shape. Obviously there is still a high potential for optimization of the PEEK powders with respect to laser sintering.

2.2 *Laser sintering apparatus*

The LS process was performed on a modified EOSINT P380 polymer laser sintering machine. The set up contains a 50 W CO_2 laser with gaussian beam (focus diameter at working surface 0,35 mm) and an integrated nitrogen generator It offers a building volume of $340 \times 340 \times 620\,mm^3$, a maximum scan speed of 5 m/s and a lateral resolution of 0,1 mm (Fig. 2). For the purpose of increasing the powder preheating temperatures we modified the machine by integrating a circular building platform (diameter of 210 mm) equipped with an additional heating device (Fig. 3).

With the modified machine we were able to generate preheating temperatures of up to 250°C. The homogeneity of the temperature distribution at the powder surface has been estimated by measuring sub surface temperatures with temperature sensitive sensors. The observed temperature differences lie within a range of $+/-3$ percent.

In order to estimate if the observed inhomogeneities of the temperature distribution have an influence on the sintering quality we determined the homogeneity of the layer thickness d_L of single layered test specimens processed at different locations within the

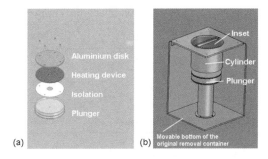
(a)

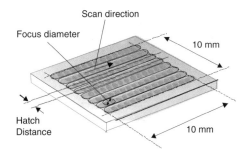

(b)

Figure 3. Modified building platform (b) with additional heating device (a).

Figure 5. Test geometry for laser sintering experiments.

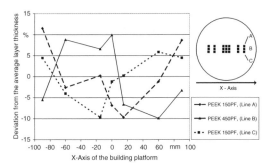

Figure 4. Distribution of layer thickness d_L within the building platform at 250°C preheating temperature.

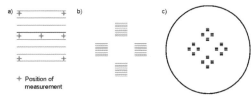

Figure 6. Position of measurement spots on sample and distribution of samples in building zone (diameter 210 mm).

building area. Figure 4 shows the values for the layer thickness measured according to the procedure described within chapter 2.3.

2.3 Sintering experiments

All experiments were carried out using single layer test samples. The samples were produced by laser irradiating as deposited polymer powders in a square shaped area with an edge length of 10 mm. The area was filled in line wise manner according to Figure 5.

Before starting any quantitative evaluation we estimated the process boundaries for the new material in a first step. Based upon existing experiences from the laser sintering of polyamide we varied the energy per unit area (deposited by the laser beam on the polymer surface) which is known to have a very decisive influence on the sintering results. A lower process border could be easily defined by looking at the mechanical stability of the test specimens. The upper limit of the process window is given by the thermal decomposition of the polymer.

In the next step we investigated the influence of input parameter variations on the sintering quality in more detail. The thickness of the sintered powder layer d_L was chosen as a comparatively easy accessible

target parameter to quantify the experimental results. d_L is a good measure for the depth of the interaction zone (the zone were the laser beam partially melts the powder). As the thickness of the powder layer after coating strongly depends on the type of powder, the coating mechanism, the required resolution and the desired productivity of the process, it is very important to be able to adjust the depth of the interaction zone in order to achieve a good adhesion of adjacent layers on one hand and a good pixel resolution on the other.

The layer thickness was measured with a mechanical ball point measurement tool. Different measurements taken within a region (1 mm²) of a specimen result in the same values with a statistical error of $+/-1\%$. However, thickness values determined at different regions of one and the same sample sometimes differ by up to 15%. Taking into account additionally the observed layer thickness variation within the building area (chapter 2.2) we decided to use an average value of 4 test samples for every given set of input parameters. Within each of the 4 samples we took the average value of 7 measurements in different regions. Figure 6 shows the geometry of the test spots on a sample (a), a set of samples with identical input parameters (b) and a typical distribution within the building zone (c).

After rough estimation of the process boundaries the parameter dependencies were identified by a "Paired Comparison Test" (Shainin, Shainin 1988). We considered the following set of input parameters: the laser power P_L, the scan velocity v_S, the hatch distance h_S,

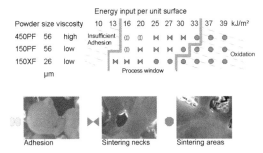

Powder size	viscosity	Energy input per unit surface 10 13 16 20 25 27 30 33 37 39 kJ/m²
450PF	56	high
150PF	56	low
150XF	26	low
µm		

Adhesion Sintering necks Sintering areas

Figure 7. Process boundaries for different PEEK precursors and achieved sintering levels (Preheating temperature T_P = 140°C nitrogen flow = 14 l/min, h_S = 0, 3 mm, 13,5 W < P_L < 52,5 W, v_S = 4,5 m/s).

the preheating temperature T_P, the nitrogen flow rate N2 and the melt viscosity η of the precursor material. A closer look was taken on the relation between the most relevant input parameters and the chosen target parameter d_L by a full factorial analysis in the last step.

3 RESULTS AND DISCUSSION

3.1 Determination of process boundaries

By varying the energy input per unit surface we were able to clearly identify the process boundaries according to the specifications given in chapter 2.3. The quality of the test samples could be further categorized according to a set of three different quality classes (Fig. 7).

For very low energy input per unit surface the test samples do not exhibit a sufficient mechanical stability for further examination. At slightly higher energy densities the powder particles start to glue together (lower process boundary). By increasing the energy density the porosity is continuously reduced. Beyond the upper process boundary a black discolouration of the laser sintered layer can be observed. Analysis has shown that this discolouration is a sign for a change in oxygen to carbon ratio which is not acceptable for a lot of applications for example in medical technology.

Especially the 150 XF powder turned out to be difficult to handle. We were not able to achieve a homogeneous powder layer with the existing coating device. Further investigations thus have been done only using the 450 PF and the 150 PF powders.

3.2 Paired Comparison

The Paired Comparison Test uses two adjustments, called stages, for each factor to estimate the impact of a single factor on a target parameter or to discover interactions between the factors. "Good" and "bad"

Table 1. "Bad" and "good" values for Paired Comparison.

	"Bad"	"Good"
P_L in W	40	50
v_S in m/s	5,0	4,0
h_S in mm	0,33	0,30
T_P in °C	140	250
N_2 in l/min	0	14
η (viscosity)	high	low

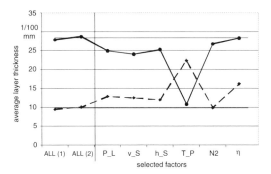

Figure 8. Results of Paired Comparison Test.

values for the factors P_L, v_S, h_S, T_P, N_2 and η were chosen based on the existing experience as as shown in Table 1.

In general the "good" values yield a higher average thickness of the sintered layer what is favourable with respect to the layer to layer adhesion. Two series of experiments (a, b) were carried out to verify the proper choice of the "good" and "bad" values: (a) two tests with all factors on "good" stages and (b) two tests with all factors on "bad" stages. The difference D between the mean values was compared to the spread d, as shown in equation (1):

$$ D = \left| \frac{G_1 + G_2}{2} - \frac{B_1 + B_2}{2} \right| \; ; \quad d = \frac{|G_1 - G_2| + |B_1 - B_2|}{2} $$

(1)

According to Shainin, D., Shainin, P. 1988 the quotient D:d must exceed 5:1 to get reasonable information from the Paired Comparison Test. In our case D:d was calculated to be 25,97. This means that the condition is clearly fulfilled. Figure 8 shows the results of the test. The lower line corresponds to experiments done with all factors – except one – on the "bad" level, the upper line shows results for samples processed with all factors – except one – on the "good" level. The most significant difference between the "good" and the "bad" line occurs for the factor TP. The lines cross each other. This means that TP has an extraordinary strong influence on the target parameter. P_L, v_S

346

Table 2. Values for Full Factorial Analysis.

| | PA 2200 | | PEEK 150PF | |
	+	−	+	−
P_L in W	57,5	40,0	47,5	33,0
v_S in m/s	3,5	5,0	3,5	5,0
h_S in mm	0,30	0,33	0,30	0,33
T_P in °C	160	60	250	140

Table 3. Comparison of impact coefficients.

Impact coefficients	Unit	PA 2200	PEEK 150PF	Deviation of impact coefficient
P_L	mm/W	0,0209	0,019	−8,6
v_S	mm/(m/s)	−0,022	−0,020	−8,6
d_S	mm/mm	−0,0067	−0,0062	−7,5
T_P	mm/°C	0,045	0,046	2,6

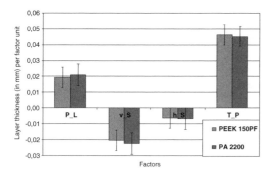

Figure 9. Impact coefficients for target parameter d_L.

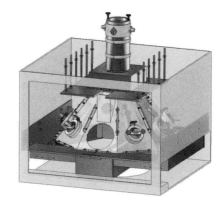

Figure 10. Additional heating element for EOSINT P380i.

and h_S also are important factors because the distance between the "good" and the "bad" line narrows nearly symmetrically for these factors.

According to Shainin it is likely that a factor has an important interaction with another factor when the target parameter varies only from the "bad" or from the "good" line. In Figure 8 this is the case for N2 and η.

3.3 Full Factorial Analysis

A Full Factorial Analysis (Montgomery, 1978) has been done only for the four most important factors T_P, P_L, v_S and h_S. The low viscosity powder PEEK 150 PF was chosen for the experiments because a higher average thickness of the sintered layer could be achieved with this material. Based on the results of the Paired Comparison the "+" and "−" values for the factors were slightly refined. As reference material PA 2200 was studied. Table 2 gives an overview of the chosen values for both materials.

Figure 9 shows the extracted impact coefficients which quantify the impact of the different factors on the target parameter d_L. The confidence interval at a level of 95%, also called significance interval, is indicated by the error bars.

Again the strongest influence is observed for T_P. P_L and v_S show impact coefficients of the same order of magnitude (0.2 mm/W, 0,2 mm/(m/s)) respectively.

Considering the sum impact factor of these two factors, the impact is nearly equal to the impact of the preheating temperature. Comparing the coefficients for both materials it has to be noticed that they basically behave in the same way. The impact coefficients for the laser parameters (P_L, v_S and h_S) however are about 8% lower for PEEK whereas the impact coefficient for the preheating temperature is about 3% higher (Tab. 3).

4 CONCLUSION AND OUTLOOK

We can conclude that the process of laser sintering PEEK does not differ essentially from the laser sintering of polyamide powder. The fact that the preheating temperature is specially important for both materials combined with the fact that the best sintering result for polyamide is achieved at TP slightly below the melting point of the precursor powder, leads to the conclusion that further modifications of the experimental set up are necessary to generate higher preheating temperatures. An additional top level heating element for integration between the focusing lens and the building platform (Fig. 10) already has been designed and is currently under construction.

REFERENCES

Hon, K.K.B., Gill, T.J. 2003. *Selective Laser Sintering of SiC/Polyamide Composites*. CIRP Annals 52:1, 173–176

Jahur-Grodzinski, J. 1999. *Review – Biomedical application of functional polymers*. Reactive and Functional Polymers 39, 99–138

Kruth, J.P. 1991. *Material Incress Manufacturing by Rapid Prototyping Techniques*. CIRP Annals 40:2, 603–614

Levy, G.N., Schindel, R., Kruth, J.P. 2003. *Rapid Manufacturing and Rapid Tooling with Layer Manufacturing (LM) Technologies, State of the Art and Future Perspectives*. CIRP Annals 52:2, 589–609

Montgomery, D.C. 1978. *Design and Analysis of Experiments*. John Wiley, New York, 1978

Shainin, D., Shainin, P. 1988. *Better than Taguchi Orthogonal Tables*. Quality and Reliability Engineering International 4, 143–149

Virtual modeling and rapid manufacturing – Bártolo (eds)
© *2005 Taylor & Francis Group, London, ISBN 0 415 39062 1*

A homemade binder for 3D printer

F.F. Aguiar, I.A. Maia, M.F. Oliveira, J.V.L. Silva
Centro de Pesquisas Renato Archer – CenPRA, Campinas-SP, Brazil

ABSTRACT: The composition of a homemade 3DP binder was proposed based on the following points: quality of the parts printed only with deionized water, the basic components of the ink contained in ink jet print head and the migration of solutions in 3DP commercial powder packed inside plastic tubes. By comparing the homemade binder to the commercial one in terms of dimensional accuracy, surface roughness, mechanical strength and depowdering efficiency it was found that both binders are similar. As a result the homemade binder, that is significantly less expensive, is been utilized in a routine basis to build biomodels.

1 INTRODUCTION

1.1 *Motivation: cost reducing for spreading medical applications*

For the last five years CenPRA, a research institution of the Brazilian Science and Technology Ministry, keeps a program called PROMED which supplies physical biomodels (Figure 1), built with Rapid Prototyping (RP) technology, for assisting an increasingly number of awkward orthopedic surgeries. The availability of these biomodels which are used in diagnosis, surgical planning and training has contributed significantly to diminish the suffering of severe injured people and giving them a better quality of life that otherwise could not be achieved.

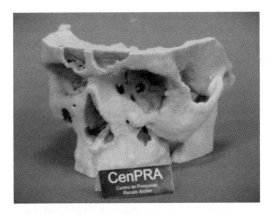

Figure 1.   Skull built with 3DP.

Thanks to these benefits PROMED has been pursuing spread RP technology in Brazil and others developing countries. Two major initiatives have been employed towards this objective. Firstly, developing and providing free software for medical image re-construction. Secondly, looking for ways to reduce the costs of physical biomodels. A first measure to achieve costs reduction was the acquisition of a 3D Printing (3DP) machine to be utilized in substitution of the most expensive Selective Laser Sintering (SLS) always as possible. Due to the high prices that the imported consumables achieve in Brazil, efforts have also been employed for substituting them. Development of a homemade binder for 3DP is a starting point towards this goal.

1.2 *Basics about 3D printing*

3D Printing is a Rapid Prototyping (RP) technology (Pham & Dimov 2001, Cooper 2001) in which a powder, after being compacted in thin layers, is selectively bound by a liquid dispensed by nozzles such as those found in the ink jet print heads. By alternating successively the processes of layering and binding the powder, a three dimensional part is built. Post process includes depowdering followed by infiltration. Depowdering is the removal of the non binded powder using vacuum cleaner, brush followed by air jets whereupon the green part which is infiltrated with a resin. By infiltration the green part acquires the strength characteristics of the infiltrating material. A major drawback of this technique is that the green part is quite fragile and consequently requires careful handling in all steps before infiltration. A major advantage is that all processes are carried out at room temperature.

1.3 ZPrinter® 310 machine – Z Corporation

This machine operates with a unique print head and utilizes the ZP® 102 plaster powder which is composed of the following dry compounds: calcium sulfate hydrate, vinyl polymer, carbohydrate, soda lime glass beads, calcium sulfate dihydrate, fatty acid ester and salts (Z Corporation 2004). The binder is water based and is dispensed by a HP thermal print head (hpc4800a). The machine automatically replaces the ink by the binder. The chemical reaction involved in the powder binding promotes a crystallization process. This process is intensified until a large amount of crystals is formed and interlock with one another to form a dense solid part.

1.4 Considerations about the binder

As the ZPrinter® 310 utilizes the same print head of the 2D printers, it was inferred that the composition of the binder should present the same set of compounds except the dye and its associate additives. A first proposal for a homemade binder composition should include, at least, water and a surfactant agent. The water plays the hole of the crystallization agent and the solvent for the powder's vinyl polymer. The surfactant has multiple functions. It helps the penetration of the binder in the print medium, avoids drop deflection from the print head nozzle plate, controls the size of the drop and finally it works as an emulsification agent. Highly viscous emulsion works as a barrier which confines the binder in the limits of the zone where the binder drop collides with the powder layer (William et al 2004).

2 EXPERIMENTAL WORK

2.1 Water as a binder

The experimental work has started by printing a part using only distilled water as a binder. Water was proposed after bench test has shown it hardened the powder. The part presented good dimensional and mechanical characteristics. The dimensions were well closed to the part which was built using the commercial binder. Moreover part was handled and undergone the depowdering step without any damage. However the morphology of the walls parallel to the z axis exhibited severe surface roughness which was clearly observable with naked eye (Figure 2).

The amount of compounds to prepare the solutions was also based on the amounts found in the ink of thermal ink jet heads (Lauw 1996). The amount of surfactant in samples 1 and 2 was chosen based on the weight percentage required to reach the CMC (Critical Micellar Concentration).

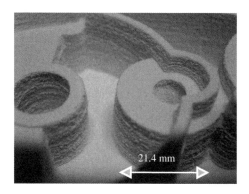

Figure 2. Surface roughness of the part built with water only as a binder.

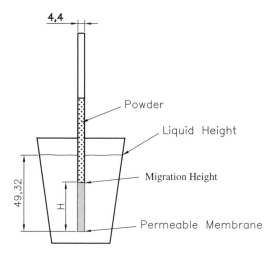

Figure 3. Schematic drawing of the experimental setup to determine the Migration Height (H) of solutions (dimensions in millimeters).

2.2 An experimental approach towards a homemade binder

The approach assumes two assumptions as follows: (1) as the commercial binder is colorless, the homemade binder should contain all compounds of the ink of a thermal print head except the die and their additives. Then the following ingredients were taken for the experiments: water, anionic surfactant, alcohol and polyalcohol; (2) the criteria to find the relative amount of these compounds in a solution to be tested as a homemade binder is the Migration Height value (H). H is the height reached by the solution in a powder column.

The experimental set up is presented in Figure 3. To prepare the powder column the powder was introduced into a plastic tube (4.4 mm diameter) which had one of its sides covered with a permeable membrane while the

Table 1. Weigth concentration of compounds in water solutions candidates as homemade binder.

Solutions	Surfactant (%)	Monoalcohol (%)	polyalcohol (%)
1	0,0999	0	0
2	0,1499	0	0
3	0,2997	0	0
4	0,4250	0	0
5	0,4246	9,9800	0
6	0,4256	9,9950	4,9975
7	0,3002	9,9967	5,2316

other remained opened to the air. In order to have the same packing degree the tubes were allowed to drop three times in the vertical position from the same height position. Then the membrane side of each tube was immersed in a corresponding solution. All tubes were left in contact to the solutions for a same period of time. The Migration Height (H) was immediately measured (Table 1) after the tubes were removed from the solutions. The water solution having both surfactant and alcohol, whose H value was the closest to that reached by the commercial binder was chosen as the homemade binder and used then in the printing tests.

2.3 Printing tests

A set of testing parts (Figure 4) was built using the commercial and the homemade binder utilizing a ZPrinter® 310 machine and ZP® 102 powder (Z Corporation®). The printing parameters were also the same, i.e., layer thickness (0.1 mm) and value saturation (100%). After depowdering, the parts were not infiltrated except the cylinders (S4) that are testing parts for porosity/infiltration evaluation.

2.4 Parts characterization

Evaluation of the surface roughness of the testing parts was made by naked eye. Dimensional measurements were made using a caliper ruler (0.001 mm precision). The mechanical characteristics were qualitatively evaluated by handling the parts after removal from the part bed and through the depowdering process.

The quantitative mechanical characterization consisted of raising the stress strain curves (EMIC DL 300 machine with 100 N transducer) from which it was obtained the maximum tension (MPa). Also the mechanical characterization was evaluated indirectly by measuring the cyanoacrylate infiltration in the cylinders. The infiltration front depends on the material porosity which is a critical factor that determines mechanical strength. The infiltration was made by immersing the cylinders in cyanoacrylate for 10 seconds. To expose the infiltration front the cylinders were

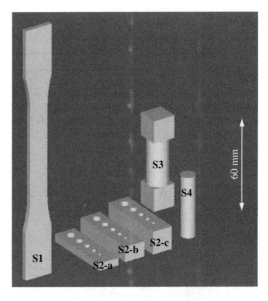

Figure 4. Schematic drawing of the testing parts built with homemade and commercial binder and applied as follows: S1 – tensile test; S2a,b,c – depowdering efficiency; S3 – dimensional characterization, S4 – cyanoacrylate infiltration.

firstly cut parallel to the circular sections. Afterwards the non infiltrated material presented in the cylinders core was thoroughly removed utilizing water jets. To give a good contrast the empty core was filled with liquid wax and then photographed with a digital camera (Sony F828, 8 megapixels resolution).

The depowdering efficiency was evaluated by trying to remove the non binded powder from the holes having different aspect ratio (S2, a, b, c) with air jets. Aspect ratio refers to the height/diameter ratio.

2.5 Results and discussion

Solution 6 was chosen as the homemade binder because it fulfills the two following conditions: (1) it contained all basic compounds of the ink utilized in ink jet print head and (2) it presented the H value closest to that of the commercial binder (Figure 5). It should be noticed that the both conditions are arbitrary and they were taken as a first approach to formulate a homemade binder. Such approach was mostly based on the feeling that the migration of the binder on the machine part bed should be similar to those of the columns. Moreover there is some question that should be answered for a better understanding of the migration mechanism, for instance the severe drop of H value (solution 5) by adding alcohol to the solution 4 which contains only water and surfactant. However a detailed investigation of the physical chemistry phenomena is beyond the scope of the present work

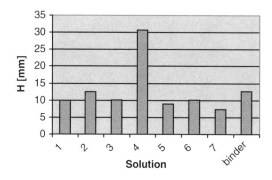

Figure 5. Migration front (H) of water solution of surfactants and alcohol. The numbers given in X axis refers to the solutions shown in Table 1.

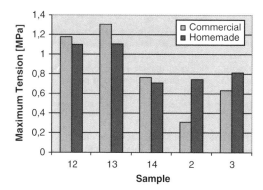

Figure 6. Maximum strength exhibited by parts built with commercial and homemade binders. Samples 12 and 13 were built perpendicularly to the Z axis and samples 14, 2.

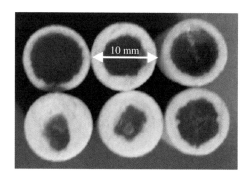

Figure 7. Cyanoacrylate infiltration in cylinder samples built parallel to the Z axis: The white contour shows the infiltration front in the cylinders built with homemade (top) and commercial (bottom) binder. The black core is wax that replaced the non binded powder.

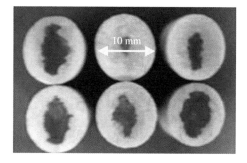

Figure 8. Cyanocrilate infiltration in cylinders samples built perpendicularly to the Z axis: The white contour shows the infiltration front in the cylinders built with homemade (top) and commercial (bottom) binder. The black core is wax that replaced the non binded powder.

which, due to this fact, should be considered as an exploratory work. In despite of that, the approach adopted gave the satisfactory results concerning roughness surface, mechanical strength, dimensional characterization and depowdering efficiency as explained as following.

Surface roughness: The parts built with homemade binder presented much less surface roughness than that found in parts printed with water and also similar surface roughness compared to that found in parts printed with commercial binder.

Mechanical strength: The maximum tension obtained from the stress-strain curves (Figure 6) of the ASTM638 samples showed that the mechanical strength of the parts built with homemade and commercial binders are similar. For both binders it is confirmed the expected behavior in which the parts built perpendicular to the z axis (samples 12, 13, 14 in figure 6) are stronger than that ones built parallel to the z axis (samples 2 and 3).

Among the three samples built parallel to the Z axis, the strength of the sample built with homemade binder is greater than that built with commercial binder. This is corroborated by the infiltration experiments. Cyanoacrilate infiltration front of the cylinder built parallel to the z axis with homemade binder is narrower, therefore less porous, than the infiltration front related to commercial binder parts (Figure 7). This could be indicating that the homemade binder migrates more efficiently towards the layers interface. This is another phenomena that should be investigated in a further work.

Regarding the cylinders built perpendicular to the z axis (Figure 8) the opposite occurs. Those built with commercial binder are stronger because present lower infiltration, therefore are less porous, than those built with commercial binder.

Dimensional characterization: The percentage difference between the parts built with homemade and commercial binder in both positions – parallel and perpendicular to the z axis – is in the 0, 2%–2.0% range. Although these differences could be considered acceptable for many applications they show that improvements can still be done in the homemade binder.

Depowdering efficiency: None noticeable difference was observed regarding the depowdering efficiency. This means that the migration of the homemade binder during the printing process was similar to the homemade binder.

3 CONCLUSIONS AND FURTHER WORK

The homemade binder presented performance similar to the commercial one and costs a fraction of the latter. As result it has been utilized to build cranium and maxilar biomodels at regular basis in the PROMED program.

A further work list includes: (a) detailed studies about the migration of the binder into de powder; (b) attempt to formulate a homemade powder; (c) build biomodels that can be sterilized in autoclaves in order to be brought to the surgery rooms.

ACKNOWLEDGEMENTS

The authors wish to thank the Division of Qualification and Analysis of Electronic Products (DAPE) of CenPRA for the tensile tests.

REFERENCES

Cooper K. G. 2001 Rapid Prototyping Technology: Selection and Application:50–63, New York. Basel, Marcel Dekker.
Lauw H. P. 1996 Specific dye set for thermal ink-jet printing, US Patent number 5534051.
Pham, D.T. & Dimov S. S.1980. Rapid Manufacturing:36–37, London, Springer-Verlag.
William C. R., Cima M. J., Pryce L., Wendy E., Monkhouse D. C., Kumar S. & Yoo J. 2004 Methods and materials for controlling migration of binder in a powder, US Patent number 20040062814.

Virtual modeling and rapid manufacturing – Bártolo (eds)
© *2005 Taylor & Francis Group, London, ISBN 0 415 39062 1*

Photopolymer material jetting in rapid prototyping

A. Liberman & H. Gothait
Objet Geometries Ltd, Rehovot, Israel

ABSTRACT: A new three-dimensional (3-D) model building technology uses inkjet printing-heads to build rapid prototypes. This technology jets 16-micron layers of acrylic-based photopolymers onto a build tray via multiple-nozzle jetting heads. The block of jetting heads moves over the tray in a raster-type sequence, much like a line printer, simultaneously jetting two different types of photopolymer material, layer by layer, to form different parts of the 3-D model being built. Each individual layer is immediately cured and hardened by use of UV light. Each layer includes both model material and support material, which is necessary to enable overhangs, cavities, holes, etc., to be jetted. After each layer is deposited and cured, the build tray moves downwards and a new layer of materials is jetted. Once all the layers of the model have been deposited and cured, there is no need for any post-curing. The support constructions are then removed using a water jet and the model is ready for use.

1   INTRODUCTION

Rapid prototyping includes many techniques used to produce 3-D objects directly from a computer. Several techniques use fluid material, which is solidified to produce the solid object. The material used to build the objet is usually referred to as build/building material or model material, while material used to support the object during build is usually referred to as support material. Object building is frequently performed by producing thin layers corresponding to computer data and solidifying the layers one upon another.

Successful technologies must find the optimal balance between high detail replication, smooth surfaces, good material properties (high flexibility, high temperature resistance, maximum dimensional stability, etc.), and easy support removal, ease of use, clean technology, good serviceability, and of course capital cost.

The inkjet-head technology described in the present paper is designed to jet multiple photopolymer materials, layer by layer, onto a build tray with simultaneous formation of both model and support constructions. Instant curing and easy removal of support material are basic features of the technology.

This enables instant handling of the finished model once the building process is completed. The materials are delivered to the jetting head via 2 kg sealed cartridges of both model and support material. Thus the user has no contact with liquid photopolymer during the build process.

The machine size is similar to that of a photocopier. Noise level is very low and there are no fumes emitted during build.

2   TECHNOLOGY FOR JETTING PHOTOPOLYMERS

The photopolymer jetting head described in the present paper, jets photopolymer resin at a resolution of 600×300 dpi, in 16 micron layers.

The printing block or bridge, comprising 8 separate and individual printing or 'jetting' heads, slides back and forth along the X-axis, much like a line printer, depositing single layers of photopolymer materials onto the build tray according to a predetermined configuration. The exact location and amount of support and model material in the entire model is automatically preprogrammed. Immediately after depositing, UV bulbs alongside the jetting bridge emit UV light, curing and hardening each layer.

The internal building tray moves downwards with extreme precision and the jetting heads begin building the next layer. This process is repeated until the model is complete. Due to the thin 16 microns layers and the repetitive UV curing after each layer, the final model is fully cured and additional postmodelling curing is unnecessary.

The jetting block uses 8 separate and individual printing heads with multiple nozzles. Sophisticated software tools enable the 8 heads to work in perfect harmony, synchronously jetting identical amounts of resin on the tray (Figure 1).

Two different types of photopolymer materials are used for building: one material is used predominantly for the actual model, while a second, gel-like photopolymer material is used for support. The construction of the support structure is pre-programmed to

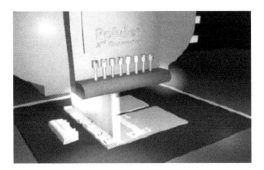

Figure 1. Jetting block with 8 jetting heads.

cope with complicated geometries, such as cavities, overhangs and undercuts, delicate features and thin-walled sections. This soft support material is only slightly linked with the model material. When this support material is wet and pressure is exerted on to this structure (by either brush or water jet), it loosens and separates from the model material. This occurs in small enough sections so that support material can be removed from gaps or holes of a few hundred microns.

2.1 Time for model construction

Input into the system is carried out using any CAD program, converted to STL files. The system's pre-processing program proposes an optimal build orientation, enabling control of the building process according to specific needs.

Once the desired build orientation is selected, the system's software "slices" the CAD model (i.e. the STL files) into 16-micron layers and the model is ready for building.

The technology uses a raster process to produce photopolymer models. This enables the machine to build several models at once in only slightly more time it would take to build one model. All this is controlled by a job management system.

To shorten build times, the shortest dimension is programmed in the Z direction, reducing the total number of layers to be created.

The system software automatically take into account physical dimensions, the geometry of the part, direction of printing, printing cell temperatures, resolution, type of interface materials, repeatable motion errors and so on to optimize the build surface. Users can also make some modifications in parameters to improve the quality of the final product.

All of the above factors, along with simultaneous construction of different object parts and their adjacent support constructions, significantly decrease total build time.

2.2 Single Head Replacement (SHR™)

The new Polyjet apparatus has multiple, separate and replaceable printing heads. The typical configuration is a printing block comprising 8 such printing heads.

The use of separate and replaceable printing heads enables efficient printing head replacement without necessitating removal and/or replacement of the entire printing block from the machine (removing at the same time non-defective, functioning heads) or recalibration. Any single head may easily be replaced with the aid of a software program with step-by-step instructions for the SHR process. After replacement, sensors attached to the printing block detect if the head and head driver have been properly inserted.

In addition, a special printing head registration facility does away with the necessity for postcalibration after SHR. The concept consists of mechanical realignment of the printing heads with the aid of alignment pins and stoppers and their software calibration, to ensure highly accurate nozzle realignment.

2.3 Material-dedicated printing heads

Each separate and replaceable printing head is 'dedicated' to jetting or dispensing a certain type of interface material. Some printing heads may be dedicated to dispensing building material alone and other to dispensing only support material or other materials. Use of different interface materials in different combinations enables variation and flexibility of types, colors, relative quantities of different interface materials jetted simultaneously to comprise any given layer of a 3-D object being printed.

2.4 Quality enhancement

Fine detail and surface smoothness are achieved by special algorithms pre-programmed according to the shape of the object to be printed, in addition to construction of the model in 16-micron layers.

2.5 Other features

- Other features include built-in part alignment properties, automatic calibration and memorization of part, material and printing parameters.
- Consistent flow of jetted materials is ensured by controlled heating, vacuum and reflux systems.
- Print resolution, printing offsets and layer thickness control mechanisms are additionally built-in to ensure the production of high-quality, accurate and finely finished models.

2.6 Jetting materials

An integral part of the 2nd generation photopolymer jetting technology is a proprietary and specialized line

Vero

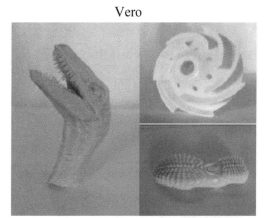

Tango

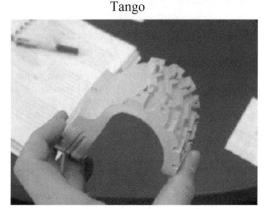

Figure 2.   Vero and Tango new resins.

FullCure

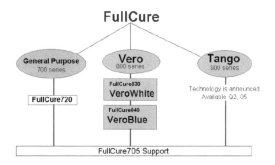

Figure 3.   FullCure new family of resins.

of photopolymer resins. This family of resins is acrylic-based.

The technology ensures full curing of the materials so models can be examined and handled as soon as the model is complete. All material use is controlled by a built-in materials management system.

Resin & Color	Characteristic highlights & Common Application
FullCure720 Transparent / yellow	General Purpose
	Liquid flow or where need to see internal details
VeroBlue FullCure 840	Detail visualization & improved technical properties
VeroWhite FullCure 830	White color and improved humidity resistance
Tango Technology Black	Rubber-like 61 Shore Shoes, Toys, Industrial
Tango Technology Grey	Rubber-like 75 Shore Shoes, Toys, Industrial

Figure 4.   FullCure characteristics.

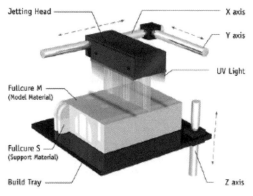

The Objet PolyJet Process

Figure 5.   PolyJet process

2.6.1   *FullCure™ resins*
- Acrylic-based photopolymers
- UV-cured – no post-cure
- Dual cartridge system
- Transparent, clear models
- Basis for additional resins

2.6.2   *New resins: Objet introduces the Vero and Tango*
See Figures 2 & 3.

2.6.3   *FullCure resins*
See Figure 4.

2.6.4   *Support material*
- Pre-programmed support for
- ƒnCavities
- ƒnOverhangs
- ƒnUndercuts
- ƒnDelicate features
- Clear definition of support-model border
- Fast support removal with WaterJet
- Clean, smooth remaining surface

357

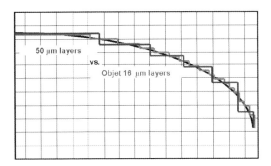

Figure 6.   Resolution graph.

2.6.5   *Best resolution: 16 μm – smoothness and details*
See Figure 6.

2.6.6   *Geometries: fine details & thin walls*
Fine details due to:

- Good separation of support from model
- Precise jetting
- Can build thin walls of 600 μm or less, depending on geometry

2.6.7   *Geometries: accuracy*
- High accuracy due to a combination of:
- Minimal model warp age due to material stability

- Software corrections to optimize model-support interface surface
- Precise electronics
- Precise mechanical movement (eg. z-tray +1 μm)
- Drop size of ~75 μm
- Accuracy of below 100 μm for most parts

2.6.8   *Geometries: smooth surface finish*
- No "step" effect due to 16 μm
- Good separation of support layer – including inside cavities
- No hard support grid leftover
- High quality surface for finishing, easy to paint

3   CONCLUSIONS

Jetting of photopolymers to build rapid prototypes offers the possibility to create accurate, detailed 3-dimensional models with any geometry required. The technology described provides successful jetting of multiple photopolymer materials, in flexible combi-nations, with easy removal of support constructions and includes easily maintainable parts, built-in trouble-shooting and automatic calibration systems. All of these ensure efficient rapid prototype construction in a time saving and cost effective system, which is suit-able for office use.

Rapid tooling and manufacturing

Virtual modeling and rapid manufacturing – Bártolo (eds)
© *2005 Taylor & Francis Group, London, ISBN 0 415 39062 1*

Statistical analysis of the effect of processing conditions on powder catchment efficiency in the Direct Laser Deposition (DLD) process

W. Syed, A.J. Pinkerton & L. Li
LPRC, School of Mechanical, Aerospace and Civil Engineering, The University of Manchester, Manchester, UK

E. Al-Eid & J. Pan
School of Mathematics, The University of Manchester, Manchester, UK

ABSTRACT: Metallic powders are used in direct laser deposition (DLD) processes to build three-dimensional objects directly from a computer aided design (CAD) model. The low powder deposition efficiency (below 40%) has been a concern for the industries using the techniques, although undamaged powders can be recycled in some applications. However, for aerospace industries, re-use of powders may not be desirable because of the very high part quality required. In addition, poor efficiency increases the proportion of misdirected, semi-molten powders, which adhere to the surface adversely affecting the surface finish. In this investigation, design of experiments and statistical modeling techniques were used to understand and identify the most influential parameters affecting the deposition efficiency in this multiple variable process. Experiments were conducted to simultaneously study the effect of 9 DLD process parameters. Using a 1.5 kW diode laser to deposit 316 L steel powder the maximum powder usage efficiency was found to be limited to 60%. After identifying the most influential factors and trends, further tests were then carried out and the deposition efficiency was raised to 75%. The work identifies the significant deposition parameters and shows the importance of targeting them.

1 INTRODUCTION

The Direct Laser Deposition process is an additive manufacturing process in which a laser creates a melt pool on a metallic substrate or previously deposited surface and additional powder is delivered to it to increase its size. Relative movement of the substrate and the laser cause rapid solidification of the pool, resulting in a raised track fused with the original substrate. A three dimensional part can be build by adding multiple tracks side by side to form layers or on top of each other to form thin walls.

The deposition process, using blown powders in conjunction with lasers, is complicated and difficult to model because it is governed by multiple different parameters. External (exogenous) process parameters that have a direct effect include: laser power, laser beam diameter, scanning speed, powder particle size, and powder delivery rate, powder carrying gas flow rate, powder delivery nozzle diameter, its offset distance and angle. Other (less well investigated) parameters that have an effect on the process include the influence of shielding gas, substrate temperature and position (Pinkerton & Li, 2004c), initial powder temperature

and other laser beam parameters (pulsed or continuous wave).

The influence of these parameters on the deposition layer thickness and width (McLean et al., 1997), microstructure (Smugeresky et al., 1997), dilution (Weerasinghe & Steen, 1983), melt pool shape (Pinkerton & Li, 2004b), thermal stress distribution (Rangaswamy et al., 2003) and surface finish (Smugeresky et al., 1997) are well documented. Some of these output parameters can be controlled to a good accuracy and some require further research (e.g. stress distribution control).

Deposition efficiency is an important factor, particularly in aerospace industry where the wastage of costly powder can negate the economic benefits of the whole process. It has received some attention, but in most cases the reported deposition efficiency remains below 40%. Many researchers have reported the trend of obtaining high deposition efficiency by increasing the laser power. Liu & Li (2005) studied the relationship between powder concentration distribution and power density and found that deposition efficiency increased with laser power. Syed et al. (2005) confirmed this relationship and also found that deposition efficiency increased with powder flow rate to an optimum value but can decrease with further increases above that value (Syed et al., 2004). Other researchers (Frenk et al., 1997; Resch et al., 2001) have also suggested that the deposition efficiency increases with powder delivery rate; Frenk et al. (1997) managed to achieve an efficiency of 69% when depositing Stellite on steel with an off-axis nozzle. Koch & Mazumder (1993) investigated the effect of laser power and beam diameter on residual stresses and deposition efficiency and found that:

$$y \propto 2Qr_l \tag{1}$$

where y is the catchment efficiency; Q the power; and r_l the laser beam radius. By utilizing a very high power CO_2 laser of 4 kW with a beam diameter of 3 mm and the powder delivery rate of 0.18 g/sec, they achieved a 90% deposition efficiency of aluminium powder with an off-axis nozzle. The maximum achievable with a coaxial nozzle was 60%. Values of this type are not the norm and are at the expense of feature resolution and commonly surface finish (Smugeresky et al., 1997).

Mclean et al. (1997) examined the effect of process velocity, laser power and powder mass flow on clad geometry. They found the catchment efficiency using a coaxial powder nozzle to be 7–8% to 7% and suggested that high powder catchment efficiency is obtained at lower process velocities. They claimed to achieve 85% catchment efficiency by changing the design of the nozzle, however this was not substantiated by any experimental results. Work by Carty et al. (1994) also showed that catchment efficiency falls with traverse speed. Contrary to these results, work with a 2.5 kW continuous wave CO_2 laser, coaxial deposition nozzle and CPM 10 V alloy powder (Hu et al., 1998) showed that for a given laser power, spot size and powder feed rate, the deposition efficiency is not affected by change in the traverse speed. That work also indicated that deposition efficiency increases with powder feed rate and laser beam diameter.

A series of work by Lin (1999, 2000) and Lin et al. (Lin & Hwang, 1999, 2001; Lin & Steen, 1997) modeled and experimentally investigated the effects of nozzle geometry and the powder stream on powder catchment efficiency in coaxial laser cladding. Modeling, based on the assumption that only powder falling on a melted surface is absorbed (Hoadley & Rappaz, 1992) and therefore powder efficiency is related to the ratio of powder stream and laser beam diameters, showed that the design of a coaxial nozzle can limit the maximum theoretical powder deposition efficiency it can achieve. The authors quoted this as 40% for the nozzle they used (Lin & Steen, 1997). Other work showed that catchment efficiency can be increased up to this maximum by reducing the particle size and velocity, reducing bonding temperature and ensuring

the powder stream is of smaller diameter than the laser beam at the deposition point (Lin, 1999). The significance of surrounding gas jet flows, particularly for the final factor, was also shown in this work (Lin, 2000).

Carty et al. (1994) compared the design of different nozzles to achieve maximum powder catchment efficiency. They used coaxial, lateral, venturi and multiple nozzles and found that the factors that affect the deposition efficiency are the supply rate of the powder, the divergence of the flow after leaving the nozzle and the velocity on hitting the substrate. Marsden et al. (1990) demonstrated that powder efficiency with an off-axis nozzle is linearly related to the angle of the nozzle to the substrate and varies non-linearly with the orientation of the nozzle axis to the traverse direction.

Despite all this work, optimization of the deposition phenomena has so far been limited to the traditional single factor approach. Because of the numerous exogenous parameters, researchers have held most values constant and assessed the effect of one (or sometimes a small set of) parameters independently. This has led to a piecemeal collection of relationships, however correct, and the lack of a unified model. By using design of experiments (DOE) and statistical tools, the effect and relative influence of each parameter can be considered together and calculated in order to determine those that are critical for the process and those that have little effect. Subsequently, the few most influential factors can be examined in further detail and manipulated to optimize the deposition process. Such a methodology has proved highly successful in the study and optimization of other laser-driven processes (Ghoreishi et al., 2002).

2 EXPERIMENTAL PROCEDURE

The laser used was a Laserline LDL 160–1500 diode laser, generating approximately equal proportions of 808 nm and 940 nm radiation up to a total power of 1.5 kW. The output beam had a nominally rectangular cross section that was focused through a lens of 300 mm focal length. The beam size at the substrate surface was experimentally measured by exposing an infrared detector card to the pilot beam and found to be 2.5 mm (slow axis) × 3.5 mm (fast axis). During deposition the laser head was fixed and vertical, and the substrate was traversed below it along the slow axis of the beam, as shown in Figure 1.

The deposition material was 316 L stainless steel (0.03% C, 2.0% Mn, 1.0% Si, 16.0–18.0% Cr, 10.0–14.0% Ni, 2.0–3.0% Mo, bal. Fe) powder with particle diameters of 53–106 μm or 107–150 μm. A FST PF-2/2 disk type powder feeder was used to deliver the powder. The substrate comprised 50 × 50 × 5 mm blocks of EN43A mild steel that were first sand blasted in a Guyson sand blaster and then degreased using ethanol. The surface roughness (Ra) so obtained was 4–5 μm, measured using a Surtronic 3+ contact probe. The optical absorptivity of the substrate at the mean wavelength of the diode laser ($\lambda \cong 874$ nm) was measured as 44% using an SD 2000 fibre optic spectrometer (Ocean Optic Ltd). The substrate was mounted onto a continuous flow water cooler, which was then mounted on a table that was manually adjustable in the z-axis (vertical). A CNC table controlled movement in the x- and y- axes (horizontal plane). The melt pool was shrouded in a 40 l/min, slow moving flow of argon gas, delivered via a lateral nozzle of 30 mm diameter. Single tracks were deposited in experiments performed according to the DOE procedure outlined below and the samples were then sectioned in a traverse plane. The cross-sectional area of deposited material was measured using optical microscopy and image analysis using the Adobe Photoshop 7.0 software package. Deposition efficiency was considered as the endogenous process variable of interest for this study; it was obtained from the process variables and measured cross-section for each track via equation 2:

$$y = A\rho v/m' \qquad (2)$$

where y is the catchment efficiency; A is the measured cross-sectional area; ρ is the material density; v is the traverse speed; and m' is the mass flow rate.

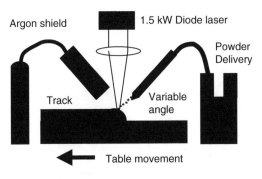

Figure 1. Experimental setup.

After statistically analyzing and studying the trend for each factor, further confirmatory experiments were carried out to endorse the trend and determine the powder catchment efficiency that could be achieved. The system was unchanged from previous tests, process parameters were set to maximize theoretical catchment efficiency and actual levels measured using the same image analysis method as previously together with equation 2.

2.1 Design of experiments

DOE is often called "active experimentation" as opposed to passive data gathering followed by analysis. Nine factors with various numbers of different levels were selected, as shown in Table (1). Varying these parameters one at a time would require more than 3,490,000 separate experimental evaluations, which practically seems to be impossible. There are several different commercial methods available for DOE and after assessing these, the orthogonal design method (OD) (Yates, 1933) was selected for this study. OD is based on application of orthogonal arrays (OA) and provides a systematic and efficient method for the optimization of a multiple parameter system. Construction of OA follow well established and clearly defined mathematical rules (Hedayat et al., 1999). They are readily expandable for required number of factors so for practically any multi-factorial experiment a suitable OA can be constructed and the conclusions that can be drawn are valid over the entire experimental spectrum. The experiments were performed in a random order; this is a part of the DOE method that eliminates the systematic problems of non-random variations by multiple other external parameters such as temperature or humidity change in the laboratory, effects of continued equipment use or powder settling, operator alertness and so forth.

A commercial software package, SPSS 11.5.1 (SPSS Inc, Chicago IL USA) was used to determine the experimental design and order. The total numbers of experiments was reduced to 81 using different configurations of the parameters.

Table 1. Parameters and their different levels used for the experiments.

	Parameter	Levels								
1	x_1 = Laser power (W)	700	800	900	1000	1100	1200	1300	1400	1500
2	x_2 = Beam diameter (mm)	3.5	3.75	4	4.25	4.5				
3	x_3 = Table speed (mm/sec)	3	4	5	6	7				
4	x_4 = Particle size less than (μm)	106	150							
5	x_5 = Powder delivery rate (g/sec)	0.13	0.19	0.25	0.32	0.37	0.44	0.55	0.68	0.75
6	x_6 = Gas delivery rate (l/min)	2	2.5	3	3.5					
7	x_7 = Nozzle angle (degree)	30	40	50	60	70	80			
8	x_8 = Nozzle diameter (mm)	1.5	2	2.5	3					
9	x_9 = Nozzle off set (mm)	4	5	6	7	8	9	10	12	14

2.2 Statistical analysis

Regression analysis (RA) was used to model the results. RA is a statistical technique that is widely used for analyzing and modeling multifactor data (Montgomery et al., 2001). The results arose from a designed experiment rather than observational study and in the regression model we must assume that the response variable (in this case catchment efficiency) was normally distributed. The collected data was first normalized using the Box-Cox transformation such that $U = F_{box-cox}(y)$ in order to minimize the residual sum of sequences in the model. The modeling was started by considering all the factors and all the possible interactions between these factors, forming a ninth-order model. The method of least squares was used to estimate the regressor coefficients and the significance of each (p-value) was calculated; this value represents the probability of an effect being due to random variations rather than a causal relationship if a normal distribution of error is assumed. The model was then stepwise developed by the linear multiple regression method. The insignificant interactions were gradually eliminated and after each elimination the significance level (p-value) of each variable and interaction calculated.

An Analysis of Variance for significance of regression test was performed on the final model. This is used to determine if there is a linear relationship between the endogenous variable (in this case deposition efficiency) and any of the regressor variables. The procedure is often thought of as an overall or global test of model adequacy (Montgomery et al., 2001). The final model was developed by checking the normal probability line of residuals.

3 RESULTS

The results obtained during the DOE tests showed a maximum deposition efficiency of 60%. The normal probability curve for the raw and normalized deposition efficiency data is shown in Figure 2; the more linear the probability plot the better the normal distribution. The need for, and effect of, the Box-Cox transformation can be clearly seen.

The final model is represented by equation 3:

$$U = 30.7 + 0.00166\,x_1 - 3.13\,x_2 - 0.0168\,x_3 - 0.0179\,x_4$$
$$- 0.296\,x_5 - 0.350\,x_6 - 0.0995\,x_7 - 13.5\,x_8 + 0.739\,x_9$$
$$+ 0.000013\,x_1 x_3 - 0.000328\,x_1 x_9 - 0.133\,x_8 x_9 + 1.81\,x_2 x_8$$
$$+ 0.146\,x_5 x_8 + 0.0109\,x_7 x_9 + 1.25\,x_8^2 - 0.0419\,_9^2 \quad (3)$$

Analysis of Variance of the estimated model gives a confidence value of 0.000 (to 4 d.p.) as shown in Table 2. This indicates that at least one coefficient is

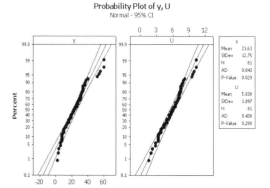

Figure 2. Probability plot for un-normalized (y) and normalized catchment efficiency (U).

Table 2. Analysis of Variance.

Source	DF	SS	MS	F	P
Regression	17	208.8	12.28	9.77	0.000
Residual error	63	79.2	1.26		
Total	80	288.0			

where DF is Degrees of Freedom; SS is Sum of Squares; MS is Mean Square; F is the ration of regression and residual MS; and P is the confidence.

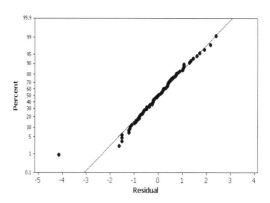

Figure 3. Normal probability plot of the residuals (response is normalized catchment efficiency, U).

different from zero and increases confidence in the model. The R- Sq value is 72.5%, indicating that the estimated model is a good fit to the experimental data.

The residual plot shown in Figure 3 gives further confirmation of the significance of the estimated model. Only one data point is outside the fitted curve, that at a lower level.

The final conclusions of the statistical modeling, showing estimated restitution coefficient (R), standard

Table 3. Regression analysis: the effect of factors and their interaction with each other.

Predictor	R Coefficient	SE Coefficient	T	P	Effect
Constant	30.727	7.354	4.18	0.000	
x_1-Laser power	0.001657	0.002267	0.73	0.468	Insignificant
x_2-Beam diameter	−3.126	1.551	−2.02	0.048	Significant
x_3-Table speed	−0.016839	0.005450	−3.09	0.003	Significant
x_4-Particle size	−0.017896	0.005310	−3.37	0.001	Significant
x_5-Powder delivery rate	−0.2955	0.1549	−1.91	0.061	Significant
x_6-Carrying gas delivery rate	−0.3498	0.2215	−1.58	0.119	Partially significant
x_7-Nozzle angle	−0.09948	0.02286	−4.35	0.000	Highly significant
x_8-Nozzle diameter	−13.534	3.678	−3.68	0.000	Highly significant
x_9-Nozzle offset	0.7389	0.3473	2.13	0.037	Significant
$x_1{}^* x_3$	0.00001304	0.00000481	2.71	0.009	Significant
$x_1{}^* x_9$	−0.0003285	0.0001639	−2.00	0.049	Significant
$x_8{}^* x_9$	−0.13260	0.07253	−1.83	0.072	Significant
$x_2{}^* x_8$	1.8089	0.6982	2.59	0.012	Significant
$x_5{}^* x_8$	0.14589	0.06927	2.11	0.039	Significant
$x_7{}^* x_9$	0.010931	0.002589	4.22	0.000	Highly significant
$x_8{}^2$	1.2527	0.5090	2.46	0.017	Significant
$x_9{}^2$	−0.04192	0.01378	−3.04	0.003	Significant

error coefficient (SE), standard error coefficient error (T) and confidence level (P) for each significant predictor are in Table 3.

After this analysis, the parameters were optimized for catchment according to the model by setting, $x_1 = 1.5\,kW$, $x_2 = 14.25\,mm$, $x_3 = 3\,mm/s$, $x_4 = 100\,\mu m$, $x_5 = 0.25\,g/s$, $x_6 = 2\,l/min$, $x_7 = 40°$, $x_8 = 1.5\,mm$ and $x_9 = 5\,mm$ (variables as defined in Table 1). With these values, the measured catchment efficiency was 72%. This was considerably higher than in any of the previous tests.

4 DISCUSSION

The first observation to be made from Table 3 is the large number of significant variables predicted by the statistical analysis, confirming the multivariable nature of the process and therefore the benefits of analyzing it using statistical methods.

The most significant factors from the table were found to be powder delivery nozzle diameter (x_8), its angle with the substrate (x_7) and the interaction between the nozzle angle and the nozzle offset distance (x_7*x_9). These are the parameters previously investigated by Marsden et al. (side feed nozzles) (Marsden et al., 1990), Lin et al. and Pinkerton & Li (coaxial nozzles) (Lin, 1999; Lin & Steen, 1997; Pinkerton & Li, 2004a). They can be considered as first-level parameters in that they directly alter the amount of powder falling within the extent of the melt pool and therefore the catchment (Hoadley & Rappaz, 1992). They also have a direct effect on the laser-powder interaction time above the melt pool and hence degree of powder pre-heating before reaching the pool;

another significant factor for all aspects of track formation layer dimension (Pinkerton, 2005). Both x_7 and x_8 have negative coefficients, indicating that increasing either powder delivery nozzle diameter or angle with the substrate will reduce the catchment efficiency. The former is intuitive as it reduces the powder concentration toward the centre of the stream; the latter is not in accordance with previous work. Assuming a horizontal melt pool, increasing the nozzle angle increases the area of it exposed to the powder stream and Marsden et al. (1990) found the opposite: that catchment efficiency increases with nozzle angle. A possible explanation is the height of tracks that was being produced. Referring to Figure 1, for nozzle angles above the inclination of the melt pool front, further increases in nozzle angle actually reduce the area of melt pool presented to the stream. Table 1 shows that angles of between 30° and 80° were tested so this may well have occurred in some cases.

The table speed and mass delivery rate were assigned negative coefficients in Table 3, indicating that deposition efficiency could be increased by reducing these factors. This is in accordance with the work done by other researchers (e.g. (Carty et al., 1994; McLean et al., 1997)); these relationships have been explored in detail in the past. The negative relationship found between particle size and catchment efficiency has been less well explored, but is in accordance with results by Lin (1999).

The coefficient for laser power (x_1) in the table is positive, suggesting that higher laser power would give high deposition efficiency. Its significance alone is low, but when interacting with table speed (x_3) or nozzle offset distance (x_9) it is significant. The negative relationship between beam diameter (x_2) and catchment efficiency can be explained by considering the fact that a diode laser with rectangular beam was being used so the diameters are in fact values chosen to give equivalent beam area and used to show variations in intensity. At smaller values the power density is higher and the actual, rectangular beam shape is more elongated along the slow axis, leading to an increased irradiated area (and therefore presumably melt pool) in conjunction with the impinging powder stream.

At the confirmation and optimization stage, working with the same system and setting the most important parameters to their optimum values increased the deposition efficiency to a maximum of 72%. However, it was clear that this was the limit of the system. For example, further reduction in the highly significant predictors of powder delivery nozzle diameter and the interaction of nozzle angle and offset caused the flow of the powder conveyance gas to become so fast that it significantly distorted or detached the melt pool. It is clear that the optimization could be carried out only within an operating envelope limited by factors inherent in the process.

5 CONCLUSION

Design of Experiments and statistical analysis using multiple linear regression and analysis of variance of techniques has been used to study the effect of nine process variables on the laser direct metal deposition process. The number of experiments was reduced to practicable levels and the data proved suitable for analysis using this method, yielding a regression model with no abnormalities.

The most significant process variables from those considered were found to be powder delivery nozzle diameter (negative relationship), its angle with the substrate (negative relationship) and the interaction between the nozzle angle and the nozzle offset distance (positive relationship). This is attributed to these being first-level effects that have a direct effect on powder falling within the limits of the melt pool possibly enhanced by their secondary effect on laser-powder interaction above the melt pool.

REFERENCES

Carty, S., Owen, I., Steen, W. M., Bastow, B., & Spencer, J. T. 1994. Catchment efficiency for novel nozzle designs used in laser cladding and alloying: 1994 NATO Advanced Study Institute Conference on Laser Processing: Surface Treatment and Film Deposition: 395–410.

Frenk, A., Vandyoussefi, M., Wagniere, J.-D., Zryd, A., & Kurz, W. 1997. Analysis of the laser-cladding process for stellite on steel. Metallurgical and Materials Transactions B: Process Metallurgy and Materials Processing Science, 28: 501–508.

Ghoreishi, M., Low, D. K. Y., & Li, L. 2002. Comparative statistical analysis of hole taper and circularity in laser percussion drilling. International Journal of Machine Tools and Manufacture, 42: 985–995.

Hedayat, A. S., Sloane, N. J. A., & Stufken, J. 1999. Orthogonal Arrays: Theory and Applications: Springer.

Hoadley, A. F. A., & Rappaz, M. 1992. A thermal model of laser cladding by powder injection. Metallurgical Transactions B, 23: 631–642.

Hu, Y., Chen, C., & Mukherjee, K. 1998. Innovative laser-aided manufacturing of patterned stamping and cutting dies: Processing parameters. Materials and Manufacturing Processes, 13: 369–387.

Koch, J. L., & Mazumder, J. 1993. Rapid prototyping by laser cladding: 12th International Congress on Applications of Lasers and Electro-Optics (ICALEO): 556–565.

Lin, J. 1999. A simple model of powder catchment in coaxial laser cladding. Optics & Laser Technology, 31: 233–238.

Lin, J. 2000. Numerical simulation of the focused powder streams in coaxial laser cladding. Journal of Materials Processing Technology, 105: 17–23.

Lin, J., & Hwang, B.-C. 1999. Coaxial laser cladding on an inclined substrate. Optics & Laser Technology, 31: 571–578.

Lin, J., & Hwang, B.-C. 2001. Clad profiles in edge welding using a coaxial powder filler nozzle. Optics & Laser Technology, 33: 267–275.

Lin, J., & Steen, W. M. 1997. Powder flow and catchment during coaxial laser cladding. Proceedings of SPIE, 3097: 517–528.

Liu, J., & Li, L. 2005. Effects of powder concentration distribution on fabrication of thin-wall parts in coaxial laser cladding. Optics and Laser Technology, 37: 287–292.

Marsden, C. F., Frenk, A., Wagniere, J. D., & Dekumbis, R. 1990. Effects of injection geometry on laser cladding: 3rd European Conference on Laser Treatment of Materials (ECLAT): 535–542.

McLean, M. A., Shannon, G. J., & Steen, W. M. 1997. Mouldless casting by laser. Proceedings of SPIE, 3102: 131–141.

Montgomery, D. C., Peck, E. A., & Vining, G. G. 2001. Introduction to Linear Regression Analysis: Wiley-Interscience.

Pinkerton, A. J. 2005. Alternative Nozzle and Metal Delivery Methods for Laser-assisted Metal Deposition (invited): UK Association of Industrial Laser Users Workshop on Laser-assisted Metal Deposition.

Pinkerton, A. J., & Li, L. 2004a. Modelling powder concentration distribution from a coaxial deposition nozzle for laser-based rapid tooling. Transactions of the ASME, Journal of Manufacturing Science and Engineering, 126: 34–42.

Pinkerton, A. J., & Li, L. 2004b. Modelling the geometry of a moving laser melt pool and deposition track via energy and mass balances. Journal of Physics D: Applied Physics, 37: 1885–1895.

Pinkerton, A. J., & Li, L. 2004c. The significance of deposition point standoff variations in multiple-layer coaxial laser cladding (coaxial cladding standoff effects). International Journal of Machine Tools and Manufacture, 44: 573–584.

Rangaswamy, P., Holden, T. M., Rogge, R. B., & Griffith, M. L. 2003. Residual stresses in components formed by the laser-engineered net shaping (LENS) process. Journal of Strain Analysis for Engineering Design, 38: 519–527.

Resch, M., Kaplan, A. F. H., & Schuocker, D. 2001. Laser-assisted generating of three-dimensional parts by the blown powder process: XIII International Symposium on Gas Flow and Chemical Lasers: 555–558.

Smugeresky, J. E., Keicher, D. M., Romero, J. A., Griffith, M. L., & Harwell, L. D. 1997. Laser Engineered Net Shaping (LENS) process: optimization of surface finish and microstructural properties: 1997 International Conference on Powder Metallurgy and Particulate Materials. Part 3 (of 3): 21–33.

Syed, W., Pinkerton, A. J., & Li, L. 2004. Combined wire and powder feeding laser direct metal deposition for rapid prototyping: 23rd International Congress on Applications of Lasers and Electro-optics (ICALEO): CD.

Syed, W., Pinkerton, A. J., & Li, L. 2005. A comparative study of wire feeding and powder feeding in direct diode laser deposition for rapid prototyping. Applied Surface Science, 247: 268–276.

Weerasinghe, V. M., & Steen, W. M. 1983. Laser cladding with pneumatic powder delivery: 4th International Conference on Lasers in Materials Processing – American Society for Metals (ASM): 166–174.

Yates, F. 1933. The analysis of replicated experiments when the field results are incomplete. The Empire Journal of Experimental Agriculture, 1: 129–142.

Virtual modeling and rapid manufacturing – Bártolo (eds)
© *2005 Taylor & Francis Group, London, ISBN 0 415 39062 1*

Rapid manufacturing and the modern mould maker

Steven Andrews
The EDM Shop affiliated Central University of Technology, South Africa

Jacques Combrink, Ludrick Barnard, Gerrie Booysen & Deon de Beer
Central University of Technology, South Africa

ABSTRACT: As delivery times and costs of dies and moulds are on a downward trend, the modern mould manufacturer is under more and more pressure to produce moulds and dies quickly, accurately and at a lower cost.

This paper deals with manufacturing techniques that enable the mould manufacturer to produce "net shape"
moulds and dies in a short period, using modern machining methods.

Whilst designers have high tech design tools, in the form of complex CAD systems, the mould maker has
being playing catch-up to manufacture these high tech designs. The traditional tool room consists of many different machines using both conventional and CNC technology, but in order to meet the new tight delivery
schedules, most clients require, much of the machining needs to be done in the "hard", the use of WEDM, high
speed machining, CNC EDM, and 5 axis machining in non-traditional roles, become the high tech tools the
modern mould maker can use to bridge the gap between designer and mould maker.

Much of this paper deals with the interface between man and machine and techniques to use "conventional"
machines in "non-conventional" roles.

1 INTRODUCTION

Whilst the advents of computers was supposed to
make our lives easier, it is a fact they have made the 1
lives of "Modern Machinists" more difficult. This
may be a large statement, but before the advent of
computers the things they had to manufacture were
simple. Now designers have many computerised tools
at their disposal and their designs have become more
pleasing to the consumers eye, but the moulds and
dies to produce these "curvaceous" products are far
more difficult to manufacture.

Many mould manufacturers have the same com-
plaint, "I have spent money on software and machinery, but I feel like I have gone backwards". The facts
of the matter are clear; one needs to understand the
mechanics of turning modern design into product.

The machinery, machinists use today, is similar to
that used 50 years ago, the only difference is the motion
is now controlled by computer and not by the artisans
hands. In order to bridge the gap between the designer
and mould maker, one needs to think outside the box.

2 TOTAL DESIGN

The tool design is the foundation of successful and
rapid metal removal. A hybrid system (one that can

deal with both solids and surfaces) is best as it allows
for both geometric and free form modeling. These
packages are relatively inexpensive (around USD
7000 – 00), very easy to use, and produce brilliant
"blue prints".

Before you even start your tool design, one needs
to have a product that is perfect in all manners. If this
is to be a progression tool then one should design the
strip, before commencing with the tool design.

The rest is simple. Place the part or parts, scale them
up for shrinkage, put blocks of steel around the product,
subtract the product, look for electrode shapes, subtract
the electrodes and print blueprints. In our research, even
more complicated tooling it took only 25–45 hours
for a complete design, including process planning.

When printing "blue prints" the modern machinist
only needs to dimension basic reference sizes as the
rest comes from the electronic data.

This data was then converted to the following
formats:

1. For 3D shapes (i.e. CNC milling):   *.iges
2. For 2D shapes (i.e. Turning and WEDM):   *.dxf
3. The rest should be printed onto paper.

Whilst it was found that a solid modeling package was
best employed to design the tooling, it was found that
the tool paths were created more quickly and easier

369

using surface machining and simple 2 axis turning and WEDM.

There are two reasons for this. Firstly, it was much easier to identify machining areas using surfaces (i.e. you can specify specific areas to machine). For turning and WEDM we found we only needed the simple tool path and decided not to "clutter" the PC's with un-needed data.

The second reason was all CAM programs normally come with a dxf and iges converter. There was no reason to spend extra money on expensive parasolid converters.

When we didn't complete the design (i.e. a total design) we found the project soon lagged behind the targets that were set. The conclusion to this is that a complete design, both 3d models and 2d blue prints are the foundations of successful machining. To start machining an incomplete design is downfall modern machinists. It is common knowledge that most machine tool shop owners do not to have the patience, and customers often put pressure on the machinists to start cutting steel, but starting to machine before the design and planning is in place is the same as building a house with poorly prepared foundations, disastrous.

3 MACHINING OUTSIDE THE ENVELOPE – AN OVERVIEW

The target delivery time for any single mould that has been made during this project was 21 working days, so during the design process the engineers were already playing with ideas and solutions to bottlenecks. The solid modeling packages were used to simulate fixturing and they were constantly trying to minimize the use of artisans and increase the use of CNC motion control.

A number of common fixtures were used for the manufacture of components, including slides, punches, ejectors, pockets and such like.

The place were most ground was gained was in figuring out how to machine materials in their hardened state and machine to as close as possible to the required surface finish.

Machining hard materials only has one drawback, there are not easy ways to put something extra in you haven't planned for. Cooling is a good example. If you forgot to drill a hole for cooling it takes a long time to spark the hole in. So the design intent was changed and cooling was machined in the hardened state, by hollowing out the material behind the cavity or punch and then using beryllium-copper baffles and cover plates to seal the hollow areas and direct the cooling.

Basically all that was done to the soft blocks was they were drilled and tapped in the appropriate places (often using a radial arm drill and a plotted out drawing!) and wire start holes were also drilled at this

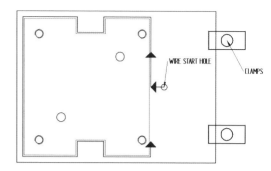

Figure 1.   Typical wire path.

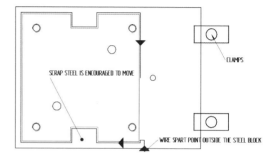

Figure 2.   Modified wire path.

point. Very often these drilled and tapped holes were only used for fixturing for machining and then these were wire cut off as the last process.

After this the mould or die components were heat-treated and tempered. Tempering is a most important part of the process as it strongly affects the accuracy of further operations. It was found that if the "growth of material" was not as expected and the material was less magnetic than expected then the material was re-tempered. Poor tempering caused a lot of movement, due to internal stresses, during high-speed wire EDM operations. The method of machining was changed from the more accepted method of cutting (see fig 1) to a method that allowed the scrap to move (see fig 2) and this produced more reliable results.

Whilst the blocks were at heat treatment the electrodes were prepared. Some simple rules of thumb were used, to determine whether or not an electrode needed to be made.

Any internal corner less than radius 0.5 mm was sparked out, as well as any feature where the depth was more than 10 times the diameter or width of the feature. An in fact during the design process the engineers tried to avoid any of these types of mould features.

The electrodes were pre-mounted on an Erowa™ system and marked as to orientation for later Ram

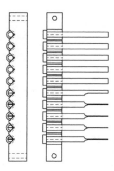

Figure 3.   Jig for holding multiple blade ejectors.

EDM operations. Blueprints were made for each electrode as well as the electrode's position according to its corresponding steel component.

Electrode material varied according to application. It was found that if the electrode had to be wire-cut at any time it was easier to manufacture the electrode from copper, but otherwise the electrodes tended to be made from graphite due to its ease of machining. Graphite grades also varied according to application. By packing the electrode with sawdust, mixed with coolant, the problem of graphite dust getting into the machine slides and spindle, during machining, was overcome.

Ejector pins were prepared at this time as well. All ejector pins were cut to length on the Wire EDM machine and although this may seem like it is not cost effective, the time saved during fitting more than made up for this cost. A V-block fixture was fixed permanently to the WEDM machines bed, with a fixed datum, on the front of the ejector pin's head. A macro program was provided so the operator only had to put in the diameter of the pin and its length to cut the pin to size. Blade ejectors were also manufactured from nitrided round ejector pins. The jig shown in fig 3 was used, together with a macro program to produce 10 pins at a time. This was normally completed overnight. These pins were then nitrided again to give a resistance to abrasion. The engineers were able to cut the exact length of the blade, thereby preventing breakages caused by the overly long blades of off the shelf items.

The cost of manufacturing these pins is around 40% of off the shelf items and the engineers could design for the maximum size, so as to prevent breakage.

The engineers always tried to think in what was eventually termed 90° design. They would look at what they could do with the part while it was fixed in a certain position and after those operations were done how they would fixture for the next operation. The solid modeling interface was used to simulate the manufacturing process and often step-by-step blueprints were produced to show the machine operators what was required.

The 90° design theory led to more fixtures being designed to speed up the clamping process and fixtures were designed to facilitate 20° fixturing. These fixtures were used to produce all the slides and moving cores on the moulds. These fixtures had common datum holes according to certain templates used by the design engineers and again these datum positions were marked on blueprints for the operators to follow. In fact all shop floor operations consisted only of clocking one axis and setting an X, Y and Z datum. These datum holes were never removed until the last operation and in fact the engineers tried to leave as many of these holes in the work pieces as possible.

4   THE SPECIFICS

4.1   *High speed machining*

There was initially a 1000×600 HSM with 18000 rpm placed at our disposal and later a 450×350 fully interpolated 5 axis HSM with 24000 rpm.

Depending on the size of the core's and cavities, they may or may not have been roughed out in the steels original state. The smaller blocks (<200 mm $\times$ 200 mm) were generally machined from a solid billet after heat treatment, but larger blocks were often "hogged out" in their soft state. The engineers found the minimum amount of material to be left on for "finish" machining was 2 mm. This is due to the extreme buckling the material underwent during heat-treatment.

Cutters with virtually no helix were used to improve tool life. The cutter library was kept very, very small. It was found that the CAM engineers learned how to use these few tools more effectively than a large range of tooling. This also helped keep costs down. In fact for die sinking the entire range of cutters were as follows.

– Ø32 r 6 button insert mill
– Ø25 r 1.2 insert mill
– Ø20 r 1.2 insert mill (common insert to Ø25)
– Ø12 r 3 button insert mill
– Ø12 minimaster with 2 different shanks and 4 insert types
– Ø12 minimaster with 5 different shanks and 4 insert types

Solid carbide cutters in various configurations were used from Ø6 to Ø1.

Programming was done on EdgeCam and PowerMill. Again there was very little or no deviation in the tool paths used. In fact the best results were obtained when only 3 cycles were used. This was probably because the engineers were familiar with these cycles and knew exactly how to make them work in various situations. However around 50% of the time, extra tool paths had to be produced during manufacture to get the required surface finish.

It was found that very shallow depths of cut at high feeds and spindle speeds produced the best results with regard to cutter life, time taken and surface finish. When it came to cutting areas with very small cutters, often the depth of cut was less than 10 microns. Although this may seem extreme, 1 mm ribs were cut to a depth of 24 mm (in 54 Rc steel), with no cutter breakage and very little wear. Again the time taken was much shorter than sparking and polishing. In fact the aim of the engineers was for polishing to begin with 800grit stones or paper. You would have needed 4 or 5 electrodes to get the same effect.

4.2 Axis machining

In the manufacture of one off parts for moulds and dies 5 Axis machining has limited success, due to the complications of programming and difficulties in fixturing. Having said that, the 5-axis machine was very useful for preparing blocks, especially slide blocks for heat treatment. All of the holes could be drilled and tapped in one operation. In mould making everything runs in one direction and therefore it was found the 90° design theory worked very well.

5 Axis machining comes into it's own, when it comes to the manufacturing of large moulds and complex split lines. Again this will be more a 3 + 2 machining situation, where the block is indexed to the correct poison and then machined using conventional 3-axis programming. Again 5 axis machining facilitated a better surface finish due to the fact that the cutter never had a zero peripheral speed.

4.3 Wire EDM operations

One of the "tools" the modern mould maker has be given to bridge the gap between designer and mould maker is the Wire EDM machine. This machine allows the modern mould maker to make extremely small or complicated parts or electrodes with the greatest of ease.

It is a fact that WEDM (Wire cutting) is a fast becoming the process of choice when it comes to manufacturing moulds, dies, electrodes and components of a mechanical nature.

Well, with the advent of affordable submerged WEDM machines with reliable AWT (Automatic Wire Threading) systems, it is possible to manufacture extremely accurate parts (±0.002 mm) of a complex nature, with little effort, in a cost effective manner.

Injection mould slides were manufactured using a combination of the 90° design theory and a WEDM machine. In the past this slide would be made using milling, grinding and possibly spark erosion. These processes would have to be done by skilled people and there is no way the slide would have been made within tolerance of 0.002 mm or the surface finish

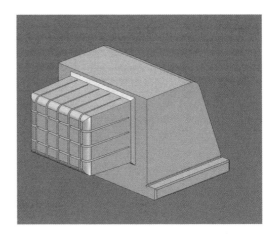

Figure 5. Typical example slide made using WEDM machine.

better than 0.6 Ra. Also the slide would have to be ground and spark eroded after heat treatment and we all know what problems distortion during heat treatment can cause.

Using a submerged WEDM with AWT, this slide (fig 5) was made with three setups (less than 20 minutes using the Erowa system) and no intervention by the operator after setup. Multiple slides were also manufactured overnight, again with no operator intervention. The wire start holes were drilled and tapped heat treatment, and the holes were wire cut afterwards to give extremely accurate fits.

Traditionally ejector pinholes are core-drilled, drilled and reamed before heat treatment on machining centers or milling machines. This necessitated using three tool holders for each size ejector pin the engineers wanted to use. The reality of this issue was forcing the engineers to use only two or three sizes of ejector pins during the design process. Also there was considerable time wasted "opening" the holes to allow the ejector pins to slide after they had "closed up" during heat treatment. The solution to this was to only drill the holes to between 1 mm and 3 mm smaller during the soft drilling phase. Then a no-core wire cutting cycle was used to open these holes to the correct H7 tolerance after heat treatment. A special post processor was written to facilitate the programming of these no-core holes. This ensured both positioning and hole-size was perfect with absolutely the minimum user intervention.

Wire EDM turning was also used to produce small round electrodes and ejectors pins. This in fact sparked the overseas supplier of the machine to start investigating and in fact build a 6th axis rotational movement into their new models. This will be particularly useful in the manufacture of helical gears and worm gears, without too much user intervention.

The engineers also used WEDM as much as possible to cut split lines, cavities, ribs and electrodes. The running costs of the WEDM machine were found to be less than USD 5-00/hr. The engineers also used WEDM where grinding operations would have been previously used.

4.4 *CNC EDM*

HSM has rapidly taken over from EDM in the modern machine shop, but there are still some places where CNC EDM operations are without an equal. All internal sharp blind corners were CNC EDM'd and often engraving was done using 3d interpolation with a sharp tungsten copper electrode after the mould had been machined using the HSM centre. Whilst often bead blasting or chemical etching were used to give a specified surface finish, it was found CNC EDM could be used to achieve better and much more controlled results. Electrodes were made using various techniques including hand carving and then EDM'd using overlapping techniques. This was found to be very useful when the mould needed to have a "combination" surface finish (i.e. Where the mould had both shiny and etched areas) Previously the only way to achieve these results were with inserts in the mould.

4.5 *Rapid prototyping*

Whilst the progress rapid tooling techniques (excluding CNC manufacturing) has been staggering, they still remain prohibitively expensive. One cannot ignore the future these techniques have and indeed spray-on layered manufacturing using multiple types of materials in a near net shape product and then machining to a finished net shape using conventional CNC techniques remains an attractive incentive to investigate these techniques. In fact the engineers used rapid prototyping techniques to manufacture inserts for a time and quality comparison to assess the future of these technologies at regular time periods. These inserts were never of the quality that could be put into a mould for production purposes.

These techniques, in the future, will give tooling engineers an increase in choice as to a production method, especially as the volumes these moulds need to produce come down in line with current manufacturing predictions.

5 CONCLUSION

This project has been very successful in terms of the following innovations.

a) Using solid modeling to predict and facilitate rapid manufacturing techniques.
b) 90° design theory to facilitate fixturing and machine setup.
c) Many new WEDM techniques and how to use WEDM to automate the manufacturing process.
d) HSM using the minimum amount of tools to get the best surface finish.
e) Using CNC EDM to get surface texture.

The following findings must also be noted:

a) Simultaneous 5 axis machining is a relatively new technique and will come to the fore in the future.
b) The process is as important as the design, and good designs are made around the processes available to manufacture the parts. For example, in China, where there is an abundance of human Resources these techniques may not be as successful, from a cost point of view, as they were in South Africa.

REFERENCES

Ke-Zhang Chen, Feng Wang, Xiang-Yang Feng, Xin-An Feng, Behavior modeling for the spraying device in the layered manufacturing process for the components made of a multiphase perfect material, RAPDASA 2004
Dr. J. Fourie, Rapid Tooling Techniques – A South African Perspective, RAPDASA 2003
Mitchell, W.P. Distributive Mould Making, RAPDASA 2003

Virtual modeling and rapid manufacturing – Bártolo (eds)
© *2005 Taylor & Francis Group, London, ISBN 0 415 39062 1*

Assembly design of hybrid rapid parts

M. Rivette, P. Mognol & J.Y. Hascoët
Institut de Recherche en Communications et Cybernétique de Nantes, Nantes Cedex, France

ABSTRACT: Currently, product development times are increasingly short but at the same time, the developed products have more and more special customization. Moreover, at each step of product evolution, prototypes are often entirely re-manufactured for testing, which increases time and cost. To reduce this time to market for new industrial products, we have developed a new methodology based on a multi-component prototype (hybrid rapid prototype) approach.

The part is decomposed into multi component prototype (MCP) instead of a part which made in one part, thus the manufacturing process is optimized and enables a larger reactivity during the development of the product.

In this paper, we present our method, which proposes, starting from features, a functional model for representation, analysed according to four different points of view:

– The feasibility analysis.
– The geometry analysis.
– The manufacturing analysis.
– The assembly analysis.

In this paper, we develop more particularly this last part. We propose a methodology of assembly for hybrid rapid part. This method on the basis of data CAD Step AP-224 must enable us to obtain an exhaustive list of solutions of gatherable module. This work is illustrated with an industrial example.

1 INTRODUCTION

1.1 Context

Modern means for the improvement of efficiency in manufacturing are quite diverse. However, the traditional cornerstones, cost, quality and time are the targets that business is managed. Thus mass customization is one of the modern means to achieve these goals, and allows customizing product to individual clients and producing those with principles of mass production. For many products, the competitive edge of the producer is dependent not merely on the price, but also on the choices or variations provided in each product line. Examples of such products range from automobiles (manifested by the increasing number of "variants" available in any base model of a car) to electronic products such as computers. The challenge is to create a variety of products from a common family without a significant trade-off in production costs or lead time.

In order to address the high cost of this practice, manufacturers develop product families from a common platform that is shared by all the products, whose variants are designed to fulfill different customer demands. The variants are created by adding specific components to the platform. The use of a small batch with an alternative is thus an increasingly frequent option.

At the same time, products have become more complex and the development times of products are increasingly long and laborious. In this context, the hybrid rapid prototype is appearing. In our work we study the rapid tooling and the prototype part.

The tools can be of evolutionary, for bridge tooling or for small series. The parts can be aspects, geometrical, functional or technological prototypes.

1.2 Related works

Hybrid rapid prototyping has been presented in some recent papers (Hur & Lee 2002, Chen & Song 2001). Their research was based on the study of two processes: CNC and Rapid Prototyping (RP) systems. CNC is used when the quality of the part is higher than what is a possible using RP system. Their methodology uses STEP AP-203 data but didn't take into account ISO specifications of the model. Likewise, concerning the manufacture of the parts, methodology that would allow the decomposition of a part into a space partitioning, was developed (Ki & Lee 2002). The weakness of these studies is that only build accessibility is considered. A choice of an adequate process is not proposed.

At the same time, the assembly of the hybrid prototypes and its repercussions on the design was never taken into account. These works do not envisage any interaction between manufacture, assembly and design.

Some research about product families (Erens & Verhulst 1997, Jiao et al. 1998), and design for assembly (Barnes, Jared & Swift 2004) related only the assembly of subset of a product (for example: a power supply) on the other hand this work does not treat assemblies of cast solid product composed of hybrid elements (for example: injection tooling).

Our concept aims to decompose the part on a Multi Component Prototype (MCP), instead of a part made in one piece. The two main reasons are to include the evolutionary requirement of the prototype regarding to the tests that are performed on it; and to optimize the manufacturing process locally, regarding to the component geometry and functional requirements. Therefore, only one or few components would have to be remanufactured separately, in order to update prototype geometry for testing purposes.

We also propose a methodology for assembly the multi component part by using the extraction and the use of entities of the CAD model.

We propose a new part decomposition for a prototype in order to guarantee the functionality requirements and to allow the evolutionary of its geometry. Furthermore each component of the new partitioned part is built with the more appropriate process. The assembly of components is design to have the same tested characteristic as the "one piece" part. This new approach is entitled "hybrid evolutionary prototypes". Our different analyses are based on CAD STEP specifications and more especially on Application Protocol (AP)-224.

2 STEP

The Standard for the Exchange of Product model data (STEP – ISO 10303) provides a neutral computer-interpretable representation of product data throughout the life cycle of a product, independent of any particular system.

STEP is organised as a series of chapters, each published separately. These chapters fall into one of the following series: description methods, integrated resources, application protocols, abstract test suites, implementation methods, and conformance testing. STEP uses application protocols (APs) to specify the representation of product information for one or more applications (Fig. 1). It is expected that several hundred APs may be developed to support the many industrials applications. STEP AP-203 is usually used for exchanging neutral format data between CAD systems. STEP

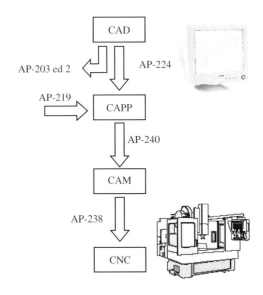

Figure 1.   STEP.

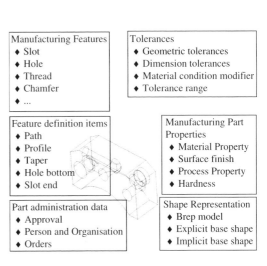

Figure 2.   Composition of the AP-224.

AP-224 is a manufacturing features oriented description. In our study we use the STEP AP-224.

2.1   AP-224: Mechanical product definition for process planning

This chapter of ISO 10303 specifies the information needed to define product data necessary for manufacturing a mechanical part. The product data is based on existing part designs that have their shapes represented by machining features.

Chapter AP224 contains all of the information and capabilities necessary to manufacture the required part (Fig. 2):

- All necessary CAD geometry and topology in a neutral format
- Machining feature information such as hole, boss, slot, groove, pocket, chamfer, and fillet (there are 20 manufacturing features)
- Dimensional and geometric tolerance information
- Part properties such as material properties, process properties, and material hardness
- Administrative information such as approval, part name and id, delivery date, and quantity
- The capability to handle both discrete parts and assemblies of parts

3 MULTI COMPONENT PROTOTYPE (MCP) CONCEPT

3.1 *Presentation*

Our MCP concept is presented in the Figure 3. The goals are to allow evolution of parts for testing purpose.

All the methodology that we have developed is based on feature analysis. For that we are not using automatic feature recognition (Tuttle et al. 1998) like in various research works made in CAPP process planning (Kang, Han & Moon 2003). We used the step-AP224 entities in the CAD model of the part because it is essential to have with the geometry several information like tolerances and part proprieties. Afterwards a 3D partitioning of the part is made in Functional Components (FC).

The result is the definition of all the components that create the Multi Components Prototype (Lesprier et al. 2003). The MCP is like a 3D puzzle of the part. After this task, the manufacturing of each component is realized. Various manufacturing processes and materials are chosen for each of them (Mognol, P., Rivette M., Jégou, L. & Lesprier, T. 2004). Then, an assembly of all components is made in order to have experimental testing on MCP. If the results of the tests do not match with the technical requirements, then only one or more components have to be re-designed and re-manufactured in order to have the prototype updated. This is the evolutionary loop. The use of the MCP concept to re-design and manufacture prototypes, allows the reduction of costs and time for each iteration in the loop. When test results match with functional requirements, design is then validated.

3.2 *MCP activities*

MCP concept involves realizing a product from a single piece part. For experimental testing, a MCP must have the same functionality as the single piece prototype from which it comes. Otherwise results could not be

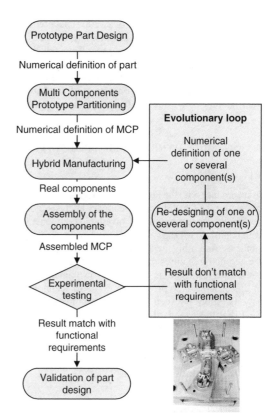

Figure 3.   MCP in design activity.

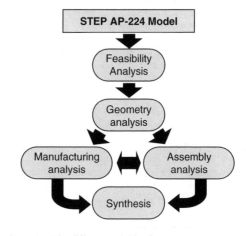

Figure 4.   The different activities for MCP concept.

interpreted, as it would be on a single piece part. Therefore all the activities shown in Figure 4 have to be perfectly analyzed.

Each analysis has is own knowledge. The synthesis allows making a MCP in conformity with initial requirements.

377

A previous paper (Lesprier et al. 2004) detailed the feasibility analysis and the geometry analysis. The goal is to group entities that participate to the same functions in the same piece of the puzzle. The manufacturing analysis (Mognol et al. 2004) proposes the best manufacturing process for each component of the studied MCP. Hereafter, we develop the Assembly analysis.

4 ASSEMBLY HYBRID RAPID PROTOTYPE

4.1 Introduction

In accordance with the Figure 4, assembly analysis takes place after the feasibility and geometry analysis. These two analyses propose a set of "available components" with regards to ISO specifications and topological possibilities. Assembly must respect initial positioning of the geometric features in the single part and technical constraints (for example tightness). The main objective of this analysis is to propose an available technologicaly assembly between the different components of MCP. For this, standard assemblies (noted CIA) have been defined. Each standard assembly is parameterized with a CIA. It is necessary to associate via fuzzy logic, for each entity a completely definite assembly in order to obtain technological solution of MCP.

4.2 CIA

A CIA (assembly identity card) is an identity card of one assembly, which gathers the general characteristics of this assembly. Each CIA has several parameters, which completely defines its geometry (Fig. 6).

Here, this example of CIA (CIA number 1), is design to contain the entity (a). Five parameters are almost independents and one (Ø4) is related to part ① and ②. This CIA is easily extractible by unscrewing.

4.3 Analysis assembly methodology

From AP-224 features of the part, and from the definitions of the functions of these entities (sealing), we obtain (N) Enhanced Entities (Fig. 5).

The analysis of evolutionary enables us to classify these enhanced entities in *i* evolutionary enhanced entities (EEE) and in *N-i* no evolutionary enhanced entities (NEEE).

At this step, each entity has some preferred manufacturing processes proposed by the manufacturing analysis.

For each EEE, the level capability of the CIA is evaluated according to the assembly criteria feasibility. Criteria have been defined in accordance to the technical possibilities of each CIA. For example global positioning and clamping by screwing are evaluated with a set of criteria. A note (0–10) is assigned at each

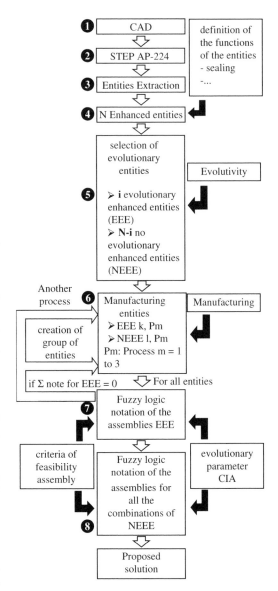

Figure 5. Method.

CIA for every entity. Note 0 corresponds to an important technical impossibility and rejects the solution. Note 10 shows a perfect adequacy with the CIA. If there is entity with no solution of CIA, another process according to manufacturing analysis and/or we make regrouping of entity is selected. Then the remaining entities and their combinations are tested. At the end of computation, we obtain a part with several components. Each component has an appropriate process and an associate CIA.

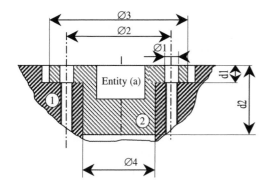

Figure 6.   Example of CIA.

5   APPLICATION

5.1   *Step 1: CAD*

In the following example (Fig. 7) there are two variants of the part for the hole 1. One with a 6.5625 mm hole {1a} and another with a 9.2 mm hole {1b}.

5.2   *Step 2 and 3: Extraction of the entities*

From the file ap-224, the information is extracted from each entity. The entities are: 7 holes located from 1 to 7 and 1 block number 0.

5.3   *Step 4: Enhanced entity*

Holes 1, 5 and 6: A liquid flow out through these holes. The designer wants to have no leak during the tests. Thus, he prefer solution with no waterproofness system for the CIA.

5.4   *Step 5: Selection of the evolutionary and no evolutionary entities.*

There are two variants of the hole 1. Diameter $\emptyset 1a = 6.5625$ mm and $\emptyset 1b = 9.2$ mm.

Thus there are 7 no evolutionary enhanced entities (NEEE) and 1 evolutionary enhanced entities (EEE), the hole 1.

5.5   *Step 6: Notation of manufacture of each entities*

Each process was evaluated for each entity according to their use by the manufacturing analysis. The manufacturing analysis proposes a level of capability process for each entity. At the present time, 3 processes were studied: EDM, HSM and DMLS. For example, the deep hole 4 is manufactured preferably with EDM (possible with DMLS, difficulty with HSM) whereas hole 7 is manufactured preferably with HSM (possible with EDM and DMLS).

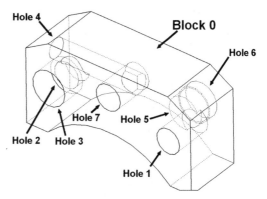

Figure 7.   Example (http://www.isg-scra.org/).

Table 1.   Notation of assembly evolutionary entities.

Entity 1	
Assembly	Note/10
CIA 1	0
CIA 2	0
CIA 3	0
CIA 4	0
...	...
CIA 20	0

Table 2.   Notation of assembly evolutionary group entities 1-5-6.

Entity 1-5-6	
Assembly	Note/10
CIA 1	10
CIA 2	5
CIA 3	6
CIA 4	0
...	...
CIA 20	3

5.6   *Step 7: Notation of the EEE assemblies*

In our example, the hole 1 have no CIA's solution (Table 1), because:

– Hole 1 have an intersection with hole 5.
– No intersection between CIA and hole 5 is authorized (waterproofness step 4).

Thus, we gather the entity in intersection with hole 1, here the holes 5 and 6. We pass by again with this group of entity by the step 7 (notation of the EEE assemblies) (Table 2).

379

Table 3. Notation of assembly evolutionary group entities 1-5-6.

Entity	$0 \cdot 2 \cdot 3 \cdot 4 \cdot 7$
Group of entity	(4-7) • (3-7) • (3-4) • (3-4-7) • (2-7) • (2-4) • (2-4-7) • (2-3) • (2-3-7) • (2-3-4) • (2-3-4-7) • (0-7) • (0-4) • (0-4-7) • (0-3) • (0-3-7) • (0-3-4) • (0-3-4-7) • (0-2) • (0-2-7) • (0-2-4) • (0-2-4-7) • (0-2-3) • (0-2-3-7) • (0-2-3-4) • (0-2-3-4-7)

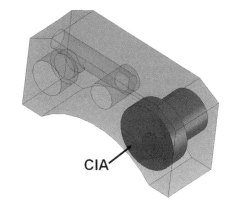

Figure 8. CIA.

The CIA number 1 (Fig. 6) is used for the assembly of the group 1-5-6.

When all evolutionary entities have CIA solutions we can note the no evolutionary entities combined between them. Only the entities having the same process of manufacture are combined. In our example we combine entities 0; 2; 3; 4 and 7 (entity 1; 5 and 6 being already gathered).

Here there are 31 solutions to test. (Table 3)

5.7 Proposed solution

Among the solutions proposed, the solution with least amount of assembly is selected. A regrouping of entities 0; 2; 3; 4 and 7 is possible, only entities 1; 5 end 6 will be manufactured separately. (Fig. 8.) We obtain two components for this example with an available assembly between them.

With the proposed method, to obtain the part with its 2 alternatives, we must manufacture only one part and 2 CIA. Without our hybrid approach we would have to manufacture 2 complete parts. We obtain a saving of time and money.

6 CONCLUSION

To optimize time scheduling and design cost or industrialization iterations, a new methodology has been presented in this paper. This methodology allows the design and manufacture real functional prototypes or dies in a 3D puzzle.

This methodology uses multiple point of view and knowledge to analyze prototypes definitions:

– Functional and evolutionary criterions,
– Feasibility and geometry to obtain 3D puzzle,
– Assembly design to re-assemble the puzzle pieces in conformity with initial requirements,
– Hybrid Manufacturing, with the choice of best processes for each Component.

Next, we have proposed a method to assemble the multi components.

The proposed method is available in an example. It can be easily automated in particular with the use of STEP Ap-224.

REFERENCES

Barnes, C.J., Jared, G.E.M. & Swift, K.G. 2004. Decision support for sequence generation in an assembly oriented design environment. *Robotics and Computer-Integrated Manufacturing 20(4)*: 289–300.

Chen, Y.H. & Song, Y. 2001. The development of layer-based machining system. *Computer-Aided Design 33*: 331–342.

Erens, F. & Verhulst, K. 1997. Architectures for product families. *Computers in Industry 33(2–3)*: 65–178.

Hur, J. & Lee, K. 2002. Hybrid rapid prototyping system using machining and deposition. *Computer-Aided Design 34*: 741–754.

Jiao, J., et al., 1998. Product family modelling for mass customization. *Computers & Industrial Engineering 35(3–4)*: 495–498.

Kang, M., Han, J. & Moon J-G. 2003. An approach for interlinking design and process planning. *Journal of Materials Processing Technology. 139*: 589–595.

Ki, D. & Lee, K. 2002. Part decomposition for die pattern machining. *Journal of Materials Processing Technology 130–131*: 599–607.

Lesprier, T., Mognol, P., Furet, B. & Hascoët, J-Y. 2004. Hybrid Manufacturing for prototypes and Dies. *Machine Engineering Vol 4 N° 1–2, ISSN 1642-6568*: 201–210.

Lesprier, T., Mognol, P., Furet, B. & Hascoët, J-Y. 2003. Hybrid Manufacturing for multi component prototype. *VRAP 2003*: 509–515.

Mognol, P., Jégou, L., Rivette, M. & Furet, B. 2004. High Speed Milling, Electro Discharge Machining and Direct Metal Laser Sintering. *The International Journal of Advanced Manufacturing Technology*.

Mognol, P., Rivette M., Jégou, L. & Lesprier, T. 2004. A first approach to choose between HSM, EDM and DMLS processes in Hybrid Rapid Tooling. AEPR-2004.

Tuttle, R., Little, G., Corney, J., Clark, D. 1998. Feature recognition for NC part programming. *Computers In Industry. Vol 35*: 275–289.

Virtual modeling and rapid manufacturing – Bártolo (eds)
© *2005 Taylor & Francis Group, London, ISBN 0 415 39062 1*

Stereolithography rapid tooling for injection moulding

S. Rahmati
Imam Hussain University, Mechanical Engineering Division

P.M. Dickens
Loughborough University, Rapid Manufacturing Research Group, UK

ABSTRACT: Increasing competition in global markets is exerting intense pressure on companies to trim their product cycles continuously. Reducing the time to produce prototypes is a key to speed up the development of new products. Rapid tooling (RT) with particular regard to injection mould fabrication using rapid prototyping (RP) technology of Stereolithography (SL) may lead to savings in cost and time. Rapid prototyping techniques are proving invaluable as aids in production planning, in performance tests conducted on plastic parts and in ensuring the rapid availability of sample parts. The design to product lead-time can be significantly shortened if the production time for tooling can be shortened.

Previous work at Nottingham University has shown that SL injection mould tooling can be used successfully in low to medium shot numbers. This paper looks at the failure mechanisms in SL injection mould tooling during injection. During injection, tool failure either occurs due to excessive flexural stresses, or because of excessive shear stresses. SL failure mechanisms have been investigated and different scenarios have been demonstrated. The failure results are investigated during injection process, and methods of improving the tool design to reduce the chances of failure are proposed.

1 INTRODUCTION

The design to production time for new components continues to decrease so that the long lead-time in producing tooling conventionally becomes more of a barrier in responding to customer demand (McDonald et al. 2001). Increase in design capabilities, increase in product variety, demand for shortened lead-time and the decrease in production quantities are the major driving forces in the development of rapid tooling technologies, where the tooling time and cost are significantly reduced. At the same time, stereolithography (SL) tooling techniques are improving and are therefore becoming increasingly popular among manufacturers (Decelles et al. 1996), (Greaves, 1997).

Previous work at Nottingham university (Rahmati et al. 1997a, 1997b, 1998) has shown that SL injection mould tooling can be used successfully in low to medium numbers, and up to 500 parts have been produced with one tool. The development of tools forms an important part of the development process. This paper describes work where the tools have been examined in detail to determine how and why the tools fail, and shows how tool failure can be reduced by suitable tool design. Section 2 outlines the experimental procedure,

Section 3 looks at failure mechanisms during injection, and the results are summarised in section 4.

2 EXPERIMENTAL METHOD

In constructing SL injection moulding tools, epoxy (SL5170) insert shells were fabricated directly from CAD data on an SL machine (SLA 250). These inserts were then fitted into steel mould bases through steel frames, and back-filled with an aluminium powder/ aluminium chip/epoxy resin mixture (Figure 1). The back-filled mixture added strength to the inserts and also allowed heat to be conducted away from the mould. The modular steel mould bases were two standard base plates machined with a cylindrical pocket to fit the steel frames and the inserts (Figure 2). The SL tools were then tested in a 50 ton Battenfeld production moulding machine to produce parts from polypropylene (PP) and Acrylonitrile Butadiene Styrene (ABS).

The fifth tool was used to study the effect of the increase in mould temperature versus tool performance; the tool successfully resisted failure up to 85°C versus the normal setting of 40°C. Although the tool didn't fail until 85°C, the moulding quality was

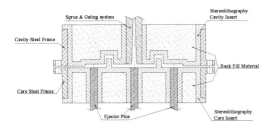

Figure 1. Cross sectional view of the SL injection moulding tool inserts.

Figure 2. Complete tool assembly components and the steel frames.

reduced due to the rubbery effect of the core. The sixth tool was built from epoxy filled material where 400 shots were made before failure. The seventh tool was aimed at investigating the effect of the increase in injection pressure versus tool performance using a load cell technique. The tool was tested at varying pressure, starting at 7600 psi, and increasing in different increments to 15660 psi, where the tool started to fail. This test led to a detailed pressure investigation at three locations inside the cavity where the pressure drop was determined. These tests enabled us to determine the tool failure mechanism using the data collected during the injection trials.

During the moulding process, the temperature and pressure of the cavity were monitored, and the melt temperature was controlled using eight thermocouples to ensure that the conditions within the cavity were as uniform as possible. Fractured samples of both the moulds and the mouldings were examined using either an optical microscope or a Scanning Electron Microscope (SEM). Fractured cubes were used to investigate the failure cross sections and fractured surfaces. Failed cubes embedded into the moulding material were mounted using a casting material and cut and polished in order to be examined, using an optical microscope. However fractured surfaces of the cubes and the core were investigated using SEM, which was a very effective technique and led to the interpretation of the failure mechanism of the SL tooling.

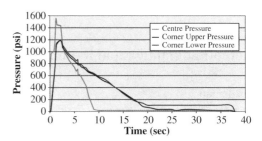

Figure 3. Pressure profile of the cavity at three locations.

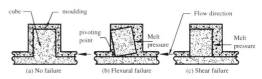

Figure 4. Schematic view of different scenarios, which may occur during injection.

3 FAILURE DURING PLASTIC INJECTION

When the plastic is injected into the cavity, there is a sudden pressure rise within the cavity which is the highest pressure reached during the moulding cycle (Figure 3). This pressure exerts a force on the core features which will give rise to plastic deformation or fracture if the ultimate tensile strength or flexural strength of the material is exceeded. Figure 4, shows the various scenarios which may arise during injection. In (a), there is no failure, in (b) there is a flexural failure and in (c) there is a shear failure. Flexural stress can lead to instant failure, or alternatively, to crack propagation and fatigue failure these are discussed in Sections 3.1 and 3.2 respectively. Failure due to shear stress is discussed in Section 3.3.

3.1 Flexural failure during injection

The majority of failures observed during this investigation were due to flexural stresses. In flexural failure the injection pressure overcomes the tool's flexural strength so that the feature rotates about its pivot point, and ultimately breaks off (Figure 4(b)). This can occur in one moulding if the injection pressure is beyond the flexural strength of the SL tool, but flexural failure is usually due to the history of the loading as explained in section 3.2 and 3.3.

The flexural stress for a cantilever beam with a uniform force F acting on it, is given in (Ives et al. 1981) as:

$$\sigma = \frac{6F.h}{at^2} \qquad (1)$$

382

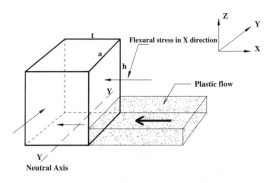

Figure 5. Schematic view of the cube's stress parameters and the approaching flow.

Figure 6. Moulding showing the attached plastic of the crack before failure.

Table 1. Flexural stresses exerted on the SL cubes.

	Moment of Inertia $I\,(\mathrm{m}^4)$	Moment $M\,(\mathrm{Nm})$	Flexural Stress $\sigma\,(\mathrm{Mpa})$	Flexural Strength at 40°C (Mpa)
Cube 1	108×10^{-12}	1.687	46.85	65.0
Cube 2	62.5×10^{-12}	1.687	67.46	65.0
Cube 3	32.0×10^{-12}	1.687	105.41	65.0
Cube 4	13.5×10^{-12}	1.687	187.40	65.0

where, h, a and t are the cube height, width and depth respectively as shown in Figure 5. Table 1 shows the theoretical calculations of flexural stresses for the SL cubes versus their flexural strength. Using this equation it can be seen from Table 1 that only the largest cube should survive the injection pressure and the rest of the cubes fail. However, in practice, the SL tools have produced hundreds of parts prior to failure, so that the theoretical model in Equation 1 overestimates the flexural stresses. There are two reasons for the discrepancy. First the flexural stress formulae assumes a minimum beam aspect ratio of ten (Douglas, 1989) while this ratio here is four. Secondly the injection pressure exerted on the cubes during injection was taken to be the pressure at front of the cubes but in reality this pressure is partly counteracted by the melt pressure behind the cubes.

In the case of the smallest cube, the net pressure is found to be 153 psi, which gives a flexural stress of 15 Mpa using Equation 1, which is less than 27 Mpa which is flexural strength of the tool at 40°C. This suggests that all of the SL cubes should survive the injection pressure. A better theoretical method for calculating the flexural stresses would be through the application of computational fluid dynamics (CFD) and finite element method (FEM), which will combine the fluid and stress analysis to model the SL tool.

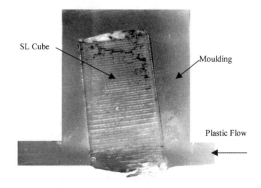

Figure 7. Flexural failure as a result of crack propagation.

3.2 Crack propagation and fatigue

Flexural stresses can also induce a "fatigue" type process, spanning a number of moulding cycles. In this situation, the cube pivots as in Figure (4-b) without being fractured but a crack is initiated at the intersection between the face of the cube in tension due to flexural stresses, and the core face perpendicular to it. During subsequent cycles, the crack propagates through the base of the cube eventually resulting in failure. Failure analysis of the SEM images has revealed that the crack propagates through the cubes prior to the ultimate failure. Microscopic pictures of mouldings numbered sequentially indicate that the crack has started well before the ultimate flexural failure. Figure 6 is a picture taken of the cross section of a moulding before the actual failure happened, where subsequent injection mouldings have exhibited a "positive" flaw corresponding to the inverse of the crack generated. The replicated crack has been curled because of its delicacy, but it was originally straight in the crack direction. Figure 7 shows the flexural failure of a similar cube to that seen in Figure 6, after a number of shots.

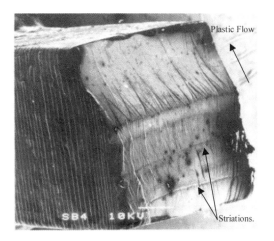

Figure 8. SEM observation revealing striation marking on the fractured surface.

Figure 9. SL cube being sheared off during injection moulding process.

Crack initiation in SL tools occurs predominately at stress concentrations, such as sharp corners or at stair steppings (an inherent property of SL parts). Crack formation may also result from flaws or microscopic defects created during photopolymerisation process due to material discontinuities (Hertzberg et al. 1980). The SL parts may contain flaws or voids, so that microscopic cracks may initiate that ultimately concentrate into a macroscopic fracture (Williams, 1984). Sharp corners, stair stepping, voids or flaws are a cause or source of crack initiation. Fatigue failure can be minimised by introducing fillets at the sharp corners in order to reduce the stress concentration and crack propagation. Evidence of the crack failure as shown in Figure 8, can be seen on the fracture surface in the form of "striations", where each one of these marks represents crack growth. At the tip of the crack and in a small region near the tip, the yield strength of the material is exceeded. In this region, plastic deformation occurs and the stresses are limited by yielding (Dally et al. 1991). After each cycle, the crack grows in the same manner until a critical crack length is reached. At this point, the crack tip can increase in velocity and spread all the way across the cube resulting in failure.

3.3 Shear failure

During shear failure, the feature is sheared off in the direction of the melt flow. Figure 9, shows the cross section of a sheared SL cube. Notice that the SL cube has been pushed across by the flow of plastic. The shear stress at a point in a section is given by (Cheng, 1997):

$$\tau = \frac{VQ}{Ia} \qquad (2)$$

Table 2. Shear stresses acting on the SL cubes.

	Shear area A_S (mm²)	Shear Force V (N)	Shear Stress τ_{ave} (Mpa)	Shear Strength at 40°C (Mpa)	T_{MAX} (°C)
Cube 1	36	421.64	11.71	24.3	65.3
Cube 2	30	421.64	14.05	24.3	61.5
Cube 3	24	421.64	17.57	24.3	55.9
Cube 4	18	421.64	23.42	24.3	46.4

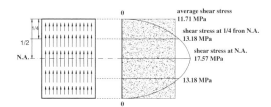

Figure 10. Distribution of the shear stresses across the largest cube base.

where V is the shear force at the given section, Q is the first moment of the area about the neutral axis, I is the moment of inertia of the cube section with respect to the neutral axis, and a is the width of the cross-section. As the shear stress calculation results show in Table 2, the maximum shear stresses produced in the SL tool during operation are below the shear strength of the SL tool. Moreover the SL tool can survive at injection temperatures beyond 40°C as shown in the last column of the Table 2. Figure 10, shows the maximum shear stresses at various points of the cube base versus the average shear stress. The plot of the maximum shear stresses at various points results in a parabolic curve.

384

4 CONCLUSIONS

Different SL tools have been successfully tested where failures were observed after 500 shots. SL failure mechanisms have been investigated and different scenarios have been demonstrated. The failure results are grouped into two main categories, first failure during injection process and second failures during part ejection.

- As experience and theoretical calculations confirm, flexural stresses during the injection process are the most probable cause of failure. Reducing the tool's aspect ratio decreases the chances of flexural failure.
- Shear stress failure during injection is less likely than flexural failure in particular when the SL tool is warmed to 40°C prior to injection.
- Failure during part ejection is strongly dependent on the freeze time. The tensile stress must be calculated and a safe release temperature must be chosen using the temperature graph. The likelihood of the failure during ejection is insignificant if the tool is injected at 40°C and the freeze time is 35 seconds.

REFERENCES

Cheng, F., 1997, Statics and Strength of Materials, 2nd edition, McGraw-Hill, ISBN 0-07-115666-6.

Dally, J. W. & William, F. R., 1991, Experimental Stress Analysis, 3rd ed., McGraw-Hill, ISBN 0-07-015218-7.

Decelles, P. & Barritt, M., 1996, Direct AIM Prototype Tooling, 3D Systems, P/N 70275/11-25-96.

Douglas, R. A., 1989, Introduction to Solid Mechanics, Sir Isaac Pitman & Sons Ltd., London.

Greaves, T. (Delphi-GM), "Case Study: Using Stereolithography to Directly Develop Rapid Injection Mold Tooling", TCT Conference, 1997.

Hertzberg, R. W. & Manson, J. A., 1980, Fatigue of Engineering Plastics, Academic, New York.

Ives, G. C., Mead, J. A. & Riley, M. M., 1981, "Handbook of Plastics Test Methods", Edited by R. P. Brown, 2nd ed., London, ISBN 0-7114-5618-6.

Jacobs, P., 1996, Recent Advances in Rapid Tooling From Stereolithography, A Rapid Prototyping Conference, Oct. University of Maryland, USA.

McDonald, J.A., Ryall, C.J. & Wimpenny, D.H., 2001, Rapid Prototyping Casebook, UK, Professional Engineering Publishing.

Rahmati, S. & Dickens, P. M., 1997a, "Stereolithography Injection Moulding Tooling", Sixth European Conference on Rapid Prototyping and Manufacturing, Nottingham, UK, ISBN 0 9519759 7 8, PP 213–224.

Rahmati, S. & Dickens, P. M., 1997b, "Stereolithography Injection Mould Tool Failure Analysis", 8th Annual Solid Freeform Fabrication, Texas, PP 295–305.

Rahmati, S., Dickens, P.M. & 1998, Pressure Effects in Stereolithography Injection Moulding tools, 7th European Conference on Rapid Prototyping and Manufacturing, Aachen, Germany, PP

Williams, J. G., 1984, Fracture Mechanics of Polymers, Ellis Horwood series in mechanical engineering, ISBN 0-85312-685-2.

Rapid Tooling using Alumide®

D.J. de Beer & G.J. Booysen
Central University of Technology, Free State, South Africa

ABSTRACT: Rapid Tooling is a well-researched and debated topic. Many companies and research institutes have moved past the subject to the Rapid Manufacturing paradigm. In previous research, different processes and associated materials such as SLA & ACES; SLS & Laserform™ ST 100/A6 Tool Steel, etc., have been evaluated and discussed in detail. Many users/researchers doubted its relevance as a competitive tooling method, whilst others were/are positive about its use. In calling it *rapid*, one has to consider the cost of the materials and process, time taken, post-processing issues, resolution, finishing time, fitments and durability of the prototype material. It then still raises the issue whether it is really rapid compared to machined tooling, and whether the time saved justifies the cost or the tool life-time. The paper takes a fresh look at rapid tooling, through the use of Alumide®, run at fairly high building speeds (compared to other plastic/nylon/epoxy/metal technologies) in an EOS P 380 system at the Centre for Rapid Prototyping and Manufacturing (CRPM) at the Central University of Technology (CUT) in South Africa. As with other Nylon/Plastic parts used with the EOS P 380 machine, no post-processing is necessary, which contributes to both speed and predictability in terms of accuracy. Cost is equal to building with other EOS plastic materials. Results are discussed in comparison with conventional tooling, reviewing its role in a concurrent product development process, with the inclusion of QFD. Both experimental results, and results obtained through real industrial case studies are compared and discussed. RT through Alumide® was the only option left for a product development company, when the time needed for conventional methods did not suffice. Results obtained, not only proved the process, but paid for.

1 INTRODUCTION

It is a very well-debated fact that manufacturers wishing to stay in business need all the possible technology edges/support to shorten time-to-market, in an attempt to stay globally competitive. Rapid Prototyping is being used more and more as a key enabling technology in reducing the time-to-market for new products, by identifying possible design flaws prior to tooling and manufacturing, and is providing the **common focus** for **multidisciplinary groups**, around which **design and development questions** can be **resolved.** Barkan and Iansti (Barkan and Iansti, 1993) present RP as a means of **rapid learning** at **every stage of the design process**. When adopting this view on the whole of the development process, one comes to the conclusion that the use of RP to enable Rapid Product Development (and Rapid Tooling to enable Rapid Manufacturing) is a fundamental challenge that must be addressed by all manufacturers to remain competitive in today's global market place (DST, 1999). Of course, the term **Rapid** is a relative term. Most prototypes require from one to seventy-two hours to build, excluding clean up and preparation time, and depending on

the size and complexity of the object. However, it enables designers to see parts of their models within a few hours, opposed to several weeks or months with conventional methods. The ability to quickly create a physical model or prototype for evaluation to assess relative merits or possible design flaws provides the possibility to inspect the object in three-dimensional form.

Rapid Prototyping can be defined as a technique in which physical models are fully created from materials, provided in various forms, completely under the control of model data created within a computer aided design environment. For the past ten years, the industry was infused by the hypothesis of not only rapidly creating prototypes of new products, but to also develop the tooling for manufacturing. Initially, this idea was hampered by the shortcoming of materials and process technologies, but as speed, accuracy, building styles, resolution, etc., are increased, the RP industry kept on coming closer to what would be acceptable for the tool-making industry. However, up to now speed, surface finish/resolution and accuracy remained a problem, as well as a generic/unified acceptance or common understanding of what constitutes Rapid – last mentioned may be a function of specific industries. From the view-point of toolmakers, the accuracy of Rapid Tooling is still below what can be achieved through CNC milling. Thus cost and time at least have to be such that it may compensate. If however, cost, quality and time-wise the process does not compete to conventional methods of tooling, there would be little motivation for a new paradigm of tooling.

1.1 Rapid Tooling – general terminology and requirements

Due to the very time-consuming, expensive and skilled task of creating/making/manufacturing tooling through conventional (subtractive) machining methods such as CNC machining or spark erosion, Rapid Tooling raised high interests in the mould making/plastic conversion industry, following a hypothesis that injection moulds can be manufactured faster and at a lower cost than possible with conventional technologies. The intention with Rapid Tooling is to provide an affordable and fast way to produce thermoplastic parts in quantities of a few hundred, for functional testing, marketing, etc, and even proceed to production quantities. Further theories were that it could provide a means of bridging the gap since the concept was finalised, till production tooling was completed – which can take anything up to six months. Releasing a product earlier, prolongs the profit cycle, and will help with its financial sustainability (De Beer *et al*, 2003). It has also been reported (Castle Island, 2005) that RT is being used to verify tool design and to test how tools will perform. A Foresight study by the South African Department of Science and Technology (DST, 1999) showed that product requirements are changing as customers are putting more accent on their own requirements, meaning more product versions may be needed, or personalised changes may need to be incorporated in production. According to the Foresight Study, growing discrimination in the tastes and desires of consumers and marketing efforts of firms, which introduce new possibilities, will drive the production of new or improved products, and sustainable consumption patterns are forcing producers to look at alternative design and manufacturing methods. The growing shortage in skilled toolmakers furthermore challenges convention mould-making techniques, as a growing quantity of tools need to be manufactured by a shrinking tool-making community.

Over and above addressing issues such as time and labour savings, the use of RP offers advantages such as the incorporation of complex conformal cooling channels, which could have an effect on thermal performance of the mould. The use of inserts into mould bolsters, or mould inserts into existing moulds, offers the best of both technologies. Rapid prototyping

injection mould fabrication methods should be considered for projects in which:

- the reduction of time to market is important,
- for prototype and short to medium volume production runs, and
- for parts which may be very hard to machine because of their geometry.

The initial investigation in conventional tooling methods highlighted the fact that parts typically suited to EDM were the candidates for rapid tooling. The time-consuming task of breaking the part up into multiple electrodes, the machining of complex shapes and the alignment of electrodes with the work piece for the sparking process – generally takes too long. Parts requiring thin slots to be produced in a tool, generally smaller than 3 mm in diameter and deeper than the cutter diameter are also fair candidates (Van der Merwe, 2005).

1.2 Current limitations

The aim/objective set for the direct production of injection moulds through RP was to offer the same level of precision and durability as CNC methods. Huge advances have been made in that direction, and important time and labour savings are being reported today by RP practitioners and service bureaus, but in comparison with CNC methods, the technology is still immature. This means that the benefits realised are not universal and must be evaluated for each case. The general limitations of RP methods compared to CNC currently are:

- production of less accurate and less durable tools;
- part size and geometry limitations,
- don't necessarily produce identical parts to hardened tooling, and
- tools may not easily be modified or corrected using typical techniques.

These limitations vary both as function of the specific RP technology used and for each individual case. The inability to modify many tools fabricated by RP technology means that making such a tool is frequently a one-shot deal; if it's not right the first time, it may be necessary to scrap the tool and start over. Rapid tooling may lose its advantage compared to conventional tooling methods under those circumstances.

2 PARALLEL PROCESSES

RP currently is used in two ways to make tooling:

- Moulds may be ***directly*** fabricated by an RP system; or
- RP-generated parts can be used as patterns for fabricating a mould through so-called ***indirect*** or ***secondary processes.***

2.1 Direct fabrication processes

The last decade saw the development specialised rapid prototyping processes, mostly in direct metal fabrication, which are available to meet specific application and material requirements necessary for moulding and casting. These may be a variant of a basic RP process, or may be a unique RP method developed to suit a specific need. A large number of technologies are in the research phase, but limited technologies are commercially available at present. Although much progress in terms of direct part production took place, RP direct metal systems remain slow and may have other limitations. It simply can't produce parts in a wide-enough material range, at a fast enough rate, to match the enormous spectrum of requirements of industry (Castle Island, 2005). Some RP OEMs also make claims about the use of polymer for injection moulds, but little success stories are available in the literature.

Conventional processes such as moulding and casting are still some of the fastest methods to produce tooling, and this often starts with an RP process. The fabrication of tooling through RP technology, however remains the most important application expected of direct manufacturing.

2.2 Indirect or secondary processes

With improvement of physical properties of RP materials and expansion in choices available, many options similar to engineering materials become available. There will however, always remains a need to reproduced RP-fabricated parts. Furthermore, tooling normally requires very specific materials to make most tools. Consequently, numerous material transfer technologies have been developed. Typically a part made by the RP system is used as a pattern or model in these processes. Similar to direct fabrication processes, many secondary processes are in various stages of development. Again, of all the methods available, currently just a few are commonly used and commercially available/applied.

2.3 Limitations of RP-generated patterns

RP-generated patterns still need finishing operations before they can be used in any indirect or secondary process. No rapid prototyping technology yet delivers surface finishes that are adequate for accurate applications such as injection mould tooling. Removal of the stair-stepping inherent in the process and other surface defects is necessary before parts will eject from a mould, and may lead to additional errors being introduced. The accuracy of most secondary processes is ultimately limited by the precision of the pattern after finishing. RP patterns are best for applications with just a few critical dimensions. For products where

many tight tolerances must be held, it's generally still faster and cheaper to use CNC.

All these texts fit in a frame which should not be changed (Width: Exactly 187 mm (7.36″); Height: Exactly 73 mm (2.87″) from top margin; Lock anchor).

2.4 Rapid tooling routes with Alumide® for the eosint P-series sintering machines

During the EuroMold 2003 (Dec 2003), EOS GmbH released Alumide®, an aluminum-filled nylon material that allows the resulting metallic-looking, non-porous components to be machined easily and to withstand high temperatures, which offers various new possibilities for both direct manufacturing, as well as direct tooling applications.

3 ALUMIDE® – GENERAL INFORMATION

A typical application for Alumide® is the manufacture of stiff parts with a metallic appearance for applications in automotive manufacture (e.g. wind tunnel tests or parts that are not safety relevant), for tool inserts for injecting and moulding small production runs, for illustrative models (metallic appearance), for education and jig manufacture, among other aspects.

Alumide® can be finished by grinding, polishing or coating. An additional advantage is that low tool-wear machining is possible, e.g., milling, drilling or turning.

Alumide® is suitable for processing on the following systems:

- EOSINT P 700 with or without powder conveying system;
- EOSINT P 380;
- EOSINT P 360 with upgrade S&P;
- EOSINT P 350/2 + upgrade 99 + upgrade S&P.

The recommended layer thickness amounts to 0.15 mm. To ensure a consistent quality of parts, it is recommended solely to use new powder. Sections 3.1–3.1 contain the Alumide® Technical Data, as per Tables 1–3.

3.1 General material properties

Table 1. General material properties for Alumide®.

Property	Standard Used	Quantity	Unit
Average grain size	Laser diffraction	60	μm
Bulk density	DIN 53466	0.64 ± 0.04	g/cm^3
Density of laser-sintered part	EOS-method	1.36 ± 0.05	g/cm^3

3.2 Thermal properties

Table 2. Thermal properties for Alumide®.

Property	Standard used	Quantity	Unit
Melting Point	DIN 53736	172–180	°C
Heat Deflection Temperature	ASTMD648 (0,45 Mpa)	177.1	°C
Vicat Softening Temperature B/50	DIN EN ISO 306	169	°C
Heat Conductivity (170°C)	Hot Wire Method	0.5–0.8	W(mK)$^{-1}$

3.3 Mechanical properties

Table 3. Mechanical Properties for Alumide®.

Property	Standard used	Quantity	Unit
Tensile Modulus	DIN EN ISO 527	3800 ± 150	N/mm^2
Tensile Strength	DIN EN ISO 527	46 ± 3	N/mm^2
Elongation at Break	DIN EN ISO 527	3.5 ± 1	%
Flexural Modulus	DIN EN ISO 178	3000 ± 150	N/mm^2
Flexural Strength	DIN EN ISO 178	74 ± 2	N/mm^2
Charpy – Impact Strength	DIN EN ISO 179	29 ± 2	kJ/m^2
Charpy – Notched Impact Strength	DIN EN ISO 179	4.6 ± 0.3	kJ/m^2
Shore D – hardness	DIN 53505	76 ± 2	

The mechanical properties depend on the $x-$, $y-$, $z-$ position of the test parts and on the exposure parameters used.

4 ALUMIDE® APPLICATIONS

4.1 General applications

Michael Shellabear (Shellabear and Hänninen, 2004) reports that Formula 1 has become an ideal application area for laser-sintering due to the extreme need for high flexibility and fast reaction times, and almost all Formula 1 teams are using this technology in one way or another. Toyota Motorsport GmbH (TMG) uses laser-sintering for building various test components for their Formula 1 cars, as well as components to be used in wind tunnel tests. The particular strength of this method is the direct manufacture of parts which offer high functionality and close-to-series properties and performance. Over and above EOSINT P systems with PA 2200 nylon powder, TMG also uses the new aluminium-filled nylon material Alumide®. This has become the material of choice for

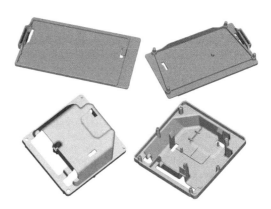

Figure 1. CAD Drawings of the 2 parts needed.

many applications which require a metallic look, good finishing properties, high stiffness or highest part quality. In one project, an on-board camera dummy for test-runs and qualifying was built using Alumide®, which provided excellent dimensional accuracy as well as easy finishing.

4.2 Tooling trials

Technimark has already successfully developed some pre-paid electricity meters. Based on their experience, they decided to tender for a new development aimed at the Southern-American market. The tender-process required the submission of injection-moulded parts with the tender documents, to prove capacity to manufacture the product. Not knowing whether they would successfully tender, they decided to keep with standard commercially available or in-house jigs, which meant they only had to introduce special jigs and fixtures to develop a risk-free new project. Figure 1 shows CAD images of both sides of the parts needed.

The parts would, amongst others, be used to hold electronic parts and PC boards. The challenge faced by our client was a huge one – 4 mould halves were needed, with less than 4 weeks available to manufacture – approximately one third of the conventional time needed, conservatively estimated. Taking the nature of the development and the tender process into account, it was very risky to commit expenses, leading to a very conservative budget. The results however, were astonishing! Figure 2 shows pictures of the moulds grown in Alumide®. The moulds required 23 hours of prototyping (one build volume), with 4 days of finishing and fitting.

This meant that the injection-moulding could start in less than a week after finalizing the design. Approximate mould costs were R23 000 (ca € 3 000), opposed to R90 000 (ca € 11 000- conservatively estimated).

Figure 2.   Mould halves grown in Alumide®.

Figure 4.   Injection moulding problem areas.

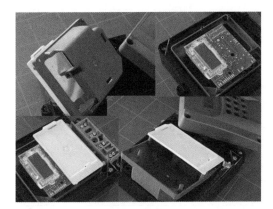

Figure 3.   Injection-moulded parts, fitted with components.

Figure 3 shows the first off-tool samples as injection-moulded in the Alumide® tools, using the required engineering materials. It can be seen that the shrinkage was acceptable, as the parts fitted (snapped) together.

Problem areas are enlarged in Figure 4, pointing out injection moulding problems. The intended features failed as the depth/cross-sectional area was to large. Having the injection-moulded parts however, did result in being able to use other methods of fixing which could be added on the existing parts, without a redesign.

Summary of results obtained:

- The company has done 30 trial samples in flame retardant ABS;
- With further tests done, 200 samples were injection-moulded – initially air-cooled;
- Final production of 600 units;

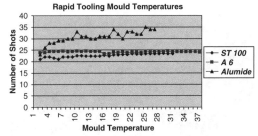

Rapid Tooling Mould Temperatures

Figure 5.   Comparative data obtained from different rapid tooling inserts.

- No visible mould damage or wear.

The following data was obtained in parallel tests on a second mould, moulding standard tensile test pieces:

- Mould temperature: 23–34°C
- Mould withstood normal working conditions and pressure of a 90 ton injection moulding machine.

Figure 5 represents comparative data for 30 injection-moulding shots in Alumide® with no cooling provided, compared to ST 100 and A6 tool steel.

Figure 6 shows the injection moulding settings as used on the Alumide® tooling.

Microscopic photos were taken of the mould surfaces both at 50 times and 100 times magnification, and no wear was shown on the photos (Figure 7). Figure 7a was taken at 50 times magnification, and 7b at 100 times magnification (De Beer and Booysen, 2005).

5   COMPARATIVE STUDIES COMPLETED

5.1   *First generation prepaid electricity meter*

The development required a combination of direct manufacturing with LS, and vacuum casting through

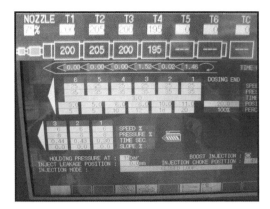

Figure 6. Injection moulding settings used on the Alumide® tooling.

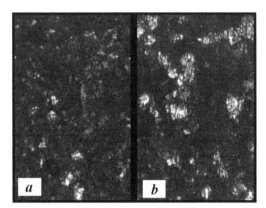

Figure 7. Microscopic images of the Alumide® Surfaces. (Figure 7 a – 50 times; Figure 7 b – 100 times)

silicone moulds made from SLA masters, de Beer *et al* reported that the total development was done at less than 10 % of the costs of injection moulding. Figure 8 shows a summary of the costs involved, as well as detail of the parts produced.

Taking the costs into account for the growing of SLA patterns, finishing, silicone mould making and casting, it was at least 4 times the cost involved than for the Alumide® case study. The time needed for last-mentioned, can in no ways be compared to that used for the first generation meter. At least one extra month of development time would have been won should the Alumide® RT route have been followed.

5.2 Body IQ development

The Compact Kiosk unit consists of a main top shell, base and a tube section made in two sections, andjoined during final fitment and fabricated via hand-laid

Total Cost of Project	R 112 000	£ 9 500	€ 13 900	$ 15 850
SLA Masters	R 8 500	£ 725	€ 1 050	$ 1 200
Finishing + Spraying	R 23 000	£ 1 960	€ 2 850	$ 3 250
Silicone Moulds	R 34 500	£ 2 940	€ 4 280	$ 4 890
Castings of Parts	R 38 000	£ 3 240	€ 4 710	$ 5 380
Finishing + Spraying	R 5 500	£ 470	€ 680	$ 780
SLS Functional Parts	R 3 500	£ 300	€ 435	$ 500
Injection Mould	R 1,3 M	£ 111K	€ 161 K	$ 184 K

Figure 8. Comparative data showing the costs involved for the 1st generation prepaid meter.

Table 4. Summary of parts and manufacturing method.

Cast in F16 (from SLS master pattern)	Grown directly by SLS RP:
• Main top enclosure	• Cuff Inner ring
• Base	• Cuff Outer ring
• Mounting Block	• Card Reader Bezel
• Puk Holder	• Card Reader Inner Sleeve

Table 5. Summary of costing breakdown.

RP/Vacuum casting:	Injection moulding:
RP via SLS for initial full set : R3 000	Tooling cost: R180 000
RP via SLS for 25 full sets: R42 100	Production cost and spray painting of 25 sets: R11 000
Cost via SLS and Casting in F16 from Silicone Mould: R19 000	
Finishing and spraying of parts for full 25 sets: R20 000	

fibreglass (hand-laid GRP). It was decided that all special design parts had to be cast from or grown by RP, and consisted of the parts indicated in Table 4 below.

The estimated costs for manufacturing the enclo-sure via injection moulding was approximately R680 000, with a lead time of 6 months, resulting in a unit price of R 3500. Table 5 shows an approximate costing comparison between direct manufacturing with RP or RP as master model for vacuum casting, measured against the cost of mouldmaking and injec-tion moulding.

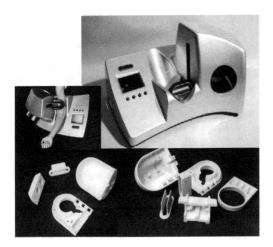

Figure 9.   Body IQ unit.

Total Cost of Project	R 46 000	£ 3 950	€ 5 750	$ 6 550
10 CAD Designs @	R 2 500	£ 220	€ 310	$ 350
6 SLA Prototypes @	R 2 500	£ 220	€ 310	$ 350
6 Silicone Moulds @	R 820	£ 70	€ 100	$ 115
26 Castings @	R 50	£ 4	€ 6	$ 7
Injection Mould	R 80 000	£ 6 815	€ 9 900	$ 11 300

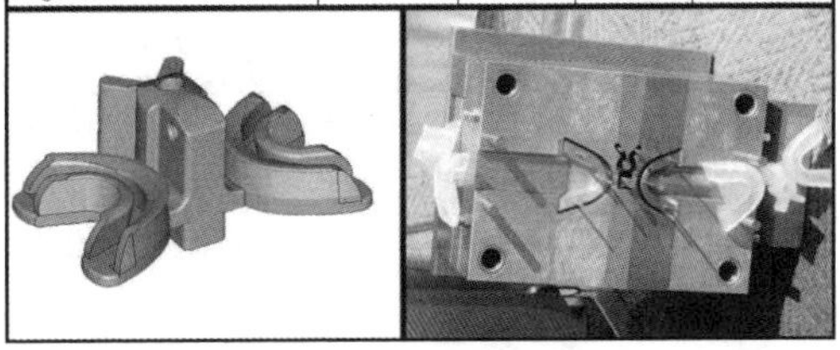

Figure 10.   Comparative data showing the costs involved for the Anaesthetic Mouthpiece Development,

Taking into account that the tool manufacturing alone would have taken at least 12 weeks, the mere fact that the 25 complete units could be developed, manufactured and tested in only 10 weeks, justified the decision on the RP route. However, the success of all processes involved became apparent in the total manufacturing cost per fully assembled and tested unit of R10 500-00 (excluding all electronic devices). Figure 9 shows images of the Body IQ with RP components.

Again, 10 fully functional prototypes needed to be produced to enable the client to physically perform the whole range of testing up front before the completion of the production tooling. At that stage, cost-wise, it was a major success, and the head start and time saving of this operation can be quantified into months.

Although the development was a big success at the time, it is again evident that the availability of Alumide® would have halved the prototype production costs, and would have lead to a much earlier release of the product (De Beer and Barnard, 2001).

5.3   Anaesthetic Mouthpiece Development

In supporting the development of a medical product, functional parts in the final engineering material were once more needed for both the clinical trials as well as for marketing of the product. Figure 10 shows comparative data for the Anaesthetic Mouthpiece Development.

Booysen et al (Booysen et al, 2004) reports that through HSM in Aluminium, the total development cost R46 000 – again twice the amount needed for the Alumide® case study. Also, last mentioned was completed in half the time. The data showed for the injection mould, represents conventional mould-making, as opposed to HSM in Aluminium.

6   CONCLUSIONS

Till recently, selecting a RT process was not easy, as a number of technologies that were available as possible solutions, were still immature, may have had significant limitations, or were still under continuous development. To make kit more complex, there were a number of routes to get to a final functional part or tool starting from a CAD definition. It is clear that the final choice would be a function of the application, the volume of parts to be produced, the final material and accuracy requirements and rapid prototyping process used.

Todd Grimm summarises it completely by stating that today's rapid tooling solutions are generally niche applications (Grimm, 2004). For applicable products, it will result in accelerated product development, especially when used in a total concurrent product development environment – as RT alone will not result in faster product development. It needs a total new paradigm. It may also result in some trade-offs to be made – the client may need to accept some limitations of the process in order to gain a time advantage. Growing in acceptance of the technology is unfortunately hampered by these niche applications and more success stories need to be published in order to gain wider recognition of Rapid Tooling as a competitive approach. As further process and material developments will take place, RT may become a standard methods in the product development and production process.

REFERENCES

ALUMIDE MATERIAL DATA SHEET: www.eos.info **(Accessed 15 April 2005)**

BARKAN, P. & IANSTI, M. *Prototyping: A Tool for Rapid Learning in Product Development*, Concurrent Engineering Research and Applications, 1, pp 125–134, 1993

BOOYSEN, G.J., DE BEER, D.J., BARNARD, L.J. & Truscott, M. 2004: *Anaesthetic Mouthpiece Development.* Proceedings of the 10th European Forum on Rapid Prototyping and Manufacture, Maison de la Mecainque, Courbevoie, Paris **AL – PARIS**

CASTLE ISLAND'S *Worldwide Guide to Rapid Prototyping.* **http//home.att.et/~castleisland/ (Accessed 22 April 2005)**

DE BEER, D.J. & BOOYSEN, G.J. *Rapid Tooling with Alumide*®. Paper presented at the EOS International Usersgroup Meeting, Bad Reichental, Germany, April 2005

DE BEER, D.J. & BARNARD, L.J. *Applying Rapid Prototyping In Concurrent Engineering – Successful SA Case Studies.* Published in the proceedings of CARS 01, University of Natal. July 2001

DE BEER, D.J., BARNARD, L.J. & BOOYSEN, G.J. *Innovation – A Function of Rapid Product Development*? Paper presented at the TCT Conference & Exhibition, Birmingham, December 2003

DST (Department of Science and Technology, South Africa) *Manufacturing & Materials Foresight Study*, **1999**

GRIMM, T. *Users Guide to Rapid Prototyping.* Published by the Society of manufacturing Engineers, Feb 01, 2004

SHELLABEAR, M. & HÄNNINEN, J. *e-Manufacturing for Automotive Applications.* Mira New Technology 2004, *http://mira.atalink.co.uk/articles/132* **(Accessed 15 April 2005)**

VAN DER MERWE, D.S. Unpublished MTech Research Report, Central University of technology, Free State, 2005

Virtual modeling and rapid manufacturing – Bártolo (eds)
© *2005 Taylor & Francis Group, London, ISBN 0 415 39062 1*

Rapid tooling for small batch production in micro injection moulding

Uwe Berger

Aalen University of Applied Sciences, Germany

ABSTRACT: Injection moulding tools consist of standard components. A frame plate in the mould contains inserts with the cavities of the negative part geometry. These inserts can be manufactured with Rapid Prototyping technology. Using this method even small batches with frequently changing parts can be produced flexibly and economically.

A flexible micro injection moulding tool was developed and tested. The inserts were manufactured using different Rapid Prototyping processes as well as High Speed Cutting. Inserts consisting of plastics and metal materials were tested. Small batches of parts were produced and assessed.

1 INTRODUCTION

Rapid tooling techniques can be classified into direct and indirect methods. Direct methods use tooling in-serts built with rapid prototyping machines. Indirect methods create the tooling inserts using master patterns built with rapid prototyping machines.

The method of removable inserts is also known as rapid insert technology. The inserts can be produced in metallic or non-metallic materials and are components of a hybrid injection mould.

The aim of this project was to investigate when epoxy resin inserts may be used for micro injection moulding.

2 TEST EQUIPMENT

An ARBURG 220 S 150-30 injection moulding machine was used for the production of micro spiral parts to test the fluidity of several kinds of plastic (Fig.1). It is the smallest machine of ARBURG. Due to its screw diameter of 12 mm it can be used for micro injection moulding.

The width of the spiral part was 1.25 mm, its thickness 0.6 mm, the gap between successive arms 1.14 mm. The part mass was 0.0478 g for POM material, with a spiral length of 50 mm and the volume of 0.0339 cm³.

To permit a fast change of cavity inserts, one of the mould halves in the injection unit had standardised dimensions. Figure 2 shows the flexible injection unit with the base plate pocketed for the inserts (Fig. 2).

Figure 1. Standardised insert with micro spiral part for fluidity tests.

3 RAPID TOOLING OF MICRO CAVITY INSERTS WITH HIGH SPEED CUTTING

At first an aluminium insert was with a high speed milling machine. High speed cutting (HSC) has some

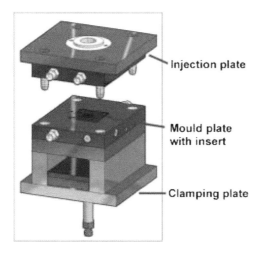

Injection plate

Mould plate with insert

Clamping plate

Figure 2. Mould injection unit with standardised insert.

Table 1.   Material properties of the resins used for testing of micro injection moulding.

	SL 5170	Accura SI 10	Accura SI 40 tempered
Tensile modulus (ASTM D-638) (MPa)	2500	3250	3300
Tensile strength (ASTM D-638) (MPa)	60	63	74
Flexural modulus (ASTM) (MPa)	3010	3100	3180
Flexural strength (MPa)	108	97	118
Glass transition temp. T_g (°C)	65	62	75
Hardness, Shore D	85	86	84

advantages for small cavities compared with Rapid Prototyping methods, as:

• hard tooling,
• short machining times,
• better surface qualities and
• closer tolerances can be realised.

Tool diameters down to 0.2 mm are available, also for the cutting of hardened steel.

Disadvantages are

• NC-programming of micro cavities requires qualified and experienced staff,
• high speed machining normally needs time consuming test and optimisation runs,
• tool aspect ratio and tool curvatures cause geometric limitations for the manufacturing of cavities.

For overcoming the geometric limitations electrical discharge machining (EDM) could be used. But this increases the amount of planning and machining time to produce a cavity insert considerably.

However, for the simple geometry used in this example HSC caused no problems. The simple geometry is convenient here for studying the injection moulding process at small injection cavities.

Tests with POM (Polyoxymethylen), PP (Polypropylene) and PEEK (Polyether Ether Ketone) were conducted successfully (Desset 2004, unpubl.).

4   RAPID TOOLING OF MICRO CAVITY INSERTS BY USE OF STEREOLITHOGRAPHY

Rapid prototyping techniques are not affected by geometric limitations as the HSC method, but by material properties.

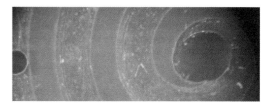

Figure 3.   Epoxy insert with sharp edge at the root of the spiral wall.

Stereolithography resins are more sensitive to mechanical stress and temperature than metals. The Stereolithography was used, because it achieves better precision and surface quality than the other Rapid Prototyping methods when micro cavities are to manufacture.

The experiments with epoxy inserts started on a 3D systems SLA 250 machine using Cibatool SL 5170 resin (Kuhnle, 2004). The next step was to apply the resins Accura SI 10 and Accura SI 40 on a Viper si2 (Bachmann, 2005).

Accura SI 10 is used as an all-round resin. Injecting POM at a temperature of 210°C with a pressure of 60 MPa the cavity of the insert was damaged at the root of its spiral wall after 40 shots, where the melt starts to flow along the spiral path. Figure 3 shows the sharp edge at the starting point of the spiral wall.

Injection at a pressure of 100 MPa sheared off the spiral root after three shots (Fig. 4).

An FEM analysis showed an equivalent stress value of about 700 MPa at this point, thus this edge had to be modified with a curvature.

Accura SI 40 can be tempered, to increase the tensile strength from about 57 MPa to 74 MPa. Its thermal resistance is also improved from about 50°C to

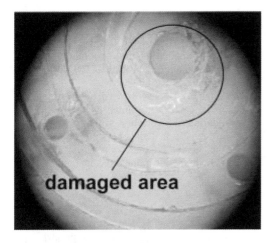

Figure 4. Accura SI 10 insert area damaged after 3 injection shots, applied pressure 100 MPa.

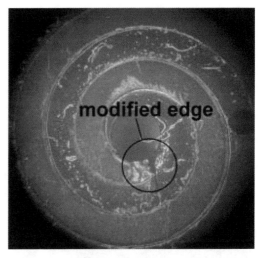

Figure 6. Accura SI 40 insert after 200 injection shots, applied pressure 60 MPa.

Figure 5. Accura SI 40 insert, POM part.

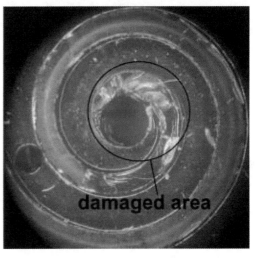

Figure 7. Accura SI 40 insert after 12 injection shots, applied pressure 150 MPa.

100°C by this, but failures due to thermal stress were not discovered during these experiments.

The next inserts were produced with it (Fig. 5). Applying an injection pressure of 60 MPa, more than 200 parts could be produced without any visible failures at the geometry of the cavity (Fig. 6).

The injection pressure of 100 MPa caused a damage at the spiral root after 56 injection shots. 12 shots were the limit when the injection pressure was increased to 150 MPa (Fig. 7).

The data given in Table 2 relate to the resins which were tested successively.

An FEM analysis was performed to simulate and visualise stresses and displacements at the insert caused by the filling process. Figure 8 shows the iso-lines of equivalent stress in a vertical cut near the middle of an insert model, which was simulated with the material properties of Accura SI 10, injecting with a pressure of 60 MPa. The resulting equivalent stress range is between 50 and 110 MPa.

397

Table 2. POM spiral parts produced without failure.

Resin/Injection pressure (MPa)	Accura SI 10	Accura SI 40 tempered
60	40	>200
100	3	56
150	–	12

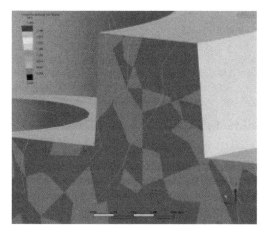

Figure 8. Model of an resin insert with iso-lines of equivalent stress in a vertical cut.

5 CONCLUSION

The experiments showed that HSC-produced metal inserts can be applied for the injection moulding of micro parts without any problems. The only limitation is the complexity of the parts.

Plastic inserts can also be used, but the miniaturisation of the cavities is limited by the mechanical stresses during the injection. Compared to other Rapid Prototyping materials epoxy resins allow the best surface quality and precision, when micro cavities are to manufacture in the direct method.

REFERENCES

Bachmann, Philipp. 2005. Anwendung von Rapid-Tooling-Verfahren zur Herstellung von Mikrospritzgießeinsätzen. Thesis of diploma: Aalen University of Applied Sciences.
Desset, Julien. 2004. Design, manufacturing and testing of a hybrid tool for micro spiral fluidy test on plastics. Final work, SOCRATES program with KaHo Ghent, Belgium: Aalen University of Applied Sciences.
Kuhnle, Martin. 2004. Development of a hybrid tool for micro injection moulding. 5th Annual International Conference on Rapid Product Development: Central University of Technology, Free State, South Africa.

Virtual modeling and rapid manufacturing – Bártolo (eds)
© *2005 Taylor & Francis Group, London, ISBN 0 415 39062 1*

Rapid manufacturing – An evaluation of Rapid Prototyping and HSC technologies in product development

J.P. Garcia & J. de Carvalho
Mechanical Engineering Department, University of S. Paulo, USP, São Carlos, Brazil

ABSTRACT: As a consequence of the demand for high quality products, low cost and short development cycles, most industries are looking for ways to improve the speed of the production processes for small and medium batch sizes. These companies are aware that they have to reduce the cost and time to market for their products in order to become more competitive. To this end it is essential that the development teams know and use the new available technologies in the product development. This work focuses on some of the new currently available technologies, such as Rapid Prototyping, Rapid Tooling and High Speed Cutting, evaluating them and indicating the main advantages, limitations and future perspectives. The usage of these technologies is presented in the development cycle of an optical mouse comparing them with respect to costs, manufacturing time and surface finishing of the parts.

1 INTRODUCTION

The globalization of markets and increasing interdependence of economic agents are reshaping the international competitive environment. These fundamental changes are imposing the organizations to re-examine and modify their competitive strategies. Within this environment of high levels of competition, the development of new products and processes plays an important role. As example, systems CAD/CAM are requested to produce in short time physical objects directly from CAD models. The concept of Rapid Manufacturing in the industrial environment encompasses tasks as prototypes design, manufacturing of small and medium batch sizes and components production directly from 3D CAD models, as well as production of dedicated tools to achieve high production levels.

A new class of manufacture processes, based on the reduction of the product development cycle and in the decreasing of life time of these products, has taken the manufacturing industry to the new frontiers in the products development area. Time reduction requirements, quality and costs associated with larger variety of products and process together with worldwide competition have included technologies of Rapid Prototyping (RP) and High Speed Cutting (HSC) as a integral part of the business processes and processes planning.

The high cost to correct errors in the advanced stages of the product development is another factor that strengthens the necessity to produce reliable prototypes.

Beyond the factors previously cited, the tools development (Molds, Dies, Foundry Patterns, etc.) is an important area in the actual manufacture processes, with raised severity according (Yarlagadda et al. 1999).

Additionally to the factors previously mentioned, new technologies of tools manufacturing (molds, dies, foundry patterns, etc.) is an important area in the manufacturing processes, with increased level of requirements. In this area, Rapid Prototyping technologies are present in tool manufacturing for small lots (Soft Tooling) as well as in tooling for high production levels (Hard Tooling). The best developments for molds through the use of Rapid Prototyping techniques are those that involve difficulties when using conventional methods.

It is important to emphasize that in the tooling production, mainly molds and dies, the High Speed Cutting (HSC) technology has an important position in the productive process chain due to the considerable Lead Time reduction.

This work focuses on the technologies of Rapid Prototyping, Rapid Tooling and High Speed Cutting, evaluating them and indicating some advantages, limitations and future perspectives.

2 THE PRODUCTS DEVELOPMENT CYCLE

The competitive market imposes severe challenges to the industries in a continuous search for improvements, technological innovations, high quality and low costs in order to keep their products competitive.

The necessary changes to follow this context imply in the effort of the organizations to increase their competitiveness. It is a new age for competition without precedents in history. This competition appears not only between traditional competitors, but also the reduction of some commercial barriers has allowed the access to isolated and protected markets.

According Krishnan & Ulrich (2001) there are at least four common perspectives in the design and development research community:

- Marketing;
- Organization;
- Engineering design;
- Operations management.

To increase the organizations competitiveness many critical success factors can be cited:

- Product positioning and pricing;
- Collecting and meeting customer needs;
- Organizational alignment;
- Team characteristics;
- Creative concept and configuration;
- Performance optimization;
- Supplier and material selection;
- Design of production sequence;
- Project Management.

The success of a company is directly linked on its capacity to understand the necessities of the customer and to develop products quickly.

3 THE RAPID PROTOTYPING TECHNOLOGY

The Rapid Prototyping techniques allow the manufacture of three-dimensional physical models from a mathematical description of the part in a CAD system. The models are manufactured by layers without the necessity of specific tooling. Complex geometry models are obtained by these processes with the same simplicity of simple parts resulting in great savings in cost and time when compared with conventional techniques.

The Rapid Prototyping technique is a powerful tool in the Product Development cycle. In an initial phase, it is indicated to test the design and the functionality of the components; in a more advanced phase it can be used in tooling production for preliminary series, prior to the production of the definitive tooling.

The Rapid Prototyping technologies are one of the most important techniques of the 90's, reducing drastically the "Time-to-Market", basic characteristic in the current competitive market. Based on the reduction of this time the Project teams can dedicate this "extra time" to other aspects in the product development as design, functionality, ergonomics and the failure reduction during the conceptual phase according (Billes et al. 1995).

In the case of small and medium production series the Rapid Prototyping is alternative technology in tools production, such as soft tooling and rubber based tooling.

Nowadays, main applications of Rapid Prototyping technologies during product development are:

- Conceptual design, where a 3D model plays an important role to check aspects such aesthetics, ergonomics, visual impact and so on.
- Functional prototypes, where precise prototypes can be used to check functional aspects such as assembly, in service performance and so on.
- Tool design, where the prototypes can be applied in soft tooling developments, or master for foundries or vacuum casting.

4 RAPID MANUFACTURING

With the increase of the users of RP technologies the applications tend to increase, not only to produce prototypes, but also production tools evolving for a process of Rapid Manufacturing (RM). The possibility to produce in a fast way a production tool, even for small batch sizes enables the validation of the final production process of the part.

4.1 Rapid tooling

The application of patterns in the manufacturing of production tools has been considered an important subject in the production of small and medium batch sizes. The cost of the production tools in these situations, represents an important parcel of the final cost of the parts and any economy in the tools cost reduces considerably the cost of the finished part.

Creating tooling for prototype and production components represents one of the most time consuming and costly phases in the development of new products. It is particularly problematic for low-volume products or rapidly changing high-volume products.

A good product development system must enable designer or design teams to consider all aspects of product design, manufacturing, selling and recycling at the early stage.

The RT methods can generally be divided into direct and indirect tooling categories, and also soft and hard tooling subgroups. Indirect RT requires some master patterns, which can be made by conventional methods like HSC or RP processes. Direct RT involves the manufacturing of a tool cavity directly in the RP or HSC machine.

Soft Tooling is associated with low costs; used for low volume of production and it uses materials with low hardness such as silicones, rubbers, etc. So an architecture of a RT system can be classified according to Figure 1.

4.2 High Speed Cutting

The High Speed Cutting is a possible technological solution for increasing the productivity with economic efficiency in the Manufacture processes. The pioneer Salomon's experience in 1931 in High Speed Cutting shows that to some cutting speeds (5–10 times larger than the conventional ones), the removal temperature of chips in the cut edge will start to reduce.

Modern researches unhappily have not been capable to verify this theory in all its extension. It has a relative reduction of the temperature in the cut edge that initiates in determined cut speeds and varies for different materials. The reduction is small for cast iron and steel and greater for non-ferrous such as aluminum and other metals.

The Salomon's experience was realized with non-ferrous materials and since then several researches have investigated other materials, tools and conditions in order to define the possibilities and limitations of the High Speed Cutting technology.

The interest of the scientific and industrial communities in the optimization of the cutting speeds has increased considerable in the last years, mainly when using the HSC as a tool in the productivity increase.

The basic strategy in the use of a HSC system is the definition of the appropriate cutting speed and the tool to be used in function of the component to be manufactured and the type of machining operation. To reach this situation it is imperative that the increase effect of the cutting speed in the part and in the cutting tool be understood, as well as the thermal effect of this action. The dynamic reactions that the machine tool will be submitted with the increment of the cutting speed is also a factor to be considered carefully.

The High Speed Cutting technology represents the integration of some concepts for manufacture such as product modeling using CAD systems, process planning, NC programs generation, and Machining involving the machine-tool, numerical command, parameters and cutting tools.

An application of the HSC technology that has become common is the manufacture of molds and dies, and also with lesser attributions techniques, in the 2D facing machining, fundamentally in the automobile industry. During some decades, the HSC has been applied in a range of metallic and non-metallic materials, including the parts production with specific necessities of superficial quality and in materials with hardness over 50 HRc.

The HSC definition is based on the material type of the part. As example cut speeds of 500 m/min is considered HSC for steel alloys, however conventional speeds

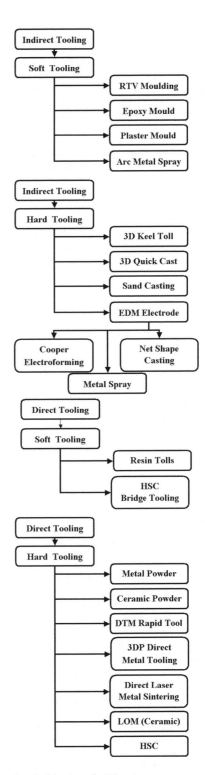

Figure 1. Architecture of a RT system.

in the aluminum machining according (Fallbohmer et al. 2000).

The molds and dies machining represents a significant area of application of the HSC technique for casting iron, casting steel and steel alloys. In the industries of the whole world the machining operations of molds and dies represent a great percentage of the productive time.

Thus, one of the main objectives of the molds and dies industry has been to reduce the productive time, in order to improve the finished part quality and also to reduce costs. In these points the industry has benefited intensively with the HSC technology.

The main advantages of the HSC technique described for several researchers like (Dagioloke et al. 1995), Yousefi & Ichida (2000), Özel & Altan (2000), Mills (1996) and (Urbanski et al. 2000) are:

- Great taxes of material removal;
- Reduction of Lead Time;
- Low cutting efforts;
- Good heat waste during the process;
- Low part distortion, due good heat waste;
- Improvement of the superficial finishing (roughness) and dimensional precision.

Although these advantages, some disadvantages or improvements opportunities are indicated:

- Necessity of special tools with coatings;
- Special Machines-Tools and high costs;
- Necessity of a continuous balancing of the spindle head;

5 CASE STUDY – OPTICAL MOUSE

5.1 Description

Independently of the technical and functional characteristics of the final product, several stages will compose the product development.

In the preliminary stage of the optical mouse project some strategies of product manufacture and the main characteristics of the product had been defined. Between these special characteristics, the product ergonomics was considered as main factor, whereas being the product an optical mouse the friction reduction is an important characteristic to be considered.

After the analysis of this factor, two strategies for the geometry definition of the optical mouse had been selected. The first one would be the creation of a conventional geometry in CAD software, based fundamentally in the ergonomic characteristic of this project. One second option would be the execution of a manual modeling instead of the definition of geometry through CAD software. In this case the strategy adopted for the definition of the more adequate geometry in point of

Figure 2. Scanning process and CAD model.

view of ergonomics was the confection of a model from the manual work of the designer.

From this model the strategy for the component production includes the surface scanning and the posterior generation of a point cloud. Through mathematical algorithms a solid model is generated in the CAD software, model that represent the most suitable surface in point of view of the ergonomics created by the designer.

Finally, from CAD file it was defined the usage of Rapid Prototyping and High Speed Cutting as strategies for the development of this product and the production the prototype of the optical mouse as well as the production o a small number of parts.

The laser scanning of the surface and the 3D CAD model are illustrated in Figure 2.

The STL files generated from the CAD system enables the manufacturing of the prototype in the Rapid Prototyping machine, in this case using the FDM (Fused Deposition Modeling) technology, elaboration of a CNC program through the use of CAM (Computer Aided Manufacturing) software for the prototype machining in the HSC Milling Center, fabrication of a Rapid Tooling through the Vacuum Casting technology, using the prototypes as master models and finally manufacturing the pre-series of parts.

5.2 Manufacturing stages using Rapid Prototyping

The first stage of the product development was the prototype manufacturing of the optical mouse using the FDM (Fused Deposition Modeling) technology.

From 3D CAD file, the prototype was produced using FDM technology, following basically the stages indicated in Figure 3.

Figure 4 presents the prototype right after its exit of the Rapid Prototyping machine.

5.3 Manufacturing stages using High Speed Cutting

The second stage of this work was the prototype manufacturing of the optical mouse using High Speed Cutting (HSC) technology to verify the effectiveness of this tool to produce prototypes in a fast way. The material used was Nylon (PA66) and in a next stage the final tooling would be produced in the same machine, as indicated in Figure 5.

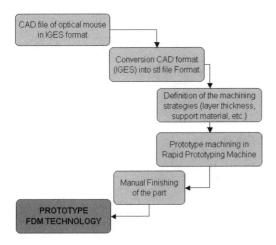

Figure 3. Rapid Prototyping machining of the optical mouse.

Figure 4. Prototype obtained in the FDM machine.

Figure 5. Prototypes obtained in the HSC machine.

Figure 6. Rubber flexible mold for pilot run.

5.4 *Pre-series production using Vacuum Casting technology*

The third stage of this work was the confection of a rubber flexible mold (Rapid Tooling), from the prototype obtained in the Rapid Prototyping machine. This technology, called Vacuum Casting, produced more than 100 parts of the optical mouse. The flexible mold in operation (producing Polyurethane prototypes) can be observed in Figure 6.

6 RESULTS ANALYSIS

It is very important to emphasize that the Rapid Prototyping (RP), Rapid Tooling (RT), are skills that allow the optimization of the Product Development cycle. The main indication of this work is that the project teams involved with development of new products should be capable to apply these technologies and choose the best one considering the product to be developed.

In the specific case of the Optical Mouse development it can be concluded that:

- The processing time to convert the CAD file in IGES format, its conversion to STL format and the definition of the best manufacturing strategy of the prototype (position, layer thickness, etc.) was lower than the strategies definition of machining product and the generation of programs through the use of CAM (Computer Aided Manufacturing) software. It was necessary, for this case, 30 minutes for Rapid Prototyping technology against 2 hours (average) for the High Speed Cutting operations.
- The lead time to the prototype manufacturing through the Rapid Prototyping technology (FDM) was of 5 hours and 20 minutes, while the lead time with the use of the High Speed Cutting technology was of 1 hour and 49 minutes to the 2 hours and 7 minutes to produce 6 parts, depending on the adopted strategy;

403

- The prototypes superficial finishing (roughness) got from the High Speed Cutting was visually superior to the one got in the Rapid Prototyping technology using the FDM (Fused Deposition Modeling) process. The prototype superficial finishing produced by FDM process needed manual finishing for posterior generation of the flexible mold, while the prototypes from HSC were finished in the machine;
- It is important to emphasize that the average time of part manufacture using the rubber mold was approximately 10 minutes representing the lowest of all processes tested.

It must be considered that each technology has its own application in distinct stages of the product development and that the results shown are applicable for the case studied.

7 CONCLUSION

With relation to the usage of Rapid Prototyping (RP), Rapid Tooling (RT) and High Speed Cutting (HSC) technologies specifically in the optical mouse case it can be concluded that:

- The Rapid Prototyping (RP) and High Speed Cutting (HSC) to produce directly the prototypes technologies had practically similar performance in the prototypes development of the optical mouse.
- The use of the Rapid Tooling (RT) technology in the production of a pilot lot had the best performance when compared with the traditional manufacturing techniques.
- The Rapid Prototyping (RP) and High Speed Cutting (HSC) technologies had similar performances in this specific case, with exception of the surface finishing, considerably better in the HSC technology than in the RP used (before manual surface finishing).
- It is important emphasize that the analyses had been done on the basis of the results got in the production of the external surface of optical mouse. The internal region of the mouse or other products with electronic interface would require several internal surfaces for components fixation which can yield to great difficulties of machining, with the necessity of the use of several tools and surfaces machining. By this way Rapid Prototyping (RP) technology emerge as an ideal tool.

For development and manufacture engineers these technologies to project and to develop new products, the right knowledge of the technologies discussed represent a competitive advantage.

The majority of the new and exciting applications of the Rapid Prototyping are showing that actually RP does not only deal with prototypes. Amongst the new applications it can be mentioned the applications in medical area and the construction of components for the aerospace industry.

As a final remark, although the results shown are valid for a particular application, it is clear that the development teams must know deeply all the characteristics of these new technologies to explore their maximum capabilities and to use them properly considering the part requirements, material, processes and application.

REFERENCES

Ahrens, C. H. et al. (2001). Considerações iniciais para confecção de protótipos rápidos. *First Brazilian Congress on Manufacturing Engineering*, Curitiba, Paraná, 2001.

Biles, W. E. et al. (1995). Computer-Aided Design and Rapid Tool Development in Injection Molding Process. *Computers ind. Engng.*, v.29, p.659–662.

Dagioloke, I. F. et al. (1995). High Speed Machining: An Approach to Process Analysis. *Journal of Materials Processing Technology*, v.54, p.82–87.

Fallbohmer, P. et al. (2000). High-speed machining of cast iron and alloy steels for die and mold manufacturing. *Journal of Materials Processing Technology*, v.98, p.104–115.

Krishnan, V.; Ulrich, K. T. (2001). Product Development Decisions: A Review of the Literature. *Management Science*, v.47, n.1, p. 1–21.

Lettice, F.; Palminder, S.; Stephen, E. (1995). A workbook-based methodology for implementing concurrent engineering. *International Journal of Industrial Ergonomics*, v.16, p.339–351.

Mieritz, B.; Dickens, P. M., (1996). A European Strategy for RP and Manufacturing. *EARP (European Action for Rapid Prototyping)*, 1996.

Mills, B. (1996). Recent Developments in Cutting Tool Materials. *Journal of Materials Processing Technology*, v.56, p.16–23.

Omokawa, R. (1999). *Utilização de sistemas PDM em ambiente de engenharia simultânea: o caso de uma implantação em uma montadora de veículos pesados*, 169p. Dissertação (Mestrado) – Escola de Engenharia de São Carlos, Universidade de São Paulo, São Carlos. 1999.

Özel, T.; Altan, T. (2000). Process simulation using finite element method – prediction of cutting forces, tool stresses and temperatures in high-speed flat end milling. *International Journal of Machine Tools & Manufacture*, v.40, p.713–738.

Urbanski, J. P. et al. (2000). High speed machining of moulds and dies for net shape manufacture. *Materials and Design*, v.21, p.395–402.

Wohlers, T. (2003). *Rapid Prototyping & Tooling State of the Industry: 2003*. Wohlers Associates, Inc. Fort Collins, Worldwide Progress Report.

Yarlagadda, P. K. D. V.; Christodoulou, P.; Subramanian, V. S., (1999). Feasibility studies on the production of electro-discharge machining electrodes with rapid prototyping and electroforming process. *Journal of Material Processing Technology*, v.90, p.231–237.

Yousefi, R.; Ichida, Y. (2000). A study on ultra – high-speed cutting of aluminium alloy: Formation of welded metal on the secondary cutting edge of the tool and its effects on the quality of finished surface. *Journal of the International Societies for Precision Engineering and Nanotechnology*, v.24, p. 371–376.

Virtual modeling and rapid manufacturing – Bártolo (eds)
© *2005 Taylor & Francis Group, London, ISBN 0 415 39062 1*

Wire EDM Performance and wire characteristics

Kamlakar P. Rajurkar & Jayakumar Narasimhan
Center for Nontraditional Manufacturing Research,
University of Nebraska-Lincoln, Lincoln, Nebraska, U.S.A.

ABSTRACT: Wire electrical discharge machining (WEDM) is used as a key machining process for cutting hard, electrically conductive materials into complex and highly precise shape in die and mold making industry. The performance of the WEDM is represented by the average discharge frequency, volume removed by single discharge, feedback voltage and overcut. The wire composition has been an important factor in improving the overall performance and application potential of WEDM. This paper presents a comprehensive experimental evaluation of 10 different commercially available wires on the basis of the wire performance, and scanning electron microscopic study of the wire surface.

1 INTRODUCTION

Wire electrical discharge machining (WEDM) produces complex two and three-dimensional shapes in hard materials with a high degree of accuracy without the use of high-cost grinding or expensive EDM electrodes. WEDM is a thermoelectric process which erodes material from the workpiece by a series of discrete sparks between the workpiece and the tool (a wire of diameter ranging from 0.05 to 0.25 mm). Deionized water is injected into the machining gap during the machining (Figure 1). The main goals of WEDM manufacturers and users are to achieve a better stability and higher productivity of the WEDM, i.e. higher machining rate with desired accuracy and minimum surface damage without wire rupture. Since the introduction of WEDM in the early seventies, a substantial improvement in its performance, reliability and machining cost has been achieved. The improvement of WEDM performance can be attributed to mainly following factors (Dauw, 1992a, b):

- More powerful generators
- New wire tool electrodes
- Better design and manufacture of basic machine tool components
- Improved machine intelligence
- Better flushing techniques

The wire composition and production methods have been very important factors in improving the overall performance and application potential of WEDM. Additionally every breakthrough based on better understanding on the mechanism of discharge has resulted into a significant achievement.

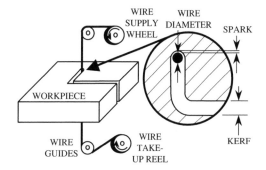

Figure 1. Wire EDM system (Wang, 1992).

2 MACHINING PERFORMANCE TESTS

A comprehensive list of general measures of "machining performance" of machine tool and/or its components includes: efficiency, stability, quality and cost. This study mainly evaluates the first three measures by testing, recording and analyzing specific parameters as shown in the Figure 2.

3 EXPERIMENTS

To understand the characteristics and performance of the various kinds of wire from different manufacturers ten type of wires were selected for comparison. Table 1 lists these ten wires and their composition. Stainless steel was used as the workpiece and all the wires were of diameter approximately 0.25 mm.

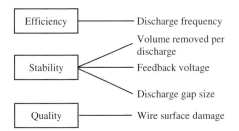

Figure 2. Relation between machining performance and parameters.

Table 1. Wire types and properties.

Wire type	Core material	Coating material	Coating thickness (μm)
Dyna X	CuZn5	CuZn50	16.67
Dyna Y	CuZn37	CuZn50	25.00
Dyna Z	CuZn20	CuZn50	17.68
Full Hard	CuZn37	None	N/A
Type X	Cu	CuZn50	17.46
Type D	CuZn20/III	CuZn50	21.67
SW25X	Cu	Zn	18.79
Super Brass	CuZn36	None	N/A
Alpha Cut	CuZn36	Zn	16.49
Stamm	CuZn37	None	N/A

4 TEST RESULTS

4.1 Average discharge frequency

Under a constant dielectric flushing condition, the machining speed is mainly determined by single spark discharge energy and discharge frequency as well as the wire flushability. The volumetric machining speed can be described as:

$$S_v = C_w \cdot E_p \cdot f \tag{1}$$

where,
S_v = machining speed (mm^3/min).
E_p = average single discharge frequency (KHz).
f = discharge frequency (KHz).
C_w = coefficient determined by the wire properties including the flushability.

The product of C_w and E_p represents the amount of metal removed from workpiece by single pulse discharge. The selection of peak current or capacitance is usually constrained by surface roughness requirement, therefore, increasing the discharge frequency seems to be a viable option for improving the machining speed. However, the risk of wire breakage also increases with the increase in discharge frequency. A wire which can sustain a higher discharge frequency without breaking leads to a higher productivity.

Table 2. Workpiece height related parameters.

Workpiece height (mm)	Charge pulse frequency f (KHZ)	Reference cutting speed S (mm/min)	Reference servo voltage Aj (%)
5	13	1.0	50
10	19	0.84	50
20	30	0.58	50
40	52	0.36	50

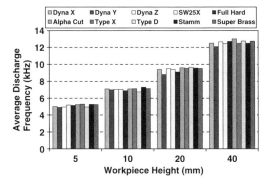

Figure 3. Comparison of average discharge frequency.

The reference discharge frequency can be set by adjusting two parameters, the Charge Frequency 'f' and the Reference Servo Voltage 'Aj' as shown in Table 2. The f value determines the pulse interval or offtime and the charge frequency for charging the capacitor which is connected to the discharge gap in parallel. Aj is a reference servo set point. The servo control system regulates the size of discharge gap and keeps the average gap voltage to a set point. Aj is set as 50% for all experiments of this study so that the higher charge frequency means higher discharge frequency.

Although discharge frequency settings were uniform, actual discharge frequencies measured on-line for each wire are not exactly equal due to the difference of wire properties and corresponding gap conditions, Results shown in Figure 3 do not indicate any significant differences in average discharge frequency values of wires for the conditions used in this study.

The volumetric machining speed can also be described geometrically as:

$$S_v = MRR \cdot (d + 2S_g) \tag{2}$$

where d is the diameter of the wire electrode (mm) and S_g is the gap size (mm).

Equating equations (1) and (2), the eroded volume per discharge ($C_w \cdot E_p$) can be obtained as:

$$C_w \cdot E_p = MRR \cdot (d + 2S_g)/f \tag{3}$$

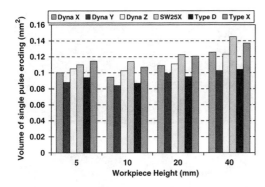

Figure 4. Comparison of volume removed by single discharge.

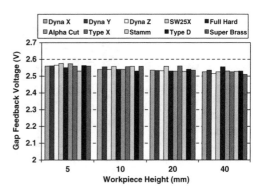

Figure 5. Gap feedback voltage.

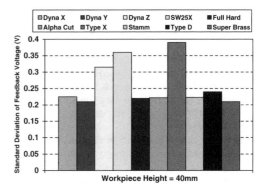

Figure 6. Comparison of Standard Deviation of feedback voltage.

According to equation (3) the eroded or removed volume per discharge, an indicator of wire flushability can easily be calculated knowing material removal rate and discharge frequency. The gap size depends on machining condition and wire properties. If machining conditions are fixed, the gap size can be assumed to remain constant for a given wire. Figure 4 shows the calculated values of eroded volume per discharge for six kinds of coated wires.

Charmilles SW25X wire displays the highest value of eroded volume per discharge for all workpiece heights except 5 mm. Bedra Bronco Cut Type X shows slightly lower value than SW25X, but it exhibits a little higher value for 5 mm thick workpiece, In Dyna wire series, Dyna Z and Dyna X wire values are better than Dyna Y wire.

4.2 Gap feedback voltage

The gap feedback voltage is an indicator of average gap size between wire electrode and workpiece. It is related to set parameters and adjustment of control system, wire tension, and other wire characteristics like tensile strength and coating material. The servo control system adjusts the table movement to compensate for the increase of gap size caused by the discharge process and maintains a proper gap size to ensure a stable spark discharge frequency. Although a single discharge erodes only a small amount of workpiece material, discharges occur at a frequency of several kHZ. The response time of the servo control system is not small enough to effectively compensate for the change in a timely manner. Results of gap feedback voltage measurement are shown in Figure 5. The Figure 5 does not indicate any significant difference among different wires. The servo control systems seem to adjust the gap size dynamically and keeps the feedback voltage to the set point. The feedback voltage slightly reduces with increasing workpiece height.

The Standard Deviation (SD) of feedback voltage indicates the variation of feedback voltage. Figure 6 illustrates results of standard deviation measured during machining for each wire. The sampling time is 100 ms and the workpiece height is 40 mm. It can be found that the coated wires display a higher SD of feedback voltage than non-coated wires. Among coated wires, wires with a higher material removal rate exhibits a greater SD value. For example, Bedra Bronco cut Type X and Charmilles SW25X have higher values of standard deviation of feedback voltage. Dyna Y wire with a lower material removal rate under the selected machining conditions shows a lower SD value per discharge. The higher eroded volume per discharge results from a more powerful discharge which leads to a stronger disturbance at the gap or a larger variation of the feedback voltage.

4.3 Overcut

Overcut is an important technical data for wire performance. The over cut is essential for setting CNC offset and ensure the machining accuracy. The over cut for five kinds of wire have been measured by

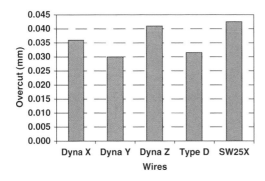

Figure 7. Comparison of overcut.

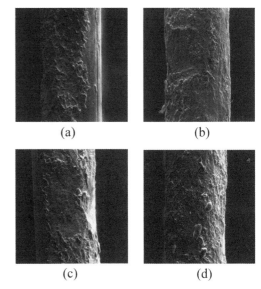

(a) (b)

(c) (d)

Figure 8. SEM of the wires.

using Mitutoyo Toolmaker Microscope. The results of over cut measurements are shown in Figure 7.

It can be seen that, Charmilles SW25X wire has the highest value of over cut and SWIL Dyna Y wire shows the lowest over cut. The difference in over cut values for different wires depends in the difference in material removal are for each wire. Thus, cutting with a wire with a higher material removal rate, results in a larger size of over cut.

The more powerful discharge also means a better flushability as more workpiece material can be ejected out and flushed away from the discharge gap. Thus, a large over cut value is obtained with wires of higher flushability.

5 METALLURGICAL ANALYSIS

5.1 Optical microscopic analysis

Previous research has shown that the presence of a coating with a low volatilization temperature leads to a higher machining speed, a lower electrode wear ratio and a higher reliability in avoiding wire breakage (Tomalin, 1989). The coated wire exhibits significant improvements over non-coated homogenous wires due to the enhancement of flushability and the ability to absorb thermal energy by vaporizing much less volume of wire coating material. Therefore the coating material of a wire and its structure in cross section is an important aspect when investigating the wire characteristics.

The optical microscopic images of the cross section of the wires showed the difference in coating thickness of wires and its variations through out the wire length.

5.2 SEM analysis of eroded wire

As an electrode in the spark discharge process, wire is also subjected to thermal energy generated by electrical discharges. The thermal energy to wire electrode

melts and vaporizes wire material similar to workpiece material. The melted and vaporized material is ejected out and flushed away by the dielectric from the discharge gap. A portion of melted material is cooled by dielectric fluid and redeposits on the wire surface. The SEM images of wire used in WEDM are shown in Figures 8a–d.

The resolidified layer can easily be seen on all 10 kinds of wires. Non-coated wires exhibit relatively flat and larger craters compared to coated wires. Coated wires show large number of smaller pits on the resolidified layers. It seems that the amount of melted material in coated wires is usually less than that of non-coated wires because a large part of thermal energy is absorbed in vaporizing a smaller volume of zinc coating.

SWIL Dyna X wire exhibits deeper craters on its surface than those of Bedra Type X. Similarly Bedra Megacut Type D wire displays a rougher surface than that of SWIL Dyna Z. In both pairs the zinc coating content is similar but the thickness varies. For example, the zinc coating on Bedra Type X has been found to be thicker than that of SWIL Dyna X (Table 1). Therefore, under similar conditions, SWIL Dyna X wire is subjected to a higher portion of the thermal energy at the core, which eventually results in large craters.

According to the catalog of electro tools products from Intech EDM (Intech EDM, 1994), Alphacut 500 has a brass core wire coated with pure zinc. Its SEM image shows that bottom of craters is flatter than other wires. It is observed that some material has been peeled off from the coating layer on the right side of the wire image. It seems that some stress remained in the coated layer. The thermal impact and losing of outer layer releases the stress it and hence some material peels off during machining.

6 SUMMARY

A comprehensive experimental evaluation of 10 different commercially available wires on the basis of the wire performance and scanning electron microscopic study of the wire surface as been presented. The results show that the wire with coating with low volatile material lasts longer and erodes higher volume for single discharge.

ACKNOWLEDGEMENTS

Authors acknowledge the support from Global Trade Network, Inc. Authors are also thankful to Professors J. Kozak and W.M. Wang for their assistance in this work.

REFERENCES

Dauw, D.F., & Albert, L. 1992a. Two Decades of Wire EDM Evolution: A Significant Improvement of Overall Performance. *The Proceedings of ISEM-10*: 300–319. Magdeburg: Germany.

Dauw, D.F., Albert, L. et al. 1992b. About the Evolution of Wire Tool Performance in WEDM. *Annals of the CIRP* 41/1: 221–225.

Wang, W.M., Rajurkar, K.P., & Boyina, S. 1992. Effect of Thermal Load on Wire Rupture in WEDM. *Transaction of NAMRI/SME*: 139–144.

Tomalin, D.S. 1989. Not all EDM Wires are Created Equal. *The Proceedings of EDM'89*:18/1–18/14.

Intech EDM. 1994. Catalog of electro tools products.

Virtual modeling and rapid manufacturing – Bártolo (eds)
© 2005 Taylor & Francis Group, London, ISBN 0 415 39062 1

Rapid Manufacturing: Reality of today or dream of tomorrow

Olivier Jay

Danish Technological Institute, Århus, Denmark

ABSTRACT: The aim of this paper is to analyze the possibilities of Rapid Manufacturing (RM) in a global perspective of low volume production. The paper should give a general approach for the companies interested in using the technology and describe the economical aspect incorporated in the product development. This paper is referring to the results of different projects that Danish Technological Institute has been working with (Studies on quality (surface, process) and a Danish Project. "Spare parts on Demand" This project was made in collaboration with Aalborg University, Institute of Production.)

1   INTRODUCTION

1.1   Definition of Rapid Manufacturing (RM)

Rapid Manufacturing (RM) has been used in many different ways and that is the reason why it is important to define it.

Rapid Manufacturing: Is the production of parts for the end user, produced directly or indirectly with the help of RP machines. The important point is that the part will be used by the end user, and will be sold as a finished part.

Terry Wohlers[1] definition could have been used as well: Rapid Manufacturing is the direct production of finished goods directly from a Rapid prototyping device. Low volume production or one of the other terms emerging on the marked could also have been used.

Production should be the right term but the name were not responding to the fact that the project was dedicated to RP machines, and customers are still only seeing the RP technologies in a prototyping way.

1.2   History behind the project

Most vendors of RPT machines are already promoting their machines for producing parts for RM. But trends of industry show that only a few companies are actually using them for that application. InvisAlign[2] manufacturing from Align Technology is certainly the best example, not only because they are producing all their parts on RP machines, but most of all because they have based their business strategy on it.

Few other companies have been following the direction but many of them are not in the direct concurrent market such as military and space industry.

How the others could become interested in working with RM?

1.3   Introduction

The Dream started a few years ago, when the first diagrams representing the production of plastics parts compare to the production of the same parts on a RP machine emerged in an EU project. The most fascinating was that the curves of that diagram were crossing about 1000 parts and at that precise moment, everybody started to believe that RP machines were going to produce millions of plastics parts. But nothing really happened afterwards. Materials were partly responsible, but the opening to the east and the reduction of the mould prices should also be in the equation. Another important point in the equation was and still is the manpower necessary to finish the parts on RP machines.

Danish Technological Institute has been involved in many projects concerning Rapid tooling and RM, and DTI history started with a success; a customer who produced for thousands of € on our Actua, then Thermojet and finally SLS. The interest was suddenly not only the price but most of all the function. The parts produced were corresponding better to the production, the reduction of the delay; reduction of stock and the continuous optimization were also mentioned. And all together, the equation was presenting the RP production as the best choice. The light was maybe coming, but one more time it was a quite narrow industry.

The solution was there anyway as well as the equation and the parameters, nothing more was necessary for us to believe in the future of it.

2 PARAMETERS

The parameters were found by interviewing customers and searching an answer to the question: What is a finished product? This way was chosen instead of making the approach of using RP machines to produce parts and then find out how to use them.

The panel is composed solely of Danish customer, even if examples from the international scene will be used later on in this paper.

As the rest of the world, Denmark is divided. Only a few customers are in front and they are developing new products produced exclusively by means of RP technologies and many of them are still using the RP technologies as prototyping only.

The leaders are most of all in the medical domain, and to be more precise, the hearing aids industry. Jewelry should be mentioned as well for an increasing demand. The analyses of those two examples present the reality about RM today.

RM is applied for medical application (Invisalign[2], Phonaks, Oticon, etc.) and the FDA material on a Stratasys is confirming it. And it is applied to high technologic and/or precision products.

But why are the other companies refusing to use it? Some of them are, of course, producing too many parts, and it is only realistic to say that RM is unable to compete with a multi-cavity mould producing thousands of parts per minute. But here are some of the other answers from them:

- How is the repeatability of the production of parts?
- Can the material be certified for a minimum of 20 years?
- Can the material be available for the next 10 years?
- How is the stress in the part compared to the layer production form?
- How is the material compared to ABS, PE?

Four parameters can be extracted from all this.

2.1 The material

It is maybe one of the most sensible parameters. Many research centers have been trying to learn more about the material. Of course, the results are depending on the orientation of the layer, and somehow as long as the process is a layer manufacturing, it will be a big parameter in the strength of the part. Material manufacturers are producing new materials every year, but it is also a weakness due to the fact that engineers do not have the time to follow up on them. An example is given at the end of that paper to understand a little more about the complexity of the materials, the example is coming from projects report.

The conclusion about the material is that they are maybe not ready to compete directly with plastic materials, but the point is that it is maybe even more important to stop competing with them. They are different and it could be more interesting to use the material when it is conforming to the specification of products instead of using our time trying to demonstrate that the materials are getting closer.

2.2 RP technologies as a production

The repeatability is almost impossible to obtain. The position of the part on the platform, the power of the laser, and many other parameters are against it. The point is that the parts are in the dimensions respecting the specifications of the constructors and that what should be presented.

But there is a much more important point to it; the finishing of the parts. Most of the RP machines have been developed to produce prototyping, this involves that the manual work has been accepted as the only solution to finish the part. The point is not a problem in itself but it will become a problem when the economical equations are made. A solution could be to develop machines to clean and finish the parts automatically.

You have, of course, many advantages from using RP machines as production. Moulds can be forgotten, stock can approach zero and products can always be better in the next series.

The freedom of fabrication is also an important point. This means that you can produce almost the form you would like to produce without thinking about the production process. The only problem is that engineers do not think of the possibilities of RM until they are almost ready to order the moulds.

That was one reason mentioned by many of the people we interviewed. When they approached deadline and the time when the project should go into production, which way would they choose? The secure way or the risky way? Even if the risky way is 10 times cheaper, most of them will take the traditional way, the one they have been using for the last 20 years, and somehow it is easy to understand that they are not taking the risk. This reason leads us to the next parameter and the worst enemy of RM today: Ourselves!

2.3 Cultures of engineer

Culture and the locked process of product development are the biggest difficulties when introducing a new process. Of course few companies have started with the process, but in most cases it has involved a strategic decision about the product.

The question about using RM is involving much more than just deciding to use it. It involves getting to learn new materials, new ways of designing parts, and new ways of thinking as well. As I mentioned before, the decision of using RM technologies should come in the stages of specifications, before the first drawing.

But to do so, companies will need more information about how to decide the right way to produce parts, including economical and real material specifications.

But culture is the one that takes longest to change, so RM needs to be patient and show what it can.

2.4 Economical aspects

The economical aspect is the most interesting aspect. Many examples are going around the world and some of them will be presented during my presentation. But it is more exiting to be a little more general.

The next points will present the different parameters to consider when you will try to evaluate the idea of using RM.

Product lifetime: Products have been dramatically reducing their lifetime on the market.

Do you need to have a new product on the market often? And are you bound to wait the breakeven point before introducing a new product?

Time to market: Introduction on the market is really important; and a delay of six month can bring a reduction of the gain by 33%. RM should not only be considered a substitution to more traditional technologies, it could also be used as a shortcut to market. It can be used in the period when you are waiting for the mould and the first parts to be produced.

Response to the product: Do you know the reaction of the public and are you sure about it? Are you taking a big risk if the product is failing? In this case it could also be interesting to produce the first parts in a more expensive way and earn a shorter margin on them before introducing a product that is cheaper to produce but which involves more heavy investments.

Personalization: Could you get some of your customers interested in getting a product just for them? RM is an easy way to make each of your customers a little more special.

Design: Can the product respond to the specification in a better way? By making complex geometries, could the product obtain a better productivity or maybe longevity? Can you simplify the montage and maybe the number of parts by changing the design?

Prices: Maybe the most easily to consider, because it only involves numbers. ☺ But you should not forget the next points in the equations.

Working hours: It is easy to calculate the price of a mould and the price per parts, and then on the other side of the balance the price per part on a RM machine. But it is important to consider the number of hours used to prepare the part, drawing the mould, programming the machines to make the mould as well. On the other hand, you may consider the extra hours/operations used to finish the part

after the production of RM. Remember to add possible costs due to revisions of the parts, and possible modification of the mould. It does not happen every time but it is still a factor of risk.

All these parameters should be considered during the decision period. It is also important to make the decision at an early stage as possible to give your engineer the possibility of designing for the right technology.

3 CONCLUSION

RM is ready to take the next challenge but there is still a long way to go. It is not only a dream of tomorrow, it will be used; the only question is when.

Materials producers work hard to introduce new materials, but the machine of the future is maybe not yet on the market. When waiting, it is important to make your homework and learn as much as you can in order to be ready when the technology will fully correspond to our needs.

Culture will change slowly, but a general publication and a bigger introduction in engineering schools as a production possibility is necessary in order to change it.

RP technologies are maybe not completely ready, but it is maybe time to focus on the finishing part of the process to do so.

4 EXAMPLE OF MATERIAL RESEARCH[3][4]

With RM and with silicon rubber tooling (SRT), ABS-like parts can be produced, but what are the differences between the real ABS from a "classic injection" and ABS-like produced by these new technologies.

The materials have been used for a long time due to their mechanical properties because it allows customers to simulate ABS-like parts but, as we can see from the table below, there are in fact still differences between these materials.

	UNIT	Properties		
		SL 7560	ABS (SRT)	ABS
Tensile Modulus	MPa	2500	2500	2225,5
Flexural Modulus	MPa	2500	2500	2415
Tensile strength	MPa	52	70	43
Flexural strength	MPa	93,5	105	76,5
Izod impact	J/m	36	70	225
Elongation at break	%	10,5	15	24
Hardness	-	D86		R110
Density	g/cm^3	1,22	1,2	1,05

The tensile and flexural performances are quite similar but the biggest difference comes from the impact strength (525% of difference) and from the elongation at break (129% of difference).

This comparison shows us that the SL 7560 has interesting properties for certain applications but stay limited for others.

The whole of these mechanical properties is very important for the material choice but we mustn't forget that a SLA part is built layer by layer and can also be built in the three directions X, Y, Z. So the layer orientation influence can be a anisotropic factor.

All the parts built for tests in this report respect the dimensions of the standard ISO 527-2(E) named "Test conditions for moulding and extrusion plastics". Currently, there is no standard concerning the SLA materials.

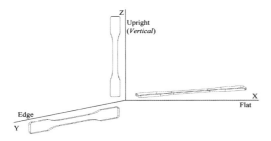

Results of the layer orientation tests

	Flat	Edge	Vertical	45°
Young's Modulus (MPa)	2972	3160	3130	3232
Maximum load (kN)	3,11	3,07	2,32	2,8
Tensile strength at maximum load (MPa)	71,83	76,36	59,8	70,29

Young's Modulus is very closed (8,5% of difference between flat and 45° parts) so the SLA material SL7560 has the same elastic modulus in each direction meaning that layer orientations does not have any influence on it. But the lowest value should be used for product development.

With regards to the maximum load, the difference is more important, about 25% between vertical and flat parts. For the vertical parts, the tensile effort is applied in the same direction as the layer accumulation. This means that the resistance between each layer can be a weak point of this process or material.

To complete the test, it is possible to study some of the results of the Loughborough University[5] (United Kingdom) where they made tests in order to know the influence of the post curing effect. They have compared the material properties after a normal curing (NC) of 30 minutes and after thermal post curing (TPC) which is normal curing of 30 minutes followed by two hours curing at 80°C.

To offer as many possibilities to our customers as possible, a test on a coating technology has been chosen. The experience of the company *Repliform Inc – USA*[6] has been involved to reduce the number of unknown parameters.

	SL7560	
Mechanical properties	without coating	with a metal coating thickness of 0,05 mm
Young Modulus (MPa)	2971,9	7997,1
Maximum load (kN)	3,11	4,47
Tensile strength (MPa) at maximum load	71,83	97,2

The metal coating had been realized on flat parts, so to analyze the metal coating effect, we are going to compare it with the properties obtained on flat parts.

The result analysis is easy because all properties increased with the metal coating: 169% for the Young's modulus, 43,4% for the maximum load and 35,3% for the tensile strength at maximum load. The origin of these properties comes from the fact that with this kind of finishing, parts are composed by a bi-material (resin + metal). Even if the metal thickness is thin, 0,05 mm, it completely modifies the behavior of parts.

Another influencing parameter is the humidity and UV exposition; tests are run actually with UV painting to research the changes of the materials over time.

REFERENCES

1. Terry Wohlers, Wohlers Report 2003
2. Align Technology's: http://www.invisalign.com
3. Lionel Bonnet, Final engineering Project, ISTN
4. Sebastien Chaussin, Low volume manufacturing of Metal Parts, Final engineering Project, UTBM
5. Loughborough University: http://www.lboro.ac.uk/
6. Repliform Inc – USA: http://www.repliforminc.com/

Virtual modeling and rapid manufacturing – Bártolo (eds)
© 2005 Taylor & Francis Group, London, ISBN 0 415 39062 1

RSP Tooling: Advancements and case studies

James R. Knirsch

President RSP Tooling, LLC, Solon, Ohio, United States of America

ABSTRACT: RSP Tooling is an indirect spray form additive process that can produce production tooling for virtually any forming process and from virtually any metal. In the past 24 months a significant amount of research and development has been performed. This resulted in an increase in the basic metallurgical understanding of what transpires during the rapid solidification of the metal, significant improvements in the production machine up time, ceramic developments that have improved finish and process changes that have resulted in a shorter lead time for tool delivery.

RSP Tooling, LLC has been shipping production tooling since the middle of 2004. There are now numerous case studies documenting the benefits of the process related to delivery, cost, and tool life. These case studies include prototype and production tooling for forging, die casting, and plastic injection operations. Customers markets include, military, electronics, automotive, medical, to name but a few.

RSP Tooling, LLC is expanding the machine capabilities. New machines are being designed that increase the machine output and size capabilities. New exotic materials have been used in making tools and bimetallic tools using two different metals have been developed.

1 INTRODUCTION

The recent developments in Rapid Production Tooling have all but made the need for prototype tooling disappear. There are several approaches that are now as fast and inexpensive as prototype tooling but can continue after part approval to run high volume production. The best of these approaches is an indirect spray forming process invented by Dr. Kevin McHugh of the Idaho National Laboratories (INL). The advantages can be found in its accuracy, finish, cost and speed compared to the other Rapid Tooling processes (Wholers 2003).

The commercialization effort for this spray forming process started in February of 2002. The beta production machine was operational in November of 2003 and started to produce production tooling in February of 2004. Since that time tooling has been produced and put into production in many forming industries and a significant number of case studies now exist. In all but the simplest tools the process has proven to be less expensive and faster than standard machining of tools or any other rapid production tooling process. Research and development of the process has continued making the process faster, more accurate and less expensive to operate. Also a better understanding of the underlying metallurgy has been obtained.

2 THE METHOD

RSP Tooling is a spray forming technology that was developed by Dr. Kevin McHugh at INEEL for producing molds and dies (Knirsch 2003, McHugh 2002, Folkestad 2002, McHugh 2001). The general concept involves converting a mold design described by a CAD file to a tooling master using a suitable rapid prototyping (RP) technology such as stereolithography (SLA). A pattern transfer is made to a castable ceramic, typically alumina or fused silica (Fig. 1). This is followed by spray forming a thick deposit of tool steel (or other alloy) on the ceramic pattern to capture the desired shape, surface texture, and detail. The deposit is built up to the desired thickness at a rate of about 500 lb./hr. Thus, the spray time for a $7'' \times 7'' \times 4''$ thick insert is only 9 minutes. The resultant metal block is cooled to room temperature and separated from the pattern. Typically, the deposit's exterior walls are machined using a wire EDM, and bolt holes and water lines are added.

The turnaround time for cavity or insert is unaffected by complexity. From receipt of a CAD solid model to shipment of the cavity is about 8 days. Molds and dies produced in this way have been used for prototype and production runs in plastic injection molding, die casting, and forging operations.

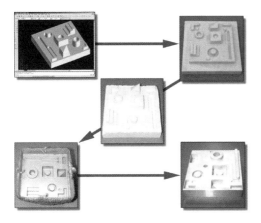

Figure 1.   Process sketch.

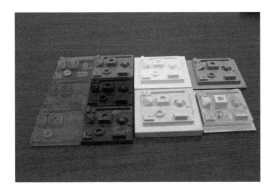

Figure 2.   Various model methods.

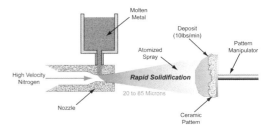

Figure 3.   Machine process.

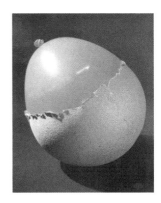

Figure 4.   Rapid solidification in action.

Generation of the physical model or "master" is straightforward. A number of rapid prototyping (RP) approaches are available commercially to accomplish this, but they differ widely in terms of cost, accuracy, surface finish. As part of an R&D study conducted with Colorado State University and an industry team (Folkestad 2002), the suitability of various RP-generated physical models as well as physical models machined from aluminum and various tooling boards, was assessed for use with RSP Tooling. Examples are shown in Figure 2.

Ceramic patterns are made by slip casting or freeze casting ceramic slurry, typically made of alumina or fused silica on to the tool master. Ease of casting, material cost, surface roughness, strength, thermal shock resistance, maximum use temperature, flatness, and dimensional accuracy are assessed. With the right equipment and procedures, very accurate and reproducible ceramics are easily made.

The ceramic is one of the most important steps of the process. It represents a significant variable in the cost, timing, finish, and accuracy of the final tool. The ceramic is the main time element in the cycle representing 40% of the entire turn around time for a tool. Research using a machined graphite master or machined ceramics have been done and several tests have shown these to be a feasible solutions that could eliminate the majority of the time consumed by this step. This approach however is only feasible for one "of a kind" tools since the master gets destroyed in the process.

The spray forming step is at the heart of the RSP Tooling process. Spray forming involves atomizing, i.e., breaking up a molten metal stream into small droplets, using a high velocity gas jet. Aerodynamic forces overcome surface tension forces producing an array of droplet sizes that are entrained by the jet and deposited onto the pattern, as shown in Figure 3.

As the droplets traverse the distance separating the atomizer and ceramic tool pattern, they cool at very high rates that vary depending on size. As a result, a combination of liquid, solid, and "slushy" droplets impact the ceramic, and "weld" together to form a coherent deposit. Figure 4 demonstrates the effect of rapid solidification with molten tin sprayed on a party balloon.

The high cooling rate of the deposit greatly impedes atomic diffusion, so segregation is very limited compared to cast metal. It also minimizes the erosive

416

Figure 5. H13 as deposited at 500×.

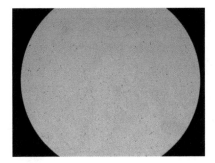

Figure 6. A2 as deposited at 50×.

interaction of the metal and ceramic tool pattern, allowing the deposited metal to accurately capture surface details of the ceramic that would not be possible if the metal was cast onto the ceramic.

The rapid solidification rate also results in non-equilibrium solidification, extended solid solubility, and very limited segregation as can be seen in Figures 5, 6.

3 ACCURACY

Dimensional accuracy and repeatability of all processing steps have been analyzed by Colorado State University personnel and industry partners using coordinate measuring machines (CMM). This has helped to identify suitable materials and processing conditions. Several conclusions have been drawn from the study.

Molds made from the same master but different ceramic patterns were essentially identical which is of major importance in multiple cavity dies or replacement inserts. It also means that accuracy can be increased by making a test tool and then modifying the model to the dimensional data.

Most of the shrinkage comes from casting and firing of the ceramic. Some ceramic formulations nearly eliminate this shrinkage. It has also been demonstrated that modest variations in binder and firing temperatures have no effect on this shrink.

Figure 7. Finger print.

The overall conclusion of the dimensional accuracy study is that the accuracy of molds made by the RSP Tooling method is comparable to the conventional practice of machining, benching, polishing and heat treating. Detailed algorithms are under development that will automatically apply scaling factors to a CAD drawing of an insert for various processing sequences.

Accuracy has been improving by making a significant reduction in the ceramic shrink rate and by use of the more accurate rapid prototyping model methods. Further improvements can be expected if machined ceramic or graphite is used.

4 REPLICATION

The process can replicate very small features. When sprayed on quartz glass the process reproduced a fingerprint in steel that was accidentally left on the pattern (Fig. 7). In tests making a small stamping die with engraved details features as small as .003 inches could be transferred to the ceramic and then to the sprayed steel tool. All of the detail of a laser burned model was transferred to a tool (Fig. 8). This is of even more significance now that the latest SLA machines can achieve producing details as small as .005 inches in width.

Work with Complete Surface Technologies has resulted in a process in which grained surfaces can be applied to the SLA and then sprayed directly into the tool which eliminates the need to etch the tool after machining. This again will time and cost of getting a tool into production (Fig. 9).

5 TECHNICAL AND ECONOMIC BENEFITS

The main benefits of RSP Tooling involve cost and turnaround time reductions without sacrificing quality

Figure 8.   Leather grain surface.

Figure 9.   Leather detail.

Figure 10.   Tool detail.

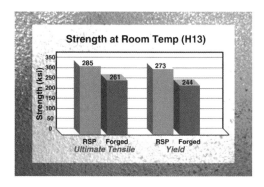

Figure 11.   H13 after artificial aging.

6   DIE MATERIALS

or accuracy. When the atomized spray covers the surface of a ceramic tool pattern, it replicates the features very accurately, regardless of the complexity as seen in Figure 10. By so doing, it eliminates many steps in normal mold making practices such as milling, EDM, benching, polishing, and engraving.

Since cost and timing of the spray form process are not affected by complexity, the savings achieved by using this process is variable based on the existing cost. On recent projects the cost savings varied from $0 to $3000 per insert. Savings in time is proportional.

The timing for multiple cavities is always significantly improved. The first cavity takes five days but each subsequent cavity follows in 3 hour intervals. This means 32 virtually identical cavities can be shipped in 10 days.

The cost for a RSP tool is also a constant except for the solid model cost and the material used. The master pattern varies depending on the method used, the size, and the material. With the improvements in the model manufactured discussed earlier both the cost and timing will be dramatically effected. With the right ceramic process it is conceivable that the timing for delivery of an insert can reduce to 8 hours.

The RSP Tooling machine is designed so each tool can be made from a different alloy. Because of the rapid solidification of the metal the quality of the tool is the same or in some cases better than machined tools of the same alloy. P-20 when sprayed typically has a better grain and density than the standard machined tool but has the same strength and hardness. Tool life is generally the same.

Spray formed H-13 tool steel results in significantly better properties than can be achieved by standard methods. As sprayed the tool has a Rockwell C hardness of 56. This can, through artificial aging, be increased to 62Rc. When aged the strength of the material is significantly higher than heat treated H13. Figure 11 compares standard and sprayed H13 tool steel.

This means if high wear materials are used in plastic injection molding or if the process is die casting tool life can be extended by 25% or more over a standard H13 tool. Since the machine can use any metal with a melt point below 1500 degrees centigrade tool life can be further improved by using new or existing hard to machine alloys.

The density of a sprayed tool using this technique is as dense as a tool using forged machined steel of

Figure 12. Polished to three finishes.

Figure 13. Reflection of a penny.

the same specification. Tests with H13 have shown densities of 99.7%. This means that not only can water lines be machined with out fear of leaking but the material can be hand polished with out fear of opening up porosity (Fig. 12).

A major distinction between this process and other RT processes is that the material used is the same as in a normal machined tool. Thus all standard operational practices can be continued unchanged. Welding, stress reliving, polishing are all performed identically although not as often.

7 SURFACE FINISH

The surface finish that can be achieved is dependent on the ceramic and the initial model. The spray system replicates the ceramic with extreme accuracy and can pick up details as small as .0001 inches. Using the standard process now in use a surface finish of 45 micro inches can be achieved. Several ceramics now being examined have the potential of significantly improving this finish. In rare cases a fused quartz glass pattern could be sprayed which can result in a mirror finish as shown in Figure 13.

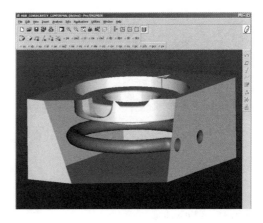

Figure 14. Conformal cooling.

8 CYCLE TIME IMPROVEMENT

An additional potential benefit of the spray forming approach involves the ability to add conformal cooling lines which rapidly cool the molding surface of a mold or die. Mold cooling accounts for about two thirds of the total cycle time in plastic injection molding, die casting and most other metal mold casting operations. Ideally, cooling lines would be placed near the surface of a die, and would conform to the geometry of the die surface. This is referred to as "conformal cooling" and is viewed by molders as very beneficial because it provides better thermal management of the tool and reduced part cycle time. In plastic injection molding, for example, conformal cooling has been shown to reduce part cycle time 15–50% compared to standard cooling practices.

The incorporation of cooling lines has traditionally involved machining straight-bore holes into the back of the die insert. Unfortunately, conformal cooling lines cannot normally be incorporated into machined dies due to their complex geometries.

Two approaches to solve this problem are being investigated. Dr. McHugh at INL is working under a DOE grant to perfect a dissolvable core that would be inserted into the spray automatically and after the tool was finished the core would be removed resulting in conformal cooling lines (Fig. 14).

The second approach is to make a clad tool with standard tool steel at the surface and a high conductivity metal as the back up material. Preliminary results indicate that spray forming is well suited for producing these clad tools. In Figure 15, a copper/steel clad was formed by depositing a high conductivity copper backing onto an H13 die insert. This allows for a simple water line to be added through machining into the copper cladding yet the cooling will be uniform over the entire surface of the die. Copper backed tools have been

Figure 15.   Clad tooling.

manufactured and have shown to have an extremely high mechanical bond due to the rough surface that is left when only a thin layer of steel is sprayed. Tests are now being run to determine the amount of additional heat that can be removed in this manner and to determine how long the bond holds up to the thermal cycling.

9   LIMITATIONS

There are limitations to the size of molds and dies that can be produced with the current equipment. The original laboratory-scale equipment at INL can produce inserts that are about $3'' \times 3'' \times 2''$ thick. Commercial beta equipment located at RSP Tooling, LLC has increased this to $7'' \times 7'' \times 4''$ thick. However, the process has no inherent size limitation, and machines with larger capacity are being planned. Preliminary research has been done to show that multiple spray heads can be used. This will be the approach used in going to larger machines. The next proposed size will be $20 \times 20 \times 10$ and the machine should be able to produce this size tool in less than eight hours.

The second limitation is the aspect ratio for standing features of the mold. Cavity features on the mold surface do not present problems. However, boss features on the mold surface do. Recent projects have shown that this limitation is more significant than originally thought. For small features the process can now make features with aspect ratios of about 4 to 1 but for larger features ($2'' \times 2''$) the ratio is closer to 1 to 1. This is because, when spraying molten metal down into a cavity in the ceramic the metal will tend to bridge across the hole before it is entirely filled. While R&D in this area continues, it is currently recommended these types of features be inserted.

10   CONCLUSIONS

Spray forming has demonstrated great potential for reducing the cost and lead time for tooling by eliminating many of the machining, benching and heat treatment unit operations. In addition, spray forming provides a powerful means to control segregation of alloying elements during solidification, carbide formation and growth, and the ability to create beneficial metastable phases in many tool steels. As a result, relatively low temperature artificial aging heat treatment can be used to tailor properties such as hardness, toughness, thermal fatigue resistance, and strength which will increase tool life.

Clad tooling with high heat conductivity metals will substantially decrease cycle times in plastic injection molding and die cast operations.

REFERENCES

Folkstad, J. E., Knirsch, J. R., and McHugh, K. M., "Die Casting and Rapid Solidification Process (RSP) Tooling – An Applied Research Project," *Proceedings of the 2002 NADCA Congress,* Paper T02-051, NADCA, Rosemont, IL, October, 2002.

Knirsch, J. R., McHugh, K. M., and Folkestad, J. E., "RSP® Tooling – A Revolutionary New Process to Manufacture Die Cast Production Tooling in Prototype Timing," *Die Casting Engineer 46* (3), May 2002, P. 56.

McHugh, K. M., and Knirsch, J. R., "Producing Production Level Tooling in Prototype Timing," *Time Compression Technologies 6* (2), March 2001, P. 23.

McHugh, K. M., and Knirsch, J. R., "Producing Production Level Tooling in Prototype Timing – An Update," *Moldmaking Technology 5* (10), October 2002, P. 42.

Wohlers, T., "The Rapid Prototyping/Manufacturing Industry," *Adv. Mat. Process. 161*(1), 35 (2003).

Virtual modeling and rapid manufacturing – Bártolo (eds)
© *2005 Taylor & Francis Group, London, ISBN 0 415 39062 1*

Hybrid moulds: The use of combined techniques for the rapid manufacturing of injection moulds

P. Martinho & P.J. Bártolo
Institute for Polymers and Composites, Polytechnic Institute of Leiria, Portugal

M.P. Queirós, A.J. Pontes & A.S. Pouzada
Institute for Polymers and Composites, University of Minho, Portugal

ABSTRACT: Injection moulding is the major polymer processing technique. The mould is a key element of this process and its characteristics have a key importance on the parts produced. The fabrication of products with higher added value, especially in terms of production time and costs, is essential for the success of the mouldmaking industry within a global competitive environment.

This paper presents the concept of hybrid mould, a recent concept for injection moulding of prototype and short series. Direct and indirect rapid tooling technologies can be used for the fabrication of these moulds. Hence, new design guidelines need to be defined to produce these moulds as well to assess its performance.

Epoxy tooling and selective laser sintering are the two rapid tooling technologies explored to produce inserts for hybrid-moulds. This work studies the fabrication of hybrid-moulds through these rapid tooling technologies, as well its mechanical and thermal behaviour.

1 INTRODUCTION

In today's highly competitive marketplace with short life-cycles of products, to develop a new product to meet consumers' needs in a shorter lead-time is crucial for an enterprise. Facing this global environment, the strategy of developing a product is transformed from "product-push" type to "market-pull". To improve competitiveness, a product should not only satisfy consumers' physical requirements but also their needs, increasing the product complexity and reducing its lifetime. Apart from this, market segmentation as a demand of individualistic consumers has led to the concept of "niche markets", increasing product choice. Additionally, companies must meet increasing customer expectations in terms of products' quality and cost. These new strategies adopted by modern companies lead to a tremendous change in their internal flexibility. As a consequence, the current industrial trend is moving from mass production, *i.e.* high-volume and small-range of manufacturing products, to a small-volume and a wide range of products. This has resulted into a remarkable industrial transformation, described by Rosochowski and Matuszak (2000) as the world class manufacturing concept, characterised by: (1) total quality, (2) concurrent/simultaneous engineering, (3) short manufacturing lead-times, (4) high

flexibility to enable rapid changes in the product development cycle, and (4) all employees are engaged in a continuous product and process improvement.

The search for alternative methodologies for the design and fabrication of moulds, for short runs of plastic components, nowadays has a great industrial demand, possible through the use of both rapid tooling technologies and new and different materials for the moulding inserts.

The idea of building moulds with moulding inserts, fabricated either in alternative metallic materials or in synthetic materials is the basis for the hybrid-mould concept (Figure 1), which can have a significant impact in the mouldmaking industry. This concept was investigated under a research project, named *Hibridmoulds*, supported by the Portuguese Agency for Innovation. This project involved the Polytechnic Institute of Leiria (Department of Mechanical Engineering), the University of Minho (Department of Polymer Engineering), the Technological Centre for Mouldmaking and DISTRIM a mouldmaking company.

This project had several objectives:

– to investigate the use of distinct rapid tooling approaches to produce inserts for hybrid-moulds
– to investigate the structural and thermal behaviour optimisation of hybrid-moulds

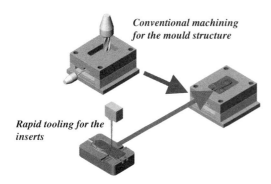

Conventional machining for the mould structure

Rapid tooling for the inserts

Figure 1.   The hybrid-mould concept.

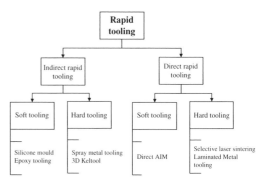

Figure 2.   Classification of rapid tooling technologies.

– to study the viability of existing rapid tooling technologies for hybrid-moulds
– the development of appropriate methodologies and project rules for the optimised manufacturing of hybrid-moulds.

This paper describes the use of two rapid tooling technologies to produce inserts for hybridmoulds: epoxy tooling and selective laser sintering.

2   RAPID TOOLING

Chua *et al* (1999) describes rapid tooling technology as the technology that adopts rapid prototyping techniques to tool and die making.

There are two approaches for rapid tooling (Figure 2): the indirect approach that uses rapid prototyping master patterns to produce moulds, and the direct approach that uses directly rapid prototyping systems to produce tooling inserts (Rosochowski and Matuszak 2000, Chua *et al* 1999). Tools for short manufacturing runs are known as soft tools, while tools for longer manufacturing are known as hard tools.

2.1   *Epoxy tooling*

This process, also called as aluminium-filled epoxy tooling, composite tooling or cast resin tooling is a process for tooling manufacturing using an epoxy resin as the tool material.

A model pattern is prepared onto a model board to define the negative of the first half of the tool. Epoxy resin reinforced with aluminium is then cast onto the pattern directly into a supporting frame, and cured. The process is repeated for the second half of the mould. When both halves have been cast and cured, the tool can then be split and prepared for moulding. Ejectors and runners are typically milled into the tool to ensure total functionality, and the tool support frames are integrated into a mould base for assembly onto the moulding machine.

The cost of this process is typically less than 40% of the conventional prototype tooling cost, typical lead-times lie between two and four weeks and fine detail, such as graining or print, can easily be reproduced. However, flash can occur more often than with conventional tooling, leading to an increased effort in trimming the mouldings.

2.2   *Selective laser sintering*

Selective laser sintering (SLS) uses a high-power laser to selectively heat powder material just beyond its melting point (Dimov *et al* 2001). The laser traces the shape of each cross-section of the model to be built, sintering powder into a thin layer. It also supplies energy, which not only fuses neighbouring powder particles but bonds each new layer to those previously sintered. The sintering takes place into a sealed heated chamber at a temperature near the powder melting point, and is filled with an inert gas to reduce oxidation of the powder being sintered. Maintaining the hot environment in the build chamber also reduces both the power required from the laser and the thermal shrinkage of the model.

After each new layer is solidified, the piston over which the model is built retracts to a new position and a new layer of powder is supplied. The powder that remains unaffected by the laser acts as a natural support for the model and remains in place until the model is complete. After all the layers have been defined, the non-sintered powder is discarded to reveal the solid object formed inside. Commercial machines differ in: the way the powder is deposited (roller or scraper), the atmosphere (argon or nitrogen) and the type of laser they use (fibre, CO_2 or Nd:YAG laser).

The building cycle consists of three main stages:

• Warm up – the chamber is filled with inert gas and both the chamber and feed material are warmed up
• Part build
• Cool down – the chamber is returned to ambient temperature in a controlled way.

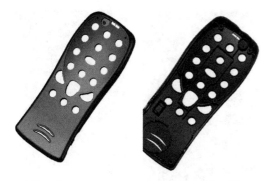

Figure 3. The front panel of a cell phone selected as case study.

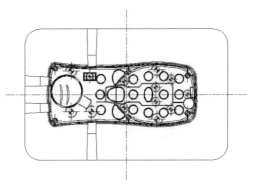

Figure 4. Position of the ejectors.

There are two types of SLS processes (Majewski and Hopkinson 2004, Kruth *et al* 2003, Simchi *et al* 2003): indirect and direct processes. In the indirect one, low temperature polymers are used as binders. The polymer binders are either mixed with the primary powder at a high melting temperature or coated on the surface of the primary powder. The direct SLS process involves the direct melting and consolidation of selected regions of the powder bed. Direct SLS enables to produce high or full density parts.

Many types of materials, such as polyamide (PA), polystyrene (PS), polycarbonate (PC), acrylonitrile butadiene styrene (ABS), metals, ceramics and composites can be used in laser sintering applications.

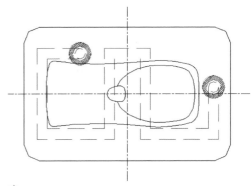

a)

3 EXPERIMENTAL WORK

3.1 *Part design*

The front panel of a cell phone, indicated in Figure 3 was chosen as a case study. This part was selected due to its shape and geometrical features, which pose some difficulties with respect to the moulds' insert fabrication. The front panel has numerous holes in the front face and extensive vertical ribs in the back. Its nominal thickness is 1,5 mm and the total volume is 11,47 cm^3.

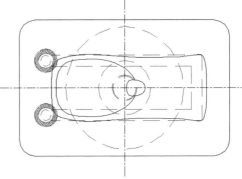

b)

Figure 5. Position of the cooling channels for a) core and b) cavity.

3.2 *Mould fabrication*

Moulding inserts were fabricated using epoxy tooling and laser sintering.

Figure 4 and 5 show, respectively, both the ejection and cooling systems considered on this work.

3.2.1 *Epoxy tooling*

Polyjet technology from Object was used to fabricate the master pattern withObject FullCure 700 photocurable resin. After fabrication, the model was subjected to several post-processing operations like support

removal and post-curing, to fully solidify the model this way increasing its mechanical properties.

Mould inserts (Figure 6) were made using Neukadur VG SP 5 from Sika and the procedure previously described. The resin was cured during 4 hours at 70°C.

Due to processing limitations, some of the part features were suppressed, namely the larger vertical ribs

Figure 6.   Core and cavity made by epoxy tooling.

Figure 8.   Core and cavity produced by selective laser sintering.

Table 2.   Main properties of Laser Form ST100.

Material properties	Values
Density	$7.7\,g/cm^3$
Coefficient thermal expansion	12.4×10^{-6}
Thermal conductivity (100°C ASTM E457)	49 W/mk
Hardness	87 Rockwell B
Young Modulus	137 GPa
Tensile-yield strength	305 MPa
Elongation	10%

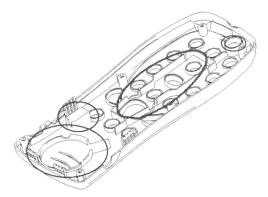

Figure 7.   Suppressed geometrical patterns.

Table 1.   Mechanical and thermal properties of Neukadur VG SP 5.

Material properties	Values
Density	$2.7\,g/cm^3$
Elasticity module	$9.800\,N/mm^2$
Poisson Coefficient	~0,42
Hardness	90 Shore D
Linear Expansion Coefficient	$30{\sim}35\ 10^{-6}/K$
Specific heat	1300 J/kg°C
Thermal conductivity	0,7 W/mK

on the middle of the back face, the snap fits and a big boss on the top of the back face (Figure 7). Table 1 shows the main properties of Neukadur resin.

Preliminary tests showed that obtaining those features was quite difficult, due to problems in separating the mould insert from the master pattern. The top edges of the bosses, corresponding to the holes in the front face of the part, were particularly degraded. In the major central boss, this was particularly evident.

The surface quality was not quantified, but it could clearly be described as rough. The production of cooling channels in the blocks was also difficult.

3.2.1   Laser sintering

Inserts were produced using metallic powder Laser Form ST100 (3D Systems, USA), infiltrated with bronze, resulting in composite structures containing 40% of bronze and 60% stainless steel alloy (Figure 8).

Core and cavity were identically oriented to minimize adjustment errors during the fabrication process. There was no need of part simplification.

Table 2 shows the main properties of Laser Form ST100.

3.3   Mould instrumentation

Produced hybridmoulds were instrumented with temperature and pressure sensors as shown in Figure 9. Due to geometric limitations related with both the water circuit and the ejection systems, an additional sensor (T1) was located in the core, just opposite to T2.

A Priamus system was used to get temperature and pressure data. Figure 10 shows the sensor signal processing system used in this research study.

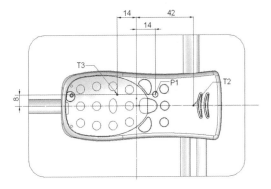

Figure 9. Sensors position within the cavity insert.

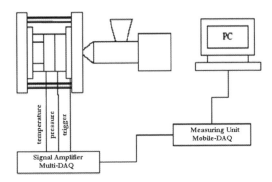

Figure 10. Sensor signal processing system.

Figure 11. Injection moulding equipment, Engel 200/45.

3.4 *Injection moulding*

Parts were obtained through injection moulding, using an Engel 200/45 machine with 450 kN of clamping force (Figure 11).

The material used was polypropylene, Hifax BA238 from Montel. Material data is shown in Table 3.

Table 3. Typical processing conditions of polypropylene.

Properties	Values
Density	900 Kg/m^3
Injection Temperature	180–290°C
Mould Temperature	20–60°C
Máx. shear stress	0,26 MPa
Máx. shear rate	24001/s

Table 4. Experimental injection moulding processing conditions used for both epoxy and laser sintered inserts.

Processing conditions	Resin inserts	SLS inserts
Melt temperature	265°C	230°C
Mould temperature	~70°C	50°C
Holding pressure	10 MPa	15 MPa
Holding time	10 s	10 s
Cooling time	40 s	10 s

Processing conditions used for both inserts made through epoxy tooling and selective laser sintering are shown in Table 4.

4 RESULTS

4.1 *Epoxy tooling*

Due to the thermal characteristics of the epoxy material, processing parameters are different from those conventionally used. Thus, cycle times are longer than with conventional tooling. Additionally, as a consequence of its low mechanical properties, pressure values are typically lower, which can lead to problems like sink marks. The major difficulties in establishing a suitable set up are related to the injection pressure, which should not be too high to prevent damage of the moulding block details, and the coolant temperature. The occurrence of "short shots" was frequently observed during the first tests.

Extraction was difficult, mainly owing to both the roughness of the surface and the small draft angle used. Accuracy is dependent on a multi-step process, including the accuracy of the process used to produce the pattern.

Figure 12 shows the variation of the temperature at the moulding surface. It can be observed that the temperature stabilises after 40 s and that a decrease of temperature is obtaining by increasing the distance from the gate. The higher temperature occurs at the position T3, closer to the gate, which is near 30°C, higher than at the other positions, T1 in the core and T2 in the cavity. A 10°C temperature difference is reached between the two moulding surfaces is also evident.

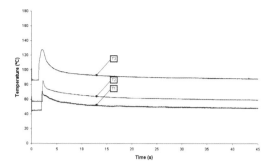

Figure 12. Temperature variation (°C) over the cycle time for the inserts made by epoxy tooling.

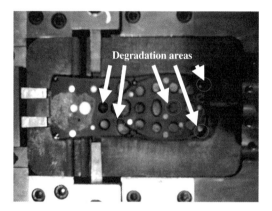

Figure 13. Degradation areas within the core made by epoxy tooling.

A low filling pressure was used. The insert degradation started after 14 cycles, when eroded pieces of the cavity insert were extracted with the injected part. The parts showed degradation, especially at the boss edges (Figure 13). Some small pits also appeared either in the core surface or in the cavity surface. In the areas of the inserts, where contact happened with the lateral movements, no evidence of wear was observed. The injection run was suspended at shot number 81.

4.2 Selective laser sintering

Beyond gaining in cost and time, rapid tooling also enables the creation of conformal cooling within the tooling. Conformal cooling can reduce the cycle time by 15–20% improving part quality.

However, in this case study it was too difficult to create conformal cooling channels, due to problems associated with the extraction of non-sintered powder from the internal cavity. At the infiltration stage, the material gets consolidates and blocks the cooling circuit.

To prevent blocking of the cooling system, the inserts were constructed in two separate halves that

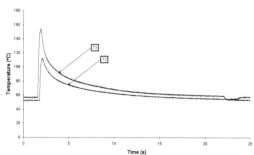

Figure 14. Temperature variation over the cycle time for laser sintered inserts.

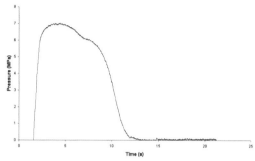

Figure 15. Pressure variation through the cycle time for selective sintered inserts.

were glued together and infiltrated. However, this solution does not enable to build the *real* conformal channels and is prone to fluid leakage during operation.

Figure 14 shows the temperature variation at mould surface, for the laser sintered inserts. Results suggest that the temperature stabilises after 20 s.

No pressure restrictions were considered due to the mechanical properties of the laser sintered inserts. Figure 15 shows the pressure variation through the cycle time. This figure enables to observe that from the instant the material reaches the pressure sensor, pressure rapidly increases to about 7 MPa and slowly drops to a minimal value after 12 s. This instant of time corresponds to the onset of thickness shrinkage.

5 CONCLUSIONS

Epoxy tooling and laser sintering were used to produce inserts for hybrid-mould produced for a complex shape.

The results suggest that moulding inserts made with the Neukadur resin are not adequate to produce parts, containing extensive ribs and deep bosses. These resins pose some problems during the preparation of

Table 5. Comparison between SLS and epoxy tooling as technologies for hybrid-moulds.

		SLS	Epoxy tooling
Initial information	Physical model		•
	Virtual model	•	•
Number of parts	>10.000	•	
	>1000	•	
	>100	•	•
	>10	•	•
Major dimension	<250 mm	•	•
	<500 mm		•
	>500 mm		
Shape complexity	High	•	
	Simple	•	•
Accuracy	<0,1 mm	•	
	<0,2 mm	•	
	<0,5 mm	•	•
Surface quality	Good		•
	Medium	•	•

moulding blocks and their fitting to the mould base. Manipulation and assembly must be carried out with particular care, as inserts are easily damaged. Injection moulding processing conditions is different from those used at conventional injection moulding. The set up of the injection process requires analysis and experimentation as it is quite different from standard moulding. Thermal properties of these inserts are key aspects of its performance. Thus, it is essential to obtain a proper cooling of the moulding blocks.

Epoxy tooling is a versatile technique to produce mould inserts of simple geometry. This rapid tooling strategy reproduces accurately small details of the master pattern. These small details will be prone to faster degradation. Therefore, epoxy tooling is more suitable to fabricate inserts with open and simple geometries.

Selective laser sintering is a direct rapid tooling technique adequate to fabricate moulding inserts, though conformal cooling channels are difficult to produce. Experimental results show that Laser Form does not affect the standard process setup. Establishing the processing set up can be followed by a similar procedure to that used for conventional moulds.

A comparison between the two techniques is provided by Table 5.

Further research is needed to evaluate the use of computer-aided engineering tools (structural and rheological tools) in order to improve hybrid-moulds design. Its main focus will be the assessment of software results through online data collection.

ACKNOWLEDGEMENTS

The authors acknowledge the financial support given by the POCTI to the project *Hibridmolde*, and by FCT – Portuguese Foundation for Science and Technology, through the POCTI and FEDER programmes.

REFERENCES

Chua, C.K., Hong, K.H. and Ho, S.L. 1999. Rapid tooling technology. Part 1. A comparative study. *Int J Adv Manuf Technol* 15: 604–608.

Dimov, S.S, Pham, D.T., Lacan, F. and Dotchev, K.D. 2001. Rapid tooling applications of the selective laser sintering process. *Assembly Automation* 21: 296–302.

Ding, Y., Lan, H., Hong, J. and Wu, D. 2004. An integrated manufacturing system for rapid tooling based on rapid prototyping. *Robotics and Computer-Integrated manufacturing* 20: 281–288.

Kruth, J.P., Mercelis, P., Vaerenbergh, J.V., Froyen, L. and Rombouts, M. 2003. Advances in selective Laser Sintering. *Proceedings of the Advanced Research in Virtual and Rapid Prototyping, VRAP2003*, Edited by Bartolo *et al*, ESTG Leiria, 59–70.

Majewski, C. and Hopkinson, N. 2004. Reducing ejection forces for parts moulded into direct metal laser sintered tools. *Int J Adv Manuf Technol* 24: 16–23.

Rosochowski, A. and Matuszak, A. 2000. Rapid tooling: the state of the art. *Journal of Materials Processing Technology* 106: 191–198.

Simchi, A., Petzoldt, F. and Pohl, H. 2003. On the development of direct metal laser sintering for rapid tooling. *Journal of Materials Processing Technology* 14: 319–328.

Advanced rapid prototyping technologies and nanofabrication

Virtual modeling and rapid manufacturing – Bártolo (eds)
© 2005 Taylor & Francis Group, London, ISBN 0 415 39062 1

Manufacture of stainless steel parts by selective laser melting process

G. Jandin & J.-M. Bertin
MB Proto, Blangy-sur-Bresle, France

L. Dembinski & C. Coddet
LERMPS, Sévenans, France

ABSTRACT: MB Proto is the first French company to develop on the Selective Laser Melting (SLM) process. Through this process, parts can be manufactured directly from CAD files. This emerging technology can meet the ever-growing customer demand for the highest level of quality. However some process developments are needed to control parts manufacturing and improve their structure and their characteristics. In addition this direct laser sintering process depends on the powder characteristics used. Concerning the development of this technology, MB Proto and LERMPS Laboratory work together in effectively studying and elaborating optimized stainless steel 316L/1.4404 powder to manufacture parts with improved both mechanical properties and material density. The aim of this paper is first to show the process used to elaborate stainless steel 316L/1.4404 powders and to manufacture parts with the Selective Laser Melting process. Then, part samples and industrial component microstructures are shown using microscope observations and characterization. Recent results have shown that the elaboration of components using the selective micro-laser melting is promising due to the high quality level of the components.

1   INTRODUCTION

The direct manufacture of good material metal parts, directly elaborated from powder material, is an emerging process in the field of rapid prototyping. It involves the use of specific laser and metal powders to create end-use products through a layer-by-layer process. As a result of a melting process, the part can be free of pores and cracks; the bond between layers is extremely strong. One key point of the development is the ability to use powders having good flowability and shapes. Other development activities are directed towards the elaboration of fully dense parts by optimising process parameters.

2   METAL POWDER ELABORATION THROUGH A GAS ATOMISATION PROCESS

2.1   *Process description*

This process uses an atomisation tower (Figure 1). It consists in melting a metallic ingot, which composition is identical as the one expected for the powder. The material is melted in a crucible (induction heaters),

Figure 1.   Picture of the tower implantation.

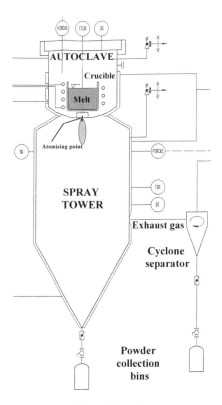

Figure 2.   Schematic description of the process.

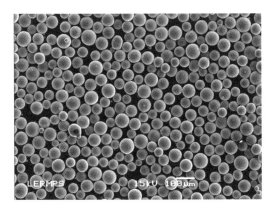

Figure 3.   SEM observation of atomised powder.

itself comprises into an oxygen free chamber. When the material reaches the liquid state, the molten liquid metal flows through a nozzle into a neutral atmosphere chamber (Figure 2). An Argon flow circulates into the chamber and cools down the material droplets. These droplets solidify during their flight in the atomizing chamber to be collected either at the bottom or in vacuum separators. All powders produced with this technology are spherical. The pressure is up to 15 bars inside the chamber.

With an adapted gas flow, spherical powder can be obtained, as shown on Figure 3. The specific system installed at LERMPS laboratory enables to produce powder with fine size distribution, with mean values generally comprises between 1 and 100 μm. The amount of powder that generally can be obtained is around 5 and 70 kg.

2.2   Materials and applications

This technology enables to develop specific alloys with a metallic base: Cu, Ni, Fe, Al that cannot be found on the market or difficult to obtain, such as shape memory alloys. It also permits to produce more standard materials such as: pure copper, bronze or stainless

steel. Elaborated powder is mainly dedicated to some applications: Metal Injection Moulding (MIM), thermal spraying, Selective Laser Sintering (SLS) or Selective Laser Melting (SLM) processes.

3   ELABORATION OF DENSE PARTS WITH THE SLM PROCESS

3.1   Description of the SLM process

MB Proto company possesses the MCP-HEK SLM Realizer machine. This process permits to build parts directly from Computer Aided Design (CAD) files. These file have to be prepared specifically to achieve the fabrication. The main operations are: an automatic vertical slicing of the model, with a thickness that is the production thickness; and a support preparation that will maintain the part in a stable position during the production.

During this production, a laser spot (wave length 1065 nm) with a typical spot size diametre of 100 μm melts the powder. This spot follows the contour of the current slice as well as the hatch of the contour inner. After the laser illumination, the melted part solidified (see Figure 4). This solidified surface is the profile of the part for the current slice.

In the following step, the building platform goes down, an amount of powder is deposited on the building surface and a mechanical system levels the surface. Then, the spot trajectory starts again.

This procedure is repeated until the ultimate layer to finally get the complete part.

3.2   Optimisation of the SLM process

Tests have been achieved on specimen elaborated with the powder built from LERMPS on the Realizer machine. This machine has a working zone of

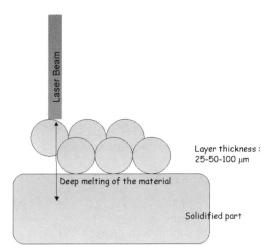

Figure 4.   Schematic description of the melting of one layer.

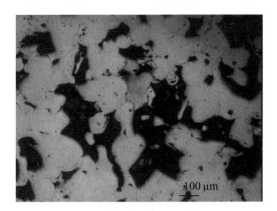

Figure 5a.   Low power and high laser scanning velocity.

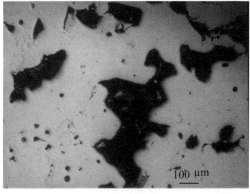

Figure 5b.   High power and high laser velocity.

Figure 5c.   Low power and low scanning laser velocity.

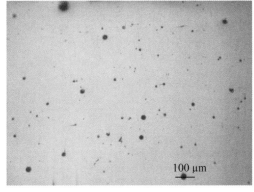

Figure 5d.   High power and low scanning velocity.

$250 \times 250 \times 240\,mm^3$, it can contain more than 100 kg of powder, and an Ytterbium fibre laser with a power of 100 W. The fabrication productivity of this machine is between 3 and 5 cm³/h.

Experimental tests have mainly focused on machine parameters. Validation tests have mainly consisted in studying fabrication parameters influence to be able to control the density of produced parts. Tested parameters are laser scanning velocity and laser power. Presented pictures (Figures 5 – a, b, c and d) show the microstructure of elaborated specimens. Dark zones represent the porosity and light zones are the metallic structure. Results clearly show that the higher the energy, the more melted the material and the denser the parts.

Measured Vickers hardness gives values between 190–200 Hv300 on the densest parts.

3.3   Construction of parts

MB Proto company mainly focused its activities with the SLM process on the construction of stainless steel

433

Figure 6. Sheet metal parts (stainless steel) elaborated with SLM process.

Figure 7. Stainless steel part elaborated with SLM process.

technical parts having complex shapes and thin walls. Most of the time, the number of parts to produce is low (1–5 parts) and for which the construction of a tooling should be prohibitive and a machining would remove more than 80% of material to get the part.

Part 1: This part (Figure 6) is a prototype for sheet metal manufacturing in stainless steel. Its dimensions are: $200 \times 160 \times 40\,mm^3$. The construction time of this part is 32 hours.

Part 2: It has the same application as part one. The dimensions are $120 \times 150 \times 150\,mm^3$ with a mean thickness of 1.2 mm. This part has been elaborated into 60 hours.

4 CONCLUSION

The powder quality is a main point to reach the ability of building parts with the SLM process. Powder developed by LERMPS has enabled MB Proto to build parts. Main SLM process parameters have been studied to be able to reach the building of nearly fully dense parts. As expected, the highest exposure sequences with highest power lead to the densest parts.

Nevertheless, such parameters partially lead to an overheating of the material and also to higher residual stresses. These ones have not been studied yet and some studies should be led in this direction. Some solutions can be also explored, such as preheating the material. Thus, the thermal shock should be degreased and the necessary power amount needed to reach the melting temperature should be degreasing. Further investigation will be led in this direction.

Nevertheless, it is now possible to build parts directly from powder, only in using a 100 W laser. Application fields have to be explored but it is expected that medical applications as well as aeronautics can be interested in building stainless steel parts without machining. Further more if, in the future, it will be possible to build titanium parts with this process.

Virtual modeling and rapid manufacturing – Bártolo (eds)
© *2005 Taylor & Francis Group, London, ISBN 0 415 39062 1*

New technology for the manufacture of sheet metal parts directly from CAD files

B. Cavallini & L. Puigpinos
ASCAMM Technology Centre, Barcelona, Spain

ABSTRACT: Since 2004, ASCAMM Technology Center is investigating the Dieless forming technology. This paper describes the technology and its principles. Some experimental results of thickness reduction and spring-back are developed and multi stage forming strategies are presented. Finally future developments at ASCAMM are mentioned.

1 INTRODUCTION

"Dieless incremental forming" is a new sheet metal forming technology allowing for the easy and rapid manufacture of sheet metal parts of complicated profile, with the advantage of requiring none or greatly reduced tooling efforts compared to traditional processes[1].

The technology is based on:

- the "incremental deformation technique": the part is formed as a result of small deformations applied successively in localised areas of the sheet
- rapid manufacturing technologies, from which the part can be formed directly from a 3D CAD model.

2 PRINCIPLES OF DIELESS FORMING

Complex sheet components are produced by the NC movement of a forming tool (spherically tipped or of more complex shape and construction) actuating locally on the sheet material, in combination with simplified and stationary dies.

This simple process configuration invests it with the flexibility necessary to fill the niche of small batch production and prototyping, in addition to short time to market as well as good reproducibility.

Simple die   Forming tool

Metal sheet

Figure 1. Principle of Dieless forming.

In a conventional drawing process, the material is more or less free to flow within the die, acquiring the desired shape. There is little and distributed thickness variation. On the contrary, during Dieless forming the sheet is held in its entire contour, and can not flow. Moreover, the very local pressure applied by the forming tool creates localized strains: the material stretches locally which results in thickness reduction[2].

The volume constancy of the material V1 = V2 and the fact that the initial area of the material is smaller than its final area $(S_2 > S_1)$ dictates that: $e_2 < e_1$, where $V = e * S$.

In fact the final sheet thickness can be related to the initial thickness thanks to a Sine law[3]:

$$e_2 = e_1 \times \sin(90 - \alpha) \qquad (1)$$

where α is the angle of the sheet with respect to the horizontal.

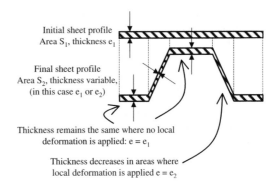

Initial sheet profile Area S_1, thickness e_1

Final sheet profile Area S_2, thickness variable, (in this case e_1 or e_2)

Thickness remains the same where no local deformation is applied: $e = e_1$

Thickness decreases in areas where local deformation is applied $e = e_2$

Figure 2. Principle of Dieless forming.

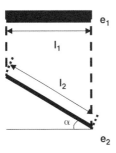

Figure 3. Consequence of the Sine law.

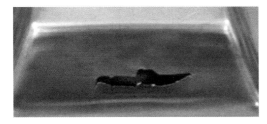

Figure 4. Fracture on AL 5754 blank.

Therefore the amount of strain is directly dependant on the parts' geometry when using single stage strategies, or more precisely on the steepness of the parts' wall.

However it is possible to overcome the Sine law thanks to multiple step strategies[4].

Equation [1] can be used for all kind of metals. So, the final thickness is only influenced by the process steps and not by other parameters such as the type of metal.

Testing quite different sheet materials from different family or groups such as pure and magnesium alloyed aluminium, stainless steel, low carbon steel and deep drawing steel, it was found that the maximum angle inclination achievable without the occurrence of fracture is in the range of 60°–65°. The above picture shows a typical fracture occurring at 63° on a Al 5754 sheet.

3 INFLUENCE OF TOOLING

1. Forming tool: The forming tool tip has a fixed diameter, which is the limiting factor in forming inward radii. The forming tool tip is of course inter-changeable, and different size and shapes of tips may be used.

2. Support tool: It is possible to work with or without a support tool, depending on the complexity of the

geometry and the desired accuracy. The following summarizes the 3 different cases.

*   Dieless
    *   – No tooling required
    *   – Very high flexibility
    *   – Reduced accuracy/simple parts

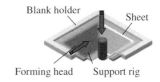

*   Using a partial die
    *   – Intermediate
    *   – Good compromise

*   Using a dedicated die
    *   – High accuracy
    *   – enables complex geometries
    *   – Good flexibility
    *   – Needs moveable blank holder

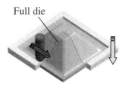

The stationary die or support tool mounted on the X-Y table, can be manufactured from low cost materials such as plastics, hardwood, resin, cast iron, or a combination of those which allows to reduce costs and weight.

4 EXPERIMENTAL

4.1 *Technological needs and objectives*

Due to the time necessary to form a part, there is up to now no simulation tool suitable for the Dieless forming process. Therefore during the development of a part all the efforts must be concentrated in the design of the process, focusing on the consecution of the difficult areas of the part such as steep flanges,

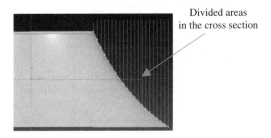

Divided areas in the cross section

Figure 5.   Pilot geometry to verify Sine Law.

Experimental and Theoretical thickness mild steel AP01

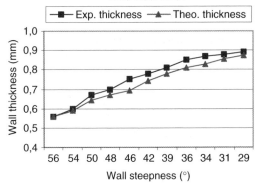

Figure 6.   Experimental vs theoretical thickness.

Major strains

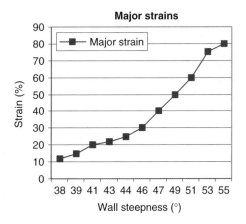

Figure 7.   Major and minor strains in the hyperbola.

The graph here before shows a good correspondence between theoretical and experimental measurements in the range of angles studied (25°–55°).

Using the same pilot geometry the relation between angle and strain has been observed. A circle mesh was marked on the sheet by laser. When the sheet is deformed the mesh also deforms, and the circles appear as ellipses. Major strains have been measured and plotted vs. the part's wall inclination angle. It is observed that the strains obtained are much greater than with conventional processes (Major strain of 80% is obtained at 55°).

It is clear that the strain increases with the forming angle.

4.3   Spring-back effect

It has been empirically observed that the spring-back effect greatly depends on the sheets characteristics:

– Material type (hardness, hardening exponent)
– Sheet thickness (lower spring-back with greater thickness)

But it also depends on process characteristics:

– In the case that no support tool is used, the material experiences a high amount of spring-back therefore affecting part's accuracy
– In the case of using a complete support tool, the sheet stretching is better controlled, and therefore the spring-back is limited

The method followed to compensate spring-back is an empirical trial-error method.

A first part is manufactured directly from the original CAD data. This part is analyzed[5] and spring-back is localized and quantified. Based on this analysis, the process CAD of the part is modified compensating

and intending to avoid excessive sheet thinning, while ensuring a correct dimensional accuracy.

Therefore the prediction of the final sheet thickness, as well as the development of strategies and procedures to ensure parts integrity are essential in the optic of development of the technology.

4.2   Strain distribution and thickness reduction

ASCAMM has designed an experiment to verify the "sine law". A pilot geometry has been designed, consisting of a truncated hyperbola, ideal for the experiment due to its variable inclination. Its characteristics are:

Thickness = 1 mm
Material = DC 01
Maximum angle = 55°
Minimum angle = 25°

The experiment consists, firstly, to evaluate the corresponding inclination in the cross-sections areas. Then to obtain the thickness with the theoretical expression of the sine law. Finally to compare the measured thickness with that obtained theoretically.

Optical digitizing of the part   Analysis of dimensional accuracy

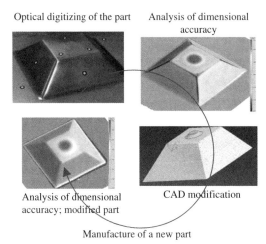

Analysis of dimensional accuracy; modified part   CAD modification

Manufacture of a new part

Figure 8.   Spring-back compensation loop.

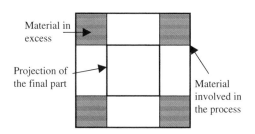

Figure 10.   Material projection in the case of a cube; 81° steep walls.

5   FUTURE INVESTIGATIONS

- Use standardised pre-forms
- Use of hot treatments in between process steps
- Materials characterization after processing; this investigation is already proceeding: The test samples were cut out of a "quadratic-based pyramid", of which each side has a different inclination. Therefore from each side are obtained test samples of different characteristics. The materials are currently tested under tensile test
- Develop FLC curves for Dieless forming.

6   CONCLUSIONS

Dieless forming is an emergent technology for the manufacture of short series or prototypes of sheet metal parts[6,7]. ASCAMM Technology Centre is leading investigations in several fields from material characterisation to process development, in order to create a greater knowledge in the field of Dieless forming.

In this paper an experimental verification of the Sine Law as well as an empirical method to compensate spring-back have been presented. Then the case of a multi-stage strategy for a cube has been shown, with the design modification added to complete a part of good aspect. All the work presented, as well as the other investigation lines of ASCAMM in the field of Dieless forming are oriented at simplifying the process design, and at the manufacture of parts within industrial requirements in terms of quality and accuracy.

Through its close links with industry, ASCAMM has also the opportunity to apply this technology in an early development stage to industrial applications.

Figure 9.   Spring-back compensation loop.

the experimentally measured spring-back, and the part is manufactured again.

This compensating loop can be repeated until acceptable results are obtained.

4.4   *Multi-stage forming strategies: "The case of a cube"*

To solve the limitation of the maximum forming angle (60°) multiple stage strategies have been developed.

The following example is the case of a cube with 81° steep walls. To reach 81° without failure it was necessary to start "pre-forming" a cube with shallower walls (60°) and then go step by step increasing the angle until the final desired angle of 81°.

Every step consists in an increase of 5° (60-65-70-75-81) in the wall steepness. Five steps have been necessary to achieve the 81° and the process time is superior to 90 minutes.

As this geometry was designed for investigation, design changes were possible. Some conic shapes have been introduced in the corners, in order to absorb the excess of material generated during the intermediate steps.

This design change avoided wrinkles in the corners which could not be eliminated otherwise. Finally, with these design modifications, the final part was achieved easily, with no fracture.

REFERENCES

[1]Amino H., Makita K. Maki T.: Sheet Fluid Forming And Sheet Dieless NC Forming, International Conference "New Developments in Sheet Metal Forming", Fellbach, Germany, 2000

[2]Kitazawa K.: Incremental Sheet Metal Stretch-Expanding With CNC Machine Tools, Advanced Technology of Plasticity 1993, 1993

[3]Hirt G., Junk S., Witulski N.: Surface Quality, Geometric Precision and Sheet Thinning in Incremental Sheet Forming, Materials Week, Munich, PP. 1–8, 2001

[4]Junk S., Hirt G., Chouvalova I.: Forming Strategies and Tools in Incremental Sheet Forming, 10th International Conference on Sheet Metal, Belfast, 2003

[5]Sawada T., Matsubara S., Sakamoto M., Fukuhara G.: Deformation Analysis for Stretch Forming of Sheet Metal with CNC Machine Tools, Advanced Technology of Plasticity, Vol. 2, PP. 1500–1504; Proc. of the 6th ICTP, 1999

[6]Jeswiet J., Hagan E.: Rapid Proto-Typing of a Headlight with Sheet Metal, 9th International Conference on Sheet Metal, Leuven, PP. 165–170, 2001

[7]Amino H., Lu Y., Ozawa S., Fukuda K., Maki T.: Dieless NC Forming of Automotive Service Parts, Advanced Technology of Plasticity, Proceedings of the 7th ICTP, Yokohama, Japan, 2002

Virtual modeling and rapid manufacturing – Bártolo (eds)
© 2005 Taylor & Francis Group, London, ISBN 0 415 39062 1

Process planning method for curing accurate microparts using Mask Projection Micro Stereolithography

A.S. Limaye & D.W. Rosen
George W. Woodruff School of Mechanical Engineering, Georgia Institute of Technology, Atlanta GA, USA

ABSTRACT: Mask Projection Microstereolithography (MPμSLA) is an additive manufacturing process capable for fabricating true three-dimensional microparts and hence, holds promise as a potential 3D MEMS fabrication process. In this paper, a process planning method for curing accurate microparts is presented. A MPμSLA system was designed and assembled. The process of curing a single layer in resin using this system was modeled was the Layer cure model. The Layer cure model was validated by curing test layers and parts with simple geometries. This model is used to formulate a process planning method to cure dimensionally accurate layers. This method has been found to be accurate within 3% error. The process planning method has been extended to cure layers larger than the field of irradiance of the imaging module of the MPμSLA system. The method has further been extended to cure layers with edges that conform to the part boundary. The errors occurring in the vertical dimension of a multi-layered micropart built by curing layers upon each other have been identified and a process planning method to avoid these errors has been proposed.

1 INTRODUCTION

Mask Projection Micro Stereolithography (MPμSLA) is a micro-fabrication process capable of fabricating 3D microparts. The principle of the MPμSLA process is illustrated in (Bertsch et al. 2000). The CAD model of an object to be built is sliced at different heights by a computer and the slices are stored as bitmaps. These bitmaps are displayed on a dynamic mask. Radiation from a light source is patterned by the bitmap displayed on the mask. This patterned radiation is focused onto the resin surface to cure a layer. The cured layer is coated with a fresh layer of resin and the next layer of the required shape is cured over it. The micropart is thus built by stacking layers one over the other.

Unlike the conventional microfabrication processes like silicon etching and LIGA, MPμSLA is able to fabricate 3D structures and is therefore a potential supplement to the MEMS fabrication processes. Since this technology is only about a decade old, it is still at the proof of concept stage. In order to mature this technology into a MEMS fabrication process, it is essential to be able to obtain dimensionally accurate microparts using the process. For this, the part dimensions have to be modeled in terms of the process parameters used to cure it. At the Rapid Prototyping and Manufacturing Institute (RPMI) at Georgia Tech., we have realized a MPμSLA system. In this paper, we have presented a

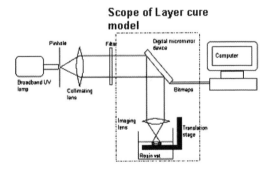

Figure 1. Schematic of the MPμSLA system realized.

process planning method for curing dimensionally accurate micro parts using our MPμSLA system.

2 MPμSLA SYSTEM

The schematic of the system realized at RPMI is shown in Figure 1.

UV light from a broadband UV lamp is collimated, and the required radiation is obtained by passing the beam through a filter. The bitmap corresponding the layer to be cured is displayed on a Digital Micro Mirror

Device (DMD™). The DMD™ is an array of individually addressable, bistable micromirrors, which can be selectively oriented to display any bitmap. This bitmap is irradiated by light from the UV lamp and is imaged onto a platform, which translates inside a vat holding the photopolymer resin. The irradiated portion of the resin cures into a layer. The platform lowers the cured layer into the resin-vat so that it is coated with a fresh layer of resin. The next layer is cured on top of the already cured layer. The platform is mounted onto an XY stage so that parts with larger cross-section can be built by scanning the platform horizontally underneath the imaging optics.

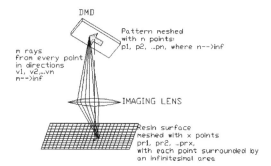

Figure 2. Nomenclature used in theoretical derivations.

3 PROCESS PLANNING METHOD TO CURE ACCURATE MICROPARTS

A micropart is built by curing layers one over the other. The process planning method to cure accurate microparts using MPμSLA is formulated in two steps. First, a process planning method to cure single layered microparts accurately is formulated in Section 3.1. Then, in Section 3.2, this method is extended to multi-layered microparts.

3.1 *Process planning method for single layers*

The field of irradiance of the imaging lens of the MPμSLA system is limited and so, it is not possible to cure layers with large cross-sections in one exposure. Large cross-section layers are built in an additive fashion by dividing the desired layer into areas that would be smaller than the field of irradiation of the imaging lens. By curing these areas next to each other, the entire layer is cured. We refer to this method of curing layers lager than the field of irradiation of the imaging lens as the "Layer stitching method". The layer stitching is implemented by mounting the platform on an XY stage and translating it horizontally underneath the imaging lens. An overlap between adjacent cured areas is deliberately allowed to ensure their bonding. In Section 3.1.1, we present a process planning method to cure layers with cross sections smaller than the field of irradiation. This method is extended in Section 3.1.2 to cure layers with larger cross-sections. In Section 3.1.3, we extend this method further to cure layers with thickness profiles that conform with the micropart geometry.

3.1.1 *Process planning method to cure layers with small cross-sections*

When a bitmap displayed on the DMD is imaged onto the resin surface, all rays emanating from all points on the pattern are directed onto the resin surface by the imaging lens. Every ray irradiates the infinitesimal area centered at the point where it intersects the resin surface. The pattern can be assumed to be composed of n number of points: $p_1, p_2, \ldots, p_n$, where $n \to \infty$. A cone of rays is emitted from each pattern point. The directions in which the rays are emitted from a bitmap-point can be represented by direction vectors $v_1, v_2, \ldots, v_m$, where $m \to \infty$. The resin surface can be assumed to be composed of x number of points $pr_1, pr_2, \ldots, pr_x$. Refer to Figure 2.

The normalized irradiance distribution across the beam incident on the DMD has been experimentally measured as given by equation (1)

$$I_{irr} = 1 - 0.00086\,p - 0.00883 p^2 \qquad (1)$$

where p is the distance from the center of beam in mm.

The irradiance received by the resin surface is derived in (Limaye 2004) as

$$H(pr_i) = (H_{av}x / \sum_{j=1}^{n} w_j m) \sum_{j=1}^{n} \sum_{k=1}^{m} w_j \delta(p_j, v_k, pr_i) \qquad (2)$$

where $H(pr_i)$ is the irradiance received by point pr_i on the resin surface; H_{av} is the average irradiance across the resin surface; w_j is the weight function given by equation (1) to account for the irradiance distribution across the beam incident on the DMD; $\delta(p_j, v_k, pr_i)$ is a function which determines whether a ray from the point p_j on the bitmap in the direction of vector v_k will strike an infinitesimal area centered on point pr_i on the resin. The function δ is evaluated by following the ray tracing procedure as explained in (Smith 1996).

The resin used with our system is the DSM SOMOS 10120 resin. It has been characterized and its working curve is plotted. The experimentally measured values of critical exposure (E_c) and depth of penetration (D_p) of the resin have been found to be $9.6\,\text{mJ/cm}^2$ and $0.056\,\text{mm}$ respectively. Using the working curve and the irradiance at the surface given by equation (2), it is possible to compute the dimensions of the cured layer.

The Layer cure model has been validated on our system and found to be accurate within 3% error.

A process planning method is formulated to generate the bitmap to be displayed on the DMD and the time for which it should be imaged onto the resin surface to cure a layer of the required dimensions. A bitmap displayed on the DMD is nothing but a cluster of micromirrors oriented in a particular direction. Using the Layer cure model, a one to one correspondence between the location of an "ON" micromirror on the DMD and the location of the pixel irradiated by it on the resin surface can be established. This correspondence is stored in the form of the "Pixel-micromirror mapping database". The database can be applied to map the pixels of the intended layer onto the micromirrors that need to be turned "ON" to cure that layer. The spatial location of "ON" and "OFF" mirrors on the DMD allows us to generate the bitmap that is needed to be supplied to the DMD. By running the bitmap through the Layer cure model, the irradiance distribution across the aerial image formed on the resin surface can be obtained. By applying the Beer Lambert's law, the time for which the bitmap needs to be imaged onto the resin surface for the entire aerial image to cure to a depth of one layer thickness or more can be computed. Thus, the inputs to the process planning method are the lateral dimensions of the layer to be cured and the outputs are the bitmap to be displayed on the DMD and the time of exposure (TOE) for which it should be imaged onto the resin surface.

The process planning method is tested on the following case study. Problem statement: A solid circular layer of diameter 1 mm and of thickness 30 μm is to be cured in DSM SOMOS 10120 resin by displaying a bitmap on the MPμSLA system. Generate the bitmap and compute the time for which it should be imaged onto the resin surface.

Solution: The required layer was cured by executing the following steps:

1. Pixel-micromirror mapping database was generated.
2. Circle was meshed with points.
3. Each of these points was snapped to its closest point on the resin surface in the Pixel-micromirror mapping database. These points were mapped onto the micromirrors on the DMD. This gave the spatial location of ON and OFF micromirrors on the DMD.
4. Based on the spatial location of ON and OFF micromirrors, bitmap was generated.
5. Bitmap was run through Irradiance model. Minimum irradiance was computed at the edges of the aerial image = 0.17 mW/cm².
6. Using the experimentally determined values of E_c and D_p in the formula:

$$TOE = (E_c/\min H(pr_i)) \, e^{(LT/Dp)} \qquad (3)$$

the time of exposure was calculated to be 56s.

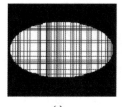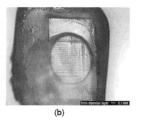

(a)   (b)

Figure 3. (a) Bitmap displayed on the DMD; (b) Disc shaped layer cured by imaging the bitmap onto the resin surface for 56 s.

The bitmap created in Step 4 and the disc shaped layer cured by imaging this bitmap onto the resin surface for 56 s are shown in Figure 3.

It can be seen in Figure 3(a) that the bitmap is elliptical in shape. In the bitmap, there are certain black lines passing through the white ellipse. These black lines exist because only a finite meshing density was used to populate the solid circular layer. These black lines result in ridges on the cured layer, visible in Figure 3 (b). It can be seen that the cured layer is slightly distorted. The maximum diameter of the disc is measured to be 1 mm and the minimum diameter to be 975 μm. Thus, there is an error of 2.5% in the process.

3.1.2 Process planning method to cure layers with large cross-sections

In this section, the process planning method presented in Section 3.1.1 is extended to cure layers with cross-sections larger than the field of irradiation of the imaging lens. As shown in Figure 4, a large layer is segmented into four smaller areas, each of which can be irradiated in one exposure step. There is an overlap between adjacent areas to ensure their bonding to each other. The overlapped area (shaded region) receives irradiation from all of its adjacent areas. This over-irradiance can be computed using the Layer cure model, which can then be used to formulate the process planning method for curing large layers with accurate lateral dimensions.

3.1.3 Process planning method for curing conformal layers

Any micropart built layer by layer would have a stair-step appearance. This causes inaccuracies in the surface definition of microparts. The stair stepping can be reduced by curing conformal layers. Conformal layers are layers whose edges conform to the geometry of the micropart. In Figure 5, an example of conformal layers is shown. Edges of the layers that constitute the slanting face of the micropart are slanting themselves so the part has a better surface definition.

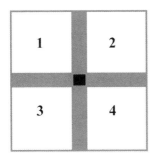

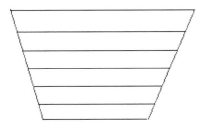

Figure 4.   Curing large cross sections by Layer stitching method.

Figure 5.   Conformal layers giving better approximation to part geometry.

Conformal layers don't have a constant thickness. They have thickness as a function of the lateral coordinates given as LT(x,y). The process planning method is Section 3.1.1 computes the time of exposure (TOE) as the time required for the least irradiated area to cure down to a depth of one layer thickness (LT). Conformal layers can be cured by controlling the TOE such that the layer will get cured to a depth given by LT(x,y). Thus, the time of exposure should be such that the exposure E(x,y) should cure the layer down to a depth profile given by LT(x,y). This can be written mathematically as

$$E(x,y) = E_c e^{LT(x,y)/D_p} \qquad (4)$$

With the irradiance distribution H(x,y) obtained by ray-tracing, the time of exposure can be given by the function TOE(x,y), such that

$$TOE(x,y) = [E_c / H(x,y)]e^{LT(x,y)/D_p} \qquad (5)$$

The time of exposure can be controlled by a method we call as "Differential exposure of bitmap". A bitmap displayed on the DMD consists of a collection of micromirrors that are "ON". By controlling the time for which each of these micromirrors constituting the bitmap displayed on the DMD is "ON", the exposure supplied to any layer can be accurately controlled.

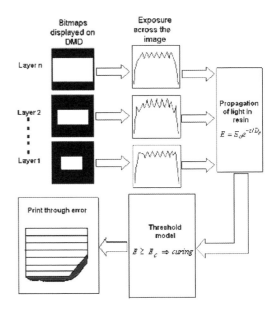

Figure 6.   Print through error in a multi-layered micropart.

3.2   Process planning method for multi-layered microparts

In this section, the process planning method for curing single layers is extended to multi-layered microparts. When a micropart is built by curing layers over each other, the vertical dimension of the micropart is slightly greater than the algebraic sum of the thickness of all layers. This error occurs because when any layer is cured, radiation penetrates beyond its intended thickness. The resin underneath a micropart receives radiation from all the layers cured above it. When the exposure received by this resin reaches its threshold exposure for polymerization (E_c), it undergoes curing. This unwanted curing causes an increase in the Z dimension of the micropart. This error is called as the "Print through error". Thus, when dimensionally accurate layers are cured by following the Process planning method presented in Section 3.1, the resulting part has an error in the vertical dimension. The source of Print through error is pictorially shown in Figure 6.

3.2.1   Zone of compensation

A deliberate reduction in the Z dimension of the micropart can be used to compensate for the increase in dimension due to Print through. In this section, a process planning method is proposed to cure microparts with accurate vertical dimension by using the "Zone of compensation" approach. As shown in Figure 7, a volume is subtracted from the bottom of the desired micropart. We refer to this volume as the "Zone of

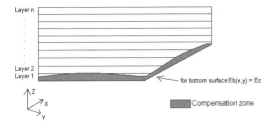

Figure 7. Zone of compensation to avoid Print through errors.

compensation". The thickness of the zone of compensation is a function of the lateral coordinates and is given by $Z_c(x,y)$.

In Figure 7, a compensation zone of thickness $Z_c(x,y)$ is subtracted from the bottom of the desired micropart. The remaining micropart is cured with n layers of thickness $LT_1(x,y)$, $LT_2(x,y)$, ..., $LT_n(x,y)$ respectively. Note that the layer thickness is not constant but a function of the lateral coordinates. Let the exposure received by the bottom surface of the micropart shown in Figure 7 be given by $Eb(x,y)$. In order for the micropart to have accurate Z-dimensions, the exposure received by its bottom surface should be just equal to the critical exposure required for polymerization (E_c).

$$E_b(x,y) = E_c \qquad (6)$$

The bottom surface will receive radiation from all the layers cured above it. Let the exposure received at the surface of the kth layer be given by the function $E_k(x,y)$. Let the height of the top of the kth layer from the bottom surface be $Z_k(x,y)$. The exposure that the bottom surface of the micropart would receive due this radiation penetrating the resin a distance $Z_k(x,y)$ would be:

$$E_k(x,y)e^{-Z_k(x,y)/D_p} \qquad (7)$$

The height Z_k would be equal to the sum of the thickness of all the layers from 1 to k as well as the thickness of the zone of compensation. So, Z_k can be given as:

$$Z_k(x,y) = Z_c(x,y) + \sum_{m=1}^{k} LT_m(x,y) \qquad (8)$$

Substituting the value of $Z_k(x,y)$ from equation (8) into equation (7), we get the exposure received by the bottom due to irradiation coming from the kth layer as:

$$E_k(x,y)e^{(-Z_c(x,y)-\sum_{m=1}^{k} LT_m(x,y))/D_p} \qquad (9)$$

Likewise, the bottom surface will receive exposure from all the n layers cured above it. The total exposure would be a sum of all these exposures.

$$E_b(x,y) = \sum_{k=1}^{n} E_k(x,y)e^{(-Z_c(x,y)-\sum_{m=1}^{k} LT_m(x,y))/D_p} \qquad (10)$$

From equations (1) and (5), the first constraint of the process planning method can be identified as:

$$\sum_{k=1}^{n} E_k(x,y)e^{(-Z_c(x,y)-\sum_{m=1}^{k} LT_m(x,y))/D_p} = E_c \qquad (11)$$

As the micropart is built layer by layer, we need to ensure that every layer cures down to a depth of 1 layer thickness. If it doesn't it will float away. So, the cure depth of the kth layer has to be equal to LT_k. This is the second constraint for the process.

$$E_k(x,y) = E_c e^{(LT_k(x,y)/D_p)} \qquad (12)$$

The two constraints to be satisfied are given by equation (11) and equation (12). The variables under user control are:

1. Thickness of layers $LT_k(x,y)$
2. Thickness of zone of compensation $Z_c(x,y)$
3. Exposure used to cure every layer $E_k(x,y)$.

Any combination of values of variables that satisfies the two constraints given by equations (11) and (12) would enable us cure the micropart accurately.

The process of curing a dimensionally accurate micropart would start with dividing the CAD model into conformal layers. The layer thickness can be chosen according to the resolution required in the vertical direction. The finer the resolution, the smaller would be the values of layer thicknesses. This selects $LT_k(x,y)$. Using equation (12), $E_k(x,y)$ can be determined. Using the values of $LT_k(x,y)$ and $E_k(x,y)$ in equation (11), the function $Z_c(x,y)$ can be determined. The Print-through errors can be avoided by subtracting the zone of compensation from the micropart and then, building the remaining micropart layer by layer by giving each layer an exposure given by equation (12).

4 SOME MICROPARTS CURED ON OUR SYSTEM

In Figure 8 (a), the four wheels and axle of a micro-SUV is shown. This is a nine-layer part. The axle is 57 μm in width and is overhanging. Figure 8 (b) is the close-up of the teeth of a micro spur gear. The thickness of the teeth at the pitch circle diameter of the gear is measured to be 40 μm. Figure 8 (c) the RPMI logo. It is a single layer, 30 μm thick part. Figure 8 (d)

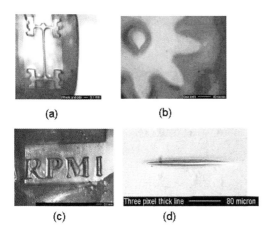

(a)   (b)

(c)   (d)

Figure 8. Microparts cured on our system. (a) Four wheels and axle of a SUV (b) Micro spur gear (c) RPMI logo (d) 6 μm wide line.

shows a two layered, three pixel wide rib, which was cured to validate the Layer cure model for very small features. The experimentally measured width of the line is 6 μm while the width computed by the Layer cure model is 6.2 μm. This experiment shows that the Layer cure model is valid for very small features too.

5 CONCLUSIONS

The process planning method for curing dimensionally accurate layers has been formulated by modeling the dimensions of a cured layer in terms of the process parameters used to cure it. A "Layer Stitching method" is proposed to cure layers with cross-sections larger than the field of irradiation of the imaging system. A scheme of differential exposure of bitmap on the DMD is proposed to cure layers with edges that confirm to the part geometry. Finally, the process planning method which involves modification of the part geometry by "Zone of compensation" method to cure dimensionally accurate multi-layered microparts is presented.

6 FUTURE WORK

The process planning method proposed in this paper to cure accurate 3D microparts will be implemented. The process planning method to cure microparts with accurate vertical dimensions shall be tested on various geometries. The accuracy of the process shall be improved by calibration (if needed) and the precision of the process shall be determined by experimentation.

ACKNOWLEDGEMENTS

We gratefully acknowledge the financial support from the RPMI member companies and the Manufacturing Research Center at Georgia Institute of Technology.

REFERENCES

Beluze, L., Bertsch, A., Renaud, P. 1999. Microstereolithography: a new process to build complex 3D objects. *SPIE Symposium on design, test and microfabrication of MEMS/MOEMS* 3680(2): 808–17.

Bertsch, A., Zissi, S., Jezequel, J., Corbel, S., Andre, J. 1997. Microstereolithography using liquid crystal display as dynamic mask-generator. *Microsystems Technologies* 3(2): 42–47.

Bertsch, A., Lorenz, H., Renaud, P. 1999. 3D microfabrication by combining microstereolithography and thick resist UV lithography. *Sensors and Actuators* 73(1): 14–23.

Bertsch, A., Bernhard, P., Vogt, C., Renaud, P. 2000. Rapid prototyping of small size objects. *Rapid Prototyping Journal* 6(4): 259–266.

Bertsch, A., Bernhard, P., Renaud, P. 2001. Microstereolithography: Concepts and applications. *8th International IEEE Conference on Emerging Technologies and Factory Automation* 2: 289–299.

Chatwin, C., Farsari, M., Huang, S., Heywood, M., Birch, P., Young, R., Richardson, J. 1998. UV microstereolithography system that uses spatial light modulator technology. *Applied Optics* 37(32): 7514–22.

Chatwin, C., Farsari, M., Huang, S., Heywood, M., Young, R., Birch, P., Claret-Tournier, F., Richardson, J. 1999. Characterisation of epoxy resins for microstereolithographic rapid prototyping. *International Journal of Advanced Manufacturing Technologies* 15(4): 281–6.

Dudley, D., Duncan, W., Slaughter, J. 2003. Emerging Digital Micromirror Device (DMD) applications *SPIE proceedings* 4985: 14–25.

Farsari, M., Huang, S., Birch, P., Claret-Tournier, F., Young, R., Budgett, D., Bradfield, C., Chatwin, C. 1999. Microfabrication by use of spatial light modulator in the ultraviolet: experimental results. *Optics Letters* 24(8): 549–50.

Farsari, M., Claret-Tournier, F., Huang, S., Chatwin, C., Budgett, D., Birch, P., Young, R., Richardson, J. 2000. A novel high-accuracy microstereolithography method employing an adaptive electro-optic mask. *Journal of Material Processing Technology* 107: 167–172.

Fujimasa, I. 1996. Micromachines: A new era in Mechanical Engineering. *Oxford University Press*.

Hadipoespito, G., Yang, Y., Choi, H., Ning, G., Li, X. 2003. Digital Micromirror device based microstereolithography for micro structures of transparent photopolymer and nanocomposites. *Proceedings of the 14th Solid Freeform Fabrication Symposium, Austin Texas*: 13–24.

Jacobs, P. 1992. Rapid Prototyping and Manufacturing: Fundamentals of Stereolithography *Society of Manufacturing Engineers*.

Limaye, A., Rosen, D. 2004. Quantifying dimensional accuracy of a Mask Projection Micro Stereolithography

System *Proceedings of the 15th Solid Freeform Fabrication Symposium, Austin Texas*: 481–492.

Limaye, A. 2004. Design and Analysis of a Mask Projection Micro-Stereolithography System, *Masters Thesis, Georgia Institute of Technology, Atlanta GA.*

Monneret, S., Loubere, V., Corbel, S. 1999. Micro-stereolithography using dynamic mask generator and a non-coherent visible light source *Proc. SPIE* 3680: 553–561.

Monneret, S., Provin, C., Le Gall, H. 2001. Micro-scale rapid prototyping by Stereolithography *8th IEEE international conference on emerging technologies and factory automation (ETFA 2001)* 2: 299–304.

Smith, W. 1990. Modern optical engineering: the design of optical systems, *McGraw Hill.*

Virtual modeling and rapid manufacturing – Bártolo (eds)
© *2005 Taylor & Francis Group, London, ISBN 0 415 39062 1*

Rapid Prototyping: Rapid flashing for SLA layering techniques

J.R. Zyzalo, O. Diegel & J. Potgieter
Institute of Technology and Engineering, Massey University, Auckland, New Zealand

ABSTRACT: The goal of this research is to investigate a more efficient layering technique than that used by SLA machines that use liquid resins for Rapid Prototyping. This new technique aims to allow a faster layering of model slices, resulting in much shorter production times, by flashing an entire slice of the model at a time, rather than scanning the slice with a thin beam form the laser. The project will use an existing SLA machine mechanism to set the slice thickness and apply the polymer layer, but all associated hardware and software for flashing a layer at once, and suitable modifications to the polymer will be developed.

1 INTRODUCTION

The manufacturing companies of today are faced with the challenge of unpredictable, high frequency market changes in both local and international markets. These are results of shorter product development times, changes to existing products, fluctuations in product demand, and changes in processes and technology. To be successful, rapid time-to-market of new products is critical in the midst of highly competitive markets (Krar, 2003).

Rapid Prototyping (RP) is one technology that has aided in shortening product development times and costs. Its main advantage is that it speeds up product design verification. Through quick design and error debugging and elimination, rapidly prototyping parts show great cost savings over traditional methods in the total product life cycle (Wohlers, 1999). RP is a relatively recent technology that allows the creation of complex three dimensional parts or models. These models are created on computer using computer-aided design (CAD) packages. Then the RP machine's software package takes the CAD model data and slices it up into thin slices and then builds the model, one slice at a time. Rapid Prototyping enables designers to test out their ideas and concepts before tooling for moulds and fixtures are made. Having parts rapid prototyped certainly reduces the costs involved in designing products and it also reduces product time-to-market. Complete parts and assemblies can be made within a 24 hour period.

2 BRIEF OVERVIEW OF RP TECHNOLOGY

There are numerous RP systems available on the market. They are broadly categorised by the initial form of their build material, i.e. the initial material from which a prototype is formed. These categories are: solid-based, powder-based, and liquid-based.

2.1 *Solid-based Rapid Prototyping*

One solid-based technology is Fused Deposition Modelling (FDM). FDM uses the extrusion of plastic through a nozzle and builds each layer by essentially drawing the shape with the nozzle whilst extruding beads of plastic for that layer. The plastic hardens immediately after being squirted from the nozzle and bonds to the layer below. This method results in a reasonable surface finish and reasonable dimensional stability. The final part has properties similar to that of ABS plastic. Other materials available for this process include wax for investment casting models.

Laminated Object Manufacturing (LOM) is another solid-based RP technology. LOM is one of the older methods and simply cuts each layer's profile out of a sheet of paper or plastic and bonds each sheet to the previous one. This method results in a reasonable surface finish and reasonable dimensional stability. The finish is similar to that of wood, and the part can then be worked on as if it were wooden. LOM is still often used to produce patterns for sand-casting.

2.2 *Powder-based Rapid Prototyping*

Powder-based build material can be considered as solid state. However, it has its own category, as build material has intentionally been made into a powered form (Chua, 2003). One powder-based rapid prototyping is known as Selective Laser Sintering (SLS). SLS uses a powder which is spread in a layer of the desired thickness, and a laser beam is then passed over the profile of the model slice which cures the powder into a solid in that section. The build platform is then dropped down one level, and the next layer of powder is spread and the process repeated for the next slice.

SLS results in a good surface finish and excellent dimensional stability, though not quite as fine as SLA technology. SLS can produce parts for both aesthetic and functional testing. The main advantage of this technology is that it can work with a variety of materials provided that they are in powered form. The most common material used has properties similar to that of glass filled nylon, but other materials include flexible rubber materials, and even steel powders, which can be used to produce tool inserts (Krar, 2003).

3D Printing is one of the more recent solid-based technologies. It uses standard inkjet technology to print each slice of the model, one layer on top of the other until the model is built. This method results in a reasonable surface finish and reasonable dimensional stability. The most commonly used materials are starch or plaster, so the models are not all that strong. They can however be further strengthened by infusing them with wax or epoxy. Newer models of the printers can now print in colour, using different coloured materials as the model is being built.

2.3 Liquid-based Rapid Prototyping

One form of liquid-based rapid prototyping uses a machine called a Stereolithography Apparatus (SLA). SLA build objects a layer at a time by tracing a UV beam on the surface of a vat of liquid resin. This class of materials, which was originally developed for the printing and packaging industries, quickly solidifies wherever the UV beam strikes the surface of the liquid. The SLA controls the UV beam and draws one model slice or layer at a time. Once one layer is completely traced, the build platform is lowered a small distance into the vat and a second layer is traced right on top of the first. The self-adhesive property of the material causes the layers to bond to one another and eventually form a complete, three-dimensional object after many such layers are formed.

Support structures, in the form of a fine lattice structure of cured resin, are needed to maintain the structural integrity of a part. They support overhangs, as well as provide a starting point for the overhangs and for successive layers on which to be built. After the part is fully built, the support structures are removed and the part is cleaned in a bath of solvent and air dried (De Laurentis et al., 2004).

This method can produce parts for both aesthetic and functional testing. SLA is generally considered to provide the greatest accuracy and best surface finish of any rapid prototyping technology. The technology is also notable for the large object sizes that are possible with the use of "mammoth" SLA machines.

On the negative side, working with liquid materials can be messy and parts often require post-processing to remove support material and post-curing operation in a separate oven-like apparatus for complete cure and stability.

3 THE CHEMISTRY OF PHOTOPOLYMERS

3.1 Photopolymers

Liquid-based RP systems require build material that contains certain chemical properties. The build material is made from photopolymers. Liquid photopolymers contain a chemical cocktail of molecules that solidify when exposed to electromagnetic radiation. The forms of radiation include gamma rays, X-rays, UV and visible range, or electron-beam (EB). Most RP systems use photopolymers that cure in the UV range and curing technology use mainly UV or EB radiation (Jacobs, 1996). There is a small variety of these UV-curable resins available, each contains fillers and other chemicals to achieve different chemical and physical properties (Chua et al., 2003). These resins can be classified into acrylate, expoxy, or vinylether systems.

3.2 Photopolymerization

Some UV-curable resins accomplish photopolymerization by using a free-radical mechanism. These resins are acrylate-based polymers. These resins are a makeup of photoinitiators and reactive liquid monomers. When the resin is exposed to a UV light source, the photoinitiators absorb some of the radiant energy and are raised to an excited state. A fraction of these excited photoinitiators become reactive after going through various complex chemical transformations. These then react with the monomer molecules to form a polymerization initiating molecule. This is called chain initiation. Additional monomers react with polymerization initiating molecule to form a chain. This is called chain propagation. The chain terminates due to an inhibition process determined by the polymerization reaction. The longer the UV exposure time, the longer the chains become resulting in a polymer with a higher molecular weight (Chua, 2003).

An advantage of acrylate-based polymers is that they have quite a high photospeed or photosensitivity, that is, they provide faster curing (Kunjappu, 2000). However, acrylates have several disadvantages. One disadvantage is that oxygen must be kept from the acrylates environment as it retards the polymerization reaction by removing free-radicals. This results in parts that are not cured enough and still in a "tacky" state. Another disadvantage of oxygen in the acrylate environment is the possibility of producing intolerable ozone caused by the reaction with the UV source.

To overcome the disadvantages of acrylate-based polymers, cationic photopolymerization is more commonly used. In cationic photopolymerization, cationic initiators start the polymerization. Cationic polymerization is similar to the chain propagation that occurs with free-radical polymerization, but in this case the growing polymer chain has an active end that is a positive ion. The two types of cationic photopolymerizable monomers are epoxy resins and vinylether resins. These cationic resins have advantages in that they are not affected by ambient oxygen and they have a continued thermal curing after exposure to the UV source. They have some disadvantages, however. The polymerization of cationic resin is affected by bases and water requiring a low humidity environment. Also, the photospeed is relatively slower when compared to acrylate-based resins (Jacobs, 1996). This can be overcome by using a higher power light source and also varies depending on the types of chemical components used (Chua, 2003; Kunjappu, 2000).

3.3 Cross-linking

An important part of the photopolymerization process is cross-linking. Cross-linking that occurs in a resin generates an insoluble continuous network of molecules, especially a resin containing monomers with three or more reactive chemical groups. The parts that are made in these resins have strong structural strength due to the resulting chemical covalent bonds. Considerable heat is necessary to break these bonds and that is why parts do not melt upon applying heat. They soften at high temperature before thermally decomposing. Greater heat thresholds can be accomplished depending on the types of monomers used in the resin (Jacobs, 1996).

4 IMAGE PROJECTION TECHNOLOGY

4.1 Micro-electromechanical systems (MEMS)

Image projection technology has been able to advance in the form of micro-electromechanical systems (MEMS). MEMS is a technology that combines computers with micro-mechanical devices such as sensors, valves, gears, mirrors, and actuators embedded in semiconductor chips. Basically, a MEMS device contains micro-circuitry on a tiny silicon chip into which some mechanical device such as a mirror or a sensor has been manufactured. Such chips can be built in large quantities at low cost, making them cost-effective for many applications.

MEMS find applications in image projection technology as optical switching devices that can switch light signals over different paths at switching speeds up to 20-nanoseconds. Most advancement has been centred on projection TV systems.

The projectors used in current TV systems rely on two general approaches to project images. One method shines light through the image-forming element (a CRT tube or LCD panel). This is known as a transmittive projector. The other method bounces light off the image-forming element. Projectors using this method are called reflective projectors. In both types of projectors, a lens collects the image from the image-forming element, magnifies the image, and focuses it onto a screen. It is important to note that the terms "transmittive" and "reflective" refer to the optoelectronics inside the projector, not to how the projector is arranged within the projection TV system, i.e. reflective projectors with rear projection.

The reflective approach has been more progressive in recent times, though the transmittive approach has been around longer and appears in many of the small portable projectors on the market today.

MEMS have aided in a recent innovation in reflective technology. In reflective projectors, the image is formed on a small, reflective chip. MEMS have a movable or deformable reflective surface on top of a semiconductor chip. The chip generates voltages in response to digital information. The voltages change the shape of the reflective surface rapidly and in a controlled way to produce the image that was encoded by the digital information. The projected light bounces off the reflective surface and gets collected by the projector lens.

Two types of MEMS chips available are digital micromirror devices (DMD, DLP) and grating light valves (GLV).

4.1.1 Digital micromirror devices (DMD)

DMDs, also called digital light processing (DLP), were invented by Dr. Larry Hornbeck in 1987 and developed by Texas Instruments (Texas Instruments, 2005). The DMD is a chip that has anywhere from 800 to more than 1 million hinge-mount microscopic mirrors on it, depending upon the size of the array. Each mirror rests on support hinges and electrodes.

Each mirror on a DMD covers a 16-μm^2 area and consists of three physical layers and two "air gap" layers. The air gap layers separate the three physical layers and allow the mirror to tilt $+10$ or -10 degrees. The middle layer consists of two address electrodes (connected to SRAM) that are diagonally opposite each other and the torsion hinge and yoke mechanism (Figure 1, bottom right). When a voltage is applied to either of the address electrodes, the mirrors can tilt $+10$ degrees or -10 degrees, representing "on" or "off" in a digital signal.

In a projector, light shines on the DMD. Light hitting a mirror that is in the "on" state will reflect through the projection lens to the screen. Light hitting a mirror in the "off" state will reflect to a light absorber. Each mirror is individually controlled and is totally independent of all the other mirrors. When all the mirrors

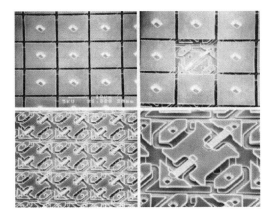

Figure 1. Enlarged view of micromirror structure. (Texas Instruments, 2005).

on a DMD chip are correctly coordinated and controlled, an all-digital image is reflected by the mirrors on to a screen or surface. The system that comprises the projectors sophisticated electronic components is called Digital Light Processing (DLP) technology. This includes the digital video or graphic signal, the light source, the projection lens, and of course the DMD chip.

4.1.2 Grating light valves (GLV)

Another MEMS device is the grating light valve (GLV). GLV technology, licensed to Sony, was developed by Professor David Bloom at Stanford University, and is now produced by Silicon Light Machines in Sunnyvale, California (Kubota, 2002).

The GLV chip consists of micro-reflective ribbons mounted over a silicon chip. The ribbons are suspended over the chip with a small air gap in between. When a voltage is applied to the chip below a ribbon, the ribbon moves toward the chip by a fraction of the wavelength of the illuminating light. The deformed ribbons form a diffraction grating, and the various orders of light can be combined to form the pixel of an image. The shape of the ribbons, and therefore the image information, can be changed in as little as 20 billionths of a second.

To make a projector, the GLV pixels are arranged in a vertical line that is 1,080 pixels long. Light from three lasers, one red, one green and one blue, shines on the GLV and is rapidly scanned across the display screen at 60 frames per second to form the image.

A major advantage of GLV technology is that GLV chips can make high-resolution images at a relatively low cost. For example, because a $1,920 \times 1,080$ pixel image can be achieved by scanning a 1,080 pixel linear array, a GLV chip can be manufactured to achieve this resolution with only 1,080 pixels, instead of the 2 million needed for other technology, such as DMD.

Also, because the ribbons are aligned vertically, there are no horizontal gaps in the image, providing a much better quality image.

5 DISCUSSION

SLA technology has established itself as a rapid prototyping technology that offers designers increased resolution of prototyped models. With SLA technology, the slice thickness can be chosen to be generally ranging from around 0.01 mm to 0.25 mm per slice, and this determines the resolution (quality/surface finish) of models.

As all of the RP systems available can produce physical objects, there are many overlapping strengths and limitations. Some are faster and better at certain aspects of model creation than others. Therefore, to determine which process is the best depends on what is intending to be accomplished. For example, designers may use an FDM RP machine purely for aesthetics of small to medium size parts, while the use of SLA is better for more detailed aesthetics and functional testing.

One problem with the SLA process happens at the layering stage. Models require long production times which have only been reduced by a minimal amount with the use of efficient lithography tracing algorithms. Yang et al. applied an equidistant hatching algorithm to an FDM RP system that improved part processing and quality (Yang et al., 2001). Often models are set up and left to run during the day when the operator is present in the factory. The operator can monitor the model's build progress throughout the day. However, for parts that require longer build times, say more than 12 hours, the operator needs to be able to leave the SLA machine running unattended, i.e. overnight. Problems can occur during this time that causes the model to be ruined requiring a completely new model to be made from scratch. It is important to note that this problem does not make the overall process of rapid prototyping slow. The overall process is indeed quite fast in terms of product idea right through to time-to-market.

One way models can be built efficiently is by processing many parts on the build platform at the same time. This is faster and more efficient than building parts separately. Efficient batching of parts is managed by the operator of the SLA machine. However, the same problem may occur as mentioned previously. Some parts may not have enough supports to build correctly and delamination occurs. Delamination is when a small section of a layer (usually on edges or points) does not bond to the previous layer and so floats freely in the resin. The result is that delaminated fragments float into neighbouring parts which in turn ruins those parts. Most of the batch must be built again from scratch.

6 RAPID FLASHING

One way to obtain more efficiency in part build times is by changing the layer scanning method. The most common technique used by current SLA machines is a single solid-state laser beam guided by mirrors to trace out the 2D slice on the surface of the resin. As it uses as raster pattern of going back and forth until the entire slice has been traced, this can be time consuming due to the resolution of the laser beam having to trace out such comparatively large slices.

The technique proposed by this research is to use DLP technology to flash an entire layer at once. By using DLP technology, the expensive solid-state lasers can be done away with making this both a lower cost and faster method. Most data projectors using DLP technology can be purchased at a reasonable price. Some systems come ready to plug into a computer monitor port and use.

This research will use such a projector but with a high UV light source, such as a xenon tube, to take a single slice created by the computer software, and to project the entire slice onto the photopolymer until it is cured to the appropriate depth.

The software will take the image of each slice, convert it to a negative black and white image, in which the slice itself will be white, thus allowing UV light to be projected in the shape of the slice. The software will also add a structural lattice pattern around the slice to give adequate support for undercuts as required. It will then repeat this process for each subsequent part until the model is built.

Though the principle of rapid flashing is relatively simple, several areas that require investigation and research present themselves.

The DMD chip needs to be subjected to high intensity UV radiation. An investigation into how much radiant power can be absorb by the DMD chip is necessary. Power absorption limits need to be specified for the DMD chip and a suitable UV light source acquired. In addition to this, it may be necessary to filter out any infra-red light (heat) from the beam of light, with a narrow band dichroic infra-red filter. Any excessive heat, that serves no purpose in the flashing process, would only decrease the amount of power that the DMD device can be subjected to.

Most standard optical systems are specifically coated to block UV light coming through, as for most applications UV radiation is seen as undesirable. As this application requires UV light, quartz lenses that pass UV light will need to be made to replace those in the data projector.

A liquid polymer that photopolymerizes at the wavelength of the UV light source must be procured or formulated. This will of course need to be adapted to suit the amount of UV that the DMD device will be able to sustain. The amount of curing catalyst present in the resin will need to be carefully tuned to allow the polymer to harden under the amount of UV radiation delivered by the system within a chosen time frame of, ideally, a few seconds per model slice.

In addition to this, as the curing catalyst in the resin will be constant, a method of curing to different depths will need to be found. This will allow the resolution of each slice to be set. The most likely methods for achieving this will be to change the amount of UV radiation hitting the DMD device, or by adjusting the distance from the UV source and DMD device closer to photopolymer layer. If this second method is chosen, the optical system will, of course, have to be adjusted to keep the component size accurate at any chosen resolution.

Work will also need to be done to ensure that the layering technique does not affect the structural integrity of the part. Current SLA layering incorporates hatching of layers during the scanning of each layer. This gives rise to the better mechanical properties of the model. Flashing whole layers at once will, in theory, not have this inherent problem but this will need to be further investigated.

The project will use the mechanism of an existing SLA machine to control the step-by-step lowering of the model into the polymer vat, and to apply the next layer of polymer to the top of the model. All other hardware involved in the DMD system and UV light source system, as well as a complete UV-transmissive optical system will be developed as part of the project. The software package that slice the model up into slices of the desired thickness, converts them into the appropriate form for single layer rapid flashing and add the required support structures will also be developed.

7 CONCLUSIONS

A possible new technique for SLA machines to rapidly flash layers has been presented in this paper. Through the use of DMD technology and a suitable photopolymer, it may be possible to minimise build times of SLA parts. This rapid flashing technique would consist in exposing an entire slice of the model to a UV source, rather than the current thin-beam raster tracing method currently employed by most systems. This will allow for even more rapid creation of rapid prototyped parts.

There are several problems that need be investigated and research further. These include the power absorption of DMD chips, the ability of the optical system to transmit UV light, obtaining a polymer which photopolymerizes at the right wavelength and of which the curing depth can be controlled, and the structural strength of the part being built.

Though an ambitious project, this represents the potential to turn conventional rapid prototyping processes into rapid manufacturing processes.

REFERENCES

Chua, Leong, & Lim (2003), *Rapid Prototyping: Principles and Applications.* (2nd ed). Singapore: World Scientific Publishing Co.

De Laurentis, Mavroidis, & Kong (2004), Rapid Robot Reproduction. *IEEE Robotics & Automation Magazine, June, 2004*: 86–92.

Jocobs, P. F. (1996), *Stereolithography and other RP&M Technologies.* Michigan: Society of Manufacturing Engineers.

Krar, S. & Gill, A. (2003) *Exploring Advance Manufacturing Technology.* New York: Industrial Press Inc.

Kubota, S. R. (2002), The Grating Light Valve Projector. *Optics & Photonic News, September, 2002*: 50–53.

Kunjappu, J. T. (2000), Radiation Chemistry in EB- and UV-Light-Cured Inks. Retrieved 21st Mar, 2005, from *Paint & Coatings Industry website*, http://www.pcimag.com/CDA/ArticleInformation/features/BNP__Features__Item/0,1846,11435,00

Texas Instruments Incorporated. (2005), How DLP™ Technology Works. Retrieved 21st Mar, 2005, from *Texas Instruments Incorporated website*, http://www.dlp.com/dlp_technology/dlp_technology_overview.asp

Wohlers, T. (1999), *Rapid Prototyping and Tooling State of the Industry: 1999 Worldwide Progress Report.* Wohlers Associates.

Yang, Fuh, & Loh (2001), An Efficient Scanning Pattern for Layered Manufacturing Processes. *IEEE International Conference on Robotics & Automation Proceedings*: 1340–1345. Seoul: Korea.

Virtual modeling and rapid manufacturing – Bártolo (eds)
© 2005 Taylor & Francis Group, London, ISBN 0 415 39062 1

Sculpting of expanded foam plastics for rapid prototyping applications

D.R. Aitchison, T.S. Germann & M. Taylor
University of Canterbury, Christchruch, Canterbury, New Zealand

B. Bouillard
INSA, Rouen, France

ABSTRACT: The paper presents the findings of on-going research ultimately aimed at producing a rapid manufacturing and prototyping facility capable of cutting polystyrene geometric and free form surfaces using a novel hotwire technique. Hotwire foam cutters are frequently used when working with rigid foams like low density expanded polystyrene. Most of the commercially available production hotwire cutters utilize computer control technology to manipulate the movement of the tensioned hotwire or the foam to perform profile cutting in 2D; though specialist multi-axis and highbred machines do exist. Despite these reported developments little work has been published on foam cutting mechanics and performance. To establish a general understanding of the foam cutting process performance a series of tests were undertaken for a range of wire temperatures and feed-rates. The findings have provided the researchers with a clearer insight into the associated hot-wire cutting mechanics and system performance.

1 INTRODUCTION

This paper presents the findings of an on-going research program which is aimed at producing a rapid manufacturing and prototyping machine capable of cutting polystyrene geometric and free form surfaces using a novel hotwire technique. One of the fundamental project tasks was to determine the optimal cutting conditions for the hotwire and work-piece system. This objective was deemed necessary for the reliable production of cut surfaces with superior accuracy and form.

An introduction to existing foam cutting facilities is presented below alongside an overview of reported work on foam cutting mechanics. This backdrop paves the way for subsequent sections on the conducted cutting trials and the consequent findings.

Hotwire foam cutters are frequently used when working with rigid foams like low density expanded polystyrene (EPS). They make neat, straight and smooth cuts that simply cannot be duplicated with any other type of saw or cutter (Ahn 2002). Most of the commercially available hotwire cutters are basic hand operated devices with an electrically heated pre-tensioned wire. More sophisticated models utilize computer control technology to manipulate the movement of the tensioned hotwire or the foam work-piece to perform profile cutting in 2D (Svecko 2000). An extension to this configuration has seen the number of axes

increased to five in some specialist machines, but the straight hotwire arrangement still limits the generated faces to ruled surfaces. Assorted highbred machines have recently been reported (Ahn 2002). Some invariably permit the creation of complex 3D facetted surfaces and concave features. However this is achieved through the machining of discrete laminae which are then manually recombined with predetermined registration to produce a solid artifact: similar to a conventional rapid prototyping process like that employed by a Fused Deposition Modeler (FDM). Surprisingly, despite these reported developments little work has been published on foam cutting mechanics and performance.

Whilst cutting expanded polystyrene, high wire temperatures are known to vaporize the foam and as a consequence create poor surface texture and accuracy (Aitchison 2003a, b). Likewise, temperatures that are too low cause the foam to tear with even worse accuracy and surface finish effects. Feed-rate is known to exhibit similar effects at sub-optimal values. Therefore, the temperature and speed of the cutting operation must be controlled and optimized to avoid these unfavorable conditions and produced surfaces with tight accuracy and form control.

To establish a general understanding of polystyrene foam cutting performance a series of tests were undertaken for a range of wire temperatures and feed-rates. Cutting temperatures ranging from 200 to 400°C were applied whilst carving standard grade polystyrene slabs of various thicknesses (20, 30 and 40 mm) at feed-rates up to 2500 mm/min. The electrical power absorption, feed force, wire temperature and a notional measure for cut surface texture were recorded. Suitable combinations of feed-rate and wire temperature were identified for the production of smooth surfaces.

Stable cutting conditions were seen to develop for some of the test cases but a three stage settling-in process was noted to occur. At higher feed-rates pronounced surface striations were consistently evident which possessed an apparent relationship with the oscillating feed force under the given conditions. General relationships between feed force, feed-rate and cut surface texture was also investigated as part of the study.

The overall findings have provided the researchers with a clearer insight into the associated hotwire cutting mechanics and performance which are presented and discussed in the following sections.

2 EXPERIMENTAL PROCEDURE

2.1 Apparatus

An OKUMA CNC milling machine was used as the basis for the apparatus to perform the polystyrene cutting tests. The test samples consisted of expanded polystyrene slabs fixed to solid base plates. The base plates were used to rigidly clamp the test samples to the bed of the milling machine. The hotwire cutting tool (Fig. 1) was fixed to the spindle housing of the mill. This arrangement made it possible to use the mill's ability to traverse the hotwire cutting tool through the samples, at a range of feed-rates with consistency and accuracy. The wire carrying frame was mounted onto the mill via a proprietary load cell. Eccentric loading produced by the vertically offset wire was eliminated through the specific configuration of the full-bridge load cell.

The frame used to support the hotwire was constructed with a tensioning axle at one end of the wire. This axle was turned and locked with a spanner to pretension the wire. Tensioning was done at the beginning of each test session and periodically throughout testing. The pre-tension load was not officially standardized or measured. PTFE plastic bushes were incorporated to electrically insulate the wire from the support frame.

Wire heating was achieved through the use of a direct current electrical power supply. To control the wire temperature the power supply current was initially varied then fixed once the desired wire temperature was attained. The current setting remained unchanged while the cutting tests were conducted at the desired pre-set temperature. A thermocouple and digital thermometer were used to establish the wire temperature.

A data-logging PC was adopted to capture the designated data during the trials.

2.2 Calibration

Before the cutting tests were commenced, each of the four signals, to be measured by a dedicated data-logging PC, were calibrated (Wire: force, current, voltage and temperature). This was done for every testing session to ensure that changes to signal gains and/or operating conditions that may have occurred when the apparatus was unattended were essentially eliminated.

Pre-testing calibration of the temperature channel was done by comparing the converted thermocouple signal, received by the PC, with the temperature measured by a second thermocouple linked to a hand-held digital thermometer. Five PC based readings were taken over the temperature range concerned (200°C to 400°C) and a linear conversion factor, between the input signal and actual wire temperature, was found.

Calibration of the load cell was achieved by mounting the cutting tool horizontally and hanging weights vertically from the wire mounts. Again five PC based measurements were recorded within the expected range of wire force values (up to 50 N). A linear conversion factor was then established between the load cell signal and actual load force on the wire.

To calibrate the current and voltage drop measurements a digital multi-meter was used to take voltage and current readings where the electrical power supply was connected to the cutting wire. Measurements were taken at various wire temperatures within the range to be used during testing.

2.3 Test procedure

A procedure for conducting the test sessions was developed to make full use of the time available on the CNC mill. For this reason, the variables which took the longest time to alter were changed as infrequently

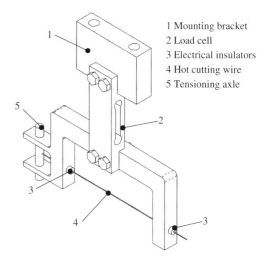

1 Mounting bracket
2 Load cell
3 Electrical insulators
4 Hot cutting wire
5 Tensioning axle

Figure 1. Hotwire cutting tool.

as possible. Due to the thermal inertia of the temperature measuring apparatus and power source, it took two to three minutes to obtain a stable wire temperature close to the desired value. Changing the expanded polystyrene test samples also took several minutes, while changing the cutting feed-rate only took a matter of seconds. Consequently the general test cutting procedure was conducted as follows:

– First test sample was fixed in the milling machine vice.
– Wire temperature was set.
– Cutting run conducted at lowest feed-rate.
– Next cutting run conducted at increased feed-rate on same test sample at same temperature.
– Cutting runs were repeated with increasing feed-rate until maximum feed-rate was reached.
– Test sample was replaced with sample of different thickness.
– Cutting runs were repeated as before with each test sample thickness.
– Wire temperature was changed and procedure repeated at different temperature.

2.4 Data collection

Data from each cutting run was recorded by a dedicated data-logging PC. The four channels of data were recorded at a sampling rate of 1000 Hz. The four parameters recorded were:

– Force on the cutting wire
– Temperature of the cutting wire
– Current through the cutting wire
– Voltage drop over the cutting wire

The data collection procedure for each test cut was made very simple through the use of proprietary software. A simple start/stop button was adopted to record the data. Data collection was started a few seconds before the wire began cutting the test sample, and stopped a few seconds after cutting had finished. The recorded binary data was immediately filtered, pre-processed and plotted by an Excel Macro. This was done between test cuts to ensure that none of the input signals were saturating, and there were no obvious anomalies with the data. Satisfactory data sets were saved for subsequent processing and analysis. Once the data set had been swiftly verified and saved the next test was commenced.

2.5 Data processing

Data collected by the PC was recorded and saved as binary files at 1000 data points per second. These binary files were then filtered to 100 data points per second and saved in ASCII text format for data processing. Once in this format, the majority of the data processing was accomplished using Matlab software. A schematic diagram indicating the method of data processing is shown in Figure 2.

The most difficult part of processing the gathered data was determining the points at which the hotwire entered and exited the test sample. It was established that the best way to find these points was to trace where the largest step-change in cutting force occurred. This location, in the data, was known to directly correspond with the instant when cutting finished and the hotwire exited the test sample. Once located, the starting point of the cut was simply derived as both the cut length and feed-rate were known. The entry and exit points were then used to truncate the data and remove irrelevant information collected before and after cutting. However, at low feed-rates, the force exerted on the cutting wire was negligible. Because of this, the collected force data gave no indication of where cutting commenced or concluded. Due to the immeasurable scale of the force this situation was obviously irrelevant for force related assessments. This situation occurred in only a few of the test cases.

Due to the 'noisy' nature of the collected data, it was first smoothed with a moving average function (the number of points was dependent upon the feed-rate) so that the step change in the cutting force, at the end of the cut, and other features could be clearly identified. Once the start and end points of the cut had been located, the calibrated raw data was truncated

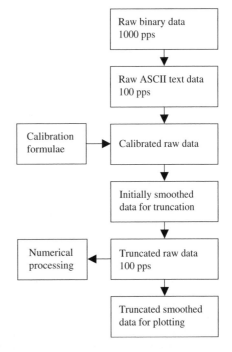

Figure 2. Schematic diagram for method of data processing.

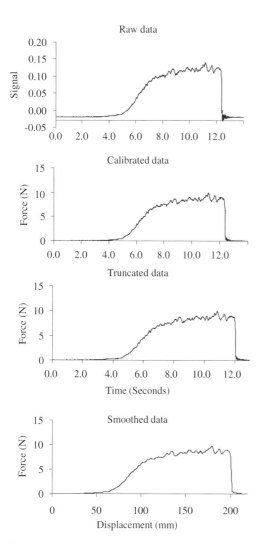

Figure 3. Stages of data processing for cutting data.

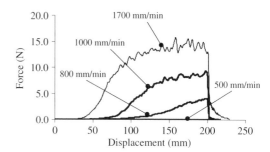

Figure 4. Force-displacement graphs for several wire feed-rates.

and a corresponding displacement vector formed so that the raw data could be plotted against its relative position, in millimeters, along the test sample. Figure 3 shows cutting force data at all stages of data processing for a sample test run.

3 RESULTS

A total of 117 test runs were completed at three nominal wire temperatures of 200, 300 and 400°C. Polystyrene test samples of 20 mm, 30 mm and 40 mm thick were cut at each of the three nominal temperatures. The ranges of feed-rates tested were slower for the cooler 200°C cutting temperature: ranging from

200 to 900 mm/min. At the higher cutting temperature of 400°C, the range of feed-rates was generally higher: from 800 to 2500 mm/min. The force on the cutting wire at higher temperatures was less for the same feed-rate at lower temperatures and so higher feed-rates were possible with the hotter wire settings.

Figure 4 shows the processed force-displacement data for the four data sets of the 30 mm thick test samples cut with a 300°C hotwire. This figure clearly shows that for some settings (feed-rates of 800 and 1000 mm/min in this case), stable cutting conditions did not develop prior to the cessation of cut. For this reason, longer samples were subsequently tested to ensure stable cutting conditions did in fact develop. Therefore, in addition to the 200 mm long standard test samples, 500 mm long polystyrene test samples were prepared and cut with a wire temperature of 300°C and otherwise identical conditions. Stable cutting conditions were reached.

4 DISCUSSION

4.1 Force curve characteristics

The most distinctive data collected during each test run was that of the force-displacement of the cutting wire. The general form of the force curve consists of three distinct sections or phases as shown in Figure 5. The initial section of the cut can be seen to have negligible cutting force exerted on the wire. Following this, if the feed-rate is high enough, there is a period of increasing force on the cutting wire, until a point where quasi-equilibrium cutting conditions are developed.

The most apparent and obvious trend, found from the testing, was that the force on the cutting wire increased as the wire feed-rate was increased or conversely the wire temperature was reduced. Figure 4 clearly illustrates this situation with no apparent cutting force at a feed-rate of 500 mm/min, but a peak force of approximately 15 N is evident at a feed-rate of 1700 mm/min.

Plots of force-distance curves at various feed-rates (Fig. 4) show a trend in the quasi-equilibrium cutting

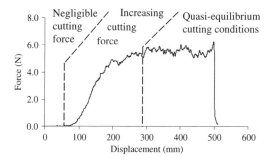

Figure 5.   The three stages of a typical force curve.

conditions. This figure also shows that at higher feed-rates, the magnitude of the equilibrium state cutting force is higher, and furthermore that the variation or noise about the mean increases.

4.2   Surface finish

To aid in the descriptions and discussion of the generated surfaces the following definition are made:

– Surface Texture: surface features that are visible with the naked eye and follow regular or irregular patterns. (Associated descriptions could include granular, ridged, nodular etc.).
– Surface Finish: the visual effect presented by fine surface detail which on its own is not apparent to the naked eye. (Associated descriptions could include dull, glossy, etc.).

The way in which the cutting force develops as shown in Figure 5, is also reflected in the resulting surface characteristics of the cut test samples. Test cuts done with very low or negligible cutting force produced a very flat surface with minimal texture and a smooth but dull finish. In contrast, striated or pimpled textured surfaces with a smooth and glassy finish are associated with higher cutting forces.

From all the generated samples, the surface texture of the polystyrene was seen to exhibit three distinctly different classes. The first class of surface texture observed was smooth and even; typically produced at slow feed-rates and at the beginning of most test cuts: the surface finishes were essentially dull. Fine whiskers laid flat on the surface and aligned with the direction of wire travel were visible on some surfaces. The second class of surface texture was seen to comprise of wavy striations but a smooth and satin-like finish. These striations occurred for cutting forces of at least 4 N. At even higher cutting forces, the surface texture deteriorated further, resulting in not only large striations but a distinctly lumpy (nodular) surface which was deemed to be a third class of surface texture. A glassy finish was evident with this classification of surface.

The first class of surface texture is hypothesised to be created due to the fact that the polystyrene is being cut purely by melting and/or smearing, hence the whiskers. The striations observed in the second class of surface texture only occur with elevated cutting forces, and are therefore thought to be created due to cutting by mechanical means as well as thermal. Surface smearing is visibly apparent. The third class of surface texture was notably rougher than the previous two and was produced during the development of high and fluctuating cutting forces. It is assumed that the wire was partly tearing through the test sample, lifting and partially releasing polystyrene beads from the surface as it passed. Melting still occurred and as a consequence the resultant striations and bumps exhibited smooth form and a glassy surface finish. Hence cutting was reliant upon a degree of mechanical slicing with the wire in addition to melting the samples. Small surface pockets and hollows suggest that the polystyrene beads are also forced away from the advancing wire and then subsequently relax to produce the nodular but glassy surface.

The form of the surface striations appeared irregular but from preliminary observation they are of similar frequency to the variation or noise seen in the recorded cutting force data. If there is a direct correlation between the fluctuation in cutting force and the creation of striations, there is no immediately obvious reason for this phenomenon. The striations are essentially parallel and generally uniform across the width of the cut samples, essentially forming a ruled surface. Therefore they are known to result from a vertical oscillation of the cutting wire. The wire oscillatory movement was at right-angles to the direction of cut but it appears that full width contact (with the EPS sample) has been maintained.

Investigation into possible quantitative methods of measuring the surface roughness of the polystyrene revealed there was no available apparatus to collect this data. A widely available Tallysurf apparatus would be ideal but due to the material softness and level of texture a stylus based apparatus was known to be impractical. Surprisingly, further investigations revealed that no suitable surface texture/finish measuring equipment were available. A new device may need to be developed in the future as a means of assessing the performance of prototype hotwire sculpting devices from a surface finish viewpoint.

4.3   Electrical and thermal dynamics

From the data collected during testing, it was seen that the current and voltage values for the cutting wire remained constant throughout each of the individual cutting trials. This revealed that the difference in electrical power consumption between cutting and not cutting was indiscernible at the recorded level of measurement.

459

By slicing polystyrene of different thicknesses, it was expected that some variation in power consumption would be noticed, as more energy would be required to make the thicker cut. However, from the tests, it is evident that the energy requirements to melt through the polystyrene is almost negligible compared to the energy lost to the surrounding air; as there was no discernable difference in energy consumption between thin and thick cuts.

The actual dynamic conditions at the cutting wire–polystyrene interface are unfortunately the most difficult to observe. It was hoped that recording the temperature of the wire would give some indication of how conditions changed as a cut was made. However, no significant temperature changes were observed during the test run. The temperatures recorded by the thermocouples were seen to remain almost constant; essentially within one degree Centigrade.

The way in which the cutting force on the wire changed during a cut indicates that the wire temperature, inside the polystyrene, must have also changed. To confirm this hypothesis, thermo-graphic pictures were taken during a test run. The results showed a definite temperature gradient along the cutting wire from the thermocouple attachment point, near the extremity of the exposed wire, to the point where the wire entered the test sample: with close to 100°C of cooling where cutting was taking place.

The fact that the cutting wire was cooling towards the polystyrene during cutting helps to explain some of the observed cutting force dynamics. However, at this stage it is not known why this temperature variation did not register on the thermocouple.

5 CONCLUSIONS

The general form of the force curve consists of three distinct sections or phases. The initial section of the cut can be seen to have negligible cutting force exerted on the wire. Following this, if the feed-rate is high enough, there is a period of increasing force on the cutting wire, until a point where quasi-equilibrium cutting conditions are developed.

Plots of force-distance curves show that at higher feed-rates, the magnitude of the equilibrium state cutting force is higher, and also that the variation or noise about the mean increases.

Test cuts done with very low or negligible cutting force produced a very flat surface with smooth finish. For mid-range feed-rates the cut surfaces exhibit a pronounced waviness of varying frequency and amplitude. In contrast, striated and pimpled textured surfaces with a smooth and glassy finish are associated with higher cutting forces.

- It is thought that the smooth low feed-rate produced surfaces are created due to the fact that the polystyrene is being cut purely by melting and/or smearing.
- The striations observed in the mid feed-rate range of cutting only occur with elevated cutting forces, and are therefore known to be created due to cutting by mechanical means as well as thermal.
- In the high feed-rate tests it is assumed that the wire was partly tearing through the test samples, lifting and partially releasing polystyrene beads from the surface as it passed. Hence cutting was reliant upon a degree of mechanical slicing with the wire in addition to melting the samples.

From initial cursory observations a relationship between the cutting force oscillations and surface striations is thought to exist.

The drawing of additional power during cutting (manifested as cooling in wire) was not evident during the tests. This unexpected situation was supported by the lack of any discernible temperature variation between the cutting and idle conditions, as recorded by the adopted thermocouples. However, thermographic camera results have indicated the presence of a thermal gradient in the wire at close proximity to the active cutting region; results which will receive further attention in due course.

Further investigations into the data characteristics and interrelationships are on-going. It is recognized that the ultimate findings will provide solid intellectual foundations and much improved understanding of foam cutting mechanics and foam machining optimization. Though the authors have a specific Rapid Prototyping and Manufacturing (RP&M) application in mind the results will benefit others who are working in the field of cutting rigid and flexible foams: a market sector which has witnessed considerable demand and seen strong growth in recent years.

REFERENCES

Ahn, D.G. Lee, S.H. & Yang, D.Y. 2002. Investigation into thermal characteristics of linear hotwire cutting system for Variable Lamination Manufacturing (VLM) process by using expandable polystyrene foam. *Int. Journal of Mach. Tools and Manf.* 42: 427–439.

Aitchison, D. & Sulaiman, R. 2003a. Determining the surface form of polystyrene through the co-ordinate measurement machine. *Journal of Mech. Eng. Sci.* 217(Part C): 839–844.

Aitchison, D. & Sulaiman, R. 2003b. Cutting expanded polystyrene: Feed-rate and temperature effects on surface roughness. Proc. Int. Conf. on MSO, Banff, 2–4 July 2003, Canada.

Svecko, R. Chowdhury, A. Bolcina, M. & Tusek, R. 2000. Computer controlled machine for cutting expanded polystyrene. *Electrotechnical Review* 67: 145–151.

Virtual modeling and rapid manufacturing – Bártolo (eds)
© *2005 Taylor & Francis Group, London, ISBN 0 415 39062 1*

Enhancement of surface finish in Fused Deposition Modelling

N.V. Reddy
Department of Mechanical Engineering, Indian Institute of Technology, Kanpur, India

Pulak M. Pandey
Department of Mechanical Engineering, Indian Institute of Technology, Delhi, India

ABSTRACT: A surface roughness model based on the measured surface profiles of Fused Deposition Modeled part is used as a key to adaptively slice a tessellated (STL) CAD geometry. It is ensured while slicing that at any location of the part the surface roughness does not exceed the specified value. Hot cutter machining is used to machine the build edges (staircase) of ABS material based on the previous work. Experimental study is carried out by hot cutter machining of build edges of a freeform FDM part to enhance its surface finish. A virtual simulation of surface roughness before and after hot cutter machining along with adaptive slicing (using limited *Ra* value as a criterion) is implemented to understand the enhancement of part surface quality.

1   INTRODUCTION

Layered Manufacturing (LM) has emerged an important technology as it supports design of the product and produces prototype to analyze it for customer specifications in a short span of time. However, the functionality of the LM parts is severely affected by poor surface finish due to staircase effect resulting from layer-by-layer deposition. Staircase formation is a geometric constraint of the layered manufacturing, which can not be eliminated. Presence of staircase on the surface of prototype detracts surface finish and hence restricts functionality of prototypes. Surface roughness value on an LM component is dependent on geometry of the enclosing surface, the layer thickness and the deposition orientation of the part. In LM, the enclosing surfaces of the solid model are approximated by series of triangles in 3D space, which results in chordal error. This chordal error also leads to non-smooth surface. In addition to staircase and chordal error, road width, air gap between roads and polymer melt temperature also contribute to the surface roughness in case of Fused Deposition Modelling (FDM).

Issues related to poor surface finish of LM parts have been addressed by many researchers. Kruth et al. [1998] reported that accuracy and surface finish are major handicaps of LM parts than their strength. Bharath et al. [2000] reported that the best possible surface finish on a FDM part could be obtained by choosing the optimal FDM process parameters. They carried out fractional factorial experiments considering layer thickness, road width, air gap, build orientation

and polymer melt temperature as process parameters. Using analysis of variance (ANOVA), they concluded that layer thickness, build orientation and interaction of these two parameters are factors with significance index 23.8828, 41.5151 and 18.4001% respectively, which affect the surface finish. Gautham et al. [unknown] studied the surface roughness of FDM parts considering slice thickness, build orientation, edge profile (assumed elliptical without measuring), layer composition and sub-perimeter composition as the parameters. They concluded that the rate of variation of surface roughness in the range 40 to 90 degree build orientation is less as compared to 0 to 40 degree region. There have been some attempts to study the effect of process parameters on part quality produced by SL using statistical techniques. Diane et al. [1997a,b] measured *Ra* values of SL parts and concluded that layer thickness and part orientation are important parameters. Perez et al. [2001a,b] developed surface roughness model for SL parts by assuming the edge profiles as filleted for values of φ between 30° and 85° and rounded for values of φ between 85° and 90°. For $\varphi = 0°$, a constant value of surface roughness (*Ra*) is proposed based on experimental results. Here, φ is angle between horizontal and surface tangent. They concluded that theoretical models proposed for surface roughness agree well with experimental values. They also explored the possibilities of manufacturing SL parts with constant slice thickness or with surface roughness confined within given values of surface roughness with variable layer thickness. Campbell et al. [2002] presented a computer graphics based visualization system for surface roughness of LM parts (SL 250, Actua 2100, FDM 1650, LOM1015 and Z 402). A part named "Truncheon" is fabricated using different LM processes with constant slice thickness. Surface roughness value is measured using *contact Talysurf* system. Experimentally obtained surface roughness values were used to generate the graphical output.

Even though the surface roughness can be reduced, by reducing the slice thickness but the build time of the prototype increases drastically. This problem can be handled by adaptive slicing of CAD models in which variable slice thickness is used (slice thickness is determined based on geometry of part and LM machine specifications) instead of using constant slice thickness. Many attempts have been made to reduce the build time of RP parts by the modification of slicing algorithms. These modifications are mainly aimed at using variable slice thickness instead of constant thickness. The layer thickness in these methods is decided based on the local geometry of the component. Dolenc and Makela [1994] introduced the concept of cusp height (figure 1) and implemented adaptive slicing for polyhedral models. Pandey et al. [2003a] implemented slicing of tessellated CAD models using real time edge profile, i.e., parabolic for FDM parts and used a constraint on *Ra* value, which is used as a design specification in place of cusp height. Pandey et al. [2003b] carried out a critical review of slicing algorithms used in LM.

A suitable part deposition orientation can improve part accuracy and surface finish and reduce the production time and support structures needed for building the part. Many attempts have been made to decide a suitable part deposition orientation using different criteria like part accuracy, surface quality, build time, volume of support structure and cost and their details can be seen in Pandey et al. [2004] and Trimuthulu et al. [2004].

Literature survey reveals that not much work has been done to enhance surface finish through staircase machining. Kulkarni and Dutta [2000] attempted to

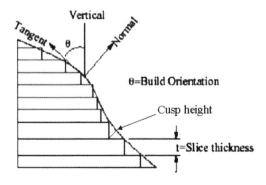

Figure 1.   Build orientation and cusp height.

enhance the accuracy and surface finish of layered manufactured part through CNC machining of staircase with ball end mill cutter, once the part is deposited completely. Contour crafting has been suggested by a parabolic trowel in place of a planar trowel for sculpting. This work is mainly aimed at the development of CNC cutter path. The developed procedures are tested on standard simulation softwares for CNC milling. Values of the surface finish that can be achieved was not reported in their work. The major limitation of this work is the inaccessibility of intricate features and details to the ball end mill cutter after the component is completely deposited. Abrasive Flow Machining (AFM) has been used by Williams and Melton [1998] to enhance surface finish of Stereolithography (SL) parts. A cuboidal workpiece produced by SL is finished by AFM and concluded that media pressure, grit size of the abrasive particles and build orientation are the significant parameters. The major limitation of using AFM for enhancing surface finish of a RP components is that there is no control on the pressure distribution in the viscous media, which can lead to uneven material removal from the different location of RP part and hence it is not possible to control the mathematically defined surface profile of the component. In addition, intricate details, blind holes, slots etc. can not be machined as the flow of media is not possible in AFM. Freeform thick layer object manufacturing technology for concept modelling of large prototypes is developed by Broek et al. [2002]. In this method, the slice edges are cut by a flexible cutting tool with higher order approximation. This method is reported to produce physical prototypes with or without little finishing. Recently Hur et al. [2002] introduced eclipse RP system by decomposing machining and deposition features of a component. This system utilizes the advantages of material deposition and CNC machining at single station.

It can be concluded from the above discussion that the surface finish in RP can not be enhanced up to a level obtainable through precision machining by adjusting RP process parameters or by modifying slicing and orientation algorithms only. Conventional or CNC machining processes can not be directly applied to finish RP parts because of high cutting forces and inaccessibility of intricate features of the part once it is completely deposited. In the present work a surface roughness model for FDM processed parts developed by Pandey et al. [2003a] is used which is based on the actual surface profile measured using *Surfanalyzer5000*. Surface roughness model developed during the present work is validated by comparing its predictions with the results published in the literature as well as the experimentally measured value in the present work. Experimental study is carried out by machining of build edges of a freeform axisymmetric FDM processed part on a lathe machine to enhance its surface finish. Part surface roughness before and after machining with adaptive slicing is simulated. Average part surface roughness is calculated to understand the overall quality of a product.

2 SURFACE ROUGHNESS MODEL

Pandey et al. [2003a] has considered layer thickness and build orientation, as the two most significant process variables that effect surface finish, based on the previous literature. To study the surface roughness of FDM part surface, a part with different build orientation (angle between surface tangent and vertical direction as shown in figure 1) was fabricated on FDM-1650 machine with 0.254 mm slice thickness, 270°C polymer melt temperature of part material i.e. ABS, 265°C polymer melt temperature of support material, 70°C envelope temperature and zero air gap. Surface roughness (Ra) value of the different faces has been measured by them using *Surf-Analyzer5000*. It is observed from measured surface profiles (between $0° \leqslant \theta \leqslant 70°$) that there is no gap between deposited roads while there is gap between deposited roads (between $70° \leqslant \theta < 90°$), where θ is build orientation. The measured surface profiles clearly indicate that the geometry of build edge profiles can be approximated as a parabola. Stochastic model is developed in the range $0° \leqslant \theta \leqslant 70°$ by approximating the layer edge profile by a parabola with base length ($t/\cos \theta$) and height as η percentage of base length, where t is slice thickness [Pandey et al., 2003a]. Constant η is established to follow normal distribution and hence for a 99% of confidence level the expression for centerline average surface roughness (Ra) is obtained as

$$Ra(\mu m) = (69.28 - 72.36)\frac{t\ (mm)}{\cos \theta} \quad 0° \leq \theta \leq 70° \quad (1)$$

Surface profile for $\theta = 90°$ (horizontal surface or surface parallel to FDM platform) is idealized as semicircle (instead of parabola) with base length t and height $0.5 \times t$. Surface roughness value for $\theta = 90°$ is thus obtained as

$$Ra(\mu m) = 112.6 \times t(mm) \quad for \quad \theta = 90° \quad (2)$$

Surface roughness for $70° \leqslant \theta < 70°$ is calculated by assuming linear variation of surface roughness between $Ra_{70°}$ and $Ra_{90°}$.

3 ADAPTIVE SLICING

Limited cusp height and limited area deviation are the two criteria used by researchers to slice a CAD model

in most of the works. These two criteria have limitations, as a standard manufacturing measure of surface quality is not considered in these works. Pandey et al. [2003b] developed stochastic model of surface roughness (expression 1) and used it as a key to slice a tessellated CAD model. In the present work, slice thickness is also calculated by the following expression as used by Pandey et al. [2003b]

$$t(mm) = \frac{Ra(\mu m) \times \cos \theta}{70.82} \quad (3)$$

where Ra is maximum possible value of surface roughness (bound kept on Ra) permitted on part. The average part surface roughness (surface quality) has been calculated using following expression

$$Ra_{av} = \frac{\sum Ra_i A_i}{\sum A_i} \quad (4)$$

where Ra_{av} is average surface roughness of the part, Ra_i is the roughness and A_i is the area of the ith triangular facet of STL file. Slicing of the tessellated CAD model is performed by the procedure proposed by Pandey et al. [2003b] and the slice thickness and corresponding height of slicing plane (which are parallel to FDM platform, xy planes) are stored.

4 VALIDATION OF SURFACE ROUGHNESS MODEL

In the present work, a freeform part shown in figure 2 is fabricated on FDM-1650 machine with 0.254 mm layer thickness, 270°C polymer extrusion temperature; 265°C support material extrusion temperature, 70°C envelope temperature and zero air gap. Surface roughness is measured at five different locations of the part as shown in the figure 2 and are compared with the

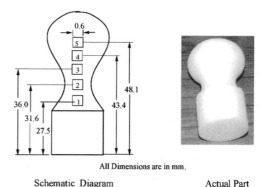

All Dimensions are in mm.

Schematic Diagram   Actual Part

Figure 2. Freeform part used in the experiment.

Table 1. Comparison of surface roughness at different locations before machining.

Location	Ra Predicted	Ra Measured	% Difference
1	24.2	25.3	4.3
2	18.2	19.1	4.7
3	18.1	18.2	0.5
4	41.1	42.0	2.1
5	19.1	25.0	23.6

predicted values of surface roughness using stochastic model by taking average value of the multiplier (70.82 t/cos θ). The comparison of the experimentally obtained and predicted values of surface roughness is presented in table 1. It can be seen from the table 1 that there is good agreement between the experimental and the predicted values. The discrepancy in the values may be due to the effect of FDM process variables like polymer extrusion temperature, viscosity of polymer, voids present between the roads, effect of support structure and radius of curvature of the part which are not considered while modelling. At fifth location there is comparatively large difference between two values. The presence of support structure at this location may be the major cause for this difference. Alexander et al. [1998] proposed to weight the cusp height in order to incorporate the effect of support structure. Additional cusp height of 2.0 mm has been taken in their work for FDM process. At fifth location the difference in the experimental and predicted value with respect to experimental value is approximately 23.6%. If Ra value at this location is weighted by 25% (assumption based on measurements) then it increases to 23.8 micrometer from 19.1 micrometer. Thus, the difference between experimental value and predict value is 4.8%.

5 MACHINING OF BUILD EDGES

Surface finish of an RP component can be improved by machining of staircase. The problem of high cutting forces and accessibility of intricate locations by the cutting tool motivated Pandey et al. [2003a] to attempt hot cutter machining as the other conventional machining methods are unsuitable to machine staircase during deposition of the part on FDM machine. The schematic diagram of a Hot Cutter used by Pandey et al. [2003a] to show that this process can be used to enhance the surface finish of FDM processed parts is shown in figure 3. In HCM, ABS (material of FDM processed part) gets softened in the vicinity of hot cutting edge of the cutter hence the build edges can be machined/ploughed without application of high cutting forces. Experiments have been conducted just above

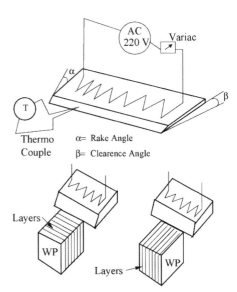

α= Rake Angle
β= Clearence Angle

Figure 3. Schematic diagram of HCM modes. [Pandey et al., 2003a].

glass transition temperature of ABS (T_g to T_g + 5 Celsius) with 15 degree relief angle of hot cutter. A minimum of 0.5 micrometer and maximum of 4.4 micrometer center line average value was obtained over a wide range of experiments.

Fractional factorial experiment design with two levels and four process variables ($2^{(4-1)}$) has been applied to understand the effect of three important machining parameters namely rake angle of the cutter, feed rate and direction of cut with respect to layers. In addition to the above mentioned machining parameters; build orientation, i.e., workpiece parameter related to part geometry (θ), has also been considered as a parameter because it affects surface roughness. Standard statistical software STATISTICA was used to find the regression equations and application of ANOVA and adequacy tests reduced these expressions to

$$Ra_0 = 2 + .425x_1 + .625x_2 + .275x_3 + .375x_4 \qquad (5)$$

$$Ra_{90} = 2.525 + .275(x_1 + x_2 - x_1x_2) + .425x_3 + .675x_4 - .175x_2x_3 \qquad (6)$$

where

$$x_1 = (V_c - 22.5)/7.5 \qquad x_2 = (\theta_c - 27.5)/17.5$$

$$x_3 = (a - 50)/10 \qquad x_4 = (\Phi - 45)/45$$

here V_c is cutting speed in mm/min, θ is build orientation in degree, a is rake angle of hot cutter in degree and Φ is cutter movement direction with respect to the layers in degree. *Rao* is surface roughness value in the direction of machining and Ra_{90} is surface roughness value perpendicular to the direction of machining.

6 MACHINING OF FREEFORM

Surface finish improvement for planar (flat) surfaces of FDM processed parts by machining staircases using straight edge hot cutter has been studied [Pandey et al., 2003a]. Surface finish of a freeform surface can be enhanced if it is machined by curved edge cutting edge, which conforms with the geometry of freeform surface. Pandey et al. [2003a] developed surface finish prediction models after HCM for planar surfaces. In order to have confidence whether the model of surface roughness prediction for planar surfaces (expression 4) can be used to predict surface roughness of freeform surfaces, experimental study is carried out by machining an axisymmetric FDM part using profiled hot cutter. For this purpose, an axisymmetric part is fabricated on FDM-1650 machine using 0.254 mm slice thickness. The material of the part is ABS and it is built at 270°C extrusion temperature, 265°C support material temperature and 70°C envelope temperature.

A hot cutter is designed and fabricated with cutting edge geometry similar to the geometrical profile of the axisymmetric part. Cutter is provided with a heating coil, which is connected through a variac to 220 Volt AC mains. Variac is used to control the temperature of the cutter, which is measured by a digital thermocouple. The hot cutter is fixed on KISTLER dynamometer (insulation plate is kept in between the two) and this assembly is mounted on tool-post of lathe to measure the cutting forces. Cutting edge is maintained at the horizontal level of axis of axisymmetric part which is mounted on the chuck of lathe machine. The voltage signals of dynamometer are fed to a computer through a data acquisition card and displayed by Labview software. The actual photograph of the experimental setup is given in figure 4.

The cutting edge of the hot cutter is heated up to glass transition temperature of ABS material, i.e., 220°C. This temperature of the cutter is maintained in the range of 220° to 220 + 5°C by adjusting the coil voltage of the heater by variac. The cutting edge of the tool is given a depth of cut of 100 micrometer (i.e., average peak to valley height) based on the previous work [Pandey et al., 2003a].

Surface roughness of the machined part (on lathe machine with hot cutter) is measured at different locations (these locations are given in figure 8) on the part, before and after machining, and is presented in table 2. It can be seen from the table 2 that good improvement in surface finish as compared to initial surface finish (before machining, Table 1) is achieved. It is also clear

Figure 4. Photograph of experimental setup.

Table 2. Comparison of surface roughness at different locations after machining.

Location	Surface roughness after machining by Simulation	Surface roughness after machining by Experiment
1	2.7	4.4
2	1.9	3.2
3	1.9	2.6
4	2.1	2.2
5	2.0	2.3

from the table that there is reasonably good agreement between the predicted and experimentally obtained values of surface roughness after HCM. The discrepancy in the values may be due to the reasons that similar machining conditions could not be maintained in the present experimental work as in the case of machining with straight cutter [Pandey et al., 2003a]. In the present experimental work the clearance angle is also not provided to the cutter which may be a reason that caused comparatively rougher surface.

7 SURFACE ROUGHNESS SIMULATION

Simulation of the part surface roughness (sliced adaptively using tessellated CAD model) is implemented to have an idea of variation of part surface roughness over the part. The adaptive slicing algorithm proposed by Pandey et al. [2003b] is used in the present work.

Adaptive slicing of tessellated CAD model gives variable slice thicknesses. A triangular facet (of STL file) may be common to more than one slice, therefore average slice thickness for a triangular facet is considered for calculation of average *Ra* on a facet. Build orientation of each triangle is calculated by the information of surface normal form STL file of the

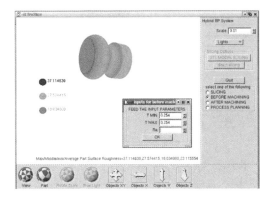

Figure 5. Distribution of surface roughness before machining.

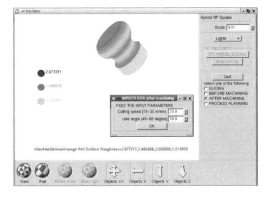

Figure 6. Distribution of surface roughness after machining.

part. Surface roughness corresponding to each triangle is calculated using expressions (1) to (3) and average part surface roughness is calculated using expression (4). Surface roughness values over the part are displayed in graphical form by rendering the triangular facets with different colors according to their Ra values. The surface roughness variation of a typical part (this part is used for experiments) before machining for uniform slice thickness 0.254 mm (adaptive slicing is not used here) is shown in figure 5. Maximum, minimum and average surface roughnesses are predicted as 37.11, 18.03 and 23.11 micrometer respectively.

The surface roughness after HCM is predicted using the expressions (5) and (6), which are developed for plane and freeform surfaces. The average surface roughness of the part after machining is calculated using expression (4). The simulation of surface roughness is carried out by rendering the triangular facets with appropriate colors. The surface roughness simulation is presented in figure 6 for the component used in experimental study, for 20 mm/min cutting

speed and 50° rake angle of cutter. Maximum, minimum and average part surface roughnesses after machining are obtained as 2.67, 0.30 and 1.21 micrometer respectively.

8 CONCLUSIONS

A surface roughness model for freeform FDM processed parts are presented and validated. Experimental study is carried out by machining of staircase (build edges) of freeform FDM part to enhance surface finish. It is concluded that there is good improvement in surface finish. A surface roughness model proposed by Pandey et al. [2003a] can be used to predict surface roughness of freeform as well as flat surfaces. Surface roughness simulation system to predict the surface finish before (with adaptive slicing) and after HCM is implemented. Further work is in progress to modify FDM hardware so that build edges can be machined on FDM platform for enhancing surface finish.

ACKNOWLEDGEMENTS

The authors are thankful to Professor S. G. Dhande, Director, IIT Kanpur, for his valuable suggestions while carrying out this work. One of the authors (NVR) gratefully acknowledges CSIR New Delhi, India, for the financial assistance provided.

REFERENCES

Alexander, P., Allen, S., Dutta, D.,1998. Part orientation and build cost determination in layered manufacturing. *Computer Aided Design* 30(5): 343–356.

Bharath, V., Dharam, P. N., Henderson, M. 2000. Sensitivity of RP Surface Finish to Process Parameters Variation. *Solid Free-form Fabrication Proceedings*: 251–258.

Broek, J. J., Horvath, I., Smit, B. de, Lennings, A. F., Rusak, Z., Vergeest, J. S. M. 2002. Free-Form Thick Layer Object Manufacturing Technology for Large-Sized Physical Models. *Automation in Construction* 11: 335–347.

Campbell, R. I., Martorelli, M., Lee, H. S. 2002. Surface Roughness Visualisation for Rapid Prototyping Models. *Computer Aided Design* 34: 717–725.

Diane, A. S., Kou-Rey Chu, Montgomery, D. C. 1997a. Optimizing Stereolithography Throughout. *Journal of Manufacturing Systems* 16(4): 290–303.

Diane, A. S., Mongomery, D. C. 1997b. Using Experimental Design to Optimize the Stereo-lithography Process. *Quality Engineering* 9(4): 575–585.

Dolenc, A., Makela, I. 1994. Slicing Procedure for Layered Manufacturing Techniques. *Computer Aided Design* 1(2): 4–12.

Gautham, K., Novi, M. I., Henderson, M. *A Design Tool to Control Surface Roughness in Rapid Fabrication*, http://prism.asu.edu/publication/Manufacturing/abs_design.html.

Hur, J., Lee, K., Hu, Z., Kim, J. 2002. Hybrid Rapid Prototyping using Machining and Deposition. *Computer Aided Design* 34: 741–754.

Kruth, J. P., Leu, M. C., Nakagawa, T. 1998. Progress in Additive Manufacturing and Rapid Prototyping. *Annals of the CIRP* 47(2): 525–540.

Kulkarni, P., Dutta, D. 2000. On the Integration of Layered Manufacturing and Material Removal Process. *International Journal of Machining Science and Engineering* 122: 100–108.

Pandey, P. M., Reddy, N. V., Dhande, S. G. 2003a. Improvement of Surface Finish by Staircase Machining in Fused Deposition Modeling. *Journal of Material Processing Technology* 132(1): 323–331.

Pandey, P. M., Reddy, N. V., Dhande, S.G. 2003b. A Real time adaptive slicing for fused deposition modeling. *International Journal of Machine tools and Manufacture.* 43(1): 61–71.

Pandey, P. M., Reddy, N. V., Dhande, S. G. 2003c. Slicing Procedures in Layered Manufacturing: A Review. *Rapid Prototyping Journal* 9(5): 274–288.

Pandey, P. M., Thrimurtullu, K., Reddy, N. V. 2004. Optimal part deposition orientation in FDM using a multi-criterion genetic algorithm. *International Journal of Production Research*, 42(9): 4069–4089.

Perez, C. J. L., Calvet, J. V., Perez, M. A. S. 2001a. Geometric Roughness Analysis in Solid Free-Form Manufacturing Processes. *Journal of Material Processing Technology* 119: 52–57.

Perez, C. J. L., Vivancos, J., Sebastian, M. A. 2001b. Surface Roughness Analysis in Layered Forming Processes. *Precision Engineering* 25: 1–12 .

Thrimurtullu, K., Pandey, P. M., Reddy, N. V. 2004. Optimal part deposition orientation in fused deposition modeling. *International Journal of Machine Tools and Manufacture* 44: 585–594.

Williams, R. E., Melton, V. L. 1998. Abrasive Flow Finishing of Stereo-lithography Prototypes. *Rapid Prototyping Journal* 4(2): 56–67.

Virtual modeling and rapid manufacturing – Bártolo (eds)
© *2005 Taylor & Francis Group, London, ISBN 0 415 39062 1*

Effect of scan mode on the strength of selective laser sintering part

Yusheng Shi, Wenxian Zhang, Yanjie Cheng & Shuhuai Huang
State Key Laboratory of Plastic Forming Simulation and Die & Mould Technology, School of Material Science and Engineering, Huazhong University of Science and Technology, Wuhan, P.R.China

ABSTRACT: The strength of selective laser sintering (SLS) part is one of the most important indexes, which judge the quality of SLS part. Factors affecting the strength of SLS part are numerous, being the scan mode one of them. Some examples are taken to study the effect of laser scan modes on the tensile strength, flexural strength and shock strength of SLS polymer testing parts. The mechanism of the effect of scan modes on the strength of SLS part is analyzed, helping to develop better scan modes suitable for SLS process.

1 INTRODUCTION

SLS parts are fabricated with laser beam scan powder material through a layer by layer strategy. During the process of accumulating materials, layer by layer, from a point to a line, from a line to a plane, and from two dimensions to three dimensions, scan system has to do much scan work. Therefore, it is very important to improve the quality of SLS parts, optimising the scan mode.

The aim of optimising scan mode is to properly design scan paths regarding precision, strength and fabrication efficiency of SLS process, as well as its easier implementation.

Owing to the importance of this field the research has been significant, performed by several authors. They have found all kinds of scan modes, for instance, parallel scan mode where the laser beam moves parallel to axis x and axis y [1], scan mode with changing scan direction and scan mode with changing scan direction in divided areas [1,2,3], dissipating scan mode with the shape of star polygon and dissipating scan mode with the shape of angled star polygon [4], fraction scan mode [5], scan mode with the spire rails and equidistant path [7,8], the algorithm of forming scan path based on figure Voronoi [9,10], and scan mode using Pythagorean hodograph as filling line [11]. Besides, there are other slice-filling modes, which are presented by researchers, such as Bertoldi [12], Wasser [13], Tiller and Hanson [14], Ganesan and Fadel [15], Pham [16] and Takashi [17,18]. The aims of studying scan modes are as follows: improving the precision of SLS part, increasing the fabrication speed of SLS part, reducing the number of turning on or off laser, and prolonging the life of laser. Until recently, nobody has studied how

scan modes affect the strength of SLS parts. For this reason, a changing direction scan mode, a subarea and changing direction scan mode, an improved subarea and changing direction scan mode are taken for some examples in order to study the effect of laser scan modes on the strength of SLS polymer part, which can help to improve the quality of SLS part.

2 THREE KINDS OF SCAN MODES

2.1 *Changing direction scan mode*

This scan mode makes laser beam directly fill the solid area of the slicing plane of a three-dimensional (3D) model according to the storing orders of intersecting points. Considering the scan efficiency of laser beam, different filling directions are adopted between neighbouring two scan lines. For example (Figure 1(a)), for scan line i and scan line i + 1, if scan line i is filled according to the orders from lines[i][0] to lines[i][1], from lines[i][2] to lines[i][3], and so on, the next scan line is filled from the last point to the forward point, namely from lines[i + 1][j end] to lines[i + 1][j end −1], from lines[i + 1][j end −2] to lines[i + 1][j end −3], and so on. J end is the last intersecting point of the scan line i + 1.

In order to increase the binding power between scan layers and to prevent SLS part from producing anisotropy in the same scan direction. The scan directions between neighbouring two scan layers should have an angle which generally is 90° (Figure 1(b)). During layering, the scan lines in even layers are set as the direction of axis x, while those in odd layers are set as the direction of axis y [2].

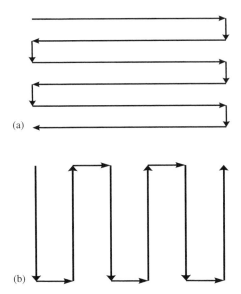

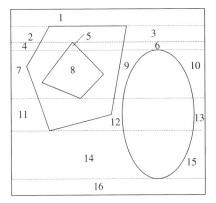

(a)

(b)

Figure 1.   Changing direction scanning mode.

Figure 2.   Subarea and changing direction scan mode.

2.2 *Subarea and changing direction scan mode*

Though the changing direction scan mode is simple and practical, it has such shortcomings that SLS parts are easy to warp and distort and laser scan efficiency is low, and so on. For this reason, a subarea and changing direction scan mode is presented in Reference [2]. Its implementation ideas are as follows:

(1) In order to increase the scan efficiency during SLS and to reduce the warp and distortion of SLS part, the scan plane of a 3D model is divided into some scan subareas without holes shown in Figure 2.
(2) Principle that the scan plane is divided into subareas is determined according to the intersecting points between the scan lines in the horizontal direction and the figure contour of the scan plane. Figure 2 is taken for an example to explain the principle that the subareas form. The scan lines form two intersecting points with the figure contour of the scan plane when they are within area 1. Intersecting points among the scan lines and the figure contour of the scan plane become into four when they are within area 2. In the same way, the intersecting points in area 4 are six, the intersecting points in area 7 are eight, and the intersecting points in area 11 are six. The whole scan area is divided into sixteen subareas according to the conditions of intersecting points.
(3) Scan order in subareas is from top to bottom, and from left to right. For example, the scan order in Figure 2 is:

area 1 → area 2 → area 3 → area 4 → area 5 → area 6 → area 7...

(4) The scan subareas are filled with lines, but there is a α angle between neighbouring two scan layers to reduce shrinkage stress of SLS part in the same direction. Generally, α = 90°.

The scan mode shown in Figure 2 is simple and reliable. It is not only beneficial to form quickly scan paths, but also beneficial to improve greatly the fabrication efficiency of SLS because a connected region is divided into some subareas to avoid the holes and grooves of scan section.

2.3 *Improved subarea and changing direction scan mode*

It is seen from the subarea and changing direction scan mode shown in Figure 2 that area 1, area 2, area 4, area 7, area 11 and so on compose a connected region without inner holes. Area 5 and area 8 compose a whole convex polygon, which is divided into two scan areas by the scan mode in Figure 2. The subareas of this scan mode are too many not to be good enough for powder material to absorb heat power because there is time interval among scanned subareas. In order to increase the scan efficiency of laser scan system, improved scan mode shown in Figure 3 is adopted.

If the scan order shown in Figure 2 is changed into: area 1 → area 2 → area 4 → area 7 → area 11 → area 14 → area 5 → area 8 → area 3 → area 6 → area 9 → area 12 → area 10 → area 13 → area 15, the number of laser scan system spanning cavity can be not only reduced, the number of subareas can be but also reduced. For example, Figure 2 can be united into four subareas: area 1, area 2, area 4, area 7, area 11, area 14 and area 16 can be united into a subarea; area 5 and area 8 can be united into one; and area 3, area 6, area 9 and area 12 can be united into one; and area 10, area 13 and

470

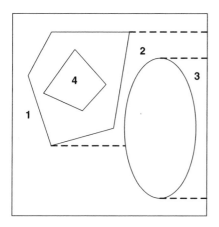

Figure 3.   Improved subarea and changing direction scan mode.

area 15 can be united into one. Improved subarea and changing direction scan mode is shown in Figure 3, which can overcomes such disadvantage that a connected region is divided into some subareas and reduce the melted seams on the surface of SLS part.

3   EXPERIMENTS ON THE EFFECT OF SCAN MODE ON THE STRENGTH OF SLS PART

3.1   *Experimental equipment*

Experimental equipment is HRPS-IIIA type SLS machine (Figure 4) made at Huazhong University of Science and Technology (HUST), Wuhan, P.R.China.

3.2   *Experimental methods*

For different scan paths, single SLS test part can not reflect the effect of different scan modes on the strength of SLS part. For example, the effects of different scan modes, such as the changing direction scan mode, the subarea and changing direction scan mode, as well as the improved subarea and changing direction scan mode, on the strengths of single shock test piece, single flexural test piece and single tensile test piece, have little difference. Different scan modes can not reflect their characteristics until the test piece has more complex structure.

For this reason, in order to simulate the strength of complex SLS parts, standard test pieces, which are sintered according to the arranging modes shown in Figure 5~Figure 7, are selected to measure the strengths of SLS parts under scan modes in the light of the experimental standards of the strength of plastic part.

Figure 4.   HRPS-IIIA machine made at Huazhong University of Science and Technology (HUST).

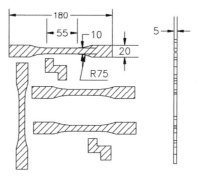

Figure 5.   SLS test pieces for tensile strength.

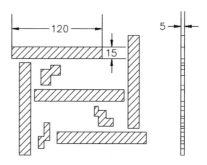

Figure 6.   SLS test pieces for shock strength.

3.3   *Experimental conditions*

Laser power: 14 W
Scan spacing : 0.1 mm
Layer thickness: 0.2 mm
Scan speed: 2000 mm/s
Preheating temperature: 95°C~110°C

SLS powder material and its particle size: polystyrene (PS) powder with the particle size of less than 74 μ.

471

3.4 The test of the strength of SLS part

3.4.1 Test principles
(1) The test principle of tensile strength

$$\sigma_t = P/(B \times D) \qquad (1)$$

Where,
σ_t: tensile strength (Mpa)
P: ultimate tensile load (N)
B: thickness of test pieces (mm)
D: thickness of test piece (mm).

(2) The test principle of flexural strength

$$\sigma_t = 3 \times P \times L/(2 \times B \times D^2) \qquad (2)$$

Where,
σ_t: strength of flexural strength (Mpa)
P: flexural failure load (N)
L: test span (mm)
B: width of test pieces (mm)
D: thickness of test pieces (mm).

(3) The test principle of shock strength

$$\sigma_t = 10^6 * E/(B \times D) \qquad (3)$$

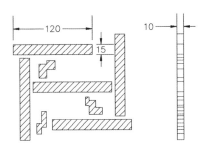

Figure 7. SLS test pieces for flexural strength.

Where,
σ_t: strength of shock (J/ m^2)
P: loss in energy of shock test (J)
B: width of test pieces (mm)
D: thickness of test pieces (mm).

3.4.2 Test conditions and test instruments
(1) The experiments of tensile test and flexural test are carried out in tensile testing machine with the speed of 10.00 mm/min, environment temperature of 20°C and the humidity of 60%. Same experiment is carried out three times with SLS standard test pieces.

(2) The experiments of the shock test is carried out in the testing machine of freely supported beam.

Same experiment is carried out four times with SLS standard test pieces.

3.4.3 Test results
The test results of tensile, flexural and shock experiments are shown in Table 1, Table 2 and Table 3.

Table 1. The test results of tensile experiment.

	Tensile strengths (MPa)			Mean tensile strengths
Scan modes	Test 1	Test 2	Test 3	(MPa)
Changing direction scanning mode	1.881	1.508	1.608	1.67
Subarea and changing direction scan mode	0.921	0.811	0.840	0.857
Improved subarea and changing direction scan mode	3.057	2.862	2.667	2.862

Table 2. The test results of flexural experiment.

	Flexural strengths (MPa)			
Scan modes	Test 1	Test 2	Test 3	Mean flexural strengths (MPa)
Changing direction scanning mode	0.207757	0.231136	0.136659	0.192
Subarea and changing direction scan mode	0.187223	0.155589	0.301715	0.215
Improved subarea and changing direction scan mode	0.453963	0.454774	0.492338	0.467

Table 3. The test results of shock experiment.

	Shock strengths (J/m^2)				
Scan modes	Test 1	Test 2	Test 3	Test 4	Mean shock strengths (J/m^2)
Changing direction scanning mode	733.33	866.67	866.67	800.00	816.67
Subarea and changing direction scan mode	600.00	533.33	533.33	533.33	550.00
Improved subarea and changing direction scan mode	933.33	1000.00	933.33	1000.00	966.67

4 DISCUSSION AND ANALYSIS OF EXPERIMENTAL RESULTS

(1) The strength of SLS part is related with the energy absorbed by SLS powder [20]. Within a range of variation, the more energy SLS powder absorbs, the more is the strength of SLS parts. The energy absorbed by SLS powder depends mainly on two main factors: the energy directly supplied to the SLS powder and the dissipated energy of the SLS powder. The supplied energy depends on laser energy and preheating temperature. The dissipated energy depends on the sintered time of each layer. Tables 1, 2 and 3 show that the strengths of tensile, flexural and shock test pieces of sintered parts with the improved subarea and changing direction scan mode have the highest values, which indicates that the energy absorbed by SLS powder under this kind of scan mode is the highest.

(2) The strength of SLS part is related to the energy density, defined by the following equation [20]:

$$e_d = \frac{P_L}{\pi V d_{sc}} \ln(e + \frac{t_p}{t} - 1) \qquad (4)$$

Where,
P_L: laser power
V: laser scan speed
d_{sc}: scan spacing
t_p: laser heating time (usually, $t_p = 0.001$ s)
t: time interval.

The above mentioned equation shows that the main factors affecting the energy density (ed) are laser power, scan speed, scan spacing, and scan time interval. Under the same scan speed and scan spacing, the effect of different scan modes on the time interval is great. Among the three kinds of scan modes for the experiments, the mean time interval of changing direction scanning mode is the longest, the mean time interval of improved subarea and changing direction scan mode is the shortest. This is the reason why the strengths of SLS parts fabricated with improved subarea and changing direction scan mode are better than those of SLS parts fabricated with the other two scan modes.

(3) The strength of SLS part is related with the melted seam. For some SLS parts with complex structure, the connected traces (called the melted seam) among subareas must happen when the parts are sintered with the subarea and changing direction scan mode.

The strength of melt seam is the weakest in the SLS part. Therefore, the strength of the SLS part depends on the strength of the melt seam for the thinnest part. The melted seam may appear in the thicker part of the SLS part by the improved subarea and changing direction scan mode, so it has less influence on the strength of the part. The melted seam may occur in the parts which long scan traces pass by the changing direction scanning mode, which is the contribution of longer time interval between two adjacent scan routes, and it causes the strength of the SLS part much weaker.

ACKNOWLEDGEMENTS

This work is financially supported by the key project under the 863 programme of China (2002AA6Z3083), the Hubei Natural Science Foundation (2004ABC001), the key R&D programme of Hubei Province of China (2001A107B02).

REFERENCES

[1] WANG Junjie, LI Zhanli & LU Bingheng. Mar 1997. The Study of Scan Path in Rapid Prototyping. MECHANICAL SCIENCE AND TECHNOLOGY in China 16 (2): 303–305.
[2] SHI Yusheng, ZHONG Qing & CHEN Xuebin. Feb 2002. Research and Implementation of a New Kind of Scan Mode for Selective Laser Sintering. CHINESE JOURNAL OF MACHANICAL ENGINEER in China 38 (2): 35–39.
[3] CAI Daosheng, SHI Yusheng, CHEN Gongju & HUANG Shuhuai. Feb 2002. The Study of the optimization of SLS Rapid Prototyping Scan Path. METAL FORMING MACHINERY in China: 18–20.
[4] Onuh, S.O. & Hon, K.K.B. 1998. Application of the Taguchi method and new hatch styles for quality improvement in stereolithography. Proceedings of the Institution of Mechanical Engineers 212: 461–471.
[5] LIU Zhengyu, BIN Hongzan, ZHANG Xiaobo & GAO Hongqing. Aug 1998. The Influence of the Fractal Scan Path on the Temperature Field of the Layer in Material Increase Manufacturing. JOURNAL OF HUAZHONG UNIVERSITY OF SCIENCE AND TECHNOLOGY in China 26(8): 32–34.
[6] ZHANG Renji, SHAN Zhongde, SU Guanghua & YANG Yongnian. Oct 1999. Study on Laser Scan Process of Selective Laser Sintering for Powder Material. APPLIED LASER in China 19(5): 299–302.
[7] CHEN Hong, ZHANG Zhigang & CHENG Jun. Sep 2001. Laser scan path stratagems on SLS rapid prototyping craft. JOURNAL OF BASIC SCIENCE AND ENGINEERING in China 9 (2–3): 202–207.
[8] Yang, Y. Loh, H.T. Fuh, F.Y.H & Wang. Y.G. 2002. Equidistant path generation for improving scan efficiency in layered manufacturing. Rapid Prototyping Journal. 8(1): 30–37.
[9] Kim, D. 1998. Polygon offsetting using a Voronoi diagram and two stacks, Computer Aided Design 30(14): 1069–1075.
[10] CHEN Jianhong, MA Pengju, TIAN Jiemo & LIU Zhenkai. Sep 2003. Scan Path Algorithm for Rapid

Prototyping (RP) Basedon Voronoi Diagrams. MECHANICAL SCIENCE AND TECHNOLOGY in China 22(5): 728–731.

[11] Farouki, R.T. & Sakkalis, T. 1990. Pythagorean hodographs, IBM Journals of Research and Development 34(5): 736–752.

[12] Bertoldi, M. & Yardimci, C.M. 1998. Domain decomposition and surface filling curves in toolpath planning and generation, Solid Freeform Fabrication Symposium Proceedings: 267–274. Austin, TX.

[13] Wasser, T., Dhar, A., Jayal & Pistor, C. 1999. Implementation and evaluation of novel buildstyles in fused deposition modeling (FDM), Solid Freeform Fabrication Symposium Proceedings: 95–102. Austin, TX.

[14] Tiller, W. & Hanson, E.G. 1984. Offsets of two-dimensional profiles: 36–46. IEEE Computer Graphics and Application.

[15] Ganesan, M. & Fadel, G. Hollowing rapid prototyping parts using offsetting techniques, Proceedings of the Fifth International Conference on Rapid Prototyping, University of Dayton, June, 1994, pp. 241–251.

[16] Pham, B. 1992. Offset curves and surfaces: a brief survey, Computer Aided Design 24(4): 223–229.

[17] Takashi, M. 1999. An overview of offset curves and surfaces, Computer Aided Design 31(4): 165–173.

[18] Takashi, M. Jun. 1999. An overview of offset curves and surfaces, Computer Aided Design: 130–132.

[19] BAI Junsheng, ZHAO Jianfeng, TANG Yaxin & YU Chengye. 1997. Study on the strength of SLS rapid prototyping test-pieces, AVIATION PRECISION MANUFACTURING TECHNOLOGY in China 33(5): pp 29–31.

[20] Elizabeth Goode. Jan 2003. Selective laser sintering system & materials. Advanced Materials & Processes. Metals Park 161(1): 66–67.

Rapid prototyping by reactive extrusion: Mixing process

A. Rocha, S. Marques, A.J. Mateus & P.J. Bártolo
Institute for Polymers and Composites, Polytechnic Institute of Leiria, Portugal

ABSTRACT: Rapid prototyping by reactive extrusion corresponds to a novel concept of rapid prototyping that produces thermosetting parts through a low-pressure extrusion process at room temperature of a mixture of polyol and isocyanate. One of this process key aspect is an appropriate mixture of these chemical compounds, performed at appropriate chambers (mixing-heads) with high pressure conditions, to create high turbulence. This paper presents a computer simulation study, developed to select the most suited configuration of the mixing head. Its position and orientation of the entrance channels is crucial to create appropriate mixing conditions that will enable to improve the curing process, so producing more homogeneous parts with better properties.

1 INTRODUCTION

In 1989 the extrusion technique, known as Fused Deposition Modelling (FDM), was developed by Crump. By this process, thin thermoplastic filaments are melted by heating and guided by a robotic device (extruder) controlled by a computer, to form the three-dimensional objects as shown in Figure 1. The material leaves the extruder in a liquid form and hardens immediately. The previously formed layer, which is the substrate for the next layer, must be maintained at a temperature just below the solidification point of the thermoplastic material, to assure good interlayer adhesion (Pham and Gault 1998). This process only works with thermoplastic materials, which properties are strongly dependent on processing operation conditions, such as extrusion temperature, environmental temperature and solidification time. As the process requires the melting of solid material it is a disadvantageous one from an energetic point of view.

One of the major problems associated with models produced by FDM is the "curl" effect. "Curl" corresponds to the curvilinear geometric distortion induced during the cooling phase (Crump *et al.* 1994). Such distortion is due to changes in density of the material as it is transformed from a melting state to a solid state. To reduce the impact of curl distortions different processes have been proposed. One involves the heating of the building chamber, thus reducing possible temperature differences. Another one involves extruding the build material at the lowest possible temperature.

An alternative method of liquefying the material is using a metal inert gas (MIG) welding process. Experimental machines have also been developed using multiple axis robots to build steel or aluminium

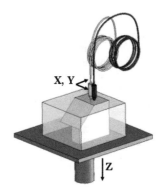

Figure 1. The FDM process.

models into a process called shape welding or 3D welding (Zhang *et al.* 2003 and 2002).

The main advantages and disadvantages of the extrusion process are indicated in Table 1.

In order to solve some of the main disadvantages of this FDM process, a novel strategy called Rapid Prototyping by Reactive Extrusion (*Rapid PRE*) has been developed. This process is mainly based on the mixing reaction of polyurethane (PU) materials. Other thermosetting materials can also be used, like epoxy and unsaturated polyester resins curing at room temperature or reinforced thermosetting material.

Polyurethanes (PU) are step-addition polymers formed by the reaction of di- or poly-isocyanates with diols or polyols. They represent a unique class of polymers whose properties can be easily tailored by the variation of its components.

Table 1. Main advantages and disadvantages of the FDM process.

Advantages
Easy to change materials
Variety of materials available
No exposure to toxic materials or lasers
Good mechanical strength

Disadvantages
Supports are required
The process is slow on models with a large mass
The models have poor strength in the vertical direction
Temperature fluctuations during production could lead to de-lamination
Limited accuracy due to the diameter of the filament used
The need of filament material
Narrow processing window

The fabrication of models through *Rapid PRE* involves the following steps (Figure 2):

i)   CAD model generation
ii) Model decomposition into sub-components or sub-parts using appropriate algorithms
iii) Evaluation of the extruders' paths for both deposition of soluble wax material (support structures or soluble mould) and casting polyurethane material
iv) Deposition of wax material to create walls for casting the thermosetting material
v)   High pressure mixture of polyol and isocyanate (impingement stage) and low pressure extrusion of the obtained low viscosity mixture (casting stage)
vi) High speed curing of the polymeric system at room temperature
vii) "Demoulding" the part through water jet.

This process enables the production of high quality parts in which materials show low shrinkage and the fabrication of the part is not made through the conventional layer-by-layer approach. Additionally, the combination of different types of polyols and isocyanates, without changing processing conditions, enables to produce parts with a wider range of properties. It can also produce reinforced polyurethane parts or polyurethane foams. Other important advantage of this technology is the cost associated with the fabrication stage, materials and machine.

Rapid prototyping by reactive extrusion involves the mixture of low viscosity compounds, which are extruded at low pressure and cured at room temperature.

In order to produce better products it is necessary to ensure the appropriate mixing, which strongly depends on the mixing head configuration (Trautmann and Piesche 2001). Thus, a simulation study was developed

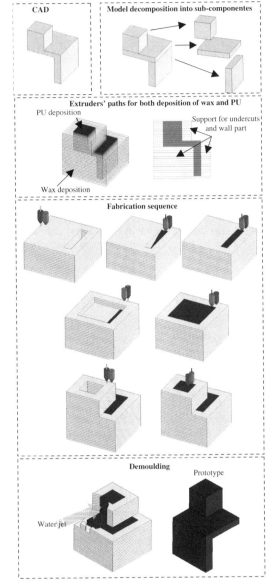

Figure 2. *Rapid PRE* system – main steps.

to determine the appropriate configuration of the mixing head used in this *Rapid PRE* system.

2 MIXING HEADS

The mixing head is one of the most important components of the *Rapid PRE* system. Curing is activated by contact between reactants (polyol and isocyanate) in the impingement mixing stage, as previously mentioned. This operation is performed by request, *i.e*, it is not a

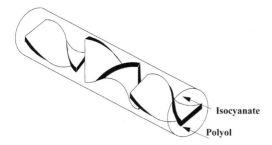

Figure 3.   Static mixer for low-pressure systems.

Recirculation   **Shot position**

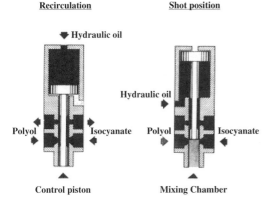

Figure 4.   Schematic representation of the functional principle of impingement mixers (Courtesy of Krauss-Maffei).

continuous operation. It is necessary a mechanism controlled by computer to both start and stop the mixing and keep the mixing chamber clean.

According to the type of metering pump the mixing heads are classified into: (1) low pressure systems, and (2) high pressure systems.

Low-pressure systems use a static mixer (Figure 3). Valves on each stream start and stop the flow while a solvent or air is used to flush out the reactants before they can gel. The cleaning cycles, which are necessary at regular intervals, are disadvantageous.

High-pressure systems do not require these cleaning steps. On high-pressure systems the chemical compounds are mixed by the kinetic energy of the fluid jets. Figure 4 shows the functional principle of an impingement mixer.

High-pressure mixing heads are usually characterized by (Schulte and Ersfeld 1981, Makosco and McIntyre 1984):

– recirculation of reactants near the mixing chamber to maintain both temperature and dispersion and permit the rapid start up of mixing
– rapid opening valves to control stoichiometry right from the start up of mixing

– inlet nozzles, which accelerate reactant jets to high velocity
– a chamber where streams impinge and initial mixing occurs.

3   COMPUTER SIMULATIONS OF THE MIXING PROCESS

3.1   Mathematical model

To study the mixing process there is a need to consider mass equations, as well momentum and energy conservation equations, which are written as follows:

• The continuity equation:

$$\frac{\partial \rho}{\partial t} + \nabla \bullet (\rho U) = 0 \tag{1}$$

• The momentum equations:

$$\frac{\partial \rho U}{\partial t} + \nabla \bullet (\rho U \otimes U) = \nabla \bullet \left(- p\delta + \mu \left(\nabla U + (\nabla U)^T\right)\right) + S_M \tag{2}$$

• The energy equation:

$$\frac{\partial \rho h_{tot}}{\partial t} - \frac{\partial p}{\partial t} + \nabla \bullet (\rho U h_{tot}) = \nabla \bullet (\lambda \nabla T)$$

$$+ \nabla \bullet \left(\mu \nabla U + \nabla U^T - \frac{2}{3} \nabla \bullet U \delta U\right) + S_E \tag{3}$$

where ρ is the density, U is the velocity, p is the pressure, μ the dynamic viscosity, T is the temperature, S_M is the momentum source and S_E is the energy source. The specific total enthalpy, h_{tot}, is expressed in terms of the specific enthalpy, h, by the following equation:

$$h_{tot} = h + \frac{1}{2} U^2 \tag{4}$$

3.2   Viscosity turbulence model

Turbulence consists of small eddies continuously forming and dissipating, in which Reynolds stresses are assumed to be proportional to mean velocity gradients. This defines the so-called 'eddy viscosity model' (Craft et al. 1993, Behnia et al. 1999, Chen et al. 2004).

This model assumes that Reynolds' stresses can be related to the mean velocity gradients and eddy (turbulent) viscosity by the gradient diffusion, in a manner analogous to the relationship between the stress and strain tensors in a laminar ideal or Newtonian flow:

$$- \rho \overline{u \otimes u} = -\frac{2}{3} \rho k \delta - \frac{2}{3} \mu_t \nabla \bullet U \delta + \mu \left(\nabla U + (\nabla U)^T\right) \tag{5}$$

where, μ_t is the eddy viscosity or turbulent viscosity, κ is the turbulence kinetic energy, δ is the kronecker matrix.

According to this model, the Reynolds fluxes of a scalar are linearly related to the mean scalar gradient (diffusivity hypothesis):

$$- \rho \overline{u \phi} = \Gamma_t \nabla \phi \qquad (6)$$

with ϕ being a general scalar variable and Γ_t the eddy diffusivity, given by the following equation:

$$\Gamma_t = \frac{\mu_t}{Pr_t} \qquad (7)$$

where Pr_t is the turbulent Prandtl number.

The above equations only express the turbulent fluctuation terms of functions of the mean variables when the turbulent viscosity, μ_t, is known.

According to the abovementioned hypotheses, the Reynolds average momentum and scalar transport equations become:

$$\frac{\partial \rho U}{\partial t} + \nabla \bullet (\rho U \otimes U) = B - \nabla p' + \nabla \bullet \left(\mu_{eff} \left(\nabla U + (\nabla U)^T \right) \right) \qquad (8)$$

$$\frac{\partial \rho \phi}{\partial t} + \nabla \bullet \left(\rho U \phi - \Gamma_{eff} \nabla \phi \right) = S \qquad (9)$$

where B is the sum of the body forces, μ_{eff} is the effective viscosity, and Γ_{eff} is the effective diffusivity, defined respectively by,

$$\mu_{eff} = \mu + \mu_t \qquad (10)$$

and

$$\Gamma_{eff} = \Gamma + \Gamma_t \qquad (11)$$

p' is a modified pressure, defined by:

$$p' = p + \frac{2}{3} \rho k + \nabla \bullet U \left(\frac{2}{3} \mu_{eff} - \zeta \right) \qquad (12)$$

where ζ is the bulk viscosity.

The Reynolds average energy equation becomes:

$$\frac{\partial (\rho h_{tot})}{\partial t} - \frac{\partial P}{\partial t} + \nabla \bullet (\rho U h_{tot}) = \nabla \bullet \left(\lambda \nabla T + \frac{\mu_t}{Pr_t} \nabla h \right) + S_E \qquad (13)$$

- *The $k - \varepsilon$ model*

The $k - \varepsilon$ model uses the gradient diffusion hypothesis to relate the Reynolds stresses to the mean velocity gradients and the turbulent viscosity. The turbulent viscosity is modelled as the product of a turbulent velocity and turbulent length scale.

In this model, the turbulence kinetic energy is defined as the variance of the fluctuations in velocity, while ε is the turbulence eddy dissipation (the rate at which the velocity fluctuations dissipate). Therefore, the $k - \varepsilon$ model introduces two new variables into the system of equations. The continuity equation is then:

$$\frac{\partial \rho}{\partial t} + \nabla \bullet (\rho U) = 0 \qquad (14)$$

and the momentum equation becomes:

$$\frac{\partial \rho U}{\partial t} + \nabla \bullet (\rho U \otimes U) - \nabla \bullet (\mu_{eff} \nabla U) = \nabla p' + \nabla \bullet (\mu_{eff} \nabla U)^T + B \qquad (15)$$

The $k - \varepsilon$ model assumes that the turbulence viscosity is linked to the turbulence kinetic energy and dissipation through the following equation:

$$\mu_t = C_\mu \rho \frac{k^2}{\varepsilon} \qquad (16)$$

where C_μ is a constant.

The values of k and ε come directly from the differential transport equations of the turbulence kinetic energy and turbulence dissipation rate:

$$\frac{\partial (\rho k)}{\partial t} + \nabla \bullet (\rho U k) = \nabla \bullet \left[\left(\mu + \frac{\mu_t}{\sigma_k} \right) \nabla k \right] + P_K - \rho \varepsilon \qquad (17)$$

$$\frac{\partial (\rho \varepsilon)}{\partial t} + \nabla \bullet (\rho U \varepsilon) = \nabla \bullet \left[\left(\mu + \frac{\mu_t}{\sigma_\varepsilon} \right) \nabla \varepsilon \right] + \frac{\varepsilon}{k} (C_{\varepsilon 1} P_k - C_{\varepsilon 2} \rho \varepsilon) \qquad (18)$$

where $C_{\varepsilon 1}$, $C_{\varepsilon 2}$, σ_k and σ_e are constants. P_k is the turbulence generation, due to viscous and buoyancy forces, represented by:

$$P_k = \mu_t \nabla U \bullet (\nabla U + \nabla U^T) - \frac{2}{3} \nabla \bullet U (3 \mu_t \nabla \bullet U + \rho k) + P_{kb} \qquad (19)$$

3.3. Computer implementation

Distinct approaches were devised to obtain numerical approximation of viscous fluxes. They can be

478

classified into three main categories: (1) finite difference, (2) finite element, and (3) finite volume, all methods relying on some kind of mesh in order to discretise the governing equations.

The mathematical models previously described were numerically implemented using the finite element method (FEM) with tetrahedral elements. This method was originally developed for structural analysis, though on the 1990s it gained popularity in flow problems, particularly non-Newtonian fluids. The use of FEM to study flow problems is described by Pironneau (1989), Hassan et al. (1993) and Reddy and Gartling (1994).

4 FINDINGS

A commercial polyol plus isocyanate were used in order to select the most suitable geometry of the mixing head for the *Rapid PRE* machine. The main characteristics of these materials for flow simulation are indicated in Table 2.

Four different mixing heads configurations were tested (Figure 5).

Figures 6 to 8 shows the variation of mass fraction of the mixture, which must be near 1, the velocity of the material flow and the turbulence inside the mixing head. The configuration A was selected for the *Rapid PRE* mixing head from these findings. The mixing behaviour enabled by this configuration allows a good mixture ($\approx$94%) without any reaction. This can be observed in Figure 9, which shows the variation of temperature. Temperature values are closed to room temperature, so no curing occurred, as the curing reaction is highly exothermic. Highest turbulence inside the chamber is also produced with configuration A.

Figure 10 shows the assembling of the mixing head designed for the Rapid PRE equipment.

Table 2. Flow characteristics of the reactants.

Polyol	
Density [kg/m³]	1110
Dynamic Viscosity [Pa.s]	1.35
Flow rate [Kg/s]	0.2035
Specific heat [J/g.K]	1.84
Thermal conductivity [W/m.K]	0.0017
Inlet temperature [°C]	29

Isocyanate	
Density [kg/m³]	1200
Dynamic Viscosity [Pa.s]	1.2
Flow rate [Kg/s]	0.2
Specific heat [J/g.K]	1.84
Thermal conductivity [W/m.K]	0.0017
Inlet temperature [°C]	29
Mixing ratio of reactants	1:1

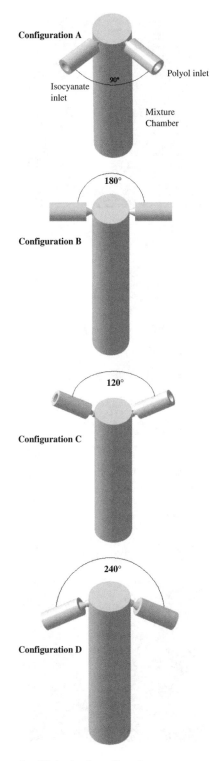

Figure 5. Mixing heads configurations.

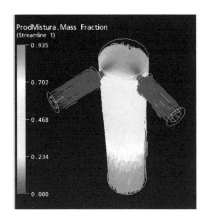

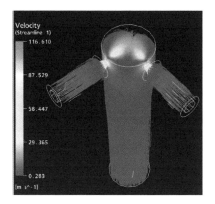

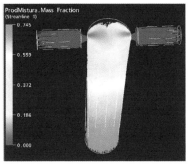

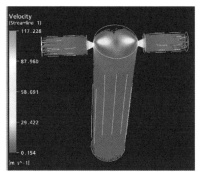

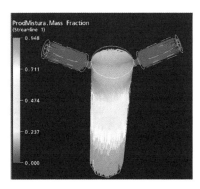

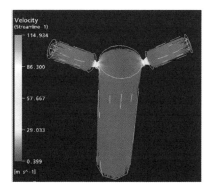

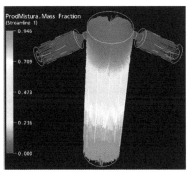

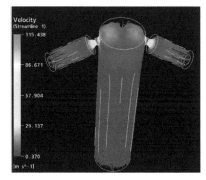

Figure 6.   Variation of the mass fraction of the mixture.

Figure 7.   Velocity of the material flow.

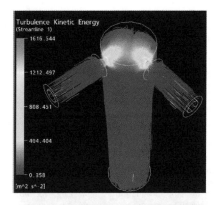

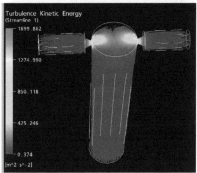

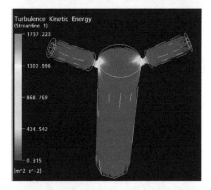

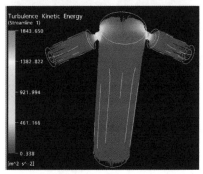

Figure 8.   Turbulence within the mixing head.

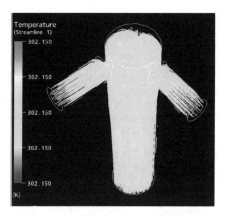

Figure 9.   Temperature profile for configuration A.

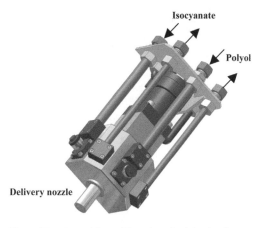

Figure 10.   Assembling of the selected mixing head.

5   CONCLUSIONS

The mixing behaviour of an impingement mixer is determined by flow conditions within the mixing chamber. The non-steady vortex motions induced by the jets are responsible for the mixing of the low viscosity reactants.

Despite the widespread use of multiple impinging fluid jets for cooling, heating and drying in various areas of engineering, there are still reliable and sufficiently general correlations that can serve for optimal design of jet configurations for specific applications.

A proper configuration for the mixing head was determined based on computer simulations of the flow behaviour of a polyol plus a isocyanate. Depending on the jets configuration, complex flow patterns may develop enabling proper mixing. Appropriate mixing conditions will enable to improve the curing process, producing more homogeneous parts with better mechanical properties.

The mixing head is a key component of the Rapid PRE system, representing a novel concept of a rapid prototyping process.

REFERENCES

Behnia, M., Parneix, S., Shabany, Y. and Durbi, P. 1999. Numerical study of turbulent heat transfer in confined and unconfined impinging jets. *Int. J. Heat Fluid Flow* 20: 1–9.

Chen, W.L., Guo, Z. and Chen, C.K. 2004. A numerical study on the flow over a novel tube for heat-transfer enhancement with linear Eddy-viscosity model. *Int. J. of Heat and Mass Transfer* 47: 3431–3439.

Craft, T., Graham, L. and Launder, B. 1993. Impinging jet studies for turbulence model assessment – II. An examination of the performance of four turbulence models. *Int. J. Heat Mass Transfer* 36: 2685–2697.

Crump, S.S. 1989. US Pat. 5121329.

Crump, S.S., Comb, J.W., Priedeman, W.R. and Zinniel, R.L. 1994. US Pat. 5503785.

Hassan, O., Probert, E.J., Morgan, K. and Peraire, J. 1993. *Adaptive finite element methods for transient compressible fluid problems.* In: Brebbia, C.A. and Aliabadi, M.H. (eds) Adaptive finite and boundary element methods, Elsevier, London.

Makosco, C. and McIntyre, D. 1984. US Pat. 4473531.

Pham, D.T. and Gault, R.S. 1998. A comparison of rapid prototyping technologies. *Int. J. of Machine Tools and Manufacturing* 38: 1257–1287.

Pironneau, O. 1989. *Finite element methods for fluids,* Chichester, John Wiley.

Reddy, J.N. and Gartling, D.K. 1994. *The finite element method in heat transfer and fluid dynamics.* CRC Press, Boca Raton.

Schulte, K. and Ersfeld, H. 1981. US Pat. 4291991.

Trautmann, P., Piesche, M. 2001. Experimental investigations on mixing behaviour of impingement mixers for polyurethane production. *Chem. Eng. Technolo.,* 24: 1193–1197.

Zhang, Y.M., Chen, Y., Li, P. and Ad Male, A.T. 2003. Weld deposition-based rapid prototyping: a preliminary study. *Journal of Materials Processing Technology* 135: 347–357.

Zhang, Y.M., Li, P., Chen, Y. and Male, A.T. 2002. Automated system for welding-based rapid prototyping. *Mechatronics* 12: 37–53.

Virtual modeling and rapid manufacturing – Bártolo (eds)
© 2005 Taylor & Francis Group, London, ISBN 0 415 39062 1

Micro-fabrication: The state-of-the-art

J. Vasco, N. André & P. Bártolo
Polytechnic Institute of Leiria, Portugal

ABSTRACT: In the last two decades micro applications have been increasing due to the huge evolution of micro-fabrication techniques. The increasing knowledge obtained from the application of these techniques and the miniaturisation required provided new solutions for the manufacture of micro components and/or micro devices, suitable for distinct applications like biomedical, microelectronics, optical, communication and automotive. An overview of the state-of-the-art of micro-fabrication techniques is carried out emphasising its main advantages and disadvantages.

1 INTRODUCTION

1.1 *Micro components*

The common definition of micro components or microstructures refers to those structures presenting sub-millimetric dimensions and tolerances within a micron range domain (Evans *et al.* 2001). An overview of the state-of-the-art of micro-fabrication processes developed to produce either micro-components or micro tools for micro replication processes. This work is part of a major research programme on micro and nanofabrication in which the Polytechnic Institute of Leiria is a partner. It aims at the establishment of a National Centre for Excellence on Microfabrication, also involving the University of Minho, the University of Coimbra, the Technical University of Lisbon, the Portuguese Centre for Computer Graphics, the Technological Centre for the Mouldmaking Industry and the Technological Centre for the Glass and Ceramic Industry.

1.2 *Micro-fabrication processes*

The processes used to produce micro components differ from each other in terms of its capabilities, like dimension of obtained components, achievable aspect ratio (height/width), use and type of resist, etc.

Micro-fabrication processes, according to the adopted fabrication strategy, can be classified into: (1) material removal processes, and (2) material addition processes (Figure 1). Material removal processes comprises both lithographic and non-lithographic processes like laser milling, micro milling, micro turning, micro EDM among others. Material addition processes are based on the photon-polymerisation of

liquid resins, which may occur through either a single-photon or a two-photon excitation process.

Micro-fabrication techniques can also be classified according to their ability to produce tools for micro replication processes (Figure 2).

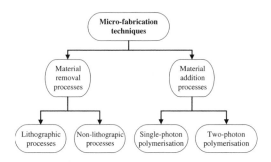

Figure 1.   Classification of micro-fabrication techniques according to the adopted fabrication strategy.

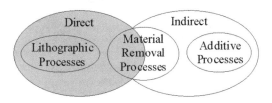

Figure 2.   Classification of micro-fabrication processes according to the ability to produce components suitable for micro replication.

2 LITHOGRAPHIC PROCESSES

The most successful of these micro-fabrication processes is the so-called LIGA (Lithograph Galvanoformung und Abformung) process developed by the Research Centre FZK at Karlsruhe, Germany nearly 20 years ago. LIGA is a German abbreviation meaning lithography, galvanoforming and plastic moulding, the main phases of the process. The LIGA process has been used to produce microsensors, micro-optical devices, and electrical and optical micro-connectors.

2.1 X-ray LIGA

This is the original LIGA process and the only lithographic technique currently available, which can produce microstructures with aspect ratio above 100:1. The X-ray LIGA uses X-rays generated by a synchrotron and masks made by exotic and expensive materials. As a result of their high energy, X-rays are capable of deeply penetrating thick resists (typically over 200 μm). The short wavelengths of these rays allow low diffraction enabling high resolution (Hirata 2003). However, from a practical point of view, synchrotron X-rays pose limitations mainly due to the size, cost and availability constraints (Kupka et al. 2000).

The basic X-ray LIGA process (Malek and Saile 2004) is described in Figure 3. Firstly, an X-ray photoresist, usually polymethylmethacrylate (PMMA) is coated onto a conductive substrate. A pattern from a mask is then transferred into the resist using X-rays from a synchrotron radiation source. After the irradiation process, selective dissolution of the irradiated parts of the resist in a chemical developer results in a relief replica of the mask pattern. Different fabrication routes can subsequently be chosen.

Specialised equipments are commercially available from companies like Technotrans (www.technotrans.de), ANKA (hikwww1.fzk.de/anka) and Dover Industrial Chrome (www.dovertvj.com).

Synchrotron direct radiation (Figure 4) is a maskless variation of X-ray LIGA that enables the production of real 3D structures, with aspect ratios up to 50:1 and high etching rates (Katoh and Zhang 1998).

2.2 UV LIGA

UV LIGA is a technology which allows the fabrication of electroformed micro-components using UV irradiation onto a photo-resist, and the subsequent electroplating to form the metal-plated components. The process shows lower resolution than X-ray LIGA and lower aspect ratio, up to 20:1(Hsieh et al. 2002). However, it is a highly cost effective process (Bischofberger et al. 1997).

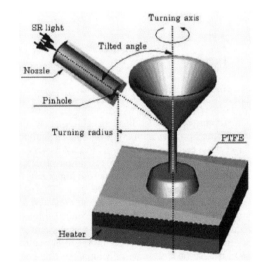

Figure 4. Synchrotron direct radiation of PTFE (Nishi et al. 2002).

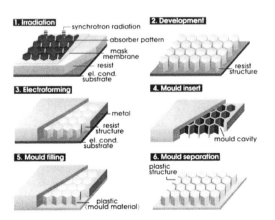

Figure 3. Basic LIGA process sequence (Hornes et al. 2003).

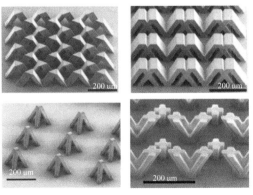

Figure 5. SEM photographs of microstructures produced with several inclined UV exposures (Han et al. 2004).

The introduction of the SU-8 resin, patented by IBM in 1989 and commercialised, among others, by Shell Chemicals under the name EPON-SU-8, has created a breakthrough in the fabrication of low-cost, large structural height micro-systems. This is an epoxy-based resin containing glycidyl ether derivative of bisphenol-A novolac, photo-initiator like triaryl sulphonium and an organic solvent. The percentage in weight of resin determines the viscosity, so the maximum thickness can be achieved.

Recently, Han *et al.* (2004) proposed a novel process to produce 3D microstructures with inclined/rotated UV lithography using negative thick photoresist, SU-8. In the exposure process, a photo-mask and a SU-8 coated substrate are fixed together, and either tilted or tilted and rotated to a UV source. Several microstructures were fabricated through this process, such as oblique cylinders, embedded channels, bridges, V-grooves and truncated cones (Figure 5).

2.3 *Ion beam LIGA*

Ion beam LIGA (IB-LIGA) is a technique that uses high-energy (0.5–3.5 MeV) light ions (protons or helium ions) for the irradiation of a photo-resist material (Munnik *et al.* 2003, Schrempel and Witthuhn 1998). Masks can be used to obtain 2D patterns, though the ion beam can also be used as a mask-less tool. IB LIGA enables to produce components until up to 40:1 of aspect ratio. This technique is a very promising one, although improvements and further developments are still needed (Lindeberg *et al.* 2000, Lawes *et al.* 1996).

3 NON-LITHOGRAPHIC

3.1 *Laser milling*

Laser milling (Figure 6) is a new technology suitable for machining a wide range of materials like metals, glass, ceramics and plastics (Pham *et al.* 2004). This technique involves the heating of a material and subsequent melt and material removal. It is a temperature dependent process, so problems like micro-cracks can commonly appear at heat affect zones (HAZ).

Whenever metals are used, the laser beam heats, melts and vaporizes the metal (metal sublimation), while in the case of polymers the process is based on the rupture of molecular chains (laser ablation). In spite of polymers exhibit low thermal conductivity the problem of heat affected zones are still present.

Typically a laser milling system consists of four main components: a laser source, a system for positioning the workpiece, a system for checking the work surface, a computer to control and synchronise the laser beam and the workpiece positioning (Pham *et al.* 2004).

There is a wide range of laser sources available for laser milling. The two most common lasers are the CO_2

one in the far infrared (IR) region and the Nd:YAG in the near IR region. The laser source used allows to determine the wavelength, spot size, average beam intensity and pulse duration. The process of laser milling is only possible if the material can absorb the laser radiation.

In this process, the laser beam scans the surface to be processed performing a hatch closely spaced using trajectories similar to CNC strategies. A collaborative research project between the Institute Polytechnic of Leiria and Hasco Portugal is currently analysing the effect of operating parameters such as laser power, wavelength, scanning strategy, etc., on the surface finishing. The evaluation of the heat affected zones appearing on higher material removal rates is also under investigation. Research findings are being used to produce moulds for micro-injection (Figure 7).

Laser milling allows the production of components with aspect ratios until up to 10:1 (Kim *et al.* 2004).

3.2 *Micro milling*

Micro milling (Figure 8) uses a physical tool to remove material. Like in conventional milling, aspects related to cutting conditions, vibrations and part fixation are important issues. Tools are very small, with a diameter between 20 and 100 μm, and extremely sharp. The use of a physical tool poses some problems regarding the geometries to be obtained. Aspect ratios of approximately 10:1 can be achievable (Dimov *et al.* 2004).

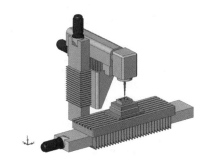

Figure 6.   Typical setup for a laser milling equipment with three motion axes. Rotational axes can also be added to improve its capacities.

Figure 7.   Mould for micro-injection produced by Laser Milling.

Figure 8. Photo of a micro Milling industrial equipment (Courtesy of Kern Micro und Feinwerktechnik GmbH & Co.).

Figure 9. Photo of a micro EDM industrial equipment (Courtesy of AGIE Ltd.).

3.3 *Micro EDM*

Micro-electro-discharge machining (micro EDM or μ-EDM) is based on material removal by electric discharges, through an electrode onto the part that cause erosion on its surface, creating high-aspect ratio micro-components. It is a similar process to conventional EDM, whose main difference is the size of the tool, the power supply of discharge energy and the resolution of the x-y-z movements (Wong *et al*. 2003). Similarly to what happens in conventional EDM, the gap depends on both the voltage/current ratio used and the type of removal operation, e.g., rough or finishing. In micro EDM, gaps are extremely small (under 3 μm) which may cause problems on chip removal because the circulation of dielectric under these circumstances is difficult (Benavides *et al*. 2002).

The main advantages of micro EDM are its low set-up cost, high accuracy and large design freedom. It is a non-contact process, so there is no machining force to produce deformation on either the electrode or the workpiece. However, apart from environmental problems associated to the use of dielectric materials, there are many parameters that affect the process, like voltage, current, discharge energy, pulse time, pulse duration and gap (Lim *et al*. 2003).

The efficiency of a micro EDM process can be improved with the use of ultrasound vibrations (Murali and Yeo 2004), which optimises the flow of the dielectric and the quality of the eroded surface.

A complementary technique to micro EDM is the micro wire-EDM (micro WEDM), which uses a wire continuously circulating. This process eliminates the wear on the electrode, and facilitates eroded particles evacuation.

Figures 9 and 10 show some commercial available micro EDM and micro WEDM machines, while some micro-components are shown in Figure 11.

Figure 10. Photo of a micro WEDM industrial equipment (Courtesy of AGIE Ltd.).

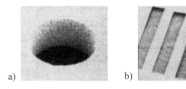

a)   b)

Figure 11. Micro EDM applications: a) 140 μm Platinum hole, photo magnified 500X, b) 200 μm Air Slit in Stainless Steel, photo magnified 70X (Courtesy of Optimation, Inc.).

3.4 *Micro turning*

According to Rahman *et al*. (2005) and Lu and Yoneyama (1999), micro turning is a conventional material removal process that has been miniaturised (Figure 12). Thus, it is possible to perform almost the same operations like turning, facing, grooving or threading (Lu and Yoneyama 1999).

Figure 12. Micro turning equipment and examples of micro turned components (Courtesy of Mikrotool Pte Ltd.).

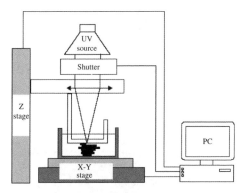

Figure 13. Constraint surface technique: microstereolithography through a glass window (Bertsch *et al.* 1997).

High aspect ratios can be achieved. In complex operations, like grooving or threading, where the cutting conditions are critical, aspect ratios can vary from 5 to 10:1. However, when turning a shaft, it may be possible to obtain components with much higher aspect ratio (Sharon *et al.* 2003).

The major limitation of the micro turning process is the machining force, which influences machining accuracy and the limit of machinable size. Consequently, control of the reacting force all along cutting is one of the important factors to improve machining accuracy. The cutting force must be lower than the force that causes plastic deformation of the workpiece as shown by Lu and Yoneyama (1999).

4 ADDITIVE PROCESSES

4.1 Single-photon polymerisation: microstereolitography

Conventional stereolithography builds shapes using ultraviolet light to selectively solidify photo-sensitive resins. The system, firstly proposed by Hull (1984), consists of a computer, a vat containing a photosensitive polymer, a moveable platform on which the model is built, a laser to irradiate and cure the polymer, and a dynamic mirror system to direct the laser beam. The computer uses the sliced model information to control the mirrors, which direct the laser beam over the polymer surface, "writing" the cross-section of one slice of the model through the polymerisation of a set of elementary volumes called *voxels*. After drawing a layer, the platform dips into the polymer vat, leaving a thin film from which the next layer will be formed. The next layer is drawn after a waiting period to re-coat the surface of the previous layer.

The basic principle of all microstereolithography is very similar to the stereolithography technique. However, to get a better resolution, the beam is focused

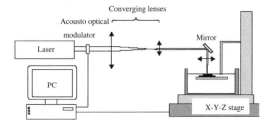

Figure 14. Free surface technique.

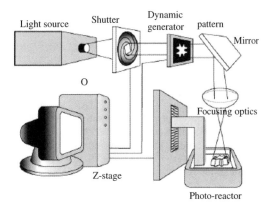

Figure 15. Integral microstereolithography.

more precisely in order to reduce the spot size to a few micrometers of diameter. Several strategies have been proposed (Bertsch *et al.* 2003): constrained surface techniques (Figure 13), free surface techniques (Figure 14), and integral processes (Figure 15).

Integral microstereolithography represents the most recent advancement in this field, enabling the solidification of each layer in one irradiation step by projecting the corresponding image onto the surface

Figure 16. a) SEM image of a micro gear (Maruo and Ikuta Chung 2003), and b) SEM image of a micro-wheel (Farsari 2000).

of the photo-polymerisable resin through either a liquid crystal display or a digital micro mirror device.

Figure 16 shows some examples of micro-parts produced by microstereolithography.

4.2 *Two-photon polymerisation*

Two-photon-initiated polymerisation curing processes represents a useful stereolithographic strategy to produce micro/nanoscale structures by using femtosecond infrared laser without photo-masks (Lemercier *et al.* 2005, Tormen *et al.* 2004, Kawata and Sun 2003). The molecule simultaneously absorbs two photons instead of one, being excited to an higher singlet state. The use of two-photon-initiated polymerisation allows a submicron 3D resolution, on top of enabling both 3D fabrication at greater depth and an ultra-fast fabrication. It is a true 3D fabrication process.

One of the most important characteristics of two-photon excitation is the proportionality of the square of the incident light intensity, whereas in single-photon absorption it is directly proportional to the light intensity (Nguyen *et al.* 2005). Hence, photo-polymerisation can be confined to volumes with dimensions in the same range of the light wavelength.

Figure 17 represents activation and deactivation pathways of initiator molecules in two-photon polymerisation reactions, while Figure 18 shows some examples of micro-structures produced by two-photon polymerisation.

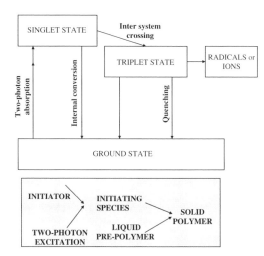

Figure 17. Various deactivation pathways of a two-photon excited initiator molecule.

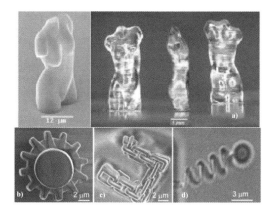

Figure 18. Images of a) a micro-scale Venus (Stute *et al.* 2003), b) a micro-gearwheel (Kawata and Sun 2003), c) a micro-chain (Kawata and Sun 2003), and d) a micro-spring (Kawata and Sun 2003), produced by two-photon polymerization.

This process was recently applied to inorganic-organic hybrid materials.

5 COMPARISON AND CONCLUSIONS

A wide range of research work on micro-fabrication technologies has been summarised. These technologies have made a significant impact in many areas of engineering, electronics, biology and medicine.

Micro-fabrication can present several types of advantages like a higher surface-area-to-volume ratio, a smaller size, a better geometrical control, facilitating batch processing, high throughput and integration.

Table 1. Comparison of micro-fabrication technologies.

Process	Aspect ratio	Geometry	Time	Availability
X-Ray LIGA	>100:1	2½D	Fast	Limited
UV LIGA	20:1	2½D	Slow	Good
IB LIGA	40:1	2½D	Medium	Medium
Laser Milling	10:1	3D	Fast	Good
Micro Milling	10:1	3D	Medium	Good
Micro EDM	>10:1	3D	Medium	Good
Micro WEDM	100:1	2½D	Medium	Good
Micro Turning	>10:1	2D	Medium	Good
Micro stereo-olithography	–	3D	Medium	Limited
Two-Photon polymerisation	–	3D	Medium	Limited

Table 1 indicates some of the most relevant characteristics of the micro-fabrication processes described in this paper.

REFERENCES

Benavides, G.L., Bieg, L.F., Saavedra, M.P. and Bryce, E.A. 2002. High Aspect Ratio Meso-Scale Parts enabled by Wire Micro-EDM. *Microsystem Technologies* 8: 395–401.

Bertsch, A., Jiguet, S., Bernhards, P. and Renaud, P. 2003. Microstereolithography: a review. *Mat. Res. Soc. Symp. Proc.* 758: LL.1.1.1–13.

Bertsch, A., Zissi, S., Jézéquel, J.Y., Corbel, S. and André, J.C. 1997. Microstereolithography using a liquid crystal display as dynamic mask-generator. *Microsystem Technologies*, 42–47.

Bischofberger, R., Zimmermann, H. and Staufert, G. 1997. Low-Cost HARMS Process. *Sensors and Actuators* A61: 392–399.

Dimov, S., Pham, D.T., Ivanov, A., Popov, K. and Fansen, K. Micromilling Strategies: Optimization Issues, *Proceedings of the Institution of Mechanical Engineers* Volume, 218, Part B, 2004.

Evans, B. and Mehalso, R. 2001. How Small is Small?: A Guide to the New Microfabrication Design and Process Techniques. *Medical Device & Diagnostic Industry*, 11.

Farsari, M., Claret-Tournier, F., Huang, S., Chatwin, C.R., Budgett, D.M., Birch, P.M., Young, R.C.D. and Richardson, J.D. 2000. A novel high-accuracy microstereolithography method employing an adaptive electro-optic mask. *Journal of Materials Processing Technology* 107: 167–172.

Han, M., Lee, W., Lee S.K. and Lee S.S. 2004. 3D microfabrication with inclined/rotated UV lithography. *Sensors and Actuators*, 111: 14–20.

Hirata, Y. 2003. LIGA process – micromachining technique using synchrotron radiation lithography and some industrial applications. *Nuclear Instruments and Methods in Physics Research B*, 208: 21–26.

Hornes, J., Gottert, J., Lian, K., Desta, Y. and Jian, L. 2003. Materials for LIGA and LIGA-based Microsystems. *Nuclear Instruments and Methods in Physics Research*, 199: 332–341.

Hsieh, G.-W., Hsieh, Y.-S., Yang, C.-R. and Lee, Y.-D. 2002. Novel Positive-Tone Thick Photoresist for High Aspect Ratio Microsystem Technology. *Microsystem Technologies* 8: 326–329.

Hull, C.W. 1984. US Pat. 4575330.

Katoh, T. and Zhang, Y. 1998. High Aspect Ratio Micromachining by Synchrotron Radiation Direct Photo-Etching. *Microsystem Technologies* 4: 135–138.

Kim, J.T., Kim, B.C., Jeong, M.Y. and Lee, M.S. 2004. Fabrication of a Micro-Optical Coupling Structure by Laser Ablation. *Journal of Materials Processing Technology* 146: 163–166.

Kowata, S. and Sun, H.B. 2003. Two-photon photopolymerization as a tool for making micro-devices. *Applied Surface Science* 208/209: 153–158.

Kupka, R.K., Bouamrane, F., Cremers, C. and Megtert, S. 2000. Microfabrication: LIGA-X and Applications. *Applied Surface Science* 164: 97–110.

Lawes, R.A., Holmes, A.S. and Goodall, F.N. 1996. The Formation of Moulds for 3D Microstructures using Excimer Laser Ablation. *Microsystem Technologies* 3: 17–19.

Lemercier, G., Mulatier, J.C., Martineau, C., Anémian, R., Andraud, C., Wang, I., Stéphan, O., Amari, N. and Baldeck, P. Two-photon absorption: from optical limiting to 3D microfabrication. *Comptes Rendus Chimie* 8: 1308–1316.

Lim, H.S., Wong, Y.S., Rahman, M. and Lee, M.K. 2003. A study on the machining of high-aspect ratio microstructures using micro-EDM. *Journal of Materials Processing Technology*, 140: 318–325.

Lindeberg, M., Buckley, J., Possnert, G. and Hjort, K. 2000. Deep Ion Projection Lithography in PMMA: Substrate Heating and Ion Energy Concerns. *Microsystem Technologies* 6: 135–140.

Lu, Z. and Yoneyama, T. 1999. Micro cutting in the micro lathe turning system. *International Journal of Machine Tools&Manufacture* 39: 1171–1183.

Malek, C.K. and Saile, V. 2004. Applications of LIGA technology for precision manufacturing of high-aspect-ratio micro-components and systems: a review. *Microelectronics Journal* 35: 131–143.

Maruo, S. and Ikuta, K. 2002. Submicron stereolithography for the production of freely movable mechanisms by using single-photon polymerization. *Sensors and Actuators* A 100: 70–76.

Munnik, F., Benninger, F., Mikhailov, S., Bertsch, A., Renaud, P., Lorenz, H. and Gmür. 2003. High aspect ratio, 3D structuring of photoresist materials by ion beam LIGA. *Microelectronic Engineering*, 67–68: 96–103.

Murali, M. and Yeo, S.H., 2004. Rapid Biocompatible Micro Device Fabrication by Micro Electro-Discharge Machining. *Biomedical Microdevices* 6: 41–45.

Nguyen, L.H., Straub, M. and Gu, M. 2005. Acrylate-based photopolymer for two-photon microfabrication and photonic applications. *Advanced Functional Materials*, 15: 209–215.

Pham, D.T., Dimov, S.S., Ji, C., Petkov, P.V. and Dobrev, D. 2004. Laser milling as a "rapid" micromanufacturing process. *Proc. Instn. Mech. Engrs.* 218: 1–7.

Rahman, M.A., Rahman, M., Kumar, A.S. and Lim, H.S. 2005. CNC microturning: an application to miniaturization. *International Journal of Machine Tools&Manufacture* 45: 631–639.

Schrempel, F. and Witthun, W. 1998. P-LIGA: 3D-integration of microstructures with curved surfaces by deep ion irradiation. Nuclear Instruments and Methods in Physics Research, 139: 363–371.

Sharon, A., Bilsing, A., Lewis, G. and Zhang, X. 2003. Manufacturing of 3D Microstructures using Novel UPSAMS Process for MEMS Applications. *Material Research Society Symposium* 741: 151–156.

Stute, U., Serbin, J., Kulik, C. and Chichkov, B.N. 2003. Three Dimensional Nanostructure Fabricated by Two-Photon Polymerization of Hybrid Polymers. *Proceedings of The 1st Conference on Advanced Research in Virtual and Rapid Prototyping* 399–403.

Tormen, M., Businaro, L., Altissimo, M., Romanato, F., Cabrini, S., Perennes, F., Proitti, R., Sun, H.B., Kawata, S., Fabrizio, E.D. 2004. *Microelectronic Engineering* 73/74: 535–541.

Wong, Y.S., Rahman, M., Lim, H.S., Han, H. and Ravi, N. 2003. Investigation of micro-EDM material removal characteristics using single RC-pulse discharges. *Journal of Materials Processing Technology*, 104: 303–307.

Virtual environments and concurrent engineering

Virtual modeling and rapid manufacturing – Bártolo (eds)
© 2005 Taylor & Francis Group, London, ISBN 0 415 39062 1

Development and use of analytical cure models to improve stereolithography surface finish

Benay Sager & David W. Rosen

George W. Woodruff School of Mechanical Engineering, Georgia Institute of Technology, Atlanta, USA

ABSTRACT: To facilitate the transition of Stereolithography (SLA) into the manufacturing domain and to increase its appeal to the micro manufacturing industry, the ability to manufacture repeatable parts with improved surface finish is needed. In order to truly improve surface finish of SLA technology, a more systematic approach that will incorporate process parameters is needed. In this paper, an analytical model for determining surface finish is presented. Based on the irradiance profile of a Gaussian laser beam, this analytical model is used as a computational tool to simulate SLA surface finish. Simulated surface finish is compared to SLA parts' surface finish and results are discussed. How this analytical model can be used in SLA process planning to improve surface finish is presented.

1 INTRODUCTION

With the growing interest in applying Stereolithography (SLA) technology to the microfabrication and rapid manufacturing domains comes the need to build repeatable parts with improved surface finish. In order to improve the surface finish of SLA parts, the limits of SLA process resolution, both theoretical and empirical, need to be established. In other words, the resolution of the SLA process needs to be characterized and quantified.

While the traditional SLA cure model (Jacobs 1992; Jacobs 1996) is a simplified form of the laser beam characteristics, it does not reflect the dynamic nature of the SLA build process, i.e. changing laser beam angle, beam diameter, and irradiance, and hence resulting cured parts. In order to quantify SLA surface finish, the effect of these changing parameters on the build surfaces needs to be captured in an analytical cure model.

Quantification of SLA surface finish can be extended to the process planning domain for the SLA process in order to improve the surface finish. Traditional SLA machines have been designed with a laser beam scanning a particular cross-section where the laser beam angle with the build surface deviates only slightly from the vertical. Such a configuration not only limits attainable resolution due to stair steps, but also results in particular areas of the build getting more exposure than needed; thus producing inefficient builds and overexposure of parts. Furthermore, it has been shown that generating better upfacing surface finish is possible

by changing the angle of the build surface while keeping the laser stationary (Reeves and Cobb 1997; Reeves 1998; West 1999; Kataria and Rosen 2000).

Therefore, it is possible to improve SLA surface finish by using a combination of cure profiles with varying shapes and sizes via adaptive control of the laser beam angle, scan speed and other process parameters (Sager and Rosen 2004). In this paper, an analytical cure model based on dynamic laser beam characteristics is presented as a first step towards achieving SLA surface finish improvement.

2 ANALYTICAL SLA CURE MODEL

At the center of the SLA process is the UV laser or light source. The irradiance profile of this light source is the determining factor of cured SLA profile shapes. Traditional SLA systems use a stationary UV laser with galvanometer-driven mirrors to scan a particular cross-section on the build surface. For modeling purposes, some general behavior of the laser beam has been widely assumed.

In order to simulate exact build shapes, an irradiance model was developed at Georgia Tech by taking into account the dynamic parameters of the SLA process (Sager and Rosen 2004). The assumptions about the behavior of the laser beam during the SLA build process, as presented by (Jacobs 1996) in traditional models, and the assumptions of the developed model are compared in Table 1.

Table 1. Model assumption comparison for SLA laser beam.

Parameter	Assumption about behavior	
	Traditional SLA model	Dynamic model developed at GT
Beam diameter	Circular shape & constant diameter	Shape & size change according to location on vat
Beam angle with vat surface	Perpendicular	Changes during scan due to location
Beam irradiance profile	Constant	Changes depending on focus location & location of beam
Refraction	Does not affect cure profile, not taken into account	Changes size, shape and location of cure profile

As shown in Table 1, the changes in several parameters during the SLA process, which have a significant effect on the resolution of the SLA process, are adequately captured in the developed model.

2.1 Development of analytical irradiance model

In order to develop an SLA cure model, first the irradiance behavior of the laser beam used in the process needs to be modeled. The irradiance distribution of an SLA laser is modeled as Gaussian. Irradiance is the radiant power of the laser per unit area (mW/cm^2), and is often denoted by $H(x, y, z)$. A Gaussian beam always either diverges from or converges to a point. However, because of diffraction, converging to a single point does not occur [O'Shea, 1985 #12]. Instead, the beam does reach a minimum value, d_o, the beam waist diameter. For a Gaussian laser beam, the maximum irradiance H_o occurs at the beam waist. The irradiance at any given point along a laser beam depends on the radial and perpendicular distances d and z from the beam waist, as shown in Figure 1.

The maximum irradiance is determined by integrating the irradiance function over the area covered by the beam at this point. This integral is equal to the laser power used in the SLA machine, P_L:

$$H_o = \frac{8P_L}{\pi d_o^2} \tag{1}$$

The extent of the beam waist region can be characterized with a parameter called the Rayleigh range, z_R, which is the distance from the beam waist where the diameter has increased to $\sqrt{2}d_0$. The collimated region

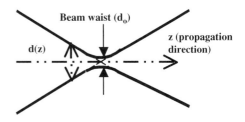

Figure 1. Gaussian laser beam waist.

of a Gaussian beam waist is equal to $2z_R$ (O'Shea 1985). The Rayleigh range is:

$$z_R = \frac{d_o}{\theta_b} = \frac{4\lambda}{\pi\theta_b^2} = \frac{\pi d_o^2}{4\lambda} \tag{2}$$

The irradiance along the laser beam is given by:

$$H(d,z) = \frac{H_o \exp\left(\frac{-\left(\frac{2d^2}{d_o^2}\right)}{1+\left(\frac{z^2}{z_R^2}\right)}\right)}{1+\left(\frac{z^2}{z_R^2}\right)} \tag{3}$$

What is important for characterization of the laser beam is how irradiance changes along the Stereolithography vat. The irradiance is affected by two main factors: distance away from beam waist, and the angle between laser beam and vertical. In order to take these two factors into account, a more convenient spherical coordinate system was used.

SLA parts are manufactured by scanning the laser beam in the x and y directions using two galvanometer-driven low inertia mirrors. Even though the mirrors rotate, the center point of the second mirror does not change significantly. Therefore, by choosing the second mirror as the origin point of the laser beam, the origin of the spherical coordinate axis is selected at a point with height h above the center of the build surface. The conventional and the spherical coordinate systems are shown in Figure 2.

In Figure 2, the origin of the Cartesian coordinates is the center of the vat surface, (0, 0, 0). The spherical coordinate system is defined by ρ, β, ψ in Figure 2, where ρ is the spherical distance away from point (0, 0, h). The axis X' makes an angle β with axis X. The angle ψ is between the galvanometer mirror and axis X', and angle $\gamma = 90 - \psi$ from basic geometry. The focus depth f in Figure 2 into the resin is the location of the theoretical beam waist, and is typically 3 mm (Partanen 2002). For any point P, the coordinates (x_p, y_p, z_p) in the vat are converted into spherical coordinates

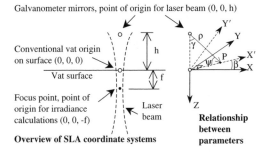

Galvanometer mirrors, point of origin for laser beam (0, 0, h)

Conventional vat origin on surface (0, 0, 0)

Vat surface

Focus point, point of origin for irradiance calculations (0, 0, -f)

Laser beam

Overview of SLA coordinate systems

Relationship between parameters

Figure 2.   Coordinate systems used in SLA analytical model.

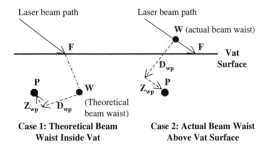

Case 1: Theoretical Beam Waist Inside Vat

Case 2: Actual Beam Waist Above Vat Surface

Figure 3.   The two cases for beam waist location determination.

ρ, β, γ. The location of the beam waist along the laser beam is constant, and it is at a radial distance $\rho_f = (h + f)$. However, the angle that the focus point makes with the vertical, γ_F, will change accordingly, as the beam moves around the build surface. The center of the laser beam intersects the vat surface at F (x_F, y_F, 0) , which forms the basis for several irradiance calculations.

In the SLA process, the path of the laser beam is refracted inside the resin, resulting in a change of the location of the theoretical beam waist within the resin (it is assumed that the maximum angle made with the surface of the resin is not large enough to result in a significant percentage of the laser beam to be reflected). Using Snell's law for refraction, the path of refracted laser beam in the vat is calculated. For SLA resins, the refractive index, φ, is estimated to be 1.5. It is assumed that the refractive index of SLA resins does not change significantly over time and refractive indices of cured and uncured resins are similar (Narahara and Saito 1994).

The Beer-Lambert absorption law is considered valid for the SLA process (Flach and Chartoff 1995). The Beer-Lambert law is:

$$H = H_{surface} \exp\left(- g / D_p\right) \tag{4}$$

where $g = \|z_p/\cos\varphi\|$ is the length of the laser beam attenuation path along the refracted ray, z_p is the vertical distance of travel, and D_p is the depth of penetration, a resin constant.

Since SLA laser beams are typically focused underneath the build surface, there are 2 distinct cases for calculating SLA irradiance within the vat. For scans close to the center of the vat, the theoretical beam waist is within the vat. This is Case 1. The reason why the beam waist is theoretical in Case 1 is because the focus depth (3 mm) in SLA is much larger than the typical depth of penetration (0.25 mm). Therefore, the laser beam never penetrates deep enough into the resin as a result of absorption. For scans that are at the corners of the vat, the actual beam waist is either on or above the vat surface, constituting Case 2.

Calculating the location of the beam waist, W, is essential in determining the irradiance at any given point inside the vat. Our model determines the location of W (x_w, y_w, z_w) relative to the point F for both cases in Figure 3. To calculate irradiance at any arbitrary point using Equation 3, the distances between P and W that are parallel and perpendicular to the beam propagation path must be found.

In short, the distances Z_{wp} and D_{wp} in Figure 3 need to be calculated. In this context, the distance D_{wp} refers to the shortest distance between the line FW and point P (d in Equation 3), whereas the distance Z_{wp} (z in Equation 3) refers to the distance between points W and P parallel to the beam path of $\overline{FW}$. These parameters are computed as follows (Edwards and Penney 1994):

$$\vec{M} = \vec{F} - \vec{W} \tag{5}$$

$$t = \frac{\vec{M} \bullet \left(\vec{P} - \vec{F}\right)}{\vec{M} \bullet \vec{M}} \tag{6}$$

$$\vec{D}_{wp} = \vec{P} - \left(\vec{F} + t\vec{M}\right) \tag{7}$$

$$Z_{wp} = \left\|\left(\vec{P} - \vec{W}\right) - \vec{D}_{wp}\right\| \tag{8}$$

Inserting the calculated parameters into Equation 3 yields the irradiance equation for any arbitrary point P in the vat:

$$H(P) = \frac{H_O \exp\left(\dfrac{-\left(\dfrac{2\|D_{wp}\|^2}{d_o^2}\right)}{1 + \left(\dfrac{Z_{wp}^2}{z_R^2}\right)}\right) \exp\left(-\dfrac{g}{D_p}\right)}{1 + \left(\dfrac{Z_{wp}^2}{z_R^2}\right)} \tag{9}$$

495

2.2 Development of analytical cure model

The irradiance model was developed to predict cure profiles. To do so, the exposure at any point in the vat needs to be known. Exposure is defined as the integral of irradiance at a point over time:

$$E(x, y, z) = \int_{t_{start}}^{t_{end}} H[x(t), y(t), z(t)] dt \qquad (10)$$

Direct integration of Equation 9 is not possible using Equation 10 since irradiance is not only a function of time, but space as well. Therefore, a suitable numerical integration method must be chosen to evaluate this integral. In this research, Simpson's 3/8 rule was used for numerical integration of exposure.

Using Simpson's 3/8 rule, the resulting exposure at point P from each particular laser beam scan is calculated. After adding up the cumulative effect of exposure from several laser beam scans (within one layer and from additional layers), the exposure at a particular point is calculated and compared to the critical exposure value, E_c, above which a point is cured (Hull 1986). In doing so, E_c, which is a resin constant, is taken as a meaningful threshold and exposure is considered to be additive.

3 APPLICATION OF CURE MODEL: SLA SURFACE FINISH CHARACTERIZATION

The developed cure model was implemented in MAT-LAB to simulate the resulting cure shape of parts built in a SLA250/50 machine. It should be noted that the analytical model can simulate a build at any location on the build platform and it can be extended to any SLA machine.

3.1 Parameters affecting SLA surface finish

During the SLA process, several parallel laser beam scans per layer are used to scan a part. The spacing between these scans is called hatch spacing (hs). Hatch spacing affects not only the surface finish, but also the resulting cure depth of the downfacing surface, as shown in Figure 4.

In Figure 4, the gap between each individual bullet shaped cure profile is the cusp height, δ. Cusp height is the primary metric that is used for determining the surface finish in freeform-fabricated parts (Lu, Fuh et al. 2001). As the hatch spacing is decreased, the resulting cure depth increases while the cusp height decreases up to a certain value.

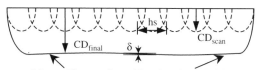

Figure 4. Cross section of parallel scans showing cumulative effect on cure depth.

3.2 SLA part build simulation

SLA part builds with varying hatch spacing values were simulated using the developed analytical cure model. A 10-layer build of a 5 mm by 7.5 mm area consisting of the same parallel laser beam scans on each layer was simulated using hatch spacing values of 75 μm, 100 μm (default SLA build value), 150 μm, and 190 μm. In the simulation, the layer thickness was 100 μm with a laser power of 33 mW and scan speed of 250 mm/s. A 10-layer part was chosen in order to level off the effect of SLA build errors such as over-cure and print through on surface finish, which level off after the first 5–7 layers of a build.

The spacing was varied in order to quantify the change in surface finish due to the different builds. At the 75 μm hatch spacing, the build has many overlapping laser beam scans so its surface is expected to be very smooth. However, when the hatch spacing is 190 μm, the cured outline of each individual laser beam scan should be visible with large cusp height values. The cross-section of downfacing surface of the 190 μm hatch spacing simulation is shown in Figure 5.

In Figure 5, the cross-section of the simulated cure profile is shown in dark. Quantification of the cusp height between cure profiles on the downfacing surface, δ_y and δ_z in Figure 5, is important for surface finish characterization. In addition, the surface roughness average, R_a, and root-mean-square average, R_q, were used to quantify the surface roughness of the overall downfacing surface:

$$R_a = \frac{1}{n} \sum_{i=1}^{n} z_i = \frac{1}{l} \int_0^l |z| dy \qquad (11)$$

$$R_q = \sqrt{\frac{1}{n} \sum_{i=1}^{n} z_i^2} = \left[\frac{1}{l} \int_0^l z^2 dy \right]^{1/2} \qquad (12)$$

where l is the cutoff length and z is the vertical height of the profile (Kalpakjian and Schmid 2003).

3.3 SLA part build and model comparison

The simulated SLA parts were built using an SLA 250/50 machine and DSM Somos 7110 resin (E_c = 8.2 mJ/cm², D_p = 0.14 mm) with same laser and

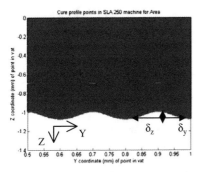

Figure 5. Cross-section of simulated 190 μm hatch spacing part.

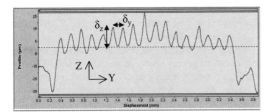

Figure 6. Inverted surface profile of SLA part from Talysurf.

process parameters. The SLA parts were cleaned using industrial grade (99%) isopropyl alcohol in an ultrasound machine for 5 minutes, dried, and post-cured in an UV oven for 30 minutes. During the drying process, the amount of light and moisture received by the parts were minimized by keeping them in metal containers. After post-curing, the surfaces of the parts were measured along the center by a Taylor-Hobson Form Talysurf Profilometer. The inverted profile of the measured surface of the 190 μm hatch spacing part is shown in Figure 6.

In Figure 6, each bump corresponds to the cross-section of a laser beam scan line. Qualitatively, the downfacing surface profile in Figure 6 is similar to the simulated profile in Figure 5.

Three other SLA parts with hatch spacing values of 75, 100, and 150 μm were also built on the build platform. The average δ_y, δ_z, R_a, and R_q values from each part are compared to the simulated parameters from the analytical model in Table 2.

In Table 2, it is seen that for the horizontal spacing between cured profiles, δ_y, the simulated values agree well with the measured values. However, for δ_z, R_a, and R_q, the simulated and measured values are significantly different. The analytical model predicts a change in R_a from 0.2 to 20.6 μm when the hatch spacing is doubled from 100 to 190 μm. However, for

Table 2. Comparison of simulated and measured surface finish.

Hatch spacing of build		Surface roughness parameter (μm)			
		δ_y	δ_z	R_a	R_q
75 μm	Simulated	75	0.08	0.05	0.05
	Measured	114	5.0	1.6	2.2
100 μm	Simulated	100	0.3	0.2	0.2
	Measured	101	14.8	4.4	5.5
150 μm	Simulated	150	17.9	5.9	6.5
	Measured	150	27.3	7.8	9.3
190 μm	Simulated	190	60.0	19.8	20.9
	Measured	154	24.8	7.3	8.9

the same change in hatch spacing, the measured results are between 1.6 and 7.3 μm.

4 DISCUSSION OF RESULTS

The focus of this study is to use an analytical irradiance model to predict cure shapes in the SLA process. In doing so, the exposure threshold model is applied to determine whether a point is solidified. The threshold model is only a first order approximation of resin behavior. In reality, due to the thermal curing and the dark reaction in the resin, the level of cure in a cross-section varies. During cleaning, along with partially cured areas, some cured areas are eliminated. In fact, up to 20% of the tip of the cured profile could be removed during cleaning (Tang 2005).

The actual irradiance profile of the HeCd laser used in the SLA 250/50 machine is not Gaussian and changes over time as much as 10%. The Gaussian beam assumption used in the analytical model is an approximation, and changes in the irradiance profile of the laser beam over time are difficult to model. However, the irradiance profile of the SLA 250/50 laser was recorded for a particular point in time, and used to simulate the resulting cure shape of two overlapping lines. This cure shape is compared to the Gaussian profile cure shape in Figure 7.

From Figure 7, it is seen that the resulting cured height in the SLA 250/50 machine is different from the Gaussian cure profile at some places by about 30%. In addition, the cure shape is asymmetric about the center of the laser beam, creating valleys in cured profile that are as wide as 40 μm and deep as 60 μm. These valleys trap resin as much as 10–20% of the cusp height. Furthermore, the amount of trapped resin increases as hatch spacing becomes larger, between 10–30%, due to the exponential relationship between the surface exposure and hatch spacing as shown in Figure 8. In Figure 8, the measured R_a values with error bars of 100% due to the compounding of the

497

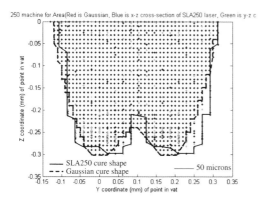

Figure 7. Comparison between Gaussian model and actual SLA 250/50 machine cure profile.

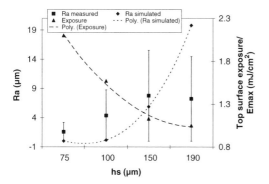

Figure 8. Relationship between hatch spacing, exposure, & R_a.

aforementioned effects, and the simulated R_a values are plotted.

In Figure 8, all of the simulated R_a values fall within the error bars of the measured R_a values, except for the 190 μm hatch spacing case. This means that the trapped resin due to increased partial curing in this case is very severe and contributes to smoothing out of the surface more than the other 3 cases.

The results show that the compounded effects of trapped resin, changes in irradiance profile and post-processing can cause significant variations to the final surface roughness of a part and are very difficult to capture in an analytical model.

5 FUTURE WORK

Surface finish trends have been accurately detected with the analytical model, including larger values of R_a, R_q, δ_y and δ_z with increased hatch spacing. The differences between the simulated and measured

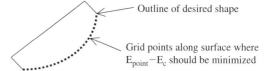

Figure 9. Minimizing deviation of exposure from E_c.

values can be attributed to partially cured and trapped resin, changes in the irradiance profile of the SLA 250/50 machine, and post-processing. As a next step, SLA parts will be simulated and built at different locations in the SLA machine with different hatch spacing values in order to quantify how well the analytical model predicts surface finish when the laser beam is not perpendicular to the vat surface. Also, design of experiments approach will be applied to post-processing to quantify the effects of different steps on surface finish.

The analytical cure model can be applied in SLA process planning to improve surface finish. During the SLA process, many areas of a build receive excess exposure due to the use of horizontal layers. In addition, the laser beam is almost always vertical in the current SLA machines, which results in large stair-steps. If the surface of the SLA part is represented as a grid whose points should receive just enough exposure to solidify, then by maximizing the number of grid points, uniform exposure across the grid can be obtained, as shown in Figure 9.

By minimizing $E_{point} - E_c$ at each grid point on a desired surface, the shape of the cure profile can be controlled. If $E_{point} = E$ (hs, γ, x, y, z, t...) for each grid point, then values of process parameters, which will conform to the desired surface and hence improve surface finish, can be determined by parameter estimation. Inverse design techniques (Özisik and Orlande 2000) present a suitable solution method for the SLA parameter estimation formulation.

ACKNOWLEDGEMENTS

We gratefully acknowledge the support from the Rapid Prototyping and Manufacturing Institute member companies and the George W. Woodruff School of Mechanical Engineering at Georgia Tech.

REFERENCES

Edwards, C. H. and D. E. Penney (1994). Calculus with Analytic Geometry. Englewood Cliffs, NJ, Prentice-Hall Inc.

Flach, L. and R. P. Chartoff (1995). "A Process Model for Nonisothermal Photopolymerization with a Laser Light Source I: Basic Model Development." Polymer Engineering and Science **35**(6): 483–492.

Hull, C. (1986). Apparatus for production of three dimensional objects by Stereolithography. U.S.A, Hull/UVP Inc.

Jacobs, P. F. (1992). Rapid Prototyping & Manufacturing: Fundamentals of Stereolithography, Society of Manufacturing Engineers.

Jacobs, P. F. (1996). Stereolithography and other RP&M Technologies: from Rapid Prototyping to Rapid Tooling, Society of Manufacturing Engineers.

Kalpakjian, S. and S. R. Schmid (2003). Manufacturing Processes for Engineering Materials. Upper Saddle River, NJ, Prentice Hall.

Kataria, A. and D. W. Rosen (2000). Building Around Inserts: Methods for Fabricating Complex Devices in Stereolithography. ASME DETC, Baltimore, MD, ASME.

Lu, L. J. and Y. H. Fuh et al. (2001). Laser-Induced Materials and Processes for Rapid Prototyping, Kluwer Academic Publishers.

Narahara, H. and K. Saito (1994). Fundamental Analysis of Single Layer Created by Three Dimensional Photofabrication. International Conference on Rapid Prototyping, Dayton, OH, University of Dayton.

O'Shea, D. C. (1985). Elements of Modern Optical Design, John Wiley and Sons.

Özisik, N. M. and H. R. B. Orlande (2000). Inverse Heat Transfer: Fundamentals and Applications. New York, Taylor & Francis.

Partanen, J. (2002). Personal Communication.

Reeves, P. (1998). Reducing Surface Deviation of Stereolithography. Manufacture Engineering and Operations Management, Ph.D. Dissertation University of Nottingham.

Reeves, P. E. and R. C. Cobb (1997). "Reducing the surface deviation of stereolithography using in-process techniques." Rapid Prototyping Journal 3(1): 20–31.

Tang, Y. (2005). Stereolithography Cure Process Modeling. Chemical Engineering, Ph.D. Dissertation Georgia Institute of Technology. 133.

West, A. (1999). A decision support system for fabrication process planning in stereolithography. Mechanical Engineering, Master's Thesis Georgia Institute of Technology.

Virtual modeling and rapid manufacturing – Bártolo (eds)
© *2005 Taylor & Francis Group, London, ISBN 0 415 39062 1*

Application of simulated annealing in improving the performance of stereolithography

V. Canellidis, V. Dedoussis & S. Sofianopoulou

Center for Product Development and Rapid Prototyping, University of Piraeus, Piraeus, Greece

ABSTRACT: Effective utilization of Stereolithography (SL) mainly relies on orienting and packing parts optimally on the fabrication platform of the machine, so to achieve maximum space utilization and minimum build time, without of course compromising surface quality. The present work focuses on an effective way to pack parts optimally on the fabrication platform of SL machine. Due to technical constrains set by SL technology, the original 3-D packing problem is simplified by one dimension by projecting each one of the parts on the build platform (x-y plane) and packing their projections instead of the actual parts themselves. In order to solve the resulting 2-D packing problem a heuristic method has been adopted. The heuristic method consists of a Simulated Annealing algorithm employing a polynomial-time cooling schedule and a new improved placement rule.

1 INTRODUCTION

Stereolithography is one the most widely used techniques belonging to the group of Layer Manufacturing Technologies (LMT) or else Rapid Prototyping Technologies (RPT). Rapid prototyping technologies involve fabrication of parts directly from 3D-CAD models by successive addition of layers of various kinds of materials like polymers, metals, ceramics or paper. In stereolithography parts are fabricated by successive solidification (polymerization) of thin resin layers. For the solidification of the layers an UV laser beam, focused on the surface of the resin, is used. The laser beam, guided by a system of galvanometer mirrors, scans selected areas at the surface of the resin, thereby solidifying thin resin layers. Next, the platform, on which the part is attached, lowers a small distance (0.10–0.25 mm) and a recoating blade (sweeper) moves across the resin surface applying and leveling a new layer of liquid resin on top of the already fabricated part of the part. The process continues in the same manner, gradually building the part from the bottom-up, until the entire part is fabricated.

In the past few years researchers have investigated extensively the time needed for a part to be built by stereolithography systems (e.g. Giannatsis et al. 2001). Thus the build-time has been expressed as:

$$\sum_{i=1}^{n} TLayer(i) \tag{1}$$

where TLayer is the time required for the addition of the ith layer and n is the total number of layers. The addition of a layer consists of recoating and scanning. Thus the time needed for the fabrication of a single layer may be expressed as the sum of recoating and scanning time. The recoating time now can be calculated as the sum of the time required for the lowering of the platform, the time required for the recoater movement and the pre- and post-scan delay periods.

The time required for the lowering of the platform, the time required for the recoater movement and the pre- and post-scan delay periods have been found to be independent of the number of parts being fabricated at a time. So, if we proceed in fabricating one part at a time instead of fabricating as many parts as possible, the total sum of build-time will be significantly larger. Noting that the total cost of ownership (TCO) increases as the build time increases, it becomes obvious that effective utilization of stereolithography relies on packing parts optimally on the fabrication platform of the machine.

Quite a few researchers have examined the problem of packing several parts on the fabrication platform of the machine. The methodologies and algorithms employed tackle the relevant 3D bin packing problem mainly for the case of Selective Laser Sintering technology. In most of the cases, heuristic methods have been used in conjunction with a placement rule (e.g. Hur et al. 2001) to produce near optimal solutions. The corresponding work in stereolithography technology, however, is rather limited. Decisions regarding packing rely mainly on the skill and experience of the stereolithography apparatus operator. Thus the purpose of this work is to present a methodology that "automates" the process of making these decisions for the stereolithography technology and provides the operator with effective packing solutions.

Due to the existence of support structures in SL technology, packing solutions/fabrication layouts where some parts are built on top of others are unacceptable. Thus, the original 3-D packing problem is simplified by one dimension, by projecting each one of the parts on the build platform (x-y plane) and packing their projections instead of the actual parts themselves. The parts are being projected on the build platform only after the calculation of the desirable orientation for each part. Then the minimum bounding rectangle is being defined for each one of the projections. Finally the minimum bounding rectangles are being used as input in the packing algorithm.

The packing algorithm consists of a simulated annealing procedure, which utilizes a polynomial time cooling schedule, in conjunction with a new improved placement rule.

2 SIMULATED ANNEALING OVERVIEW

Simulated annealing is a generalization of the Monte Carlo method, suitable for large scale optimization problems, especially those where a desired global extremum is hidden among many, poorer, local extrema. It was motivated by an analogy to the thermodynamics of annealing in solids. In an annealing process, material is being heated to a temperature that permits many molecules to move freely with respect to each other. Then it is cooled in a slow manner, until the material freezes into a crystal, which is completely ordered, and thus the system is at the state of minimum energy. The temperature of the material must be gradually lowered so that at each temperature the atoms can move enough to begin adopting the most stable orientation. Noting that the target in a combinatorial optimization problem is to find an optimal solution or else the "minimum energy state of the problem", simulated annealing technique uses an analogous cooling operation for transforming a poor, unordered solution into an ordered, desirable solution, so as to optimise the objective function.

By analogy to the physical annealing, simulated annealing uses the control parameters that follow:

Cost function: An objective function that measures how well the system performs when a certain configuration is given.
Move schemes: A generator of random changes in the configuration so as to create new candidate solutions.
Temperature: A control parameter analogous to the temperature in the annealing process of solids. This artificial parameter acts as a source of stochasticity, which is convenient for eventually detrapping from local minima, and it represents the willingness of a system to accept a state/solution that is worse than the current.

Cooling schedule: A definition of the cooling speed to anneal the problem from a random solution to a good, frozen one. In its details, it must provide a starting temperature, together with the rules to determine when and how much the temperature should be reduced and when annealing should be terminated.
Metropolis acceptance criterion: As it was pointed out simulated annealing is suitable for problems where the desired global optimum is hidden among many, poorer, local optima. Thus the method should be capable of escaping local optima. In order to achieve the previous mentioned goal, simulated annealing takes not only downhill moves, but also permits uphill moves with an assigned probability of $P(\Delta C) = \exp(-\Delta C/T)$ where ΔC is the change in the objection function value and T is the temperature parameter. Thus a new candidate solution is accepted, if it has the same or smaller objective function value. However, if the new solution yields an increase in cost of the system, then the new solution is judged for suitability probabilistically according to the probability $P(\Delta C) = \exp(-\Delta C/T)$. Thus the new solution is being accepted if a random number, generated in the interval $(0,1)$, is less than or equal to $P(\Delta C) = \exp(-\Delta C/T)$. Otherwise, it is rejected.

A simulated annealing optimization starts with a Metropolis Monte Carlo simulation at an initial high temperature T. This means that a relatively large percentage of the random steps that result in an increase of the cost of the objection function will be accepted with probability $P(\Delta C) = \exp(-\Delta C/T)$. Metropolis Monte Carlo simulation is a randomization technique being used for optimizing a function using a random sampling of the solution space. After a sufficient number of Monte Carlo steps, the temperature is decreased. The Metropolis Monte Carlo simulation is then continued. This process is repeated until the final temperature is reached. Actually a simulated annealing algorithm consists of a pair of nested loops. The outer loop controls the temperature parameter and the inner loop (or else Markov chain) runs a Metropolis Monte Carlo simulation at that temperature. The way in which the temperature is decreased has already been referred as the *cooling schedule*.

3 PROBLEM FORMULATION

In this section, the basic concepts of the methodology utilized for solving the packing problem are described in detail. As it was mentioned before, our target is to pack as many parts as possible on the platform of a stereolithography system in order to reduce the TCO or else to minimize the unoccupied area (trim loss) on the platform. Thus, we are dealing with a minimization problem and the cost function F may be defined as the percentage of the area of the platform that is unused by the, say n, rectangles:

$$F=1-\frac{\sum_{i}^{i=n} l_i w_i}{lw} \qquad (2)$$

where l and w are the length and the width of the fabrication platform respectively, and l_i, w_i are the length and width of the ith rectangle.

A packing pattern can be represented by a permutation $\pi = (i_1, ..., i_n)$ where i is the index of the r_i rectangle. Actually, a permutation represents only the sequence in which the rectangles will be packed by a placement rule. The advantage of this data structure is the facile creation of new permutation-solution by simply changing the sequence. For the assignment of each permutation to an unique packing pattern a placement rule is being utilized which will be discussed later in detail. After the creation of the packing pattern we are able to calculate its "cost" through the cost function.

3.1 Cooling schedule

The cooling schedule is being considered to be the most important factor in a simulated annealing algorithm. It is composed by the starting temperature and the rules to determine when and how much the temperature should be reduced and when annealing should be terminated. The cooling schedules may be divided into two broad groups, static and adaptive. Static schedules are those schedules that follow a predetermined course of decrement. Static schedules generally disregard the dynamic behaviour of the problem at hand. On the other hand, dynamic schedules use statistical analysis to control the temperature decrement, while they are computational expensive.

It is apparent that if the temperature parameter is kept high or it is decreased too quickly then the process will not tend toward globally optimal configurations. Therefore, an efficient cooling schedule must be designed in order to successfully implement the simulated annealing algorithm. Thus, a dynamic polynomial-time cooling schedule has been adopted as the most promising approach.

By using the polynomial-time cooling schedule suggested by Van Laarhoven (1988), the initial temperature is being determined by generating candidate configurations/solutions and evaluating their suitability according to the Metropolis Monte Carlo algorithm. At the beginning of the annealing process the acceptance rate of new configurations, disregarding the objection function values, should be high enough

in order not to get trapped easily in a local minimum. The acceptance rate may be expressed as:

$$x \cong \frac{m_1 + m_2 \exp(\frac{-\overline{\Delta C^+}}{T})}{m_1 + m_2} \qquad (3)$$

in Eq. (3), C is defined as the cost of the objection function, m_1 as the number of transitions occurred resulting in decrease of the objection function ($\Delta C \leq 0$), m_2 as the number of the transitions occurred resulting in increase of the objection function ($\Delta C > 0$) and finally $\overline{\Delta C^+}$ is defined as the average increase in cost over m_2 transitions. Eq. (3) can now be rewritten as:

$$T=\frac{\overline{\Delta C^+}}{\ln\left(\dfrac{m_2}{m_2 \cdot x - m_1 \cdot (1-x)}\right)} \qquad (4)$$

By generating a fixed number of transitions and accepting all configurations with increased objective function values an initial temperature can be obtain using Eq. (4).

After determining the initial temperature a decrement rule must be established. Keeping a record of the cost values of the configurations $\pi_1, ..., \pi_j$ that occur during the generation of the kth Markov Chain (the inner loop), where j is the length of the kth Markov Chain, we are able to approximate the probability distribution of the cost values of the kth Markov Chain by a normal distribution with mean μ_k given by

$$\mu_k=\frac{1}{j}\sum_{i=1}^{i=j} C(\pi_i) \qquad (5)$$

and variance

$$\sigma_K^2 =\frac{1}{j}\sum_{i=1}^{i=j} C^2(\pi_i)-\mu_K^2 \qquad (6)$$

Therefore, the decrement rule for the temperature may be expressed as

$$T_{k+1}=\frac{T_k}{1+\dfrac{T_k \cdot \ln(1+\delta)}{3\sigma_K}} \qquad (7)$$

where δ is called the distance parameter. The choice of δ determines how closely the algorithm will approximate a globally minimum state. Moreover, δ controls the computational effort that is going to be needed to reach an approximate globally minimum configuration.

503

Finally the simulated annealing algorithm ends when temperature reaches zero.

3.2 Move schemes

Move schemes in the annealing process denote the generation of candidate configurations/solutions for the system. A new configuration/solution may or may not be accepted as the next state of the system depending upon the Metropolis criterion. Typically, modifying the current state of a system in some way is creating a new solution. In the presented methodology three move schemes have been adopted. The first one simply selects two rectangles from a permutation $\pi = (i_1, ..., i_n)$ and swaps their locations. For instance say that the current state of the system corresponds to the permutation $\pi = (1, 2, 3, 4, 5, 6, 7)$ where each integer is the index of a rectangle. This move scheme would randomly select two indices, e.g. 3 and 5, and swap their locations. Thus, the new solution would be $\pi = (1, 2, 5, 4, 3, 6, 7)$.

The second move scheme was inspired by the SJX operator, introduced by Jakobs (1996) as a crossover operator in genetic algorithms. Given a permutation $\pi = (i_1, ..., i_n)$, as well as two integers b and c, the scheme takes c consecutive indices starting from the b-th index of the permutation and place them at the beginning of the new permutation. Then it fills the remaining unallocated slots of the new permutation by the remaining indices. Suppose, for example, that the initial configuration is $\pi1 = \{1, 2, 3, 4, 5, 6, 7\}$, $b = 2$ and $c = 3$. Then 2, 3, 4 are the first three indices of the new configuration, and the other unused indices 1, 5, 6, 7 will fill the rest of the new configuration. Thus the new solution is $\pi2 = \{2, 3, 4, 1, 5, 6, 7\}$.

The third move scheme randomly selects the 40% of the indices of a permutation $\pi = (i_1, ..., i_n)$ and rotates by 90 degrees the rectangles that are being represented by those indices.

3.3 Placement rule

As it was mentioned before, a permutation represents only the sequence in which the rectangles will be packed. Thus, a placement rule is needed in order to perform the actual packing. Taking into account the nature of our problem, we can consider it as a 2-D bin packing problem, which is commonly encountered in the field of operational research. In order to solve this kind of problems the majority of the researchers use the standard bottom-left (BL) placement policy which attempts to minimize the total length required by placing each part's bounding rectangle (BR) as far to the left as possible and by favouring positions near the bottom-lower and side of the container/platform. The disadvantage of standard bottom-left algorithm is that groups of polygons exist for which an optimal packing

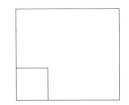

Figure 1. Placement of first rectangle.

pattern cannot be generated. Thus, a new improved placement rule should be established in order to overcome the previously mentioned drawbacks.

Before describing the algorithm the terms vertical insertion lines and horizontal insertion lines should be explained. Vertical insertion lines correspond to the lines, normal to the lower side of the platform, which are adjacent to the rightmost side of each rectangle that has been placed on the fabrication platform. Horizontal insertion lines are the lowest, in the y-wise sense, lines parallel to the x-axis of the platform that "prevent" the overlapping of the rectangle to be packed with the rectangles already placed on the platform. Where a vertical and a horizontal insertion line intersect there is an insertion point where a new rectangle can be placed without overlapping with the rectangles already placed on the platform. Thus, there are as many insertion points as vertical insertion lines.

For clarity reasons, the algorithm of placing-packing the BR on the SLA fabrication platform will be described with reference to Figures 1–5. The algorithm starts by placing the first BR on the platform employing a simple BL rule as it is shown in Figure 1.

After the placement of the first rectangle there are two vertical insertion lines in order to pack the second rectangle. The first vertical insertion line coincides with the left side of the platform, while the second is adjacent to the rightmost side of the rectangle that has already been packed. Moreover, there are two horizontal insertion lines respectively to the vertical insertion lines. A horizontal insertion line that is adjacent to the upper side (in the y-wise sense) of the rectangle already packed corresponds to the first vertical line. A horizontal line that coincides with the lowest side of the fabrication platform corresponds to the second vertical line. As it is obvious, a horizontal line is being used simply as a mean to prevent overlaps. Thus, there are two insertion points where the second rectangle may be placed. In Figures 2 and 3 one can observe the two insertion points with their vertical and horizontal insertion lines as well as the actual packing of the second rectangle in both positions. The hatched rectangle is the rectangle that we are trying to pack. The algorithm chooses that insertion point that corresponds to the lowest horizontal insertion line.

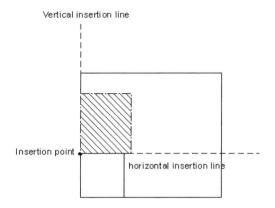

Figure 2. First possible placement of second object.

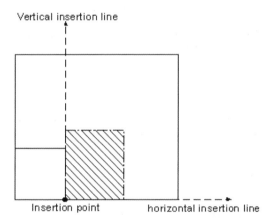

Figure 3. Second possible placement of second object.

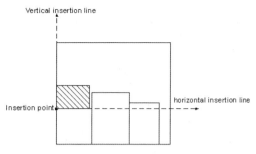

Figure 4. First possible insertion point, given a vertical insertion line.

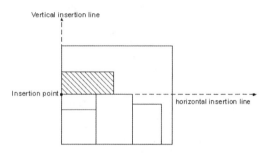

Figure 5. Second possible insertion point, given a vertical insertion line.

In that way the second and the third rectangle are placed. For the packing of the fourth rectangle there are four vertical and horizontal insertion lines indicating four insertion points. The first vertical insertion line coincides with the left side of the platform. Now the horizontal insertion line for this vertical line depends on the length, in the x-wise sense, of the rectangle to be placed. For instance if the length of the rectangle to be placed is smaller than the length of the first rectangle then the horizontal insertion line is adjacent to the upper size (in the y-wise sense) of the first rectangle, as it is shown in Figure 4. If the length of the rectangle to be placed is bigger than the length of the first rectangle then the horizontal insertion line is set at the top of an already packed BR for which there would be no overlaps (e.g. Figure 5). Using the same methodology the other three vertical and horizontal insertion lines are being calculated resulting in four insertion points. Finally, the rectangle is being placed at that insertion point that owns the lowest horizontal insertion line. If an insertion point results in a layout that is unacceptable, e.g. the part is packed partially outside the platform, then the algorithm simply disregards it. Finally, if two or more insertion points share a common horizontal insertion line, then the algorithm selects the one that owns a vertical line nearest to the left side of the platform.

3.4 Comparison with a genetic algorithm

The proposed methodology has been coded in visual C++ allowing the user to define the length of the internal Markov chain, the initial temperature (in case there is a disagreement with the initial temperature provided by the algorithm) and the distance parameter. Moreover, the user can define the initial acceptance percentage and the length of the trial run in order to obtain an estimate of a good initial temperature. Computational results were then compared with those produced by a methodology proposed by Mantzouratos et al. (2002) which utilizes a genetic algorithm and an improved BL placement rule in order to solve the 2-D bin packing problem. Genetic algorithms are also

Table 1. Test problems.

Size of fabrication platform		Number of rectangles	Known optimal solution
Width	Height		
400	200	10	Yes
400	400	15	Yes
70	80	20	Yes
70	80	25	Yes
50	50	25	No
50	50	30	No

Table 2. Trim losses per test problem and algorithm.

Number of rectangles	Known optimal solution	Trim losses by simulated annealing algorithm	Trim losses by genetic algorithm
		% of the platform area	% of the platform area
10	Yes	0	0.045
15	Yes	1.25	1.25
20	Yes	4.0179	3.571426
25	Yes	3.2143	10.714287
25	No	0	0
30	No	4.24	7.49

heuristic algorithms, which are based on the principles of natural selection and survival of the fittest. A genetic algorithm attempts to evolve a solution using a population of potential solutions, appropriately called chromosomes. New chromosomes/solutions are created through a process of breeding, where promising genetic material is passed from one chromosome to the other.

In order to compare the two heuristic procedures, four test problems proposed by Lai & Chan (1997) and Jakobs (1996) were used, as well as two others of our own. For each test problem the simulated annealing algorithm and the genetic algorithm were run 15 times and the average values where taken into account. Only the four test problems proposed by Lai & Chan (1997) and Jakobs (1996) have a known optimal solution of zero trim losses. Table 1 presents the dimensions of the fabrication platform as well as the number of rectangles to be packed for the six test problems. Finally in Table 2 the trim losses for each test and each algorithm are presented.

Though simulated annealing algorithm took much more time than the genetic algorithm to converge in a

"good" solution, we observe that in most cases performed very satisfactory indicating the efficiency of the proposed methodology. Moreover, it must be pointed out that part of the success of the methodology is due to the use of a new improved placement rule which results in much more dense packing layouts than the one employed by the genetic algorithm.

4 CONCLUSIONS

In this paper, a methodology that "automates" the process of making decisions regarding packing for the stereolithography technology and provides the operator with effective packing solutions was presented. The methodology consists of a simulated annealing algorithm and new improved placement rule. A software tool has been developed that can be used by the SLA operator at a pre-processing phase. Finally, the method has been found to be quite robust through the comparison with packing solutions provided by a genetic algorithm.

ACKNOWLEDGEMENT

The financial support of the University of Piraeus Research Center is gratefully acknowledged.

REFERENCE

Dowsland, K.A. & Dowsland, W.B. 1995. Solution approaches to irregular nesting problems. European Journal of Operational Research 84: 506–521.

Giannatsis, J., Dedoussis, V. & Laios, L. 2001. A study of the build time estimation problem for stereolithography systems. Robotics & Computer Integrated Manufacturing 17: 295–304.

Hur, S.M., Choi, K.H., Lee, S.H. & Chang, P.K. 2001. Determination of fabricating orientation and packing in SLS process. Journal of Materials Processing Technology 112: 236–243.

Hopper, E. & Turton, B.C.H. 2001. An empirical investigation of metaheuristic and heuristic algorithms for a 2D packing problem. European Journal of Operational Research 128: 34–57.

Jakobs, S. 1996. On genetic algorithms for the packing of polygons. European Journal of Operational Research 88: 165–181.

Lai, K.K. & Chan, W.M. 1997. Developing a simulated annealing algorithm for the cutting stock problem. Computers And Industrial Engineering 32(1): 115–127.

Leung, T.W., Chan. Chin Kin, & Troutt, M.D. 2002. Application of a mixed simulated annealing-genetic algorithm heuristic for the two dimensional orthogonal packing problem. European Journal of Operational Research 145: 530–542.

Mantzouratos, N., Canellidis, V., Zouzoulis, T., Dedoussis, V. & Sofianopoulou, S. 2002. Pre-processing software tool for optimizing stereolithography apparatus build performance. *Proceedings Euro-uRapid 2002*, International Users Conference on Rapid Prototyping & Rapid Tooling & Rapid Manufacturing, editor R. Meyer, Frankfurt, Dec. 2002, paper B-5/1.

Theodoracatos, V.E. & Grimsley, J.L. 1994. The optimal packing of arbitrarily shaped polygons using simulated annealing and polynomial time cooling schedules. *Computers Methods In Applied Mechanics And Engineering* 125: 53–70.

Van Laarhoven, P.J.M. 1988. Theoretical and computational aspects of simulated annealing. *CWI Tracts*, Amsterdam.

Tian-Peng, Ma-Jian, Zhang-Dong-Mo. 1998. Application of simulated annealing to the combinatorial optimization proble with permutation property: An investigation of the generation mechanism *European Journal of Operational Research* 118: 81–9

Virtual modeling and rapid manufacturing – Bártolo (eds)
© *2005 Taylor & Francis Group, London, ISBN 0 415 39062 1*

Modeling of selective laser sintering heat transfer phenomena for powder-based biomaterials

Florencia Edith Wiria, Chee-Kai Chua & Kah-Fai Leong
School of Mechanical & Aerospace Engineering, Nanyang Technological University, Singapore

Margam Chandrasekaran & Mun-Wai Lee
Forming Technology Group, Singapore Institute of Manufacturing Technology, Singapore

ABSTRACT: Tissue engineering is a fast growing field that combines engineering and medical expertise. Providing good scaffolds, as temporary support during new tissue growth, is one of the engineering challenges in this field. The potential of selective laser sintering (SLS) for tissue-engineering scaffolds fabrication using powder-based biomaterials has been investigated and preliminary results have been favorable. Scaffold specimen fabricated by SLS were mostly conducted by empirical method, hence it is beneficial to have a theoretical model for the heat transfer phenomenon. As SLS is using a heat source to bind the powder together, it has a particular challenge that it might only be used by thermally stable biomaterials. This model aims to relate the thermal properties of the biomaterials and the SLS laser beam properties. By understanding the important processing parameters, it can then reduce the empirical trial-and-error sintering trials at SLS to obtain suitable parameters.

1   INTRODUCTION

1.1   *Tissue engineering*

Tissue engineering is a growing field which combines the expertise of medical, biology and engineering. Tissue engineering offers many advantages, including providing an advanced alternative solution for tissue or organ replacements. Growing patients' own cells and tissues may no longer be impossible. This option reduces patients' waiting time for the implant from donors and also eliminates the chances of immunological rejection between donors and recipients.

However, the tissue engineering field presents a number of engineering challenges. Some of the challenges are to provide good scaffolds for the new tissues or cells to grow from suitable biomaterials (Griffith et al. 2002).

Tissue engineering scaffolds are 3-dimensional structures that function as temporary mechanical supports for the new tissues or cells to grow. As the tissues regain their natural strength, the scaffolds should naturally degrade.

Requirements of tissue engineering scaffolds include porous structures with the interconnected pores. A number of conventional methods have been proposed to make the scaffolds, yet they have many restrictions, such as the usage of toxic solvents in the process and the incapability of the techniques to produce scaffolds with controlled micro- and macroporosity (Yang et al. 2001).

1.2   *Selective laser sintering (SLS)*

Rapid prototyping (RP), a layer-by-layer manufacturing process, is the preferred alternative to fabricate tissue engineering scaffolds. RP has demonstrated the ability to build 3-dimensional intricate models with controlled micro- and macroporosity and high accuracy. Thus, this technology has been investigated for scaffolds fabrication to replace tissues and organs (Leong et al. 2003).

Selective laser sintering (SLS), one of the most versatile RP techniques, is being investigated in this research project. As with any other RP systems, SLS reads the 3-dimensional closed-volume models in .STL (StereoLithography) format. The computerized models were then sliced into very thin layers. SLS then build the objects from powder particles. A CO_2 laser scans and sinters each layer according to its cross-sectional area. After each layer is successfully sintered, the build platform is lowered for a new layer of powder to be refilled into the platform. The laser then sinters this new layer of powder and the activity continues until the part is fully fabricated (Chua et al. 2003).

Scaffold fabrication of biomaterials using SLS has been reported with favorable results (Chua et al. 2004a, b, Tan et al. 2003a, b).

1.3 Objective

Biomaterials are relatively sensitive materials that might require delicate processing treatment in order to avoid their properties degradation. Yet, until today, most of SLS scaffold fabrication research is heavily relying on trial-and-error experimental sintering work to get the optimum sintering condition for the biomaterials. This might lead to thermal degradation of the biomaterial, if the temperature distribution is not optimized.

This project aims to research on the relationship between the thermal properties of the biomaterials and the SLS laser beam properties. By recognizing the most influential parameters, SLS experiments could be set according to those parameters to prevent further thermal degradation.

2 THERMAL MODELING

2.1 Theoretical thermodynamics phenomena at selective laser sintering

Sintering is a thermal treatment to bond loose particles into a coherent, predominantly solid structure via mass transport events. These mass transport events normally occur in atomic scale. Particles fuse together when heated to a relatively high temperature, which is normally half of the absolute melting temperature (German 1996).

During sintering, particles absorb energy in the form of heat. Temperature of the particles rises and the particles softened. Necks are formed between particles, earliest at the initial contact point. The neck grows and the particles coalesce to one another to form a bigger piece of particle.

Sintering in SLS is one that results from a moving heat source, which is the laser beam. As such, there would be an overlap region of heat received by the powder particles, as shown in Figure 1.

The regions in gray and dark gray show the area that receives overlapping of heat as the laser moves, with scanning speed of a and $(a + x)$, respectively.

2.2 Theoretical formulae of heat transfer process

Taking a single powder particle as the control volume, the single particle would experience heat transfer process in terms of conduction, radiation and energy from the laser source. The powder particles are assumed as having spherical shape.

For a moving Gaussian laser beam, the energy balance at the control volume could be expressed as follows,

$$\frac{2k}{r}\frac{\partial T}{\partial r}+\frac{\partial}{\partial z}\left(k\frac{\partial T}{\partial z}\right)-\varepsilon\sigma\left(T_s^4-T_\infty^4\right)+q\left(x,y,z,t\right)=\rho c_p\frac{\partial T}{\partial t}$$

(1)

where $k =$ thermal conductivity of biomaterial; $r =$ radius of the spherical powder particle; $T =$ instantaneous temperature of the particle, $T_s =$ temperature of the surface; $T_\infty =$ temperature of the fluid; $=$ emissivity; $\sigma =$ Stefan-Boltzmann constant $= 5.67 \times 10^{-8}$; $\rho =$ density of powder bed; $c_p =$ specific heat of the biomaterial.

The function $q(x, y, z, t)$ represents the heat source function of the laser beam,

$$q(x,y,z,t)=(1-R)\beta I_0\exp\left[-\frac{\left(x-x_0-v_x t\right)^2+\left(y-y_0-v_y t\right)^2}{w^2}-\beta z_{ct}\right]$$

(2)

where $I_0 =$ initial light intensity; $R =$ surface reflectivity; $(x_0, y_0) =$ initial position of laser beam; $(x, y) =$ instantaneous position of laser beam; $(v_x, v_y) =$ laser beam scan speed in x- and y-directions; $\beta =$ absorption coefficient; $z_{ct} =$ depth of z-axis in the Cartesian coordinate system; $w =$ laser beam radius.

2.3 Experimental setup

An experiment was performed to get the temperature distribution profile from the laser beam, as shown in Figure 2.

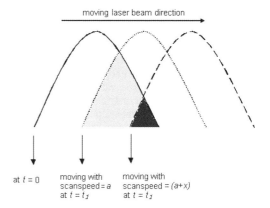

Figure 1. Sintering condition with a moving laser beam.

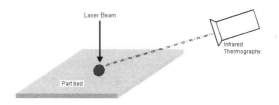

Figure 2. Temperature distribution experimental setup.

510

3 RESULTS AND DISCUSSIONS

3.1 *Temperature distribution analysis*

Temperature distribution analysis of the SLS was conducted using the experimental setup based as shown in Figure 2.

As the laser beam is shot onto the powder bed, the temperature distribution was captured using an infrared thermography. Initial measurement was to capture the temperature distribution when laser power was set to 3 W and 4 W, as shown in Figures 3 and 4.

From Figures 3 and 4 it can be seen that temperature distribution produced are different. Structures of the temperature distribution of both laser powers are nearly similar, as seen from the shapes of each circle that signifies the temperature range. However, the temperature difference between the two laser powers is very large even for 1 W difference. This difference is most significant at the innermost circle, which is exactly where the laser first hits before the heat is distributed to its surrounding.

Closed observation of the temperature maps reveals subtle differences in the distribution. The temperature rise is predominantly due to the laser power while the differences in the shape of the temperature map in the innermost region and surrounding are due to secondary terms of conduction and radiation heat transfer across the powder bed. Thus the temperature map justifies the thermal energy balance equation.

3.2 *Sintering trials of poly(vinyl alcohol) biomaterial powder*

Sintering trials of poly(vinyl alcohol) (PVA) biomaterial powder are used to verify the theoretical heat transfer study previously derived and the temperature distribution maps. PVA is a biodegradable polymer.

Figures 5 and 6 show sintering results of keeping scan speed constant at 1778 mm/s and varying the laser power at 13 and 14 W.

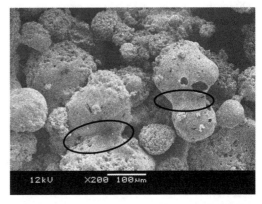

Figure 5. Sintering results of PVA at scan speed 1778 mm/s and laser power 13 W.

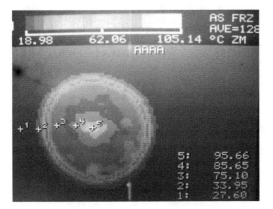

Figure 3. Temperature distribution at laser power of 3 W.

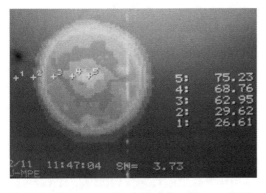

Figure 4. Temperature distribution at laser power of 4 W.

Figure 6. Sintering results of PVA at scan speed 1778 mm/s and laser power 14 W.

511

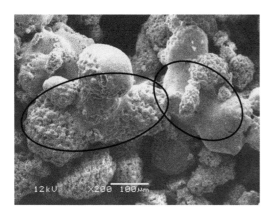

Figure 7. Sintering results of PVA at scan speed 1270 mm/s and laser power 13 W.

The black ellipses and rectangle in Figures 5 and 6 indicate the sintering effect, shown by the necking between powder particles. It can be shown that there is much difference in the sintering results. This shows that sintering laser power of 1 W higher gives larger necking among the powder particles, shown by the thicker necking when sintered at laser power 14 W. Some powder particles in Figure 6 have even clumped together into a bigger mass.

Figure 7 shows PVA powders sintered at scan speed 1270 mm/s and laser power 13 W.

Comparing Figures 5 and 7, it is seen that by lowering down the scan speed, thicker necking is obtained and hence larger sintering effect is produced.

4 CONCLUSION

A model based on energy balance relates the biomaterials' properties, namely thermal conductivity of the biomaterials and the laser properties, namely laser power and scan speed.

Greater sintering conditions were obtained by lowering scan speed or increasing laser power. This condition correlated with the energy balance equation.

Verification of sintering poly(vinyl alcohol) biomaterial powders also showed that scanning speed and laser power were indeed most influential to the sintering results.

Further experimental and theoretical studies should give more attention to scanning speed and laser power to obtain the optimal sintering conditions.

REFERENCES

Chua, C.K., Leong, K.F., Lim, C.S. 2003. *Rapid Prototyping: Principles and Applications*, 2nd ed. Singapore: World Scientific.

Chua, C.K., Leong, K.F., Tan, K.H., Wiria, F.E., Cheah, C.M. 2004a. Development of tissue scaffolds using selective laser sintering of polyvinyl alcohol/hydroxyapatite biocomposite for craniofacial and joint defects. *Journal of Materials Science: Materials in Medicine* 15: 1113–1121.

Chua, C.K., Leong, K.F., Wiria, F.E., Tan, K.H., Chandrasekaran, M. 2004b. Fabrication of poly(vinyl alcohol)/hydroxyapatite biocomposites utilizing rapid prototyping for implementation in tissue engineering. *Proc. International Conference on Competitive Manufacturing, Stellenbosch, 4–6 February 2004.*

German, R.M. 1996. *Sintering Theory and Practices*. New York: Wiley.

Griffith, L.G. & Naughton, G. 2002. Tissue Engineering – Current Challenges and Expanding Opportunities. *Science* 295: 1009–1014.

Leong, K.F., Cheah, C.M., Chua, C.K. 2003. Solid freeform fabrication of three-dimensional scaffolds for engineering replacement tissues and organs. *Biomaterials* 24: 2363–2378.

Tan, K.H., Chua, C.K., Leong, K.F., Cheah, C.M., Cheang, P., Abu Bakar, M.S., Cha, S.W. Scaffold development using selective laser sintering of polyetheretherketone-hydroxyapatite biocomposite blends. 2003. *Biomaterials* 24: 3115–3123.

Tan, K.H., Chua, C.K., Leong, K.F., Cheah, C.M., Gui, W.S., Tan, W.S., Wiria, F.E. 2003b. Selective laser sintering of biocompatible polymers for application in tissue engineerings. *Proc. Second International Conference on New Biomedical Materials, Cardiff, Wales, 5–8 April 2003.*

Yang, S., Leong, K.F., Du, Z.H., Chua, C.K. 2001. Review: The Design of Scaffolds for Use in Tissue Engineering. Part I. Traditional Factors. *Tissue Engineering* 7(6): 679–689.

Virtual modeling and rapid manufacturing – Bártolo (eds)
© 2005 Taylor & Francis Group, London, ISBN 0 415 39062 1

A predictive gas assist atomisation nozzle freeze-up model for micro spray metal deposition

Martin Jenkins & Richard Everson

School of Engineering, Computer Science and Mathematics, University of Exeter, Exeter, UK

ABSTRACT: Combining the low energy process of gas assist atomisation with a suitable RP methodology, micro spraying of a low melting point alloy (LMPA) offers a potential route for generating low cost one-off pre-production mould tools for the injection moulding of thermoplastic components. We describe and analyse the particular problem of atomiser nozzle freeze-up in micro spraying of a molten LMPA (MCP58: Bi 49%, Pb 18%, Sn 12%, In 21%) using gas assist atomisation. We discuss the profile of the rate of spray deposition decay that leads to eventual nozzle freeze-up, and further discuss how this phenomenon depends not only on the physical and thermal properties of the sprayed LMPA but also on the geometry and thermal properties of the atomiser nozzle. A simple theoretical model has been derived to describe the freeze-up process. It takes into account the time dependent heat transfer processes at the nozzle tip as well as the relative physical and thermal properties of the sprayed material and nozzle material. Simulations of time to nozzle freeze-up using the theoretical model show good agreement with experiment. We conclude with a brief analysis of the process of nozzle freeze-up, and its relevance to a practical RP system.

1 INTRODUCTION

One of the earliest references to the problem of nozzle freeze-up relating to metal spraying was that by Singer [1970] in his paper dealing with the production of metal sheet directly from sprayed molten metal. Shibani and Ozisik [1977] modelled the freeze-up phenomenon for low Prandtl number liquid metals flowing in pipes, but made assumptions that the pipe wall was at constant temperature, the flow was fully developed, and that factors such as thermal conductivity were constant. The model also indicated that the further the flow is along the pipe when measured from the axial entry point, the greater becomes the heat transfer rate, an effect compounded for a typical conical atomiser nozzle which acts as a very efficient spine cooling fin whose greatest heat transfer coefficient will occur at its tip. However, constant values are inappropriate as nozzle freeze-up is a time related process whereby temperatures and thus physical properties such as viscosity, thermal conductivity, and surface tension are changing. Shibani and Ozisik [1977] defined a freezing parameter which was related to the thermal conductivities of the solid and liquid phases of the metal indicating that the greater this ratio the greater was the cooling rate of the liquid metal. In the case of low melting point alloys (LMPA) which have a complex relationship between thermal conductivity and

temperature, the freeze factor developed by Shibani and Ozisik [1977] may well be significantly influenced not only by the temperatures of inlet, wall, and freeze point but also by the values of the thermal conductivities at those temperatures. A later work by Cho and Ozisik [1979] investigated the freezing of liquids flowing in turbulent flow inside circular tubes under transient conditions. They determined that the solid-liquid interface got closer to the central axis of the pipe with time but more significantly that its rate of translation increased at a faster rate for lower Prandtl numbers. So molten metals, which have very low Prandtl number have a faster pipe close up rate than materials having larger Prandtl numbers such as water. There is however a problem as the viscosity profile for an LMPA is such that the Prandtl number for the alloy will be high, in fact at very low laminar flow velocities extremely high, indicating the freeze-up process to be slow in comparison with other materials but experiment shows the freeze-up process occurs quite rapidly, so it does not appear as though the Prandtl number taken in isolation is sufficient as a guide to the freeze-up phenomenon, it must also be looked at in conjunction with the viscosity prevailing under the flow conditions. Shibani and Ozisik [1977] also used the freezing parameter $\bar{\theta}$ to show that the higher its value the greater was the rate of freezing and the thicker was the freeze layer at any given position along a pipe. Sampson and Gibson [1981] presented a mathematical model to not only describe the build up of the solidification layer but to also predict the conditions under which eventual blockage would occur. The model calculates the change in solidification layer radial co-ordinates for changes in temperature and time, and plots the results in a phase-plane plot. The model was dependent upon two dimensionless factors P_p and B_f, the pressure parameter and freeze parameter respectively. The pressure parameter is related to the Peclet number that contains a velocity function, thus P_p will decrease for increased flow and therefore a greater wall temperature θ_1 is required to stave off blockage. Conversely, and importantly, if no flow occurs blockage cannot occur even at a very low wall temperature. If however the pipe length l_0 is increased P_p is increased and blockage will occur at a lower wall temperature but again this must also take into account the flow. Phase-plane plots were then developed for a range of freeze parameter values and the curves used to indicate critical values of B_f leading to regions of blockage or non-blockage. The mathematical model developed for generation of the $P_p B_f$ curve was then tested against actual experimental data from work on freezing of hydraulic systems by Des Ruisseaux and Zerkle [1969] and showed good agreement. This analysis concurred with our findings where raising the temperature of the nozzle tip reduced the spray decay rate and extended the time to blockage.

A more recent analysis of the freeze-up phenomenon during gas atomisation of metals is that by Liu et al. [1995] who developed a numerical model to describe the flow and heat transfer behaviour of molten metals in the delivery tube of a gas atomiser. The model is a quantitative correlation obtained by regression analysis of numerical results which expresses nozzle freeze-up as being primarily caused by inadequate melt superheat, excessive residence time of melt in the delivery tube, and recirculation of under cooled droplets at the tube exit. The model assumes an overall heat transfer coefficient between the melt at the tube wall and ambience, and that nozzle freeze-up is only dependent upon processing parameters and material properties, and is not transient in nature. The model also indicated that making the nozzle tube from a material having a low thermal conductivity, coupled with a thick wall, would reduce the melt superheat required and thereby prevent delivery tube freeze-up. However, our experimentation demonstrates that nozzle freeze-up is a transient phenomenon, and a nozzle material of low thermal conductivity only accelerates the freeze-up process, see Figure 2.

The preceding theoretical models were all based on the premise that the wall temperature of the tube was constant and below the freeze temperature of the liquid and that there was no temperature gradient axially along the tube at time zero. More importantly

however, was that the heat flow into, or out of, the tube was not considered, only referring to the heat flow from the liquid flow within the tube.

2   SPRAY DECAY

Figure 1 illustrates a typical decay rate curve when spraying a molten LMPA.

Allowing for convective and radiation heat losses from the outer surfaces of the nozzle, the combined heat flow into the nozzle from alloy flow and nozzle conduction was insufficient to maintain the tip temperature above the solidus temperature of the alloy and thus prevent nozzle freeze-up. This indicated that the temperature profile of the nozzle inner surface and tip was changing with time even though the nozzle unit base temperature remained constant, and that the heat flow into the nozzle, particularly the tip region, was slower than the heat flow from the nozzle even allowing for the additional heat input due to the metal flow. Modelling the freeze-up process assuming a constant wall temperature cannot therefore be correct, as the process is transient in nature and the wall temperature profile is not constant. There must therefore be a changing heat flow into the nozzle even assuming a constant melt flow temperature through it.

Thermal conductivity of the nozzle material governs the rate of heat flow through the nozzle, as evidenced from Fourier's law;

$$qx'' = -k\frac{d\theta}{dx}$$

where qx'' is the heat transfer rate in the x direction per unit area, and $d\theta/dx$ is the temperature gradient in

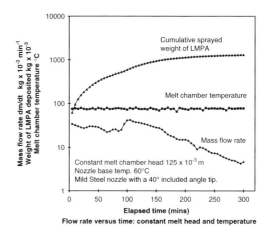

Figure 1.   Flow rate with constant nozzle base temperature.

514

the direction x, with k the thermal conductivity of the material.

The simplest way to vary the heat flow into the nozzle was to make a set of geometrically identical nozzles from three materials of distinctly different values of thermal conductivity, namely Acetal, Mild steel, and Duralumin aluminium alloy which have thermal conductivities of 0.23, 50, and 236 w/mK respectively.

Figure 2 illustrates that thermal conductivity of the nozzle material had a very significant effect on the spray rate and nozzle freeze-up rate, and therefore the characteristics of the heat flow into the nozzle cannot be ignored in the development of a theoretical model. The theoretical effect that nozzle angle might have on tip temperature, for the three nozzle materials, for various average heat transfer coefficients h, was also considered (see Figure 3).

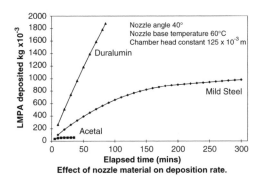

Effect of nozzle material on deposition rate.

Figure 2.   Effect of nozzle material on spray rate.

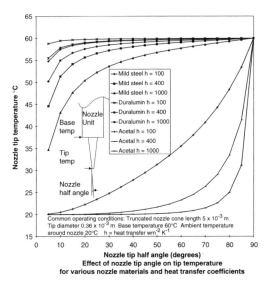

Common operating conditions: Truncated nozzle cone length 5 x 10⁻³ m Tip diameter 0.36 x 10⁻³ m Base temperature 60°C Ambient temperature around nozzle 20°C h = heat transfer wm⁻² K⁻¹

Effect of nozzle tip half angle on tip temperature for various nozzle materials and heat transfer coefficients

Figure 3.   Effect of nozzle cone angle, material, and heat transfer coefficient on tip temperature.

The general expression for the temperature profile of a conical spine fin:

$$\theta_i = (\theta_0 - \theta_a)\frac{\cosh m(l - x)}{\cosh ml} + \theta_a$$

where m is the general differential solution for conical spine fin efficiency:

$$m = \sqrt{\frac{2h}{k(l - x)\tan \phi}}$$

establishes a relationship between the nozzle material thermal conductivity k, the average surface heat transfer coefficient h, the nozzle cone angle ϕ and the tip temperature θ_i.

The modelling illustrates that a nozzle of a material having a low thermal conductivity, irrespective of the heat transfer coefficient, would give a low tip temperature and thereby be unsuitable for reducing the problem of nozzle freeze-up.

Epstein [1976] did consider the transient heat flow in the wall. The model took account of the movement of the phase conversion front within the freezing liquid and the transient conduction in the wall and demonstrated that the growth of the frozen layer within the pipe came to a stop when the conduction heat flux into the wall balanced convection from the liquid. At that instant, the frozen layer would then begin to melt and ultimately disappear. If values for Acetal, Duralumin, Mild Steel, and the LMPA were used in the Epstein model it showed that the solidification thickness-time history approached the conditions of an isothermal wall where a steady state heat balance occurs and the frozen layer never melts, i.e. nozzle freeze-up.

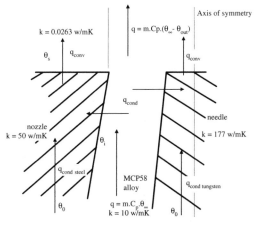

Figure 4.   Thermal processes taking place in the tip region of a spray nozzle unit.

It is evident that the problem of nozzle freeze-up is an extremely complex one and that as yet a general model that satisfactorily describes the freeze-up process has not been derived. Although various models predict freeze-up for very specific theoretical cases none are specific to the case of a spray nozzle configuration, and in particular what takes place at the very tip of a nozzle. The models discussed have been developed around a macro scale approach with undefined tube length and provide dimensionless proofs, and do not take into account the possible micro scale heat transfer effects that apply more specifically to very micro scale geometric aspects of a spray nozzle.

3 HEAT FLOW MODELLING

Analysis of the transient nature of the heat flow processes in the nozzle is limited by the unknown nature of the temperature profile for any tip region of the nozzle linked to unknown heat transfer coefficients, whether average contour values or localised values.

A spray nozzle can be considered to be a blunt ended, base heated conical spine fin, which because of its small size does not enable a temperature profile along its length to be obtained with any certainty, and so its tip temperature cannot be known with any certainty. All that can be stated is that for an internally heated truncated cone the tip face temperature would be above the ambient temperature, and have a temperature profile across it [Cho et al., 1997].

If the average heat transfer coefficient were known it would at least go some way to assisting with the derivation of a reasonably appropriate temperature profile for the nozzle.

From Fourier's law of heat conduction the flow of heat energy into the nozzle by conduction must be equal to the energy leaving the nozzle by convection to the surroundings before flow through the nozzle starts. Thus:

$$q_{cond} = q_{conv} = q_{ideal} = h\pi d_0 l \frac{\theta_0}{2}$$

where h is the average heat transfer coefficient for the nozzle surface, d_0 is the cone base diameter, and θ_0 is the base temperature excess [Kern and Kraus, 1972].

Unfortunately h cannot be found directly from this expression, as without knowledge of the temperature profile along the fin, q_{cond} cannot be derived. At best, the nozzle can be considered to be a very stumpy fin whose characteristic fin efficiency will be typified by $ml < 1$ [Lienhard, 1981], which concurs with Bejan [1993] who showed that in the case of tapered fins the heat transfer coefficient was proportional to the reciprocal of the taper constant, so for large angles of taper

the heat transfer coefficient will be low resulting in the maintenance of a high tip temperature. It would appear therefore that if the temperature excess were great enough i.e. a high base temperature, then a point would be reached where the tip temperature was greater than the freeze temperature. Unfortunately the argument is based upon an average heat transfer coefficient for the whole nozzle, i.e. a macro approach, whereas a micro approach shows the problem to be more complex.

3.1 Mass and heat transfer through small holes

Cho et al. [1997] used a naphthalene sublimation technique in order to measure the mass transfer taking place throughout different regions of a short hole. Direct measurement of the mass transfer enabled the mass transfer coefficient h_m to be derived. Using this with known values for the hole diameter and mass diffusion coefficient of naphthalene allowed the mass transfer profile throughout the hole to be expressed as a Sherwood Number. Given the analogy between heat and mass transfer the Sherwood number could be considered to also be an indicator of the heat transfer taking place across the profile of the hole. Their research showed that, as in fully developed tube flow, the transfer coefficient was constant along the fully developed flow region of the hole but that there was a sudden increase in the transfer coefficient at the very exit point of the hole. The indication was that the transfer coefficient at a sharp edge is high. Also the mass transfer across the end face of the nozzle was very low as a result of the low entrainment velocity on the surface compared to the exit velocity of the flow. In the context of spray nozzle development it would indicate that the heat transfer rate at the very inner edge of the nozzle hole is high but that it decreases rapidly within a short distance from this edge across the nozzle tip face. As a consequence the nozzle tip temperature would most likely be at a minimum just inside the very edge of the nozzle orifice and would be the location at which nozzle freeze-up would commence.

3.2 Nozzle freeze-up

Although the exact temperature profile for the region around the micro geometric features of the nozzle tip was unknown, it was clear that the nozzle freeze-up process started somewhere in the tip region. For the purpose of modelling the assumption was made that for a very short length dl the tip region could be considered as part of a thick walled, small bore tube of constant inner and outer diameters (Figure 5).

Assumptions were made that the process was steady state, all conduction and convection occurred radially, there was no axial temperature gradient, that physical properties were constant and independent of

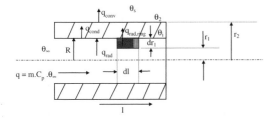

Figure 5. Nozzle schematic.

temperature change, and that representative theoretical values such as the inner nozzle tip heat transfer coefficient were valid. In addition it was assumed that the thermal conductivities of fluid and wall material were equal at the solidification boundary at incipient freeze-up, and that fluid flow was fully developed.

Radial heat flow $q_{rad} = q_{cond}$ = heat lost by convection q_{conv}:

$$q_{rad} = \frac{2\pi k}{\ln(r_2/r_1)}.dl.(\theta_1 - \theta_2) = q_{conv} = 2\pi r_2 h dl(\theta_2 - \theta_0) \tag{1}$$

Heat given up by the fluid must equal the radial heat loss:

$$\dot{m}C_p d\theta_1 = \frac{k(\theta_1 - \theta_2)}{\ln(r_2/r_1)} \tag{2}$$

and by association

$$q_{rad} = \frac{2\pi.dl.kr_2 h(\theta_1 - \theta_2)}{k + \ln(r_2/r_1)r_2 h} \tag{3}$$

As metal alloy flowing in the tube solidifies and starts to form a ring within the tube, then the volume of the frozen ring can be considered to be:

$$v = 2\pi.r_1.dr_1.dl$$

For a very thin ring radially the mass of the ring will thus be:

$$m = 2\pi.r_1.dr_1.dl.\rho$$

As further freeze-up takes place then the change in radius of the ring with time will be expressed by dr_1/dt and the radial heat loss from the ring will be:

$$q_{rad,ring} = 2\pi.r_1.\frac{dr_1}{dt}\rho.l.dl \quad \text{(For all } l) \tag{4}$$

The heat lost by the ring must equal the radial heat flow, so equating expressions (3) and (4), rearranging and integrating with respect to r_1:

$$\frac{k.r_2.h(\theta_1 - \theta_0).dt}{\rho l} = [k + r_2 h \ln(r_2)] r_1^2/2$$

$$-r_2 h r_1^2 \left[\ln(r_1) - \frac{1}{2}\right] + C \tag{5}$$

At time $t = 0$, r_1 will equal the total tube radius R:

$$C = r_2.h.\frac{R^2}{2}\ln(R - \frac{1}{2}) - [k + r_2.h.\ln(r_2)]\frac{R^2}{2} \tag{6}$$

So at any time t when the tube radius equals $R - r_1$,

$$t = \frac{\rho l}{kr_2 h(\theta_1 - \theta_0)}\left[(k + r_2.h.\ln(r_2)/2)\left(r_1^2 - R^2\right) - \right.$$

$$\left. \frac{r_2.h}{2}\left\{R^2\left(\ln(R) - 1/2\right) - r_1^2\left(\ln(r_1) - 1/2\right)\right\}\right] \tag{7}$$

The model (7) relates the mass flow rate to time as a result of the change in orifice radius with time. For a very thin wall tube r_2 can be considered to equal r_1 at time zero. The model can be further refined by linking the relative thermal properties of fluid and wall material [Epstein, 1976],

$$\epsilon = \sqrt{\frac{k_2\rho_2 C_{p2}}{k_1\rho_1 C_{p1}}} \tag{8}$$

where k, ρ, and C_p are the thermal conductivity, density, and specific heat respectively, and can also be used to compare the freeze-up in tubular flow with that of annular flow when the nozzle contains a flow control needle by relating r_1 to the appropriate Poiseuille mass flow rate expression (Figure 6).
For tube flow:

$$\dot{m} \propto r_1^4 \tag{9}$$

For annular flow:

$$\dot{m} \propto r_1^4 - r_3^4 - \frac{(r_1^2 - r_3^2)^2}{\ln(r_1/r_3)} \tag{10}$$

The model predicts the times to nozzle closure in accordance with the nozzle material in a manner that concurs with experiment. However, the model is imperfect as a result of the process assumptions made and thus still predicts times to freeze-up greater than

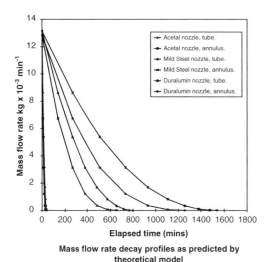

Mass flow rate decay profiles as predicted by theoretical model

Figure 6.   Nozzle freeze-up predicted by theoretical model.

experimental results. In addition the model cannot take into account other factors that may influence nozzle closure such as nozzle hole surface finish, nozzle efflux geometry, the build up of droplet particles around the orifice due to recirculation zones, and static attraction impeding efflux of sprayed material from the nozzle.

4   CONCLUSIONS

A relatively simple predictive model has been derived to predict time to freeze-up in a micro spray gas assist atomizer nozzle unit and shows encouraging agreement with experiment. Our model does take into consideration the heat flow into, and out of, the nozzle via base heating as well as the heat flow from the liquid flow within the nozzle. Although more refined models have been developed by others they have not proved to be any more useful than the simple approach developed here in terms of being a guide to the tip design of the nozzle. It appears that when metal flow is initiated, the nozzle behaves as an open tube with a tip temperature below the solidus of the metal now flowing through it. The molten metal thus starts to solidify at the cold boundary region close to the tip, and this reduction in orifice area starts to reduce the flow rate.

The heat flow rate into the tip from the metal flow is insufficient to raise the tip temperature and so the boundary stays below the solidus and further solidification build up takes place. This process continues until total blockage occurs. Once freeze-up occurs the nozzle becomes a conical spine fin whose tip temperature is below the solidus of the metal and heat flow into the tip from the combined heat flows through the nozzle

material and melt material are still insufficient to raise the tip temperature to a level sufficient for a resumption of flow [Epstein, 1976]. The only way flow can be induced under these conditions is by raising the temperature of the nozzle tip by raising the temperature of its surroundings. If the mass flow rate of melt through the nozzle is increased then the rate of heat loss from the melt at the cold boundary is reduced and the time to freeze-up is increased. Even if the overall temperature of the nozzle tip was increased beyond the solidus the high heat transfer rates of micro geometric forms such as sharp edges will still result in those features having temperatures below the solidus with the eventual effect of slow solidification build up of the melt leading to freeze-up, albeit over a greatly extended time period.

It would appear that the only sure way to eliminate nozzle freeze-up is by operating the spray nozzle in an environment whose temperature is above the solidus of the sprayed metal, which would negate its use as a process for RP. So although combining the low energy process of micro nozzle atomisation of a LMPA with a suitable RP methodology offers attractive possibilities for the generation of low cost one-off pre-production mould tools for the injection moulding of thermoplastic components, it is unlikely to be a practical proposition.

REFERENCES

A. Bejan. Heat Transfer as a Design-Oriented Course: Mechanical Supports as Thermal Insulators. *Int. J. of Mechanical Engineering Education*, 21(1), 1993.

C. Cho and M.N. Ozisik. Transient Freezing of Liquids in Turbulent Flow Inside Tubes. *ASME J. Heat Transfer*, 101:465–468, 1979.

H.H. Cho, M.Y. Jabbari and R.J. Goldstein. Experimental Mass (Heat) Transfer In and Near a Circular Hole in a Flat Plate. *Int. J. Heat Mass Transfer*, 40 (10):2431–2443, 1997.

N. Des Ruisseaux and R.D. Zerkle. Freezing of Hydraulic Systems. *Canadian J. Chem. Eng.*, 47:233–237, 1969.

M. Epstein. The Growth and Decay of a Frozen Layer in Forced Flow. *Int. J. Heat Mass Transfer*, 19:1281–1288, 1976.

D.Q. Kern and A.D. Kraus. *Extended Surface Heat Transfer*. McGraw-Hill Book Company, New York, USA, 1972.

J.H. Lienhard. *A Heat Transfer Textbook*. Prentice-Hall Inc, New Jersey, USA, 1981.

H. Liu, E.J. Lavernia and R.H. Rangel. An Analysis of Freeze-up Phenomena During Gas Atomisation of Metals. *Int. J. Heat Mass Transfer*, 38(12):2183–2193, 1995.

P. Sampson and R.D. Gibson. A Mathematical Model of Nozzle Blockage by Freezing. *Int. J. Heat Mass Transfer*, 24:231–241, 1981.

A.A. Shibani and M.N. Ozisik. A Solution of Freezing of Liquids of Low Prandtl Number in Turbulent Flow Between Parallel Plates. *ASME J. Heat Transfer*, pages 20–24, February 1977.

A.R.E. Singer. The Principles of Spray Rolling of Metals. *Metals and Materials*, 4:246–250, 1970.

Virtual modeling and rapid manufacturing – Bártolo (eds)
© 2005 Taylor & Francis Group, London, ISBN 0 415 39062 1

Heat sink silica effect: A new approach using ANSYS program simulation in Thermal Stereolithography (TSTL) process

S.R. Andrade, A.L. Jardini, R.A. Rezende, R. Maciel Filho, & M.A.F. Scarparo
Faculty of Chemical Engineering, State University of Campinas, S.P., Brazil

ABSTRACT: This paper introduces the concept of combining both form (CAD models) and behavior (simulation models) integrated on Stereolithography system. By connecting these tools, designers can create both a system level design description and a virtual prototype of the system. This virtual prototype, in turn, can provide immediate feedback about design decisions by evaluating whether the functional requirements are met in simulation. A finite element model in ANSYS program has been used for the simulation of the curing process during Thermal Stereolithography (TSTL) process. The model takes into account both the thermal and silica the heat sink phenomena involved in the process. Thermal Stereolithography is a manufacturing technique which uses a CO_2 laser beam to cure (solidify) thermoset resin in a selective way to produce three dimensional parts. This technique was initially developed at the State University of Campinas at Brazil. The materials employed are epoxy resin, diethylenetriamine as curing agent and silica powder. The physical process associated to this technology includes heat sink and localized curing of silica powder. In this analysis, the temperature distribution at each point of materials heated by a laser irradiation, and the thermal properties of thermoset resin are investigated. The results are in agreement with the experimental previous. The ANSYS program used in this work could also be integrated in the strategy to virtual prototyping, where the CAD models can be simulated directly trough this interface program.

1 INTRODUCTION

In modern engineering process, designers, engineers, and technicians commonly use various computer assisted technologies to evaluate their products at each stage in the product development cycle. These computer assisted technologies comprise: Computer Aided Design (CAD), Computer Aided Engineering (CAE) and Computer Aided Manufacturing (CAM).

CAE systems, most widen are based on the Finite Elements Method (FEM), analyses the CAD design in much little pieces, solving a set of algebraic equations in order to get desired results in function of loading and the boundary conditions. CAE softwares based on the analyses by finite elements are used to calculate tensions, shifts, vibration, heat transfer, fluid flowage, industrial installations and other applications. In this work, as a CAE tool, the ANSYS program is used to build a numerical model able to simulate the heat transfer and the heat sink phenomena during localized cure and to analyze the curing profile of thermosensitive material in function of its physical properties and of laser parameters. Taking into account previous experimental results it has been carried out the system optimization using ANSYS analyses.

A different approach to the production of three dimensional models, which can be called Thermo Stereolithography (TSTL) process, is presented. This novel process is a Rapid Prototyping (RP) technique, which uses a CO_2 laser beam to cure, or solidify, thermosensitive polymeric materials to produce three dimensional parts (Chen et al, 1992). The main advantage and need of Stereolithography process is to obtain a highly localized curing. In this work, it has been examined some of the physics and chemistry of the process, with an emphasis on understanding the parameters which most affect our ability to restrict or localize curing to a small region (Scarparo et al., 1997). Results are encouraging because they show that it is effectively possible confine the heating of the material and, therefore, its curing occurs into small regions about the order of the laser beam size, in three dimensions. To be able to localize curing in the sample, a specific ratio of epoxy, diethylenetriamine, and silica powder is required. Improper ratios of the reactants lead to the formation of low-molecular-weight molecules and it results in a sample which is either poorly cured or completely uncured. The amount of silica, relative to the amounts of epoxy and diethylenetriamine, is critical to confining the curing process to a localized volume.

Because the thermal conductivity of silica is greater than that of uncured epoxy, it appears that the silica acts as a heat sink to localize laser heating and limit the reaction volume.

Simple model to describe the flow of energy in laser-induced curing had been constructed as a function of laser parameters such as power, beam diameter, and scanning speed. By dividing the beam diameter 2ω by the scanning speed υ, one obtains the dwell time τ_d,

$$\tau_d = \frac{2\omega}{\upsilon} \tag{1}$$

which is the average time that any spot on the laser scanning path is irradiated.

In order to predict the effect that the laser has on curing it is important to determine how much energy was delivered to the sample and over what volume that energy was distributed. Because the sample is highly absorptive at the CO_2 laser wavelength, nearly all of the energy in the beam during the dwell time is absorbed by the sample within a distance from the surface equal to the absorption depth δ. The absorption depth was determined by measuring the transmittance of sample of uncured material over a wavelength range which included that of the CO_2 laser at 10.6 microns. Small cylindrical volume V, was assumed, that a absorbed an energy E during the dwell time, where V is given by (Equation 2)

$$V = \pi\omega^2\delta \tag{2}$$

and ω is the radius of the laser beam.

The energy deposited in volume V is the product of the laser power and the dwell time (Equation 3)

$$E = P\tau_d \tag{3}$$

Once the curing process is described as a function of temperature and time, the curing behavior can be predict as a function of laser irradiation conditions.

2 THERMO STEREOLITHOGRAPHY (TSTL)

Figure 1 illustrates a Thermo Stereolithography (TSTL) apparatus composed by a CO_2 laser with a laser beam focused and directed through an optoelectronic system, creating a cured layer.

In previous work (Jardini et al, 2003), the diagnostic tests of the laser scanning parameters to obtain the localized curing in thermosensitive process has been performed using the apparatus shown in Figure 1, as well as the laser parameters, involved in the curing

Figure 1. Thermo Stereolithography apparatus.

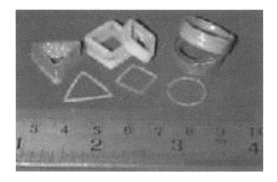

Figure 2. Multilayer parts constructed to different geometries.

process. A CW CO_2 laser (10.6 microns) with a beam waist of about 0.8 mm was used. A control system is used to scanning laser over several different trajectories on the sample surface, varying the laser speed and the laser revolution time.

Figure 2 illustrates the parts with different geometries obtained. Note that the parts obtained do not experience appreciable dimensional contraction and that the final parts are firm and stable.

2.1 Materials and characterization

Thermosensitive material consists of an epoxy resin mixed with silica powder as curing agent and diethylenetriamine as filling material. The epoxy resins were chosen based on their viscosity, thermosensitive and stability during the cure process. In order to follow the experimental model, the appropriate amount of reactants compatible with the desirable localized curing properties has to be determined. The sample behavior in terms of silica absorption of energy is of great use for study the heat sink effects and curing profiles. To study the infrared absorption of samples with different amount of silica, an optical characterization has been obtained. The transmittance of samples with

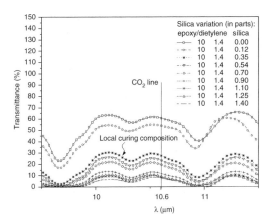

Figure 3. Transmittance in terms of silica variation in the composition.

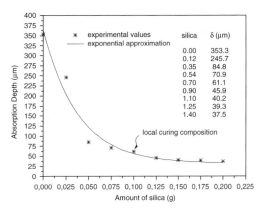

Figure 4. Depth of optical absorption in terms of silica variation.

different amounts of silica powder was measured using infrared spectrometer.

In order to determine the absorption depth in each sample, the semi-liquid sample was inserted between two KBr crystals (transparent to infrared light). An analytical solution by applying Lambert-Beer Law ($I_T = I_0 \exp^{-x/\delta}$), a satisfactory estimate was established. Considering that the intensity is partly reflected and attenuated by the crystals it is possible to determine the depth of energy absorption by the sample. The determination of δ is given by (Equation 4):

$$I_T = I_0 (1-R)^2 \exp^{-\alpha_c (x_{c1}+x_{c2})} \exp^{\frac{x_s}{\delta s}} \qquad (4)$$

where: I_T/I_0 = sample transmittance, R = KBr crystal reflectance, x_c = crystal thickness, x_s = sample thickness, α_c = optical absorption coefficient of the crystal, δ_s = sample absorption depth.

Figure 3 shows different transmittance behavior as silica is added to the sample, along the CO_2 line (at 10.6 microns).

From this experiment we could establish the relationship between the amount of silica variation and the depth of the optical absorption (Figure 4).

Considering the importance of silica in the formation of the appropriate composition of the thermosensitive resin, it is necessary to analyze the effect since it is considered to be one of the main parameters in the physical interpretation and development of the process to obtain local curing. This work intends to show a new numerical simulation, since in previous model (Scarparo, 1995) silica had not been considered in the process. As it has been mentioned before, silica plays an important role in thermal process.

3 THE NEW NUMERICAL MODEL

To the best of our knowledge, in order to optimize the thermal curing process, it has been developed a numerical model to simulate the temperature evolution at samples produced by a repetition rate pulsed laser. The numerical model has been developed using ANSYS computer code, which apply the Finite Element Method (FEM) to solve general problems in engineering area. The FEM has been used for several authors since 1967 (Zienkiewicz et al, 1977) to solve problems of structural and fluids mechanics, diffusion problems, chemical transport, heat transfer and phase change analysis (Andrade, 2003). In this work, the ANSYS code is used as tool and a numerical model is developed in order to analyze the heat transfer at thermosensitive materials during cure by laser exposure. The equation governing the conduction of heat can be derived from the principle of energy conservation and the laser radiation is tacked into account as a heat source and it can be written as (Equation 5):

$$\rho c_p \left(\frac{\partial T}{\partial t} + V_x \frac{\partial T}{\partial x} + V_y \frac{\partial T}{\partial y} + V_z \frac{\partial T}{\partial z} \right) =$$
$$\ddot{q} + \frac{\partial}{\partial x}\left(k_x \frac{\partial T}{\partial x} \right) + \frac{\partial}{\partial y}\left(k_y \frac{\partial T}{\partial y} \right) + \frac{\partial}{\partial z}\left(k_z \frac{\partial T}{\partial z} \right) \qquad (5)$$

where ρ is the density, c_p is the specific heat, k is the material conductivity, T is the temperature of sample, V is the velocity for transport of heat, q is the heat generation rate per unit volume. The amount of heat in the sample due to the laser beam scanning is taking into account in this heat generation term. The value of heat generation, is this model, is consider depending on laser power and volume irradiated as shown in Equation 6:

$$q = \frac{P}{\pi \varpi^2 \delta} \tag{6}$$

where P is the laser power, ω is radius of the laser beam and δ is the absorption depth. It has already been discussed that this parameter (δ) depends on amount of silica and its values has been showed by Figure 4.

4 RESULTS

The numerical model has been developed to calculate the temporal and spatial variation of temperature, during cure taking into account a repetition of laser beam pulse. In the model developed, the thermal physical properties vary according to silica amount in the sample.

It has beam analyzed the thermal behavior of samples with 3 different amounts of silica. Table 1 shows the physical properties of samples and the absorption depth (δ) used in each simulation. The physical properties of air are also listed in it. The geometry considered in the simulations represents a transversal cut of a piece of sample, in contact with air, under laser beam exposure. Figure 5 shows a representation of the geometry considered.

The results show the temperature distribution at two different locations at sample: at the center of laser beam and at 0.4 mm distant from the center (beam

Table 1. Physical properties of samples.

Samples	k (W/m.K)	ρ (Kg/m³)	c_p (J/g.K)	δ (m)
1 (without silica)	0.036	1155	1410	0.00035
2 (3,5% silica)	0.039	2005	1396	0.00008
3 (7% silica)	0.043	2799	1384	0.00006
Air	0.026	1.16	1007	

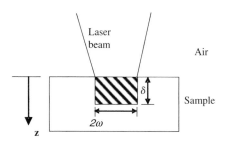

Figure 5. Representation of geometry considered in numerical model.

waist). In order to analyze the heat diffusion of the samples with different amounts of silica, the figures show the difference of temperature (ΔT) between the two considered locations in samples. The simulations represent the temperature variation at sample and in the air during 10 lasers pulses. The dwell time and the interval between pulses depend on scan speed, two different speeds have been considered. Table 2 shows the scan speeds and the respective dwell time and interval between pulses (repetition rate). The time relative of the interval between pulses is related to the trajectory of laser beam, in this case, it is considered to be a circle in order to produce a cylindrical shape object as one showed previously in the Figure 2. Figures 6 to 11 show the results obtained

Table 2. Scan speeds and respective parameters of laser beam used in simulations.

	Scan Speed (m/s)	Dwell time (s)	Repetition rate (s)
Case 1	1.6	0.000502	0.035
Case 2	2.39	0.000334	0.020

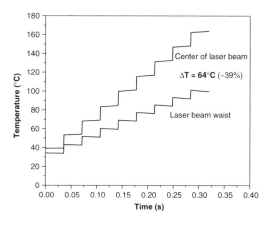

Figure 6. Time evolution of surface temperature (sample 1, case 1).

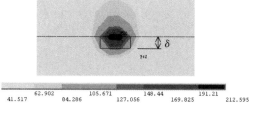

Figure 7. Cross section isothermal profile: lateral position on surface relative to the beam center (mm) vs normal distance from surface (sample 1, case 1).

from simulations of the thermal curing process of the samples 1, 2 and 3 respectively, using the parameters relatives to laser scan speed 1 (case 1). Figures 12 to 17 show the similar results using the parameters for laser scan speed 2 (case 2).

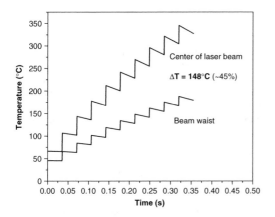

Figure 8. Time evolution of surface temperature (sample 2, case 1).

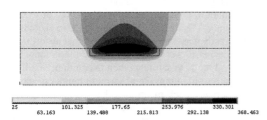

Figure 9. Cross section isothermal profile (sample 2, case1).

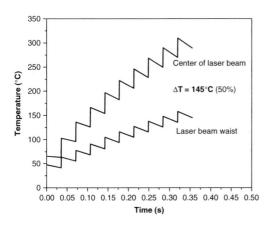

Figure 10. Time evolution of surface temperature (sample 3, case 1).

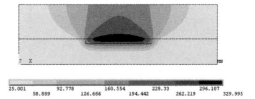

Figure 11. Cross section isothermal profile (sample 3, case 1).

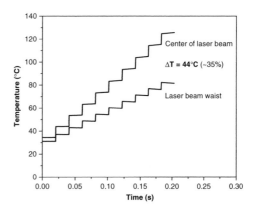

Figure 12. Time evolution of surface temperature (sample 1, case 2).

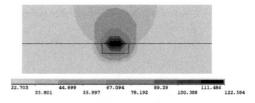

Figure 13. Cross section isothermal profile (sample 1, case 2).

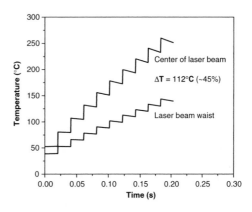

Figure 14. Time evolution of surface temperature (sample 2, case 2).

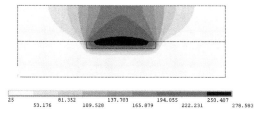

25   81.352   137.703   194.055   250.407
  53.176   109.528   165.879   222.231   278.583

Figure 15.   Cross section isothermal profile (sample 2, case 2).

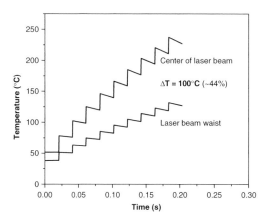

Figure 16.   Time evolution of surface temperature (sample 3, case 2).

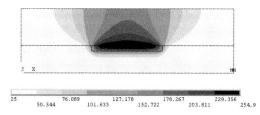

25   76.089   127.178   178.267   229.356
  50.544   101.633   152.722   203.811   254.9

Figure 17.   Cross section isothermal profile (sample 3, case 2).

5   CONCLUSIONS

This new numerical simulation of the thermal stereo-lithography process using infrared irradiation seems to be an important tool to prove the efficiency of silica powder in the localized cure process. The temperature variation for each tested case demonstrated that silica works as a heat sink. Because of the low thermal conductivity of the sample, each radiation of the site

results is essentially adiabatic, rise in the temperature which does not decay to ambient between irradiations. The temporal behavior of the temperature of an irradiated site would be a series of temperature spikes on a constant background with a decay time equal to the cycle time (repetition rate). Because of the simplifying assumption of temperature independent thermal properties, each adiabatic temperature spikes is about the same, in each case.

These results shows that the confined laser beam depend upon of the amount of the silica in the sample. By looking Figures 6 to 11 it is possible compare the laser beam waist with the center of the laser beam using different percentages of silica, according to Tables 1 and 2. For the same number of pulses we may detect different temperature variation.

In the samples 1, 2 and 3, the temperature variation ΔT in each case is showed and it shows the fact that the temperature difference increase is confined in region where silica acts as a heat sink. If the amount of silica is too small, curing is not localized. Conversely, if the amount of silica is too large, the sample does not cure properly. Its interesting to observe that when the amount of silica increases from 0 to 3.5%, the ΔT temperature between the center of the laser beam and the laser beam waist increase. If the amount of silica, increase by the double of value, for instance, 7%, the ΔT temperature starts to decrease.

It already was observed in Figure 4 that absorption depth decreases very fast as increase the amount of silica and tends to saturate when the amount of silica is enough to maintain the absorption depth constant. It had already been mentioned that the experimental results obtained through the variation of depth of optical absorption clearly indicate the necessity of controlling the thickness of the cured part in function of variation of silica amount in the sample.

It is necessary to have a very good stoichiometric composition to obtain the localized cure and amount of silica has to be chosen to successfully confine heat laterally within the diameter of the laser beam and vertically the prototype layers are mainly dependent on the absorption depth. The numerical model developed seems to be an important tool to study this influence of silica, as well, it is possible to continue the investigation of influence of other parameters in order to optimize the Stereolithography process.

REFERENCES

Andrade, S.R. 2003. "Simulação Numérica do Processo de Solidificação de Placas Finas com Redução da Espessura com Núcleo Líquido", *Tese de Doutarado,* Faculdade de Engenharia Mecânica, UNICAMP, Campinas, SP.
Jardini, A.L., Maciel Filho, R., Scarparo, M.A.F., Andrade, S.R., & Moura, L.F.M. 2004. "Infrared Laser

Stereolithography: Prototype Construction Using Special Combination of Compounds and Laser Parameters in Localized Curing Process", *Journal of Materials and Product Technology*, (21): 241–245.

Chen, H., Corbel, S., Allanic, A.L., & Andre, J.C. 1992. "Solid fabrication induced by thermal IR lasers", *Proceeding of 3rd International Conference on Rapid Prototyping*: 3–13.

Scarparo, M.A.F., Barros, M.L., Kiel, A., Gerk, E., & Hurtack, J.J. 1996. "Mechanisms of carbon dioxide laser stereolithography", *Applied Surface Science,* 106: 275–281.

Scarparo, M.A.F., Kiel, A., Zhiyao, Z., Ferrari, C.A., Chen, Q.J., Miller, J.H. & Allen, S.D. 1997. "Study of resin based materials using CO_2 laser stereolithography", *Polymer,* 38 (9): 2175.

Zienkiewicz, O.C., Parekh, C.J., & Wills, A.J. 1676 The application of finite element method in heat conduction problems involving latent heat. Rock Mechanics, 5: 65–76.

Virtual modeling and rapid manufacturing – Bártolo (eds)
© 2005 Taylor & Francis Group, London, ISBN 0 415 39062 1

Wrapping algorithms for multi-axis additive rapid prototyping

Neil Sewell, Richard Everson & Martin Jenkins
School of Engineering, Computer Science and Mathematics, University of Exeter, Exeter, UK

ABSTRACT: The 'bottom-up' layered manufacturing approach to rapid prototyping of free-form artifacts has gained prominence in recent years. However, the method is generally implemented on specialised, purpose-built machinery and requires special support structures wherever overhangs are present. Vertical machining centres provide a platform allowing 4 or 5 axis positioning of an artifact and deposition heads may be straightforwardly added to them. In this paper we investigate methods for additive rapid prototyping taking advantage of this ability to deposit material from a range of directions.

We develop algorithms to define a build path to be followed by a deposition system which is able to rotate the artifact during deposition. In outline these algorithms follow the artifact's surface, while 'wrapping' it with a bead or ribbon of material. Criteria for optimum build quality cannot be simultaneously satisfied for many surfaces. The problem is most severe when the corner between two non-coplanar facets is being rounded. We therefore define measures of the build error and develop straightforward optimisation schemes to minimise the error. Speedy optimisation is facilitated by specialised data-structures permitting rapid location of a facet and its neighbours. The method is demonstrated on simple artifacts, where we show that the build quality is effectively enhanced by the optimisation.

1 INTRODUCTION

Rapid Prototyping (RP) is now an integral part of the design and testing phases of manufacture. Solid Freeform Fabrication from CAD-generated descriptions can be achieved with a variety of technologies such as selective laser sintering, stereo-lithography, 3-D printing, laminated object manufacturing and fused deposition modelling; see, for example, Beaman, Barlow, Bourell, Crawford, Marcus, and McAlea (1997). The majority of these deposition systems use the layered manufacturing or 'bottom-up' approach: the artifact to be manufactured is 'sliced' into a series of horizontal layers, which are then built successively, deposited or attached, one upon another.

The beauty of the layered manufacturing approach lies in its simplicity: any artifact can be sliced into layers, each of which is built independently of the others. However, parts of a layer that overhang previously built layers must be supported. In current systems support is provided either by purpose built support structures (e.g., in stereo-lithography) which must be removed during hand finishing or by the rapid prototyping material in its raw state, such as unsintered material in selective laser sintering systems. An alternative to these methods is to rotate the artifact during construction so that new material is deposited directly upon previously built material, thus obviating the need for support structures. Although it might appear that the machinery necessary would be prohibitively expensive it is, in fact, integral to many Vertical Machining Centres (VMCs) whose machine bed can be rotated about an additional axis (4-axis machines) or pair of axes (5-axis machines). VMCs are already owned by many engineering businesses, and the promise of being able to augment a VMC with a RP facility has led us to investigate RP using hot-melt adhesive, for example, (Bostik 2002, Sealock 2002).

In our prototype system artifacts are constructed on the VMC bed which is moved under a hot-melt glue depositor mounted on the vertical axis. A bead of constant diameter, h, can thus be deposited. Hot-melt adhesive is used as a low cost alternative to casting wax for cores for investment casting. The adhesive has a short open-time, good resistance to aging degradation, is non-soluble, and non-toxic. Sewell (2005) gives full details of the deposition system and its use for bottom-up slicing and multi-axis RP; see also (Sewell, Jenkins, and Everson 2003).

RP utilising VMCs is attractive for a variety of reasons. Many small and medium sized engineering businesses already own VMCs and the cost of adding a hot-melt adhesive deposition unit is small, far less than the cost of a full RP system. In addition multiaxis deposition offers the possibility of building models that cannot be built by traditional RP methods and it is feasible to attach multiple depositors to a VMC tool carousel permitting artifacts to be built from several materials.

In this paper we examine the use of the fourth and fifth axes available on VMCs to avoid the construction of support structures. In outline we develop an algorithm that follows the artifact's surface, while 'wrapping' it with a bead or ribbon of material. By rotating the artifact under the deposition nozzle we ensure that material is always deposited directly upon already built material: in effect we rotate away the overhang.

In section 2 we describe criteria to ensure good build quality and describe the algorithm in more detail. It is impossible, for many surfaces described by triangular tessellations, to simultaneously satisfy all the criteria. The problem is most severe when the corner between two non-coplanar facets is being rounded. We therefore define measures of the build error in section 3 and develop straightforward optimisation schemes to minimise the error. We conclude in section 4 with a discussion on the feasibility of wrapping methods for rapid prototyping and enhancements to the algorithm possible if the diameter of the deposited bead can be accurately controlled.

2 WRAPPING

The STL format, which describes an artifact in terms of a tessellation of its surface into triangular facets, is by far the most common format for the description and interchange of CAD data, and we therefore consider the construction of a surface described in terms of a collection of connected triangular facets. As explained above, our deposition system deposits a bead or filament of material from a vertically oriented nozzle onto previously deposited material or, at the start of a build, onto the machine bed. The process can be likened to tightly wrapping the object with a cord or ribbon (the deposited material) so that successive loops of the cord always touch the previous loop.

As illustrated in Figure 1, within each facet the material is deposited along a straight line segment. Experiments and observations with bottom-up building show that the following three criteria ensure a build with high surface quality and structural properties:

1. The deposition nozzle should always be in the plane of the facet across which it is depositing.
2. The deposition nozzle should always be perpendicular to the bead.
   This ensures maximum clearance between the deposited material and the nozzle.
3. The current bead should be deposited exactly one bead diameter, h, from the previously deposited material or the build bed. Holes will be left in the

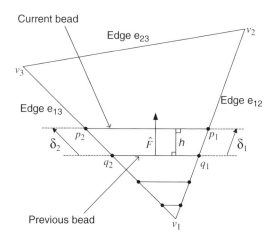

Figure 1.   Depositing successive beads of material across a facet defined by vertices $\mathbf{v}_1$, $\mathbf{v}_2$ and $\mathbf{v}_3$. The previously laid bead is $\mathbf{q}_2 - \mathbf{q}_1$; the current bead is $\mathbf{p}_2 - \mathbf{p}_1$.

surface if the bead is deposited too high, whereas there will be unwanted accumulation of material if the bead is laid too low.

These criteria straightforwardly determine the path for a bead within each facet. If, as shown in Figure 1, the vertices of the facet are $\mathbf{v}_1$, $\mathbf{v}_2$ and $\mathbf{v}_3$ and the previous bead has been laid from $\mathbf{q}_1$ (on edge $\mathbf{e}_{12}$) to $\mathbf{q}_2$ (on edge $\mathbf{e}_{13}$), then the coordinates of the ends of the line segment defining the current bead are calculated as:

$$\mathbf{p}_1 = \mathbf{q}_1 + \boldsymbol{\delta}_1 \tag{1}$$

$$\mathbf{p}_2 = \mathbf{q}_2 + \boldsymbol{\delta}_2 \tag{2}$$

where

$$\boldsymbol{\delta}_i = \frac{h}{\mathbf{e}_{1i}^T \hat{\mathbf{F}}} \mathbf{e}_{1i} \tag{3}$$

and $\hat{\mathbf{F}}$ is the unit vector perpendicular to the previous bead, $\mathbf{q}_2 - \mathbf{q}_1$ in the plane of the facet:

$$\hat{\mathbf{F}} = \frac{(\mathbf{q}_2 - \mathbf{q}_1) \times \hat{\mathbf{n}}}{\| (\mathbf{q}_2 - \mathbf{q}_1) \times \hat{\mathbf{n}} \|} \tag{4}$$

where $\hat{\mathbf{n}}$ is the normal to the facet. Note that the $\boldsymbol{\delta}_i$ need only be calculated once for each facet: on successive loops the entry ($\mathbf{p}_1$) and exit ($\mathbf{p}_2$) points of the bead are efficiently updated using (1) and (2). When the bead reaches a corner (e.g., $\mathbf{p}_2$ reaches $\mathbf{v}_3$ in Figure 1) before the facet has been completely constructed, a new exit or entry point and $\boldsymbol{\delta}_i$ must be recalculated for the new edge ($\mathbf{e}_{23}$ in Figure 1).

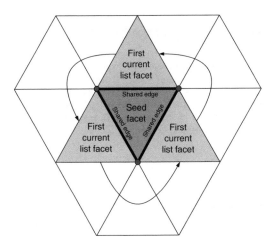

Figure 2.   Facets sharing a vertex with the seed facet(s) (dark grey) comprise the initial loop.

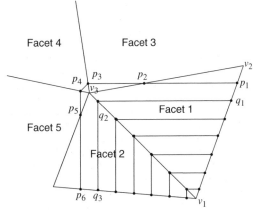

Figure 3.   Incorporation of new facets (facets 3, 4 & 5) into the current loop as the current loop moves past $\mathbf{v}_3$.

Once the line segment $\mathbf{p}_{12} = \mathbf{p}_2 - \mathbf{p}_1$ is determined, the criteria 1 and 2 determine the angle through which the artifact must be rotated to ensure that material is deposited directly onto existing material. If $\hat{\mathbf{u}} = \hat{\mathbf{n}} \times \mathbf{p}_{12} / \| \hat{\mathbf{n}} \times \mathbf{p}_{12} \|$ and the part is rotated so that $\hat{\mathbf{u}}$ is vertical then the criteria that the nozzle is in the plane of the facet and is perpendicular to the bead are both satisfied. If R_θ and R_ϕ are rotation matrices describing rotations of the VMC bed about the x and y axes by angles θ and ϕ respectively, then solving $R_\theta R_\phi \hat{\mathbf{u}} = (0, 0, 1)^T$ yields solutions for the necessary rotations:

$$\theta = \arctan(-\hat{u}_y / \hat{u}_z) \pm \pi \qquad (5)$$

$$\phi = \arctan(-\hat{u}_x / (\hat{u}_y \sin\theta - \hat{u}_z \cos\theta)) \pm \pi. \qquad (6)$$

The algorithm is initialised by choosing a facet or group of edge-connected, coplanar facets whose boundaries form the initial loop. The plane of these seed facets is deemed to be coincident with the plane of the VMC bed and determines the initial orientation of the artifact. The facets themselves are filled simply using a single sweep of a conventional layered filling algorithm (Sewell 2005). As illustrated in Figure 2, the initial loop passes through the facets that share a vertex with the seed facets. A list of the facets through which the current loop passes is initially constructed by examining the neighbours of the seed facet(s); subsequently it is updated as facets are either removed, because they have been completely filled (the current line segment in Figure 1 moves past $\mathbf{v}_3$), or added because the current loop has moved past a vertex. The addition of facets is illustrated in Figure 3. Here the loop $\mathbf{q}_1$, $\mathbf{q}_2$, $\mathbf{q}_3$ involves only facets 2 and 3; however,

$\mathbf{q}_2 + \boldsymbol{\delta}_2$ lies beyond $\mathbf{v}_3$ and it is clear that the subsequent loop must also pass through facets 3, 4 and 5.

The choice of an optimal path within these facets is not immediately clear as we now discuss.

3   ROUNDING CORNERS

A simple example shows that all of the criteria for good build quality cannot be maintained as new facets are incorporated into the build path. Figure 4 shows the extension of the bead paths $\mathbf{p}_1 \rightarrow \mathbf{p}_2$ in facet 1 and $\mathbf{p}_{n+1} \rightarrow \mathbf{p}_n$ in facet 2 into facets 3, 4 and 5. This path maintains a constant bead diameter from the previously deposited path, fulfilling criterion 3, and it is possible to rotate the artifact to maintain the other two criteria. However, it is clear from the diagram that if all the facets are coplanar the path turns at the intersection point labelled $\mathbf{p}_v$ in the interior of facet 4, rather than on the facet edges. Although it would be possible to accommodate this in the coplanar case by introducing 'virtual' edges extending the $\mathbf{v}_1 \rightarrow \mathbf{v}_2$ edge into facet 4, it is likely to lead to the proliferation of virtual edges and poor surface finish as the deposited material turns in the middle of facets. A more serious and common difficulty, however, arises when the facets are not coplanar. In this case, laying a bead to $\mathbf{p}_v$ violates criterion 2 because $\mathbf{p}_v$ would not be in the plane of facet 4. The physical artifact produced would therefore have material deposited outside the planes defining the surface and would thus be a poor representation of the STL data.

It is possible to devise heuristic methods for determining the path in new segments (Sewell 2005). The 'constant radius' method determines intersections on the edges of new facets by insisting that the intersections are all one bead diameter from the intersection on the last loop that triggered the incorporation of new facets into the current loop ($\mathbf{q}_1$ in Figure 4). While this leads to good build quality in the vicinity of the vertex, it can produce deposition paths that are substantially closer than one bead diameter more distant from the vertex; physically, during part construction, this leads to the deposition of excess material and the 'dragging' of deposited material by the nozzle. Alternatively, the 'parallel alignment method' extends the bead paths in new facets parallel with the bead paths in the previous loop. This maintains criterion 3 except close to the apex of acute corners where it tends to produce paths that overshoot the corner leading to gaps between the current and previous loops; further details are given by Sewell (2005).

The constant radius and parallel alignment methods can be combined so that parallel alignment is used for intersections on edges that are shared between newly added facets and facets on the previous loop, while the constant radius method is used for edges which are not shared. This is illustrated in Figure 5 where the deposited material for the current loop is shown in dark grey and that for previous loop in light grey. Intersections $\mathbf{p}_2$ and $\mathbf{p}_5$ are determined by parallel alignment with $\mathbf{q}_1 \rightarrow \mathbf{q}_2$ and $\mathbf{q}_2 \rightarrow \mathbf{q}_3$ respectively, while $\mathbf{p}_3$ and $\mathbf{p}_4$ are at a constant diameter from $\mathbf{q}_2$. As can be seen from the figure there is a slight overlap of deposited material in the vicinity of the corner.

3.1   Error measure

In order to quantify the quality of the build we define an error measure as follows. Let t be a variable measuring distance around the loop and let $d(t)$ be the shortest distance from the current loop at t to the previous

loop. Since the ideal distance to the previous loop is h an error measure is the loop integral:

$$E = \frac{1}{L} \oint [d(t) - h]^2 \, dt \qquad (7)$$

where L is the length of the loop. In practice the error was calculated by evaluating $d(t)$ at points $t_n = n\delta t$ around the loop and computing

$$E \approx \frac{1}{L} \sum_{n=1}^{N} [d(t_n) - h]^2 \qquad (8)$$

We remark that an alternative measure is the integrated absolute error, which is less sensitive to extreme

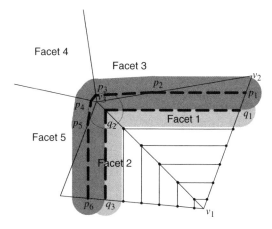

Figure 5.   Combination of constant radius and parallel alignment methods.

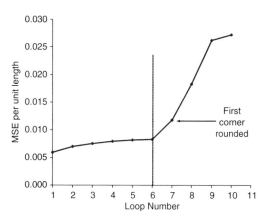

Figure 6.   Error (equation 7) for successive loops shown in Figure 7. Paths calculated using combination of constant radius and parallel alignment methods.

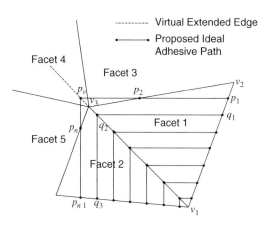

Figure 4.   Intersection maintaining criteria cannot be achieved with non-coplanar facets.

values of $d(t) - h$. However, in practice it gives similar results to the integrated squared error.

Figure 6 shows the error for successive loops during building of a sphere; the corresponding deposition paths are shown in Figure 7. No new facets are introduced during the first six loops, but corresponding to the incorporation of new facets on the seventh loop there is a sharp increase in the error. Note that in subsequent loops this error is propagated, and indeed magnified, although it may be reduced as acute corners are smoothed.

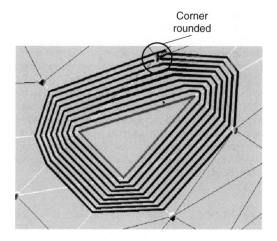

Figure 7.   Deposition paths for eight successive loops starting from a single seed facet. Paths calculated using combination of constant radius and parallel alignment methods. Beads are shown thinner than h for clarity.

3.2   Optimisation

Rather than merely measuring the loop error, we may use it as an objective to be minimised with respect to intersection locations. If the loop contains M line segments we regard the error as a function $E(\mathbf{p}_1, ..., \mathbf{p}_M)$ of the coordinates of the intersections. The function error function is continuous, but its derivatives may be discontinuous so we use a downhill simplex optimiser (Nelder and Mead 1965; Press, Flannery, Teukolsky, and Vetterling 1996). Since there may be many local minima we also investigated simulated annealers (Kirkpatrick, Gelatt, and Vecchi 1983) which are able to avoid local minima. However, although simulated annealing provided some improvement (up to $\approx 25\%$) over the simplex method, it was at the expense of many more function evaluations which are costly. The simplex optimiser was initialised with the intersections determined by the combination of constant radius and parallel alignment methods. Optimisation continued for 125 evaluations of E or until the error changed by less than 0.01 between evaluations.

Figure 8 shows the loop error obtained by optimising the intersection locations together with the error obtained using the constant radius and parallel alignment method. Figure 9 shows the optimised deposition paths. As the figures show, the optimisation is able to reduce the error on introducing new facets by a factor of three to approximately the level associated with the initial loops. Indeed optimisation is able to reduce the loop error even before new facets are incorporated.

As the length of the loop grows to incorporate many facets the computational burden of recalculating

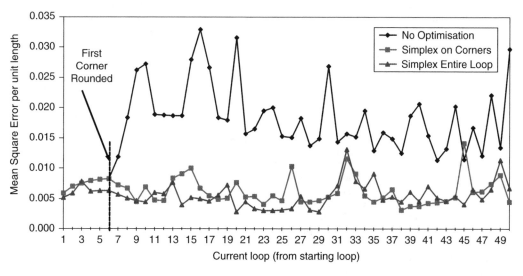

Figure 8.   Comparison of error for successive loops using no optimisation, optimisation of all intersection coordinates and optimisation of corners.

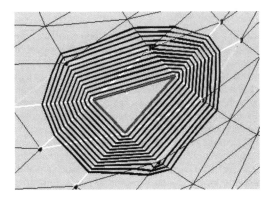

Figure 9. Optimised deposition paths for starting from a single seed facet. Beads are shown thinner than h for clarity.

the loop error for every trial solution for the simplex method becomes prohibitive. In addition the dimension of the space in which the search is conducted is equal to the number of line segments comprising the loop, so the optimisation becomes more difficult for longer loops. However, intersections distant from each other interact only weakly so the optimisation may be considerably simplified by breaking it into multiple smaller optimisations, one for each corner. By optimising the intersections of only the intersections surrounding a corner the number of parameters to be optimised is reduced and only the contribution to the error of the deposition paths around the corner need be calculated. Figure 8 also shows the error for loops optimised in this manner. The error is generally slightly larger than that obtained by optimising the full loop, however, the increase in computational speed outweighs the small increase in error. We remark that significant gains in speed can be accrued by using specially designed data structures to enable rapid reconstruction of topological and connectivity information from facets and edges (Sewell 2005; Baumgart 1975).

4 DISCUSSION

In this paper we have described a novel method for deposition based rapid prototyping that utilises the rotational ability of VMCs to ensure a good build quality without the need for support structures. Although the technique has been described in the context of hot-melt adhesive deposition on VMCs it is applicable to other additive deposition systems and materials.

The method has a number of disadvantages. Although the method is in principle capable of following arbitrary surfaces, in practice surfaces that close upon themselves are problematic because the

deposition nozzle is likely to collide with parts of the surface that have already been constructed. Also the build quality of a planar surface may be worse than is obtained with a bottom-up layered approach because the path across the planar region may undulate as a consequence of corners rounded earlier in the build. In contrast, the layered manufacturing approach produces excellent surface finishes for regions that are self-supporting. A final disadvantage is that filling regions between surfaces built with the wrapping technique is more complex than with a layered method because the loops are rarely planar. A similar difficulty has been described by Norman and Dickens (1995) in the context of 3D welding fabrication.

Much of the complexity of the algorithms discussed here arises from the fact that a bead of constant width is deposited. In practise hot-melt adhesive flows slightly before solidifying which compensates for some of the imperfections in the wrapping algorithm. However, if the bead width is continuously variable the rounding of corners is vastly simplified and build quality can be enhanced. Current work is focused on enhancements to the deposition hardware to permit bottom-up layered RP and deposition with variable width beads.

REFERENCES

Baumgart, B. (1975). A Polyhedron Representation for Computer Vision. In *National Computer Conference*, pp. 589–596.

Beaman, J., J. Barlow, D. Bourell, R. Crawford, H. Marcus, and K. McAlea (1997). *Solid Freeform Fabrication: A New Direction in Manufacturing*. Kluwer Academic Publishers, Dordrecht/Boston/London.

Bostik (2002). Bostik Thermogrip 9677 General Purpose Adhesive, Product Datasheet B412/5. Bostik Findley UK Limited.

Kirkpatrick, S., C. Gelatt, and M. Vecchi (1983). Optimization by Simulated Annealing. *Science 220*, 671–680.

Nelder, J. and R. Mead (1965). Downhill Simplex Method. *Computer Journal 7*, 380.

Norman, R. and P. Dickens (1995). 3D Welding Fabrication: Strategies for Sloped Surfaces. In *National Conference on Rapid Prototyping and Tooling Research*, pp. 51–66.

Press, W., B. Flannery, S. Teukolsky, and W. Vetterling (1996). *Numerical Recipes in Pascal: The Art of Scientific Computing*. Cambridge University Press.

Sealock (2002). Sealock H115 Carton Adhesive, Product Datasheet H115, Sealock UK Limited.

Sewell, N. (2005). *Algorithms for multi-axis additive rapid prototyping*. Ph. D. thesis, The University of Exeter, Exeter, UK.

Sewell, N., M. Jenkins, and R. Everson (2003). Algorithms for Multi-Axis Additive Rapid Prototyping. In *Advanced Research in Virtual and Rapid Prototyping*, pp. 117–122. Escola Superior de Tecnologia e Gestão de Leiria.

Virtual modeling and rapid manufacturing – Bártolo (eds)
© *2005 Taylor & Francis Group, London, ISBN 0 415 39062 1*

Selection of Rapid Prototyping systems based on the Axiomatic Design theory

A.M. Gonçalves-Coelho & António J.F. Mourão
UNIDEMI, Department of Mechanical and Industrial Engineering, Faculty of Science and Technology, The New University of Lisbon, Portugal

ABSTRACT: The type of available Rapid Prototyping (RP) systems has significantly increased since the late eighties. The best RP process for each application depends on several factors and constraints, including cost, building speed, accuracy, operating environment characteristics, and material type and properties. The RP user must evaluate all these factors and constraints to select the correct system, but there are not methodologies to support a consistent selection. This paper uses published information of a benchmarking study where RP systems were compared to allow comprehensive analysis of application requirements versus process capabilities. The authors use Axiomatic Design theory, specifically the information axiom, to select the more adequate RP system. The procedure allows considering all the system factors to appraise the more or less probability of obtaining the specified functional requirements with each system.

1 INTRODUCTION

The type of available Rapid Prototyping (RP) systems has significantly increased since the late eighties, after the introduction of Stereolithography.

None of them is universal, and designers are facing a problem that is common to all manufacturing technologies: which is the best RP technology to use for each specific application?

In fact, the best RP technology for each application depends on several factors and constraints, including costs, building speed, accuracy, operating environment characteristics, and building materials properties.

Moreover, product development in concurrent engineering environments makes the selection of manufacturing systems a critical task, as it is ascertained by Steele *et al.* (2002).

Under these circumstances, the current decision-making methodologies reveal some limitations that are hard to overtake, such as the psychological bias that is embedded in the so-called weighting factors.

The present paper uses published information of a benchmarking study where RP systems were analysed to allow the comprehensive assessment of the application requirements versus process capabilities.

The authors use Axiomatic Design theory (AD), specifically the AD's information axiom, to set up a method for selecting the best RP system for specific applications.

The proposed procedure allows considering all the system factors to appraise the more or less likelihood of obtaining the specified functional requirements with each system.

2 THE AXIOMATIC DESIGN THEORY

Axiomatic Design was developed in the late 1970s, to be used as a systematic model for engineering education and practice. Its underlying hypothesis is that there exist certain fundamental principles that govern good design practice, and its key components are design domains, axioms, hierarchies and zigzagging (Suh 1990, Suh 2001).

AD provides a general framework to assist in the decision making of any design process. Moreover, AD offers a means to show in detail and to assess the interplay between the design goals and the tentative solutions at any point in the design process. AD key components are *design domains, axioms, hierarchies* and *zigzagging*.

2.1 *The design domains*

On the AD point of view, the design outcomes belong to one of the four distinct domains: the customer domain, the conceptual domain, the physical domain and the process domain (see Figure 1).

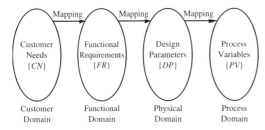

Figure 1. The design process as a mapping (Suh 2001).

The design process begins in the customer domain with the settlement of the customer needs (CNs), i.e. the features that the customers are looking for in a technical system, being it a product, a process, or any other tangible or intangible system.

Mapping between the customer and the conceptual domains – a process that is currently known as "conceptual design" – is used to find out the functional requirements (FRs) of the technical system.

Once this is done, another mapping – the "product design" – allows for the translation of the FRs into design parameters (DPs), which are the set of properties that describe the design object in the physical domain.

At last, mapping from the physical domain to the process domain – the "process design" – leads to the process variables (PVs), which outline how the product is to be made.

Each one of the above-mentioned mapping steps is not unique, and may lead to either "good design solutions" or "poor design solutions", depending on the way they are performed. In mathematical terminology, those mapping operations are represented by a design equation. For the product design, for instance, we have

$$\{FR\} = [A]\{DP\},\tag{1}$$

where the generic element of the design matrix, $[A]$, is given by

$$A_{ij} = \frac{\partial FR_i}{\partial DP_j}.\tag{2}$$

2.2   The design axioms

Distinct sets of DPs (or design solutions) may be settled to fulfil any specific set of FRs (or design problem). Consequently, the main job of the designers is to generate several possible solutions for the design problem, and try to adopt the best one for further development.

According to AD, good design solutions are those that comply with the design constraints and that conform to the Independence Axiom.

The Independence Axiom: Maintain the independence of the functional requirements. (*Alternative statement*: In an acceptable design, mapping between the FRs and the DPs is such that each FR can be satisfied without affecting the other FRs.)

The kind of the existing relationships between FRs and DPs is of major importance in the assessment of design solutions.

As regards to the independence axiom, there are three basic design types: *uncoupled, decoupled* and *coupled.*

The uncoupled solutions are the best ones and are identified by diagonal design matrices.

The decoupled solutions, which are characterized by triangular design matrices, are still acceptable. However, their FRs must be tuned according to a certain sequence, in order to avoid a trial and error process.

At last, the coupled solutions are considered "poor design". Their design matrices are populated in any other manner and the tuning process must be performed iteratively.

Several examples of uncoupled, decoupled and coupled designs can be found in the references (Suh 1990, Suh 2001, Gonçalves-Coelho & Mourão 2003, Gonçalves-Coelho *et al.* 2003).

If two or more design solutions conform to the Independence Axiom, then the choice should be made based on the Information Axiom.

The Information Axiom: Minimise the information content of the design. (*Alternative statement*: From a set of designs that satisfy the same FRs and conform to the Independence Axiom, the best is the one with the minimum information content.)

The information content is a means to quantify the design's complexity (Suh 1990), and it is worth to note that the design components' count do not estimate complexity. In fact, a design with a small number of components may be complex, and this do not necessarily happens to a design with a large number of components.

For the simple case of a one-FR, one-DP design, the information content is defined as being the logarithm of the inverse of the probability of achieving the desired value for the FR:

$$I = \log_x \frac{(\text{area of the system range})}{(\text{area of the common range})}\tag{3}$$

where the area of the system range is computed from the FR's probability density function, and the area of the common range is the portion of the above-mentioned area that is located inside the design range limits (see Figure 2).

Usually, logarithms of base 2 are used, case of which the information unit is the *bit*.

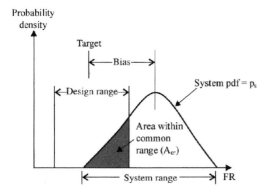

Figure 2. The probability of success for a one-FR, one-DP design solution (Suh 2001).

For any uncoupled design with n FRs, the total information content, I_t, can be computed through

$$I_t = \sum_{i=1}^{n} -\log_2 p_i = \sum_{i=1}^{n} I_i , \qquad (4)$$

where p_i is the probability of FR_i being satisfied by DP_i.

If all the probabilities p_i, are equal to one, then the design's information content is zero. Conversely, the information content is infinite when at least one of the p_i is equal to zero, which means that the design solution will never fulfil the FR_i.

According to Equation 4, large systems are not necessarily complex if their probability of success is high. On the other hand, systems with a small number of components are complex if their probability of success is low.

2.3 Hierarchies and zigzagging

According to Axiomatic Design, the design of compound objects is to be developed in a top-down manner, beginning at the *system level* and progressing through levels of more detail, until a point that is enough to clearly describe the entire design object.

A unique procedure is used to achieve the design object synthesis, for it is quite different to analyse an already existing entity or to synthesise an object that still does not exist. This procedure is named *hierarchical synthesis*, and its outcome is depicted by a *tree-model* in each one of the four design domains.

To synthesise a "not yet existing object" one has to go forth and back between at least two contiguous design domains along a zigzagging path (Suh 1990). Figure 3 depicts the tree-models resulting from zigzagging between the functional and the physical domains.

One begins at the system level, finding a design parameter $DP_{1.1}$ that satisfies $FR_{1.1}$. Once an acceptable

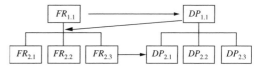

Figure 3. Hierarchy and zigzagging.

$DP_{1.1}$ is found, one comes back to the functional domain to decompose $FR_{1.1}$ into the lower-level $FR_{2.1}$, $FR_{2.2}$ and $FR_{2.3}$. The next step is to find out a set of suitable $DP_{2.j}$, in order do satisfy $FR_{2.j}$. The assessment of candidate sets of $DP_{2.j}$ is done at the light of the independence and the information axioms, and the process continues until sufficing detail is achieved.

One can find several examples about the zigzagging process at the references (Suh 1990, Suh 2001, Gonçalves-Coelho & Mourão 2003, Gonçalves-Coelho et al. 2003, Ion & Gonçalves-Coelho 2003).

3 SELECTING A RP SYSTEM

3.1 The traditional approach

Let us suppose that one needs to select a RP system on a basis of distinct criteria, as for example capital costs, building speed and acceptable raw materials.

The oldest way to deal with multi-criteria decision-making is to apply the pertinent criteria, one by one, to all the alternative solutions. Every system not complying with any of the selected criteria is to be rejected. This simple technique, however, has a remarkable shortcoming: it does not allow for setting up an overall ranking for the systems that comply with all the decision criteria.

Therefore, to set up an overall ranking is common to use weighing functions.

The following weighing function gives the overall capability of each one of the candidate RP systems:

$$C_S\left(c_{S1},..,c_{Sn}\right) = \frac{\sum_{i=1}^{n} w_i c_{Si}}{\sum_{i=1}^{n} w_{Si}} , \qquad (5)$$

where c_{Si} are specific capabilities of the system, and w_i are weighing factors, for which

$$\sum_{i=1}^{n} w_i = 1 . \qquad (6)$$

Notice that the weighing factors are always defined in a subjective manner, for they usually express someone's belief on the relative importance of the considered

Table 1. The capability matrix.

Capability	SLA	SLS	FDM	LOM	3DP	MJM	Object	DoD
	1–10 relative scale (dimensionless)							
Accuracy	8	7	6	6	4	6	8	10
Robustness	5	10	8	6	6	2	7	2
Building speed	6	6	5	5	9	8	10	2
Capital costs	2	3	5	5	8	9	8	9
Running costs	1	1	4	3	9	7	7	3
Surface texture	10	4	6	5	5	10	10	10
Texture of downward facing surfaces	10	4	4	5	5	3	10	10
Material range	5	10	5	5	7	1	3	1

criteria. In addition, it is worth to stress that weighing functions do not flag solutions not complying with all the criteria.

We argue that a good way to overtake the above-mentioned difficulties is to use AD's information axiom, provided that the functional requirements of each RP system can be attained in an independent mode.

In this case, Equation 4 can be used to compute the information content of each candidate solution. Subsection 3.2 below contains a numerical example that explains how to proceed.

3.2 The axiomatic design approach

Extensive benchmarking studies about different RP systems were carried out at the Rapid Prototyping Research Centre, Glasgow, UK. Based on those studies, Ryder et al. (2002) published the specific capabilities of the following RP systems:

Stereolithography (SLA),
Selective Laser Sintering (SLS),
Fused Deposition Modelling (FDM),
Laminated Object Modelling (LOM),
3-Dimensional Printing (3DP),
Multi Jet Modelling (MJM),
Object Quadra (Object), and
Droplet on Demand (DoD).

Eleven specific capabilities were assessed, and Table 1 summarises the eight quantitative ones. A 0–10 dimensionless scale was used, and the grade 10 was given to the best system on the market for each technology.

In an AD standpoint, the figures of Table 1 are related to the RP system ranges for all the specific capabilities of the equipments under analysis.

To begin with, we need to define the selection criteria that are to be used and these will be the functional requirements of our problem. We have selected the following:

FR_1 = accuracy,
FR_2 = robustness,

FR_3 = building speed,
FR_4 = capital costs, and
FR_5 = running costs.

More criteria could be adopted, but we will keep only these ones for a matter of simplicity.

The next step is to verify if the selected FRs can be accomplished in an independent mode. Equation 7 shows a possible design equation for each RP system that complies with the independence axiom.

$$
\begin{Bmatrix} Accuracy \\ Robustness \\ Building\ speed \\ Capital\ costs \\ Running\ costs \end{Bmatrix} = \begin{bmatrix} x & 0 & 0 & 0 & 0 \\ 0 & x & 0 & 0 & 0 \\ 0 & 0 & x & 0 & 0 \\ 0 & 0 & 0 & x & 0 \\ 0 & 0 & 0 & 0 & x \end{bmatrix} = \begin{Bmatrix} System\ topology \\ Building\ conditions \\ Power\ rating \\ Equipment\ size \\ Raw\ materials \end{Bmatrix}
$$

(7)

One can ascertain that running costs depend on the raw materials and on the equipment's power rating. However, raw materials are much more expensive than energy. Therefore, one can neglect this dependency.

Therefore, Equation 4 can be used to compute the total information content.

At this point, we have to characterize the system ranges and the design ranges for each FR of every RP system.

Using a conservative approach, we have interpreted the figures of Table 1 as being the system range upper limit, except for capital and running costs. Concerning to the costs, we have taken the table's figures as the central point of the system range, and a breath of ±5% of the dimensionless scale's full-range was adopted. Table 2 summarises the system ranges for all the RP systems. As regards to the analyzed design ranges, they are expanded in Table 3.

Case A depicts a typical general purpose RP system: the design ranges for accuracy, robustness and building speed are centred at the mid point of the 0–10 dimensionless scale, and have a breath of ±15% of the scale's full-range. Concerning to the costs, they should not exceed the mid-scale value.

Table 2. System ranges for the considered cases.

		SLA	SLS	FDM	LOM	3DP	MJM	Object	DoD
Capability		1–10 relative scale (dimensionless)							
Accuracy	min	0.0	0.0	0.0	0.0	0.0	0.0	0.0	0.0
	max	8.0	7.0	6.0	6.0	4.0	6.0	8.0	10.0
Robust.	min	0.0	0.0	0.0	0.0	0.0	0.0	0.0	0.0
	max	5.0	10.0	8.0	6.0	6.0	2.0	7.0	2.0
B. speed	min	0.0	0.0	0.0	0.0	0.0	0.0	0.0	0.0
	max	6.0	6.0	5.0	5.0	9.9	8.8	10.0	2.0
C. costs	min	1.5	2.5	4.5	4.5	7.5	8.5	7.5	8.5
	max	2.5	3.5	5.5	5.5	8.5	9.5	8.5	9.5
R. costs	min	0.5	0.5	3.5	2.5	8.5	6.5	6.5	2.5
	max	1.5	1.5	4.5	3.5	9.5	7.5	7.5	3.5

Table 3. Design ranges for the considered cases.

	Case A		Case B		Case C		Case D	
	min	max	min	max	min	max	min	max
Functional requirement	0–10 relative scale (dimensionless)							
Accuracy	3.5	6.5	7.0	10.0	3.5	6.5	3.5	6.5
Robustness	3.5	6.5	3.5	6.5	3.5	6.5	3.5	6.5
Building speed	3.5	6.5	3.5	6.5	7.0	10.0	0.0	10.0
Capital costs	0.0	5.0	0.0	10.0	0.0	10.0	0.0	3.0
Running costs	0.0	5.0	0.0	10.0	0.0	10.0	0.0	3.0

Table 4. The total information content for the considered cases.

	SLA	SLS	FDM	LOM	3DP	MJM	Object	DOD
Case	information content (bit)							
A	4.42	4.22	4.42	4.26	∞	∞	∞	∞
B	6.00	∞	∞	∞	∞	∞	5.56	∞
C	∞	∞	∞	∞	6.40	∞	4.37	∞
D	3.15	2.96	∞	∞	∞	∞	∞	∞

Case B is based on Case A, but high accuracy is required and the incurred costs are not crucial. It represents the typical needs for a system to produce high accuracy single components.

Case C is based on Case A as well, but now the building speed of the RP system is of paramount importance. As for capital and running costs, they are not vital. Case C describes a situation related to the production of small series of components.

Case D is also a modification of Case A, where low costs are needed and building speed is not important. It portrays a common situation in production of prototypes.

Table 4 contains the computed information content for all the candidate RP systems for all the four case studies, as computed through Equation 4. Uniform probability density functions (pdf) were used for all the computation, in the lack of more accurate data.

3.3 Discussion

Table 4 above depicts the total information content related to all the analyzed RP systems for the four case studies.

As one can see, there are a number of situations where the information content equates to the infinity. These situations correspond to a zero probability of achieving one FR at least.

The table shows how important the selection criteria can be: only four RP systems (out of eight) are suitable for a situation that can be considered "general purpose" (Case A). In addition, the number of suitable RP systems decreases when the FRs became more challenging (Cases B, C and D).

The same table allows for setting up a ranking of the suitable systems for each case study based on the information content: the less information content, the better.

For Case A, for instance, the ranking is: 1) SLS (4.22 bit); 2) LOM (4.26 bit); 3) SLA and FDM (4.42 bit) (*ex aequo*).

4 CONCLUSIONS

An Axiomatic Design based procedure to select RP systems for any specific purpose was proposed in the present paper.

The focal point of the procedure is the quantitative evaluation of the information content of the candidate systems.

In order to achieve this, one has to precisely define both the system and the design ranges. This is a remarkable advantage of the method, since it contributes for the understanding of the nature of the problem under analysis. It is worth to note, however, that the method is difficult to use without previous training.

As for the deployed case studies, it was made clear that the proposed method does not make use of subjective considerations, such as weighting factors. Notwithstanding, it allows for easily eliminate the systems not complying with any of the selection criteria, and to set up a ranking of the suitable systems.

ACKNOWLEDGEMENTS

The present authors would like to thank the Faculty of Science and Technology of The New University of Lisbon and the Portuguese Foundation for Science and Technology (FCT). Their valuable support is helping to make possible our research work.

REFERENCES

Ion, S., Gonçalves-Coelho, A.M. 2003. *Hierarchic Design Decomposition Through Zigzagging*, Proc. X International Science and Engineering Conference, Tome 4, pp. 95–99, Sevastopol, Ukraine, Sept. 8–13.

Gonçalves-Coelho, A.M., Mourão, A.J.F. 2003. *The Axiomatic Design as a framework for Concurrent Engineering*. In: Bártolo, P.J., Mitchell, G., Mateus, A., Batista, F., Vasco, J., Correia, M., André, N., Lima, P., Novo, P., Custódio, P., Martinho, P. (Eds.), Advanced Research in Virtual and Rapid Prototyping, pp. 593–600, Leiria: ESTG.

Gonçalves-Coelho, A.M., Mourão, A.J.F., Pamies-Teixeira, J.J. 2003. *Axiomatic design as a background for Concurrent Engineering Education and Practice*, In: Cha, J., Jardim-Gonçalves, R., Steiger-Garção, A. (Eds.), Concurrent Engineering: Advanced Design, Production and Management Systems, pp. 419–427, Lisse: A.A. Balkema Pub.

Ryder, G.J., Harrison, D.K., Green, G., Ion, W.J., Wood, B.M. 2002. *Benchmarking the rapid design and manufacturing process*. In: Rennie, A.E.W., Bocking, C.E., Jacobson, D.M. Third National Conference on Rapid Prototyping, Tooling and Manufacturing, London: Professional Engineering Pub. Ltd.

Steele, K.A.M., Ryder, G.J., Ion, W.J., Thomson, A.I. 2002. *Reducing the uncertainty of the prototyping decision*. In: Rennie, A.E.W., Bocking, C.E., Jacobson, D.M. Third National Conference on Rapid Prototyping, Tooling and Manufacturing, London: Professional Engineering Pub. Ltd.

Suh, N.P. 1990. *The Principles of Design*, New York: Oxford University Press.

Suh, N.P. 2001. *Axiomatic Design: Advances and Applications*, New York: Oxford University Press.

Virtual modeling and rapid manufacturing – Bártolo (eds)
© *2005 Taylor & Francis Group, London, ISBN 0 415 39062 1*

Development of an internet-based collaborative design system in a context of concurrent engineering

A.J. Álvares
Universidade de Brasília, Departamento de Engenharia Mecânica e Mecatrônica, GRACO – Grupo de Automação e Controle, Brasília, DF, Brazil

J.C.E. Ferreira
Universidade Federal de Santa Catarina, Departamento de Engenharia Mecânica, GRIMA-GRUCON, Caixa, Florianópolis, SC, Brazil

ABSTRACT: This work presents a description of the implementation of the WebMachining methodology and system (http://WebMachining.AlvaresTech.com) developed in a context of e-Mfg and Concurrent Engineering, aiming at integrating CAD/CAPP/CAM for the remote manufacturing of cylindrical components through the Internet. The methodology and its implementation are conceived starting from the modelling paradigm based on synthesis of design features, in order to allow the integration among collaborative design (WebCADbyFeatures), generative process planning (WebCAPP) and manufacturing (WebTurning). The system is implemented in a distributed environment of agents (agents' community), and Knowledge Query and Manipulation Language (KQML) is adopted as the language for message exchange among the design, process planning and manufacturing agents.

1 INTRODUCTION

It is in course a new revolution in the work system adopted in the manufacture companies, migrating from the activities aided by the computer (CAD, CAPP, CAM, CAP, etc), to activities based on e-Work (electronic-work), which characterize the beginning of work in the information era, with intensive use of Information Technology (IT).

IT, especially the technology of communication networks and the convergence of technologies based on networks and the Internet, is opening a new domain for constructing the future manufacturing environment called e-Mfg (electronic-Manufacturing), using work methods based on Collaborative e-Work, especially for the activities developed during the product development cycle in integrated and collaborative CAD/CAPP/CAM environments. This will allow the product developers and planners to have easier communication, enabling design sharing during the development of the product, as well as the teleoperation and monitoring of the manufacturing devices.

This paper describes the implementation of the WebMachining system and methodology (http://WebMachining.AlvaresTech.com) developed in a context of e-Mfg, aiming at integrating CAD/ CAPP/CAM for the remote manufacture of cylindrical components through the Internet. The methodology and its implementation begins with the modelling based on synthesis of design features, in order to allow the integration among collaborative design (WebCADbyFeatures), generative process planning (WebCAPP) and manufacturing (WebTurning). The system is implemented in a distributed environment of agents (agents' community), and Knowledge Query and Manipulation Language (KQML) is adopted as the language for message exchange among the design, process planning and manufacturing agents.

2 COLLABORATIVE CAD AND RELATED SYSTEMS

In design engineering practice, many activities associated with the several manufacturing aspects are being considered during the design phase. Feature-based modelling has been used in the integration of engineering activities, from design to manufacturing. Thus, the concept of features has been used in a wide range of applications such as component design and assembly, design for manufacturing, process planning and other countless applications. These applications are migrating to heterogeneous and distributed computer environments to give support to the design and manufacturing processes, which will be distributed both in space and in time.

It should be noted that it is undesirable and frequently unlikely to require that all participants in product development activities use the same hardware and software systems. Consequently, the system components should be modular and communicate with one another through a communication network, for effective collaboration.

Many research efforts have been made in the development of design environments oriented to computers networks, usually called network-centred. Shah et al. (1997) developed an architecture for standardization of communication between the kernel of a geometric modelling system and the applications. Han and Requicha (1998) proposed a similar approach that enables transparent access to several solid modelers.

Smith and Wright (2001) described a distributed manufacturing service called Cybercut (http://cybercut.berkeley.edu), which makes it possible the conception of a prismatic component that will be machined using a CAD/CAM system developed in Java in a context of remote manufacturing.

Shao et al. (2004) described a process-oriented intelligent collaborative product design system based on the Analysis-Synthesis-Evaluation (ASE) design paradigm and the parameterization of product design, using agents.

Hardwick et al. (1996) proposed an infrastructure that allows the collaboration among companies in the design and manufacture of new products. This architecture uses the Internet for information sharing, and the STEP standard for product modelling. Martino et al. (1998) proposed an approach to integrate the design activities with the other manufacturing activities based on features, which supports both feature-based design and feature-recognition.

Collaborative modelling systems typically have a client/server architecture, differing in the functionality and data distribution between clients and servers. A common problem in client/server systems is associated with the conflict between the limitation of the complexity of the client application and the minimization of the network load. A commitment solution can be conceived between the two ends, the so-called thin and fat clients. A pure thin client architecture typically places all the functionality in the server, which sends an image of its user interface to be shown on the client.

On the other hand, a pure fat client offers total interaction and local modelling, maintaining its own local model. Communication with the server is required when it is necessary to synchronize the data modifications in the local model with the other clients.

Lee et al. (1999) presented the architecture of a network-centred modelling system based on features,

in a distributed design environment, called NetFeature. This approach combines feature-based modelling techniques with communication and distributed computing technologies in order to support product modelling and cooperative design activities in a computer network.

The WebSpiff system (Bidarra et al., 2001) is based on a client/server architecture consisting of two main components on the server side: (i) The SPIFF Modelling System that supplies all the functionality for feature-based modelling, using the ACIS modelling kernel (Corney e Lim, 2001); (ii) The Session Manager that supplies functionality to start, associate, finish and logout a modelling session, as well as the management of all communication between the SPIFF system and the clients.

Li et al. (2004) and Fuh and Li (2004) mention several distributed and integrated collaborative design systems and Concurrent Engineering, and none of such systems implements collaborative design activities integrated with process planning and remote manufacturing systems via Web for the cylindrical components domain, with symmetrical and asymmetrical features. Most of those systems consider prismatic components, like WebCAD 2000 of the Cybercut system, which does not implement collaborative design.

The development of the WebCADbyFeatures collaborative design system differs from the above systems because it models cylindrical components, based on synthesis of design features (symmetrical and asymmetrical), having as motivation the development of an integrated CAD/CAPP/CAM system that allows the collaborative design through the web, in a context of Concurrent Engineering.

3 METHODOLOGY OF THE WEBMACHINING SYSTEM

Figure 1 presents part of the IDEF0 model of the proposed system, called WebMachining. The proposed methodology is divided into three basic activities: collaborative product modelling (WebCADbyFeatures), Generative CAPP (WebCAPP) and CAM (WebTurning).

The client, WebCADbyFeatures interface agent (Fig. 2) is connected to the neutral feature modeler via Web, and it begins the instantiation of a new component to be modeled from a database, using a library of standardized form features.

Then, the data about the component are sent to the server. Since the component is cylindrical, the user models the component in two dimensions, and may visualize it in 3D through VRML. A database was implemented in MySQL that stores the information on the product modeled by features, containing information associated with the form features, material features, tolerance features and technological features

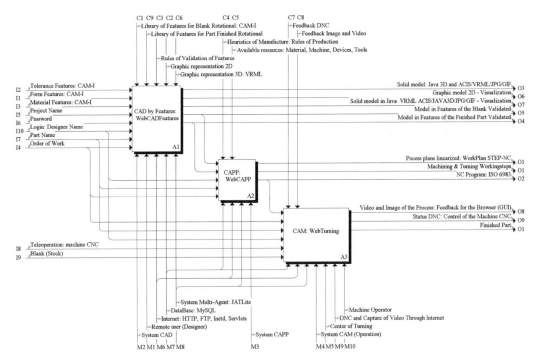

Figure 1.   Modelling of the WebMachining system using the IDEF0 methodology.

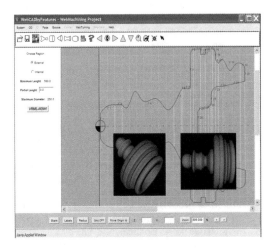

Figure 2.   Main window of the WebCADbyFeatures system, showing the profile of a cylindrical component.

(which include surface treatment, thermal treatment and production data). This combined information allows the mapping of design features into machining features, which is fundamental for process planning.

After completing and validating the model, the designed component is stored and made available to the CAPP module to generate the process plan, and the representation of the linearized process plan is based on STEP-NC (ISO 14649 – Part 12) (ISO, 2003). Then, the NC program is generated for a specific CNC lathe, in the case the Romi Galaxy 15 M turning centre (http://video.graco.unb.br).

The communication with the Romi Galaxy turning centre, with CNC Fanuc 18i-Ta (Fig. 2) is accomplished through an Ethernet connection (physical and data link layers of the ISO/OSI standard), using the TCP/IP protocols (network and transport layers of the ISO/OSI standard) associated with the application protocol Focas1/Ethernet libraries of Fanuc. Focas1 is an API for the development of applications using a standardized data structure to access 300 CNC functions (http://webdnc.graco. unb.br).

Then, the work order is sent through the Internet to the remote manufacturing system, which provides the necessary resources for producing the component (i.e. NC programs, cutting tools, fixtures and blank).

The teleoperation system of the Romi Galaxy turning centre, called WebTurning, is based on a client/server architecture, composed by two modules (Figs. 1 and 3): WebCam and WebDNC Servers, interfaced with the programs in a computer under the Linux operating system, connected logically via TCP-IP sockets and Ethernet to the machine tool and the clients, being responsible for capturing images

Figure 3. WebTurning: teleoperation and remote monitoring of the CNC turning centre.

(http://video.graco.unb.br) and supervisory control of the CNC lathe. The clients are interfaced through Java applets and HTML pages.

The WebTurning teleoperation servers are composed by the video and teleoperation servers of the machine, which provide command services, program execution, download and upload of programs, troubleshooting and other functions associated with the DNC1 communication protocol, available in the CNC Fanuc 18i-TA, accomplishing the remote supervision of the machine. All control action is executed locally, as a function of the delay of the TCP/IP protocol.

The video server is responsible for video and image capture (with four cameras), and for its distribution through the TCP/IP protocol. The other servers associated with the teleoperation services work in a bi-directional way, receiving commands through the Internet and sending status data about the machine.

The implementation of the three computer modules of the WebMachining system is presented in IDEF0 and UML diagrams, and for system development the Java, JavaScript, HTML, C and C++ programming languages were used. A case study is presented in this paper, showing the collaborative modelling of a component, the generation of the process plan with alternatives and its linearization associated with a NC program, and finally the teleoperation and monitoring of the CNC turning centre.

4 MULTI-AGENT ARCHITECTURE FOR THE WEBMACHINING METHODOLOGY

The proposed architecture for the multi-agent system (MAS) can be characterized by the agents' behavior

as being Deliberative, in the internal organization as being of the Blackboard type, and in the architecture itself as being of the Federated type using the Facilitator approach (Shen et al., 2001).

The use of an architecture based on a multi-agent approach is the most attractive currently, mainly due to the evolution of operating systems, especially Unix for personal computers, and the use of network communication based on TCP/IP in a client/server architecture (Shen e Norrie, 1999). In this way several types of agents working cooperatively and in a distributed way can be used in order to solve many problems associated with CAD/CAPP/CAM integration in a context of a community of agents.

The JATLite (Java Agent Template Lite) software tool is used for implementing the collaborative product design system. JATLite (http://java.stanford.edu/index.html) is a software written in Java that allows the users to create agents that communicate in a robust way over the Internet. JATLite offers a basic infrastructure in which agents that are registered in an Agent Message Router (AMR), usually called facilitator or mediator, use a name and a password, being connected and disconnected through the Internet, sending and receiving messages, transferring files via FTP and usually exchanging information with other agents, through the many computers in which they run. These means are used for developing the management system of the collaborative design sessions, where a design agent interface provides its design for the other participant agents of the product modelling session.

The proposed architecture is composed by six groups of agents (Fig. 4): facilitator (1) database manager (2) collaborative design (3 and 4) VRML manager (5) process planning (6) and manufacturing (7, 8, 9 and 10):

1. FACILITATOR AGENT (FA): it performs communication management among the agents, managing the routing of the messages among the agents, system safety and agents registration. It is implemented through the Message Router Agent, which is part of the JATLite architecture. It will be necessary more than one FA due to the amount of agents present in the system, in order to improve the its performance. The AMR is very important in the JATLite environment, because the agents always communicate with other agents through AMR.

2. DATABASE MANAGER AGENT (DMA): this agent performs the interaction with the MySQL database. Any agent that wants some information from the database (through the SQL language) makes a request to DMA, and it sends the answer to the agent that requested the information. The Facilitator Agent accomplishes the routing of the messages among these agents.

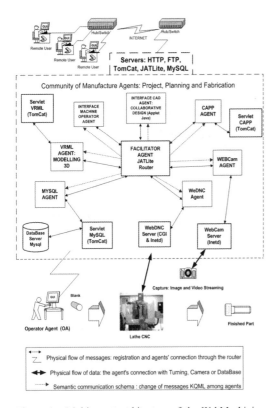

Figure 5. WebCam GUI : capture video (http://video. graco.unb.br).

Figure 4. Multi-agent architecture of the WebMachining system.

3. COLLABORATIVE DESIGN (CADIA): this is a GUI for feature-based design, implemented through a Java applet. This GUI is executed by a remote client aiming at defining the model and geometry of the raw material and of the finished component based on features. This agent will communicate with the community of agents through a connection to FA, and this will perform message routing to the correct agent. Messages are sent to the other modules of the system and users, containing the data regarding the design underway (i.e. the product model), which includes information such as: user, component name, design name, etc.; this will allow the identification of the product model that the client is creating. The creation of the component by features and the verification of the feature library, for which it is necessary a connection with the MySQL database, are accomplished directly through PHP (Personal Home Page), in order to improve the execution of the system. The 3D visualization of the product model is managed through this GUI, which communicates with the agent for 3D modelling. Figure 2 shows a prototype of the developed

GUI, which is a Java applet, and the 3D visualization of a component through VRML.

4. REMOTE USER CAM INTERFACE AGENT (WebDNC): every GUI associated with CAM that is executed by a remote client and used to teleoperate the CNC machine, has WebDNC embedded in the interface. This agent communicates with the community of agents through a connection to FA, performing message routing to the corresponding agent.

5. 3D VRML BASED MODELLING AGENT (VRML): it is responsible for 3D modelling using VRML (Virtual Reality Markup Language). It receives messages from CADIA for building 3D component models based on features.

6. PROCESS PLANNING AGENT (WebCAPP): it is responsible for process planning.

7. WebCam AGENT (WebCam): it is responsible for the video and image capture of the teleoperation system, sending the captured images directly to the GUI associated with CAM. It receives messages from FA regarding the user's identification, login and password, to allow the execution of the WebCam server (Fig. 5).

8. WebCNC AGENT (WebCNC): it is responsible for the remote control of the CNC machine, receiving commands and sending the machine status to the GUI associated with CAM, i.e. WebDNC. It receives messages from FA regarding the user identification, login and password, filename with the NC program and process planning data (fixtures, tools and raw material), being responsible for implementing the Distributed

Numeric Control (DNC) protocol through the Web (Fig. 3) via Focas1/DNC1.

9. MACHINE OPERATOR INTERFACE AGENT (MOIA): This agent gives the instructions to the shop-floor operator about fixturing the raw material, tools setup, machine setup, production scheduling, among others.

10. OPERATOR AGENT (OA): this agent corresponds to the machine-tool operator, which receives instructions for fixturing the raw material, tool setup, machine setup, production scheduling of a component and other data associated with process planning and that can only be treated by a human operator.

5 WEBCADBYFEATURES IMPLEMENTATION

The inputs for the WebCADbyFeatures system (fig. 5) are the feature model and other necessary information, and it outputs the feature model of the raw material and the finished component, which becomes an input to the CAPP module, which in turn is responsible for generating the process plan and the NC program for the corresponding component (fig. 1).

On the client side the GUI is represented by applets, and two servers (VRML and JATLite) are implemented (Fig. 6):

- VRML server based on servlets (TomCat): used for generating the 3D model of the component in VRML, from the design feature model of the component;
- JATLite Router/Facilitator server: allows the management of many sessions of collaborative product modelling, performing the coordination of communication between the many WebCADbyFeatures interface agents, as well as the other agents in the system, managing the routing of messages between the agents, system security and agent registration. It is implemented through the Agent Message Router (AMR) of the JATLite architecture. The AMR is very important in the JATLite development environment, because the agents always communicate between each other via AMR, performing activities such as sending an e-mail message (e.g. via SMTP) or a file with the feature model in 2D (e.g. via FTP).

WebCADbyFeatures allows the creation and manipulation of the feature model for the raw material and finished component in a collaborative way, the storage of that information in a MySQL database, the validation of the model and the visualization of the geometric model in 2D and 3D (via VRML).

It is composed by a GUI that has menus, visualization options, error messages, feature manipulation, communication with the JATLite session manager for

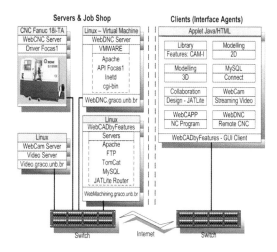

Figure 6.   Architecture of WebCADbyFeatures.

collaborative modelling, communication with the database server, communication with VRML server, monitoring of the shop floor (Web-Cam), teleoperation of the CNC turning centre, among other functions.

The main components of WebCADbyFeatures are: GUI as a Java applet, the feature library, 2D Graphical Interface, Collaborative Design IPLayer Router Client, components for 2D visualization (graphical primitives such as straight lines and arcs) and components for 3D visualization (VRML). The information regarding the features is handled through a database management system.

The modelling of the component begins with the access by the client to the web page for running the CAD Java applet. If the user is registered, an access to the database is made in order to verify the user login and password. Then, the applet is called, downloaded via web, and automatically the local Java machine runs the applet. The AWT standard (Abstract Windowing Toolkit) is used in order to allow a better performance and compatibility with any Java machine version 1.1, which is implemented in a native way in most browsers (Netscape, Iexplorer, among others), without need of a specific plug-in for a certain Java version, facilitating its execution by the user.

The first window to appear shows the initial options (Fig. 7a). For a non-registered user it is only possible to create a new project.

Then, a new window opens up (Fig. 7b) that gathers the design information. The system guides the user, asking for the relevant information for component modelling and process planning.

If the solid bar raw material is chosen, a new window appears (Fig. 7c), requesting the geometric information about a solid bar, which are its diameter and length.

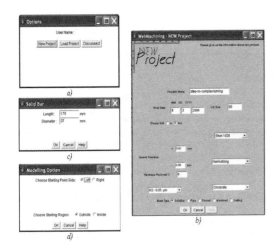

a)

c)

d)

b)

Figure 7. Stages in the design of a component: (a) options window; (b) window with data about a new project; (c) window with data about the raw material (solid bar); (d) window with modelling options.

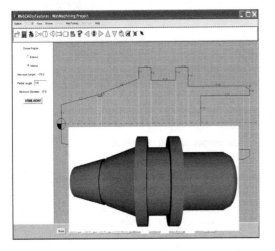

Figure 8. WebCADbyFeatures GUI with the STEPNC component: 2D and VRML views.

The last window in this stage (Fig. 7d) provides the options for selecting the floating zero (left or right) and if the user prefers to begin modeling with the external or internal portion of the part. The default is the modeling from the left-hand side, and beginning with the external features, which is the most common procedure among designers.

Proceeding with collaborative modelling, a drawing window opens up (Figs. 8 and 9), where the desired component is modeled, and the available form features in the feature library are selected. Initially, the component is modeled using the feature union

method, in which features are used as blocks for building the component geometry (like bricks), based on the CAM-I taxonomy (CAM-I, 1986). After finishing the union phase, features are subtracted from the component, including the features associated with the C-axis of the CNC turning center, which include keyways, eccentric holes, radial holes, etc.

The user has the option of zooming the drawing in 2D, move it on the screen, and also generate the VRML representation at any moment for 3D visualization. When selecting the "VRML-NOW!" button, the component model is sent to the WebMachining server through servlets, which saves the file in the server, and it is sent to the client's browser via FTP, which calls the available VRML plug-in.

The visualization in 3D is made through a plug-in for VRML, previously installed by the user in the browser. When selecting the button "VRMLNOW!" (Fig. 8), the component model is sent to the WebMachining server through servlets, which saves the file in the server, and it is sent to the client's browser via FTP. Then, it calls the available VRML plug-in, allowing the visualization of the component. There are the options for saving the geometric model locally in 2D and 3D (.wrl extension) or through features (.ftr extension), and this can be done because the security policy of the local Java machine is changed, allowing reading and writing files to the client computer. The Java machine is configured in a safe way, preventing applets from having access to the local resources of the machine. In Figures 2, 8 and 9 an example component is shown in 2D and the 3D solid in VRML.

6 CAPP/CAM SYSTEM IMPLEMENTATION

The CAPP/CAM modules can be characterized according to the four dimensions defined by Shah and Mantyla (1995): in the Planning dimension it generates non-linear process plans for cylindrical components machined in CNC lathes, containing definition of operations, fixtures, setup, sequences of operations, cutting tools, cutting parameters, NC code generation and cost estimation, in other words, micro-planning; in the Planning dimension related to time, it can be applied both on the tactical level and on the operational level of a Manufacturing Management System; in the Planning Method dimension, it is a generative CAPP system; in the Planning Depth dimension, it is defined as dynamic, in other words, plans can be changed dynamically during manufacturing depending on the dynamic characteristics of the manufacturing system.

The CAPP system, called WebCAPP (http://WebMachining.AlvaresTech.com), is composed by ten activities, based on STEP-NC:

1. Mapping of design features into manufacturing features (machining): it accomplishes the mapping

545

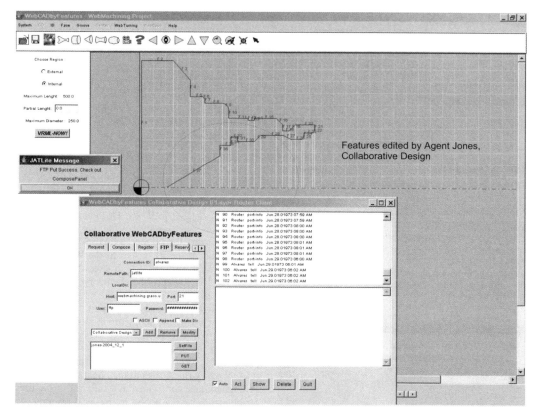

Figure 9.   Collaborative modelling between two design agents: Alvares and Jones.

of design features into manufacturing features, including machining operations such as internal and external cylindrical turning, facing, boring, parting-off, threading, etc.

2. Determination of the machining operations with alternatives, associated with the machining features (workingsteps): it selects the machining processes for the identified features, and it also considers the constraints associated with the dimensions, tolerances, material of the component, among others.

3. Determination of the machining sequence with alternatives (nonlinear workplan): it determines the machining sequence with alternatives and setup for fixturing the component.

4. Strategies for generating tool paths: it determines the strategies for generating tool paths based on STEP-NC.

5. Determination of the cutting tools (inserts and supports): it selects the cutting tool considering the machine-tool, the component material and geometry, tool life, etc.

6. Determination of the time model and calculation of the time and cost standards for each workingstep.

7. Determination of the technological conditions of optimized machining using genetic algorithms: it determines the cutting conditions considering the tool parameters and material, and subject to the following constraints: the tool life criteria used, machine power, machine capacity, among others.

8. Linearization of the nonlinear process plan using genetic algorithms.

9. Generation of the NC program (ISO 6983): the NC program (ISO 6983) is generated in such a way that collisions are avoided.

10. Generation of reports and process plan: it sets up the document regarding the process plan, including information on alternative plans and cost estimates.

7   CONCLUSION

In this article the implementation of a system for of collaborative product development was described, based on modelling for synthesis of design features for cylindrical components (symmetrical and asymmetrical features) through Internet. This software, called

WebCADbyFeatures, is one of the modules of the WebMachining system, which proposes a framework for CAD/CAPP/CAM integration.

A multi-agent system that enables collaborative design was developed, being implemented in a client/server architecture, composed by servers, HTML pages and Java applets, which allow the remote user to carry out the collaborative modelling of the component in 2D and its visualization in 2D and 3D, through VRML.

Some of the characteristics of the WebCADby-Features system are pointed out below:

- It uses multi-platform servers based on Servlets, JATLite, HTTP, MySQL and FTP; implemented in Java, HTML, Javascript and PHP. The servers were developed in the Linux platform;
- The client is based on Java applets, using AWT, not being necessary the Java plug-in, enabling total compatibility with the browsers; the user will just have to install a plug-in for visualization of the component in VRML;
- Multi-user and multi-task system, based on threads, both on the server side and on the client side;
- Remote communication among people, eliminating the geographical and temporal barriers for product development, allowing the implementation of Concurrent Engineering;
- The user can model parts with splines for "general_revolution" type features, and he/she can use eccentric features (C-axis) in the design, moving beyond STEP NC-Part 12 (ISO 14649, 2003).
- Speed and safety in the communication among the agents.

The WebMachining system can be accessed via web through the following URL: http://WebMachining. AlvaresTech.com. Many of its modules are available for use, thus providing a remote laboratory and a machining rapid prototyping system, in a context of e-Mfg (TeleManufacturing), allowing collaborative modeling, process planning and remote machining through web.

REFERENCES

Bidarra, R., Van den Berg, E., Bronsvoort, W. F., 2001, "Collaborative Modelling with Features", Proceedings of DET'01, ASME Design Engineering Technical Conferences, Pittsburgh, USA.

CAM-I, Deere & Company, 1986, Part Features for Process Planning, Moline Illinois, CAM-I.

Corney, J., Lim, T., 2001, "3D Modelling with ACIS", Saxe-Coburg Publications, 2a Edition.

Fuh, J. Y. H. Li. W. D., 2004, "Advances in collaborative CAD: the-state-of-the art", Computer-Aided Design xx 1–11.

Han, J. H., Requicha, A. A. G., 1998, "Modelerindependent Feature Recognition in a Distributed Environment". Computer-Aided Design, Vol 30, No. 6, pp 453–463.

Hardwick, M., Spooner, D. L., Rando, T, Morrir, K. C., 1996, "Sharing Manufacturing Information in Virtual Enterprises", Communications of the ACM, Vol 39, No. 2, pp 46–54.

ISO 14649, 2003, Data model for Computerized Numerical Controlers – Part 12: Process Data for Turning, Draft International Standard, V09.

Lee, J. Y., Han, S. B., Kim, H., Park, S. B., 1999, "Network-centric Feature-based Modelling", Pacific Graphics.

Li, W. D., Ong, S. K., Fuh, J. Y. H., Wong, Y. S., Lu, YQ, Nee AYC., 2004, "Feature-based design in a collaborative and distributed environment", Computer-Aided Design, Vol 36, No. 9, pp 775–97.

Martino, T. D., Falcidieno, B., Hasinger, S., 1998, "Design and Engineering Process Integration Through a Multiple View Intermediate Modeler in a Distributed Object-oriented System Environment", Computer-Aided Design, Vol 30, No. 6, pp 437–452.

Shah, J. J., Dedhia H., Pherwani, V., Solkhan, S., 1997, "Dynamic Interfacing of Applications to Geometric Modelling Services Via Modeler Neutral Protocol", Computer-Aided Design, Vol 29, pp 811–824.

Shah, J. J., Mäntylä, M., 1995, "Parametric and Feature-Based CAD/CAM: Concepts, Techniques, and Applications", John Wiley & Sons, New York.

Shao, X., Li, Y., Li, P., Liu, Q., 2004, "Design and implementation of a process-oriented intelligent collaborative product design system", Computers in Industry, Vol 53, No. 2, February, pp 205–229.

Shen, W., Norrie D. N., Barthés J. P. A, 2001, "Multi-Agent Systems for Concurrent Intelligent Design and Manufacturing", Taylor & Francis, New York.

Shen W., Norrie D. H., 1999, "Agent-Based Systems for Intelligent Manufacturing: A State-of-the-Art Survey" Knowledge and Information Systems, an International Journal, Vol 1, No. 2, pp 129–156.

Smith, C. S., Wright, P. K., 2001, "Cybercut: An Internet-based CAD/CAM System", ASME Journal of Computing and Information Science in Engineering, Vol. 1, No. 1, pp 1–33.

Applications

Virtual modeling and rapid manufacturing – Bártolo (eds)
© 2005 Taylor & Francis Group, London, ISBN 0 415 39062 1

Customer-centred product development

R.I. Campbell
Loughborough University, UK

D.J. de Beer
Central University of Technology, Free state, South Africa

ABSTRACT: The process of new product development (NPD) should begin with determining customer require-ments and conclude with a product that is delivered on-time, on-cost and on-spec. Throughout the process, design outcomes must be presented to customers in a way they can understand. Research conducted by the authors has shown that there is one presentation format that is readily understood by most customers, i.e. a physical model. Using RP models, customer feedback can be actively sought and incorporated into design iterations throughout NPD. This paper begins by arguing the case for increased usage of physical models, in combination with other media, to gen-erate customer input. It describes the research that has been undertaken in ascertaining what kind of information can be gained from customers during NPD and several case studies are presented to illustrate this. Conclusions are derived as to the methods and tools required for product development to become truly customer-centred.

1 INTRODUCTION

The whole process of new product development (NPD) should begin and end with the customer, i.e. it starts with determining customer requirements and concludes with a product that is delivered to the cus-tomer, on-time, on-cost and on-spec. However, there can be a tendency, even within companies who practice concurrent engineering, to limit customer involvement only to the beginning and end of the NPD process. This opens up the possibility that the customers' opin-ions will be lost or at least diluted by the time the product is launched. Techniques such as quality func-tion deployment (QFD) are aimed at propagating the "voice of the customer" throughout the entire NPD process and have been shown to produce tangible bene-fits. At each stage of the process, the customer require-ments are re-interpreted into desirable outcomes related to the specific tasks at hand. For example, the need to have a certain textured finish on the surface of the product may be interpreted into a particular spark-erosion technique during the tool manufacture stage. To ensure that the outcomes at any particular stage still coincide with the customers' requirements, it is nec-essary to present the outcomes to them in a way they can readily understand. This is often problematic since the outcomes may be in the form of a highly technical text document, numerical calculations, computer-generated images, test results or some other format that is beyond the understanding of the lay person.

Research conducted at Loughborough University and the Central University of Technology, Free State has shown that there is one presentation format that is readily understood by most customers, i.e. a physical model. Traditionally, physical models have been used by companies to present design concepts both inter-nally and externally, e.g. clay model mock-ups of automobiles. They have also been used internally for a range of other activities, e.g. assembly checks, func-tional testing, production operations training, etc. However, as more of these activities have come into the realm of virtual prototyping, the requirement for physical modelling has decreased. This means that the opportunity to receive customer feedback at key stages in the PDP has also decreased since a virtual prototype can not yet replicate all the characteristics of interest to a customer. However, with the advent of rapid prototyping, it has now become relatively quick and easy to convert a virtual prototype (in the form of a CAD model) to a physical model. This enables cus-tomer feedback to be actively sought and incorpo-rated into design iterations throughout the PDP.

This paper begins by arguing the case for increased usage of physical models, in combination with other media, to generate customer input throughout the PDP. It describes the research that has been under-taken in ascertaining what kind of information can be gained from customers during the PDP and what types of prototypes they need to have access to (both virtual and physical). Several case studies are presented which show the value-adding potential of involving customers directly in the PDP. Finally, conclusions are derived as to the methods and tools required if product development is to become truly customer-centred.

2 USING PHYSICAL MODELS TO GENERATE CUSTOMER INPUT

In recent years, virtual prototyping (VP) has made a great contribution to the NPD process both in terms of supporting individual design analysis tasks, e.g. finite element analysis (FEA), computational fluid dynamics (CFD), etc. and also in enabling closer integration of the design team within a concurrent engineering approach, e.g. through the use of product data management (PDM). Many engineers today would consider themselves to be computer literate and would not hesitate to adopt a virtual prototyping strategy if the benefits could be clearly shown to them. Other professionals within the NPD team may not be so familiar with graphics-based packages but with some familiarisation should be able to understand and use the output from virtual prototype "tests", e.g. sales and marketing using photo-rendered images within product brochures.

However, when one moves outside of the NPD environment into the general public, computer literacy and computer acceptance vary dramatically. Even with the younger generation, who are typically very computer literate, trying to use specialised VP software to present product appearance or functionality can be fraught with problems. For example, using an FEA fringe plot to prove to a prospective customer that a skate-board deck will withstand their weight (even when landing from a jump) is likely to leave them nonplussed. There is no real interaction with the prototype and hence little sense of realism. As VP develops in future, e.g. making use of virtual reality (VR) techniques such as immersion and haptic feedback, this may change. However, at present it is not suited to determining customer opinion on design concepts (or at least, not on its own). For this reason, many companies will resort to using physical prototypes when dealing with customer input.

Over the past decade or so, rapid prototyping (RP) models have brought a new dimension to capturing customer input. Prior to the use of RP, designers would normally have to produce two different types of physical prototype for a new design (Evans & Campbell 2003). A "block model" would be made using manual modelling techniques that would closely resemble the appearance of the final product. However, it would be non-functional and could be too fragile for the customer to handle. Additionally, a working prototype would be created to demonstrate how the product would function.

Typically, this would bear little resemblance to the final product. Customer evaluations would be hindered by the fact that it was impossible to obtain a holistic view of the product from either prototype. This would severely limit the quality of the opinions produced. Using RP it is possible to create accurate geometric models directly from CAD data which will include all the internal details. If necessary, secondary processing, such as investment casting, can be used to convert the model into a wider range of materials. The functional components created can then be assembled into a fully working "appearance prototype". This enables potential customers to use the prototype as they would the final product in the normal environment and without fear of damaging it. This will yield more representative customer input compared to traditional modelling methods and has made the use of physical models an extremely accurate way of generating customer input.

3 INFORMATION OBTAINED USING PROTOTYPES

The aim of a customer-centred product development process should be to firstly capture customer requirements into the product design specification (PDS) and then to continually verify that the evolving design matches these requirements. Customer requirements can range from "hard" functionality such as power output or load-bearing capability to much "softer" qualities like "I want this product to make me feel successful". Some requirements might be shared by a large number of people in society, e.g. "must be safe to operate", whereas others may be specific to an individual, e.g. "must fit my foot". Obviously, the wide diversity of requirements that can be captured will require a range of different tools and techniques. The problem is that it is the more difficult-to-capture qualities that will often have the most impact upon product success. The main categories of input are discussed below.

3.1 Functional requirements

These refer to what the product will actually do, i.e. its primary purpose. This will include immediate performance targets but also longer term aspects such as reliability, serviceability and the life in service that customers will expect.

3.2 Environmental requirements

These are concerned with the impact that the manufacture, use and disposal of the product will have upon the environment. Customers may have a view on manufacturing processes or materials that they regard as "environmentally unfriendly". They would not buy a

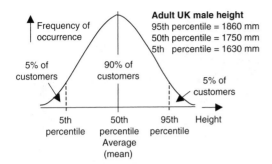

Figure 1. Anthropometric data for adult UK male height.

product made in this way. Likewise, they may only wish to own a product that has lower energy consumption and that is easy to recycle.

3.3 *Ergonomic requirements*

These are best understood as the product features that lead to ease-of-use. Therefore, they include all aspects of the human-product interface, e.g. size of a handle, comfort of seat cushions, overall weight, font used in decal lettering, clarity of instruction booklets, etc. In terms of the physical size aspect, ergonomists often make use of standard anthropometric data that gives various human dimensions in terms of percentile values (see figure 1 for an example). These can be used to estimate what percentage of customers will be accommodated by the product.

3.4 *User-fit requirements*

This could possibly be seen as a subset of ergonomic requirements. It refers to the physical interface between the product and the customer's body. However, it has the extra connotation that the product must be fitted to the individual, not merely a percentile range. Example products where this is crucial are dentures, spectacles and orthopaedic devices.

3.5 *Aesthetic requirements*

The term aesthetics refers to the impact that a product will have upon the human senses, most notably the sense of vision. This will include the overall form of the product but also the colour scheme and surface textures. Other important sensual requirements may include how the product feels to touch and how it sounds when used.

3.6 *Emotional requirements*

This is perhaps the most difficult category to define because it could possibly encompass all of the above.

For example, think of the various requirements that would be necessary for a product to provide a feeling of luxury. However, the unique aspect of emotional requirements is that they are normally not directly measurable in the product. Rather, it is the effect that the product creates within the customer. This also tends to be the most difficult area to predict and to measure.

An essential question for a product development team to answer is "how can customer input be captured?" This will obviously depend heavily upon the nature of the input. In most cases, the potential customers are exposed to an existing product design or a new design concept and asked for their opinions. It is preferable if the customers are allowed to actually use the product (or prototype) and even more desirable if they can do so in the normal operating environment. Various aspects of the design will be evaluated and sometimes rated against some sort of benchmark, for example a competitor's product. Questionnaires can be used to assist with this. The results from many users will be compiled, analysed and then translated into specific requirements (often the most problematic aspect of the process).

Techniques available to help with the process of obtaining customer input are well-covered elsewhere (Langford & McDonagh 2002), (Akao 1990). This paper concentrates on the role that physical prototypes can play in capturing customer requirements. Referring to the preferable situation for customer involvement that has just been described, it can be seen that a physical prototype should enable the customer to use it naturally and in its entirety. That is, the ideal physical prototype to be used for customer involvement must be virtually indistinguishable from the final manufactured product. This is not always possible, primarily because of expense, but the advent of RP has enabled high quality appearance prototypes to be created for many products. However, this is not the case for every product and besides, just using a prototype will not enable the product development team to capture all types of customer input. Consequently, the use of physical prototypes should be supported by other techniques also. Some of these are computer-based and link well into the same digital format that is used by RP. Examples of these include CAD rendering, virtual reality, reverse engineering, virtual sculpting and CAD modelling. The precise combination of technologies required for the development of any specific product is the subject of ongoing research at Loughborough University (Cain 2005).

4 CASE STUDIES

A number of product development case studies have been undertaken at Loughborough University and the Central University of Technology in order to

Figure 2.   Original gardening fork design.

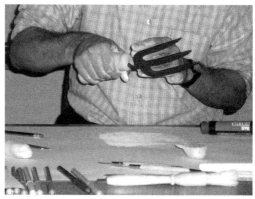

Figure 3.   New handle design being modelled.

Figure 4.   Scanning of a fork handle clay model.

demonstrate the effectiveness of customer-centred product development as well as to identify some of the current limitations. Three of these case studies have been selected and are presented below. The first demonstrates the combined use of physical models with other techniques, the second highlights the feasibility of using rapid manufacturing to obtain full-functional prototypes and the third illustrates the way physical prototypes can support a highly iterative product development process which can lead to a highly optimised design.

4.1   Case study one – customised fork handles

As a demonstration of how customer input can be captured within the NPD process using computer-based technologies, a customer-centred case study was undertaken. The product chosen was a small, hand-held gardening fork (original design is shown in figure 2). The product was selected because it embodied several of the requirement types listed in section 3. The aim was to create customised handles for four individual users according to their specific requirements. The following process was followed:

1. Discuss handle requirements through a semi-structured interview.
2. Evaluate the original handle design against a set of pre-determined criteria such as grip, aesthetics, usability, etc.
3. Generate an improved user-fit design, recording ideas in verbal, sketch and written format.
4. Capture user-fit and other ergonomic requirements using modelling clay.
5. Translate into CAD model (using reverse engineering if necessary).
6. Capture and verify aesthetic requirements using CAD rendering.
7. Verify functional requirements using RP model.

The modelling and prototyping stages are now discussed in turn.

4.1.1   Customer input using modelling clay
Customers were shown the original fork design and then provided with the metal element (shaft and prongs) together with an air-drying modelling clay. They were asked to model their own design of handle that would fit their hand as they desired, incorporate other ergonomic aspects such as finger grips or wrist supports and functional aspects such as hanging holes. They were encouraged to attach the clay to the metal element during this process to give a representative feel of weight and balance. An image of one of the new designs being modelled is shown in Figure 3.

4.1.2   Translation into CAD model
Some of the handle designs created were relatively simple in shape and it was possible to model them in CAD through direct observation. However, some of them were more complex and reverse engineering had to be used. 3D laser scanning was undertaken with 3D Scanners' ModelMaker and a FARO arm system (see figure 4). The point clouds of data were imported into Geomagics Studio software where it was merged and

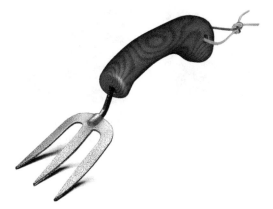

Figure 5.   CAD Rendering of wooden handled fork.

Figure 6.   Verification of fork handle functionality using RP model.

refined before being used to create NURBS surfaces. If necessary, these surfaces were further refined using the Freeform virtual sculpting system and then used as the basis for building a solid CAD model within Solidworks.

4.1.3   CAD rendering
Once the CAD model was completed, it could then be used as the basis for high-quality rendered images that were used to convey alternative colours and surface textures that the finished product could have. The users were shown different versions of these until they were happy with the aesthetic appearance of the design. The example shown in Figure 5 is a representation of what a wooden handled fork would look like.

4.1.4   Verification of functionality
The final stage in the process was to verify that the new design of handle met the customer's requirements for functionality. This was achieved by creating an RP model of the handle using an FDM 2000 machine with ABS material. The handle was hand-finished and attached to the metal element of the fork ready for functional testing (see figure 6).

The outcome of this case study was proof that customers can become closely involved in the design process. Designers can work directly with them to capture requirements that can then be used to produce new product designs. The case study also demonstrated the role that computer-based technologies can play in this interaction. A simple product was deliberately chosen for the case study but the principles followed are applicable to more complex products also.

4.2   Case study two – self-tensioning unit for display system

The Central University of Technology, Free State worked with Technimark, a Cape Town based product development company, to develop a portable display system that could display banners in a bowed form. The banners were to be tensioned via a device which would, upon application, tension the banner to approximately 80 mm. Having identified the device as a crucial part of the solution, the next agreement between the designers and the customer was that this would be a separate self-tensioning unit (STU), which opened up numerous deployment possibilities. The STU was envisaged as a free-standing unit that could also be mounted on walls, structures and frames. It should be able to hold banners of up to six meters in length, which when mounted to a wall, using six or more units to tension and display the banner.

The design brief for the STU was as follows:

- The unit must have some form of tensioning mechanism or system capable of being adjusted through approximately 80 mm of linear travel
- Each unit needs to be able to provide a tension force of approximately 10 kg.

An iterative design process was then followed using physical prototypes for each version to gain customer input.

4.2.1   Design iteration MK1
The following design characteristics were applied in the MK1 concept:

- Tensioning was achieved through a spring-loaded slider moving vertically within the main body of the unit (see figure 7a);
- Variable mounting-angles were achieved by using a cylindrical hinged pin, located in a slider (see figure 7b);
- The base and cover were assembled using screw-fixing (see figure 7c);
- Design curves and lines were defined and experimented with to achieve an acceptable level of aesthetics.

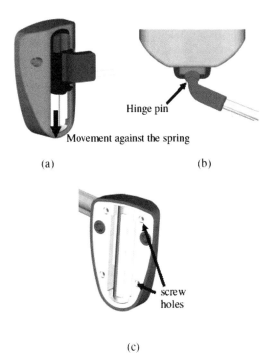

Hinge pin

Movement against the spring

(a)   (b)

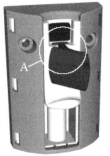

Enlargement of area
inside circle A

A

(a)

screw
holes

(c)

Figure 7.   Design characteristics of the MK1 concept.

(b)

Figure 8.   Design characteristics of the Mk2 concept.

Prototypes were produced using laser sintering (LS) and tested by means of putting up a 1 m × 2 m banner on a simple structure and evaluating the design. The main outcome of the MK1 functional design evaluation was that the curvature of the flexible carbon fibre batten used between two units caused the hinge pin to be dislodged out of the slider.

4.2.2   Design iteration MK2
The design concept was re-thought, which lead to the following alternative design improvements being applied to the MK2 concept:

- The round hinged pin part was replaced with a T-Profile mount-block with a fixed angle (see figure 8a). The angle's variation will be achieved with dedicated parts designed for required/preset angles (see figure 8b). The "square" button seen above the mount-blocks was for locking them in position.
- Snap-fitting of the cover and base to eliminate screws.

Prototypes were again produced by LS and once more tested by means of suspending a banner of 1 m × 2 m, as a functional evaluation of the design. This time, the twist force transferred by the flexible carbon fibre batten caused the slider to jam. Springs with a different stiffness value were used, but without any real improvement of the situation.

4.2.3   Design Iteration MK3
Once more, the concept was re-thought, with further improvements suggested and applied in the MK3 concept:

- The only change to the main body of the Self-Tensioning Unit was that the locking button was omitted, allowing the unit to be made shorter. This feature was judged unnecessary based on the evaluation of the previous prototype grown.
- The flexible carbon fibre baton was replaced by an off-the-shelf aluminium extrusion (see figure 9a). The extrusion could be bent to achieve the desire bow profile. This resulted in no twist force being exerted on the units.
- The T-profile mount-block part was replaced by two parts. The first part locates in the unit's T-slot and is connected to the second part via a hinge pin (see figure 9b). The second part is screw fixed to the aluminium extrusion.
- With the rail mount-block now made from two parts connected by a hinge pin, any angle can be accommodated, due to the swivel action (see figure 9c).

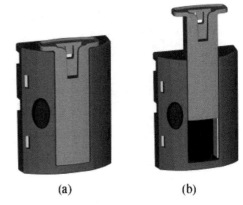

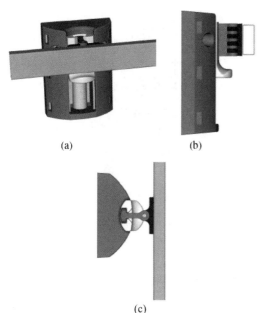

(a)   (b)

Figure 10.   Bottom (a) and top position (b) of slider.

(a)   (b)

(c)

Figure 9.   Design characteristics of the MK3 concept.

The same evaluation method of using LS functional prototypes and a banner of 1 m × 2 m on a structure was used to test the concept. During the functional evaluation of the MK3 concept, mounting of the four units on a wall proved to be very difficult, especially when trying to achieve horizontal and vertical alignment. The rail mount-block was bulky, which required that the sock of the banner to be large enough to feed the Aluminium rail through it, with the mount-block already attached. The pockets of the banner-sock needed to have holes cut in them in order for the rail mount-block's swivel end to have access to the slider on the STU. With the rail mount-block fixed to the aluminium rails, it could not be guaranteed that if the prints needed to be changed within the store, the same rail will end up at the same set of mounted self-tensioning units. Inter-operability could not therefore be guaranteed. The entire installation, assembly of the print and changing it were major concerns identified, and alternative solutions had to be found. The customer identified the following further needs:

• It would be beneficial not to have to drill any holes in order to install a unit. If standard shelving rails could be utilised, this would be of great practical benefit. (The same standard is used throughout the specific chain of stores in question).
• The unit itself must not be visible to customers, only the print on the wall must be seen. The unit must be hidden behind the print at all times.

The original design brief was amended and the following requirements were added:

(a) Only the banner must be displayed. The STU must be hidden behind the banner.
(b) The unit must be able to mount on the standard railing system in the specific customer's stores.
(c) The initial installation must be simplified as well as the process of changing prints.

4.2.4   Design iteration MK4

Once more, the concept was re-thought and further improvements incorporated into the MK4 concept:

• The slider was adjusted to extend beyond the unit so that the banner will cover the unit at all times even if the slider is in its lowest position (see figure 10a).
• A push button mechanism was incorporated that locks the slider in either the top or bottom position and which needs to be pressed to move the slider from the bottom to the top position. When the button is pressed, the spring will shoot the slider to the top position (see figure 10b) where it will again lock in place. If the slider needs to be pushed to the bottom position, the press bottom will be pressed and the slider will be pushed manually against the spring force to the bottom stop. This feature addressed the issue of simplifying the banner display process tremendously.
• Allowance was made for a slide-in mounting bracket in the base part of the STU (see figure 11). This feature helped to address the design requirement that the fixing of the unit must be able to utilise the existing standard structures.
• A mounting adaptor plate (see figure 12a) could now be developed for each specific customer requirements, since it can be implemented into the standard system of such a customer. The images below and following are specific to the specific customers

Figure 11. Slide-in mounting bracket in the base part of the self-tensioning unit.

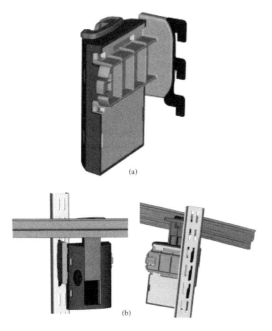

(a)

(b)

Figure 12. (a) Adaptor plate. (b) Attachment to customer's rail.

internal mount rail that is installed in all their stores at standard distances (see figure 12b).
• The rectangular aluminium extrusion was replaced by a specially designed extrusion. This adoption of a slide and guide method for the rail mount-block meant that there was no need for screw fixing and also the pitch spacing of the STU was not critical. The rail mount-block could be simply slid along inside the rail to the point where it matched up with the slide rail block location slot. This also contributed to easing the assembly and putting up of the unit.

Figure 13. Method for securing the print to the plastic extrusion.

• The method used to secure the print (see figure 13) was now a readily available commercial system, with the following advantages: (a) the 18 mm × 3 mm flexible plastic extrusion, required a much smaller banner sock, (b) feeding the extrusion through the sock was very easy and (c) changing the prints was also made very easy since all preparation could be done in the store areas, without inconveniencing consumers. Changing prints could be done in a matter of minutes.

4.2.5 The final product design
Prototypes were produced from the final design iteration (see figure 14) and they were tested and evaluated. The performance result and feedback obtained gave the entire design team a high level of confidence. A full set of ten STUs were prototyped using LS. The units were finished-off and sprayed in the specific colour required by the customer (similar to the standard colour of their rails and shelving systems). The units were couriered to the UK and installed in the customer's marketing store for various departments to evaluate the system.

Due to the design solution being tailored around the customer's specific requirements and the fact that the solution could be tested and evaluated upfront, the customer considered themselves to be a part of the development process and also felt confident that their specific requirements had been met. Based on these factors, a provisional order of 45 000 self tensioning Units were placed subsequent to the approval of the final injection-moulded components and units.

4.3 Case study three – car dashboard console

Another of the case studies conducted at Loughborough was the redesign a vehicle dashboard console to produce

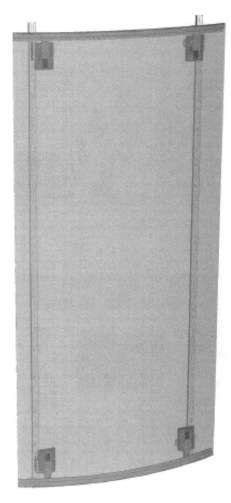

Figure 14. Final design shown in fully assembled configuration.

a new concept design to be produced using rapid manufacturing (RM). The existing product, manufactured via injection moulding, was a multiple part assembly that needed to be manually assembled with additional mechanical fixings. The original dashboard console was designed to house the vehicle's electrical instrumentation switches along with a standard automotive audio entertainment system, such as an AM/FM radio or CD player. However, it was the customer's wish that this audio equipment be replaced with an electronic GPS navigation and display system. A picture of the existing dashboard console design and the new GPS system may be seen in figure 15.

The new electronic equipment, which was bought from an external supplier as a sealed unit, needed to be fitted intact and in such a way that it could be operated without invalidating the manufacturer's warranty.

Figure 15. The current dashboard console and new electronic GPS unit.

Other similar projects had previously been undertaken by the customer, using conventional injection moulding methods to produce housings for similar electronic equipment. However, it was decided that a comparative investigation would be made into the possibility of using RM as a feasible alternative. Therefore, the method used was as follows:

1. Reverse engineer the existing part and replicate generic part geometries
2. Redesign the part so as to meet the customer's new requirements (i.e. inclusion of GPS)
3. Provide a product that complied with the customer's standards and testing procedures.

Design flexibility throughout the project was restricted by a number of factors, which included the following.

- It was necessary to accommodate all of the standard electrical controls and maintain the same instrumentation layout as used on the regular console.
- Secure "packaging" of the GPS unit was to be achieved in such a way that spatial conflict between it and the consoles reverse elements did not occur.
- Necessity existed to provide user access to the GPS control systems, which were in the form of three push buttons and an infrared sensor.
- Aesthetically, the overall appearance of the console was to remain similar to the original.

Obviously, design for RM was a key element within the project and all design concepts were developed in line with the principles of a new designers tool being produced at Loughborough (Burton 2005).

During the project, few "conventional" paper based drawings were made. Instead, the use of parametric 3D CAD software enabled generation of a "fuzzy" model. Using this generic model it was relatively easy to apply conceptual features and design revisions in a format

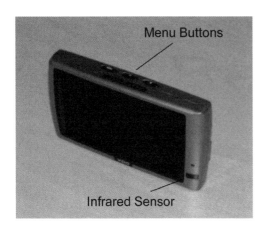

Figure 16.   GPS unit control systems.

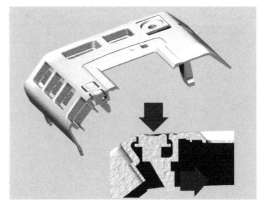

Figure 17.   A section view of the integral cantilever switch mechanism.

that could be physically reproduced and assessed using accurate RP parts. Whilst unseen by all but the designer, a number of paper based "scratch pad" sketches were made and used to roughly visualise and evaluate features prior to modelling. Likened to "envelope" or "napkin" sketches, this is a common activity within such CAD based design methodologies.

User access and control of the GPS unit was to be achieved via both infrared remote control and direct physical contact with the three push buttons on top of the unit. A picture of the unit, showing the location of these two control systems, may be seen in figure 16.

For infrared activation a simple aperture was created so as to provide a clear line of sight from the remote control to the sensor. However, access to the push button controls once the device is installed required a more elaborate solution. When mounted within the console, the three control buttons were perpendicular to the display screen area and inaccessible from any other plane due to the position of other instrumentation. In order to operate these controls it was necessary to devise a switch mechanism which allowed the transmission of force in a direction perpendicular to that which it was originally applied. To resolve this issue a purpose designed cantilever switch was employed. Hinged from the bounding box used to position the GPS unit, the switch acts as a rocker type device through which force may be transferred. When force is applied to the exterior part of this switch, the mechanism flexes about its cantilever axis and transmits force through a push rod, which actuates the GPS unit control button. The picture in figure 17 shows a section view of this cantilever switch mechanism and how it transfers user-applied force to operate the GPS unit.

Designing this mechanism in a way that it remained an integral part of the console removed the need for any additional components and the issues commonly related to them, such as manufacture, assembly and

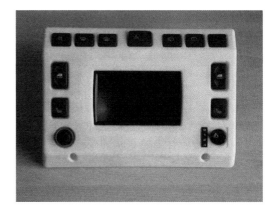

Figure 18.   RM product assembled and ready for testing.

their attached costs. The complexity of this feature and its geometry would make its production a near impossibility with conventional methods of manufacture. However, undercut features and impossible lines of draw are of little or no consequence when producing parts via RM, allowing the inclusion of such features within components.

The console design was produced using SLS and went through several iterations for fine tuning using direct customer input. The final version was assembled with existing switches (see figure 18) and successfully tested by the customer. It proved to be an excellent example of how the design freedom afforded by RM could be used to produce a high added-value product that would otherwise have been virtually impossible to produce. The success of the project had depended greatly upon the design flexibility afforded by RM and the ability for the RM components to be functionally tested by the customer. The process

560

followed had created a direct link from industrial design through to manufacture.

5 CONCLUSIONS

This paper has demonstrated how the utilisation of functional prototypes can enable complete design iterations (involving analysis of all design criteria) to be repeatedly undertaken until an optimised design is reached. This is possible because RP/RM prototypes are sufficiently similar to the final product for all functional testing and customer evaluation to be completed with a high degree of confidence. In addition, the provision of fully functional prototypes can also act as the catalyst for stimulation of further ideas and development.

A particular benefit of RP/RM prototypes is that they are readily understood by industrial designers, engineering designers, manufacturing engineers and even the target customer. They can perform an analogous role to virtual prototypes within the domain of PDM (product data management). In PDM, members of the product development team have remote access to the virtual prototype as shown in figure 19a and can perform their own specific analysis upon it. The PDM network can also act as a medium for communication between team members. As such, PDM/virtual prototyping is promoted as a means of breaking down the barrier between engineering design and manufacturing. A functional RP/RM prototype can act in a similar way (see figure 19b) but has the advantages of being portable, more widely understood and accessible to several people at the same time and place (rarely the case with virtual prototypes). This is indicated by the "cluster" members in fig 19b being shown as more numerous, and in actual contact with the prototype and with each other.

The role of the RP/RM prototypes can also be similar to that for which sketches and mock-up models are used within industrial design, i.e. for rapid and copious design iterations and as a communication medium amongst designers and between the design team and the customer. However, sketches can only be used to evaluate aesthetics whereas functional models can be used to evaluate aesthetics, ergonomics, performance, manufacturability, etc. In addition, when RP/RM is used in combination with other technologies, it enables customers to be brought more directly into the NPD process, even to the extent of geometry creation. Current technologies cannot fully support the "customer as designer" paradigm as they are often too complex to use. However, future CAD, reverse engineering and design analysis systems may indeed become accessible to the customer/user.

Customers are becoming much more discerning and selective, so for any product to be successful, it

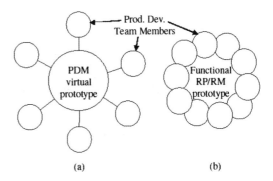

Figure 19.   Analogy between PDM network and functional prototype cluster.

must be highly attractive to the customer. Therefore, designers are required that have specific skills to analyse customer wishes, needs and requirements and understand how these can be translated to an end experience for the customer. Designers need to "get inside the heads" of customers by observing them using the product in the real-world. They need to increase their insight into the lives of potential customers. This will help expose them to the often unspoken needs and attitudes of users. This already happens in many smaller, innovation-led companies but is quite rare in larger, more organised companies (Burns & Evans 2000). The common practice of designers receiving customer preferences via a separate marketing or market research department is no longer sufficient. Designers must be trained to communicate directly with customers. Therefore, to fully implement customer-centred product development, it is firstly necessary to produce "customer-centred designers".

ACKNOWLEDGEMENTS

The authors would like to acknowledge the design inputs of Miss Rebecca Cain to case study one, the Technimark design team (Mr Alphons du Toit, Mr Dietmar Renner and Mr Hanno van Riet), to case study two (all designs have been made by Technimark) and Mr Mike Burton to case study three.

Acknowledgement for financial support goes to the National Research Foundation (NRF), the Department of Trade and Industry (DTI) and the programme for Technology and Human Resources for Industry (THRIP).

We would further like to acknowledge inputs of the management, staff and students from Loughborough University's Design Practice and Rapid Manufacturing Research Groups and the CUT's Centre for Rapid Prototyping and Manufacturing (CRPM). Particular thanks go to Mr G Booysen and M L Barnard as well as the students who were involved.

REFERENCES

Akao, Y. (Ed), (1990). Quality Function Deployment: Integrating Customer Requirements Into Product Design, Productivity Press.

Burns, A.D., Evans, S., (2000). Insights into customer delight In Scrivener, S.A.R., Ball, L.J. and Woodcock, A. (eds) Collaborative Design, Springer, London.

Burton, M.J., (2005). Design for Rapid Manufacture – Development of a Designers' Tool, PhD Thesis, Loughborough University, to be published in 2005.

Cain, R., (2005). Involving users in designing: A framework based on understanding product representations, PhD Thesis, Loughborough University, to be published in 2005.

Evans, M.A., Campbell, R.I., (2003). A comparative evaluation of industrial design models produced using rapid prototyping and workshop-based fabrication techniques, Rapid Prototyping Journal, 9(5), 2003, pp 344–351.

Langford, J., McDonagh, D., (2002). Focus Groups: Supporting Effective Product Development, Taylor & Francis, London.

Virtual modeling and rapid manufacturing – Bártolo (eds)
© *2005 Taylor & Francis Group, London, ISBN 0 415 39062 1*

New approaches in tooling design and manufacture for the packaging industry

D. Dimitrov & A. Bester
Laboratory for Rapid Product Development, University of Stellenbosch

ABSTRACT: During the different plastic conversion processes a large amount of heat must be extracted from the plastic in the mould. The rate at which this heat is removed has a large impact on the cycle time. The temperature distribution in the mould influences the heat extraction, causing some parts of the product to cool slower than others, which in turn has an influence on the dimensional accuracy of the final product. Modern simulation software has a large role to play in creating virtual representations of the heat distribution and conduction in moulds. These tools can help the designer to develop moulds and products, which are optimised both for aesthetic purposes and production. This paper discusses some experiences gained in a comparative case study of conventional cooling vs. conformal cooling. Some pitfalls are also pointed out related to the virtual model as well as to the manufacturing process of the experimental mould using rapid manufacturing techniques.

1   INTRODUCTION

The development of innovative products and their realisation by means of advanced manufacturing methods and process combinations is becoming more and more a key issue in international competitiveness (Bernard et al. 2002). Shorter product life cycle times demand rapid response and high levels of flexibility in the production of dies and moulds. Typical manufacturing sequences, however, including pre-milling, heat treatment, EDM, and manual finishing are usually associated with high lead times, high machining costs, and limited flexibility. Die and mould makers are therefore increasingly being compelled to utilise modern technologies in their production chains to satisfy their customers.

A prime example of vast possibilities for substantial performance improvement is the plastic conversion in the packaging industry. According to the common practice between 60% and 90% of the manufacturing cycle time of moulded plastic objects are used to cool the product in the mould to a temperature where enough mechanical strength has been gained to release it from the mould without any substantial distortion (Raennar 2003).

The main productivity increase, therefore, can be achieved by optimising the cooling cycle. The optimal cooling layout of a mould is usually presumed by the designer. He is, however, restricted to conventional manufacturing processes, such as drilling and EDM. The cooling design is normally influenced by

the product geometry and most often is not considered during the design process of the product itself.

Cavities are formed however, differently, and the same type of cooling cannot be used on every mould. There are certain areas of the mould, which need more cooling than others. Those are, for example, spots, where the plastic material is particularly concentrated due to the specifics of the moulding process. They become very hot during the operation and are actually the areas where adequate cooling must be assured. In bottles the typical areas of concern are the neck and the base. This is why it becomes very important that the layout of the cooling is designed in accordance with the actual conditions of heat generation and formation.

The paper discusses the use of Virtual Reality (VR) in the design process of a blow moulding tool. It reflects the experiences gained in the investigation of the cooling behaviour on blow moulds and the development of an optimised cooling layout using graphic simulations of the heat transfer patterns in the mould. For that purpose a new design of an existing mould was modelled and the cooling conditions investigated. The results were then compared with the simulated cooling behaviour in the production (existing) mould.

2   MOULD DESIGN

2.1   *General considerations*

A suitable approach to improve the cooling behaviour of a mould is the implementation of conformal cooling, which follows as much as possible the shape of the cavity. This can however, not be realised with conventional manufacturing processes. An alternative is the use of Layer Manufacturing (LM) methods, which are characterised by a high degree of geometric independence. The development of a conformal cooling layout requires from the designer to use powerful CAD/CAM software. A reliable interface between the different software tools is important to help the process to run efficiently and prevent design delays. The development of these capabilities has to assure the viability of the following concept points:

- The prediction of heat formation in the mould during the moulding process, enabling the optimal mould design from a cooling point of view
- The acceleration of the mould manufacturing process and then
- The combination of these steps into an integral solution for mould design and manufacture.

The advantages and potentials the conformal cooling could provide are demonstrated on a hypothetical model (Fig. 1a) of a blow moulding object. The simulated heat distribution can be seen when using conventional (drilled) cooling channels (Fig.1b), as well as applying the conformal cooling approach. Figure 1c shows clearly the equal distribution of the heat following exactly the contour of the cavity as a result of an optimised cooling as opposite to the happening shown on Figure 1b.

The heat flow in a mould is a very complex mathematical problem with many different variables. An important variable is the thickness of the material between the cooling water and the plastic (Rohsenow et al. 1998). It becomes, however, obvious that this will be simplified by the use of a mould with conformal cooling. The wall thickness of the mould stays in this case constant throughout. The ideal for the heat transfer in a mould would be that the plastic part is of uniform wall thickness. This combined with an uniform wall thickness in the mould together with a constant cooling fluid speed will reduce the heat flow problem in a mould to a trivial problem, which could be solved using simple methods. To achieve this, however, is not that simple, due to different features in the plastic articles such as handles, indents and others, which require different inserts in the mould.

To explain the heat flow in a mould the assumption will be made that the plastic and the mould are of uniform thickness and that the water flows at a constant speed (Fig. 2).

No heat is generated in the plastic or in the mould, the all heat present in the parison is a result of the extrusion process. When the parison expands to touch the wall of the mould, the major heat conductions start. There will be some other heat losses but all this is negligible compared to the conduction between the

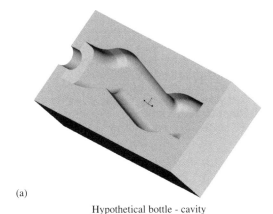

(a) Hypothetical bottle - cavity

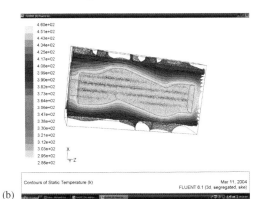

(b) Heat distribution with conventional cooling layout

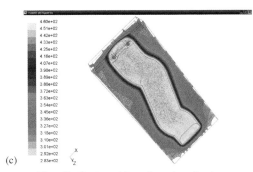

(c) Heat distribution with conformal cooling layout

Figure 1. Hypothetical blow moulding object.

parison and the mould. The mould acts as a heat sink, resulting of its excellent conductivity compared to plastic as well as its superior mass. The energy, which is now flowing from the plastic to the mould, is lost to the water. The water temperature will stay virtually unchanged if the mass through flow is sufficient. In the specific case, for example, of the container as illustrated

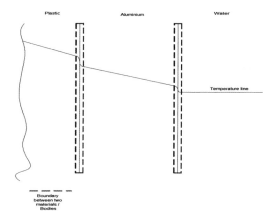

Figure 2. Heat flow in a mould (simplified).

below with a mass (shot weight) of 85 g, a mass flow of 10 kg water for the cooling cycle will ensure that the average cooling water temperature increase will not be more than 1 Kelvin if the bottle is cooled from 180°C to 50°C.

The major uncertainties in the heat flow problem between the plastic and the water are the boundary conditions between the plastic and the mould, and the mould and the water. It can, however, be calculated from textbook data and formulas (Rohsenow et al. 1998, Bejan 1993), but it is general knowledge that these calculations will have a certain percentage of error depending on the assumptions made. The only way to quantify these values correctly will be by determining them experimentally.

A method of determining these values is by inserting thermocouples in the mould wall at certain depths. This will allow the temperature gradient in the mould to be measured, from which the heat flow can be calculated. The placement of these thermocouples will be determined by analysing the simulation results.

2.2 Creating a simulation model

The next step contains the modelling of the production mould for the container as used by the particular packaging manufacturer. For simulation purposes a 3D solid model of the object is needed. This model of the mould was translated with the use of a STEP translator and imported into the relevant simulation package.

At this stage it has to be made clear that the usability of a model file depends mainly on the capability of the CAD package selected. It is true that an experienced CAD operator can positively influence the simulation results. To minimise the problems possibly caused by the CAD system a proper solid modelling package, which is set to tight tolerances, should be used (instead of a surface modeller) when the simulation model is developed.

There is also a question about the reliability of the interface, which is employed to translate the data between CAD – systems and simulations packages. The well known IGES translator, for example, seems to be totally unsuitable for simulations purposes. It exports the geometry as surfaces. These surfaces need to be joined again in the meshing package in order to recreate the solid model, which is very time consuming and frustrating. Good results were achieved with the Unigraphics translator, which exports Parasolids. It works relatively well, although some difficulties were encountered especially with the surface of the solids. Problems such as lines not matching up appeared and this in turn made the surface meshing, which is necessary before the solid meshing can be done, very difficult and in some cases impossible. The best interface turned out to be the STEP translator. It translates the solid as a solid, thus the surface geometry of the model is of a better quality than this obtained through a Parasolid. The STEP translation was still not perfect. However, a well designed solid could eliminate most of the potential problems.

Figure 3 shows the virtual model of the existing production mould as used by the packaging manufacturer. While Figure 3a reflects the solid model of the mould, the picture right – Fig. 3b, shows the solid model of the water jacket related to the conventional cooling channels. Figure 3c on the other hand illustrates the velocity conditions in the three sections of the mould. Subsequently, Fig. 3d shows the heat distribution throughout the mould. The inequality of this distribution is more than clear.

Based on empiric approach a prototype mould for the same object was constructed. The cooling was performed using a flooded cavity in the back of the mould. The cavity was created by offsetting the mould cavity by 6 mm, which formed the conformal cooling layout. The cooling simulation of the prototype mould was carried out using the same input parameters as applied for the production mould. The results can be seen in Figure 4.

An observation of the cooling conditions on the prototype mould shows much stronger presence and equal appearance of the cooling fluid surrounding the mould cavity (Figure 4a & b). The velocity diagrams in the three sections can be interpreted as follows: the water enters the top cavity at a relatively high speed. It follows the shape of the bottle neck, which ensures relatively good cooling. The cooling is still not fully uniform, due to the fact that the water speed is not completely uniform over this section of the mould. The water speed in the central cavity is relatively low, but it is uniform over the mould surface, which ensures uniform cooling. In the third mould section the water flows into the bottom of the mould cavity and hits the mould bottom at a relatively high speed, which will assure a good cooling in this area. In the rest of the bottom cavity the water

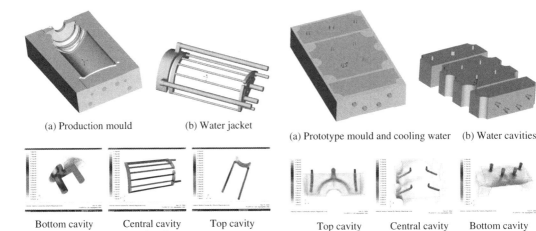

(a) Production mould   (b) Water jacket

(a) Prototype mould and cooling water   (b) Water cavities

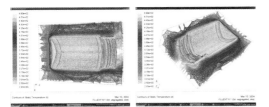

Bottom cavity   Central cavity   Top cavity

(c) Velocity diagrams

Top cavity   Central cavity   Bottom cavity

(c) Velocity diagrams of the cooling water
in the different cavities

(d) Heat distribution of bottle cavity from
different view angles

Figure 3.   Solid models, velocity diagrams and head distribution on the conventional production mould.

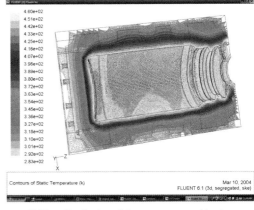

(d) Heat distribution of the prototype mould using
conformal cooling

Figure 4.   Solid models, velocity diagrams and head distribution on the prototype mould.

flow velocity is low, but uniform and will therefore cool the mould section equally. The heat distribution as shown on Figure 4(d) demonstrates the improved cooling as shown by the temperature contours which follow the cavity much closer.

There are, however, substantial possibilities for further improvement as shown on the pictures below (Fig. 5). Once again the pictures in (a) are reflecting the solid models of mould, water layout and possible inserts for better regulation of the water velocity. The velocity diagrams in section 5b show that the water flow can be manipulated to give an uniform surface velocity over the mould. In Figure 5c the temperature contours can again be seen. If Figure 5c is compared to Figure 4d a further improvement of the cooling can be clearly established.

3   IMPROVED MOULD MANUFACTURE

The original mould was designed in a surfacing package. Due to the loose tolerance setting of this package the model was unusable for simulations and prototyping

purposes. The basic cavity design is relatively simple. It was therefore possible to create a complete new solid model using appropriate software package such as Pro/Engineer. This formed the foundation for the conformal cooling layout. Thereafter the cooling side of the mould was divided into three water pockets. These water pockets were placed in such a way that it separated the three critical cooling areas in the mould.

An ".stl" file was created as an input for a layer manufacturing device using the 3D Printing process. In this way time consuming milling and EDM operations were avoided. The two halves of the mould, fabricated on the 3D printing machine in plaster based powder, were used as pattern for sand moulds. The sand castings of the mould cavities were subsequently machined to

Mould   Water and inserts   Inserts

(a) Possible cooling layout of prototype mould
for next iteration

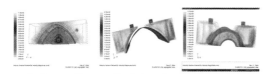

Bottom cavity   Central cavity   Top cavity

(b) Velocity diagrams of water in proposed cooling layout

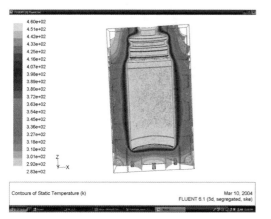

(c) Heat distribution of the bottle mould using the
proposed cooling layout

Figure 5.   A possible further design iteration of the cooling
layout.

final tolerance and finished by hand as required. The
cooling water was supplied through the back plate
which helped to keep the mould geometry simple.
This also allowed for cooling layout flexibility.

4   CONCLUSIONS

The conclusions can be summarised as follows:

1. The utilisation of Virtual Reality tools demonstrates clearly how mould design changes can drastically influence the cooling parameters of a mould
and therefore the reduction of cycle times.

2. The simulation of the cooling behaviour of both
the production and the prototype moulds, whereby
in the later some conformal cooling design ideas
based on empiric assumptions were implemented,
shows that the prototype mould has a better cooling than the production tool currently used. This
can be clearly seen when looking at the figures
showing the temperature distribution contours.
The physical comparative tests between the production mould and the prototype confirmed qualitatively this observation through the fact that the
bases of the bottles produced in the prototype
mould deformed less. None of the bottles from
either mould showed signs of any side deformation, but this was to be expected due to the simple
geometry of this part of the product. The duration
of the tests, however, was only a few hours, which
was extremely insufficient with regards to gaining
reliable data for comparison and further conclusions. Therefore proper experiment procedures
have to be designed and sufficient time for comprehensive tests – in the region of 72 hrs, allocated.

3. The use of VR tools showed good possibilities for
further substantial improvements. It is recommended that this or a similar version of the mould
is produced in order to carry out proper comparative tests, and to get calibration data regarding
extrusion, solidification, and ejection temperatures
of mould and parison. Only in this way a prediction
of the cycle times and their quantitative variations
will be possible. On the other hand, some steps of
the manufacturing process could be tested using
new materials available in layer manufacturing and
more specifically in 3D Printing.

4. Based on these experiments and trials a generalized
heat flow prediction methodology can be developed.

REFERENCES

Bejan A. 1993, *Heat Transfer*, Toronto, John Wiley & Sons,
Inc.
Bernard A., Fischer A.: August 2002, *New Trends in Rapid
Product Development,* keynote paper CIRP-General
Assembly, San Sebastian. CIRP Annals 2002, Volume
51/2/2002.
Raennar L-E. 1–4 October 2003, "Efficient cooling of FFF
injection moulding tools with conformal cooling chan-
nels – an introductory analysis", *Proceedings of the 1st
International Conference on Advanced Research in
Virtual and Rapid Prototyping,* , Leiria, Portugal.
Rohsenow Warren M., Hartnett James P., Cho Young I. 1998,
Handbook of Heat Transfer, Third Edition, Milan,
McGraw – Hill.

Virtual modeling and rapid manufacturing – Bártolo (eds)
© *2005 Taylor & Francis Group, London, ISBN 0 415 39062 1*

Flexible manufacturing concepts/new business models like mass customization – Advantages and new possibilities by using laser-sintering technologies

Markus Glaßer

EOS GmbH, Electro Optical Systems, Germany

ABSTRACT: This paper focus on the concept of flexible manufacturing highlightening its advantages and new possibilities through laser-sintering technologies. Several examples are used to illustrate the concept e-Manufacturing™, which means fast, flexible and cost-effective production of high-quality products directly from electronic data, is since a long time a common term on the RP and RM market. Laser-sintering technologies allow you to implement your development-/ manufacturing ideas better and completely different in every phase of the product life cycle than by using conventional technologies. Flexible manufacturing concepts and new business models are getting more and more of importance. Like mass customization which is meanwhile often used in the medical-, dental- and fashion industry.

1 INTRODUCTION

"The mass market is dead," Kotler proclaimed to describe the evolution from mass marketing to customer segment marketing, that finally arrived at personalized or "one-to-one" marketing today. The requirements that result from this paradigm shift are twofold:

- First, you need to better know than your competitor what customers want and to what extent they are willing to pay for.
- You need to transfer this knowledge into physical products and services to offer.

This results in the ability to offer individualized products and services in the sense of mass customization. This both requires restructuring efforts in sales and manufacturing. While the efforts in sales focus around minimizing the customer perceived uncertainty when acquiring customized products, research in manufacturing revolves around the issue of producing customized products at costs comparable to series production. In this paper we will investigate what contribution e-Manufacturing can make in this sense. From our point of view, e-Manufacturing means the direct, flexible and cost effective production directly from 3D CAD data files. In this way laser-sintering is a very efficient way of making e-Manufacturing reality.

As a rule of thumb we can claim that e-Manufacturing has a good chance of unlashing its potential if …

- the customer demands a high number of varieties which culminates in the extreme in the demand for individual products.
- demanded products are of high complexity.
- customer demand is subject to sudden and unpredictable change.
- product life-cycles ever shorten.

the sheer number of identical products sold is low.

In the following we will give a number of examples that proof that e-Manufacturing with laser-sintering can be a viable solution for industries that fit the above defined rule of thumb.

2 FLEXIBLE MANUFACTURING CONCEPTS AND NEW BUSINESS MODELS

2.1 *e-Manufacturing of direct plastic parts*

2.1.1 *Sunglasses*

Implementing e-Manufacturing also can open new ways in sales. In this case it very much could be possible that the customer enters a kind of "mini-factory" where he could design his own sunglasses at a suitable "tool-kit" providing him with support in doing so.

Once his design efforts result in a sunglasses he will be happy to buy, the created 3D data is directly transferred to an EOSINT system where the sunglasses are built. In order to transfer the crude sintered

Figure 1.   Laser-sintered sunglasses (Tecnologia & Design, Crabbi Sunliving, EOS).

Figure 2.   Laser-sintered hearing aids (Phonak).

part to a posh fashion product it still has to undergo an AutoFinish process that results in sunglasses as shown in Figure 1.

2.1.2   Hearing aids

Another industry that holds true for our defined rules of thumb are hearing aids. Since the success of hearing aids very much depends on its ability to adapt to the anatomy of the auditory canal. Figure 2 illustrates a hearing-aid with a laser-sintered shell.

The process of e-Manufacturing a hearing aid is as following:

- Take a copy of the auditory canal anatomy by creating a wax cast.
- Scan the wax cast with a scanner to create 3D data.
- Integrate an identification number in the 3D data that helps identify the product after the laser-sintering process.

Figure 3.   Laser-sintered skiboot buckles (Tecnologia & Design).

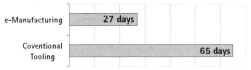

Figure 4.   Comparing time-to-market of a laser-sintered skiboot buckle with traditional manufacturing methods.

- Laser-sinter the shell.
- Combine the laser-sintered shell with the electronics components.

2.1.3   Skiboot buckles

Figure 3 illustrates the result of a laser-sintered and auto finished skiboot buckle that was painted in the last process steps.

Whether skiboot buckles, inline skaters, ice-hockey boots or any other trendy sports shoe, the buckle can become a vital part in determining both individual fit and aesthetic preferences. The major benefit of e-Manufacturing buckles like these with laser-sintering equipment is in a significantly reduced time-to-market. Because of omitting tooling that is necessary in traditionally manufacturing (e.g. blowmolding) the entire manufacturing process is speed up. Figure 4 illustrates the time saving effects of up to 58%.

A T&D research program in 2004 on this particular product indicates that laser-sintering leads blowmolding from an economic point of view up to 630 units.

For many consumer goods industries this number is indeed very hard to succeed, making e-Manufacturing with laser-sintering the manufacturing method of choice.

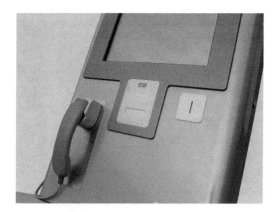

Figure 5.   Laser-sintered public phone handset holder.

Figure 6.   Fitting 120 handset holders into build window (Model Shop Vienna).

2.1.4   *Public phone handset holder*

Leaving consumer goods, e-Manufacturing can be even more valuable for certain B2B applications. A very good example is the e-Manufacturing case of Model Shop Vienna, using laser-sintering technology to e-manufacture a holder for a public phone handset.

The issue for Model Shop Vienna was to produce a lot size of 250 of the described cradles. Since this lot size is quite limited conventional manufacturing

methods were not suitable. Instead heavy investments in tooling would have increased cost per unit dramatically. On top, once tooling was fixed design changes would be prohibitive. Since Model Shop could not say for sure how many units of this cradle they would have to build, they decided to go for laser-sintering with Alumide (which is an EOS trademark). On the one hand this material combines all the properties the customer Telekom Austria demanded (metallic appearance, high stiffness, etc.). On the other laser-sintering made Model Shop Vienna flexible to unpredictable design changes the customer might want to add in a second order. Since laser-sintering results in a linear cost curve Model Shop Vienna did not have to calculate the break-even-point where the heavy investments in tooling would pay off. Model Shop Vienna estimates that laser-sintering leads the costs compared to conventional manufacturing up to 500 identical pieces.

2.2   *e-Manufacturing of tools for plastic parts*

2.2.1   *Hammtronic joystick "HI-drive"*

Another impressive example for e-Manufacturing is the manufacturing of a joystick navigating a Hammtronic tandem roller. By means of an automatic translation the vehicle always moves in the direction of the joystick deflection whichever way the driver is sitting. All in all this joystick offers 16 different functions and consists of 15 plastic parts plus parts for electronics, mechanics and switches.

The way e-Manufacturing by laser-sintering helped implement this product was not by laser-sintering the joystick itself, but laser-sintering the tools required for injection molding the required parts. With lot size of ca. 5000 the advantages of e-Manufacturing were not in laser-sintering the components itself but in laser-sintering the metal tools for injection molding these component. The first step was to divide the parts into four mother tools for optimum utilization of the injection molding machine. Then minimization of the volume of the laser-sintered inserts was required for minimum building times and material consumption. This resulted in "onserts" instead of inserts that were attached to a conventional manufactured tool base. The total volume of laser-sintered material was 3.1 liters for all tooling for the 14 plastic parts. Figure 7 illustrates some of these inserts.

The major value add in e-Manufacturing these inserts with laser-sintering was in a dramatically reduced time to market. In this particular case, the Hammtronic was about to be presented at a trade show just a few months ahead. Since the injection molding process was not yet optimized the tools underwent certain iterations until the molding process.

2.2.2   *Multi-component molding*

The superior quality of laser-sintered tools can very well be demonstrated by a example of a two color key

571

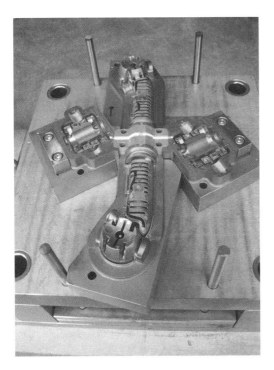

Figure 7. Some of the insert geometries laser-sintered on EOSINT M (Fruth Innovative Technologien – FIT).

Figure 8. Process of molding a two color key ring (Arberg).

ring for motor scooter. Figure 8 illustrates the process of getting a twin colored key ring using the DirectTool method employing direct metal laser-sintering.

The parting surfaces fitted perfectly, only reaming of ejector holes and polishing of molding surfaces was necessary after the laser-sintering process. The entire process from 3D CAD data to a finished tool took less than 2 days. In the manufacturing process the tools withstood 10,000 + high quality parts molded. This is

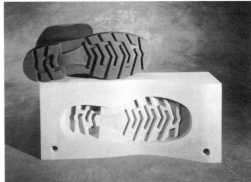

Figure 9. DirectTool injection mold and resulting rubber boot sole (Tecnologia & Design).

shows that e-Manufacturing with laser-sintering is very well a suitable application even for these large production batches. The quality of the tools can be estimated from the fact that a twin color mold requires very well finished tools as otherwise the two colors would mix.

2.2.3 Rubber boots
A very interesting example showing the potential of direct tooling in plastics laser-sintering is the injection molding of a rubber boot sole as illustrated in figure 9.

The advantages of employing laser-sintering in this case mainly is the reduced time-to-market. Following the mantra of e-Manufacturing this means the direct way from 3D CAD data to a tool featuring the sole of the rubber boot or any other trendy shoe. Using EOS Alumide further has the advantage that it can be used as a proper tool that can be fixed to the injection molding machine as it is. This is mainly because of the material properties resembling that of metal. This also results in a tool endurance that is sufficient for producing several hundred shots of shoe soles. This number is very well sufficient in trendy market like fashion shoes where design aspects are integrated into a trendy sole design. Laser-sintering the tooling for this kind of shoes can help this industry adapt sole design to sudden changes in customer preferences.

2.3 e-Manufacturing of direct metal parts

2.3.1 Dental camera
A good example for a metal part directly laser-sintered from 3D CAD data is the Planmeca dental camera as illustrated in figure 10.

The market for dental cameras is on the one hand relatively small and on the other can offer substantial value add by customizing to the anatomy of a dentists hands. This makes the customized economic point of view. By doing so, manufactures of such equipment

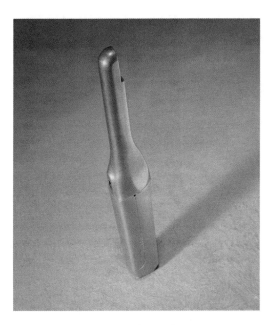

Figure 10.   Dental camera (Planmeca).

Figure 11.   Imaginary example for a spare pump housing.

can both address new markets by offering customization and gain an image boost by leveraging innovative technology manufacturing of dental cameras as one-offs feasible from an economic point of view. By doing so, manufactures of such equipment can both address new markets by offering customization and gain an image boost by leveraging innovative technology.

2.3.2   Metal spare parts

Another prominent example were e-Manufacturing of metal parts with laser-sintering offers a promising potential is the manufacturing of spare parts. Gallagher; Mitchke, and Rogers quantify the supply of after-market parts as a $400 billion business. With the traditional manufacturing of spare parts there are several problems. These are described in the following on the surface:

- Spare parts have to be provided for a long time even after the active product life-cycle ended.
- Tooling, documentation and the like have to be stored for a substantial amount of time.
- Restarting the original manufacturing process (e. g. blow molding) is prone to problems for on-off production of spare parts.
- There are spare parts where no original tooling nor documentation exists (e. g. ancient automobiles, museum artifacts, etc.).

One viable solution to all of these problems is e-Manufacturing of spare parts with laser-sintering. This becomes especially a viable solution for the time

period after the active product life cycle ended after 40 years. Imaging the fleet of trucks that serve as precious vehicles in developing countries after they were junked in industrialized nations. If you were the African operator of a MAN 15.215 DHS truck, build year 1968 you certainly would experience great difficulties replacing an aging pump housing with original spare parts such as illustrated in figure 11.

Employing reverse engineering tools that generate 3D data from the pump housing to be replaced. This data serves as the base for e-Manufacturing on laser-sintering systems making up for a spare part that meets the requirements.

The idea of e-Manufacturing spare parts also becomes a very interesting solution for military purposes. In this field researchers evaluate the feasibility of a mobile parts hospital. Here the vision goes that required spare parts are directly manufactured in the field. This vision may also be tempting for civil purposes. Imaging for example an ocean cruiser that e-manufactures required spare parts while on the sea. The same might hold true for aerospace applications if spare parts could be e-Manufactured on site at remote airports.

2.3.3   Computer locks

The requirement in this case study was the production of 300 sets of computer locking parts, each comprising assembly of 3 components as illustrated in figure 12.

The solution in this case was a DirectPart solution on an EOSINT M system using DirectMetal 20 powder. The required 300 sets were built in 5 jobs, 60 sets per job. The built time per job was 32 hours. This means that the required 300 sets were built in approximately 7 days. Although the time requirement of 7 days seems long on first sight it has to be considered that the first 60 sets were built immediately after the 3D CAD phase was over. With any conventional

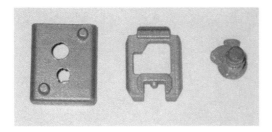

Figure 12.   Computer lock: the three laser-sintered assembly components (Rapid Product Innovations).

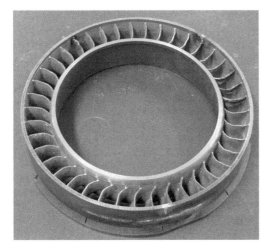

Figure 13.   Helicopter Stator ring (Pôle Européen de Plasturgie).

method of manufacturing any tools would have been necessary. The design phase of those would easily have consumed these 7 days.

2.3.4   *Helicopter turbine parts*

This case study was aimed at investigating e-Manufacturing of rotors and stators for test rigs at 30,000 rpm and 250°C (500 F). This stator as illustrated in figure 13 is for use in a MAKILA 2A helicopter engine. Its dimensions are 180 mm (7 inch) in diameter and 52 mm (2 inch) height. The blade thickness is between 0.5–1.8 mm (0.02–0.07 inch).

The solution to this case was direct metal laser-sintering with DirectSteel 20. This resulted in a reduction of both time and costs by 50% compared to a conventional manufacturing method.

Although this is not an example for an endproduct it still demonstrates the magnitude of both e-Manufacturing in the product design phase and the method of direct metal laser-sintering.

Figure 14.   Laser-sintered sand core and end-product (ACTech).

2.4   *e-Manufacturing of patterns/casts for metal parts*

2.4.1   *Hydraulic valve for rail vehicle*

When designing a new hydraulic valve for a rail vehicle the company approached ACTech, an EOS customer. The initial goal was to manufacture two prototypes to see whether the design was sufficient. The product is illustrated in figure 14.

To produce these prototypes ACTech laser-sintered the cores that were used in a follow on casting process. The unconventional way of producing cores however, had the positive side effect that the mold draft was not necessary anymore. This had the effect that the switching properties of the prototype valve exceeded expectations that far, that the rail car company insisted to produce the entire batch size of 200 with laser-sintered sand cores. This order motivated ACTech to further quench laser-sintering technology and succeeded in integrating the immanent product complexity in the laser-sintered single piece sand core thus dramatically simplifying the outer casting molds. The result not only was a hydraulic valve with unprecedented switching properties but also an e-Manufacturing process that lead the costs.

2.4.2   *Racing gearbox housing*

Many interesting examples for e-Manufacturing stem from the racing industry, helmed by formula one racing teams. This not only is because of an abundance of money but also because of an inherent drive for innovation that can give the lead of the tenth second needed to win the race.

The example presented here for a DirectPattern application was the need to produce a newly designed motor cycle gear box in titanium. Figure 15 shows both the sacrificed model laser-sintered as a DirectPattern with EOS PrimeCast 100 and the final titanium cast product.

In this application the DirectPattern was used as a sacrificed model in investment casting. Therefore the

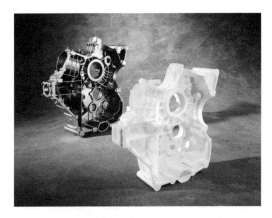

Figure 15. Laser-sintered PrimeCast sacrificed model and final gearbox housing (Poggipolini).

DirectPattern was coated with a ceramics based material that later formed the cast. After finishing the DirectPattern was burned and melted leaving the empty cast ready to be filled with the titanium alloy the gear box was intended to consist of. The entire process was that fast that the originator Poggipolini Titanium coined the term "Racing parts in racing speed".

3 STATE OF THE ART OF LASER-SINTERING

The cases described above show that there has been a trend from prototyping towards flexible manufacturing concepts and new business models. This trend has been enabled and supported by continuous innovation to improve the performance of laser-sintering technologies in several key areas.

An important success factor for providing flexible manufacturing concepts and new business models with laser-sintering are the material properties. Over the years both the plastic and metal materials have been significantly improved so that, for example, parts built in today's PA 2200 polyamide material have mechanical properties comparable to and in some cases better than injection moulded PA12 or ABS. Metal parts in DirectSteel H20 can have a tensile strength up to 1100 MPa. At the same time the accuracy and surface finish of the parts has been greatly improved. This is also important for the DMLS tooling application. The introduction in 1999 of fine metal powders for building in 20 micron layer thickness represented a real breakthrough for this application.

There has also been a trend towards larger build envelopes. Larger build envelope means not only the possibility to build larger parts, but also that more parts can be built in one job. In a full load, an EOSINT P 700 can manufacture about 150 pieces with a size of about 100 mm × 100 mm × 100 mm. The same load can hold almost 20,000 pieces with a size of 20 mm × 20 mm × 20 mm, or more than 150,000 pieces with a size of 10 mm × 10 mm × 10 mm.

Especially when considering larger parts or batch production, the system productivity is a critical factor, and this has been improved in various ways. The EOSINT P 385 system achieves 120% or more productivity improvement compared to the EOSINT P 350 by using optimized thermal management and laser exposure strategies. Also on the EOSINT M product line the brand new EOSINT M 270 is about 50% to 100% faster (in dependance of the used material type) than the EOSINT M 250Xtended and designed to be able to run also light alloy materials.

The time and cost of producing parts depends not only on the build speed in the machine, but also on the efficiency of the other steps in the process chain. EOS has developed Integrated Process Chain Management (IPCM) systems for EOSINT P and S systems which include peripheral devices for removing laser-sintered parts from the powder bed, recycling unused powder back into the machine and other handling steps. Both these systems also build the parts in an exchangeable frame to enable fast turnaround time between jobs, which also increases total productivity.

4 CONCLUSIONS

Laser-sintering technologies are today used for a wide range of applications, which can collectively be referred to as e-Manufacturing. Continuous technical innovation has enabled new applications by improving both the technical and economical performance, and this trend will continue. There are also a number of general market trends which support the increasing use of laser-sintering, especially for low volume production of end-use parts. Particularly due to the available material properties, laser-sintering is and will be the key technology for e-Manufacturing.

Virtual modeling and rapid manufacturing – Bártolo (eds)
© *2005 Taylor & Francis Group, London, ISBN 0 415 39062 1*

Rapid manufacturing through direct rapid prototyping in the South African context

L.J. Barnard & D.J. De Beer
Central University of Technology, Bloemfontien, South Africa

ABSTRACT: The South African market is very small if it is compared with international markets. South Africa is one of the top ten countries with new inventions but ranks in the bottom ten of new products released into the market. One of the problems that add to this is the fact that tooling is very expensive and a small entrepreneur can very often not afford to develop a product, manufacture the tooling and manufacture a 1st batch to release into the market.

Rapid manufacturing through direct rapid prototyping may offer a solution to the problem. A limited quantity of products can be ordered and grown to have enough stock to release the product, on a just-in-time-manufacturing concept. This will not have the huge initial cost of tool-making and manufacture as traditionally. Time is also saved since growing the parts will take less than a week, where the traditional way can take anything from eight to twelve weeks before a 1st off-tool sample is delivered. It can thus be seen that the company can much faster get a return on investment in this way, which will ensure the survival of the company. The paper will review case studies where rapid manufacture, by means of Laser Sintering, was used. The case studies compare the cost of traditional manufacturing vs. rapid manufacture and highlight advantages and disadvantages of the different processes. Issues like finishing and design for manufacture will be touched on.

1 INTRODUCTION

South Africa has a population of 46.6 Million people of which 11.2 Million stay in normal houses (http://www.statssa.gov.za). 4.31 Million of these households earn less than R 1 600 (€195) per month for the total household. The rest are staying in low cost and informal housing, and are daily fighting to survive and have enough to eat and cloth themselves. Of the 46.6 Million people, 61.2% are older than 18 years and can potentially earn an income. Taking all of these factors into account, it is evident that the South African markets are cannot be compared with international markets.

The South African population consist of 46.6% male and 53.4% female. These possible clients also consist of different ethnic groups and cultures. It therefore also reduces the market tremendously since different culture- and first and third world economies will buy different products. This once more re-iterates the difficulty that a product designer and innovator face, and development decisions that need to be taken in terms of product acceptance, market share and upfront-investment. If the product is a need to have, some risks are eliminated. However, survivors cannot be taken into account to support any product over and above basic living necessities.

Production cost in South Africa is also more expensive than for its international competitors, since the majority of production-equipment used is either imported from Europe or the USA. The exchange-rate of the Rand is in the ratio of R 8 to €1 and R 6.5 to 1$. Prime interest rate at which the banks get their money from the Reserve Bank or international banks is approximately 9%. Financing from an accredited financial institution may vary between interest rates of 9% to 14%. A production house therefore needs to do its pricing in such a way that the investment in equipment can be repaid. It thus has a direct influence on the individual or company's potential to successfully develop a new product. New ways methods for the development and manufacturing of products are needed to increase the percentage of inventions that are registered and successfully commercialised. Currently only 10% of new inventions successfully make it into the South African market and generate a profit.

New products fail at an alarming rate. The figures vary a great deal, because different definitions of what constitutes a new product and what constitutes success are used. It is however widely accepted that for every 10 ideas for new products, 3 will be developed and 1.3 will be launched and only 1 will generate a profit (De Beer 2002). This is a very disturbing piece of statistics since the cost of product development is very high.

2 CONVENTIONAL PLASTIC PRODUCT IS DEVELOPMENT IN SA

A new idea is normally put on paper and patented if it is patentable. After the provisional patent was lodged and granted, the person either designs the product themselves or takes it to a service bureau to do the development on their behalf. When the 3D CAD design is completed, the prototype development can start. This is most of the time an iterative process which can be repeated a few times.

In most of the cases after a first prototype is produced some things which are missed through the design phase or all the features were not put into the design by the designer are realized. When the client got a prototype in their hands they see things that will work better in another way. The prototype can also be used to ensure that the manufacturing, assembly or filling process will work as anticipated. Some inventors try to skip the CAD design- and the prototype phase to save money. They go ahead and immediately place an order for the manufacturing of the moulding, if it is a plastic product. This is a very risky way of developing a plastic product since the idea of the inventor and the conceptualization of the mouldmaker most of the time differs, resulting in a different final product than the original idea.

The mistake are only realized when the first of tool samples are delivered. This implies that time and money is lost due to the necessary changes to correct the product. The product henceforth, has a late market-entry and the residual time in which the product can generate profit is shortened. Figure 1 (Neel 1996) clearly indicates if the product development time and hence, the release are extended, the net-profit period of the product becomes shorter.

Out of the figure above it can be seen that it is very critical to get the product released as fast as possible, at the lowest possible cost, as it only starts to make money after the release. The product needs to be supported or subsidised by funding from other sources than the income of the product until the brake even point. Therefore it is critical to get past the release point with the least amount of funding. It can be seen that a different way of manufacturing is needed to ensure that the product gets to the market as fast as possible and at the lowest cost, without sacrificing the quality or the lifecycle of the product. This new way of manufacturing is called Rapid Manufacturing (RM).

3 INTERNATIONAL SUCCESS IN RAPID MANUFACTURING

According to Todd Grimm (Grimm 2004) RM is the production of end-use parts directly or indirectly – using rapid prototyping (RP) technology. RP is a toolless process that manufactures production parts directly from a digital product definition, without machining, moulding, bending or forming. Using the additive process can decrease time and cost while creating innovative process and part designs. This way of manufacturing was already used for the creation of parts for aircrafts (3D Systems 2004), the space shuttle and the space station (3D Systems 2004). Fully qualified for flight, the rapid manufactured parts have yielded time and cost savings. For the limited number of units in production, conventional tooling and moulding would have been much more expensive and time-consuming. Internationally, with a demand for smaller batch sizes, Rapid Prototyping technologies are used to produce final parts (Venuvinod). In the context of this paper RM is the manufacture of end use product by growing it by means of Selective Laser Sintering.

Phonak is currently using RM to produce Hearing Aid Shells (©EOS GmbH 10/03). Manufacturing the

Figure 2. Laser Sintering Hearing Aid Shells.

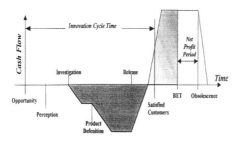

Figure 1. New product development cash flow (Neel 1996).

hearing aids by means of Laser-Sintering with PA 2200 material, enables the company to make shells that fit exactly in the patient's ear. Each person's ear profile differs. Already 50 000 hearing aid shells were laser sintered at a rate of 1400 per week. There are 600 000 devices sold per year. The aim is that in 2 to 3 years all shells are laser sintered. Figure 2 show a picture of the Hearing Aids as laser sintered.

Another example of RM is the laser sintering of low batch size manufacturing of sunglasses. The sunglass' frames were manufactured and market ready within 16 days against 47 days with normal toolmaking and injection moulding. The sintered frames passed all break value and optic impressions tests (©EOS GmbH 10/03). The great advantage of the use of RM is that the size and shapes can be changed as the market require. The picture in figure 3 shows a lot of different grown Sunglasses.

Design optimisation could be achieved by making use of RM. The requirement was to develop a better

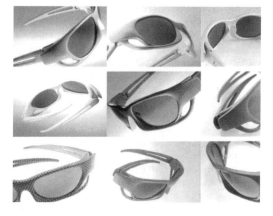

Figure 3.   A vast selection of glasses.

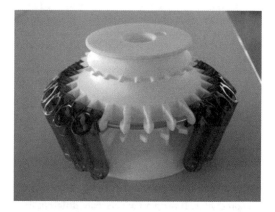

Figure 4.   Rotor for Rotolavit.

washing rotor for Rotolavit (©EOS GmbH May 03). The design was changed in such a way that the result was a high functional integration of thirty two parts into three parts. The finish and assembly effort was substantially reduced. Traditionally the thirty two single components was injection moulded in several tools. The test tube holder need perfect finish and created high effort after moulding. Figure 4 shows a photo of the Rotolavit rotor.

It is evident that manufacturing quantities play an important role in the decision whether RM should be used or not. If thousands of parts are needed, the cost per unit comes down tremendously and RM will not be used for the manufacture of such parts (Grimm 2004). The initial cost for the moulding is very expensive but the unit cost is low. If RM is used the unit cost becomes much higher. Other factors should also be looked at when a decision is made on the manufacturing method, such as the size of the market, is the design finalised, is there an advantage in changing the design slightly to be able to be more versatile and advantages in combining some parts to decrease the total number. All these factors should be weighted against the production time before a decision is taken.

4   RAPID MANUFACTURING IN THE SA CONTEXT

The South African market is very small, as indicated earlier. RM can be applied fast and cost effective to produce end use parts at low or medium production runs. In most cases this low or medium production is enough for the total lifetime of the product. Another big advantage of RM is that design limitations imposed by the moulding process are not applicable anymore. No need to pay attention to release of parts out of mould or draft angles. Undercuts are no problem since the way of manufacture is done by depositing layers of powder.

A further advantage of RM is that the design can be changed when the product is already launched. This will not mean a loss of a total mould or a lot of production time due to changes that need to be done to the mould. Constant development of the product can be done while the market dictates what they want and are willing to pay for. When a standard mould is manufactured the inventor does not have this advantage.

5   RAPID MANUFACTURING SUCCESS IN SA

5.1   *Case Study 1*

The Body IQ Compact Kiosk (figure 5) concept was developed and funded by Winning Wellness Pty (Ltd), for fitness club members to be issued with a Body IQ smart-card once they have joined and taken

Figure 6.   Full set of parts grown on the SLS.

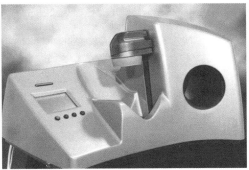

Figure 5.   Pictures of the Futrex device.

up Body IQ membership. The card will be used to store the last twelve groups of weight, blood pressure and body fat measurements (all generated through the BODY IQ and stored through the system's card reader), to build a complete history of each member's physical and medical condition.

The kiosk had to accommodate a power supply, an Adamslab Weigh measurement head section (Electronic boards and display boards), a Futrex mechanism, an Udex blood pressure measurement system, a USB hub, and a display and card reader. As can be seen out of the pictures below the unit consist out of a big housing with a few smaller components.

The client needed an initial 25 units to start marketing the concept, and there after 20–30 units per year. For the manufacturing of the housing, hand laid fibreglass (hand laid GRP) was chosen, as the size and quantity did not justify any other process. The nine special parts had to be manufactured through injection moulding, for which the tooling would have cost R 180 000. Tooling production (in parallel) would have taken 6–8 weeks, which would further slow down the release of the product.

Taking the development stage into account, as well as the quantities needed by the company, this would have priced the system out of the market. A full set of

parts (figure 6) were then grown on the SLS for R 3 000 to demonstrate and test the device. Based on excellent functionality of the SLS-produced parts, together with its complexity, it was decided to directly manufacturing them with SLS. Full production of 25 sets of parts would have cost R 42 100. This was calculated as full batches, and not spaced over time with other parts, which could have resulted in further time-saving. Measured against the estimated tooling costs only (without taking injection moulding part costs into account), this was a much less risky development option for the company (De Beer & Barnard 2001).

5.2   Case Study 2

A set of moulds for a security pin punch pad housing, measuring 150 mm in length, 50 mm in width and 30 mm thick, will cost approximately R 80 000 (€10 000). The housing consists of three parts. It will take in-between eight to twelve weeks to manufacture this mould dependent on the efficiency of the mould shop. For the price of the mould, 1000 remote housings can be grown on the EOS P380 in PA 22000. The client, who needed this product to test their new system, only needed 30 sets to release the product into the market. The cost to grow the 30 sets was R 13 835 (€1 525), which included the finishing of the grown parts. The product can be guaranteed to be eight weeks earlier on the market than if it was manufactured in the traditional way.

The company who handled this development is a very small company with limited resources. If all the company's resources are tied up in the manufacturing of the injecting moulding tool, not enough funds will be available for the marketing and testing of the new product. By making use of RM the company can place an order for a limited quantity as the need arise and the product is introduced into the market, which will not tie up a lot of their resources. In this way the company can also see what the market reaction on this product is. If the market needs are different from what the product provides, the design can be changed

Figure 7.   Pin Punching Pad.

Figure 8.   Densito Meter.

economically and efficiently. When the demand rises for the product a decision can be taken if a tool should be manufactured. The Pin Punching Pad as described in the case study is shown in figure 7.

5.3   Case Study 3

This product measures grey-scales, used in the calibration of x–ray machines. Legislation requires that such a product be used by all facilities using x-ray machines. It consists of five parts if manufactured in the traditional way. By planning to manufacture the product with RM the design can be changed so that it only consists of three parts. The injection moulding tools for the product will cost R 235 000. The injection moulded parts will cost R 63.00 a set. The initial demand is calculated at a maximum of 100 per year for the first three years.

The company is a start-up company that cannot afford the tooling costs, which will take 12 weeks to complete. If the company borrows the money and pay for the mould it will take them in the order of 16 weeks to get the product ready for the market. The money will be tied up for a long period before they will get a return on their investment. It can close down the company before any profits are generated, and totally remove the product from the market. With RM the company can sell their product within a week from the time they place the order for the manufacturing of the parts. Figure 8 shows the assembly of the Densito Meter.

5.4   Case Study 4

Freedom of Creation, a company founded by Materialise to develop decorative items, showed that custom-designed functional artworks, which are difficult to manufacture with conventional methods, can be manufactured by means of RM. These special designs are exclusively manufactured for interior designers or architects wanting to create special effects or themes with their designs. Some of these designs are impossible to manufacture in one piece with traditional moulding methods, and may even be difficult to assemble. If this is intended for an exclusive design, quantities do not justify the manufacturing of a mould, as the design would not be for general use, and thus not for mass-production.

This opens up new possibilities for South African companies to for the first time be able to produce impossible functional artworks. In this way of manufacture the design trend can be used in different sizes throughout the house or building it is intended for. In using e.g. an African theme, this method of manufacturing is very useful. As a trial, a design for a custom-designed lamp-shade was made. Should this be grown, the cost will be R 3 200. If however, the quantity is increased to 9, it will be R 4 800 for all 9. For a custom-designed item, a client may be willing to pay such an amount for the opportunity of owning a unique art-piece. Planning by the designer should just be done so that it can be done in combination with other builds. Figure 9 shows the design for such a lampshade.

6   FURTHER RESEARCH ON RAPID MANUFACTURING

The case studies reported forms the foundation for further research to establish the break even point for RM, to replace traditional manufacturing processes such as injection moulding. The study will concentrate on case studies to determine the advantage of RM in the South African context. Parameters such as design complexity, part geometry and material volume, delivery time and production volume over time should be measured against price.

Figure 9.   Special designed lampshades.

Advantages such as design freedom, combination of a number of parts in smaller assemblies, financial advantages and time to market will be focused upon and highlighted. Finishing of the parts to compete with the quality available from other conventional manufacturing processes will also form part of the study.

7   CONCLUSION

It is evident that the production of plastic parts through direct growing with Laser Sintering offers numerous advantages. The parts will give the company the advantage of testing their product in the market without investing all their finances in very costly tooling. For the South African market, with a very low success rate of commercialisation of products, it is an added advantage to use RM to produce the product until such time that a clear decision can be taken or when enough money is earned to invest in hard tooling. Finances normally tied up in tooling can now be used to do proper marketing of the product. In it also offers the company the opportunity to enter into niche markets, which have higher value and higher profit margins. Products that are normally not possible to manufacture due to economies of scale, or for which limited productions runs will be to expensive, can now be manufactured. RM is currently used internationally in a wide range of applications. For South Africa it can make a huge difference and open new possibilities. It may also help South African SMMEs to escape from bankruptcy or may help to save a potentially good product that just need to enter the market at the required pace.

REFERENCES

De Beer, D.J. 2002. The Role of Rapid Prototyping to Support Concurrent Engineering in South Africa, DTech Thesis, Mechanical Engineering, Central University of Technology, Free State.

Grimm, T. 2004. User's Guide to Rapid Prototyping.

De Beer, D.J., Barnard, L.J. 2001. Applying Rapid Prototyping in Concurrent Engineering – Successful SA Case Studies. Paper published in the proceedings of CARS 01, University of Natal.

Neel, R.M. 1996. Accelerated Product Development as a Strategic Weapon, Proceedings of the 2nd Asia Pacific Conference on Rapid product Development, Brisbane.

Venuvinod, P.K. Rapid Prototyping- Laser-based and other Technologies.

©EOS GmbH, 10/03, UW, CS_P_Phonak_en.ppt

©EOS GmbH, 10/03, UW, CS_P_Sunglasses_en.ppt

©EOS GmbH, May 03, UW, CS_P_Hettich-Rotolavit_en.ppt

http://www.statssa.gov.za

3D Systems Case Studies. 2004. Boeing sees growing value and Versatility in SLS System and Duraform Materials, http;//www.3dsystems.com

3D Systems Case Studies. 2004. NASA Sends Duraform parts into Space, http;//www.3dsystems.com

Virtual modeling and rapid manufacturing – Bártolo (eds)
© 2005 Taylor & Francis Group, London, ISBN 0 415 39062 1

Rapid manufacturing: A path to new markets

S. Killi

Oslo School of Architecture and Design, Norway

ABSTRACT: Over the last three years collaboration between a hip surgeon and the rapid prototyping laboratory at the Oslo School of Architecture and Design in Norway has resulted in a system for improving the hip replacement surgery. The method is to use custom fitted measuring devices during surgery, so called anteversionheads, these devices could have hundreds of different designs. By using rapid prototyping techniques like SLS it is possible to produce these parts directly, no tools or storage, production on demand. The report goes through the different aspects of the process, pinpointing problems and solutions. The conclusions are that Rapid manufacturing, using standard rapid prototyping processes definitely has its advantages when it comes to freedom of form, economy and adaptability. The problems encountered are typical for the rapid prototyping industry; low accuracy, few materials to choose from and mainly for this project, cleaning of the parts.

1 INTRODUCTION

Rapid manufacturing (RM) is a legitimate child of the Rapid prototyping (RP) technology developed during the late 80's and 90's. Especially the layer-by-layer method can show an escalating quality performance since the early "3D printers". Early, the idea to use this technology to produce spare parts on demand was introduced. One problem was to come up with processes that were fast and accurate enough, and materials with sufficient qualities. Since 2000 this problem has been addressed and some projects using this technology have been realized, the most famous being customized hearing aids from Siemens and Widex, presented in Wohlers report 2004[1]. Another problem was to come up with products that were suited for the current technology; i.e. small and fetching a high price. In the last two years a number of interesting projects have developed products that meet these criteria.

2 BACKGROUND

The first commercial layer manufacturing system was presented at the AUTOFACT show in Detroit, MI in November 1987 by the 3D Systems company[2] and intended primarily for rapid prototyping application. Several other processes were subsequently developed through the 1980s and 1990s[1]. The technologies now available include a variety of different processes, such as Stereo lithography (SLA)[2], Selective Laser Sintering (SLS)[2], Fused Deposition Manufacturing (FDM)[2], OBJET[2], Laminated Object Manufacturing (LOM)[1].

All these systems are based on a three dimensional CAD file presented as a triangulated polygon mesh, stl file[2]. The digital object is sliced into thin layers and then manufactured by producing these layers one at a time.

Over the last decade, the SLA and the SLS technologies, has increased the build speed more than 1500%[5] and developed materials with much better properties – both mechanical and chemical. The price of the materials has remained relatively stable while the investment cost has dropped some 20–25%[1].

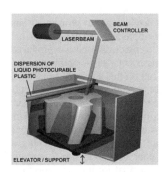

Picture 1. Schematics of SLA process[3]. Stereo lithography was the first RP system, launched by 3D systems[4]. It is based on an epoxy resin which is cured (solidified) with a laser. This gives accurate parts with high finish, however it needs support structure and can not stack parts in several layers since it is a liquid bath. Several materials are available, lately also transparent versions. It is an expensive system, both in use and to purchase.

Picture 2. Schematics of the SLS process[3]. Selective laser sintering was developed at the University of Texas at the end of the 80s. The first commercial machine came in 1992. It is based on a crystalline polymer (often Polyamide) which is sintered (close to melting temperature) with a laser. This gives strong, ductile parts. The surface is somewhat porous and loose powder is hard to remove completely. Since the powder gives support it is possible to stack the whole build envelope with parts, and also to add parts during production. The build envelope is for 3D systems machines $330 \times 380 \times 450$ mm.

As mentioned earlier, the primary use for these machines was producing prototypes and visualizing new designs. This both reduced the design time and the lead-time for tooling, since the need for test-running the tools is less. The obvious next step would be to produce small series for consumer testing. This would make it possible to stop a project before big investments in tools are made. As the need for investments in production tools is one of the biggest problems getting your new product out to the market.

3 CUSTOMIZATION

Car companies were some of the first to introduce delivery on demand and to provide an opportunity to customize their products. They no longer had a need for big storage facilities as each car produced is custom made to some extent: colors, interior, car stereo, etc. Car companies can do this since the price level is so high and the lead-time necessary for production does not increase due to computer-controlled production lines.

Could customized production also be implemented on a much smaller scale? The technology is there and both old and new markets are opening up but how would it be done? The following case study illustrates just such a potential for this new form of production on a smaller scaled product and what it would entail. Problems with these methods will also be addressed.

4 RAPID MANUFACTURING FOR SURGERY

A hip replacement surgeon In Oslo, Dr Bjørn Iversen had been doing hip replacement for several years

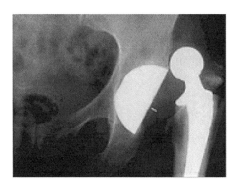

Picture 3. A picture of a misplaced hip joint (picture courtesy of Bjørn Iversen) There is two parts that needs to be correctly aligned. The bolt through the femur and the cup in the pelvis. If the bolt is not in the right angle it will jump out of the cup when the patient moves his legs (crossing their legs etc.).

when he saw the need to address one of the biggest problems with this form of surgery, misplacement. Studies done, had shown that a staggering 78% of all hip replacement was misplaced[6], leading to painful lives for the patients and huge insurance payments, 1.5 billion dollar just in the US[6]. As a result, the USA has witnessed a decrease in the number of hip surgeons of more than 20% over the last 3 years in states like Louisiana[6].

A method that would produce a prosthetic hip that would decrease the number of hip misplacements would certainly be in great demand. The problem Dr Iversen found was that there were some universal angles in the mounting of the replacement hip joint and the femur, even though size and form of humans vary greatly. To correct his problem, Dr Iversen and his company, Orthometer, has patented[7] an interface (called an anteversion head) to use during surgery. This system gives the surgeon the possibility to place the artificial hip joint at the exact correct position. The interface is just used a couple of minutes during the operation and then disposed.

The hip replacement market is today approx one million operations per year in the Western world[6]. The Orthometer Company's biggest problem was that there is close to 100 different types of artificial hip joints, a left and a right side and a grinder for each of these types, bringing the number of potential hips up to many hundreds of different types.

In each case the interface had to match. Even if only one type was employed, differences in cup sizes still meant that the number of possibilities for one type of hip necessitated 12 different designs. With tooling cost in the order of 30 000 euros for each part and a lead-time of 10 weeks it would be impossible to realize this idea.

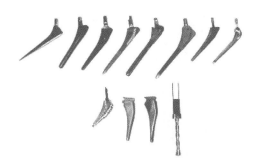

Picture 4. A small assortment of different artificial hip joints. All have different shape and interface on top. Actually there is not one that has the same system to mount the pelvis ball! Photo: Steinar Killi.

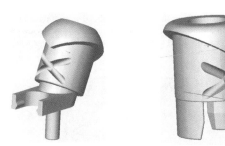

Picture 5. Two examples of measuring interfaces (anteversion heads), designed by Bjørn Iversen and Steinar Killi. The rounded top should fit in to the pelvis cup and will vary for different cup sizes. The cross on the middle is used for positioning the hip prosthesis (30 degrees angle), at the bottom we see the different ways to mount the interface to the prosthesis. To cover all possibilities the number of different designs will exceed 1000. Photo: Steinar Killi.

In the summer of 2002 Dr Iversen approached Oslo School of Design and Architecture with his project. The technology we are running, SLS, could make it possible to actually produce these parts.

The polyamide powder had been approved by NAMSA (North American Medical Surgery Association) in 1997 for contact with blood, tissues and flesh for up to 24 hours.

Because of the porosity of the powder the anteversion head would suck blood during the operation and had to be disposed afterwards, a key element to make this an economical success.

For each operation it would be necessary with four different interface parts. In the machines at AHO it is possible to run 250 of this packages at a time, making delivery possible eight days after ordering. So, no stock of parts, "just in time" production. The price for this kind of surgery could justify the relatively high price on the parts. New variants of hip replacements

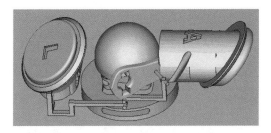

Picture 6. For each operation there is a need for one of these four form quadruplets. All parts in the quadruplet can vary depending of the prosthesis used. Drawing: Steinar Killi.

Picture 7. A production package consisting of 240 quadruplets. In theory, each of them could be different. All quadruplets have a tag with design number and a powder batch lot number. Drawing: Steinar Killi.

could be provided in matter of weeks, with no investment cost in tooling.

5 PROBLEMS AND CHALLENGES

During the project three main problems had to be addressed.

1) Accuracy
2) Cleaning of the parts
3) Economically sound production logistics.

5.1 *Accuracy*

The SLS technology has an accuracy level of ± 0.15 mm. With variations due to positions in the production chamber (picture 7) the accuracy could be sometimes lower. The prosthesis is made of polished steel, or some times in titanium, with very few possibilities to cause enough friction between the RM produced plastic part and the prosthesis. This means that it would be very difficult to get a perfect fit, easy to mount and remove the plastic part, but tight enough to not fall off during surgery.

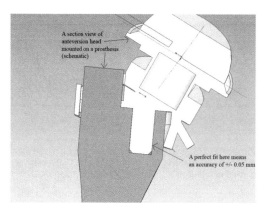

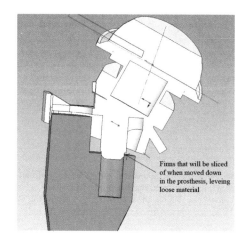

Picture 8.　A schematic section view of an anteversion head mounted on a prosthesis. Anteversion head and prosthesis are sectioned in the midplane, showing how the anteversion head are attached to the prosthesis (Prosthesis in green and anteversion head in white).

Picture 10a.　Finns that would be sliced into correct size during mounting. Schematic picture show a section view of the anteversion head just before entering the prosthesis. Drawing: Steinar Killi.

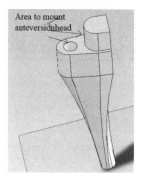

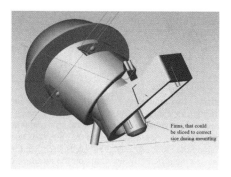

Picture 9.　Drawing of one of the prosthesis used. The anteversion heads are to be mounted on the top section. The drawing is a "dummy" for easier presentation, the actual prosthesis comes with more details and are hard to get good pictures of. This prosthesis "dummy" is being used in the following example. Drawing: Steinar Killi.

Picture 10b.　Anteversion head with fins. Drawing: Steinar Killi.

It would be easy to obtain this in an injection molded design, but with parts that could vary within 0.3 mm, resulting in either too tight or too loose mounting, we had a problem.

One of the prosthesis could serve as an example. This prosthesis has a cylindrical hole were the anteversion heads could be mounted (schematic sketch in pictures 8 and 9).

The biggest problem here is that the hole is not conical and is polished; we need a perfect fit and an accuracy deviation of no more than ±0.05 mm. This is impossible to obtain. In a batch of 100 pieces, maybe 10% would succeed in fitting. So, we had to make the design in a way that could adapt to various accuracy deviations. We tested out a long range of

designs, one with fins (pictures 10a and 10b) that would cut off unnecessary material during mounting and give a one time perfect fit. However this would leave loose material that could lead to infections.

Another possibility would be to make the pin in the anteversion head collapsible (see pictures 11a and 11b), but this would lead to problems in the cleaning process: powder would be hard to remove from the hollow area.

The solution in this case was to mix the two methods: instead of cutting the material during mounting, the material was allowed to deform; see pictures 12a and 12b.

Other prosthesis had to find similar individual solutions. The positive aspect is of course that you don't need to think "production" (draft angles, parting lines, side pulls and so on). The only obstacle would be the possibility to clean the parts afterwards, which leads to the second problem.

Picture 11a.   Shafts that will allow flex to compensate for inaccuracies. Drawing: Steinar Killi.

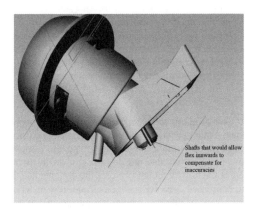

Picture 11b.   Anteversion head with shafts. Drawing: Steinar Killi.

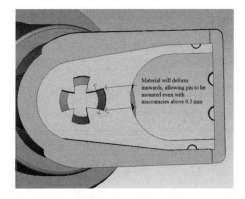

Picture 12a.   Heart- or club-shaped ribs that will allow material deformation. Drawing: Steinar Killi.

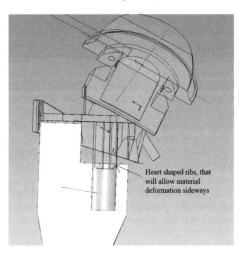

Picture 12b.   Anteversion head with heart- or club-shaped ribs mounted on the prosthesis. Anteversion head pictured translucent for better understanding. Drawing: Steinar Killi.

5.2   Cleaning of the parts

Parts that would come in direct contact with blood, tissues etc will demand an extensive cleaning process, Not only need the parts be free for bacterias, but no loose powder can be left anywhere on the surface. In the SLS process the parts are produced in a "powder cake" and loose powder has a tendency to stick to the part, partly due to the heat during production, but also because the porosity create cavities for the powder to "hide" in.

The bacteria problem could easily be solved, by using new powder for each build and using gamma radiation after packaging, no bacteria will survive. Cleaning of all loose powder was, and is, a far more difficult task. The standard procedure for cleaning sls parts are glass bead blasting. However this means human interaction, and the following problems with contaminated pressured air and glass beads. Some of the things that

were tested was tumbling the parts with small ceramic beads and distilled water. This improved the surface quality to a great extent. However, holes, shafts and hidden cavities were not powder free. Using ultra sonic sound showed some interesting improvements, but still not a fully satisfactory result.

The working solution had to be to go back to bead blasting, but improve the quality of the compressed air and glass beads, followed by a tumbling. This will require a individual handling of each part to ensure that all powder are gone. This is still not an optimized process and should be improved.

5.3   Economically sound production logistics

The products dealt with in this paper are in a typically "high cost" area, tedious cleaning processes and quality insurances makes it still a challenge to make money.

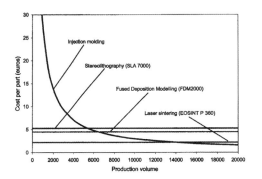

Picture 13. Illustration showing the break-even cost analysis of a small but complicated part, done at Loughborough University. As depicted in Wohlers Report 2004[1].

Since the powder used always has to be new, a system for handling the one time used powder is needed. Either sold on a second hand marked or for production of other parts.

Secondly, a large scrap amount would easily decrease the earning. Problems due to machine error resulting in discarding a full batch would cost thousands of euros and increased delivery time. A well maintained machine park is an absolute must.

Working in this market also demands a fully operational and strict quality system, there is no room for errors. A weak or inaccurate part could have devastating results both during surgery and after.

Of course, if one type of prosthesis should become some kind of a standard, (15 000 units or more) an injection molding manufacturing method could be cost effective. Picture 13 shows graphically the break-even point for producing a small, complicated part using traditionally manufacturing methods and different RP processes. For quantities less than a couple of thousands, the cost for traditional methods are literally going out of the charts. The chart in picture 13 is not accurate for our case; the cost pr unit will be around 25–35 euros. But the amount of each design could be as few as one. Then the cost for the design work would be the really expensive part of the production line.

6 CONCLUSIONS

Rapid manufacturing are showing great possibilities in a future customized marked. A very low threshold for getting your product out to the marked because of no tooling costs, storage cost and so on will definitely be a booster for innovation.

The case discussed above give some interesting guidelines for getting full advantage of this way of manufacturing products.

1) The products should be in a high cost market, the cost using this technology is still high.

2) There are still rather few materials available; colors, texture, and finish are to a large extent set. Preferably, RM projects should be in areas were these things are less important.
3) Complexity of the product is a freebee, while size cost.
4) Cleaning could be a problem. Traditionally manufacturing methods like injection molding usually need little post processing.
5) Accuracy is still a problem. The best RP systems have tolerances down to some hundreds of a millimeter, but this is not for a large area. Big parts, or lots of parts will still give deviations in accuracy.
6) Most systems on the marked today have their positive and negative sides. The SLS technology has a great advantage because of the possibility to build large quantities of fairly strong, ductile parts. Poor tolerances and loose powder is a challenge. Other systems with better accuracies and easy cleaning, like SLS, cannot build larger quantities (at least you cannot stack parts on top of each other) and you also need to build support structure.
7) Designing products for RM needs a different approach than designing for traditionally manufacturing methods. On one hand, you don't need to think about draft angles, complex tooling, and such, on the other hand, accuracy and handling of the parts will be an issue.

Other products for other applications will probably generate other problems and other solutions. The whole design process will be different, giving the designer more freedom but also more responsibility. In theory, there could be a direct link between the designer and the end consumer. New RP techniques and new materials are on its way, giving a whole new perspective to how we will design and manufacture products in the future.

For further info about the RM future, Wohlers annual reports[1] are a very good source.

The Orthometer Company has performed the first test surgeries with good results, further optimization and fine tuning are necessary, but the system works. At this moment a further optimizing of the process is being developed.

REFERENCES

1) Wohlers Report 2004. Rapid Prototyping, Tooling & Manufacturing State of the Industry Annual Worldwide Progress Report
2) "Rapid manufacturing" by D.T Pham & S.S Dimov ISBN 1-85233-360-X
3) http://www.designinsite.dk June 10th 2005
4) http://www.3dsystems.com June 10th 2005
5) Rapid prototyping laboratory at AHO, speed test done summer 2003, SLSUG proceedings 2003
6) J. Arthroplasty 2002 April; 17(3): 359–64
7) European Patent nr 1212016

Virtual modeling and rapid manufacturing – Bártolo (eds)
© 2005 Taylor & Francis Group, London, ISBN 0 415 39062 1

Conception and fabrication by stereolithography of a monolithic photocatalyst for air cleaning

M. Furman, S. Corbel, H. Le Gall, O. Zahraa & M. Bouchy
Département de Chimie Physique des Réactions, Nancy Cedex, France

ABSTRACT: The aim of the study is the fabrication by stereolithography of a monolithic photocatalyst and to use it in photocatalysis for air decontamination. The efficiency of the monolithic photocatalyst is controlled by both the kinetic rate and the external mass transfer rate. The global rate depends on a competition between the two regimes. The purpose of this paper is to quantify the influence of the geometry of the monolithic support on these two rates using methanol as a model pollutant. Three different geometries are tested: a honeycomb geometry, a static mixer geometry and a helix geometry. It appears that geometry has practically no influence on the external mass transfer rate since depends exclusively on the flow rate. However, the chemical kinetics is very influenced by the configuration of the support.

1 INTRODUCTION

Photocatalysis using TiO_2 as catalyst is an advanced oxidation process which enables the complete degradation of many volatile organic compounds (VOCs) (Avila et al. 1997, Gao et al. 2001). In this process, the catalyst, generally TiO_2 Degussa P25 (Wu et al. 2004) is activated by near UV light (λ around 365 nm) and becomes a redox system capable to react with organic molecules (Orlov et al. 2004). TiO_2 catalyst can be used in the form of slurry or immobilized over suitable supports. More and more photocatalytic reactors are built up immobilizing catalyst as they provide advantages compared with slurry reactors (no need to separate the catalyst particles from treated water, due to the strong absorption of UV light by TiO_2 only a thin layer near the surface is activated…). However, immobilization of TiO_2 on supports also creates its own problems: the accessibility of the catalytic surface to the photons and reactants is reduced and the external mass transfer plays a significant role, particularly at low fluid flow rate, due to the increasing diffusional layer from bulk flow to the catalyst surface (Dingwang et al. 2001).

From this, our chemical engineering research lab aims to make a monolithic photocatalyst by stereolithography used for air cleaning. The stereolithography is a rapid prototyping process which enables the fabrication of many three dimensional supports some times very complex (Dufaud & Corbel 2003). Besides, thanks to the CAO, we are able to create lots of geometries quickly and modify them as many times as necessary.

The criteria to have a good geometry of support are:

– a great accessibility to the catalytic surface to the photons,
– a great ratio surface over volume,
– a geometry able to promote a good external mass transfer.

2 EXPERIMENTAL DETAILS

Three monolithic supports with completely different geometry (Fig. 1) have been made by a classic stereolithographic apparatus in epoxy resin (SI30) using a Nd-YAG laser.

Geometry 1 (Fig. 1a):

This is a honeycomb support with vertical and horizontal cylindrical channels. The vertical channels enable the circulation of the gas flow, whereas the horizontal channels let the light pass. Every channel has a diameter of 4 mm. Thus, the porosity of the structure is 0.3.

Geometry 2 (Fig. 1b):

This second geometry is based on the geometry of a static mixer therefore it is named like that.

Geometry 3 (Fig. 1c):

This structure looks like the DNA helix with a 1 cm helicoidal step.

These supports primary fixed on a rotary motor (Fig. 2) have been coated thanks to a syringe containing a well known volume of an aqueous solution of TiO_2 (Degussa P25) (6 g/l).

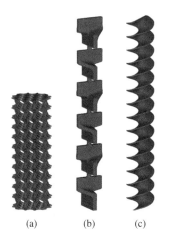

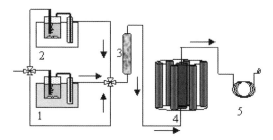

Figure 3. Experimental setup of photocatalysis with circulation of gas. (1) Cooling bath for methanol; (2) Cooling bath for water; (3) Static mixing; (4) Photocatalytic reactor; (5) Gas chromatography.

(a)   (b)   (c)

Figure 1. Three geometries used as monolithic support; (a) Honeycomb geometry; (b) Static mixer geometry; (c) Helix geometry.

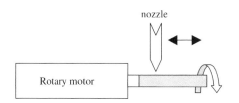

Figure 2. Schematic diagram of the impregnation system.

Table 1. Characteristics of the three monolithic photocatalysts.

	Length cm	Width cm	Surface area cm^2	Specific surface cm^{-1}	Catalyst weight mg
Honeycomb	7.5	2	120	23	15
Static mixer	14	2	80	8	12
Helix	14	2	105	17	13

This experimental setup enables to control the mass of the coated catalyst and obtain a uniform layer on the entire surface of the supports. Eventually, the characteristics of the monolithic photocatalysts are shown in Table 1.

The experimental setup used for the photocatalytic tests is shown in Figure 3. A pure air flow is splitted into three paths for reactor feed preparation. The first path is meant for pollutant and the second one for water. Both of them use cooling bath with adjustable temperature in order to obtain the expected vapor pressure of each compound. The third one is used to dilute both water and pollutant with a view to get a concentration range from 50 to 1000 ppm of pollutant and flow rate from 100 to 400 cm^3/min.

The monolithic photocatalysts are introduced in a Pyrex tubular reactor (2.1 cm diameter, 70 cm in length) with 6 UV lamps around (Mazda TFWN, electric power 18 W, light power 3 W, λ_{max} = 365 nm). The reactor is 4 cm far from each lamp so that the incident light power is 14 mW/cm^2 at the surface of the catalyst.

3 RESULTS AND DISCUSSION

3.1 *Photocatalytic activity*

The efficiency of the three monolithic photocatalysts was carried out by estimating the average degradation rate in methanol per square centimetre of support R_{av} (mg/min/cm^2) at different flow rates and inlet concentrations (see Equation 1 below):

$$R_{av} = \frac{Q_e C_{in} X}{S} \qquad (1)$$

where Q_e = flow rate in mL/min; C_{in} = inlet concentration in mg/L; X = experimental conversion of methanol and S = apparent surface area of the support in cm^2.

In order to understand the behaviour and the influence of the geometry on the photocatalytic activity, a model which takes into account the chemical kinetics and the mass transfer limitations was proposed.

The relationship among the global degradation rate R, the external mass transfer rate $R_{m, ext}$ and the kinetic reaction rate R_{kin} is given by the following expression:

$$\frac{1}{R} = \frac{1}{R_{kin}} + \frac{1}{R_{m,ext}} \qquad (2)$$

The rate R depends on the concentration in the bulk flow and the corresponding average rate is R_{av} mentioned above.

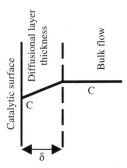

Figure 4. Schematic diagram of the profiles of concentration near the catalyst.

As most photocatalytic reactions take place on the surface of the catalyst, the intrinsic kinetic rate can be given by the model of Langmuir-Hinshelwood (Meng et al. 2002):

$$R_{kin} = \frac{kKC}{1+KC} \qquad (3)$$

where C = concentration in the bulk flow; k = intrinsic kinetic constant proportional to the absorbed light; K = constant of adsorption, depends on the affinity of the pollutant for TiO$_2$, and on the geometry.

However, when external mass transfer limitations occur (Fig. 4), the rate can be expressed as:

$$R_{m,ext} = k_{m,ext}(C - Cs) \qquad (4)$$

with k$_{m,ext}$ = external mass transfer constant; C$_s$ = concentration in pollutant on the surface of the catalyst. It can be assumed to be negligible compared to the concentration in the bulk flow.

The Equation 4 above becomes:

$$R_{m,ext} = k_{m,ext}C \qquad (5)$$

The constant k$_{m,ext}$ depends on the diffusional layer thickness δ as follows:

$$k_{m,ext} = \frac{D}{\delta} \qquad (6)$$

with D = molecular diffusional coefficient of methanol in air.

The greater k$_{m,ext}$ the smaller δ and the smaller the mass transfer limitations. Taking in consideration the above limitations, the expression of the rate can be rewritten as:

$$R = \frac{\dfrac{kKC}{1+KC} \times k_{m,ext}C}{\dfrac{kKC}{1+KC} + k_{m,ext}C} \qquad (7)$$

Table 2. Value of constant k and α for the three geometries.

Flow rates (mL/min)	k (mg/min/cm^2)	α(min.cm^2/L)			
		100	200	300	400
Honeycomb	4.1×10^{-4}	343.1	144.6	126.3	55.8
Static mixer	1.5×10^{-3}	187	107	96.6	57.6
Helix	3×10^{-4}	599.1	251.9	239.9	218.9

Now, the constants k and k$_{m, ext}$ have to be determined.

The reactor is considered as being integral, and the mass balance on a fraction of the reactor can be written as follows:

$$-Q_e\frac{dC}{dS} = R = \frac{\dfrac{kKC}{1+KC} \times k_{m,ext}C}{\dfrac{kKC}{1+KC} + k_{m,ext}C} \qquad (8)$$

The integration of Equation 8 gives:

$$\left(\frac{1}{k_{m,ext}} + \frac{1}{kK}\right)\ln(1 - X_{mod}) - \frac{1}{k}(X_{mod}C_{in}) = -\frac{S}{Q_e} \qquad (9)$$

We define a constant α which takes into account of the mass transfer (k$_{m, ext}$) and the kinetic (k and K).

$$\alpha = \frac{1}{k_{m,ext}} + \frac{1}{kK} \qquad (10)$$

There is only one solution X$_{mod}$ to this equation for every couple of constant k and α. The best couple (k/α) is determined after minimization of the following criterion (see Table 2):

$$\sum_i \frac{R_{avi} - R_{modi}}{R_{avi}} \qquad (11)$$

where: $R_{mod} = \dfrac{Q_e C_{in} X_{mod}}{S} \qquad (12)$

and i = index of an experimental point; X$_{mod}$ = modelled conversion; R$_{mod}$ = average rate corresponding to the modelled rate.

When α remains constant in function of the flow rate it means there is no limitation due to the mass transfer.

We can estimate the constant K for each geometry because α = 1/kK. For example, for the helix geometry with k = 3.10^{-4} mg/L/min: K = 15.2 L/mg.

The Figures 5–7 show the experimental (symbolised by points) and modeled rates (solid line) for the three different geometries.

591

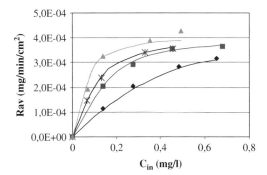

Figure 5. Average photocatalytic degradation rate of methanol against the inlet concentration with honeycomb geometry: ◆ Q_e = 100 ml/min, ■ Q_e = 200 ml/min, × Q_e = 300 ml/min, ▲ Q_e = 400 ml/min, — average modeled rate R_{mod}.

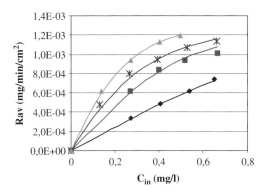

Figure 6. Average photocatalytic degradation rate of methanol against the inlet concentration with static mixer geometry: ◆ Q_e = 100 ml/min, ■ Q_e = 200 ml/min, × Q_e = 300 ml/min, ▲ Qe = 400 ml/min, — average modeled rate R_{mod}.

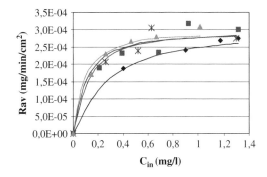

Figure 7. Average photocatalytic degradation rate of methanol against the inlet concentration with helix geometry: ◆ Q_e = 100 ml/min, ■ Q_e = 200 ml/min, × Q_e = 300 ml/min, ▲ Q_e = 400 ml/min, — average modeled rate R_{mod}.

The knowledge of the experimental kinetic constant k and external mass transfer constant $k_{m, ext}$ is necessary to estimate the relative contribution of the two regimes. Two limiting cases can be discussed. In the first case:

$$\frac{kKC}{1 + KC} > k_{m, ext} C$$

This case is quite complex because limitations depend on the concentration in the bulk flow. This is what is observed with the static mixer geometry (Fig. 6) at a flow rate of 100 mL/min: the concentration in the bulk flow is low enough so that the global rate is controlled by the external mass transfer limitations. As a consequence, it is proportional to concentration. However, the external mass transfer rate increases with increasing concentrations and kinetic dependency is then expected for higher concentrations.

In the second case:

$$\frac{kKC}{1 + KC} < k_{m, ext} C$$

In this case, the kinetic dependency occurs. The phenomena becomes more significant at high flow rate as the constant $k_{m, ext}$ is also higher. This is what happens for the honeycomb geometry and the helix geometry (Figs. 5, 7) and also for the static mixer geometry when the flow rate is higher than 100 ml/min.

When kK and $k_{m, ext}$ are in the same order of magnitude both the kinetic rate and mass transfer rate have the same contribution which is difficult to settle.

Besides, the experimental results show that $k_{m,ext}$ is not very influenced by the geometry. The most likely explanation is the relative low velocity of the gas flow which do not allow to promote turbulence.

However, the kinetic constant k clearly depends on the geometry. The static mixer geometry is the one which has the highest value of k. The most likely explanation is a lost of light due to the absorption by the support in the honeycomb and helix geometries. Therefore the catalyst is not as much activated as for the static mixer geometry. This assumption will be verified with further experiments.

3.2 Optical properties of the support material

It has been interesting to know the optical properties of the polymer used as monolithic support in order to quantify the lost of light. The Figure 8 shows the absorption of light for three thicknesses of polymer against wavelengths. The experiment has been carried out using an integrating sphere. An absorption greater than 90% at 365 nm has been observed for thicknesses of 1 mm and 2 mm. These thicknesses are in the same order of magnitude as for the three supports. The strong

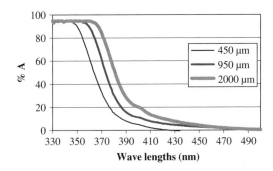

Figure 8. Percentage of absorption of the polymer used as support against wave lengths.

absorption is due to the presence of photoinitiators in the resin used in stereolithography which absorbs at the same wavelength as the maximal emission of the UV lamps. As a consequence, the light passing through the support is completely absorbed and lost. The result corroborates the explanation given above, about the accessibility to the catalytic surface to the light.

4 CONCLUSION

The photocatalytic activity of three supports with completely different geometries has been tested in this study. The experiments have been performed in order to know how the geometry influences the chemical rate and mass transfer rate and finally the photocatalytic efficiency. It has been shown that the geometry influences the kinetic constant k, but not significantly the mass transfer constant $k_{m, ext}$ since it depends exclusively on the flow rate.

As a consequence, the perspective is the development of a geometry having a great ratio surface to volume. At the same time, a very good accessibility of the photons to the catalytic surface must be achieved.

Besides, the optical properties of the polymer show that the support is not inert since it absorbs the light and probably leads to a decrease of the reactivity. That is why the use of another material transparent to the UV light, and able to promote favourable interaction with the pollutant will be developed in the future.

REFERENCES

Avila P., Bahamonde A., Blanco J., Sanchez B., Cardona A.I., Romero M. 1998. Gas-phase photo-assisted mineralization of volatile organic compounds by monolithic titania catalysts, *Applied Catalysis: Environmental*, 17:75–88.

Dingwang C., Fengmei L., Ajay K. Ray 2001. External and internal mass transfer effect on photocatalytic degradation, *Catalysis Today*, 66:475–485.

Dufaud O., Corbel S. 2003. Oxygen diffusion in ceramic suspensions for stereolithography, *Chem. Eng. J.*, 92:55–62.

Gao R., Stark J., Bahnemann D.W., Rabani J. 2002. Quantum yields of hydroxyl radicals in illuminated TiO₂ nanocrystallite layers, *J. Photochem. Photobio. A: Chem.*, 148: 387–391.

Meng Y., Huang X., Wu Y., Wang X., Qian Y. 2002. Kinetic study and modelling on photocatalytic degradation of para-chlorobenzoate at different light intensities, *Environmental Pollution*, 117:307–313.

Orlov A., Jefferson D.A., Macleod N., Lambert R.M. 2004. Photocatalytic properties of TiO₂ modified with gold nanoparticles in the degradation of 4-chlorophenol in aqueous solution, *Catalysis Letters* 92:41–47.

Wu C., Yue Y., Deng X., Hua W., GAO Z. 2004. Investigation on the synergetic effect between anatase and rutile nanoparticles in gas-phase photocatalytic oxidations, *Catalysis Today*.

Virtual modeling and rapid manufacturing – Bártolo (eds)
© 2005 Taylor & Francis Group, London, ISBN 0 415 39062 1

Porous cooling passageway for rapid plastic injection moulding

K.M. Au & K.M. Yu

Department of Industrial and Systems Engineering, The Hong Kong Polytechnic University,
Hung Hom, Hong Kong

ABSTRACT: The application of solid freeform fabrication (SFF) techniques to the design of plastic injection moulds has presented a new challenge to mould design and analysis strategy. Injection mould cooling process is important from the viewpoint of quality and productivity of tooling part. Efficient injection mould cooling systems can reduce the cooling time required and hence improve the productivity. Recently, many researchers have made attempts to design conformal cooling channel by various rapid tooling technologies. However, the cooling performance cannot turn up to mould engineer's expectation. Undesired injection moulded defects such as thermal stress, warpage or shrinkage are unavoidable. The paper, thus, proposes a novel approach using porous structure cooling passageway for the design of more uniform cooling performance in rapid plastic injection mould. Cooling performance of the proposed porous structure cooling passageway is investigated with mechanical feature. Computer-aided engineering (CAE) analysis is utilized for linear static analysis. The results of the proposed porous cooling will be compared with the traditional injection mould cooling model.

1 INTRODUCTION

1.1 Injection moulded cooling consideration

In contemporary plastic product manufacturing, injection moulding of polymeric component faces growing demands for several fancy plastic items. The quality requirement for these items is increasing. Proper control of injection moulding cycle is important for managing the standard of high-quality items. The injection moulding cycle can be divided into four stages. Figure 1 shows the injection moulding cycle. They are mould opening time, injection time, cooling time and ejection time. The cooling time (Rees 2002) includes two-thirds of the whole cycle period of injection moulding process.

Traditional injection mould cooling system is formed by conventional machining processes, such as straight-line drilling. Several straight-line cooling channels are formed around the mould cavity. In general, the performance of traditional mould cooling system, to a great extent, is highly depended on the experiences and machining techniques of mould engineers. The fabrication techniques place severe limitations in geometric considerations of cooling system design. Figure 2a illustrates the traditional cooling channels in which coolant flows through the mould inserts. The plugs will ensure the coolant flow circulation to be in a single circuit. Figures 2b and 2c show the traditional cooling circuit.

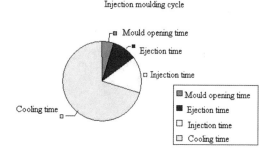

Figure 1. Injection mould cycle.

Non-uniform cooling will arise through conventional material removal (MR) methods. Sometimes, diverse problematic issues are inevitable and will be extended to injection mould defects formation, such as warpage, thermal stress distribution or shrinkage.

Recently, many researchers have made attempts to design conformal cooling system by diverse rapid prototyping (RP) and rapid tooling (RT) technologies (Hilton & Jacobs 2000, Grimm 2004, Rosochowski & Matuszak 2000). Some complex geometric devices or kits such as conformal cooling channel can be feasibly produced.

Conformal cooling channel (CCC) (Venuvinod & Ma 2004) can be defined as a water passageway in

(a)

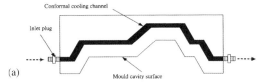

(b)

Figure 3. a) Layout of ideal conformal cooling channel (CCC), and b) CCC formed by copper duct bending (Direct AIM™ prototype tooling.).

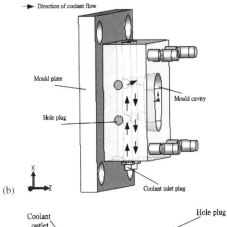

(b)

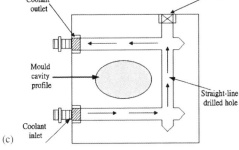

(c)

Figure 2. Traditional cooling circuit in plastic injection mould, a) inlet of traditional cooling channel, b) pathway of coolant flow, and c) layout of traditional cooling channel.

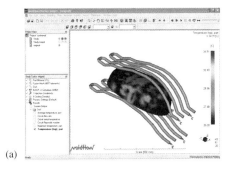

(a)

(b)

Figure 4. Melt flow simulation analysis, a) mould temperature distribution by CAE of MoldFlow, and b) thermal stress analysis by FEA package COSMOS/Works.

tooling that follows the contours of the part to be formed, i.e., form of cavity surface. However, this cannot be produced by contemporary MR processes. The geometric complexity of the cooling system around the cavity cannot be accomplished by conventional machine tools and manufacturing processes. Theoretically, conformal cooling covers entirely the whole surface of the part, the cooling rate can be increased by reducing the injection moulded cycle time. Figure 3a illustrates the layout of CCC. A common way to achieve conformal cooling in contemporary

RT is to bend copper ducts. For example in direct AIM prototype tooling (3D Systems) 1997 (Decelles 1997), Figure 3b shows the layout of copper duct bending as the CCC. Figure 4a shows the CAE melt flow analysis of mould temperature distribution by MoldFlow (MoldFlow Corp. 2005). It indicates that the mould cavity temperature around the mouse surface is uniformly distributed and the temperature maintains around 37.6°C and 39.8°C. The melt temperature is approximately at 220°C. Figure 4b demonstrates the finite element method of thermal stress analysis of the mouse model by FEA package COSMOS/Works (Structural Research & Analysis Corp. 2005). It shows that the corners and edges of the mouse experience potential risk for thermal stress formation without proper cooling. These corners or edges cannot obtain an effective cooling by localized CCC'Ss using bending copper ducts. However, the geometry of the copper duct can only partially follow the shape of the mould cavity surface. It cannot provide a totally uniform temperature distribution. Injection moulded defects such as hot spot or warpage cannot be avoided in the corner or undercut regions. The size and geometry of the mould cavity surface will influence the bending of copper ducts and hence performance and effectiveness of the cooling process such as coolant flow rate.

1.2 Importance of uniform cooling for rapid injection mould

Contemporary RT technologies present a range of methods by which CCC's for rapid mould manufacturing can be produced. The achievement of CCC's in rapid mould production is to achieve uniform heat transfer during the cooling stage of injection moulding process. CCC's in injection mould can improve the cooling performances better than straight-drilled channels which located at the side of the mould cavity and core. However, the number of CCC located is limited within the mould. Theoretically, the distance between adjacent channels is about 1.5 D (1.5 times of diameter) (Dym 1987). Uniform cooling is restricted to regions where the coolant is flowing through inside the mould. Risks for defects formation will exist. The wet area of the coolant flow is the decisive factor for controlling the cooling time and defect formations. Typically, the heat transfer during the injection moulded cooling includes heat exchange originated from polymeric melt to the mould material by conduction. Then the heat is conducted from the mould material to the coolant in the cooling passageway, where the coolant is heated up by convection and flows out.

2 PROPOSED POROUS STRUCTURE FOR CONFORMAL COOLING

In this paper, an original approach is proposed to achieve conformal cooling by employing porous cooling

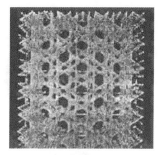

Figure 5. Scaffolding structure in QuickCast v1.1.

Figure 6. Natural structure of honeycomb.

passageway (Gibson & Ashby 1997, Ashby et al. 2000, Kalita et al. 2003, Fang et al. 2005.) A porous solid is one made up of an inter-connected network structure configuration of several three-dimensional elements. They can be assembled effectively with face-to-face or surface-to-surface connectivity. Figure 5 illustrates the QuickCast™ v1.1 (Lee 1999) that is mainly used as scaffolding support of rapid investment casting pattern. Figure 6 illustrates the natural structure of honeycomb. The inter-connectivity of these polyhedral elements can form continuous passageways for fluid flow. Theoretically, larger surface area of coolant flow can increase the rate of heat transfer from the injection mould to the coolant. Equations 1 and 2 show the rates of conductive and convective heat transfer.

$$q_{cond} \ or \ \dot{Q}_{cond} = -\frac{kA}{l}(T_s - T_\infty) \ \text{(W/m)} \qquad (1)$$

$$q_{conv} \ or \ \dot{Q}_{conv} = hA(T_s - T_\infty) \qquad \text{(W)} \qquad (2)$$

where k = thermal conductivity, W/m · K; h = convective heat transfer coefficient, W/m² · K; A = heat transfer surface area, m²; T_s = temperature of the

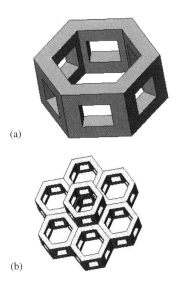

(a)

(b)

Figure 7. Porous structures, a) hollow hexagonal prism element, and b) hexagonal prism assembly.

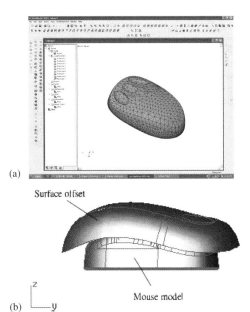

(a)

Surface offset

Mouse model

$\overset{z}{\llcorner\!\_y}$

(b)

Figure 8. a) Mouse model for FEA and melt flow analysis, and b) surface offset of mouse model.

mould surface, K; T_∞ = temperature of the coolant sufficiently away from the mould surface, K; and l = length of cooling passageway, m. The rate of heat transfer is observed to be proportional to the temperature difference and its surface area. Increase in length of cooling passageway will also reduce the rate of heat transfer. The injection moulded temperature should be well controlled by reducing the temperature difference between diverse areas during the mould closing stage.

For the purpose of plastic injection moulding, the mould materials are mainly stainless steels or any other metallic alloys. As their strong mechanical strengths, both the functions of these porous structures are mainly acted as mechanical support to withstand heavy compressive force or tensile strength. Figures 7a and 7b show the proposed porous structure using hollow hexagonal prisms and its assembly for developing the conformal cooling passageway.

3 MODEL DEVELOPMENT OF POROUS COOLING PASSAGEWAY

The use of RT technologies offers a compact fabrication of a complex 3D model with integrated production of conformal cooling passageway. This section outlines the method for the approximation of automatic design of topological and geometrical design of cooling passageway with porous structure configuration. The proposed method includes three main steps. The first step is to formulate the porous cooling passageway by surface offsetting. A 3D real life object is

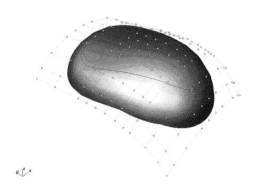

Figure 9. Spatial enumeration of surface offset for cooling passageway location.

employed as an example with melt flow analysis in Figures 8a and 8b. The second step is to undergo spatial enumeration of the conformal cooling passageway by porous structural pattern approximation in Figure 9. The final step is topological unionization of the approximated porous structural pattern to form the cooling passageway of the injection mould. Figures 10a and 10b show connectivity of polyhedral elements in 3D Euclidean spaces. Figure 11 shows the model with porous cooling passageway which covered the part's boundary profile within the mould insert.

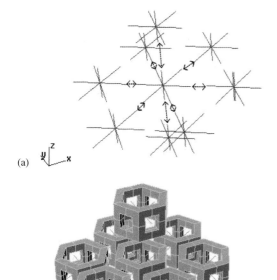

(a)

(b)

Figure 10. a) Wireframe skeleton in 3D Euclidean spaces, and b) skeleton fleshed-out into porous structure.

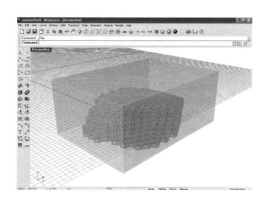

Figure 11. Formation of porous cooling passageway.

4 MELT FLOW SIMULATION AND ANALYSIS OF VARIOUS INJECTION MOULDED MATERIALS

In order to determine the mould temperature distribution during the cooling process, finite element technique is employed to investigate the thermal stress distribution and deformation for the proposed three-dimensional elements. Thermal and mechanical

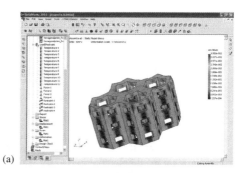

(a)

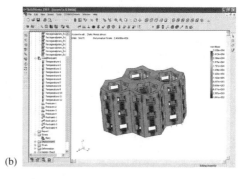

(b)

(c)

Figure 12. FEA simulation by COMSOS/Works for assembly porous structure various AISI mould steel, a) P-20, b) H-13, and c) A6.

Table 1. Thermal stress and thermal strain for various mould materials from FEA.

Mould steel (AISI)	P20	H13	A6
Thermal stress (MPa)	0.1251	0.1047	0.0983
Thermal strain (mm)(e^{-10})	2.121	1.948	1.643

simulations have been done by using commercial FEA package COSMOS/Works. Figures 12a to 12c show the result of FEA simulations of porous structural element assembly with various common mould materials AISI (P20, H13 and A6). Table 1 shows the thermal stress and thermal strain from FEA for various mould materials.

5 RESULTS AND DISCUSSIONS

Typically, an injection mould with proper thermal management is important for enhancing part quality, reducing injection mould process cycle time and in that way increasing productivity. The porous metal is envisaged as a system that heat is transferred from a high temperature mould material into lower temperature coolant. The proposed porous structure of conformal cooling passageway has large accessible surface area and thus high heat transfer rate. It can develop a uniform cooling system with higher effective heat transfer coefficient. It can have acceptable pressure drops and withstand the intense clamping force. To determine the mechanical behavior, three types of common mould materials are examined, such as AISI P20, H13 and A6. From the simulation results, A6 tool steel can provide the best mechanical strength to withstand the injection mould clamping force.

Not all steel materials can withstand the high clamping force or stress and the injection mould may crack. Comparing with traditional cooling channel using A6 tool steel as the injection mould material, the mechanical strength of A6 tool steel for the proposed conformal cooling passageway can have a better performance in thermal stress and thermal strain analysis. The productivity can thus be increased with decreasing cooling time. Heat transfers faster from injection moulded materials to the coolant. Effective and uniform cooling can be obtained by the large surface area porous structure materials. Thus efficiencies of heat conduction and convection of the injection mould are increased. Part quality can be improved by reducing the residual thermal stress in the parts being moulded. Obviously, some parts with complex geometries, holes or sharpened edges can experience an effective heat transfer by the proposed porous structure cooling system.

6 CONCLUSIONS

This paper has demonstrated a novel approach to further improve the performance of current CCC in thermal management for rapid plastic injection moulding.

A porous-based approach of cooling passageway is proposed to accomplish uniform cooling of the polymeric melt in the mould cavity. The designed porous configuration cooling passageway increases the rate of heat transfer as well as providing the necessary mechanical strength. Short cycle time and good product quality can thus be ensured.

ACKNOWLEDGEMENTS

The work described in this paper was supported by a grant from The Hong Kong Polytechnic University.

REFERENCES

Ashby, M. F., Evans, A., Fleck, N. A., Gibson, L. J., Hutchinson, J. W. & Wadley, H. N. G. 2000. *Metal Foams: A Design Guide*. USA: Butterworth-Heinemann.
Decelles, P. 1997. *Direct AIM™ prototype tooling procedural guide*. Valencia Calif.: 3D Systems.
Dym, J. B. 1987. *Injection molds and molding: a practical manual, 2nd edition*. New York: Van Nostrand Reinhold.
Fang, Z., Starly, B. & Sun, W. 2005. Computer-aided characterization for effective mechanical properties of porous tissue scaffolds, *Computer-Aided Design*, 37(1): 65–72.
Gibson, L. J. & Ashby, M. F. 1997. *Cellular solids, Structure and properties, 2nd edition*. Cambridge: Cambridge University Press.
Grimm, T. 2004. *User's Guide to Rapid Prototyping*. Michigan: Society of Manufacturing Engineers.
Hilton, P. D. & Jacobs, P. F. 2000. *Rapid Tooling: Technologies and Industrial Applications*. New York: Marcel Dekker, Inc.
Kalita, S. J., Bose, S., Hosick, H. L. & Bandyopadhyay, A. 2003. Development of controlled porosity polymer-ceramic composite scaffolds via fused deposition modeling, *Materials Science & Engineering*, C 23(5): 611–620.
Lee, K. W. 1999. *Principles of CAD / CAM / CAE Systems*. Reading: Addison Wesley Longman, Inc.
MoldFlow Corp. 2005 (http://www.moldflow.com/stp)
Rees, H. 2002. *Mold Engineering, 2nd edition*. Munich: Hanser.
Rosochowski, A. & Matuszak, A. 2000. Rapid tooling: the state of the art, *Journal of Materials Processing Technology* 106: 191–198.
Structural Research & Analysis Corp. 2005 (http://www.cosmosm.com)
Venuvinod, P. K. & Ma, W. 2004. Rapid prototyping: *Laser-based and other technologies*. Boston: Kluwer Academic.

Virtual modeling and rapid manufacturing – Bártolo (eds)
© 2005 Taylor & Francis Group, London, ISBN 0 415 39062 1

The GlucoFridge: A case study in "Rapid Prototype as Design"

O. Diegel, W.L. Xu & J. Potgieter
Institute of Technology and Engineering, Massey University, Auckland, New Zealand

ABSTRACT: The GlucoFridge is a pocket-sized portable battery powered insulin refrigerator. It is a technologically complex product involving a combination of thermo-electric cooling, electronic design, mechanical hardware design, heat-sink design, Injection molding design, and large amounts of computer simulation. The entire product was developed to a stage where it was ready to take into production within a time-frame of three months. This was achieved through efficient use of available Computer Aided Design, Finite Element Analysis and Rapid Prototyping technologies. It was the tight integrated use of these technologies that allowed for an extremely fast re-iterative design approach and allowed for a short development time. This paper presents a case study on how Rapid-Prototype as Design was used to help stimulate the fast and successful innovative outcomes of the project.

1 INTRODUCTION

1.1 Project background

The latest WHO estimate for the number of people with diabetes, (world-wide, in 2000) is 177 million. This will increase to at least 300 million by 2025. It has long been known that the number of deaths related to diabetes is considerably underestimated. A plausible figure is likely to be around 4 million deaths per year related to the presence of the disorder. Overall, direct health care costs of diabetes range from 2.5% to 15% annual health care budgets, depending on local diabetes prevalence and the sophistication of the treatment available (W.H.O., 1985).

When traveling, diabetics currently have to keep their insulin cool by putting it in a hotel refrigerator (if there is one available) and by carrying cooling devices such as icepacks when they travel away from their hotel.

Insulin should be kept at temperatures below 25°C. Many countries have temperatures ranging from 20°C to 40°C, which rapidly spoils insulin. The American Diabetes Association recommends that insulin be stored in a refrigerator (American Diabetes Association, 2005) Many hotels also do not have room refrigerators. These factors make it troublesome for diabetics to travel as they have a constant worry about whether there insulin is safe.

There are currently 3 commonly available cooling methods for insulin-dependant patients as well as a number of derivative solutions. They are as following:

1) Ice packs: This still seems to be the most commonly used means of refrigeration. Ice packs are frozen and put in a cooler box in which the insulin is transported over short trips.
   Negatives: This is very bad for the insulin as it often comes in direct contact with the ice. Chilled insulin vials should be inspected for crystals and particulates before use. Freezing may cause crystallization, resulting in variable potency.

2) MED-Ice Ice Packs: These are re-freezable ice packs filled with a non-toxic gel that last 30% longer than ice cubes. Travel Organizers with MED-Ice have a compartment in which the MED-Ice Ice Packs are inserted so that there is no direct contact with the insulin.
   Negatives: Must be frozen for 4–6 hours before use. Does not last more than a few hours. Gel packs often leak with change of atmospheric pressure.

3) FRIO Wallets: This device came on the market in 1999 and it is designed to keep insulin cool and safe for 48 hours. The main advantages are that there are no bulky ice packs, you do not have to worry about finding a freezer to get supplies of ice and the wallet is light to carry.

   It is activated by immersing it in cold water for 5–15 minutes. The panels of the wallet contain crystals and these expand into gel with the immersion in water. The wallet remains at a cool temperature for 48 hours, according to the prevailing conditions. The system relies on the evaporation process for cooling.
   Negatives: The inner bag stays very damp despite drying it with a towel as recommended and so the labels on the insulin vials start to disintegrate. Even though the unit is damp, the instructions say

Figure 1.   Glucofridge final pre-production prototype.

that the cooler pouch should not be put in a plastic bag because it does not work so well. This makes it inconvenient to keep with other items that may be damaged by the dampness.

1.2 *The GlucoFridge portable insulin refrigerator*

The GlucoFridge, as shown in figure 1, is a portable, battery powered, pocket sized refrigerator. This "worlds smallest" refrigerator is designed for carrying insulin (or other medication), which needs to be kept at a constant and cool temperature.

The applications for this technology are numerous, from transporting insulin, blood and sperm samples to anti-venom vaccines. It consists of a refrigeration unit that has drawers made for different size injection devices or medical samples. The unit comes equipped as standard with drawers for the most popular insulin-injection devices.

The refrigerator is powered by rechargeable Lithium ion batteries and is of a size that allows it to fit in a jacket pocket. As its cooling technology it uses a Peltier device (heat pump).

The novelty of this product lies in its portability, its uniquely small and compact size, and in its application of Peltier device technology. There is currently no other such product available.

The GlucoFridge is designed to be plugged into mains power at night, thus cooling the insulin and simultaneously charging the battery. Away from the hotel, it is powered by the batteries. As batteries it uses two lithium ion rechargeable batteries giving it a life 12 to 24 hours depending on how many times it is opened during the day.

Although many people enjoy traveling, patients with diabetes often fear or avoid it. This product gives diabetics both freedom of mind and freedom of movement when traveling.

2   PROTOTYPE AS DESIGN

Is traditional project management up to the task when it comes to managing new product development projects? According to Frame (2002), "…traditional project management is broken." as it has failed to adapt to meet changing times and technologies.

Whitney (1989) notes that: "In many large companies, design has become a bureaucratic tangle, a process confounded by fragmentation, overspecialization, power struggles, and delays."

Traditional project management tends to focus on what is often called the "holy-triangle" of project management: Cost, Time, and Quality (which is usually defined by the technical requirements of a project).

With global commerce supported by technology and communication made possible by the internet, time is now 7/24 and cost and technical challenges are addressed on a global basis using team members in different countries, with different cultures and time zones, languages, and methods. In this changing world, New Product Development time is rapidly becoming the most critical factor. High-tech products that come to market six months late but on budget will earn 33% less profit over 5 years. In contrast, coming out on time and 50% over budget cuts profit by only 4% (McKinsey and Company, 1989). If companies develop products on budget, but in shorter times, they develop a commercial advantage and increased flexibility.

Design-by-Prototype is a technique in use by such organizations as the National Aeronautics and Space Administration's Ames Research Center (Mulenburg, 2004). It shows significant success in simplifying and speeding up the development of research hardware with large cost savings. Design-by-Prototype is a means of using the old artisan's technique of prototyping as a modern design tool.

Prototyping is probably the oldest product development technique in the world and has been used by artisans for centuries. These artisans created prototypes of their ideas, to ensure that they worked, before making the primary artifact they were planning. Traditional Design-by-Prototype is useful in creating hardware for projects by eliminating much of the formal engineering design process.

It is often impossible to precisely specify requirements at the fuzzy front end of a project. Even when possible, it may be undesirable to do so (Frame, 2002). This often makes Design-by-Prototype critical to projects.

Design-by-Prototype is a highly interactive, integrated process that allows multiple iterations of complex aspects of a desired R&D product to be quickly evaluated and adapted into a properly functioning whole (Mulenburg, 2004). This 'whole' almost always meets the users' needs, as they actively participate in the design as it evolves during development. It gets their buy-in with each further improved iteration of the prototype.

The need for using this new/old process in new product development companies is largely due to the proliferation of highly functional and easy to use computer-aided-design (CAD) design tools to highly skilled and versatile engineers.

One of the problems with CAD is that it does not always reflect reality accurately. In a review of 72 development projects in the computer industry (Eisenhardt and Tabrizi, 1995) it was found that the common perception that Computer Aided Design greatly enhanced product development time, was often not the reality. Further anecdotal experience also shows that the extensive use of computer design tools can result in both excessive time expended in design, and a lack of imbedded reality in the final product. A design may look pretty on the computer screen, but will it meet the users' needs and can it be efficiently made as designed? Often many design changes occur during the manufacture of these pretty designs that increase both schedule and cost to the project without a commensurate increase in product usability or quality. Beautiful three-dimensional computer models and detailed CAD drawings can result in difficult to manufacture hardware that requires expensive fabrication processes that add cost and/or increase schedule.

Prior to computers, designers who often were not engineers, converted engineering sketches into finished drawings for manufacture. While doing so, much design detail was added to not only meet manufacturing's needs, but also to ensure the end user's satisfaction. Computers have gradually eliminated the designer's role, leaving a gap that engineers are often not trained to fill: making the design manufacturable and optimizing its desired usefulness. For many high technology products, much design time can be saved and expensive rework eliminated during fabrication by using design-as-prototype.

Barkan & Insanti (1992) advocate prototyping as a core development process for a way out of this dilemma. Mulenburg (2004) sees this is a major contributing factor in the 70–80% of projects that never make it through complete development, or fail in the marketplace because of compromises made during development that reduce content to save cost and schedule.

Mulenburg sees one of the major contributors to problems during the traditional linear design process as being an attempt to make every part as effective as possible. Trained in design, many engineers try to optimize every portion of a product in trying to create an optimized whole, which is exactly the opposite needed of what is required for both speed and parsimony in design.

The result is sub-optimization adding both time and cost to the design process without optimizing the final product. An old Zen proverb captures this problem in a few effective words; Perfection comes not when there is nothing more to add, but when there is nothing more to subtract.

The desired product must, of course, meet the basic needs of the intended user, and these needs must be agreed upon and defined as early and as clearly into the project as possible. Reality is that things are often optimized simply because they can be; not because they need to be. As an example, when only a few units of a product will be built, does a lengthy comparison of which fasteners to use need to optimize the highest quality with the lowest cost when only a minimum order quantity will be purchased anyway? If the functional requirements can be adequately met by an early choice, it is much more important to make the selection and move on to more complex aspects of the design that may need extra time to ensure they meet the desired needs. In new product development, time truly is money.

3 RAPID PROTOTYPE AS DESIGN

The recent advent of the latest computer aided design (CAD), computer aided engineering (CAE) and computer aided manufacturing (CAM) technologies has added a new twist to the traditional prototype as design process. It is now transforming from a 'prototype as design' process into a 'Rapid Prototype as Design' process.

This new generation of computer tools now allows engineers to, for example, relatively easily perform complex finite element analysis (FEA) calculations on their products, to test for any thermal or structural problems, and even to simulate how plastic may flow through an injection molding tool during manufacturing.

The latest generation of rapid prototyping technologies, such as stereolithography (SLA), Selective Laser Sintering (SLS) and Fused Deposition Modeling (FDM) now allow physical prototypes to be produced within hours rather then days. These rapid prototyping processes, which were previously only able to make plastic-like parts are now increasingly becoming able to produce metal parts. Not only is the choice of materials and processes increasing, but the last few years have seen a significant reduction in the cost of these technologies. Systems are now also available for not only simulating the behavior and performance of electronic circuits, but also for rapid prototyping circuit boards.

These technologies mean that it is now possible to construct first highly advanced virtual prototypes, and then working physical prototypes almost as fast as they are designed, thus allowing more iterations of a design within a shorter time-frame. This, in turn, potentially allows for products that are even better suited to their intended users in even shorter times.

The following case study demonstrates how using these rapid prototyping technologies can help in bringing a new product to market faster.

4 THE GLUCOFRIDGE DESIGN PROCESS

The GlucoFridge went through three major, and many minor, design reiterations, the first two of which identified the major technical problems to overcome, before arriving at the final production design.

The design process began with the inventor identifying the need for the product, and researching the market to the extent that a medical device manufacturing company was willing to take on the project, both from a manufacturing and distribution point of view, upon receiving a working prototype.

The first two weeks of the project were spent producing a conceptual CAD design of how the product might work from an engineering perspective. At the same time, a large range of Peltier devices (thermoelectric coolers) was ordered so that a series of tests could be undertaken to test the ideas that were to be at the core of the product. Peltier devices are semiconductor devices that, when current is passed though them, get hot on one side and cold on the other. These devices are commonly used on small camping refrigerators, for example.

Though it may not always be considered a true use of technology, the Internet should not be ignored as a tool to speed up the product development process. It is a rich source of technical information and an extremely effective way of rapidly sourcing components. Most electronic components, for example, can often be sourced within a few days, and the extra cost this may involve is usually insignificant when compared to the benefits of the time saved. During the first 2 weeks of the project, the Internet was used extensively, both to order a wide range of Peltier devices from several manufacturers, and to obtain much information about thermoelectric cooler theory.

From the initial CAD design, as shown in figure 2, a thermal finite element analysis was performed to calculate the size of the cold plate that would be required to keep two vials and the insulin compartment of two NovoPen III (a commonly used insulin injection device) at a temperature of approximately 10°C. The initial design was base on using a Sony Handycam style 7.6 V, 3800 mAh rechargeable battery running a 7.8 V, 5.5 W Peltier device sandwiched between a large hot plate heatsink and a cylindrical cold plate extrusion.

From within the CAD software, toolpaths for all the aluminium components were generated within a few minutes, and the aluminium components were then rapidly produced on a CNC milling center. This effectively demonstrated the tight integration now available between CAD and CAM systems.

A rapid prototype of all the plastic components was printed on a Dimension 3D printer which produced ABS plastic parts of the design which was assembled and tested.

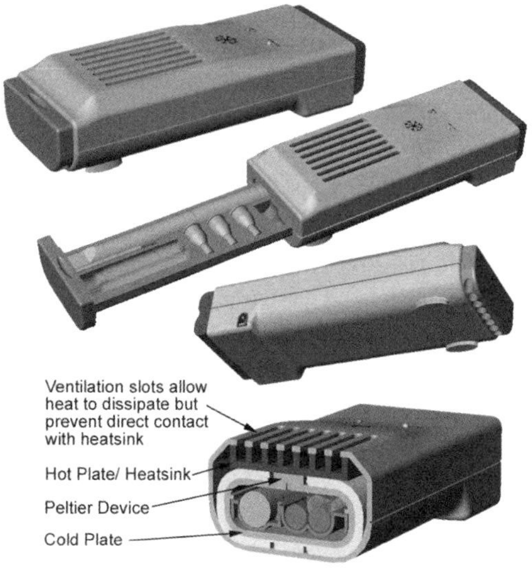

Figure 2. First iteration of GlucoFridge design.

A simple version of the temperature control circuit was prototyped on Vero-board to allow for testing, while at the same time the more complex real temperature control circuit was still being designed.

The initial tests revealed an almost immediate problem in reaching the target temperature. Though it had initially been hoped that it would be possible to reach the target temperature without the use of a fan, it was almost immediately found that, even with a large and efficient heatsink for the hot side of the Peltier device, the system was not quite efficient enough to dissipate enough heat for the temperature differential created by the Peltier to allow the desired refrigerator temperature to be reached.

The design was immediately modified to allow for the use of a 40 mm × 40 mm × 6 mm fan, and a new plastic top cover component was printed and the unit was once again tested.

In this second batch of tests on this minor revision of the first design, the cold plate easily reached the desired temperature, but a second major problem soon became apparent. The design included an electronic temperature control circuit which, when the cold plate reached the lower end of the target temperature range of 5°C, would cause the power to the Peltier to be switched off. When the temperature, after gradually increasing, then reached the upper end of the temperature range of 15°C, the Peltier would switch back on. The intention of this feature, as shown in figure 3, was to extend the battery life by making do as little work as possible.

What became almost immediately apparent was that, once the power to the Peltier was switched off, it

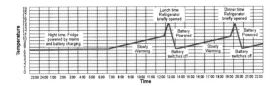

Figure 3.   Desired effect of temperature control circuit.

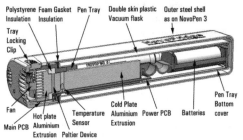

Polystyrene Foam Gasket Pen Tray   Double skin plastic   Outer steel shell
Insulation   Insulation   Vacuum flask   as on NovoPen 3

Tray
Locking
Clip

                                                              Pen Tray
          Cold Plate   Bottom
Fan   Hot plate   Temperature Aluminium Power PCB Batteries cover
Main PCB Aluminium   Sensor   Extrusion
          Extrusion   Peltier Device

Figure 4.   Second design iteration of GlucoFridge.

only took minutes for the cold plate to climb back up to the upper end of the temperature range. Even with the polystyrene insulation that was used to insulate the system from the outside ambient temperature, the temperature still climbed too rapidly, which meant that the battery had to switch on more often in order to keep the temperature in the desired range. This meant, in effect, that the refrigerator only had a battery life of approximately six hours, which was much less than the minimum requirement of 12 hours of battery life.

A solution was soon found to this problem: To build the entire cooling assembly into the inside of a vacuum flask type cooling container, as shown in figure 4. The vacuum would act as an effective insulation that would ensure that, once the cold plate reached the low end of the temperature range, the Peltier would switch off, and the cold plate would then, because of the

effective insulation, take a long time to warm back to the upper end of the temperature range.

This major change forced a complete redesign, as the vacuum flask meant that the battery could no longer be on the rear of the unit. This, in turn, meant that the size and shape of the Handycam type battery made it less than ideal for the product, so a new battery needed to be found. Some Internet research and fast tests were carried out with a variety of batteries, and it was found that four 2500 mAh, 1.2 V, AA size batteries, in combination with a 3.75 V, 5.5 W Peltier device could achieve the desired temperature levels. This appeared to be an ideal solution as the AA batteries were substantially smaller and would thus allow for a smaller product.

New CAD models were generated to see if the AA batteries could be configured in such a way as to fit within the vacuum flask together with the insulin vials and pens. This proved possible by completely changing the configuration of the hot plate, Peltier, and cold plate, and by changing to a pair of smaller 25 mm fans. Another prototype was made of the plastic and aluminium components. The lead-time to get a sample of the vacuum flask made was about 3 weeks, so test were initially carried out with an off-the-shelf vacuum flask of the same volume as that in the design.

This second prototype proved to work, and was taken to potential users for comment. From these focus groups a few minor and a potentially major problem were identified.

One of the minor problem was one of battery status and temperature status indication, which was easily solved through the use of multiple coloured LED indicators. Other minor problems identified included the need for the addition of a physical on/off switch so that the unit could be physically switched off when not in use, and the effectiveness of the locking clip that locked the medication tray into the vacuum container and provided a tight seal between the two.

The potentially major problem was to do with the batteries. Almost all the users immediately asked about the possibility of putting in regular, non-chargeable AA batteries. Though this was, in theory possible, the batteries had to be good quality high mAh rated batteries capable of providing a high continuous current output (such as the Energizer Titanium e^2 batteries, for example). This was perceived as a major problem by the users because of the perceived high chance of them using the product with unsuitable AA batteries. If the wrong type of batteries were used, they would not only tend to get very warm, thus canceling the effect of the cooling, but also ran the risk of being damaged and leaking inside the product. It was therefore indicated by the users that it would be better if the batteries were not changeable by the user, and they even indicated a preference not to use standard AA batteries in order to eliminate the risk of the wrong type being used.

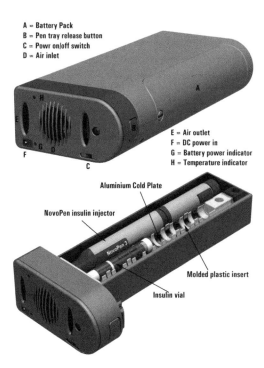

A = Battery Pack
B = Pen tray release button
C = Powr on/off switch
D = Air inlet

E = Air outlet
F = DC power in
G = Battery power indicator
H = Temperature indicator

Aluminium Cold Plate
NovoPen insulin injector
Molded plastic insert
Insulin vial

Figure 5. Final design iteration of GlucoFridge.

From this user feedback, the third major design iteration, as shown in figure 5, was entered into, and a new type of batteries was selected. The new batteries were to be two 3.7 V, 2200 mAh lithium ion batteries in parallel. These batteries, at 18.3 mm diameter and 65.2 mm in length, would not be easy to fit inside the vacuum flask unless it was made too big to fit in a pocket. The new design iteration therefore changed to a removable battery pack on the outside of the vacuum flask, which at the same time alleviated any risk of any heat produced by the batteries affecting the temperature of the insulin. The larger batteries also meant a small increase in the height of the product to allow for the batteries, but this in turn allowed for the use of a larger, and more effective, 35 mm fan which in turn made the system more efficient.

From this final rapid prototype, a few minor changes were made, such as moving the power adaptor connector to the back of the unit in order to make assembly easier, and the addition of extra air-vents to the front of the medication tray in order to make the warm air exhaust off the hot plate more effective.

At this stage, less than three months after the start of the project, it was deemed that the product was at a level where it would both meet customer expectations and be manufacturable at a cost acceptable to the manufacturing company.

The entire project, including all CAD files and data, as well as the physical working prototypes was then handed over to the medical product company for manufacturing and production.

5 CONCLUSIONS

Rapid Prototype as Design can be seen as a natural evolution of the more traditional Prototype as Design process facilitated by the advent of improving CAD, CAE and CAM technologies as well as the proliferation of low cost rapid prototyping technologies.

The ability to effectively integrate these many existing and emerging technologies into the new product development process has the potential of giving companies the ability to produce new high technology products at an increasing rate.

The GlucoFridge case-study described in this paper employed a rapid prototype as design process and successfully completed three major design iterations, each to working prototype level within a time span of less than three months.

REFERENCES

American Diabetes Association, Resource Guide 2005, USA, http://www.diabetes.org/rg2005/insulin.jsp, 2005.
Barkan, P. and Insanti, M. (1992). Prototyping as a core development process. Design for Manufacturing-Course Handout, Stanford University.
Frame, J.D. (2002). The New Project Management. Jossey-Bass, New York.
McKinsey and Company, Fortune Magazine, Feb 13, 1989.
Mulenburg, G., From NASA: Don't overlook the value of Prototype-as-Design in developing your new products, Vision Magazine, Product Development and Management Association, Vol 28, Issue 4, October 2004.
Mulenburg, G., IAMOT 2004, International Association for Management of Technology, Design by Prototype: Examples from the National Aeronautics and Space Administration, Jan 23, 2004.
Whitney, D.E. Manufacturing by design, in, Managing Projects and Programs, Harvard Business School Press, Boston MA.
World Health Organisation, Diabetes Mellitus: report of a WHO study group, World Health Organisation Technical Report Series 727, WHO, Geneva, 1985.

Virtual modeling and rapid manufacturing – Bártolo (eds)
© 2005 Taylor & Francis Group, London, ISBN 0 415 39062 1

Development of a user interface for a home automation system

A.M. Jacobs, R.G. Adank & A. Shekar
Institute of Technology and Engineering, Massey University, New Zealand

ABSTRACT: This paper highlights the importance of the user interface in the development of a home automation system. The project focused on ensuring that the interface meets the needs of the end user, and followed the product development process of identifying user needs, concept generation and detailed design, followed by commercialisation considerations. Concept screening resulted in the decision to design a user interface consisting of a web site and keypad combination.

1 INTRODUCTION

Research has shown that the home automation market in New Zealand is currently in the introductory phase of the product life cycle, with few products currently available in the medium priced segment of the market. This segment was therefore identified as having the greatest potential.

The project sponsor has formed a strategic alliance with an existing alarm company, to take advantage of their distribution and installation network. The home automation system will be integrated with the alarm, using the motion detectors as sensors allowing lights to be programmed to turn off when there is no movement.

This project focused on the development of the user interface for a system aimed at this market segment. The result of idea generation and screening was the decision to control the system with both a keypad/touch screen, which allows easy setting of the alarm part of the system and a web site, which allows detailed set-up of the system.

A number of concepts for the web site and keypad/touch screen were generated and these were screened down to one concept, based on ease of use. The results were developed and tested by users and subsequently improved in their ease of use, and their details finalised.

A marketing plan was developed based upon knowledge gained about the target market from the market research phase of the project. The key points of differentiation of the system under development are energy efficiency, ease of use and low price achieved through the strategic alliance with Castle Alarms and Integration with the Alarm technology, meaning that the majority of the system is already in place. additional factors are the systems ability to be easily retrofitted

into existing homes and the web based control of the system, allowing versatile control of the system, both from within and outside the home.

2 PROPOSAL, RESEARCH AND INITIAL DESIGNS

2.1 Project objectives and constraints

It was agreed and decided with the partners of the research that the project would focus on the new home automation systems user interface.

The development of a user interface for a home automation system

The objectives of the project were:

- Establish mode of control
- Design interface software concept
- Produce prototype
- Evaluate prototype
- Establish final design of interface
- User interaction.

The system will be based on the concept developed by the project sponsor, which consists of a control hub integrated into a household alarm system which extends the alarms wiring to send and receive signals to not only the motions sensors, but to additional lights or appliances to be controlled.

2.2 Current market situation

A variety of sources and techniques were used to establish the current state of the home automation market, internet, retail and literature searches and interviewing experts including Dr Don Barnes, a retired university professor with experience in the home automation field. In addition to this, user needs were investigated through 35 interviews and two focus groups.

As a result of this research, it was found that the home automation market in New Zealand is currently in the introductory phase of the product lifecycle, with a number of systems available, but little market penetration.

It was also discovered that there are two main segments to this market, comprehensive systems that offer a wide range of automation capabilities aimed at the top end of the market, and medium priced systems that cater to basic automation needs. The top end systems generally require a microcontroller at each node (the point on the system that is to be controlled), which means they are expensive to implement, but offer a greater range of control. The medium priced systems generally offer a basic on/off control at the node.

The majority of the systems available today are in the top end of the market. The system under development would therefore be aimed at the later segment of the market, and thus its main competitor would be the X-10 system. The X-10 system operates by sending signals over the existing power lines within the home. Modules plugged into power outlets can be controlled to an on or off state, thus controlling the appliance plugged into the module.

2.3 User information

The target market for the system is middle-income families with an interest in high-tech products. "Middle-income" was defined as families with a total household income of $50,000 to $100,000.

The needs of this market were researched with 35 interviews and a focus group. From this research, it was found that 80 per cent of respondents claiming they would be prepared to pay more for a product that offered energy savings. It was found that the end users were afraid that the system would not be easy to use. Additional information on what functions the system should be able to perform was also obtained, and used in the idea generation phase.

2.4 Idea generation and screening

Black box analysis was used to identify possible means of controlling the system. Literature searches provided information on possible technologies, with an Internet site and voice recognition being identified as possibilities (Brumitt, 2000). Experts, and existing products were also researched, and from this the possibility of a touch screen control was identified. Brainstorming and the results of user focus groups were also used to provide concepts. The most feasible results of this were

screened using a matrix with a rating of good medium or poor.

From the matrix, the combination of a keypad /touch screen and web page/web TV was selected as the preferred methods of control. This allowed easy setting of the alarm from the keypad/touch screen, and detailed management of the system from the web site.

3 DETAILED DESIGN

3.1 Outcomes of concept generation

From the initial concepts, it was established that the keypad/touch screen and web page combination was the best means of control for the system. The content of the web site was based upon research completed previously and especially from the analysis of the user needs. One key concept developed was the idea of the "total home management", where the system forms part of the home communications system as well as managing the appliances (Tweed et al. 2000).

3.2 Concept development and selection

A menu tree was developed for the web site using the content that had been identified as being necessary. From this, a number of concepts for the layout of the web site were designed and refined based upon ease of use. These were then screened down to one concept. The interface for the keypad/touch screen was then based upon this concept to ensure continuity. The result was then shown to experts and some members of the original focus group for feedback. Improvements were suggested and implemented, including the creation of a help link that opens a page containing instructions relating to what the user is attempting to do. And the layout of the set-up pages was altered so that they were less complex.

4 PROTOTYPING AND EVALUATION

4.1 Materials and manufacturing issues

The case of the keypad will be injection molded and fixed together with clips built into the case. A printed circuit board secured is secured to pins and clips formed in the case. A tamper switch will be located on the circuit board so that the alarm will sound if the case is opened while it is armed. The LCD screen is located above the circuit board, and has a touch sensitive panel located above it. A microphone and speaker will be located in the printed circuit board below the vents in the top to allow a message to be recorded and played back to the user. Further detail is given in the final product specification section.

The case will be manufactured on contract, the LCD and touch panels will be sourced from an outside supplier and for the initial run, the printed circuit boards will be assembled by the company's technician. The units themselves can also be assembled by the manufacturing company, using the specified guidelines provided by the researchers.

4.2 Technical evaluation

An alpha prototype of the website and a simulation of the keypad that allow control of the system have been produced and tested by members of the original focus group. In both cases, the menu structure proved easy to understand for the users and some improvements were made in areas where there was some confusion.

For a beta prototype to be completed, a large amount of programming must be completed before it can function properly. The same is true of the keypad/touch screen.

4.3 USER evaluation

As stated above, users from the existing focus group were contacted to evaluate the web site and touch screen simulation. In addition to this, a mail survey was sent to 100 homes in areas of Auckland, New Zealand, deemed to be in middle income suburbs. The survey asked users among other things, how much they would be prepared to pay for the system, based on price information provided about other systems currently available, how likely they would be to purchase the system and where they would expect to purchase the system.

4.4 Evaluation of product design and justification of preferred solution

The web site and keypad/touch screen work well together as a combination. The web site allows comprehensive set-up and control of the system, while the keypad allows the alarm to be set easily and quick easy access to control of the system. The inclusion of the communication aspect of the system provides a valuable differentiation point from other systems currently on the market.

The solution is based on research conducted in the market research phase and takes advantage of the relatively underdeveloped medium priced segment of the home automation market. The strategic alliance with Castle Alarms ensures that the system has a network of installers and a distribution chain already in place, and takes advantage of the brand recognition currently held by Castle Alarms.

The alpha prototype has been user tested and subsequently improved, ensuring it meets user needs.

5 FINAL PRODUCT SPECIFICATIONS

The web site will be capable of:

- Programming lighting and appliance moods, in which lights can be set to come on when motion sensors are activated or at certain times
- Controlling lights and appliances
- Setting and disarming the alarm and setting alarm zones
- Setting user details for the system, including a pin for setting the alarm and retrieving e-mails and messages
- Calendars and planners
- A home shopping list and possibly a bill manager.

They keypad/touch screen will be capable of:

- Setting and disarming the alarm
- Controlling lighting and appliances and activating preset moods
- Recording messages for other users and retrieving messages form other users.

Figure 1 shows the main menu for the web site concept. It allows access tot he four main functions of the web site; lighting, appliance and alarm control and the personal features.

Figure 2 shows the concept for the keypad/touch screen.

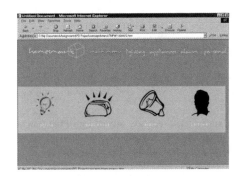

Figure 1. Main menu of web site control.

Figure 2. Keypad/touch screen concept.

609

6 COMMERCIALISATION

6.1 *Marketing and launch plans*

The product objectives are to gain a 5 per cent share of the target market within five years and a return on investment of 20 per cent per year. These objectives have been set to ensure the survival of the system and an adequate return for the company who has other projects to focus on also.

To achieve the objectives set out above, the market share can be increased by introducing add on products and services gaining competitors customers and increasing repeat purchases from existing customers. Converting existing customers can also serve to increase market share, if they can be converted to using our system. As the objectives for the marketing strategy also call for a return on investment, a balance of decreasing inputs and increasing market penetration could be used.

The core marketing strategy should take advantage of the key benefits of the system under development, which are:

- Achieves energy efficiency through "smart" use of lighting and appliances.
- Ease of use and communication features integrated into the system.
- Low price achieved.
- The strategic alliance with Castle Alarms ensuring an installation and distribution network is already established, saving a significant part of the set up cost.
- Integration with the alarm technology, meaning that the majority of the system is already in place.
- Ability to be easily retrofitted into existing homes.
- Web based control of the system, allowing versatile control of the system, both from within and outside the home.

The strategy that will take advantage of these benefits is to distribute the system as an extra with an alarm and through other lighting and electronics stores. This will reach a market that may currently be unaware of the availability of home automation systems, thus converting non-users to our customers. The cost of the system will be low enough to allow the system to be affordable to those people in the target market. The marketing should push the energy efficiency aspect of the system.

In addition to this the key points of differentiation from the X-10 system are:

- Ease of retrofitting installation with a professional alarm installer
- Ease of use
- Quality, with less of a do-it-yourself image and more of a quality hi-tech product
- Versatility of control of the system.

Another key to competing with the X-10 system is the introduction of add-on or integrated products or services. Forming alliances with companies who have complementary products or skills would increase the appeal of the system as well as keeping development costs down. Obvious industries that an alliance would be beneficial with are:

- Home Theatre Systems
- Access Control, such as automatic gates and garage doors.

In addition to this, due to the close integration of the system with the Internet, there could also be the possibility partnering with an Internet Service Provider (ISP), to allow access to the Internet through the system. Another opportunity is the idea that the ISP could subsidise part of the system to encourage the customer to join their service.

The market research phase of this project showed that one of the key factors that appealed to users was the ability of the system to save energy, with 80 per cent of respondents claiming they would be prepared to pay more for a product that offered energy savings. It also showed that users were concerned about the ease of use of the system, one of the more common comments to emerge was that "I can't even program my video". The users thought it would have to be intuitive to use. These two factors should therefore be used in promoting the product.

The launch of the product would focus around the in store displays and magazine and web banner adds detailed further in the promotional strategies section below. The main focus would be on informing users of the availability of the system, and the benefits it offers them.

6.2 *Business strategy fit*

The project sponsor's business strategy is to use its experience in cutting edge technology to diversify and thus ensure long-term growth. This product therefore is very well suited to the sponsor company. They have experience with the technology, having already developed a protocol capable of being implemented in a home automation environment. They also have experience in the programming required to make the alpha prototype web site and keypad simulation into a functioning web site capable of control and the keypad LCD menu capable of being run on the LCD screen.

6.3 *Market potential*

The home automation market is in the introductory phase of the product life cycle, and as such, it is an opportune time to be entering the market as it has a great potential. As environmental concerns grow, there will be an increasing emphasis placed on energy efficiency, something that home automation can offer. The Internet also has come into our lives, and it is the ability to act as the nerve centre for a home entertainment network that will mean that the home automation system will play an important role in the way in which the Internet is delivered in the future. In addition, there is now increased concern about raising crime rate and violent crime. The home automation system integrates with an alarm system to provide a sophisticated unit that can offer increased security.

All of these factors are working in the favour of the home automation market, ensuring that it is almost certain to grow in the future.

6.4 *Production strategies*

The initial production of the system will be small. Thus components will be sourced from outside the company and the sponsors technicians will complete the assembly. As the number of units being produced grow, it may be necessary to contract the assembly work out to a third party.

6.5 *Outline of promotional strategies*

The cost of advertising this product on television is far too prohibitive when considering the number of units that it is predicted will be sold. The main method of marketing the system will therefore be:

- Print media
- In-store literature and displays
- Internet banner advertising and web site.

In addition to this if a large chain store can be found to stock the product, then an introductory special could be run and included in a mail out circular.

To target the selected market segment, marketing of the system could include print advertisements in high tech magazines such as computer, science or Internet magazines. More mainstream magazines and newspaper advertisements could also be included to broaden the market reached. Web sites that attract the target market could also be used by the purchase of banner advertisements, and a web site promoting the product would provide interested users with more detailed information and purchase information.

7 CONCLUSIONS AND FURTHER WORK

The outcome of this project has been the development of an interface for the client company, which has the end user foremost in its consideration. The use of a systematic product development process has ensured that development of the interface has not lost focus of the end user and has been completed on schedule. The techniques used throughout the project have guaranteed that a wide base of information was used in the development and the evaluation of concepts. This provided a unique solution to the problem, and added features to the system above and beyond the physical components that make up the system, allowing differentiation from other systems that it may be competing against.

The interface should be implemented as intended, to ensure the future success of the home automation system. It should also be part of the business strategy to be actively pursuing new products that can be added to the system after its launch. These could include automatic drape controllers and access control systems. The addition of such products would strengthen the position of the system within the market and keep it competitive and updated.

The research highlighted the successful integration of actual users in the design of an interface for a home automation system using a combination of the web and a handheld device.

REFERENCES

http://www.homeautomationindex.com/
http://www.lboro.ac.uk/eusc/g_design_kiosks.html
http://www.chmsr.gatech.edu/
(Centre for human-machine systems research)
Brumitt, B., Meyers, B., Krumm, J., Kern, A., and Shafer, S. (2000). EasyLiving: Technologies for Intelligent Environments Proceedings of the International Conference on Handheld and Ubiquitous Computing 2000.
Jojic, N., Brumitt, B., Meyers, B. Harris, S., Huang, T. (2000). Detection and Estimation of Pointing Gestures in Dense Disparity Maps. Proceedings of the Fourth International Conference on Automatic Face and Gesture Recognition, 468–475.
Quigley, G. and Tweed, C. (1999). "Added-value services in the installation of assistive technologies for the elderly." Research report, School of Architecture, The Queen's University of Belfast.
Sheehy, N. (2000a). User-interface requirements for elderly people. Research report, School of Psychology, The QueenÕs University of Belfast.
Tweed, C. and Quigley, G. (2000a). The design and technological feasibility of home systems for the elderly. Research report, School of Architecture, The Queen's University of Belfast.

Virtual modeling and rapid manufacturing – Bártolo (eds)
© 2005 Taylor & Francis Group, London, ISBN 0 415 39062 1

Rapid product development of die-casting parts through virtual and rapid prototyping

J.C. Ferreira & J. Marques
Technical University of Lisbon, Portugal

P.J.S. Bártolo & N.F. Alves
Polytechnic Institute of Leiria, Portugal

ABSTRACT: The die-casting manufacturing process can be greatly improved through the integrated use of advanced virtual and rapid prototyping technologies. This paper proposes a new strategy for the rapid product development and optimisation of the manufacturing process of die-casting products, integrating conceptual design, virtual prototyping, rapid prototyping, $P\text{-}Q^2$ analysis, and numerical control (NC) connected to transducers to control in real-time the die-casting manufacturing parameters. This strategy enables to reduce the lead-time of die-casting designs and optimise the die-casting manufacturing technology parameters.

1 INTRODUCTION

The die-casting process is a well-known and established manufacturing technology (Street 1986, Herman 1992 and 1998). In pressure die-casting, the molten metal is injected into the die at high velocity, through a runner/gating system and solidifies under applied pressure (Beeley 1972). This system, composed by sprue, runners, gates, and overflows, is formed by a series of passages through which the molten metal flows into the die cavity (Herman 1992 and 1998). Casting design modifications at an early stage is crucial to produce high quality die-castings with shorter lead-times and reduced costs. An optimised casting process implies to satisfy the essential requirements of the designer and, at the same time, ensure the castability, avoiding casting defects from the manufacturing process.

This research study proposes an integrated methodology combining virtual and rapid prototyping, as well real-time control of the processing parameters to efficiently optimise both the die-casting product and manufacturing technology, allowing the reduction of the lead-time for die-casting product development.

2 METHODOLOGY

The methodology developed for the rapid product development of die-casting products is illustrated in Figure 1. It comprises four main phases. First, the: CAD

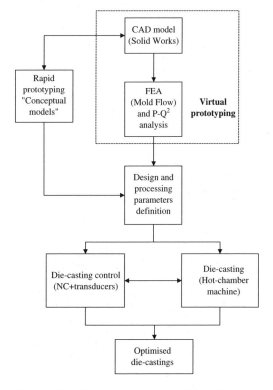

Figure 1.   Rapid product development of die-casting parts.

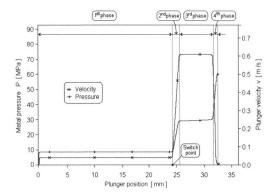

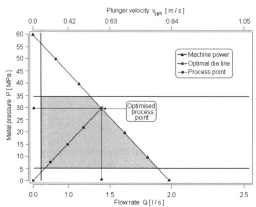

Figure 2. Pressure and velocity profiles during the die-casting filling process.

Figure 3. Example of a P-Q^2 diagram for a specific-die casting machine.

modelling and finite element analysis (FEA) for die-casting modelling and simulation, followed by the rapid prototyping of conceptual models. A third phase involves the establishment of the die-casting parameters based on P-Q^2 analysis and FEA, and the last one the die-casting control using a NC system connected to transducers integrated in the die-casting machine.

This advanced die-casting strategy starts with the CAD definition of the casting geometry. Once the first design is available, conceptual models are obtained through rapid prototyping. Physical models are then used to perform functionality and aesthetics studies, as well to analyse and optimise thickness variations, the design of the parting die planes and draft casting angles. Simultaneously, computer simulations are performed either to develop upgraded die-castings or for process optimisation. The selection of an appropriate die-casting machine for a specific cast part design is strongly dependent on the metal pressure, flow rate, and closing force.

The metal pressure and flow velocity for die-casting manufacturing processes can be improved by either computer simulation or by analysing and controlling the real parameters with a numerically controlled system linked to transducers.

2.1 Die-casting filling process phases

The die-casting filling process comprises five different stages according to Lui *et al.* (1996). Figure 2 illustrates the profiles of both metal pressure and plunger velocity as a function of the plunger position.

The first phase is called the "pre-fill stage" where the molten metal fills the shot cylinder at low velocity and pressure. The progressive velocity of this phase, $v_{1ph,}$ is calculated by the following equation:

$$v_{1ph} = k\left(\frac{100\% - f_i}{100\%}\right)\cdot\sqrt{d_p} \qquad (1)$$

where f_i is the percentage of the chamber filled, d_p is the plunger diameter, and k is a constant.

The second phase occurs after the switch point, see Figure 2 with a fast shot filling until the molten metal arrives at the gates. The casting quality can be improved analysing the position of the switch point.

At the third phase, the average pressure inside the cavity is almost constant, while the plunger speed could be controlled. This phase has a great impact on gas entrapment and porosity of the die-castings, due to very complex flow patterns occurring inside the cavity during the injection period. It can be improved through computer simulation.

The fourth phase begins when the so-called "impact pressure" occurs. At this point, the die cavity is completely filled (Herman 1998).

The fifth phase (not represented in Figure 2), is characterised by a maximum power output due to maximum pressure. The pressure is maintained at high levels until the pressure, applied by the hydraulic system, can still be transmitted to squeeze the viscous metal in the die cavity (Upton 1982). The high-pressure impact has a densifying effect on the metal surface making sounder castings.

2.2 P-Q^2 technique

In the third phase, the metal pressure can be evaluated through a P-Q^2 diagram (Figure 3).

The P-Q^2 diagram takes into account both machine and die characteristics, indicating how a die will perform on a die-casting machine (Herman 1996, Alsopp and Kennedy 1983). The P-Q^2 diagram represents the allowable working area below the die-casting machine's characteristic power line, *i.e.*, how much pressure, P, the machine applies to the metal at any given flow rate, Q. In the P-Q^2 diagram, the machine's power line, straight line with descending slope, is expressed by the following equation:

$$P = P_{max} \cdot \left[1 - \left(\frac{Q}{Q_{max}} \right)^2 \right] \qquad (2)$$

where P_{max} is the maximum metal pressure achieved by the plunger when it reaches the end of stroke and Q_{max} is the maximum flow rate when the plunger advances at dry shot speed. The plunger size and shot speed parameters are the two factors influencing the gradient of the machine's power line profile.

A die with a particular runner/gating system has its own characteristic profile, which is also represented by a straight line with ascending slope on the P-Q^2 diagram, i.e., for a given gate sectional area A_i, the relationship between metal pressure and flow rate is given by:

$$P_m = \frac{\rho}{2} \cdot \left(\frac{1}{C_d} \cdot \frac{Q}{A_i} \right)^2 \qquad (3)$$

where ρ is the metal density and C_d is the discharge resistance coefficient of the die.

The optimum operating point must lie within the allowable working area and can be obtained through the intersection point of the two profiles. The limits of the recommended working area, a grey zone, are defined by both the maximum and minimum gate velocities specifically related to the die-casting, the horizontal lines, and a maximum allowable die filling time, the vertical line. This optimised point is represented in Figure 3. This representation procedure enables to quickly improve the die-casting operating point by changing the correlated parameters.

2.3 Processing parameters

2.3.1 Metal pressure
The metal pressure P is determined through the following equation:

$$P = \frac{\rho}{2g} \left(\frac{v_i}{C_d} \right)^2 \qquad (4)$$

where g is the acceleration due to gravity and v_i is the flow velocity into the gates.

2.3.2 Metal flow rate
The flow rate Q_w is evaluated from the continuity equation:

$$Q_w = v_{3ph} A_p = v_i A_i \qquad (5)$$

The flow rate continuity is determined either through the product of the average plunger velocity of the third phase v_{3ph}, by the plunger area A_p, or by product of the flow velocity in the gate v_i, by the gate sectional area A_i.

2.3.3 Filling time
The die filling time evaluated through the casting volume, the selected alloy and the flow resistance, will determine the required flow rate in the third phase. Thus, the optimisation of the process implies the estimation of the appropriate filling time for the flow rate required for a specific die-casting machine. The characteristic wall thickness of the casting allows the definition of the ideal cavity die filling time t_f, taken at the gates through the following equation:

$$t_f = k \left(\frac{T_p - T_f + \lambda f_s}{T_f - T_d} \right) l_w \qquad (6)$$

where T_p is the pouring temperature, T_f is the minimum melt flow temperature, T_d is the die temperature, f_s is the solid fraction, l_w is the characteristic wall thickness of the casting, λ is a conversion factor, and k is an empirical constant.

2.4 Thermal casting cycles

To complete the simulation of the die-casting manufacturing process, further information is needed regarding the thermal casting cycles. The governing equation for unsteady heat conduction is:

$$\frac{\partial T}{\partial t} = \alpha \left(\frac{\partial^2 T}{\partial x^2} + \frac{\partial^2 T}{\partial y^2} + \frac{\partial^2 T}{\partial z^2} \right) \qquad (7)$$

where $\alpha = k/\rho C_p$ is the thermal diffusivity, C_p the specific heat, k the thermal conductivity, t the time, T the temperature, and x, y and z are the orthogonal coordinates. The latent heat is a major heat-transfer problem associated with phase change problems that can be determined by the following equation:

$$C_{eff} = C_p - \rho L \frac{\partial f_s}{\partial t} \qquad (8)$$

where C_{eff} is the effective capacity during the solidification period of the die-casting, L the latent heat of the casting alloy and f_s the solid fraction of the casted metal in die-castings.

The die filling temperature and solidification behavior is strongly dependent on the die's temperature gradients. The determination of the thermal casting cycles through computer simulation can provide information

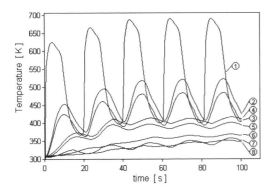

Figure 4. Simulated temperature cycles as a function of time.

on the die's transient temperature, as well the cooling temperatures function of time (Figure 4).

3 CASE STUDY

3.1 Virtual prototyping

The casting parts of a "cremone" were selected as a case study. The "cremone" was modelled by a computer-aided design software that enables virtual analysis and assembling (Figure 5a). The majority of casting parts are manufactured in zinc alloys and only the locking handle is in aluminium. Only the "cover" in zinc AC41A alloy is considered to be manufactured in a hot-chamber die-casting machine for this research work. A tessellated model of the "cover" (*.stl format) was also produced for both rapid prototyping and die-casting computer simulation (Figure 5b).

3.2 Rapid prototyping

Physical models were produced using a ThermoJet printer from 3D Systems and ThermoJet 88 (wax) build material that yields appropriate quality and surface finish for die-casting applications (Figure 6). These models enable the study of functionality and aesthetics, as well definition of the parting die planes and the castings' draft angles. These models were also used to discuss and optimise thickness variations. Thus, the preliminary geometry of the casting, which exhibited significant thickness variation, an average $l_w = 1.68^{\pm 0.44}$ mm, was redesigned into a more uniform one with an average $l_w = 1.38^{\pm 0.25}$ mm.

3.3 Die-castings alloy properties

The cast alloy selected for the "cover" is zinc AC41A with a composition specified in Table 1.

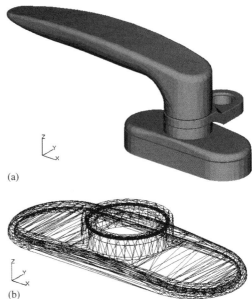

(a)

(b)

Figure 5. (a) 3D-CAD of a "cremone". (b) STL mesh.

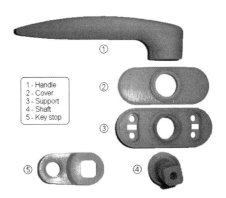

1 - Handle
2 - Cover
3 - Support
4 - Shaft
5 - Key stop

Figure 6. Physical parts of the "cremone" made by rapid prototyping.

Table 1. Zink AC41A alloy composition according to ASTM.

	Zn	Al	Cd	Cu	Fe	Mg	Pb	Sn
Wt.%	95.02	3.95	0.002	0.92	0.05	0.05	0.003	0.002

This material has an excellent balance of desirable mechanical and physical properties, outstanding castability and long-term dimensional stability, as shown in Table 2.

Table 2. Properties of Zinc AC41A alloy.

Properties	Value
Density, [g/cm^3]	6.7
Hardness, [Brinnell]	91
Ultimate tensile strength [MPa]	330
Latent heat, [kJ/kg]	113
Heat capacity, [kJ/kg-K]	0.42
Thermal conductivity, [W/m-K]	109

4 RESULTS AND DISCUSSIONS

A rheology software, with extended data properties for zinc AC41A alloy, was used for numerical simulation and analysis. The runner and gating system were then redesigned according to the results produced after several computer simulations, this way improving the injection velocity, avoiding excessive swirls and turbulence and providing homogeneous die-filling patterns. The thermo-physical data for zinc AC41A alloy was considered for these simulations (Table 2). The temperatures selected were: melting temperature $T_{melt} = 693$ K, average temperature in the hot die $T_{die} = 493$ K and average cooling temperature $T_{cool} = 343$ K.

The results of the numerical simulation of the improved casting design a presented bellow.

4.1 Die fluid dynamics

The fluid dynamics in the die cavity is an important factor influencing the casting quality in terms of both surface finish and porosity produced by gas entrapment.

It was also found that the optimal casting design consists of a compound gate with three tangential gates and a total short cross-sectional area of 9.16 mm^2. The cavity fill time is 10.5 ms and an injection average velocity of 40.25 m/s is obtained at the 3rd phase of the filling process. Figure 7 shows that with the adopted casting design, the metal flow pattern presents acceptable swirls and turbulence inside the casting.

4.2 Pressure simulation in 3rd phase

The casting is successfully filled with an injection pressure of 29.4 MPa at the 3rd phase (Figure 8).

4.3 Die filling simulation in time

The numerical simulation shows a homogeneous mould-filling pattern. The last filled areas are located at the opposite edge of the gates, where overflows and vents were conveniently attached (Figure 9).

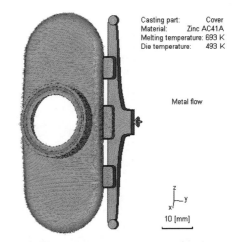

Figure 7. Die cavity filling.

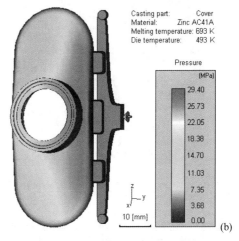

Figure 8. Pressure distribution at the end of 3rd phase.

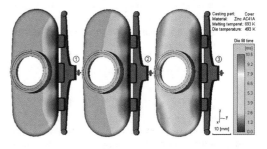

Figure 9. Die filling simulation in time: a) $t_1 = 3.15$ ms; b) $t_2 = 6.30$ ms; c) $t_3 = 9.45$ ms.

4.4 Temperature simulation during solidification

The computer simulation offers the flexibility to analyse temperature fields at various stages all along the casting cycle. For one stabilised cycle, the temperature

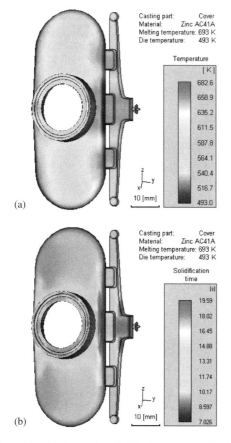

Casting part: Cover
Material: Zinc AC41A
Melting temperature: 693 K
Die temperature: 493 K

Temperature
[K]
682.6
658.9
635.2
611.5
587.8
564.1
540.4
516.7
493.0

10 [mm]

(a)

Casting part: Cover
Material: Zinc AC41A
Melting temperature: 693 K
Die temperature: 493 K

Solidification time
[s]
19.59
18.02
16.45
14.88
13.31
11.74
10.17
8.597
7.026

10 [mm]

(b)

Figure 10. (a) Temperature distribution and solidification hotspots. (b) Solidification time.

distribution during the solidification process, confirms the production of some isolated "hotspot" zones in the castings far from the gating system, which can originate some porosity and metallic contraction (Figure 10a). Figure 10a indicates that the heavy sections solidify later, so it is an important way to reduce thickness variations, as expected.

5 DIE-CASTING NUMERICAL CONTROL

The metal pressure and the die flow velocity values, established by computer simulation, were implemented and controlled by a NC system connected to both a pressure transducer and to a displacement transducer installed in the die-casting machine pumping system. The pressure and plunger velocity function of the plunger position was registered and the obtained profiles are shown in Figure 11.

The analysis of these profiles shows that during the pre-fill stage, 1st phase, the sprue is filled with molten metal at a low velocity $v_{1ph} \approx 0.04$ m/s and the pressure

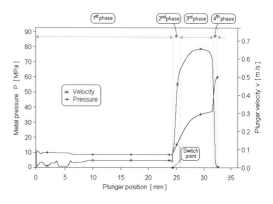

Figure 11. Diagram of pressure and velocity profiles function of the plunger position registered.

Figure 12. Casting shot manufactured with improved and controlled parameters.

of the molten metal is low $P_{1ph} \approx 9$ MPa avoiding gases entrapped in the molten metal.

A fast shot occurs through the 2nd phase, filling the runners until the molten metal arrives to the gates. The switch point was slightly moved forward to plunger position 24.19 mm. At the third phase, the plunger reaches a velocity of 0.53 m/s. It can be controlled reaching a maximum of 0.66 m/s

The highest metal pressure in the cavity is about 37 Mpa, with an average value of 29 MPa. The progressive mounting velocity avoids liquid splash owing to excessive fill velocity, on top of preventing gas entrapment and porosity enhancing the quality of the castings. The final plunger velocity and pressure are slightly reduced in this phase to avoid an excessive impact pressure at the final shot phase.

At the 4th phase, when the die cavity is completed filled, pressure attain a maximum value of 60 MPa during 2.5 s to squeeze the liquid metal in the die cavity, eliminating blisters and densifying the metal surface making sounder castings.

Good quality castings were produced, manufactured with improved and controlled parameters, as can be seen in Figure 12.

6 CONCLUSIONS

An integrated virtual and rapid prototyping methodology was proposed for advanced die-casting manufacturing with hot-chamber process and a case study was presented.

The findings of this research study can be described as follows:

- The manufacture development of die-casting products can be improved through the creation of virtual prototypes and the building up of rapid prototyping conceptual models to upgrade the castings design, considering the die-casting enhanced manufacturing parameters produced by computer simulation and the P-Q^2 technique.
- Rapid prototyping patterns allow the study of thickness variations and the design of the parting die planes and castings draft angles, a part from the study of functionality and aesthetics.
- Computer simulation of the die-casting manufacturing process enables to both improve the die-casting control parameters and produce casting products with better mechanical properties.
- The real filling parameters matched the values estimated from computer simulation can be obtained through a set of die-castings experimental shots. The metal pressure and fill velocity parameters were further improved using a NC system connected to transducers integrated in the die-casting machine. Good quality zinc-alloy castings were also produced based on the improved die-casting process control.

ACKNOWLEDGEMENTS

The authors acknowledge the contribution of Alualpha S.A. for allowing to the usage the experimental tests of their die-casting equipment.

REFERENCES

Allsop, D.F. & Kennedy, D. 1983. Pressure Die-Casting Part 2: The Technology of the Casting and the Die, Pergamon Press, Oxford, N.Y.

Beeley, P.R. 1972. Foundry Technology, The Butterworth Group (London Butterworths).

Herman, E.A. 1992. Die-casting Dies: Design, North American Die-casting Association, 15–24.

Herman, E.A. 1996. Gating Die-Casting Die, North American Die-casting Association.

Herman, E.A. 1998. Die-Casting Process Engineering and Control, Society of Die-Casting Engineers, Inc., River Grove, Illinois 60171.

Lui, Y.B., Lee, W.B., Ralph, B. 1996. A reclassification of the die-filling stages in pressure die-casting processes, Journal of Materials Processing Technology, 57, 259–265.

Street, A.C. 1986. The Die-Casting Book, 2nd Edition, Portcullis Press, 3–17.

Upton, B. 1982. Pressure die-casting, Pergamon Press, Oxford.

Virtual modeling and rapid manufacturing – Bártolo (eds)
© *2005 Taylor & Francis Group, London, ISBN 0 415 39062 1*

The need of technology transfer of RP in prototyping and casting industries of developing countries

A.B.M. Saifullah, N.P. Mahalik & B.H. Ahn
Department of Mechatronics, Gwangju Institute of Science and Technology (GIST), Gwangju, South Korea

ABSTRACT: This paper presents work on Rapid Prototyping (RP) through comparative evaluation of industrial products, which are produced by workshop-based, casting and RP. Parameters such as production time, cost, accuracy, surface finish and strength of product are used for comparative evaluation. Experimental case studies have been done to test the feasibility of RP technology implementation in developing countries (DC). Advance manufacturing system (AMS) are suggested for these DC manufacturing industries with the concept of RP.

1 INTRODUCTION

In the last fifteen years a number of educational institutions, research centres and companies have been motivated to use the computer assisted technologies in product design and manufacturing in order to reduce the production time and cost, in the aim of gaining greater market share [1]. In particular, companies experience immense pressure to provide a greater variety of complex products in shorter product development cycles, coupled with a desire to further reduce costs and improve the quality of the products [2]. One possible way to achieve this goal is through the use of state-of-the-art technology, such as RP. But still few research works has been done to validate possible application of RP technology in DC [6]. This work has been motivated to substantiate that there are many areas where RP technology can be applied with shorter production time, less cost and quality product in comparison with the conventional way of making prototype as well as final products. From the working experiences and training in some prototyping and casting companies of Bangladesh (Machine Tools Factory, Tara Pump Company, Techno-Zenith Industrial Enterprise, Laboni Enterprise, Design & Technology Centre) and also in Nepal and Pakistan, authors selects some areas where RP technology can be applied. Three case studies such as

- A part of a Hand Tube Well (HTW): A domestic used products
- Load carrying devices, like hook and chain
- Car spares parts (i.e. steering knuckle)

have been taken for experiment. There are other sectors like architectural, refinery and medical prototyping, where RP technology can be applied [4]. Research works proof that using some latest development in RP technology (3Dprinting, Z Cast System) not only prototype but also mould making can be possible with significant results. This paper also propose some mould & pattern design for single and multiple manufacturing system that can be made by using RP technology with greater economical benefit as compared with the conventional system of manufacturing. It also suggests the cost effective way of using RP technology. According to the title of the paper this research focuses on the RP application in developing countries (DC), and always considering the cost of RP technology. After comparing with all major RP systems only two or three were found compatible and cost effective in the DC economic scenario.

2 OVERVIEW OF THE RESEARCH

In this paper, RP method is suggested to apply in the prototyping and casting industries (P&CI) for DC. First, some experimental studies have been done on major RP systems, then analysing each system and comparing with the other prototyping methods. For comparative evaluation of RP systems the production cost, time and quality of the products (i.e. surface finish, accuracy, strength) have been selected. After comparing all systems, two RP methods have been selected to apply in P&CI of DC, considering the economical scenario and market demand. Comparative production process using selected RP system with conventional system of making prototyping and final product has been carried out. Finally advanced manufacturing system has been proposed with the help of RP that can be applied with significant results. Systematic overview of the research is shown in figure 1.

3 EXPERIMENTAL RESULT FOR SELECTION

Experiments have been done with the following major RP systems to test each systems production cost, time and quality of the products, such as strength, surface finish and accuracy. Experimented RP systems are:

1. Stereo lithography (SLA500)
2. Selective Laser Sintering (SLS2000)
3. Fused Deposition Modelling (FDM8000)
4. Laminated Object Manufacturing (LOM1030)
5. 3 Dimensional Printing (3DP, Z Corp.).

Products, which have been made from different prototyping methods, have different strengths, depending on the process and material used (shown in table1). Prototype from SLA shows the highest strength

(66 MPa) and the lowest is the prototype made from Z Corp., 3DP. SLA shows highest strength because it produces prototype directly from the liquid, which is turned into solid by curing reaction. Strength test has been done by universal testing machine (UTM) is shown in figure 3 (test specimen type was ASTM D638). Experimental results show that on completion of the prototype, after building with minimal post processing, SLA provides greater accuracy than the other RP processes. Accuracy and roughness tests have been done by co-ordinate measuring machine (CMM). In case of SLA, accuracy value is 100–120 micrometer and for 3DP (Z corp.) the value is around 500 micrometer. Thus SLA provides best accuracy in comparison with the other RP systems. The mechanical properties of parts made by 3D Printing are much inferior compared with SLA or FDM. However, most commercially available 3D Printer produces parts much faster than other RP technologies. Cost and finishing time have been conducted by the sample part (figure 2), which has been made by all five processes. The sample has

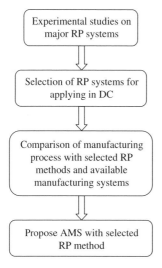

Figure 1. Systematic procedure of research.

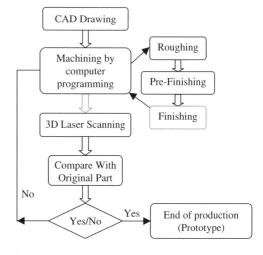

Figure 2. Process flow diagram of making prototype of Hook by CNC machine.

Table 1. Experimental data for selection of RP.

Process	SLA500	SLS2000	FDM8000	LOM1030	3DP Z Corp.
Material	SL5410	DuraformP/A	ABS	Paper	ZP100
Tensile strength (Mpa)	66	34	12	26	4
Elongation (%)	4–5	8–9	5–6	10	0.7
Accuracy (μm)	100–120	125–150	150	200	400–500
Roughness (μm)	8	12	28	30	16
Built time (Hour)	4	4	10	3	2
Part cost (Man Won*)	30	30	250	20	5

* 1 Man Won = $ 9 (Approximately).

100 mm × 100 mm × 25 mm dimension. Comparative result shows that using this 3DP (Z Corp.) RP method, 250000 mm^3 volume of prototype part can be produced by 50000 Korean Won (1200 Won = 1$) and also in 2 hours. In terms of developing countries (i.e. Bangladesh), this production cost will be 500 taka ($ 8). So the experimental result shows, Z Corp. RP System is suitable to apply in DC.

4. COMPARATIVE EXPERIMENTAL RESULT

To compare the production process of RP with conventional method, prototype of load-carrying hook (figure 4) has been made with Computer Numerical Control (CNC) machine and also with SLA500 RP method. Process flow diagram of both systems are shown in figure 2 & figure 5. Prototype has been made by CNC machine needs more time than that of RP, however, it shows better accuracy and surface finish. As in both cases, final product which is usually

 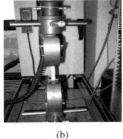

(a)   (b)

Figure 3.   (a) Sample RP part made by FDM, (b) Strength test is being done by Universal Testing Machine (UTM).

(a)   (b)

Figure 4.   Experimental manufactured part of hook, (a) Prototype made by **SLA 500** (up) final product by sand casting using the same prototype as master pattern (bottom), (b) Prototype made by CNC, then final product by casting.

made by sand casting, accuracy and surface finish does not affect much, but that do in case of production time.

Therefore, RP is preferable over CNC prototyping where prototyping does not need much precise and accurate prototype rather short production time. Other recognizable feature of making prototype by RP is the cost of production. Comparative experimental result (table 2) shows that in case of RP, production cost is 5 times less than that of CNC machining. Manufacturing system with cheaper production cost is always preferable to implement. So, RP system can implement cost effectively in DC. Hook prototype has been made local (Korea) available RP system (SLA500), which costs $ 150 and 5 hours with CAD drawing. But with the cheapest RP system (3DP Z Corp.) it can be made by cost of one-third of SLA with half production time.

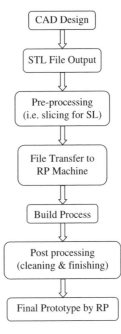

Figure 5.   Process flow diagram of making Prototype by RP (SLA500).

Table 2.   Comparative experimental data for prototyping of hook with CNC machine and RP process.

Prototype (Hook)	CNC machine	RP (SLA 500)
Production Cost ($)	400	150
Production Time (with CAD drawing)	3 days	5 hours
Accuracy (μm)	12.5	100
Roughness (μm)	2	8

623

5 PROPOSED ADVANCED MANUFACTURING SYSTEM (AMS)

5.1 *Case studies implementation with AMS*

5.1.1 *Case Study 1: Hand Tube Well*

About seventy percent of people in DC use HTW (figure 9), which is their only source of water supplier. Tube well, the term generally used to describe water well, the subcontinent including Bangladesh, Nepal, Pakistan, India, which are termed as borehole or water well in other parts of the world [3]. Prototype of HTW (figure 10) inside valve has been made with SLA500 and final product by sand casting. Comparative result with conventional method shows that production process by RP system is much better than that of available system.

5.1.1.1 Traditional Manufacturing System of HTW

HWT making by traditional method is lengthy process and quality of the product is not good. For example, the manufacturing process of HTW Inside Valve (HTW-IV) by traditional method (figure 7, TPH [7]) has been studied as comparison with that of RP. It shows that

RP process of making inside valve (IV) is much more superior than that of traditional methods. However, IV is made by RP, which needs less time and quality of the product is better.

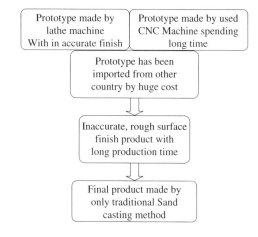

Figure 7. Flow diagram of manufacturing system of HTW by the traditional method.

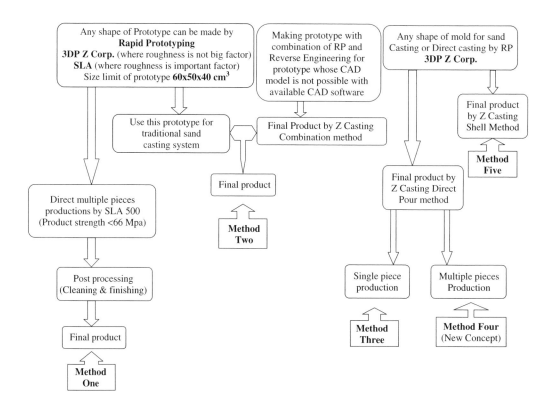

Figure 6. Advance manufacturing model by RP.

5.1.1.2 Manufacturing system of HTW by RP

In this research it has been proved that HTW can be made by RP. Experimental result shows that HTW-IV prototype can be made with almost all RP process with design flexibility, less time, better surface finish and accuracy. Then the final product can be made with sand casting using the same prototype as a master pattern.

Figure 8. Inside Valve (IV) of HTW made by traditional method.

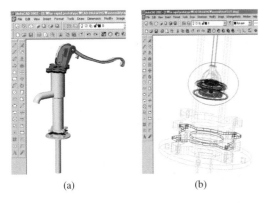

(a)   (b)

Figure 9. (a) CAD model of HTW, (b) IV in wire frame model of HTW part.

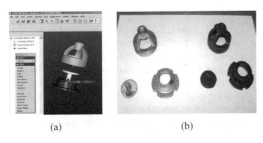

(a)   (b)

Figure 10. HTW-IV made by RP (a) CAD model by Pro-E, (b) Left: Prototype by SLA 500. Right: Final product by sand casing using same prototype as a master pattern.

5.1.2 Case study 2: Load carrying Hook

Making prototype of load carrying device, Hook, is very difficult with conventional system (i.e. by CNC machine). It needs lots of time and as well as money. So, it can introduce the RP method, which concludes better manufacturing system. By RP method not only the prototype but also mould can be made. The final product can also be made by Direct Pouring (DP) method of ZCast System [5].

Multiple pieces mould can be made in a single mould and then by DP method, mass production can be possible. Figure 12 shows conceptual CAD design for multiple pieces mould, with proper pouring and ventilation system. This design has been verified by Z Corp. USA. Using same technique multiple pieces manufacturing of HTW-IV can also be made for mass production.

5.1.3 Case study 3: Steering knuckle

One of the complex processes of making passenger car body part is steering knuckle, which is very lengthy process to design and manufacturing but with the help of RP it can be manufactured quite easily [13]. By the help of ZCast direct pour method or shell method it can be manufactured with greater accuracy, surface finish, less manufacturing time and cost. Conceptual CAD mould design has been made which is shown in figure 13.

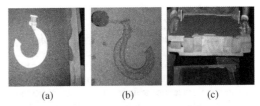

(a)   (b)   (c)

Figure 11. Experimental hook manufacturing system with traditional casting, using the RP pattern. (a) Half slice of RP hook placed in cope, (b) cope mould, (c) cope is being placed over drag (same process has been used for making HTW inside valve).

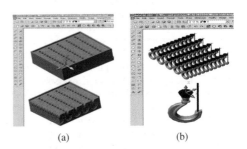

(a)   (b)

Figure 12. (a) Multiple pieces hook mould design, (b) final hook after DP method (conceptual).

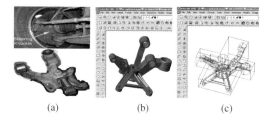 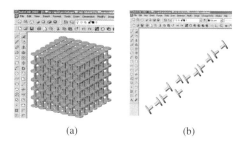

(a)	(b)	(c)

Figure 13. Steering knuckle (a) up shown in car, bottom after dismantle, (b) 3D solid CAD drawing, (c) Wire frame CAD mould design.

(a)	(b)

Figure 15. HTW IV inner part (a) the array of multiple pieces, (b) part can be separated (conceptual).

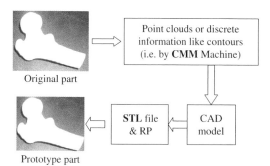

Figure 14. Flow diagram of experimental process of making prototype with combination of RP and RE.

5.2 *RP with Reverse Engineering*

With the help of Reverse Engineering (RE) prototype of very complex shape can be made. RE in light of RP refers to the process of regenerating a physical object back into a digital CAD format, then producing direct or modified copies of the original object using RP. The application of RE spans a variety of industries, and can be accomplished with about as many techniques [9, 10]. To verify its applicability the following bone part (figure 14) has been made, which shows the significant result. Final prototype part is almost same with the original product.

The process of making prototype with RE & RP is shown in figure 14. The physical object to be reproduced must be scanned electronically to provide a point cloud of co-ordinate along the various surfaces of the object. A point cloud is a set of coordinate data points in 3D space representing a solid object. This can be achieved by manual probes, which are placed against the object bay hand then the coordinate data at each point is entered into the computer by clicking a pedal or button. CMM are based on the same concept, except that the part is fixtured and the sensor tip is driven electronically to touch the surface and enter data points. Once a point cloud is generated it must be converted from simple dots in space into actual surface representations. There

are various "connect-the-dot" software applications available today that are designed for specific purpose, which usually connect the points and export either the standard STL file for RP or a standard generic CAD file interface known as IGES (International Graphics Export Standard). Finally, the CAD or STL file is created, the parts can be rapid prototyped to provide the models or hardware as needed.

5.3 *Direct manufacturing*

Along with the prototype, the final product can be made economically by direct manufacturing with SLA 500 RP system [3]. Experimental result shows that inner part of HTW-IV has tensile strength around 70 MPa. This inner part practically does not take much pressure from water as it works as one-way valve.

So, prototype of this inner part can be used as final product, no need to make it by casting. Conceptual result shows (figure 15) that at a time more than 500 pieces part can be prototyped together. This part can be directly made by RP rather than making first prototype and then final product by casting.

6 CONCLUSION

This work suggests the applicability of RP technology in various domains of product manufacturing. Essentially, it emphasizes that the RP technology can be transferred to the developing countries such as Bangladesh, Nepal, Pakistan, Indonesia and even India. RP technology has been implemented by taking several case studies such as Hand Tube Well parts, hooks and chains and a part of a car (steering knuckle). Various RP principles and methods have been described. Especially a comparative study of various manufacturing processes of products has been made, taking into account of many parameters such as production time, cost, accuracy, surface finish and strength of product. These are the important parameters as far as transfer of technology to the developing countries is concerned. This work also includes some experiment on Reverse

Engineering, as this has been an integrated method as far as RP is concerned. The combined technology (RE and RP) can be employed in the areas of interest.

ACKNOWLEDGEMENT

This research has been supported by Gwangju Institute of Science and Technology Research Fund.

REFERENCES

1 Chen, D. & Cheng, F. 2000. Integration of Product and Process Development Using Rapid Prototyping and Work cell Simulation Technology, *Journal of Industrial Technology*, vol.16, no.1

2 Evans, M. A. & Campbell, R. I. 2003. A Comparative evaluation of industrial design models produced using Rapid Prototyping and workshop-based fabrication techniques, *RP Journal*, vol. 9, no. 5

3 The Hong Kong Polytechnic University 2000. Rapid Product Development Centre

4 Z Corp. 3D printing, Industry standard three Dimensional Printers and solutions

5 CEPTECH, Z Corporation Seoul, Korea (Local Branch of Z Corp USA)

6 BANGLAPEDIA, Technological information Centre, Dhaka, Bangladesh

7 TARA Pump House Dhaka, Bangladesh

8 Gopakumar, S. 2004. RP in medicine: a case study in cranial reconstructive surgery, *Rapid Prototyping Journal*, vol. 10, no. 3

9 Lee, K. H. & Son, S. 1999. Medical Application of Rapid Prototyping Technology, *Journal of the Korean Society of Mechanical Engineers,* vol. 36, no. 9

10 Cooper, K. G. Rapid Prototyping Technology selection and application

11 Zhengying, W., Yiping, T. & Bingheng, L. 2003. A Rapid Manufacturing Method for Water-Saving Emitters for Crop Irrigation Based on Rapid Prototyping and Manufacturing. *International Journal of Advance Manufacturing Technology*, vol. 21

12 Döllner, G., Kellner, P. & Tegel, O. 2000. Digital Mock-ups and Rapid Prototyping in automotive product development, *Society for Design and Process Science*, vol. 4, no. 1

13 Kim, T. S., Toutaoui, B.W. & Le, J.H. Modern Cars With Progressive Rapid Technology Solution-Rapid Prototyping in South Korea, CAPT, KOREA

Virtual modeling and rapid manufacturing – Bártolo (eds)
© 2005 Taylor & Francis Group, London, ISBN 0 415 39062 1

Virtual engineering for decisional and operational processes enhancement in foundry

A. Bernard, N. Perry & J.C. Delplace
IRCCyN, Ecole Centrale Nantes, France

ABSTRACT: European foundries are competing against low costs countries, which are using different rules in terms of fiscals, socials and environments laws. These competitors are reducing the numbers and the proficiencies of new contracts, consequently modifying companies' financial balances. In order for the european companies to remain competitive, they need to convince new customers, by reducing their production costs; controlling the process risks and developing differentiation factors (quality, innovation, reactivity, etc.). The scientific way to amend the progresses is based on methodological propositions and deployments utilizing virtual and digital engineering technologies. Therefore, the bases of this paper are to propose methodological solutions for the integration of virtual engineering tools and methods all along the sand casting process, from the negotiations to the customer delivery. A new global approach is proposed, based on product and process definition, for costs evaluation, order strategic and logistic impact simulation, rapid product development, production performance enhancement and total digital traceability for parts and tooling. The fundaments and the innovative factors are explained, and the different new solutions that it has been necessary to create are argued, both on the conceptual and the operational levels. Finally, this approach relevance is demonstrated with a real case study, realized in SMC Colombier-Fontaine, the industrial partner of this project, where the application has been deployed and is working successfully.

1   INTRODUCTION

SMC, foundry of the AFE METAL group, uses a sand casting process for the manufacture of primary pieces in steel. The evolution of the markets drove the business to develop again technologies in partnership with research laboratories. SMC is present on many sectors of activities of which the rail one, the truck, the manutention, etc. With near of 4000 available references and about 1000 references turning on a year, the activity is based on the production of small and medium series of pieces, which have heterogenous forms and manufacturing processes. This variety, called "mix products", lets the management of this foundry become more complex than the one of factories of which the production is limited to some products.

The realization of this foundry process is characterized by the mastery lack of certain parameters the influence of which is not totally identified. The variety of the products complicates the identification of these parameters while mixing the parameters linked to the procedures to those linked to the products, which let any drift become difficult to master.

The "mix products" multiplies equally the number of orders and, consequently, weights the necessary structure dedicated to the realization of the administrative

tasks. The commercial negotiation, the planning preparation, the production scheduling, the design studies, the industrialization, the follow-up of the subcontracting, the expedition, the quality insurance are all functions that oblige the company to reinforce its operational structures. These expenses more or less are included in the sale price of the product. Thus:

– certain pieces allow covering the variable expenses, the expenses set up and assure a profitable margin: they are said contributory,
– of others on the other hand do not cover the entirety of the expenses set up: these pieces are said marginal. Consequently, according to the volume that it represents, a production absorbs a part more or less important of the expenses set up.

Events can induce changes in the organization and thus modify the financial balance defined at the time of the budget:

– if an important volume of pieces presents in its process a very long operation, it can decrease the production flow,
– at the opposite, if an important volume of pieces is corresponding to short processes, the activity can decrease on certain production stations. Consequently, the game consists then to possess industrial and

commercial plans constituted of a maximum of contributory pieces and of a minimum of marginal pieces while keeping a homogenous production plan. It is necessary not to handicap the results and the future balance of the business: the commercial offer impacting the long-term profitability of the business.

The tools and employed methods in the design office reflected equally this established fact. This service is in charge of the realization of the cost estimations for more of about one hundred parts a month and of the industrialization of about fifteen new products a month. The technicians dedicated the major part of their time to regulate the problems of perfect manufacture processes. The technicians have very little time to realize the cost estimations and the casting studies. In practice, the conception of the pieces and the casting studies represented a comparatively reduced party in the load of the technician.

Thus, SMC had itself to modify its organization, also because it is confronted to a worldwide competition. Facing this position, SMC knows anyway to emphasize its core competencies. Thus, even if the competition is very hard on the costs, the difference in terms of delays and of quality allows again winning again markets.

Despite this competition, the progresses were planned based on a unified product numerical definition. Our proposition consists in defining a global integration methodology of the design, manufacture and numerical control technologies.

The proposed approach leans on an evaluation system of profitability based on the definition of the manufacture processes of the product. This method is constituted of two applications: the conditional and variational cost estimation and the piloting of the business curve. It is based on a concept that allows simulating the impact of an order on the exploitation account. The cost estimation is the first step of the process. Its function is not to deliver a sale price but to measure directly the impact of the product on the exploitation result. This system is also dynamic, i.e. that it is possible to know instantly what influence has the modification of a criterion on the profitability of the product. These functionalities allow mastering the economical inductors of the manufacture process. The piloting curve allows next showing the balance point representing the profitability threshold of the product.

The model used for the definition of the process being common to all the applications of the numerical chain, the informations can thus be reutilized under the same forms. The conception of the products and processes and the management of the production can benefit from this information thanks to a shared database.

Our proposition continues by the suggestion of a development methodology of products. As it was previously shown, in this domain, the problematic does not concern a functionality lack of the commercial tools but in the lack of available resources for the generalization of these techniques. Consequently, we propose a methodology allowing reducing the necessary time for the realization of a study. CAD then is deployed as a tool of bureautics (like Excel or Word). Thus implemented, the visualization, the pattern making, the prototype manufacturing and the constraint analysis of the final product are as much services that easily can be proposed to the customers.

Software tools were developed to facilitate the conception of the casting cluster that will serve to flow simulation. They allow arranging quickly the pieces, the filling channels and the different shapes that compose the cluster. The simulation of the filling and solidification allow validating the chosen options in conception phase. Once validated, the placement tools used for cluster design generate automatically numerical models of the constituent entities of the process. These will be used in the different applications necessitating a numerical geometric definition.

The first one of them is the manufacture of the casting tools. The simplification of the CAM software allows generating quickly path of machining without special competences in this domain. The trajectories of machining cutting tools are thus generated from standard ranges and next are used directly on the machining center for the manufacture of the elements. Thanks to CAD modeling of the tools and the core boxes, the necessary resources for the manufacture and the assembly of the tools considerably were reduced. Thus, the reduction of the manufacture costs allowed justifying the realized investment for numerical modeling.

2 NUMERICAL INTEGRATION FOR CASTING OF STEEL PARTS

The numerical integration allowed improving considerably the development capacities of the offices of studies [Ber00]. Nevertheless, for the caster, the integration of these new technologies does not hold really its promises. Many industrial ones think to buy solutions and meet again opposite integration problems, that induces unexpected investments. These difficulties their permits with difficulty to put some places based systems on the exploitation of these technologies, that appear nevertheless comparatively accessible (prototyping, simulation, etc.).

Nevertheless, it seems that the generalization of these numerical technologies can reply to strategic needs for the business: to reduce the response time to the offer call, mastery the industrial risks linked to complex procedures, etc.

The caster needs to develop differentiation factors that allow for him to promote his commercial offer.

Among these, the reduction of the risks that takes the order giver while choosing a supplier can be the one of the levers facing the competition of the emerging countries, of course more attractive in terms of costs, but of which the fiability remains limited in terms of quality and of delays.

To succeed this objective, tools were developed to allow mastery the financial risks (calculation of the contributory margin, simulation of the impact on the exploitation result, etc.). These tools are based on a definition of the product and of his manufacture process that it suits to validate as soon as possible in order to fiabilize the decisions that are taken.

The generally used methods do not allow validating this definition before the discount of the commercial offer. Before the production phase, it is therefore necessary to go through an intermediary step for the validation of the choices technological. For that, the caster realizes the "typical pieces" that are dependent upon the customer.

This organization does not allow therefore replying to the offer call while having validated the parameters technological. It lets place to the party of inherent risks to the development of the product:

- malfunction of the product,
- inability of the procedure to realize the product,
- passing of the development delays,
- passing of the development costs,
- modification of the manufacture process,
- modification of the contributory margin.

In this framework, the integration of numerical technologies has for objective to compress sufficiently the delays and development costs to realize the "typical pieces" before the discount of the commercial offer (for judged strategic matters). Thus, the already industrialized piece uses basis for the negotiation, obliterating all the risks technological.

The integration of development technologies numerical allows equally to simulate the processes define, allowing thus to insert an intermediary step of validation. It is then possible to perfect the operational processes before their deployments, reducing thus risk them, the delays and the costs linked to the perfect operational processes. The deployment of a similar organization allows equally to the caster to validate the news using basis to the evaluation of the income costs, fiabilizing thus the strategic choices linked to the product.

Despite all, with an average delay of response to an offer call of about two weeks and a development process of about nine weeks, the placement some places such an organization constitutes an extremely ambitious tournament. The reworked one completes development process is necessary. The integration of the conception technologies and of numerical manufactures is essential.

In fact, introduction of engineering technologies numerical in foundry allows:

- to create the numerical model in three dimensions of a piece, of a cluster or of a tool,
- to communicate around the visualization of a potential piece,
- to realize quickly models or functional prototypes,
- to simulate the process of filling, of solidification, of cooling, etc.,
- to make tools quickly,
- etc.

Nevertheless, it seems impossible to organize the development process of manner to generalize the usage of these technologies within a numerical organized system.

Why? Is it a question of a structural problem or organizational?

To reply to these questions, it is important to well understand the development process used by the caster. As we saw it, this process is comparatively simple: from the client requirements and/or the drawing of the piece, the caster does an outline of the tool that it transmits to the model maker for the realization of the tools. The casting studies represent therefore, for the caster, a workload very weak (one to two hours at the maximum).

Facing these pragmatic methods, the numerical technologies are inevitably more costly. The realization of a complete study in CAD necessitates to model the piece (between 2 and 20 hours), and on the other hand to model the cluster (between 5 and 20 hours). In addition, the exploitation cost of these systems very high for two reasons: important cost acquisition and a usage rate very weak. In fact, the complexity of these tools and the necessary time for the realization of a study disgust the technician caster for that CAD is not the core trade.

The foundry technician, that needs a support to express his know-how, meets again facing a true tool of specialist, that necessitates an intensive exploitation to succeed absorbing the very important investments of which it is the subject. There is unquestionably a problem in appropriateness between the man needs and the CAD system functionalities.

The exploitation of these numerical technologies is costly, but it is supposed to bring important gains of productivity. Is it therefore possible to justify the realized investment at the time of the conception thanks to the exploitation of the applications and to the obtained gains.

The gains of productivity are in reality comparatively reduced. In fact, many numerical applications bring services without actually to be able to justify financial provisions. Among these, the 3D visualization and the rapid prototyping [Ber98] allow communicating with the different actors of the project. Without any doubt, these applications allow improving the definition process of the product.

The return on investment realized by foundry simulation is equally very difficult to justify. The study of filling, of solidification and of cooling is done from the CAD model of the cluster [Moq94]. Even if simulation allows improving unquestionably these results, the necessary financial investment for systemizing it remains widely superior to the feasible gains.

The realization of the tools is subcontracted to the pattern makers. He uses more and more CAD/CAM to model and to machine the master's model or directly model them [Ogi98].

This method modifies only very little the manufacture process of the tools and the decrease of the sale prices (about −10% in average) does not allow justifying the time and the engaged costs for the conception.

2.1 Product design

The exploitation costs and the very specialized approach of the CAD software do not allow the caster to realize the numerical integration of its means of development of products.

To reply to the needs regarding CAD, it is necessary to deploy a tool allowing the technician caster expressing simply his know-how. The specialist needs a computer tool based on common known standard. The CAD software must also become a very easy-to-use tool.

As defined by our analyses, these requirements correspond to a vision that has to be focus on the core function of the foundry technician. He has the opportunity to use it during discussions with the customer, with his laptop that disposes jointly of costs, conception and simulation evaluation systems. This type of platform enables him defining in real time and with the customer, a realistic definition of the product and of its manufacture process that is optimized in terms of costs. The CAD software has therefore to be usable on a portable personal computer and sufficiently friendly to allow modifying quickly a 3D model, to adapt the proposed process to the constraints of foundry.

After the study of the tools on the market, we proposed SolidWorks™ as 3D modeling tool for several reasons: its price, its friendliness, its performances, its development capacities, and the reactivity of its commercial organization.

The friendliness of software mainly depends on its performances. Besides modeling capacities, the easy use of the functions allows multiplying the loops in order to test different solutions (in particular for the management of the subcontracting activities).

But a CAD tool limited to the modeling of foundry pieces would not be sufficient when considering a global numerical engineering perspective. It has also to enable the development of the environment of supplementary functionalities allowing deploying the proposed engineering methodology.

In foundry, some significant developments should allow in a medium term improving software performances regarding conception oriented trades [Dou98, Mar03, Mar04]. But, based on our experience and on the company expertise, we propose first a knowledge-based methodology for the modeling of raw foundry parts. The first step consists in obtaining a skeleton in 3D piece. On this basis, the technician models next the outlines of the raw part by realizing the necessary adaptations (draft angles, fillets, etc.). Due to the possibility of managing different configurations of the parts, it is therefore very simple to create a model of the piece with access to a raw configuration of the piece and to a machined configuration of the piece. Such capability enables the optimization of the weight of material when considering the mechanical properties of the final machined part.

2.2 Process design

After defining a conception methodology of the product, we propose now a methodology for the modeling of the features describing the manufacture process. It is necessary to define the patterns, the tools, the cluster, the cores and the core boxes.

The foundry technician puts his added value while defining the industrial processes. While specifying his strategy for the casting and the filling of the piece in liquid metal, he defines the key parameters of the manufacture of the product. These aspects constitute the core of his trade.

To succeed in integrating such kind of numerical modeling tool, it is necessary to give to the designer a system at least as performant as the traditional manual method he uses. The goal is not to replace the technician while proposing a solution automatically [Bou93, Mer97, Lic91, Vex00], but well integrate him in the creation of his added value on a numerical support [Rou99].

For the creation of these models, the architecture of the CAD systems necessitates to create complex assemblies allowing merging certain entities. Thus, for the modeling of the cluster, it is necessary to merge the models, the filling system, the metal cooling channel, the masselottes, etc. Such an architecture creation enables the automatic update of the body of the "processes" entities in case of the modification of one of the components, as the piece for example.

The two main assembly models (part and cluster) have access to eight modeling configurations, which correspond to the different steps of the manufacture process. These configurations represent the following entities:

– plates (over and under),
– mold (over and under; mold empty and assembled),
– cluster (model for simulation and real topology).

The definition of new project is based on this information structure. This conception environment allows modeling the cluster thanks to:

– positioning of plates in the mold,
– positioning of the parting line,
– modification of the number of pieces present on every cluster,
– adaptation of the diameter of the flow descent and canal type of principal feeding,
– modeling of the feeding canals.

The methodology used for the definition of the masselottes is interesting. This component plays an important role at the metallurgical level, simulation allowed looking for solutions optimizing the ratio "volume of masselottes/volume of liquid metal".

In conclusion, this tool allows in some minutes to model the body of the components of a project. This methodology allows thus optimizing the casting strategy and the feeding flow.

2.3 Simulation

The conception phase must result in pieces and tools in accordance with the capacities of the production tool and compatible with the foundry procedure sand. The global objective remaining the minimization of the scraps and the direct manufacture of pieces in accordance with the production plan, the integration of numerical simulation tools gives assistance for the validation of the products and processes. We structured a classification of the simulation applications in three families:

– simulation "technological",
– simulation "functional",
– simulation "logistical".

Simulation "technological" has for objective to represent numerically the sequence of an operation for a given technology. It enables the validation of the parameters defining a process before its physical execution, avoiding thus the costly errors being able to intervene during its deployment. Available software on the market globally presents good performances, to analyze the results.

It is first of all necessary to compare similar geometric elements. In fact, the traditional method of tool manufacture allows obtaining from models a precision of about 0.1. Nevertheless, their positioning does not present any functional constraints, it is done to more or less some millimeters. If this has no effect on the physical process, the obtained results with simulation seem very uncertain. Consequently, it is important to compare similar geometric elements. Therefore, it is not possible to use the simulation as a measure tool of defects. The results have to the opposite clearly to be identified as macroscopic and analyzed with an inverse

logic, that consists of not to consider the good results as good, but rather to consider the false results as false.

The results are therefore satisfactory, even without realizing the simulation of filling. This one represents the filling of the mold in liquid metal, measuring the risk of rupture of the sand in case of too important pressure of the alloy on a part of the mold. It is not therefore determining for the validation of metallurgical health of the piece. Despite all, it is anyway preferable to simulate the process to improve the representativity of the results.

Thanks to the calculation power of the recent processors (and that does not stop improving), it is possible to integrate the simulation of filling and of solidification as being an integral part of the conception process. The filling can be simulated in masked time, while the solidification can be simulated in real time, allowing validating the technological choices in some minutes (the calculation time being configurable according to precision, a calculation of about thirty minutes allows obtaining representative results).

"Functional" simulation has for objective to represent the functioning of an object. One finds thus a multitude of applications simulables, as for example the kinematics, the efforts (constrained and deformation), the acoustics, etc. This kind of simulation concerns more the integrator than the producer (caster) of the piece.

Nevertheless, it is interesting for the caster to propose a complete service for the conception of the product. Thus, it increases the added value of the study and can thus justify the mechanical characteristics of the product in case of a competition with an alternate technology.

At the present time, simulation "logistical", that has for objective to represent numerically simultaneous or sequential sequence of several operations, does not concern directly the objectives of this approach. Mainly due to the excellent autoimmunization of the production casting line, this point has been postponed to a near future. One can imagine that in the end, this would allow simulating the impact of a new production on the flows. From this perspective, the interoperability between the processes and the production data management would allow these technologies to be integrated in the global proposed approach.

2.4 Integrated product/process/cost

As we saw it before, the design software tools were interfaced to the cost estimation system for the definition and the management of the product/project information. Consequently, we carried out an exploratory work aiming to secure the two systems. The objective of this study is to determine if it is pertinent to evaluate the profitability of the product directly in the design environment, while using the CAD models as the main source of data.

2.5 Identification of exploitable parameters

The objective of this approach is to create an integrated conception environment, allowing validating quickly the realized choices from two points of view:

- functional (the product must fit the functional needs),
- technological (the product must be feasible),
- and financial (the product must be profitable).

While having access to such a tool, it is possible to optimize the definition of the product and the processes, while fiabilizing the operational deployments and the projected results.

To satisfy this objective, it is necessary to integrate the assessment application in the conception environment trade. This integration first implies the identification of the 3D parameters usable by the assessment application.

The CAD model of the piece brings little information. In fact, the weights are directly exploitable (piece weights brute and machined). The main information is therefore present in the entities intervening in the manufacture process (clusters, cores, etc.). From these models, it is possible to identify the following parameters:

- number of pieces on the plate,
- cluster weight,
- core weight,
- number of cores for a given plate,
- number of casting accessories (manchons, filters, etc.).

Thanks to the modeling tools trade, it is possible to conceive the cluster in some minutes, this allows multiplying the loops. The interfacing with the estimate system intervenes then to evaluate directly the profitability of the chosen options and allows defining the optimum solution, validated thanks to simulation. This methodology allows thus constraining the designer to the process definition of realistic manufacture and avoids the evaluation of utopian solutions. It allows equally optimizing the load of the flow filling system.

The realization of a complete study necessitates some hours (between 2 and 10 hours, principally for modeling the piece). For comparison, the realization of a cost estimation needs very often less than 30 minutes. In fact, when a new matter represents an important sale, it is necessary to evaluate the costs more precisely possible. In this case, the obtained gains while using this methodology allow justifying its usage.

2.6 Manufacturing process of tools

Our researches have first concerned the introduction of the technologies of rapid prototyping [Ber98, Hil00, Woh04]. These can be mainly employed to make the master models. The tests, carried out with the LOM technology, allowed validating this methodology without any problem, dimensional precision being very near of the techniques of traditional pattern making.

This research brings us next to realize some economical tools. It is a matter to realize the models in an economical material (high density foams or polystyrene) and to apply a surface treatment that increases the resistance of the model.

The EOS technology of sand sintering was technically validated and allowed introducing another system for the manufacture of functional "good material" prototypes, thanks to the manufacture of models and cores in sand CRONING. Test validations during the development of a body of a turbocompressor made obvious the interest of this method for the manufacture of complex forms prototypes. The deployment is simple but the manufacture costs limit the usage of this procedure to the realization of prototypes of very strong added value parts.

These amounts take into account the casting study realized with the conception methodology. It is therefore important to separate this study of the rest of the costs to obtain an objective analysis.

The exploitation of these types of procedures of rapid prototyping still remains, at the present time, with justifiable difficulty. It is therefore necessary to have access to more realistic alternative technologies as milling.

Thanks to this global methodology of development, 3D modeling of product, that represented beforehand an unjustifiable investment in times and in money, is now totally integrated in the manufacture process of the tools and allows realizing important gains (-50% Times, -40% Costs).

Speed and flexibility of the employed methods allow a minimum time for the realization of the tools, about one week.

A tool manufacturing methodology is proposed. Its originality lies in its integration level, the body of the components and about the fact that its deployment is completely defined in the numerical environment of conception [Ber$_2$02].

The manufacturing "economical" method is based on the exploitation of CAM software easily accessible for milling preparation. CAMWorks was chosen for its integration capacity in our methodology. It works directly in SolidWorks (certified software GoldPartner) allowing thus to define, simulate and generate the milling process in the conception environment. It is not therefore necessary to carry out conversion of files. In addition, it perfectly corresponds to the technical needs.

In terms of creation of milling paths, the standardization of the process already allowed carrying out important gains. Consequently, the necessary time for the definition of the process is globally identical. The process time is improved thanks to use of algorithms of more performance (average calculation times of inferior of 80%, reduction of the global time of programming of 50%). Nevertheless, the use of software induces the increase of the global cost of programming.

Thus, the time minimum necessary for the manufacture of the tools is reduced to one day. The realized tests perfectly showed the feasibility of these delays. Globally, this is a reduction of the process time of about 25% that is noted in comparison with the "economical" method and closely of 65% in comparison with the traditional methods.

In terms of cost, one records a reduction of 20% in comparison with the "economical" method and of 50% in comparison with the traditional one.

3 CONCLUSIONS

This methodology presents also an interest for the fiabilisation of the simulation data. Thanks to the similitude of geometries, the obtained results with simulation are representative of the real deployed processes. This methodology allowed us noting an important reduction of the tool modifications (−30%).

The gains are very important but must not occult what appears as the one of the main benefit of this methodology. It allows basing the development process on the usage of the numerical conception technologies, while justifying it financially.

This is possible thanks to, on one hand, the reduction of the critical size of the systems and of the induced costs, and on the other hand, the standardization and the numerical integration. Thus, this methodology places the numerical technologies to the heart of the development process.

The complete proposed approach was deployed in an operational way on the basis of the suggestions that we carried out.

One of the key factors of this success was the integration of all these methodological and technological modifications by the operators, technicians and engineers casters that saw their trade strongly jostled by the new tools and solutions. It has therefore been necessary to unblock all the resistances to the change by successive steps and to let the method be appropriated by the users.

The methodological suggestions deployed allow now SMC evolving in a competitive market while having access to differentiation factors that allow it promoting its offer.

The deployment of the proposed solutions allowed braking the decreasing awaited. The evolution of the exploitation result has been stabilized progressively to −0,2%/year in 2004, to compare to the envisaged decreasing of about −2,5%/year. Thanks to this tendency, the exploitation result loses only 5% from 1999 to 2004 (of 11% to 6%) instead of an awaited loss of 10% without modification of the organization of the business.

These results show the coherence of the industrial political deployment and prove the existence of alternate solutions facing the delocalisation of production sites such as the one of SMC.

REFERENCES

Bernard, A., Taillandier, G., 1998, *Le prototypage rapide*, Editions Hermès, Paris, ISBN 2-86601-673-4.

Bernard, A., Taillandier, G., Desmares, A., Les enjeux du développement rapide de produit pour la fonderie, *Hommes & Fonderie*, No 301, février 2000.

Bernard, A., Delplace, J.C., Gabriel, S., CAD and rapid manufacturing techniques for design optimization in sand casting foundry, *CIRP Design seminar*, Hong-Kong, 16–18 mai 2002.

Bernard, A., Delplace, J.C., Gabriel, S., Optimisation CAD and rapid manufacturing techniques for design optimization in sand casting foundry, *IDMME*, Clermont-Ferrand, mai 2002.

Bernard, A., Delplace, J.C., Perry, N., Gabriel, S., Integration of CAD and rapid manufacturing techniques for design optimization in sand casting foundry, *Rapid Prototyping Journal*, Vol. 9, No 5, 2003, p327.

Blanco, E., L'émergence du produit dans la conception distribuée. Vers de nouveaux modes de rationalisation dans la conception de systèmes mécaniques, *Thèse de doctorat de l'Institut National Polytechnique de Grenoble*, Génie Industriel, 15 Décembre 1998.

Boujut, J.F., Un exemple d'intégration des fonctions métier dans les systèmes de CAO: la conception de pièces forgées tridimensionnelles, *Thèse de doctorat de l'Institut National Polytechnique de Grenoble*, Génie Mécanique, soutenue le 13 September 1993.

Delplace, J.C., 1999, Contribution à l'optimisation du processus de conception et d'industrialisation d'une ébauche de fonderie, *Rapport de DEA*, Spécialité Production Automatisée, UHP Nancy I.

Delplace, J.C., 2000, OPTOFORM process validation for validation for sand casting, *8èmes Assises européennes du prototypage rapide*, Prix de la meilleure étude, Paris.

Doux, V., 1998, Identification des zones caractéristiques nécessaire à la conception d'outillage consommable, *Rapport de stage ESIAL*.

Eynard, B., 1999, Modélisation du produit et des activités de conception. Contribution à la conduite et à la traçabilité du processus d'ingénierie, *Thèse de doctorat de l'Université de Bordeaux I*, Spécialité Productique.

Hilton, P., Jacobs, P., Rapid tooling, *Technologies and industrial applications*, Marcel Dekker, New York, ISBN 0-8247-8788-9.

Liccia, Y., Masselottage assisté par ordinateur, *ENSAM*, AIX-EN-PROVENCE 1991.

Martin, L., Moraru, G., Veron, P., Etude de cas et identification des modèles pour la conception de pièces intégrées de fonderie, *Int. Conf. CPI2003 Integrated design and production*, Meknes, Maroc, 22–24 October 2003.

Martin, L., Moraru, G., Veron, Ph., 2004, Structured CAD-CAM models in casting process design, *3rd int. conf. on Advances in Production Engineering APE'04*, Varsovie, Pologne.

Merrouche, A., 1997, Vers la mise en œuvre automatique de la conception optimale: architecture de systèmes et outils logiciels pour l'optimisation de forme de pièces 3D, *Thèse de doctorat de l'Université de Technologies de Compiègne*.

Moquard, E, 1994, CFAO en fonderie CASTMASTER, une solution métier attendue, *L'usine nouvelle* No 2448–17.

Ogier, G., 1998, La filière numérique en fonderie, Solutions CAO/CFAO.

Roucoules, L., 1999, Méthodes et connaissances: contribution au développement d'un environnement de conception intégrée, *Thèse de doctorat de l'Institut National Polytechnique de Grenoble*, Spécialité: Mécanique: Conception, Géomécanique et Matériaux.

Schaeffer, P., Allanic, A., 1998, Nouvelle machine de fabrication rapide par couches, *7èmes Assisses européennes du prototypage rapide*, Paris.

Tollenaere, M., 1999, Méthodes globales de conception de produits, *Ouvrage collectif PRIMECA dirigé par M. TOLLENAERE, Editions Hermes*, Paris, pp 53–75.

Vexo, F., 2000, Contribution à l'intégration de la simulation avec la CAO: application à la construction du système de remplissage en Fonderie, *Thèse de doctorat de l'Université de Reims Champagne-Ardenne*.

Wohlers, T., Wolhers report 2004, *Rapid prototyping, tooling & manufacturing state of the industry*, Annual report, ISBN 0-9754429-0-2.

Author index